ADVANCED AIR POLLUTION

Edited by **Farhad Nejadkoorki**

INTECHOPEN.COM

Advanced Air Pollution
Edited by Farhad Nejadkoorki

Published by InTech
Janeza Trdine 9, 51000 Rijeka, Croatia

Edition 2014

Publishing Process Manager Ivana Lorkovic

Technical Editor Teodora Smiljanic

Cover Designer InTech Design Team

Image Copyright Stéphane Bidouze, 2010. Used under license from Shutterstock.com

Additional hard copies can be obtained from orders@intechopen.com

Advanced Air Pollution, Edited by Farhad Nejadkoorki
p. cm.
ISBN 978-953-307-511-2

INTECH

open science | open minds

Contents

Preface

The air, as an elementary component of the biosphere, is a conveyor of many pollutants both gases and particles in the atmosphere. Air pollution is a state of the atmosphere with predominant presence of hazardous substances that are harmful to humans and animals. The air-borne pollutants degrade the air quality and constant exposure to polluted air may lead to several health problems such as cardiopulmonary disease, bronchitis, asthma, wheezing and coughing etc. Anthropogenic sources of air pollution include activities of transport (airports, highways and ship emissions), power generation, oil refineries, waste incineration, industries and agriculture. Natural emissions come from area sources (sea, ocean, countryside land, forests) and point sources (volcanoes and territorial cracks).

In writing this book, we have assumed that readers are well acquainted with the very basic concepts of air pollution. The book covers all the important areas in the science of air pollution and is designed for at least two kinds of readers: a) students of intermediate and advanced courses in air pollution; b) professional, researchers and practitioners of many areas of the field.

This book contains four main parts. There have been improvements in monitoring pollutants by providing prediction of the temporal and spatial distribution of actual pollution levels. These are described in different studies under part one both for indoor and outdoor. The second part covers several studies on air pollution modeling. Deterministic, statistical and mathematical models are explained. The third part discusses a review of the most common Geographical Information System applications in air quality management. Finally in part four the latest developments in air quality management are given through numerous studies.

June, 2011

Farhad Nejadkoorki, PhD
Yazd University,
Iran

Part 1

Air Quality Monitoring

1

Observational Study of Black Carbon in the North Suburb of Nanjing, China

Lili Tang et al.*
Nanjing University of Information Science & Technology, Nanjing, China

1. Introduction

Black carbon is a kind of amorphous carbon particles, which was formed by incomplete combustion of carbonaceous substance (Andreae et al., 1984; Bond and Bergstrom, 2006; Hansen and Rosen, 1984; Hansen et al., 1988; Penner et al., 1993). The main natural sources of black carbon are volcanic eruptions, forest fires, etc. Its shape, size and origins are extremely variable depending on the different type of fuels, combustion processes, and the age and history of air masses. It mainly resides in the submicron particle size range. This aerosol is usually separated into two main components: one is called the organic carbon (OC) and the other the black carbon (BC) component. Emissions from natural sources such as forest fires are localized and incidental (Bai and Wang, 2005), whereas anthropogenic emissions are widespread and continuous (Liu et al., 2010; Streets et al., 2001; Xu et al., 2006). The increasing demand for energy in the past century has led to a great increase in the rate of BC emissions. Black carbon particles have received increased interest recently since Hansen et al.(Hansen et al., 2000) indicate that it is perhaps more cost effective to control black carbon particles than CO_2 at the present time. Black carbon has many effects (IPCC 2001; Lary et al., 1999), such as harming to human health, changing climatic characteristics, reducing visibility, affecting the atmospheric chemical reaction process, which has become a focus of scientific research in recent years.

Although the effect of increasing BC in the atmosphere may be considerable, knowledge of BC concentrations, distributions, characteristics, and potential effects is still seriously lacking. Moreover, the surface of BC particles contains numerous adsorption sites that are capable of enhancing catalytic processes. As the result of its catalytic properties, BC may intervene in some important chemical reactions involving atmospheric sulfur dioxide (SO_2), nitrogen oxides (NO_x), ozone (O_3) and other gaseous compounds (Gundel et al., 1989). Despite the evident significances of BC in air chemistry and physics, information concerning their spatial and temporal variability is still quite limited.

---

* Shengjie Niu, Mingliang Yan, Xuwen Li, Xiangzhi Zhang, Yuan Zhu, Honglei Shen, Minjun Xu and Lei Tang
Jiangsu Environmental Monitoring Center, Nanjing, China
Jiangsu Institute of Meteorological Sciences, Nanjing, China

Considering the global distribution of black carbon, the black carbon concentrations in the northern hemisphere were significantly higher than that in the southern hemisphere, as well as there is higher black carbon concentration value in eastern China (Streets et al., 2001). Of both natural and anthropogenic origin, emissions of black carbon recently have been estimated at 24Tg per year (Penner et al., 1993). China has higher black carbon emissions in recent years. We must make observation of black carbon in order to grasp the first primary data. The research work about the systematic observation of black carbon begun from 1990s in China, which had been carried out in most areas of China after 21 century.

Nanjing is located in the center of Yangtze River Delta region, which has developed petrochemical, steel and other heavy industries. The air pollution in Nanjing is very serious, but the observations of black carbon have not been reported. The observational results of black carbon in fine particulate matter from November 2008 to April 2010 in the north suburb of Nanjing were used to study the evolution law of black carbon aerosol. The observational data of SO_2, NO_2, HNO_2 and O_3 in October 2009 was also used to analyze the correlations between black carbon aerosols and SO_2; to discuss the complex-phase photochemical reaction between black carbon aerosols and gas. Moreover, the findings may also be useful in the formulation of strategic air pollution control in China and other rapidly developing countries. To understand anthropogenic air pollutant sources, loading, and their fate in the region, it is important to measure BC for different seasons and compare their characteristics. In this paper, an intensive air pollution health effects study was conducted. The characteristics of spatial variations of BC in $PM_{2.5}$ concentrations during one more year are summarized, and the relationships between gas and BC in $PM_{2.5}$ are discussed.

2. Experimental methods

2.1 Measurement site description

The field study was conducted in north suburb area of Nanjing City which is located in Jiangsu Province, 290 km east of Shanghai, and 10 km south-southeast of the Yangtze River. There are large regional sources that can impact this site, including Yangzi Petrochemical Industries Company, Nanjing steel plant, Nanjing chemistry plant and Huaneng electric generating station (5-10 km to the east and northeast). The No. 328 national highway is just 1 km in the east of measurement site. Average daily traffic flow is over 42,000 on this highway. The dominant local sources of BC are expected to be vehicular emissions, incomplete combustion of fossil fuels and biomass burning.

Measurements were taken at a university 10 km north of the center of Nanjing city in a residential area, near the top of automatic meteorological observation station (elevation 5m). Samples were collected 4.5 m above the ground.

2.2 Sample collection and analysis

$PM_{2.5}$ aerosol samples were collected from November 2008 to April 2010. Sampling started and ended at around 8 a.m. and 8 p.m. every week; each sample was collected for 12 h. The sampler (Andersen Instruments, Smyrna, GA, USA) was equipped with a slot inlet of 2.5 μm cut size and operated at a flow rate of $1.11 m^3/min$. All the collection substrates were 20×25 cm quartz fiber filters, prebaked at 550 °C overnight to remove any absorbed organic materials. The filters were stored and transported to the field in annealed aluminum foil pouches. After the sampling was completed, the 20×25 cm filters were folded in half such

that the side with the sample on it was folded onto itself and stored in the annealed aluminum pouches. All the filter samples were stored at -4 °C in a freezer until analysis.
OC and EC concentrations were determined using a thermal/optical aerosol carbon analyzer (Sunset Laboratory, Forest Grove, OR, USA) (Birch and Cary, 1996). For the determination of OC and EC concentrations, a filter punch 1.45 cm^2 in size was removed from the 20×25 cm filter and loaded into the aerosol carbon analyzer. The thermal analysis conditions (i.e., temperature program and the type of atmosphere) employed were similar to those adopted by all research groups that participated in the carbon analysis of the ACE-Asia samples (Schauer et al., 2003). Duplicate measurements of OC and EC were made on every sample, and the average relative standard deviations were 4% and 9% for OC and EC, respectively. Laboratory and field blanks were collected at a rate of 10% to determine the limit of detection (LOD); collocated samples were collected at a rate of 10% to determine precision; replicate analysis was performed on 10% of the samples to determine analytical precision.

2.3 The aethalometer

A model AE-21 Aethalometer (Magee Scientific Inc., Berkeley CA) was used to measure aerosol black carbon (BC) in real time (Lavanchy et al., 1999; Liousse et al., 1993; Petzold and Niessner,1995; Ruellan and Cachier,2000; Sharma et al.,2002). It measures the attenuation of light transmitted through particles that accumulate on a quartz fiber filter (Hansen et al., 1984). A vacuum pump draws air through the instrument so that the particles continuously accumulate on the filter while being illuminated by visible light (incandescent light source). The effective operational wavelength of the aethalometer is 880 nm (Bodhaine 1995; Lavanchy et al., 1999). The measurement of BC by this technique is controversial since this optical technique is based on light absorption of atmospheric particles sampled on filters and Mie theory to estimate BC mass concentrations in the atmosphere. This method is similar in principle to the Coefficient of Haze monitors developed over 40 years ago (Hemeon et al., 1953), but is more sensitive and stable. Optical attenuation from BC is calculated using the decrease in light transmission through the filter and the sample volume. The Aethalometer is not designed to measure either aerosol organic carbon or atmospheric light extinction from particles. The principle of the method is described in detail elsewhere (Hansen et al., 1984; Hansen and McMurry1990). The Aethalometer sample flow rate was 4 L-min^{-1}. An 2.5μm size inlet fraction was used. The disk aging, power failure and other reason cause some data loss in the observation process, neglecting whole day data less than 18 hours per day (Lou et al., 2005), the effective data is 285 days.
This study was also conducted to determine the relationship between light absorption measurements of an aethalometer and thermally derived black carbon concentrations.

2.4 Observation of SO_2, NO_2, O_3 and HNO_2

AR500 system was used to determine the concentration of four kinds of gases, which has double optical DOAS AQM system and produced by Swedish OPSIS AB. High-pressure xenon lamp (B class) was used as launchers light. A smooth spectrum emitted by the xenon lamp can covere the wavelength. The wavelength of UV and visible light is about 200 ~ 400 nm and about 400 ~ 700 nm, respectively. The measured gases are O_3, SO_2, NO_2 and HNO_2. Observation date was from 1 October to 31 October of 2009. According to 75% of data capture rate, excluding whole hour data less than 45 min per hour and whole day date less than 18 h per day.

Meteorological and air quality data are monitored continuously at this location by the Office of Atmospheric Environment Observatory.

3. Results and discussion

3.1 Methods comparison: elemental carbon and black carbon

Comparison between the Aethalometer BC data and the EC component of the EC/OC sampler data should show good agreement, since EC is expected to be the principal (visible) light absorbing component in ambient air (Gundel et al., 1984; Hansen et al., 1984; Hansen and Rosen, 1990). Five minutes Aethalometer BC data were averaged into twelve hour values temporally matching the 61 EC/OC samples. The relationship between BC and EC is shown in Figure 1. The two methods were found to be highly correlated (R^2=0.9492). It is said that particulate black carbon measured by the Aethalometer method has been shown to be a reliable surrogate for particulate elemental carbon measured by Thermo/Optical Reflectance at this location and time of year.

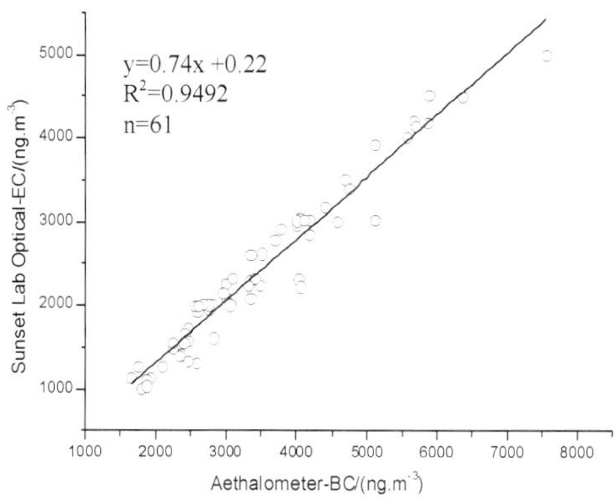

Fig. 1. Comparison between BC measurements made by EC/OC Analyzer and Aethalometer

3.2 Monthly variation of BC concentration

Fig. 2 shows the variation of monthly average black carbon aerosol concentrations during the observation period. The Black carbon aerosol concentration in autumn and winter is higher than that in spring and summer. Take the year 2009 as example, variation order (from high to low) of black carbon aerosol concentrations is October> May> November> February> December> June> April> September> August> July.

The monthly average black carbon aerosol concentration in Nanjing reaches maximum in September 2009 in the whole observation process. It is earlier comparing with other areas of the country (Qin et al., 2007; Zhao et al., 2008) The rainfall in Nanjing is very low in October 2009, and it is controlled by the high atmospheric pressure. These reasons lead to air pollutants concentrate and difficult to spread and eliminate. Moreover Nanjing was located in the transition zone between subtropical zone and temperate zone and doesn't adopt large-scale unified heating measures like north cities of China in winter. The emission inventory of black carbon aerosol won't increase sharply because of heating, which is another reason for the relative higher monthly average concentration of black carbon

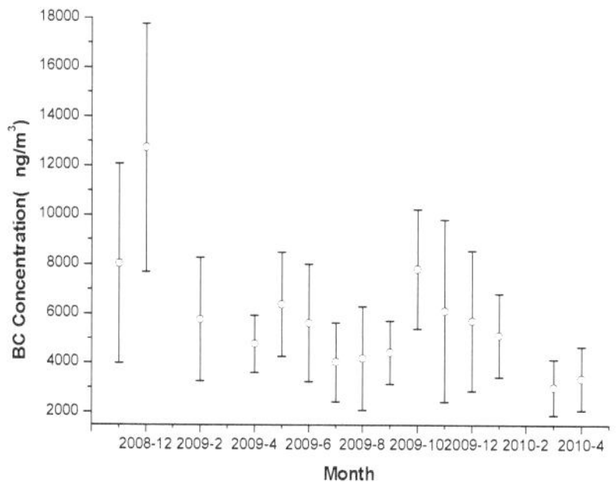

Fig. 2. Variation of BC average concentration in $PM_{2.5}$ in the air of north suburb of Nanjing

aerosol in October. By contrast, the black carbon monthly average concentrations in November and December 2009 decrease by 23.7% and 55.1% respectively over the same periods, which indicates that "on the promotion of crop straw integrated reuse decision" issued by Jiangsu Environmental Protection Administration and a series of air pollution control policies in the practical application has gained tangible effects. However, the black carbon aerosol concentration is still higher in Nanjing, the black carbon aerocolloidal concentrations in November and December 2009 are (6127 ± 3689) ng/m^3 and (5714 ± 2849) ng/m^3 respectively. It is still necessary to reduce emissions of industrial waste gas, vehicular exhaust and to prohibit straw incineration strictly.

The minimum of monthly average black carbon aerosol concentration appeared in March 2010 in the whole observation process. The precipitation in March 2010 increases by 40% over the same period, reaching 118mm, 16 precipitation days, black carbon aerosol was significantly removed by rain. Wu et al (Wu et al., 2009) found in the Pearl River Delta of China that black carbon aerosol concentrations in the rainy season were significantly lower than that in dry seasons in Panyu district of Guangzhou city. From 2004 to 2007, black carbon aerosol concentrations in the dry seasons is 10.4% higher than that in the rainy seasons (Wu et al., 2009). From the beginning of March, the amount of solar radiation reaching the ground increases with the gradual increase of sunlight hours and the atmospheric convection become strong. The black carbon aerosol concentrations decrease because of the diffusion of black carbon aerosol with atmospheric convection.

3.3 Seasonal variation of BC concentration

Nanjing is located between subtropical and temperate transition zone, which has four distinctive seasons. Therefore the criteria for the classification of astronomy is used to differentiate the Four Seasons, March, April and May are classified as spring, June, July and August are classified as summer, September, October and November are classified as fall, December and January, February in the following year are classified as winter. Fig. 3 shows the seasonal variation of black carbon aerosol concentration, where 1 to 7 in x-axis represents different seasons from the fall in 2008 to the spring in 2010. From the Fig. 3, black carbon aerosol concentrations in fall and winter were significantly higher than that in spring and summer, which is consistent with variation characterization of monthly average black carbon aerosol concentrations. The maximum of black carbon aerosol concentration appeared in the fall of 2008 during the whole observation period. It is the peak time of straw

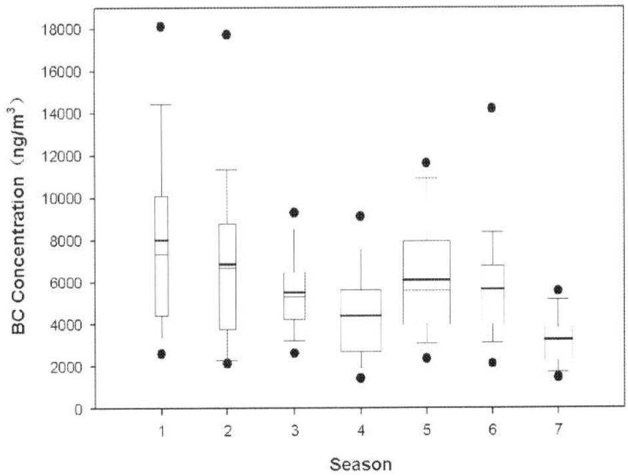

Fig. 3. Variation of seasonal average BC concentrations in $PM_{2.5}$ in the north suburb of Nanjing

incineration from the November 12 to November 31. The straw incineration attributes to the increase of the black carbon aerosol concentration. The concentrations of black carbon aerosols is relatively lower while without straw incineration. Hence the variation of daily black carbon aerosol concentration is greater in the fall of 2008 and the data has a higher dispersion degree. The concentration is 18000ng / m^3 at 95%, but less than 3000ng/ m^3 at 5% with no difference in other seasons.

3.4 Diurnal variation of BC concentration

The diurnal variation of BC concentration is characterized by a special pronounced double-peak patterns during the observation period in Fig 4. In the case of temperature inversion persisting, human activities starting from 5 a.m. lead to the increase of black carbon aerosols emissions, and centralization and no diffusion of a large number of pollutants near the ground. Black carbon aerosol concentration in the air began to increase at the same time. Auromeet et al (Auromeet and Serge 2009) found that transportation was an important

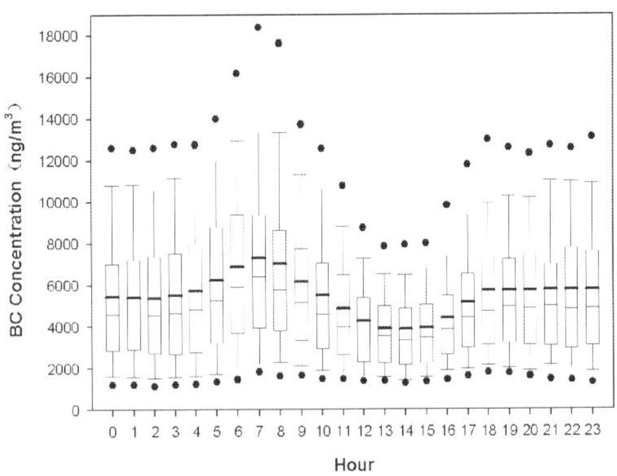

Fig. 4. Diurnal variation of BC concentration in $PM_{2.5}$ in the air of north suburb of Nanjing

source to the black carbon aerosol in Toulon region of France. In this study, the 328 national highway is just 1 km far away from observation site, so the vehicular exhaust make a greater contribution to the concentration of black carbon aerosol.

After 7 a.m., with the increase of solar radiation, temperature inversion structure was destroyed and atmospheric convection was enhanced, black carbon aerosol concentration decreases gradually. The strongest solar radiation and most intense atmospheric convection appear from 13 p.m. to 15 p.m., so the aerosol particle is not easy to concentrate. Hence the minimum of black carbon aerosol concentrations occurs at this time.

Black carbon aerosol concentration gradually increased after 16 p.m., reached maximum again at 18 p.m. Solar radiation was gradually decreased, and convection was weakened during this period. Meanwhile, a large number of black carbon aerosols were emitted to the atmosphere again by human activities; black carbon aerosol concentration begins to increase.

These changes of Black carbon aerosol concentration have similarity with other previous reports (Qin 2001; Zhao et al., 2008). But the maximum of the black carbon aerosol concentration at 18 p.m. will last until the dawn of the following day and does not decrease with the reduction of human activity during this period, which is different with other reports. It is because of the particular weather conditions in Nanjing. The book "Jiangsu weather" published by Jiangsu Province Bureau of Meteorology 90 years in the 20th century (Jiangsu Climate 1991) described that Nanjing has a high frequency of temperature inversion. The day of temperature inversion was defined that it appeared temperature inversion one time in a day, which has strong temperature inversion frequency. The structure of temperature inversion layer can be formed at toward evening in the Nanjing region. As the time passed by, strength of temperature inversion was continually increasing. The presence of strong temperature inversion layer caused that black carbon aerosols were not easy to diffuse and maintained at a stable concentrations level.

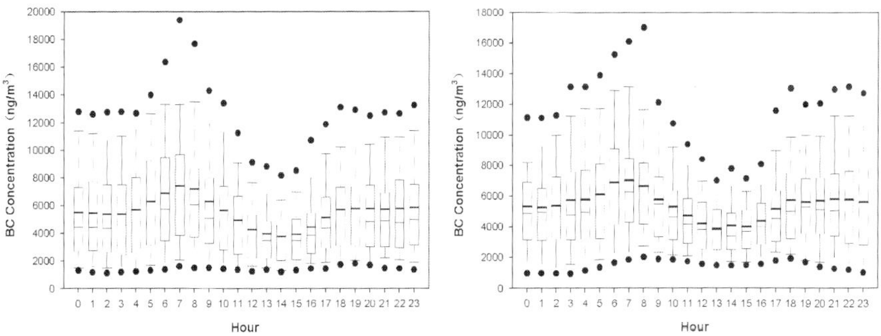

(a) (b)

Fig. 5. Variation of diurnal BC concentration in $PM_{2.5}$ on weekdays(a) and weekend(b) in the air of north suburb of Nanjing

The hourly average concentrations of black carbon aerosol on weekdays and weekend are discussed respectively. Fig. 5 shows the variation of hourly average black carbon aerosols concentration on weekdays. The tendency of black carbon aerosol concentration in Fig. 5 is consistent with the observation period, but the concentration value has more dramatic change. While the tendency of black carbon aerosol concentration given in Fig. 5 has some differences, the appearing time of maximum of black carbon aerosol concentration at 95% is put off one hour; the appearing time of maximum of black carbon aerosol concentration at 75% is still at 7 a.m. The different concentration among different numbers demonstrated that human activities are random on weekend and it is irregular about black carbon aerosols emission. However, the maximum of black carbon aerosols concentration still appears at 6 a.m. to 8 a.m. and after 5 p.m. due to the variation of temperature. The appearing time of afternoon maximum of black carbon aerosols concentration is delayed three hours on weekend, which is related that people enjoy some form of recreation and go home late.

3.5 The pollution character of BC concentration in the north of Nanjing

From Fig. 4, black carbon aerosol concentration has no significant difference with its average value, which indicates the concentration of black carbon aerosols is not a normal distribution. It is not scientific to use average value to represent the background value. Considering the request of atmospheric background, we define the concentration having maximum frequency as the background concentration of black carbon aerosols (Tang et al., 1999). Taking 100ng/m^3 as frequency statistics step, draw the frequency distribution map of hourly average concentration of black carbon aerosol. From Fig. 6, the frequency distribution of hourly average black carbon aerosol does not have normal characterization, using the log-normal distribution function to fit the frequency distribution characteristics, results show that hourly average concentration of black carbon aerosols at 2920ng/m^3 appeared maximum frequency. The observation site is located in the campus of Nanjing University of Information Science & Technology in the north suburb of Nanjing, we can conclude the background concentration of black carbon aerosol is approximately 2920ng/m^3 in the north suburb of Nanjing. The statistics data are not affected by a small number of polluted air masses and do not appear extremely high concentration of black carbon, therefore the value is most representative to the black carbon aerosols in the north suburb of Nanjing.

The background concentrations of black carbon aerosols in different areas of China have significant differences. The detailed results are listed in Tab. 1. The black carbon in $PM_{2.5}$ was studied in this paper, which is the main difference with other papers. Li Yang et al (Li et al., 2005) observed the black carbon in $PM_{2.5}$ in Xi'an, and the results showed that

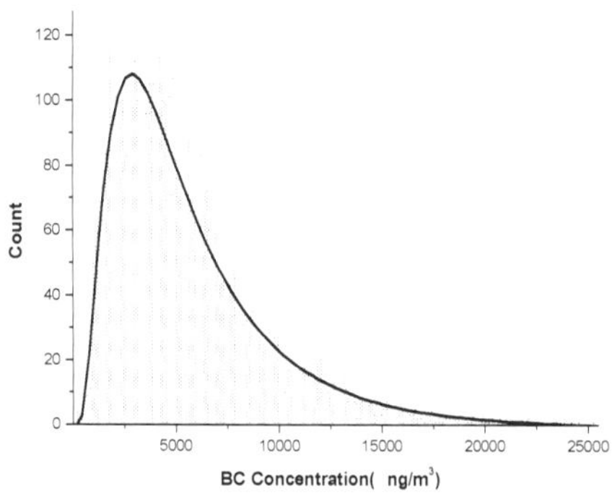

Fig. 6. Count distribution of hourly mean of BC in $PM_{2.5}$ in the air of north suburb of Nanjing

background concentration of black carbon in $PM_{2.5}$ is as high as 4500ng/m³, but the value only reflects the black carbon concentration levels in $PM_{2.5}$ of the fall in Xi'an region. There is a higher value of the black carbon aerosol concentration in the fall. It can be inferred that the whole-year background concentration in Xi'an region is lower than the value in the fall. Lou Shujuan et al (Lou et al., 2005) found that 90% of black carbon exists in the aerosol particles with the diameter less than 10μm, while 82.16% of black carbon exists in the aerosol particles with the diameter less than 2.5μm in Beijing region. Taking 80% as conversion factor between the black carbon in TSP and the black carbon in $PM_{2.5}$ in Nanjing, background concentration of black carbon in TSP is up to 3650ng/m³ in this study. The value is much higher than that in other studies. The air pollution by black carbon aerosol is very serious in Nanjing.

Site name in China	Location	Date	Position	Background concentration (ng/m³)	Literature
Wa liguan, Qinhai	Global background	September 1994-November 1995	TSP	63	Tangjie et al. (1999)
Wenjiang, Sichuan	Suburb of city	September 1999-August 2000	TSP	2890	Qin Shiguang (2001)
Lasa, Tibetan	Suburb of city	June 1998-September 1998	TSP	400	Qin Shiguang (2001)
Lin'an,Zhejiang	Region background	August 2000-February 2001	TSP	2350	Qin Shiguang (2001)
Shang dianzi, Beijing	Region background	August 1999-September 2000	TSP	222	Qin Shiguang (2001)
Xi'an, Shanxi	Suburb of city	September 2003-November 2003	$PM_{2.5}$	4500	Liyang et al. (2005)
Xining, Qinghai	Suburb of city	September 2005-February 2006	TSP	2300	Zhao Yucheng et al. (2008)
Nanjing, Jiangsu	Suburb of city	November 2008-April 2010	$PM_{2.5}$	2920	This paper

Table 1. The background concentration of BC in other areas of China

3.6 A case study

Nanjing municipal government had instituted the blue sky plan in order to control the air pollution and ameliorate the air quality since 2003. The aim of blue sky plan for 2009 is to have 320 days achieving superior rank of the atmospheric status (API ≤ 100). Considering from the whole year of 2009, there are only 14 pollution days with the help of further some weather conditions in the first six months. But from the beginning of October, due to the increasingly deterioration of atmospheric status, there are 16 pollution days within October in Nanjing. There into, it appears 15 consecutive pollution days from October 16 to October 30 in Nanjing.

The air condition of Pukou district in which observation site is located is better than whole Nanjing region, but there are nine days with air pollution index exceeding 100, slight pollution, and there are many days with the air pollution index around 95 in Fig. 7. There

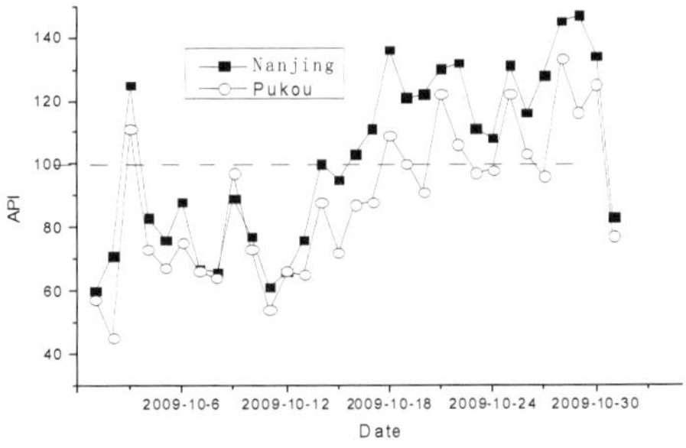

Fig. 7. The comparison of air pollution in two different site

are no large-scale infrastructural construction in Pukou district, but there are many large industries, such as Yangzi Petrochemical, Nanjing Chemical, Nanjing steel and so on in the north suburb of Nanjing. Therefore black carbon aerosols make an important contribution to air pollution in the north suburb of Nanjing.

Fig. 8 shows the evolution sequence of hourly average black carbon aerosol concentration in October, the data of October 10 and 11 were lost. From Fig. 8, the concentration of black carbon aerosol has significant changes. The air pollution is most serious in late October, the variation scope of diurnal black carbon aerosol concentrations can reach 20000ng/m^3, the maximum (26701ng/m^3) occurred at 8 p.m. on October 26. These are determined by the boundary layer structure and the emissions of pollution sources in Nanjing region. The north cold air arrived at Nanjing on October 31, black carbon aerosol concentration simultaneously decreased to 964ng/m^3 at 20 p.m. In Fig. 9, correlation analysis was used between the air pollution index and black carbon aerosol concentration. The result of correlation analysis shows that they have positive relationship (the correlation coefficient R = 0.689; $p < 0.05$), also proved that black carbon aerosol is an important component in the atmospherically particles.

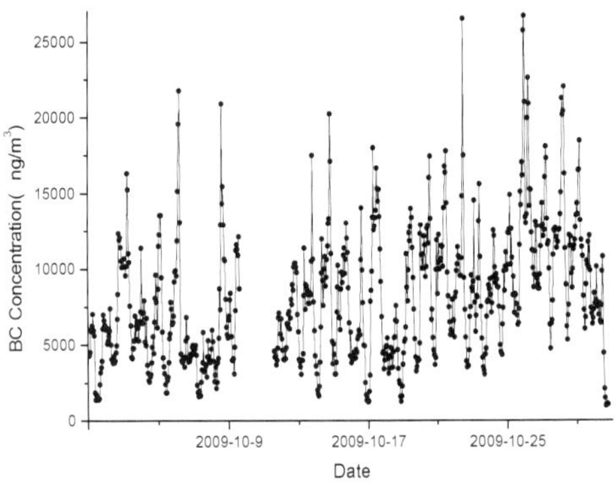

Fig. 8. Sequence of hourly average BC concentration in $PM_{2.5}$ in October 2009 in north suburb of Nanjing

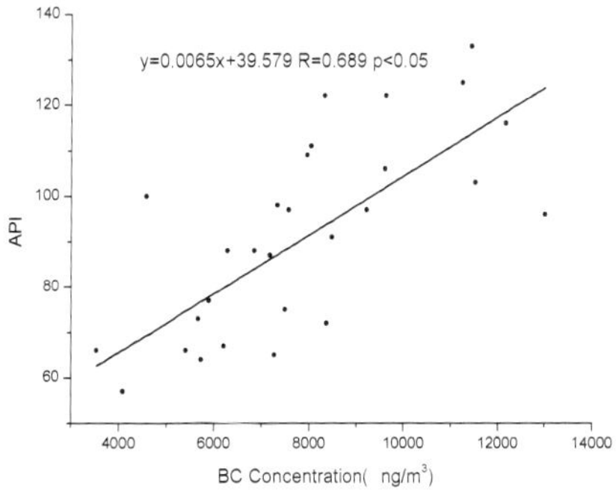

Fig. 9. Correlation between the API and diurnal average BC

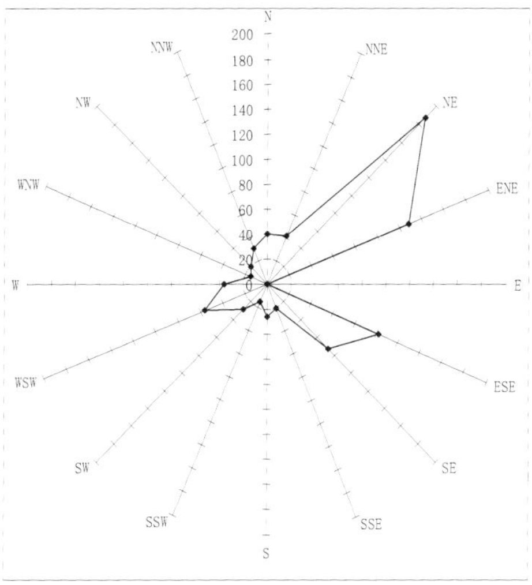

Fig. 10. Wind Rose in the north suburb of Nanjing during Oct.2009

There was no significant rainfall in October 2009 in Nanjing. The cumulative rainfall was only 8.3 mm, which decreases by 90% over the same period. Most areas of Jiangsu province have shown a meteorological drought. From Fig. 10, the dominant wind direction was northeast in October 2009 in Nanjing. There are industrial areas along the Yangtze River in this direction, where is a large emission sources of black carbon aerosol. In addition, after the harvest of crops, phenomenon of straw incineration still exists partially, and incomplete burning of biomass is also an important source of black carbon aerosols. The static wind frequency was 116 hours in October 2009 in Nanjing. The black carbon aerosol is not easy to diffuse because of the higher static wind frequency. The black carbon aerosols centralized gradually on the condition of low rainfall and higher static wind frequency, and formed a high value and serious pollution at last.

3.7 Relationship between BC and different gas

Black carbon aerosols were formed only by the incomplete combustion of carbonaceous material. Nitrogen oxides and sulfur oxides were generated with the formation of black carbon aerosols. Penner et al (Penner et al., 1993) pointed out that there are good correlation between SO_2 and black carbon aerosols in most places, and according to the correlation to calculate the global emission inventory of black carbon aerosol. Fig. 11 shows the hourly average concentration and daily average concentration of SO_2 in October 2009 and correlations between them and the Black carbon aerosol concentration. The results of correlation analysis show there is higher positive relationship between daily average concentration of SO_2 and black carbon aerosols (correlation coefficient= 0.627; $p < 0.05$), which demonstrate that it is operational to calculate the emission inventory of black carbon aerosols according to the emission inventory of SO_2 in the north suburb of Nanjing. But there is no good relationship between hourly average concentration of SO_2 and black carbon aerosols (the correlation coefficient= 0.290; $p < 0.05$), which shows that average concentration of SO_2 and black carbon aerosols concentration do not have complete consistency. So the emission inventory of SO_2 was not applied to reflecting the real emission inventory of black carbon aerosols. It is necessary to consider various factors when the emission inventory of black carbon aerosols is calculated.

Fig. 12 shows variation sequences of hourly average NO_2, HNO_2 and O_3 concentration in October 2009. Fig. 13 shows correlations between black carbon aerosol and these three kinds of gas. The results showed that black carbon aerosol has good positive relationship with NO_2 and HNO_2; correlation coefficients are 0.615 and 0.676 respectively. Nitrogen oxides in the atmosphere are mainly from fuel combustion. NO generated firstly in the combustion process was emitted into atmosphere and was oxidized to NO_2 gradually. The good relationship between NO_2 and black carbon aerosol demonstrates that air pollution in the north suburb of Nanjing presents complex features.

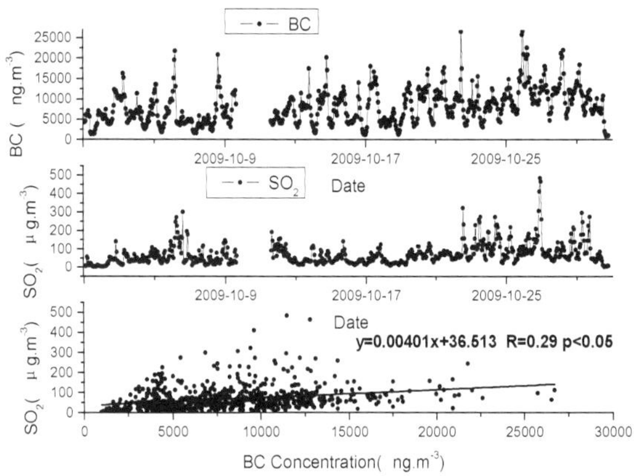

Fig. 11. Hourly mean variation series of SO_2 concentration in the north suburb of Nanjing during Oct.2009 and correlation character between SO_2 and BC

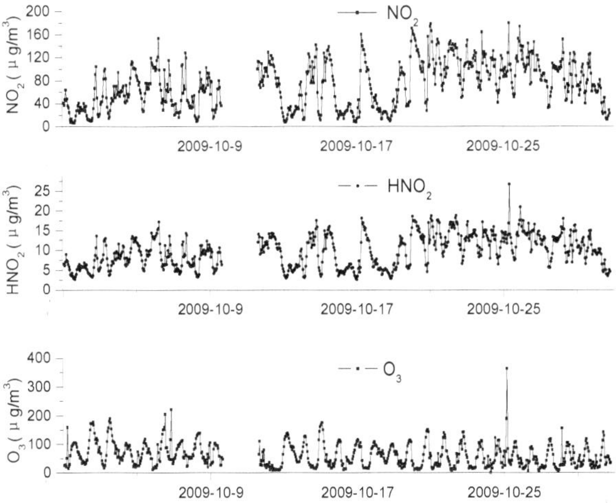

Fig. 12. Variation series of NO_2, HNO_2, O_3 concentration in the north suburb of Nanjing during Oct.2009

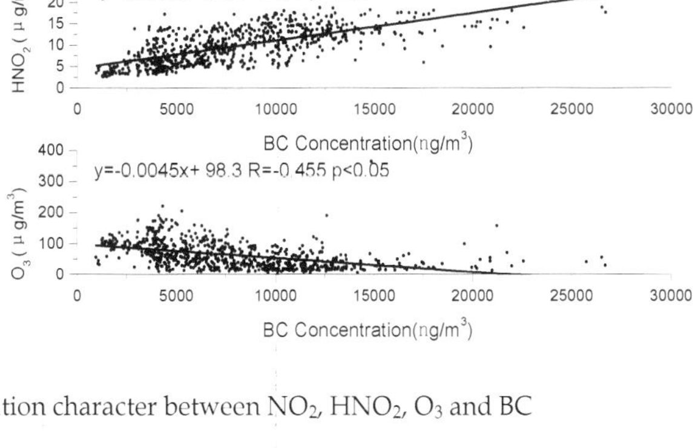

Fig. 13. Correlation character between NO_2, HNO_2, O_3 and BC

HNO_2 plays a dominated chemistry role in atmosphere. The diurnal variation tendency of HNO_2 and black carbon aerosol has good consistency and the most significant relationship. The major production of HNO_2 is from heterogeneous chemical reaction in the atmosphere, and the conversion rate of NO_2/HNO_2 was enhanced by the black Carbon Aerosol (Ammann et al., 1998; Kalberer et al., 1999; Tang et al., 2006). At the same time, HNO_2 is very unstable in the atmosphere and is accumulated only at night. The maximum concentration of HNO_2 occurs at early morning before sunrise. HNO_2 is quickly converted to OH free radicals rapidly by photolysis after sunrise. A series of photochemical reaction are initiated by the reaction of OH radicals and organic pollutants, and a large number of secondary pollutants, such as PAN, O_3 and so on, are generated in the reaction process. The migration and transformation of HNO_2 was an important reason why the concentration of HNO_2 and black carbon aerosol concentration has high correlation coefficient.

O_3 has the typical negative relationship with black carbon aerosols (correlation coefficient = -0.455; $p < 0.05$). This shows that the black carbon aerosols in the atmosphere can promote the loss of tropospheric O_3, which is consistent with observations got by Latha et al (Latha et al., 2004) in the Indian Hyderabad region. Fendel et al (Fendel et al., 1995) inferred that the black carbon aerosols contribute to break the π bond of O_3 molecule, O_2 molecules and atomic oxygen were generated in the reaction. O_2 molecules generated was released into the atmosphere directly , while atomic oxygen is able to combine with another atomic oxygen and form O_2 molecules, or react with black carbon aerosol and form carbon oxides. Meanwhile, black carbon aerosol concentration is some extent determined by the atmospheric convective activity determined by the solar radiation intensity. There is the strongest solar radiation at noon, black carbon aerosol are not easy to concentrate due to strong convective activity. But at the same time, strong solar radiation provides more energy for atmospheric photochemical reactions and promote the reactions. O_3 concentration will therefore increase. In summary, the negative relationship between O_3 concentration and black carbon aerosol concentration is determined by the two factors.

4. Conclusions

Particulate black carbon measured by the Aethalometer method has been shown to be a reliable surrogate for particulate elemental carbon measured by Thermo/Optical Reflectance at this location and time of year. The variation of diurnal black carbon aerosol concentration in Nanjing showed unique bimodal structure, high frequency of the inversion temperature lead to black carbon aerosol concentration maintaining the maximum at night.

There is a dramatic variation in the diurnal average concentrations of black carbon aerosols in Nanjing, varied from 1114 to 19408 ng/m^3 during the observation period, the concentration of black carbon aerosol in autumn and winter is higher than in spring and summer. The background concentration of black carbon in $PM_{2.5}$ is 2920ng/m^3 during observation period, air pollution is serious. The major factors which determine black carbon aerosol concentration are the structure of boundary layer temperature profiles, wind speed, wind direction, precipitation and other meteorological factors and emission intensity of sources.

The black carbon aerosols have positively correlated with SO_2, NO_2 and HNO_2. It is operational to calculate the emission inventory of black carbon aerosol according to the emission inventory of SO_2 in the north suburb of Nanjing. The black carbon aerosol in the atmosphere has certain catalysis to the formation of HNO_2. The concentration of black carbon aerosol concentration has negative relationship with O_3 concentration. Black carbon aerosol on the consumption of O_3 and meteorological factors are the main causes of this negative relationship.

5. Acknowledgments

This work was supported by the Scientific Research Foundation of Jiangsu Environmental Monitoring (NO.1016) of China, Special Scientific Fund for Non-profit Public Industry (Meteorology, No.GYHY20090607) of China and the Research Foundation of Nanjing University of Information Science & Technology, China.

6. References

Ammann M., Kalberer M., Jost D.T., Tobler L., Rossler E., Pigue D., Gaggeler H.W., Baltensperger U.(1998). Heterogeneous production of nitrous acid on soot in polluted air masses. *Nature* Vol. 395, pp. 157-160, ISBN 0028-0836

Andreae, M.O., Andreae, T.W., Ferek, R.J., Raemdonck, H. (1984).Long-range transport of soot carbon in the marine atmosphere. *Science of The Total Environment* Vol. 36, pp. 73-80 ISBN 0048-9697

Auromeet Saha, Serge Despiau. (2009). Seasonal and diurnal variations of black carbon aerosols over a Mediterranean coastal zone. *Atmospheric Research* Vol. 92, 27-41, ISBN 0-478-23223-3

Birch, M.E., Cary, R.A. (1996). Elemental carbon-based method for monitoring occupational exposures to particulate diesel exhaust. *Aerosol Science and Technology* Vol. 25, pp. 221-241, ISBN 9783527326600

Bodhaine, B. A.(1995). Aerosol absorption measurements at Barrow, Mauna Loa, and the South Pole. *J. Geophysical Research* Vol. 100, pp. 8967-8975, ISBN 0958-305X

Bond Tami C., Bergstrom Robert W.(2006). Light Absorption by Carbonaceous Particles: An Investigative Review. *Aerosol Science and Technology* Vol. 40, pp. 27-67, ISBN 9783527326600

Fendel W, Matter D, Burtscher H, Schmidt-Ott A. (1995).Interaction between Carbon or Iron Aerosol Particles and Ozone. *Atmospheric Environment* Vol. 29, pp. 967-973, ISBN 1352-2310

Gundel, L.A., Dod, R.L., Rosen, H., Novakov, T.(1984).The relationship between optical attenuation and black carbon concentration for ambient and source particles. *Science of The Total Environment* Vol. 36, pp. 197-202, ISBN 0048-9697

Gundel, L.A., Guyot-Sionnest, N.S., Novakov, T. (1989). A study of the interaction of NO2 with carbon particles. *Aerosol Science and Technology* Vol.10, pp. 343-351, ISBN 9783527326600

Hansen, A.D.A., Rosen, H. (1984).Vertical distributions of particulate carbon, sulfur, and bromine in the Arctic haze and comparison with ground-level measurements at Barrow, Alaska. *J. Geophysical Research Letter* Vol. 11, pp. 381-384, ISBN 0094-8276

Hansen, A. D. A., Rosen H., T. Novakov. (1984). The aethalometer—An instrument for the real-time measurement of optical absorption by aerosol particles. *Science of The Total Environment* Vol. 36, pp.191- 196, ISBN 0048-9697

Hansen, A.D.A., Bodhaine, B.A., Dutton, E.G., and Schnell, R.C. (1988).Aerosol black carbon measurements at the South Pole: initial results, 1986-1987. *J. Geophysical Research Letter* Vol.15, pp. 1193-1196, ISBN 0094-8276

Hansen, A.D.A., McMurry, P.H. (1990). An intercomparison of measurements of aerosol elemental carbon during the 1986 carbonaceous species method comparison study. *J. Air & Waste Management Association* Vol. 40, pp. 394-395, ISBN 1047-3289

Hansen, A.D.A., Rosen, H. (1990).Individual measurements of the emission factor of aerosol black carbon in automobile plumes. *J. Air & Waste Management Association* Vol. 40, pp.1654-1657, ISBN 1047-3289

Hansen, J., Sato M., Ruedy R., Lacis A., Oinas V. (2000). Global warming in the 21st century: An alternative scenario. *Proc. National Academy of Sciences* Vol. 97, pp. 9875–9880, ISBN 0309064279

Hemeon, W.C.L., Haines, G.F., Ide, H.M. (1953). Determination of haze and smoke concentrations by filter paper samples. *Air Repair* Vol. 3, pp. 22-28, ISBN 0830628819

Intergovernmental Panel on Climate Change. (2001). Climate Change 2001: *The Scientific Basis—Contribution of Working Group I to the Third Assessment Report of the Intergovernmental Panel on Climate Change*, edited by J. T. Houghton et al., pp. 881, Cambridge Univ. Press, New York.

Jiangsu Meteorological Administration Writing Group of "Jiangsu Climate". (1991). *Jiangsu Climate.* Beijing: China Meteorological Press (in Chinese).

Kalberer M., Ammann M., Arens F., Gaggler H.W., Baltensperger U. (1999). Heterogeneous formation of nitrous acid (HONO) on soot aerosol particles. *Journal of geophysical research* Vol. 11, pp. 13825-13832, ISBN 0148-0227

Lary, D. J., D. E. Shallcross, R. Toumi. (1999). Carbonaceous aerosols and their potential role in atmospheric chemistry. *J. Geophys. Res.,* Vol. 104(D13), 15, pp. 929- 15,940, ISBN 0148-0227

Latha K M, Badarinath K V. (2004).Correlation between black carbon aerosols, carbon monoxide and tropospheric ozone over a tropical urban site. *Atmospheric Research* Vol. 71, pp. 265-274, ISBN 0-478-23223-3

Lavanchy, V. W. H., Gaggeler H. W., Nyeki S., U. Baltensperger. (1999). Elemental carbon (EC) and black carbon (BC) measurements with a thermal method and an aethalometer at the high-alpine research station Junfraujoch. *Atmospheric Environment* Vol. 33, pp. 2759- 2769, ISBN 1352-2310

Liousse, C., Cachier H., Jennings S. G.. (1993). Optical and thermal measurements of black carbon aerosol content in different environments: Variation of the specific attenuation cross-section, sigma (s).*Atmospheric Environment Part A* Vol. 27, pp.1203- 1211, ISBN 1352-2310

Liu D., Allan J., Corris B., Flynn M., Andrews E., Ogren J., Beswick K., Bower K., Burgess R., Choularton T., Dorsey J., Morgan W., Williams P., Coe H.(2010). Carbonaceous aerosols contributed by traffic and solid fuel burning at a polluted rural site in

Northwestern England. *Atmospheric Chemistry and Physics Discussions* Vol. 10, pp. 25243-25286, ISBN 1680-7367

Li Y, Cao J J, Zhang X Y. (2005). The variability and source apportionment of black carbon aerosol in Xi' an atmosphere during the autumn of 2003. *Climatic and Environmental Research* Vol. 10, pp.229-237(in Chinese), ISBN 1006-9585

Lou S J, Mao J T, Wang M H. (2005). Observational study of black carbon aerosol in Beijing. *Acta Scientiae Circumstantiae* Vol. 25, pp.17-22(in Chinese), ISBN 0253-2468

Penner J E, Eddleman H, Novakov T. (1993).Towards the development of a global inventory for black carbon emissions.*Atmospheric Environment.Part A, General topics*27, pp.1277-1295, ISBN 1352-2310

Petzold, A., Niessner R. (1995). Method comparison study on soot selective techniques. *Microchimica Acta* Vol. 117, pp.215- 237, ISBN 0026-3672

Qin S G, Tang J, Shi G Y. (2007). Observational study of black carbon at Wenjiang, Sichuan Province. *Acta Scientiae Circumstantiae* Vol. 27, pp.1370-1376(in Chinese), ISBN 0253-2468

Qin S G. (2001). Observational Study on Black Carbon Aerosols in Rural and Remote Sites of China. Beijing: *Chinese Academy of Meteorological Sciences* (in Chinese)

Ruellan, S., Cachier H. (2000). Characterization of fresh particulate vehicular exhaust near a Paris High flow road. *Atmospheric Environment* Vol. 35, pp.453- 468, ISBN 1352-2310

Schauer, J.J., Mader, B.T., DeMinter, J.T., Heidemann, G., Bae, M.S., Seinfeld, J.H., Flagan, R.C., Cary, R.A., Smith, D., Huebert, B.J., Bertram, T., Howell, S., Kline, J.T., Quinn, P., Bates, T., Turpin, B., Lim, H.J., Yu, J.Z., Yang, H., Keywood, M.D. (2003). ACE-Asia intercomparison of a thermal-optical method for the determination of particlephase organic and elemental carbon. *Environmental Science and Technology* Vol. 37, pp.993-1001, ISBN 0013-936X

Sharma, S., Brook J. R., Cachier H., Chow J., Gaudenzi A., Lu G.. (2002). Light absorption and thermal measurements of black carbon in different regions of Canada. *J. Geophysical Research* Vol. 107(D24), pp.4771, ISBN 0148-0227

Streets David G., Gupta Shalini, Waldhoff Stephanie T., Wang Michael Q., Bond Tami C., Yiyun Bo.(2001). Black carbon emissions in China. *Atmospheric Environment* Vol. 35, pp.4281-4296, ISBN 1352-2310

Tang J, Wen Y P, Zhou L X. (1999). Observational Study of Black Carbon in Clean Air Area of Western China. *Quarterly Journal of Applied Meteorology* Vol. 10, pp.160-170(in Chinese), ISBN 1001-7313

Tang X Y, Zhang Y H, Shao M. (2006). *Atmospheric Environmental Chemistry (The Second Edition)*. Beijing: Higher Education Press (in Chinese), ISBN 9787040193619

Wu D, Mao J T, Deng X J. (2009). Black carbon aerosols and their radioactive properties in the Pearl River Delta region. *Sci China Ser D-Earth Sci* Vol. 52, 1 pp.152-1163(in Chinese), ISBN 1006-9313

Xu L, Wang Y Q, Chen Z L, Luo Y, Ren W H.(2006).Progress of black carbon aerosol research i: Emission, removal and concentration.*Advances in Earth Science* Vol. 21, pp.352-360(in Chinese), ISBN 1001-8166

Zhao Y C, De L G E, Ma Y C.(2008). Change Features of Black Carbon Concentration in Atmosphere in Xining from Autumn of 2005 to Winter of 2006. *Urban Environment & Urban Ecology* Vol. 21, pp.26-29 (in Chinese), ISBN 1002-1264

2

Air Pollution Monitoring Using Fuzzy Logic in Industries

Seyed Ebrahim Vahdat and Foroogh Mofid Nakhaee
Islamic Azad University- Ayatollah Amoli Branch
Iran

1. Introduction

In this chapter, an approach will be investigated by monitoring of industrial air pollutions by fuzzy logic. Monitoring by fuzzy logic means air pollutant risk will be evaluated on time in the specific industry and area. It will be able to identify which air pollutant has high risk and which industry is responsible for that, and also, which area is in the matter. In other word, it will evaluate air pollutant risk, related industry and related area on time. Therefore, it helps Air Pollution Control District (APCD) to identify high risk air pollutant, industry and area by numerical value.

Fuzzy logic is used in uncertainty environments so that expert's experiments can be converted into mathematical languages. On base of fuzzy logic is possible to maximize the utilization of design. In this chapter, it shows that, it used for monitoring of industrial pollutions as a case study of air pollution monitoring in iron and steel making industry and aluminum making industry.

Fuzzy logic has a real concept of natural actuation which everybody meets everyday. This logic seems to be natural phenomena in a real form. Fuzzy logic is used in uncertainty environments so that expert's experiments can be converted into mathematical languages. On the bases of fuzzy logic it is possible to maximize the utilization of decision making. It is possible to use it also for ranking of air pollutant risk, dangerous industries and high risk area by a continuous value between 0 and 1.

Fuzzy logic investigates the relative air pollutant. In order to accomplish this, fuzzy-logic approach defines a set for each property. For example, air pollutant has different criteria; industries produce different amount of air pollutant and also the mass of industry production differences in the specific area. Therefore, these matters have different membership degree in their sets. Using these sets and fuzzy logic rules, it can evaluate and compare different air pollutant risk for APCD decision making properly.

Fuzzy logic method was proposed because this method has not been used for air pollutant risk ranking. In the other word, fuzzy logic method shows the satisfactory value of air pollutant in a continuous value between 0 and 1.

In the first step, the parameters must be normalized. The second step is sets definition. These are 3 sets called air pollutant set, industry set and area set. In the final step, the air pollutant favorability degree in the specific industry at the relative area is determined; the industry favorability degree for the specific air pollutant production at the relative area is

determined and also the area favorability degree for the healthy environment in the specific industry at the relative air pollutant is determined.
Fuzzy logic explains the relativity of different features and defines a set for each feature and then determines the desirability of each variable to provide the involved feature on a scale of zero to one. This value, called the dependency degree of the involved variable in relation to the mentioned feature, will be evaluated.
Zero desirability degree means that the involved variable isn't provided at all for the related variable or the involved variable is not desirable for the correspondent feature at all; and desirability degree one implies that the target feature is completely provided for the involved variable or the involved variable is absolutely desirable for the related feature. A desirability degree of between zero and one means that the involved feature is partially provided for the target variable or the involved variable is partially desirable for the related feature.
There are different methods like AHP, Fuzzy Hierarchical TOPSIS and Fuzzy network TOPSIS for decision making. All these methods are heavily dependent on expert's opinion which does not have a scientific basis in general while fuzzy logic has a regulated pattern and is rarely influenced by expert's opinion [Astel A. 2007; Chen P.H. et al, 1998; Icaga Y., 2007; Mintz R. et al, 2005; Pokrovsky O.M. et al, 2002.; Sowlat M.H. et al, 2011]. The results of this research, in particular, are not influenced by expert's view.
After introducing fuzzy logic and comparing it to binary logic in this paper, its application in monitoring the pollutants of industries related to metallurgy in addition to a case study on monitoring the industrial pollutants, carbon dioxide, in iron and steel production industries, aluminum, zinc and lead production in some parts of the world have been stated.

2. Introduction to fuzzy logic

Fuzzy logic was first introduced by the originally Iranian professor, Mr. L. A. Zadeh, from the faculty of Electronics in the Berkeley University of America in 1965 through an article entitled "Fuzzy Sets" in Information and Control conference [Zadeh L. A., 1965].
American industrialists and scientists accepted his logic as a merely theoretical viewpoint and ignored its applicable aspects. With the arrival of this logic into the East (especially Japan), this logic was used as an applicable method to analyze processes. Since Prf. Asgharzade was a professor of electronics, the applications of this logic in electricity and electronics sciences were first taken into account but it is now used in all technical and engineering sciences [Ghafari A. et al., 1998].
We can simply say that fuzzy logic is the new view of science towards the universe. In fact, processes are analyzed as they truly are, not the way we want them to be, but how can we use them? Everything should be formulized in technical and engineering sciences; otherwise, its application will be limited and does not have a scientific basis. After all, the problem is how fuzzy logic can be formulized?
The function that yields the accuracy or inaccuracy of propositions in fuzzy logic is called the membership function. The defined membership function should be compatible with the nature of process; otherwise, the fuzzy inference will be erroneous. It is thus necessary to identify the process completely before defining the membership function.
How does fuzzy infer these figures then? There are different rules to infer fuzzy. One of the most famous ones is the fuzzy maximum-minimum rule (fuzzy peak-trough). It should be noted that the inference rule has to be compatible with the nature of process. Having

sampled inferences (especially at limit quantities) and compared them with reality (real quantities), we can make sure of the correctness of selecting the fuzzy inference rule.
The minimum rule is used when a connector i.e. the conjunction 'and' is used. In this rule, the minimum value of propositions is accepted as an inference; in other words, if there are some propositions valued a, b, c, d, ..., the minimum rule infers the minimum of these values [Ameri R., 2003; Taheri S. M., 1999].

$$\text{Fuzzy Minimum Rule} = \min (a, b, c, d, \ldots) \quad (1)$$

The maximum rule is used when a separator i.e. the conjunction 'but' is used. In this rule, the maximum value of propositions is accepted as an inference; in other words, if there are some propositions valued a, b, c, d, ..., the maximum rule infers the maximum of these values [Ameri R., 2003; Taheri S. M., 1999].

$$\text{Fuzzy Maximum Rule} = \max (a, b, c, d, \ldots) \quad (2)$$

3. Comparison between binary logic and fuzzy logic

Einstein says "where classic mathematics rules are realized, they are not reliable and where they are reliable, they cannot denote reality." Einstein's description is an imagination of the inefficiency of classical logic rules in mathematics. In natural conditions, indeed, the world of zero and one, correct and incorrect, good and bad is an abstract and imaginative world. A given subject is rarely 100 percent correct or incorrect and good or bad.
Using fuzzy logic, the precision of results increases and errors decrease. For instance, using the limited elements method, surface residual stresses resulting from grinding with an average error of 12.3 percent are predictable; however, the error level of calculation decreases to 0.037 percent by using this logic [Ali Y.M. and Zhang L.C., 1997].
We are familiar with the term 'fallacy' in binary logic. Its famous example is that if the proposition 'the fire is on the ship' is a true proposition and if the phrase 'the ship is on the water' is true too, the proposition 'the fire is on the water' should be true as well. Fire and water are two opposite things in reality and cannot be together; hence, the above conclusion is incorrect. It is obvious that the final conclusion is wrong because the value of these propositions is correct or incorrect in our view and there is no an intermediate state and also we have not considered the opposite natures of water and fire [Ghafari A. et al., 1998].
Fuzzy logic deals with this process as it is. It means that it gives a value between zero and one for the accuracy of the proposition 'the fire is on the ship'. Value one is given when fire has covered a ship completely and value zero is given when there is no fire on the ship. Generally, the degree of the accuracy of this proposition is defined according to the fraction of fire on the ship. There is a similar view for the proposition 'ship on water' i.e. a fraction of ship touching water is regarded as the degree of the accuracy of the proposition (the ship is on the water). Consequently, when the ship is completely sunk, this proposition is true with a value of one, but when the ship does not touch water at all, this proposition is true with a value of zero.
The opposite natures of water and fire are occult in fuzzy logic. Given what was said so far, if the fire has covered 30 percent of the ship, the proposition 'the fire is on the ship' is true with a value of 0.3. Also, if 20 percent of the ship is touching water, the proposition 'the ship is on water' is true with a value of 0.2. According to the fuzzy minimum rule, the proposition 'the fire is on the water' is then true with the least value of the propositions which is 0.2 percent.

$$\mathrm{Min}\,(0.3, 0.2) = 0.2$$

Notice that the defined membership functions are compatible with the nature of processes. In the above example, if water covers all the ship which means that the ship is sunk, the proposition 'the ship is on the water' would be true with a value of one. But certainly there would be no fire on the ship in this case since water and fire are opposites. The proposition 'the fire is on the ship' is then true with a value of zero and, therefore, the proposition 'the fire is on the water' is true with the least values which is zero.

$$\mathrm{Min}\,(1, 0) = 0$$

4. Monitoring of industrial pollutants

Monitoring of pollutants can be studied in three sets. The first set is the type of the metallurgical process that provokes pollution, so set M, the set of metallurgical processes {iron and steel production, aluminum production, copper production, zinc production, ... } is defined as relation 3. The second set is an area that provokes pollution; so a set of areas is formed like relation 4. Set A, is a set of areas definable all over the universe. For example, the set of world factories {Khuzestan Steel, Isfahan Iron Melt, ..., Bohler,... , Myhanaite, ...} or the set of world states or provinces {Mazandaran, Gilan, ..., Florida, Texas, ...} or the set of world countries {Iran, Japan, ...}. Also, the set of areas can be defined in a country e.g. the set of factories in Iran {Khuzestan Steel, Isfahan Iron Melt, ...} or the set of provinces in Iran {Mazandaran, Gilan, ...}. More general the selected area (the whole world) and more detailed its divisions or the members of the set (factories) are, more detailed the monitoring will be. Due to the insufficiency of data on the area set in this research, the set of countries was taken into account. The third set is the type of pollutant that led to the formation of set P, the set of pollutants {carbon dioxide, carbon monoxide, sulfur dioxide, lead, nitrogen oxides, evaporative organic compounds, arsenic, lead and copper in water, ...} according to equation number 5. Set M can have m members, set A can include a members and set P can have p members.

$$M = \{i \in N, i \in [1, m], m_i\} = \{m_i\} \tag{3}$$

$$A = \{j \in N, j \in [1, a], a_{i,j}\} = \{a_{i,j}\} \tag{4}$$

$$P = \{k \in N, k \in [1, p], P_{i,j,k}\} = \{P_{i,j,k}\} \tag{5}$$

m_i is the process i we want to study. $a_{i,j}$ is the process i in area j which will be studied and $P_{i,j,k}$ is the amount of pollutant k in process i that is in area j. It should be noted that the amount of pollutant is normalized i.e. $P_{i,j,k,Normalized}$ amount of pollutant is produced in area j for each unit of product of process i. Hence, equation 6 is used to calculate $P_{i,j,k,Normalized}$.

$$P_{i,j,k,Normalized} = \frac{\text{total k pollution in the } a_{i,j}}{\text{total mass production of} a_{i,j}} \tag{6}$$

4.1 Safety degree determination

The goal is to determine the biocompatibility or safety degree of process i in producing pollutant k in area j. Therefore, a new set named C is formed which means biocompatible processes in producing different pollutants in different areas. Each member in this set belongs to set C at a certain degree. This degree indicates the degree of biocompatibility or safety of a certain process in producing a certain pollutant in a certain area i.e. $C_{i,j,k}$. Equation 7 which is called the membership function of the members of the set C, i.e. $C_{i,j,k}$, is used to calculate $C_{i,j,k}$.

$$C = \{C_{i,j,k}\}$$

$$\text{If } C_{i,j,k} > 0 \text{ then } C_{i,j,k} = \left(\frac{\text{Min}\{P_{i,j,k,\text{Normalized}}\}}{P_{i,j,k,\text{Normalized}}} \times \left(1 - \frac{G_{k,j}}{G_{k,\text{critical}}}\right) \right) \quad \text{Else} \quad C_{i,j,k} = 0 \tag{7}$$

$G_{k,j}$ is the amount of pollutant k in area j and $G_{k,critical}$ is the critical amount of pollutant k that is reported by the Air Pollution Control District (APCD) and there is thus a fixed and certain amount for each pollutant. $G_{k,j}$ is equal in equilibrium conditions in all over the world; however, this amount is different in real conditions for different areas.

Thus, all processes are scaled continuously between 0 and 1 relatively according to the pollutants they produce and in relation to the area which is located in. As the scale 0 for the member $C_{i,j,k}$, i.e. the process i located in area j, is not biocompatible or safe at all in producing pollutant k, while scale one for member $C_{i,j,k}$, i.e. process i located in area j, is completely biocompatible or safe in producing pollutant k and a scale between 0 and 1 for member $C_{i,j,k}$, i.e. process i located in area j, is partially biocompatible or safe in producing pollutant k.

The worst conditions are considered in this research; therefore, the minimum fuzzy logic rule is used. The most dangerous process i in producing pollutant k in area j is the process that has the least biocompatibility or safety in producing pollutant k in area j. As a result, $C_{i,j,k,critical}$ is obtained from the fuzzy minimum rule like equation 8.

$$C_{i,j,k,\text{critical}} = \text{Min}\{C_{i,j,k}\} \tag{8}$$

Perhaps, it seems that monitoring of pollutants only through the set of processes is confusing. For instance, areas having advanced technologies have higher degrees of biocompatibility or safety because the normalized amount of the pollutant is low; but the total amount of the production of process at a unit of time may be so high that it practically includes a major part of the pollutant production. It is noteworthy that if the amount of the product in areas with advanced technologies reduces, with which technology would the shortage caused by the reduction in production be produced? If the involved technology is not as advanced as the intended technology, the amount of the pollutant will increase more than the past in equilibrium conditions. Hence, the total product production amount variable in an area will be placed in the second statement of equation 7 over time.

4.2 A case study on monitoring pollutants in industries

In this case study four industries including one, iron and steel production industry; second, aluminum production industry; third, zinc production industry and fourth, lead production industry were studied. It means that set M has four members (i=4). In addition, the set of countries was studied as the set of areas that included 4 members (j=4), a country in the Middle East, one in far east, one in Europe and one in America continent.

4.2.1 Monitoring pollutants in iron and steel production industry

The main pollutants in iron and steel making industries are carbon dioxide in the air, lead in the earth and sewage, nickel in the earth, cadmium in sewage and sulfur in water [Vahdat S.A. and Tohidi N., 2009]. Due to the shortage of data, the air pollutant, carbon dioxide, was only studied. It means that there is only one member (k=1) in the set of pollutants related to iron and steel production industries.

4.2.2 Monitoring pollutants in aluminium production industry

The main pollutants in aluminum production industry are carbon dioxide in the air, methane hexafluoride, methane tetra fluoride in the state of gaseous and dust, sulfur dioxide, nitrogen dioxide and mercury [International Aluminum Association, 2005]. Due to the shortage of data, the air pollutant, carbon dioxide, was only studied. It means that the set of pollutants related to aluminum production industry has only one member (k=1).

4.2.3 Monitoring pollutants in zinc production industry

The main pollutants of zinc production industry include carbon dioxide in the air, sulfur compounds and nitrogen oxides [Siegmund et al, 2010]. Due to the shortage of data, the air pollutant, carbon dioxide, was only studied. It means that the set of pollutants related to zinc production industry has only one member (k=1).

4.2.4 Monitoring pollutants in lead production industry

The main pollutants of lead production industry include carbon dioxide in the air, lead in the water and air, sulfur compounds and nitrogen oxides (International Lead Association, 2005; Siegmund et al, 2010). Due to the shortage of data, the air pollutant, carbon dioxide, was only studied. It means that there is only one member (k=1) in the set of pollutants related to lead production industry.

As the data related to the amount of pollutant in each area were not available, their amounts in different areas were considered to be the same and equal to its maximum amount in the world, i.e. 360 PPM. It means that its effect on the degree of safety was ignored.

$M = \{i \in N, i \in [1, 2, 3, 4]\}$ = {Iron and steel production, aluminum production, zinc production, lead production}

$A = \{j \in N, j \in [1,2,3,4]$, a country in America continent, a country in Europe, a country in the Middle East, one in the Far East }

$P = \{k \in N, k \in [1]\}$ = {carbon dioxide}

So, according to the equations (3) to (6) and presented data in the tables (1) and (2), we have:

$P_{1,1,1,Normalized}$= Carbon dioxide in iron and steel production industry in the Middle East country =2.5

$P_{1,2,1,Normalized}$=Carbon dioxide in iron and steel production industry in the Far East country =3.1

$P_{1,3,1,Normalized}$= Carbon dioxide in iron and steel production industry in the Western Europe country =1.5

$P_{2,4,1,Normalized}$= Carbon dioxide in aluminum production industry in an American country =3.4

$P_{3,4,1,Normalized}$= Carbon dioxide in zinc production industry in an American country =0.19

$P_{4,4,1,Normalized}$= Carbon dioxide in lead production industry in an American country =0.26

Description	Average of carbon dioxide production in ton for the production of one ton product $P_{i,j,k,Normalized}$	Current amount of carbon dioxide in the area
Iron and steel industry in the Middle East country	2.5	360 PPM
Iron and steel industry in the Far East country	3.1	
Iron and steel industry in the Europe country	1.5	
Aluminum industry in a country in Europe	1.6	
Aluminum industry in a country in America	3.4	
Zinc industry in a country in America continent	0.19	
Lead industry in a country in America continent	0.26	

Table 1. The amount of pollutant in ton for the production of one ton of product for monitoring pollutants in industry in 2005 [America Environmental Protection Agency, 2005; International Aluminum Association, 2005; International Lead Association, 2005; Geir, 2005; Siegmund et al, 2010].

Description	Allowed amount in PPM
carbon dioxide $G_{1,critical}$	300

Table 2. The allowed amount of air pollutant in PPM in 2005 [America Air Pollution Control San Diego, 2005; Geir, 2005].

Having inserted the above figures in equation seven, we have:

$$C_{1,1,1} = \frac{0.19}{2.5} \times \left(1 - \frac{360}{300}\right) = -0.01520 < 0 \Rightarrow C_{1,1,1} = 0$$

$$C_{1,2,1} = \frac{0.19}{3.1} \times \left(1 - \frac{360}{300}\right) = -0.01226 < 0 \Rightarrow C_{1,2,1} = 0$$

$$C_{1,3,1} = \frac{0.19}{1.5} \times \left(1 - \frac{360}{300}\right) = -0.02533 < 0 \Rightarrow C_{1,3,1} = 0$$

$$C_{2,4,1} = \frac{0.19}{3.4} \times \left(1 - \frac{360}{300}\right) = -0.01118 < 0 \Rightarrow C_{2,4,1} = 0$$

$$C_{3,4,1} = \frac{0.19}{0.19} \times \left(1 - \frac{360}{300}\right) = -0.20000 < 0 \Rightarrow C_{3,4,1} = 0$$

$$C_{4,4,1} = \frac{0.19}{0.26} \times \left(1 - \frac{360}{300}\right) = -0.06333 < 0 \Rightarrow C_{4,4,1} = 0$$

According to equation 8 we have:

$$C = \{0,0,0,0,0,0\} = 0$$

$$\Rightarrow C_{1,1,1} = C_{1,2,1} = C_{1,3,1} = C_{2,4,1} = C_{3,4,1} = C_{4,4,1} = 0$$

In fact, the maximum amount of carbon dioxide in the world is higher than the allowed amount. Therefore, producing carbon dioxide pollutant even at a very small amount is practically dangerous. It means that the desirability degree of all industries in all areas of the world is zero for the production of carbon dioxide pollutant which means that it is not desirable at all.
Actually, it is not possible to stop producing the mentioned products; steel, aluminum, zinc and lead because human's dependence on these products is currently inevitable. As a result, we can investigate which industry in which area in the production of carbon dioxide has a smaller role. Equation seven will thus change as follows for this particular case.

$$\text{If } C_{i,j,k} > 0 \text{ then } C_{i,j,k} = \left(\frac{\text{Min}\{P_{i,j,k,\text{Normalized}}\}}{P_{i,j,k,\text{Normalized}}}\right) \text{ else } C_{i,j,k} = 0$$

By replacing the data, we have:

$$C_{1,1,1} = \frac{0.19}{2.5} = 0.07600$$

$$C_{1,2,1} = \frac{0.19}{3.1} = 0.06129$$

$$C_{1,3,1} = \frac{0.19}{1.5} = 0.12667$$

$$C_{2,4,1} = \frac{0.19}{3.4} = 0.05589$$

$$C_{3,4,1} = \frac{0.19}{0.19} = 1.00000$$

$$C_{4,4,1} = \frac{0.19}{0.26} = 0.73077$$

According to equation 8:

$$C_{i,j,k,\text{critical}} = \text{Min}\{0.07600, 0.06129, 0.12667, 0.05589, 1, 0.73077\} = 0.05589$$

$$\Rightarrow C_{2,4,1} = 0.05589$$

5. Discussion and conclusion

At present, the amount of carbon dioxide in the world is higher than the allowed amount. Therefore, producing carbon dioxide pollutant even at a very small amount is practically dangerous. So, the safety degree of all industries in all over the world for producing carbon dioxide is zero i.e. it is not desirable at all. But, it is not possible to stop producing steel and aluminum, because at present human's dependence on these products is inevitable. As a result, we can investigate which industry in which area is in safer conditions of producing carbon dioxide pollutant.
In the set of aluminum, copper, zinc and steel industries and in the areas set, a country in the Middle East, a country in the Far East, a country in the Western Europe and a country in the America continent, Aluminum production industry located in a country in America has

the most critical conditions in producing air pollutant, carbon dioxide, with a safety degree of 0.05589. Hence, further focus should be done on controlling carbon dioxide pollutant in aluminum industry in a country in America. Thus, other areas and industries that are not safe in producing carbon dioxide, according to their significance are:

- $C_{1,2,1}$ i.e. iron and steel production industry in a country in the Far East with a safety degree of 0.06129
- $C_{1,1,1}$ i.e. iron and steel production industry in a country in the Middle East with a safety degree of 0.07600
- $C_{1,3,1}$ i.e. iron and steel production industry in a country in the Western Europe with a safety degree of 0.12667
- $C_{4,4,1}$ i.e. lead production industry in a country in the America continent with a safety degree of 0.73077
- $C_{3,4,1}$ i.e. zinc production industry in a country in the America continent with a safety degree of 1

6. Suggestions

Due to the insufficient data in this research, a limited number of industries and countries as well as a limited number of pollutants were only studied. Increasing the number of the members of the set of pollutants, industries and areas, monitoring will be carried out in a wider range and will close to the real conditions.

7. Acknowledgement

We would like to appreciate Pr. Naser Towhidi's valuable and scientific directions.

8. Variables list

$P_{i,j,k,Normalized}$, the amount of pollutant k that is produced for each unit of product of process i in area j.
$G_{k,j}$, the amount of pollutant k in area j
$G_{k,critical}$, the critical amount of pollutant k that is normalized.
$C_{i,j,k}$, the safety degree of process i in producing pollutant k in area j.

9. References

Ali Y.M. and Zhang L.C. (1997). Estimation of residual stresses induced by grinding using a fuzzy logic approach, *Journal of Material processing Technology*, Vol. 63, pp. 875-880.

Ameri R. (2003). *Introduction to Fuzzy Mathematics*, Mazandaran University publications, 401 Pages.

Astel A. (2007). Chemometrics based on Fuzzy logic principles in environmental studies, *Talanta*, Vol. 72, Issue 1, pp. 1-12.

Chen P.H., J.-H. Lai and C.-T. Lin. (1998). Application of Fuzzy control to a road tunnel ventilation system, *Fuzzy Sets and Systems*, Vol. 100, Issues 1-3, pp. 9-28.

Ghafari A., Maqsoodpoor A., Poormomtaz A. & Qasimi J. (1998). *Fuzzy thinking*, Khage Nasir Toosi University Press, 343 Pages.

Geir O. Braathen, An overview of the 2005. 2005. Antarctic Ozone Hole, WMO Global Ozone Research and Monitoring Project, Report No. 49, WMO TD No. 1312, WORLD METEOROLOGICAL ORGANIZATION, pages 76.

Icaga Y. (2007). Fuzzy evaluation of water quality classification, *Ecological Indicators*, Vol. 7, Issue 3, pp. 710-718.

Mintz R., Young B. R.& Svrcek W. Y. (2005). Fuzzy logic modeling of surface ozone concentrations, *Computers & Chemical Engineer*ing, Vol. 29, Issue 10, pp. 2049-2059.

Pokrovsky O.M., Kwok R. H. F.& Ng C. N. (2002). Fuzzy logic approach for description of meteorological impacts on urban air pollution species, A Hong Kong case study Original Research Article, *Computers & Geosciences*, Vol. 28, Issue 1, pp. 119-127.

Sowlat M.H., Gharibi H., Yunesian M., Tayefeh Mahmoudi M.& Lotfi S. (2011). A novel, fuzzy-based air quality index (FAQI) for air quality assessment, *Atmospheric Environment*, Vol. 45, Issue 12, pp. 2050-2059.

Taheri S. M. (1999). *Introduction to Fuzzy Sets* (3rd Ed.), Ferdowsi University of Mashhad Jihad publications, 212 pages.

Vahdat S.A. and Towhidi N. (2009). Sustainable development of iron and steel production in Iran of the environmental aspect by using fuzzy logic, *Journal of Environmental Studies*, Vol.51, pp. 111-122.

Zadeh L. A. (1965). Fuzzy sets, *Information and Control*, Vol. 8, Issue 3, pp. 338-353.

Siegmund A., L. Centomo, C. Greenen, N. Piret, G. Richards and R. Stephens. 2010. *Lead Zinc 2010 conference book*, A john & Sons Inc., MetSoc, COM2010, 1287 pages. ISBN 978-0-47094-315-1

America Environmental Protection Agency, 2005, available from: http://epa.gov/air/emissions/index.htm

America Air Pollution Control San Diego, 2005, available from: http://www.sdapcd.org

International Aluminum Association, 2005, available from: http://www.world-aluminium.org/ Sustainability/Environmental+Issues

International Lead Association, 2005, available from: www.lead.org

3

Indoor Nitrogen Oxides

Silvia Vilčeková
Technical University of Košice
Slovakia

1. Introduction

Nitrogen monoxide (NO) and nitrogen dioxide (NO_2) are gases produced by high-temperature combustion through reactions between nitrogen and oxygen. Both nitrogen in fuel and nitrogen in the air can participate in the reactions. Formation of NO can be followed by further oxidation into NO_2 in the atmosphere. NO_2 is the most toxic of the nitrogen oxides (grouped generically as NO_x) and it is very important from the health point of view. Emissions of nitrogen oxides have generally increased, a fact which has resulted in a lot of studies dealing with nitrogen oxides exposure, particularly over the last 15 years.

Measurements of NO_2 were carried out in residential and industrial premises over 2-day periods both in winter and summer. Personal exposures were found to vary substantially among the 43 respondents, ranging from 10.2 - 87.5 ppb with a mean of 43.7 ± 16 ppb in winter and 23.6 ± 7.8 ppb in summer. The results also showed that the season of the year, house size and outdoor occupation could significantly influence exposure as well as factors such as the type of fuel used and the ambient air quality level. Personal exposure to NO_2 was significantly greater in winter than in summer, and respondents living in smaller houses and/or having outdoor occupations were exposed to higher NO_2 levels (Kulkarni & Patil, 1998.). Another study was focused on measurements of nitrous acid (HONO) and NO_2 concentrations in a total of 119 houses. Indoor samplers were deployed in the living rooms of the houses for 6 days and outdoor samplers were placed near the houses. Average indoor HONO and NO_2 concentrations were higher than outdoor levels. The results suggest that both HONO and NO_2 are generated by indoor sources (Lee at al., 1999). Subsequent surveys focused on characterisation of the air quality in buildings and the investigation of any seasonal changes in the air quality. There was some indication of much higher concentrations of NO and NO_2 on the ground floor of the buildings and there was some evidence of a seasonal gradient when the concentrations of air contaminants obtained during the winter and summer survey were compared (Nayebzadeh et al., 1999). In one study carried out in Hong Kong (Chao & Law, 2000) a group of 60 people was selected for a study of their personal exposure to NO_2. All the participants spent most of their time at home and in an air-conditioned office environment. It was found that the average personal exposure to NO_2 was 46.0 μg/m^3, which was quite close to the average level of 47.3 μg/m^3 of the 12 participants involved in the measurement exercises in their home environment. The 12 residential premises were studied for NO_2 levels in different locations in both the indoor environment and the surrounding outdoor environment. It was found that the average NO_2 level in the indoor environment (living room, bedroom and kitchen) was 55.2

μg/m³ while the corresponding outdoor NO_2 level was 71.8 mg/m³. The study indicated that cooking activities in the kitchen had a strong impact on its NO_2 level; while cooking was going on, the average indoor NO_2 level was 59.7 μg/m³ but when it wasn't, the NO_2 level was 41.8 μg/m³. The NO_2 level in an air-conditioned office was low and the impact from other outdoor sources was not significant due to the small amount of time spent in those micro-environments. In another study researchers determined that NO_2 concentrations were higher during the heating season. NO_2 was measured in 77 homes, using Palmes tubes placed in the kitchen, living room, and outdoors. The mean outdoor NO_2 concentration for all measurements was 19 ppb, with levels significantly higher in the heating season (21 ppb vs. 17 ppb). Indoor concentrations were higher than outdoor concentrations. The kitchen concentrations were the highest with a maximum concentration of 85 ppb, suggesting the gas stove as the major source of indoor NO_2. During the heating season, 40% of the measurements made in kitchens exceeded the Environmental Protection Agency's annual average National Ambient Air Quality Standard level (Zota et al., 2005). The effectiveness of an under-floor air distribution system in controlling thermal comfort and indoor air quality in an office building was evaluated. The indoor values of NO_2 concentrations did not exceed the recommended threshold limits in buildings with under-floor air distribution systems (Yusof et al., 2006). Several studies have focused on comparison of indoor and outdoor NO_x concentrations. According to one study (Lawrence et al., 2005) the outdoor concentrations of NO_x were higher than indoor concentrations measured for an 8-h period in a day and carried out in three environments, i.e. rural, urban and roadside. The full day variations of NO_x were in an urban location. Also, for indoor concentrations two peaks dominate, one in the morning and the other during the evening/night, during which time activities such as cooking take place. For outdoor concentrations there are no such dominating peaks but regular small peaks were found which are believed to be due to the use of heavy diesel generators for supplying electricity. Other research has demonstrated that mean levels of NO_2 in houses with gas cookers are appreciably greater than levels in houses with electric cookers. The difference was shown to be greater for the kitchen, although it was also noticeably different for levels in living rooms and bedrooms. The effect of outdoor sources may be more noticeable in homes with electric cookers, as there are no obvious indoor sources of nitrogen dioxide. Furthermore, the mean levels for single and double-glazed homes have been compared to see whether window type affected indoor levels of NO_2. The results showed that in the kitchen the monitored levels in double-glazed homes were higher, although the difference was not statistically significant (Cotteril & Kingham, 1997). The NO_2 concentration in an office was measured during a 5-month period to study the horizontal flow of pollutants across the building and the indoor-outdoor ratio (IO) with respect to outdoor concentrations. The horizontal distribution pattern of measured NO_2 varied across the building. The IO ratio for NO_2 was <1 across the building demonstrating the influence of outdoor contaminants on indoor air quality (Katiyar & Khare, 2007). The study (Šenitková et. al, 2000) suggested that higher indoor nitrogen oxides concentrations occur mainly through energy conservation with its concomitant tendency to decrease the air exchange rate. The highest concentrations of nitrogen oxides were achieved during gas burning in the kitchen and showed dependence on the intensity of gas burning, the amount of gas burned and on the layout of the room(s). In another study (Kajtár & Leitner, 2007) NO_x distribution is described during the use of a gas cooker in the kitchen by computational fluid dynamics simulation. The aim of the study was to estimate the exhaust ventilation rate necessary to provide the required indoor air quality. Another paper (Kornartit et al., 2010) reports on the measurement and analysis of indoor and outdoor NO_2 concentrations and their comparison with measured personal exposure in various microenvironments during winter and summer. In winter, indoor NO_2 concentrations were found to be strongly correlated with personal exposure levels. The NO_2 concentration in houses where a gas cooker was used was higher during winter in all rooms than in houses with an electric cooker, whereas there was no significant difference noticed in summer. The average NO_2 levels in kitchens with a gas cooker were twice as high as those with an electric cooker, with no significant difference in the summer period. A time-weighted average personal exposure was calculated and compared with measured personal exposures in various indoor microenvironments. Interestingly, NO_2 personal exposure levels were found to be higher during summer than in winter.

In the present work, indoor NO_x concentrations were measured in order to understand their contribution to indoor sources and to develop a mathematical model for predicting the occurrence of indoor nitrogen oxides depending on the intensities of combustion through the use of gas stoves.

2. Formation and sources of nitrogen oxides

Nitrogen compounds form during combustion. In fact, the atmosphere itself is the source of much of the nitrogen leading to the formation of nitrogen oxides (NO_x). Molecular nitrogen (N_2) makes up most of the gases in the earth's atmosphere (79% by volume). Because N_2 is relatively nonreactive under most atmospheric conditions, it seldom enters into chemical reactions, but under pressure and at very high temperatures, it will react with O_2:

$$N_2 + O_2 \rightarrow 2NO \tag{1}$$

Approximately, 90–95% of the nitrogen oxides generated in combustion processes are in the form of nitric oxide (NO), but other nitrogen oxides can form, especially nitrogen dioxide (NO_2), so air pollution experts refer to NO and NO_2 collectively as NO_x. In fact, in the atmosphere the emitted NO is quickly converted photo-chemically into NO_2.

$$2NO + O_2 \rightarrow 2NO_2 \tag{2}$$

$$NO_2 + h\upsilon \rightarrow NO + O \tag{3}$$

$$O + O_2 + M \rightarrow O_3 + M \tag{4}$$

$$NO + O_3 \rightarrow NO_2 + O_2 \tag{5}$$

Equation (3) shows the photochemical dissociation of NO_2. Equation (4) shows the formation of ozone from the combination of O and molecular O_2 where M is any third-body molecule (principally N_2 and O_2 in the atmosphere). Equation (5) shows the oxidation of NO by O_3 to form NO_2 and molecular oxygen. These reactions represent a cyclic pathway driven by photons represented by $h\upsilon$.

Nitric oxide is a colourless, odourless gas and is essentially insoluble in water. Nitrogen dioxide has a pungent acid odour and is somewhat soluble in water. At low temperatures such as those often present in the ambient atmosphere, NO_2 can form the molecule NO_2–O_2N or simply N_2O_4 that consists of two identical simpler NO_2 molecules. This is known as a dimer. The dimer N_2O_4 is distinctly reddish-brown and contributes to the brown haze that is often associated with photochemical smog.

Both NO and NO_2 are harmful and toxic to humans, although atmospheric concentrations of nitrogen oxides are usually well below the concentrations expected to lead to adverse health effects. The low concentrations are due to the moderately rapid reactions that occur when NO and NO_2 are emitted into the atmosphere. Much of the concern about regulating NO_x emissions stems from the wish to suppress the reactions in the atmosphere that generate the highly reactive molecule ozone (O_3). Nitrogen oxides play key roles in important reactants in O_3 formation. Ozone forms photo-chemically (i.e. the reaction is caused or accelerated by light energy) in the lowest level of the atmosphere, known as the troposphere, where people live. Nitrogen dioxide is the principal gas responsible for absorbing the sunlight needed for these photochemical reactions. So, in the presence of sunlight, the NO_2 that forms from NO incrementally stimulates photochemical smog-forming reactions because nitrogen dioxide is very efficient at absorbing sunlight in the ultraviolet portion of its spectrum. This is why ozone episodes are more common during the summer and in areas with ample sunlight.

High temperature/high pressure conditions exist in internal combustion engines, like those in automobiles (known as "mobile sources"), which is why NO_x is one of the major mobile source air pollutants. These conditions of high temperature and pressure can also exist in boilers such as those in power plants, so NO_x is also commonly found in high concentrations leaving fossil fuel power generating stations. In addition to atmospheric nitrogen, other sources exist, particularly the nitrogen in fossil fuels.

The nitrogen oxides generated from atmospheric nitrogen are known as "thermal NO_x" since they form at high temperatures, such as near burner flames in combustion chambers. Nitrogen oxides that form from fuel or feedstock are called "fuel NO_x." A significant fraction of fuel nitrogen remains in the bottom ash or in unburned aerosols in the gases leaving the combustion chamber, i.e. the fly ash (Vallero, 2008). In the low-oxygen area of flames, NO can be formed via fuel radicals with nitrogen - so called "prompt NO_x" (Baumbach, 1996).

Indoor nitrogen oxides air pollution can be generated by outdoor or indoor sources. The primary indoor sources of nitrogen oxides are unvented fuel burning appliances (gas ranges, kerosene space heaters etc.), heating appliances and tobacco smoking. Increased levels of nitrogen oxides can occur in homes with gas cookers, paraffin heaters and certain portable bottled gas heaters as well as with badly maintained central heating boilers or gas fires. The rate of production is governed by three main factors: the amount of oxygen, the flame temperature, and the rate of cooling of the combustion products (since NO formation and conversion proceed more rapidly at higher temperatures). The average level in homes without combustion appliances is about half of that outdoor; in homes with kerosene heaters or unvented gas space heaters, indoor levels often exceed outdoor levels. Immediate concentrations of NO_x, occurring during cooking are still more dangerous, however, mainly during winter, when ventilation is insufficient.

3. Measurement of nitrogen oxides

Measurements of nitrogen oxides were carried out in order to determine indoor and outdoor NO_x concentrations during the year, as well as the influence of air exchange and air tightness of building structures on indoor NO_x occurrence depending on the indoor sources of NO_x. The significant part of measurements of NO_x concentrations focused on the dynamics of NO_x concentrations changes during the combustion process for various rates of gas burning in modelled conditions for the purpose of determining relevant parameters influencing NO_x occurrence. The outcome of the modelled measurements was the development of a mathematical model for predicting the occurrence of indoor nitrogen oxides.

3.1 Method

The investigation included online gas measurement for NO, NO_2 and NO_x by the 'Thermo Environmental Instruments Inc. - model 42 - Chemiluminiscence NO-NO_2-NO_x Analyzer'. This analyzer was located in the middle of the room, at the level of the breathing zone of a sitting person - 105 cm above the floor. The results showed concentration peaks during cooking, but with large differences resulting from different rates of gas burning.

The principal method used for measuring NO_2 is based on chemiluminescence (6), (7):

$$NO + O_3 \rightarrow NO_2^* + O_2 \qquad (6)$$

$$NO_2^* \rightarrow NO_2 + h\upsilon \qquad (7)$$

where $h\upsilon$ represents a photon of solar radiation of energy and NO_2^* is the NO_2 molecule in the excited state.

NO_2 concentrations are determined indirectly from the difference between the NO and NO_x (NO + NO_2) concentrations in the atmosphere. These concentrations are determined by measuring the light emitted from the chemiluminescent reaction of NO with O_3 (similar to the reaction of O_3 with ethylene noted for the measurement of O_3), except that O_3 is supplied at a high constant concentration, and the light output is proportional to the concentration of NO present in the ambient air stream (Vallero, 2008).

3.2 Indoor and outdoor NO_x occurrence during the year

Indoor sciences have been studied over the last twenty years at the Civil Engineering Faculty of the Technical University of Košice in Slovakia. Initially the monitoring of indoor pollutants (radon, particles, VOCs, odours, nitrogen oxides, etc.) was undertaken in buildings located in the east part of Slovakia. Seasonal differences in pollutant levels were confirmed. It was shown in this study that indoor NO_x concentrations were significantly higher than outdoor ones especially in the winter (Fig. 1.). All buildings studied used gas appliances, mainly in the kitchens (Šenitková, 2000).

3.3 Influence of air exchange on NO_x concentrations

The effect of air exchange by ventilation on indoor nitrogen dioxide occurrence was studied in buildings with natural and forced air replacement during and after the combustion process. Monitoring of NO_2 concentrations was carried out during the use of a gas stove with gas consumption from 0.2 to 0.4 m^3 for 30 minutes. The increase and decrease of NO_2 concentrations were monitored depending on the air exchange in the room. The measurements were performed with exhaust ventilation turned both off and on and at different adjustments of air exhaust. The maximum concentration of NO_2 with natural air exchange was 321.1 $\mu g/m^3$, and 185.9 $\mu g/m^3$ with an air exhaust rate of 130 m^3/h, i.e. the air

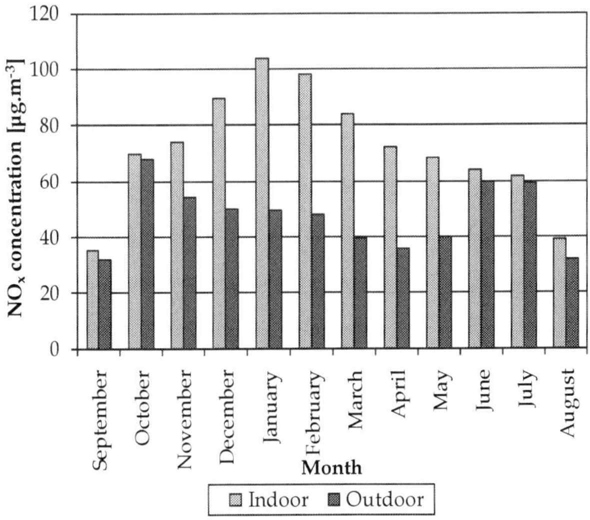

Fig. 1. Mean indoor and outdoor NO_x concentrations throughout the year

exchange rate was about 4 times per hour. Before the start of the combustion process, background concentrations of NO_2 ranging from 26.6 to 38.3 µg/m^3 were determined, as were concentrations of NO_2 at the natural ventilation of 87.8 µg/m^3 for 35 minutes after burning. In contrast, the value of NO_2 concentration reached only 20.3 µg/m^3 with forced air replacement. The dynamics of the increase and decrease of NO_2 concentrations monitored in the room with natural and forced ventilation is evident from Fig. 2.

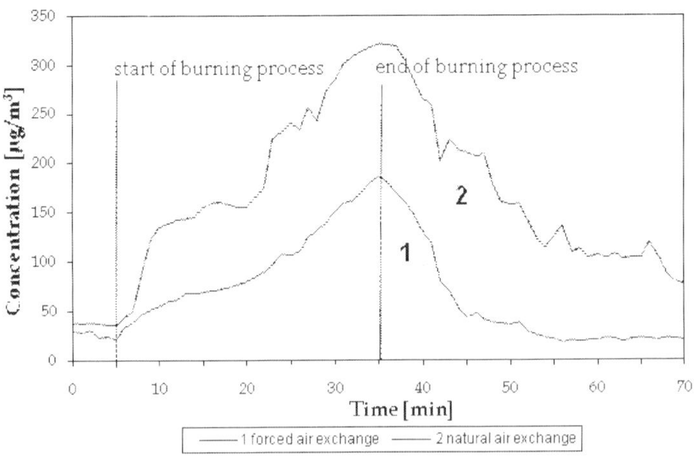

Fig. 2. NO_2 concentrations depending on air exchange

In rooms with a similar layout and volume, higher values of NO_2 concentrations were reached, especially when the combustion process did not provide sufficient air exchange (insufficient natural ventilation by infiltration). NO_2 concentrations exceeded the Slovak safety limit value of 200 μg/m³ during and after the combustion process. The measurements of NO_2 concentrations have confirmed that the level of nitrogen oxides indoors depends not only on the rate of combustion and on the volume of the room but also on the manner and rate of air exchange, and in particular on the specific distribution conditions.

3.4 Influence of building structures on NO_x concentrations

In order to verify the influence of air tightness of openings, NO_2 concentrations were measured in kitchens with both wooden and plastic windows. The gap permeability coefficient for wooden windows was $i_{lv} = 1.4 \times 10^{-4}$ m²/(s.Pa^{0,67}), the determined air exchange, n = 1.8. The gap permeability coefficient for plastic windows as determined by the manufacturer was $i_{lv} = 0.02 \times 10^{-4} - 0.4 \times 10^{-4}$ m²/(s.Pa^{0,67}), the determined air exchange, n = 0.2. Measurements of NO_2 concentrations in buildings with wooden and plastic windows were made with other parameters (rates of gas burning, time and burning conditions etc.) equal. Outdoor NO_2 concentrations, depending on the time of measurement and diffusion conditions, varied from 17.3 μg/m³ to 36.1 μg/m³. The increase of NO_2 concentrations was monitored, depending on the intensity of gas burning, for sixty minutes and the decrease of NO_2 concentrations for ninety minutes. The influence of tightness of windows on the dynamics of NO_2 concentration changes for the intensity of gas combustion from 0.1 to 0.2 m³/h is shown in Fig. 3.

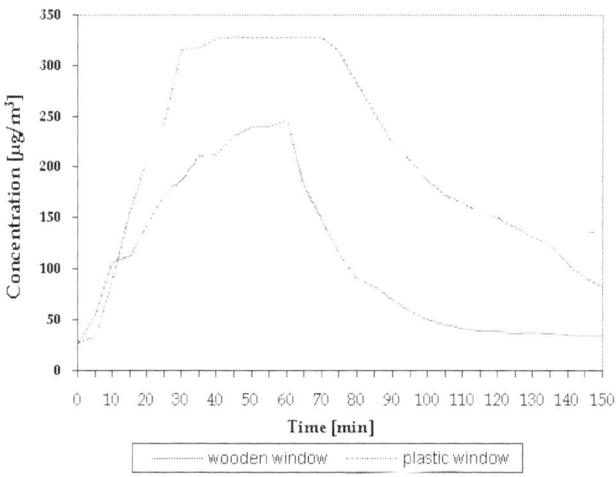

Fig. 3. Dynamics of NO_2 concentration changes

In kitchens with wooden windows a maximum NO_2 concentration of 246.1 μg/m³ was reached after 60 minutes of gas burning at a consumption rate of 0.1 to 0.2 m³/h. In kitchens with plastic windows, a significantly higher concentration of 315.2 μg/m³ was reached after 30 minutes of gas burning at a consumption rate of 0.1 to 0.2 m³/h. During the combustion process, the NO_2 concentrations increased linearly in both kitchens with wooden and plastic windows. After completion of the combustion process, the concentrations decreased and after 90 minutes a value of 33.7 μg/m³ was reached in kitchens with wooden windows. In kitchens with plastic windows the concentrations increased linearly for only the first 30 minutes, at which point they reached a value of 315.2 μg/m³. Then, for 45 minutes the concentrations steadied and values ranged from 318.7 to 328.4 μg/m³. Concentrations decreased 15 minutes after the end of the combustion process and a value of 82.4 μg/m³ was reached after 90 minutes. The dynamics of the changes in concentrations showed a need for a targeted air exchange during gas stove combustion.

3.5 Indoor NO_x concentrations in modelled conditions

Measurements were made to detail NO_x production and determine the resultant indoor concentrations while gas appliances were being used. The intensity of gas combustion varied from 0.1 $m^3.h^{-1}$ to 0.7 $m^3.h^{-1}$. As kitchens had been identified as the most polluted rooms, measurements were taken there. The levels of background concentrations of NO, NO_2 and NO_x were monitored. The increase and decrease of NO_x concentrations was found at the same time.

Changes in NO_x concentrations for the range of gas consumption intensities were studied. The changes while allowing the combustion process to continue at a maximum rate for 60 min, and for decreasing rates of gas usage, were then followed for 90 min after combustion had stopped. The mean NO_x concentrations for the intensity of gas consumption from 0.1 $m^3.h^{-1}$ to 0.7 $m^3.h^{-1}$ are presented in Table 1.

Before the start of the combustion process the average outdoor i.e. indoor background concentrations of NO_x were determined and mean values from 21.0 to 27.9 μg/m³ were recorded. The combustion process took 60 minutes and a linear increase was recorded for 5 - 60 minutes depending on the intensity of gas burning. After the completion of combustion, mean NO_x concentrations achieved high levels from 1785.9 to 2889.2 μg/m³. 90 minutes after the end of the combustion process, the mean values of NO_x were from 279.1 to 1293.6 μg/m³. The concentrations of NO_x after 1 h of gas combustion were shown to depend on the level of gas consumption ranging from 0.1 $m^3.h^{-1}$ to 0.7 $m^3.h^{-1}$ (Fig. 4).

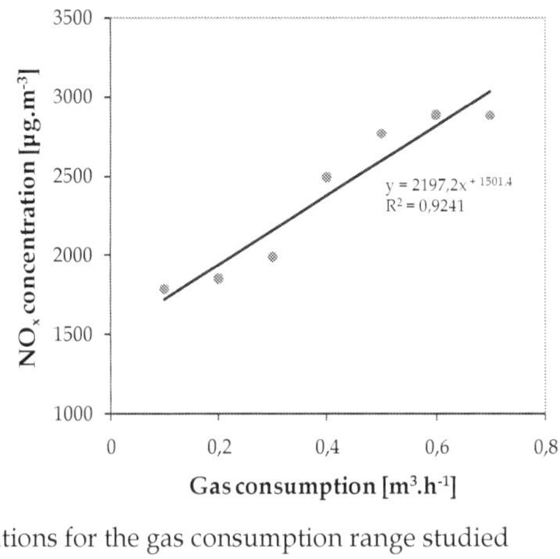

Fig. 4. NO_x concentrations for the gas consumption range studied

Time [min]	Intensity of gas consumption $[m^3.h^{-1}]$						
	0.1	0.2	0.3	0.4	0.5	0.6	0.7
0	**21.0**	**22.3**	**27.9**	**26.3**	**22.6**	**24.8**	**23.3**
5	204.7	429.2	756.8	1043.5	1819.2	3058.7	3071.2
10	489.5	829.8	1521.1	2269.8	2875.0	2886.1	2882.5
15	671.5	1140.4	2076.6	2590.6	2758.0	2887.7	2882.1
20	854.5	1366.1	2272.2	2475.9	2771.1	2900.5	2885.8
25	1013.0	1669.2	2060.8	2496.2	2773.1	2892.8	2885.8
30	1131.8	1910.1	1996.4	2497.1	2771.8	2899.5	2884.0
35	1320.8	2089.8	1994.3	2496.2	2771.8	2894.1	2882.6
40	1477.9	2092.7	1990.9	2494.9	2769.6	2892.6	2883.0
45	1565.1	1899.7	1988.2	2495.3	2770.8	2893.2	2886.2
50	1660.9	1857.3	1991.1	2493.0	2769.9	2888.2	2884.7
55	1711.7	1851.3	1990.6	2492.2	2769.2	2889.6	2883.8
60	**1785.9**	**1852.4**	**1989.0**	**2490.4**	**2769.9**	**2889.2**	**2885.1**
65	1644.4	1857.9	1990.2	2491.6	2769.4	2890.4	2883.9
70	1432.5	1851.0	1989.7	2492.7	2769.8	2890.0	2881.7
75	1282.8	1864.2	1989.0	2496.3	2769.9	2889.6	2882.1
80	1172.5	1937.9	1989.9	2493.2	2772.8	2890.3	2884.7
85	1041.1	1891.0	1987.8	2492.8	2773.6	2890.8	2885.7
90	929.6	1629.7	2028.6	2488.8	2775.4	2890.9	2885.8
95	810.7	1395.7	2142.3	2535.9	2773.1	2890.4	2886.2
100	715.3	1256.4	2003.4	2558.7	2794.8	2882.0	2873.1
105	634.9	1107.5	1759.1	2283.3	2790.3	2940.0	2944.0
110	566.5	994.4	1657.5	2034.3	2622.0	2912.1	2900.9
115	512.8	881.4	1526.9	1807.7	2383.4	2610.6	2627.4
120	468.8	800.7	1425.0	1672.8	2207.8	2404.9	2522.1
125	431.8	705.1	1315.7	1541.4	2009.2	2267.6	2359.9
130	393.5	656.8	1214.2	1420.2	1860.4	2080.5	2159.5
135	362.7	587.5	1126.9	1323.6	1712.4	1859.8	1929.1
140	340.1	538.9	1017.3	1248.1	1538.7	1650.4	1687.8
145	310.3	473.9	920.4	1128.7	1366.9	1469.2	1519.3
150	**279.1**	**441.6**	**848.8**	**1021.9**	**1202.3**	**1284.7**	**1293.6**

Table 1. Mean NO_x concentrations for various gas usage rates

The oxides formation at a gas consumption rate of 0.1 $m^3.h^{-1}$ is given in Fig. 5.
It is evident that an essentially linear increase of NO, NO_2 and NO_x concentrations occurred during the whole combustion period to a final maximum value after 1 h, when gas usage was stopped (NO 1357.4 $\mu g.m^{-3}$, NO_2 428.5 $\mu g.m^{-3}$ and NO_x 1785.9 $\mu g.m^{-3}$). A decrease in concentrations was then observed for a period of 90 min, after which the concentrations were NO 209.8 $\mu g.m^{-3}$, NO_2 69.4 $\mu g.m^{-3}$ and NO_x 279.1 $\mu g.m^{-3}$.
Importantly, it should be noted that the concentration of NO_2 after 1 h of gas combustion (428.5 $\mu g.m^{-3}$) was more than double the Slovak safety limit (200 $\mu g.m^{-3}$). The limit was exceeded 10 min after the start of combustion (270.9 $\mu g.m^{-3}$) and concentrations remained above the limit for the next 100 min (228.6 $\mu g.m^{-3}$).

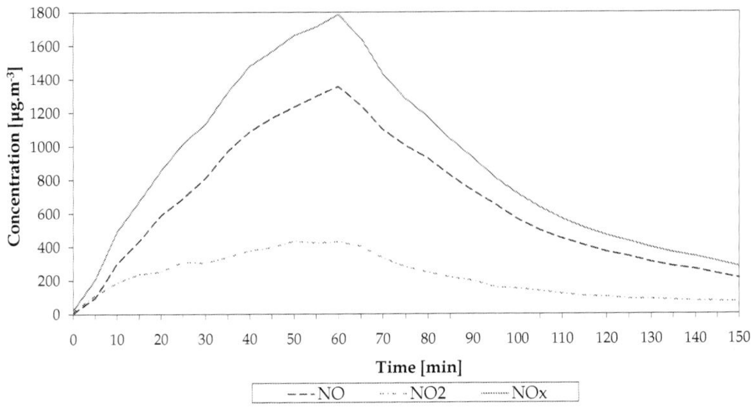

Fig. 5. Nitrogen oxides concentrations at a gas usage rate of 0.1 $m^3.h^{-1}$

Oxides formation at a gas consumption rate of 0.2 $m^3.h^{-1}$ is given in Fig. 6.

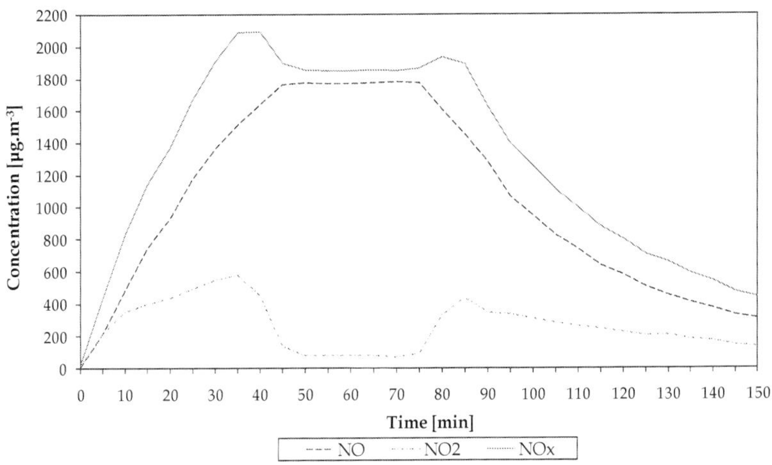

Fig. 6. Nitrogen oxides concentrations at a gas usage rate of 0.2 $m^3.h^{-1}$

The figure shows that a linear increase in concentrations of NO, NO_2 and NO_x was not sustained throughout the combustion process. Concentrations of NO_2 and NO_x rose linearly for 35 minutes, reaching an NO_2 value of 577.2 $\mu g.m^{-3}$ and an NO_x value of 2089.8 $\mu g.m^{-3}$. Subsequently the concentrations decreased and for 25 minutes were almost constant, with NO_2 values ranging from 68.8 to 88.2 $\mu g.m^{-3}$ and NO_x from 1851.0 to 1864.2 $\mu g.m^{-3}$. As a consequence of natural air exchange, an increase of NO_2 concentrations and decrease of NO concentrations occurred 15 minutes after the completion of combustion. 90 minutes after the end of combustion, the concentrations were NO 306.8 μg.m^{-3}, NO_2 134.8 μg.m^{-3}and NO_x 441.6 μg.m^{-3}.

4. Mathematical modelling of indoor nitrogen oxides

Modelling is a powerful tool that is being used extensively by scientists to investigate air movement and airflow patterns and to evaluate the performance of HVAC systems. Numerous models have been developed to study indoor air quality (IAQ). Although designers, architects, code enforcement officials, and other building scientists have been using modelling as a powerful tool to address indoor IAQ issues, more development is needed to increase the accuracy and the validity of the IAQ models. These are being used to predict the dispersion of air contaminants within the building environment. Air pollution concentration can be estimated both spatially and as a function of time. The aim of IAQ models is to provide a way to link information about sources, distribution, dynamic changes of pollutant concentrations and also building factors to estimate indoor pollutant concentrations. Many of the latest indoor air models include exposure modelling and risk modelling. The type of model used depends on the planned uses of the model. The most common uses of IAQ models are the following: (i) estimating population exposure to various indoor pollutants, (ii) estimating the impact of individual sources on pollutant concentration and (iii) estimating the impact of individual sources and IAQ control options on personal exposure (Spengler et al., 2001).

Designing a model and modelling can substantially contribute to problem solving and to obtaining new understanding of system behaviour with respect to air component dynamic changes. A mathematical description of the processes can generalise experimental results so that they may be used to interpret other processes physically similar to those introduced in the model. Although modelling doesn't achieve all the rules of similarity, it does provide overall information about a system's behaviour, which is only a little different from a real situation. Also, although it is only an approximation, the model does simplify description of the real physical measurements.

Most research institutes and universities focus on modelling and predicting outdoor and indoor pollutants. Most of the models developed to predict human exposures use empirical data on indoor/outdoor ratios (I/O) to determine indoor microenvironment concentrations as a function of time. These models are described in various studies. For example, one study (Sexton et al., 1983) describes a simple deterministic model, relating exposure to background ambient levels, indoor values, and human activities. Results suggest that indoor nitrogen dioxide concentrations in private dwellings vary primarily because of outdoor levels and the type of cooking fuel used, but are also affected by factors such as air-exchange rates and strength of indoor sources. Estimates of population exposures are obtained by combining observed distributions of nitrogen dioxide concentrations from outdoor and indoor settings with information about the number of people and amount of time spent in each microenvironment. One group of researchers (Lee et al., 1998) developed regression models for indoor NO_2 exposure. Using analytical methods to choose descriptive proxies and evaluate the tradeoffs in their implementation can help epidemiological studies improve their designs. Other researchers (Axley et al., 1994) developed and applied mathematical models to predict indoor air concentrations resulting from the transport of outdoor gas-phase nitrogen oxides, ozone and nitric acid indoors; and integral expressions that may be used to estimate model parameters from dynamic measurements of contaminant concentration. Another study (Dimitroulopoulou et al., 2001, Ashmore et al., 1999) deals with a modelling of indoor exposure to nitrogen dioxide in the UK. A dynamic multi-compartment computer model was developed to describe the physical processes determining indoor pollutant concentrations as a function of outdoor concentrations, indoor emission rates and building characteristics. The model has been parameterised for typical UK homes and workplaces and linked to a time-activity model to calculate exposures for a representative homemaker, schoolchild and office worker, with respect to NO_2. This model can be used to distinguish three types of exposures: outdoor exposure, indoor exposure resulting from penetration of outdoor air and indoor exposure resulting from indoor sources. The model predicts the above types of personal exposures for a representative homemaker, schoolchild and office worker, based on indicative activity patterns. The deterministic modelling approach applied here combines two types of models: (i) A physical model, used to calculate hourly indoor air pollutant concentrations for different microenvironments as a function of outdoor concentrations, building characteristics and indoor source emissions. This physical model, entitled INTAIR, is a simple dynamic compartment model which solves the resulting set of differential equations using a fourth-order Runge-Kutta scheme. (ii) An exposure model, used to calculate personal exposures by combining the movement of typical individuals through a series of microenvironments with the modelled ME concentrations. Another study (Baxter et al., 2007) seeks to utilize publicly available data and questionnaire responses to predict residential indoor concentrations of traffic-related air pollutants for lower socioeconomic status (SES) urban households. Indoor and outdoor 3-4 day samples of NO_2 were collected in 43 low SES residences across multiple seasons from 2003 to 2005. Home characteristics and occupant behaviours were collected via a standardized questionnaire. Additional housing information was collected through property tax records, and ambient concentrations were collected from a centrally located ambient monitor. The contributions of ambient concentrations, local traffic and indoor sources to indoor concentrations were quantified using regression analyses. Comparing models based on covariate selection using p-values or a Bayesian approach yielded similar results, with traffic density within a 50 m buffer of a home and distance from a truck route as important contributors to indoor levels of NO_2. The Bayesian approach also highlighted the uncertainty in the models.

The difficulty of predicting NO_x emissions during combustion of fossil fuels is caused mostly by the complexity of the mechanisms of NO_x formation, mechanisms which depend on many factors. Some partial mechanisms have not yet been completely clarified so it is not possible to describe them exactly. Consequently, empirical correlations between fuel properties, combustion regime and NO_x emissions cannot be properly calculated. This affects the degree to which calculations can be informative in a mathematical model of indoor NO_x production. The final result has to be assembled from the results obtained from statistically suitable collections of measurements as well as more controlled laboratory measurements.

4.1 Input parameters

The research task to describe NO_x production in the combustion process relies on the measurement of NO_x as influenced by those parameters which directly characterize the combustion process, i.e. intensity of gas combustion I ($m^3.s^{-1}$), exposure period t (s), amount of indoor air m (kg) and air change n (s^{-1}).

All the given variables are defined using basic dimensions, allowing their direct use for dimensional analysis to create a mathematical model of NO_x production. All the variables can be simply measured during combustion allowing us to then compare NO_x production under the particular working conditions of the chosen combustion device on the basis of direct measurements and by means of the created mathematical model described in the following section.

4.2 Dimensional analysis application

Dimensional analysis offers a method for reducing complex physical problems to the simplest form prior to obtaining a quantitative answer. At the heart of dimensional analysis is the concept of similarity. In physical terms, similarity refers to some equivalence between two things or phenomena that are actually different. Mathematically, similarity refers to a transformation of variables that leads to a reduction in the number of independent variables that specify a problem (Sonin, 2001). The behaviour of a physical system is defined by the complete set of dimensionless variables formed by the relevant physical variables. This fact suggests that if two systems have the same numerical values for all the defining dimensionless variables, then these two systems are dimensionally similar. Further, if there are two dimensionally similar systems, then their behaviour can be closely correlated and hence the results of measurements on either one can be "projected" to the other. This is the basis for dimensional modelling. Therefore the first step in a model experiment is to construct a complete set of dimensionless variables relevant to the system. These dimensionless variables are then made to be equal for the model and the prototype (Szirtes & Rozsa, 2006).

The mathematical model describing NO_{xi} formation is based on the formation of dimensionless arguments Π_i, from the stated variables that influence formation of the oxides of nitrogen. The valuable property of these arguments is that in all existing systems of units they have the same numerical size and they have no dimension. The formation of a mathematical model rests on derivation of functional dependence from the expressed dimensionless variables, which in general always has an exponential character. Transformation of this function into logarithmic co-ordinates corresponds to a linear character that makes the work with the model easier and enables one to determine simply the parameters of linear function (Čarnogurská, 1998, 2000).

The model given here has universal validity for all combustion devices that have at least approximately similar geometric characteristics. For every combustion device the parameters of linear function, i.e. regression coefficients, have to be determined separately.

The general relationship between the selected variables which affect NO_{xi} production can be expressed in the form:

$$\varphi\,(I,\ t,\ m,\ NO_{xi},\ n) = 0 \tag{8}$$

The dimensional matrix-relationship created (9) has the rank of a matrix where r = 3 and its lines are dimensionally independent of each other. From n = 5 independent variables a matrix r can be set up with (n-r) dimensionless arguments.

$$\begin{array}{c} \\ s \\ m \\ kg \end{array}\ \begin{array}{ccccc} t & I & m & NO_{xi} & n \\ \hline 1 & -1 & 0 & 0 & -1 \\ 0 & 3 & 0 & -3 & 0 \\ 0 & 0 & 1 & 1 & 0 \end{array} \tag{9}$$

For the general form of the argument п is valid:

$$п = t^{x_1}.I^{x_2}.m^{x_3}.NO_{xi}^{x_4}.n^{x_5} \tag{10}$$

From the condition that the left side of the identity equals one (ordinary number п), the resulting exponent of every basic dimension is equal to zero. By applying this condition for every element from r basic dimensions, we obtain a system of three linear equations with five unknowns. To solve the system, we have to select two unknowns (and this always twice) and then calculate the rest.
For the selected $x_4 = 1$, $x_5 = 0$ the exponents will be calculated from the linear equations system, thus $x_1 = 1$, $x_2 = 1$, $x_3 = -1$.
For the selected $x_4 = 0$, $x_5 = 1$ the exponents will be calculated from the linear equations system, thus $x_1 = 1$, $x_2 = 0$, $x_3 = 0$.
For these the dimensionless argument is then valid:

$$\begin{array}{c|ccccc|} & x_1 & x_2 & x_3 & x_4 & x_5 \\ п_1 & 1 & 1 & -1 & 1 & 0 \\ п_2 & 1 & 0 & 0 & 0 & 1 \end{array} \tag{11}$$

The required dimensional homogeneous function in dimensionless form is

$$\Psi(п_1, п_2) = 0 \tag{12}$$

or after adjustment and backward transformation of dimensions for the particular variables, the function will have the form:

$$\Psi\left(\frac{t.I.NOxi}{m}, t.n\right) = 0 \tag{13}$$

The real course of the dependence of dimensionless arguments $п_1$ to $п_2$ calculated according to the relationship (13) from the measured values.

$$п_1 = A.п_2^B \tag{14}$$

From this a regression straight line for the calculation of the regression coefficients can be calculated by the method of least squares.
After completing relationship (14) using relationship (13) and its modification, we obtain the relationship characterising NO_x production in the form:

$$NOxi = \frac{10^9.A.t^{B-1}.n^B.m}{I} \tag{15}$$

where A and B are regression coefficients calculated by the method of least squares.
The mathematical formula (15) presents a mathematical model of NO_{xi} production through using of gas appliances in the kitchen. The indoor NO_x is given by the formula:

$$NO_x = NO_{xi} + NO_{xo} \tag{16}$$

NO_{xi} - indoor nitrogen oxides concentration [$\mu g.m^{-3}$]
NO_{xo} - outdoor nitrogen oxides concentration [$\mu g.m^{-3}$]

5. Verification of the mathematical model

The mathematical model was verified for various amounts of gas consumption in the range of 0.1 - 0.7 $m^3.h^{-1}$. As an illustration, the verification for gas consumptions of 0.1 $m^3.h^{-1}$ and of 0.2 $m^3.h^{-1}$ using a model experimental room is given below. The amount of air was 28.28 kg and the air change rate 3.56x10-4 s^{-1}. The air change rate was determined on the basis of decreasing NO_x concentration per hour.
On the basis of measured input parameters, NO_{xi} concentrations were determined through the located constants and regression coefficients obtained by the method of least squares. These values are presented in Table 2.

Gas consumption [$m^3.h^{-1}$]	A	B
0.1	4.0000E	1.8420
0.2	2.0000E	1.7765
0.3	2.0000E	1.8179
0.4	3.0000E	1.8593
0.5	4.0000E	1.6603

Table 2. Regression coefficient values

As an illustration, the relationship for NO_{xi} production at a gas consumption rate of 0.1 $m^3.h^{-1}$ is obtained from:

$$NOxi = \frac{4.0E - 09 . t^{1.8420-1} . n^{1.8420} . m}{I} \tag{17}$$

The values of measured and calculated NO_{xi} concentrations are compared in Table 3. The differences between measured and calculated NO_{xi} concentrations are considered acceptable. The deviations reflect two things: the selection of relevant parameters is a simplification; their degree of involvement and the number of factors on which NO_{xi} production depends, which they represent, are both unknown.
The relationship for NO_{xi} production at a gas consumption rate of 0.2 $m^3.h^{-1}$ is obtained from:

$$NOxi = \frac{2.0E - 09 . t^{1.7765-1} . n^{1.7765} . m}{I} \tag{18}$$

The values of NO_{xi} concentrations measured and calculated for consumption of 0.2 $m^3.h^{-1}$ are compared in Table 4. A linear increase of NO_{xi} concentrations continued for forty minutes from the start of combustion, i.e. only this amount of time was required for complete combustion of gas burning.

t [min]	$NO_{xi,measured}$ [μg.m^{-3}]			$NO_{xi,calculated}$ [μg.m^{-3}]	Deviation [%]
	min	max	mean		
5	195.7	219.5	204.7	243.9	20.67
10	426.7	534.5	489.5	420.6	16.34
15	593.4	740.3	671.5	583.2	16.97
20	712.7	992.5	854.5	737.3	19.54
25	884.5	1161.7	1013.0	885.4	18.43
30	999.8	1262.0	1131.8	1028.9	17.1
35	1213.4	1478.5	1320.8	1168.6	13.89
40	1410.5	1566.1	1477.9	1305.1	12.15
45	1499.4	1645.3	1565.1	1439.0	10.77
50	1571.1	1747.4	1660.9	1570.5	10.02
55	1639.6	1776.3	1711.7	1700.0	11.53
60	1762.1	1803.3	1785.9	1827.7	11.25

Table 3. Measured and calculated NO_{xi} concentrations for gas consumption of 0.1 m^3.h^{-1}

t [min]	$NO_{xi,measured}$ [μg.m^{-3}]			$NO_{xi,calculated}$ [μg.m^{-3}]	Deviation [%]
	min	max	mean		
5	316.2	556.0	429.2	421.2	33.89
10	632.6	1016.2	829.8	705.6	24.98
15	1003.8	1290.2	1140.4	958.5	15.16
20	1230.8	1499.1	1366.1	1192.8	13.26
25	1527.1	1779.5	1669.2	1414.3	14.80
30	1785.4	2031.0	1910.1	1625.9	14.54
35	1960.2	2201.3	2089.8	1829.9	12.00
40	2048.6	2151.4	2092.7	2027.3	4.48

Table 4. Measured and calculated NO_{xi} concentrations for gas consumption of 0.2 m^3.h^{-1}

6. Interpretation of mathematical model

Current knowledge as well as the results of monitoring in a selected group of residential buildings in situ confirmed the significant occurrence of indoor nitrogen oxides, especially with the burning of gas in kitchens, and lead to the construction of a mathematical model for predicting the occurrence of nitrogen oxides. Subsequently, a mathematical model based on the dimensional analysis method was verified for each intensity of gas burning. The calculated values of NO_{xi} concentrations were compared with average measured concentrations of NO_{xi}. The variance between the average measured and calculated values of NO_{xi} is shown in Fig. 7. This difference was established as an absolute error which ranged from values of 4.48% - 34.05%. The average absolute error is 17.86%. This value is acceptable, and confirms that the proposed mathematical model is useful for predicting the occurrence of indoor nitrogen oxides.

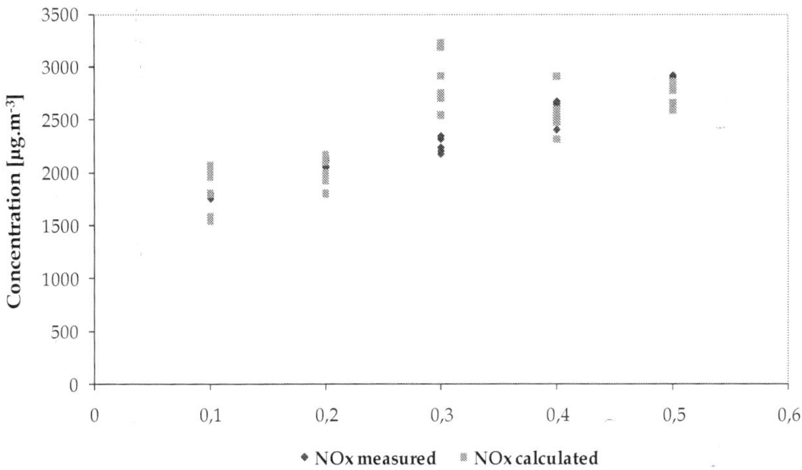

Fig. 7. Variance of measured and calculated NO_{xi} concentrations for various gas intensities

7. Conclusion

The study of indoor pollutants affecting indoor air quality is a modern problem. One solution is the use of mathematical methods to predict the level of pollutant concentrations. This chapter proposes a possible mathematical approach to indoor NO_x modelling. The correct selection and clear determination of all relevant variables at the beginning of construction of the mathematical model was important. The experimental measurements confirmed that NO_x production is mainly affected by intensity of gas combustion, exposure period, and rate of air exchange. A mathematical description of the physical processes influencing NO_x production has been derived and used to predict the occurrence of these gases. The model was verified with satisfactory results for gas consumption in the range of 0.1 - 0.7 $m^3.h^{-1}$. Differences between measured and calculated indoor NO_{xi} concentrations are probably because the selected relevant variables for construction of the mathematical model do not give a complete and fully accurate picture of the combustion processes leading to NO_{xi} production. Results from the mathematical modelling approach can be improved through better understanding of input parameters and precise verification. The dimensional analysis method can be used for modelling and prediction of indoor pollutants. The presented mathematical model can be applicable for indoor nitrogen oxides prediction within the building design process and building operation management.

8. Acknowledgment

The author is grateful to the Slovak Grant Agency for Science for supporting projects on Indoor Air Sciences (No. 1/0352/03 and 1/3342/06).

9. References

Ashmore, M.R., Byrne, M., Dimitroulopoulou, C., Kinnersley, R.: Modelling the Contribution of Indoor and Outdoor sources to Air Pollution Exposures in the UK. (1999). Proceeding of the 8th International Conference on Indoor Air Quality and Climate „Indoor Air 99",ISBN 1 86081 295 3, Edinburg, Scotland, August 1999

Axley, J.W., Hartzell, A.L., Peavey, J.B. (1994). Transport of Reactive Gas-Phase Outdoor Air pollutants Indoors. *Indoor and Built Environment*, Vol.3, No.5, (September 1994), pp. 266-273, ISSN 1420-326X

Baumbach, G. (1996). *Air Quality Control*. Springer, ISBN 3-540-57992-3, Stuttgard, Germany

Baxter, L.K., Clougherty, J.E., Paciorek, Ch.J., Wright, R.J., Levy, J.I. (2007). Predicting residential indoor concentrations of nitrogen dioxide, fine particulate matter, and elemental carbon using questionnaire and geographic information system based data. *Atmospheric Environment*, Vol.41, No.31, (October 2007), pp. 6561–6571, ISSN 1352-2310

Dimitroulopoulou, C., Ashmore, M.R., Byrne, M.A., Kinnersley, R.P. (2001). Modelling of indoor exposure to nitrogen dioxide in the UK. *Atmospheric Environment*, Vol.35, No.2, (November 2001), pp. 269-279, ISSN 1352-2310

Cotterill A., Kingham, S. (1997). Nitrogen Dioxide in the Home: Cooking, Double Glazing or Outdoor Air? Indoor and Built Environment, Vol.6, No.6, (November 1997), pp. 344-349, ISSN 1420-326X

Čarnogurská, M. (1998). *Dimensional analysis and theory of similarity and modelling in practice*. Technical University of Kosice, ISBN 80-7099-386-3, Kosice, Slovakia

Čarnogurská, M. (2000). *Fundamentals of mathematical and physical modelling in mechanics of liquid and thermodynamics*. Vienala, ISBN 80-7099-344-8, Kosice, Slovakia

Chao, Ch.Y.H., Law, A. (2000). A study of personal exposure to nitrogen dioxide using passive samplers. *Building and Environment*, Vol. 35, No.6, (August 2000), pp. 545-553, ISSN 0360-1323

Kajtár L., Leitner, A. (2007). CFD Modelling of Indoor Environment Quality Affected by Gas Stoves. *Proceeding of 9th REHVA World Congress Clima 2007 "WellBeing Indoors"*, ISBN 9789529989836, Helsinki, Finland, June 2007

Katiyar, V., Khare, M. (2007). Horizontal Measurements of SO_2, NO_2 and SPM in Indoor Air and relationship to Outdoor concentration. *Proceeding of 9th REHVA World Congress Clima 2007 "WellBeing Indoors"*, ISBN 9789529989836, Helsinki, Finland, June 2007

Kornartit, C., Sokhi, R.S., Burton, M.A., Ravindra, K. (2010). Activity pattern and personal exposure to nitrogen dioxide in indoor and outdoor microenvironments. *Environment International*, Vol.36, No.1, (January 2010), pp. 36-45, ISSN 0160-4120

Kulkarni, M.M., Patil, R.S. (1998). Factors Influencing Personal Exposure to Nitrogen Dioxide in an Indian Metropolita Region. *Indoor and Built Environment*, Vol.7, No.5-6, (September 1998), pp. 319-322, ISSN 1420-326X

Lawrence, A.J., Masih, A., Tanej, A. (2005). Indoor/outdoor relationships of carbon monoxide and oxides of nitrogen in domestic homes with roadside, urban and rural locations in a central Indian region. *Indoor Air*, Vol.15, No.2, (April 2005), pp. 76-82, ISSN 0905-6947

Lee, K., Levy, J.I., Yanigisawa, Y., Spengler, J.D., Billick, I.H. (1998). The Boston residential nitrogen dioxide characterisation study: classification and prediction of indoor NO_2 exposure. *Journal of Air and Waste Management Association*, Vol.48, No.8, pp. 736-742

Lee, K., Xue, J., Spengler, J.D., Ozkaynak, H., Leaderer, B.P.l. (1999). Residential Nitrous Acid and Nitrogen Dioxide Concentrations. *Proceeding of the 8th International Conference on Indoor Air Quality and Climate "Indoor Air '99"*, ISBN 1 86081 295 3, Edinburg, Scotland, August 1999

Nayebzadeh, A., Cragg-Elkouh, S., Rancy, R., Dufresne, A. (1999). Sources of Indoor Air Contamination on the Ground Floor of a High-Rise Commercial Building. *Indoor and Built Environment*, Vol.8, No.4, (July 1999), pp. 237-245, ISSN 1420-326X

Sexton, K., Letz, R., Spengler, J.D. (1983). Estimating human exposure to nitrogen dioxide: an indoor/outdoor modelling approach. *Environmental Research*, Vol.32, No.1, pp. 151-166.

Sonin, A.A. (2001). *The Physical Basis of dimensional analysis* (2th edition), Department of Mechanical Engineering, USA

Spengler, J.D., Samet, J.M., McCarthy, J.F. (2001). *Indoor Air Quality Handbook*, McGraw-Hill Companies, ISBN 0-07-445549-4, New York, USA

Szirtes, T., Rosa, P. (2006). *Applied Dimensional Analysis and Modeling*, Elsevier, ISBN 0123706203

Šenitková, I. (2000). Indoor Air Quality in Housing Units. *Proceeding of the XXVIII World Congress on Housing: Challenges for the 21st century*, ISBN 6713391, Abu Dhabi, U.A.E., 2000

Vallero, D. (2008). *Fundamentals of Air Pollution* (4th edition), Elsevier, ISBN 978-0-12-373615-4, California, USA

Zota, A., Adamkiewicz, G., Levy, J.I., Spengler, J.D. (2005). Ventilation in public housing: implications for indoor nitrogen dioxide concentrations. *Indoor Air*, Vol.15, No.6, (December 2005), pp. 393-401, ISSN 0905-6947

Yusof, M.Z.M., Leman, A.M., Husain, A., Shah, S.M.R. (2006). The effectiveness of Underfloor Air Distribution (UFAD) System in Controlling Thermal Comfort and Indoor Air Quality. Proceeding of 11th International Conference on Healthy Buildings 2006, ISBN 978-989-95067-0-1, Lisboa, Portugal, June 2006

4

Particularities of Formation and Transport of Arid Aerosol in Central Asia

Galina Zhamsueva[1], Alexander Zayakhanov[1], Vadim Tsydypov[1], Alexander Ayurzhanaev[1], Ayuna Dementeva[1], Dugerjav Oyunchimeg[2] and Dolgorsuren Azzaya[2]
[1]Department of Physical Problems of Buryat Science Center, Siberian Branch of Russian Academy of Sciences, Ulan-Ude,
[2]Institute of Meteorology and Hydrology of Mongolia, Ulaanbaatar,
[1]Russia
[2]Mongolia

1. Introduction

The climate change is a global problem of humanity. During last decades anthropogenic air pollution is a main cause of the content change of the atmosphere. The situation constantly is redoubled by continuing intensive growth of the air pollution in the areas located near of emission sources and also far off territories.

At the present the square of deserts increases in East Asia, and it affects on increasing of dusty days number. The total area of deserts and desertification lands is about 1.653 million km^2 in Northern China [Wang T. & Zhu Z., 2001; Wang S., 2005]. More than 70% of the pastureland area of Mongolia is under desertification [Natsagdorj et al., 2003].

Gobi Desert covers 1/3 of Mongolia territories and is the driest area in the world. The climate of Gobi Desert is dry, cold, and more continental. Soils are poor with humus, containing a lot of gravel.

Dust storm frequently occurs in arid and semi-arid territories of Central Asia. It is known that Gobi desert is one of the major sources of dust storms occurrence in the East Asia. Asian dust storms significantly influence on air quality [Chan et al., 2005; Gillette, 1986; Lee et al, 2006; Wei & Meng, 2006]. At the intense dust storms in Gobi desert the great many of sand and dust are transferred to Eastern China, Korea and Japan. Sometimes many particles of sand and dust are raised up by strong winds and transfer all over the world [Kim & Chung, 2008]. The total dust emission from sources in East Asia is estimated is 10.4×10^6 ton yr^{-1} for PM_{10}, 27.6×10^6 ton yr^{-1} for PM_{30} and 51.3×10^6 ton yr^{-1} for PM_{50}. The total dust emission from Mongolian sources is 2×10^6 ton yr^{-1} for PM_{10}, 2.9×10^6 ton yr^{-1} for PM_{30} and 8.7×10^6 ton yr^{-1} for PM_{50}. It is suggested that Southern Mongolia is an important dust source region [Xuan et al., 2004].

In recent years the intense dust storms are observed not only in Central Asia but also in different parts of the world. For example, strong dust storm was observed in Australia on 23 October 2002. The dust storm was 2400 km long, up to 400 km across and 1.5-2.5 km in height. The plume area was estimated at 840,860 km^2 and was raised 3.35-4.85 million tones of sands and dust with ground [McTainsh et al., 2005].

The researches of dust storms distribution in Mexico City [Jaurequi, 1989], spatial-temporal distribution of dust storm producing by wind erosion in USA [Gillette & Hanson, 1989], sources of dust storms formation and analysis of their far transport in China [Littmann, 1991; Sun et al., 2001; Wang S. et al., 2005; Zhou, 2001], climatology of dust storms in Mongolia [Natsagdorj et al., 2003; Zhadambaa et al., 1967a, 1967b] were carried out. However the detail researches of dust storms in Gobi Desert (Mongolia) didn't carry out to present days.
The purpose of this chapter is a study of analysis of mass and number concentrations of dust aerosol, their chemical compound, meteorological and turbulent characteristics of atmosphere and the mechanism of dust storm formation in Gobi Desert.

2. Sampling and methodology

Complex researches of aerosol and small gas components, meteorological, turbulent and radiative characteristics of atmosphere of arid territories of Mongolia have been begun by Department of physical problems of Buryat science centre of SB Russian Academy of Sciences in 2005 and continue nowadays. Measurements are conducted in the summer expeditionary periods 2005-2010 at Dornogovi aimag (station Sainshand, $44^0 54'$; $110^0 07'$, station Zamyn-Uud, $43^0 44'$; $111^0 54'$) and Sukhbaatar aimag (station Baruun-Urt, $46^0 41'$; $113^0 17'$) [Zhamsueva et al., 2008]. Observations points are situated at the Hydrometeorological centers located on considerable removal from settlements. Round-the-clock measurements of meteorological parameters are carried out by autonomous ultrasonic meteorological complexes «AMK - 03». Autonomous ultrasonic meteorological complexes measures momentary values of three orthogonal component of wind speed, air temperature, atmosphere pressure and air humidity. Momentary values of these parameters are used for calculations of turbulent air characteristics [Azbukin et al., 2009]. Fine aerosol was selected by diffusion spectrometer DSA in size range from 1.6 to 400 nm. Sampling of aerosol particulates with a size less than or equal to 10 μm was carried out on «Whatman-41" filters by high volume sampler PM-10 of General Metal Works Inc.

3. The research of air mass transport in Central Asia

3.1 The particularities of air mass circulation in Mongolia

Synoptical conditions and winds in desert regions are two main factors influenced on transport of sand, dust and atmosphere impurities. It is known that in the arid and semi-arid regions of Mongolia and Northern China located near 40^0 N the west flow of air transports great many of sand and dust particles to different parts of Earth.
For investigations of pathways of the atmospheric impurities transport in arid and semi-arid territories of Mongolia we have constructed forward and backward trajectories of air mass movement using of the reanalysis model NCEP/NCAR HYSPLIT (http://www.arl.noaa.gov/ready/hysplit4.html) [Draxler & Rolph, 2003] and archival meteorological data (archive FNL) National Oceanic and Atmospheric Administration (USA).
For construction of trajectories we set geographical coordinates of two stations of Gobi Desert (Sainshand and Baruun-Urt). Trajectories of air mass movement are calculated for 2008 with duration 120 h and step on time 6 h at heights of 1500, 2500 and 3500 m. The air mass movement of these heights mainly confirm of regional far transfer of impurities of regional

and global scales. In total 283 trajectories are constructed namely 140 forward trajectories and 143 backward trajectories for all seasons of 2008. The most often repeatable types of trajectories of air transport based on their direction and length are detailed as result of average data. It is established, that east, northeast and southeast carrying out of air mass are prevailed in this region. The episode of forward trajectories of air mass movement from stations Sainshand (13 April 2008) is presented presented on Fig. 1a for illustration. Apparently from figure, air mass from Gobi Desert passed over territories of Korea, Japan and Russia.

The results of the calculation of air mass backward trajectories show the prevalence of northwest, western, southwest and southern direction of winds to Gobi Desert. Under the southern and southwestern directions the conditions for air transport from China to the area of study are formed. This is confirmed by the construction of backward trajectories by the model HYSPLIT. On Fig. 1b is shown the most typical trajectories of air mass drift to Baruun-Urt station calculated for 26 May 2008.

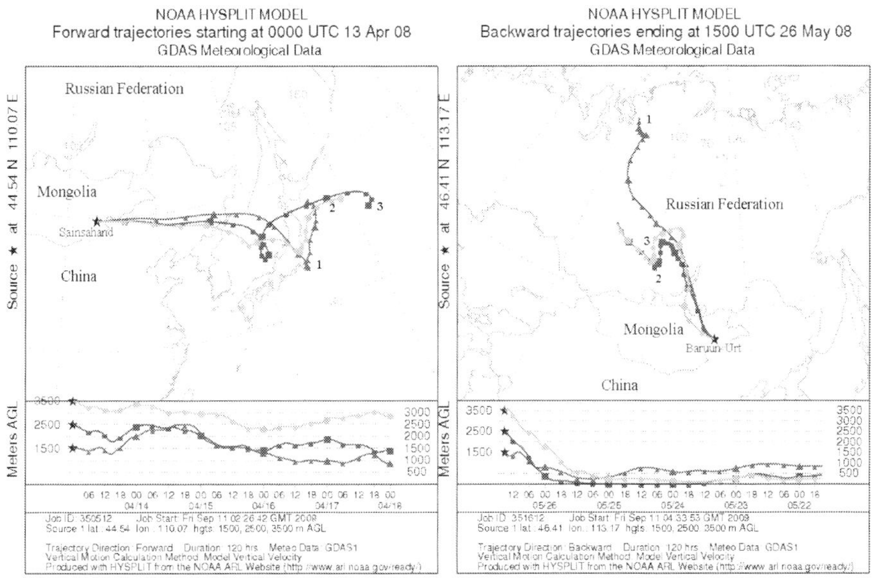

a) b)

Fig. 1. Forward and backward air masses trajectories during observation period at the Sainshand and Baruun-Urt stations by the model HYSPLIT: a) 13 April, b) 26 May

3.2 Study of wind regime in atmosphere of Gobi Desert

The variety of Mongolia relief leads to irregular distribution of air temperature over different locations of terrestrial surface (arid, semi-arid, steppe territories, mountain ecosystems and so on) and it promotes to development of local circulation of wind [Zhadambaa et al., 1967a, 1967b]. In this context great interest presents the study of wind regime in atmosphere of Gobi Desert.

The daily fund meteorological data for 2 years from 2004 to 2005 of meteorological stations (Sainshand and Baruun-Urt) have been analyzed for study of wind regime of Gobi Desert.

Roses of winds constructed on daily fund and experimental meteorological data show basically a northwest direction of wind. However in summer the formation of local atmospheric circulation is observed. Besides in summer the increasing of repeatability of southern (30%), northeast (26.4 %) directions of wind at Sainshand is noted in comparison with other seasons.

It proves to be true also experimental data of meteorological parameters received by means of acoustic meteorological complexes AMK-03 during scientific expeditions in June-July between 2005 and 2009, and also by using of meteorological data of CLIWARE system (http://cliware.meteo.ru/izomap).

4. Dust storm in East Gobi

4.1 Number of dusty days and trend

The analysis of daily fund meteorological data for 18 years between 1991 and 2009 has been carried out for detail research of frequency and duration of dust storms at Sainshand in East Gobi. Fig. 2 shows temporal trend of number of dusty days. It is revealed that last years number of dusty days is increased. For example, frequency of dust storms from 1991 to 2006 is increased in 3 times.

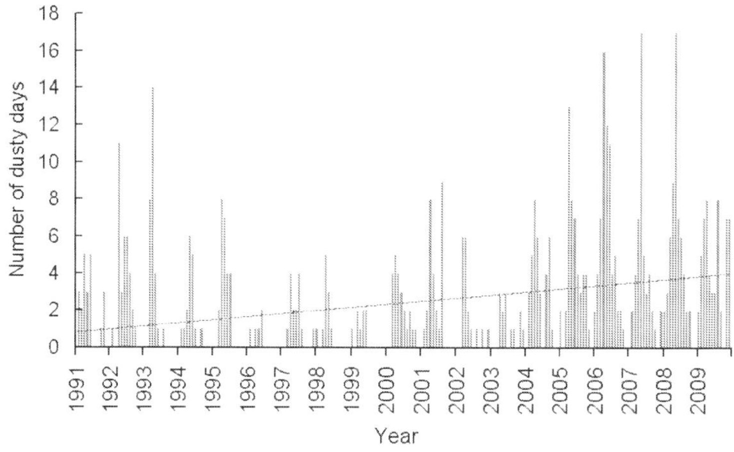

Fig. 2. Number of dusty days at Sainshand

To study the hour duration of dust storms we used daily fund data from 2003 to 2009. It is established that dust storms duration from 2003 to 2007 has increased in 40 times. The most duration of dust storms (573 hours) was observed in 2007. Analysis of number of dusty days has been allowed us to identify seasonal features. It is important to note that since 2004 dust storms began to appear in autumn and winter that it was not observed in previous years (Table 1).

The most repeatability of dust storms in atmosphere of Mongolia arid territories is observed in April-May at day and evening. The repeatability maximum of dust storms is marked in 15-18 hours of local time (Fig. 3).

	January	February	September	October	November	December
1991	0	3	0	1	3	0
1992	0	1	2	1	0	0
1993	0	0	0	0	0	0
1994	0	1	1	0	0	0
1995	0	0	0	0	0	0
1996	0	1	0	0	0	0
1997	0	0	0	0	0	1
1998	1	0	0	0	0	0
1999	1	0	0	0	0	0
2000	0	0	2	1	1	0
2001	0	1	0	0	0	0
2002	0	0	0	1	0	1
2003	0	0	1	0	2	1
2004	0	3	6	1	0	0
2005	2	0	4	4	1	0
2006	2	4	2	2	1	0
2007	0	2	2	1	0	2
2008	2	3	2	2	0	0
2009	2	5	2	0	7	7

Table 1. Number of dusty days in winter and autumn at Sainshand station

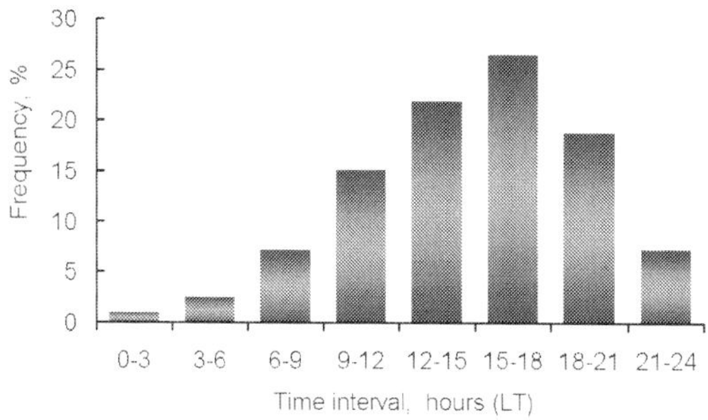

Fig. 3. The daily variation of dust storms for spring in the Gobi

4.2 Analysis of synoptical situation at severe dust storm in May 2008

The most repeatability of dust storms in arid and semi-arid territories of Central Asia is observed in spring, causing by strong winds due to passing of atmospheric fronts. For detail analysis of the episodes of strong dust storm in May 2008 in Mongolia we scrupulously considered the meteorological and weather conditions. In this time strong dust storms are mainly evidenced in East Gobi.

"Black wall" of sand and dust are accompanied by strong winds. Wind force reaches 30-40 m/s in some regions and the tones of sand and dust is transported to regions of China, Korea and Japan. The calculation of forward trajectories air mass movement using of model HYSPLIT confirms this result.

The trajectories of cyclones moving in May 2008 were investigated with help the surface and high altitude (500 hPa) meteorological maps of the Northern hemisphere (http://www.aari.nw.ru/odata/_d0010.php).
We find that the cold Arctic air masses are passed across Lake Baikal mixing with air mass of moderate latitudes in this time. Then it moved to the territories of Mongolia and China creating the significant gradients of the temperature and pressure. These meteorological and synoptical conditions were cause of intensive dust storms in arid and semi-arid territories of Mongolia and China. Therefore Lake Baikal such as huge reservoir of fresh water is a specific climatic zone which is situated on the frontier of arctic and moderate air masses. Baikal influences on pathways of air mass and dust storm occurrence in arid and semi-arid regions of Mongolia and China.
The satellite images of MODIS (http://modis.gsfc.nasa.gov) and the results of calculation of air mass backward trajectories demonstrate this fact and the episodes of severe dust storms in East Gobi on 26-27 May 2008 (Fig. 4).

Fig. 4. Satellite picture passage of cold front in Mongolia (26 May 2008)

4.3 Behaviors of meteorological parameters during dust storm episodes

The analysis of behaviors of momentary meteorological parameters during the passage of dust storms presents a special interest for study of dust storm formation.
Usually two types of dust storms are observed such as the short-term connected with passage of atmospheric front and the long-term caused by strong winds due to elevated pressure gradients [Hoffmann et al., 2008].
Two regional dust storms were observed in the evening during complex scientific expedition in Baruun-Urt (semi-arid territory) in July, 2006. Notice, that first dust storm was accompanied by short-term precipitations, second was accompanied by a strong wind, local wind soil erosion, visibility no more than 50 m [Zhamsueva et al., 2008]. High temperature (+27^{0}C) and low relative humidity of air (20%) are registered in the beginning of dust storm. Under dust storm passage the sharp decreasing of air temperature during 15-20 minutes up to 8.8 ^{o}C (July, 10) and up to 7.9 ^{o}C (July, 11) independently of atmospheric turbidity are observed. In this time the increasing of average wind speed up to 21.5 m/s (July, 11), wind flaws are achieved up to 50 m/s and sharp increasing of relative humidity of air was noticed (Fig. 5).

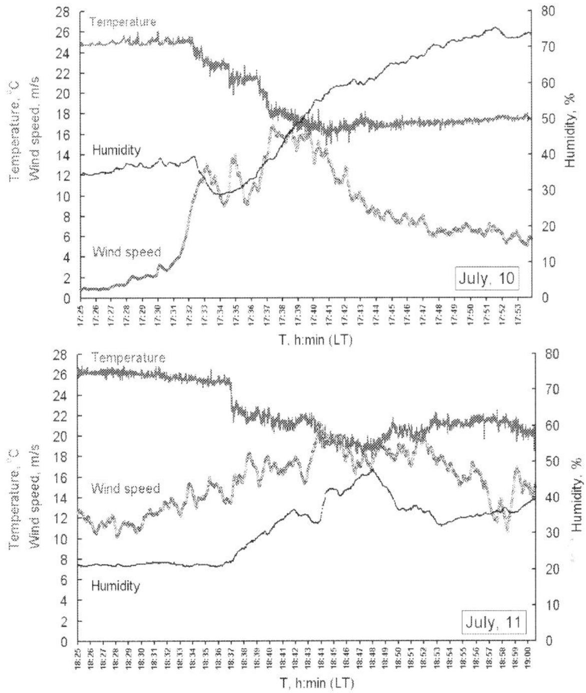

Fig. 5. Temporal variations of wind speed, temperature, humidity in atmosphere of Baruun-Urt during dust storm (10, 11 July 2006)

Under the weakening of dust storm the temperature of air stayed without change. After local dust storm we observed the decreasing of temperature and wind speed and increasing of relative humidity due to cold atmosphere front passage.

4.4 Turbulent characteristics of Gobi Desert atmosphere

The research of turbulent characteristics of atmosphere during dust storm is a great importance.

The turbulence promotes to heat and moisture exchange in the atmosphere of the Earth. Turbulence is caused by topographical heterogeneity and non-uniform heating of the Earth surface. The increasing of atmospheric turbulence conduces to the far transport of dust aerosol [Tverskoi, 1951; Bezuglaya & Berlyand, 1983].

According [Park et al., 2010; Bezuglaya & Berlyand, 1983] the dust storms usually occur under certain critical values of wind speed depending on terrain features and soil properties, which are irregular for different regions.

To get a complete picture of the atmospheric dynamics we analyzed around-the-clock daily measurements of turbulent characteristics obtained using of autonomous ultrasonic meteorological complex AMK-03 in Gobi Desert from 2005 to 2009 in summer. The turbulent characteristics of atmosphere are calculated based on the theory of Monin-Obukhov [Monin & Yaglom, (1965); Obukhov, (1988)]:

$$H = c_p \rho < T^{/} \cdot w^{/} > \quad (1)$$

$$K_h = v * T * / (\partial \theta / \partial z) \quad (2)$$

By results of the analysis it is revealed that in days without dust storms average vertical turbulent heat flux (H) in the surface layer in the day is H = 50 W/m^2. The maximal average daily value of the turbulent exchange coefficient (kh) is kh = 0.8 m^2/s. Turbulence data analysis during dust storms showed that the vertical heat flux was directed downward from atmosphere to surface due to a sharp temperature fall. Maximum H value was -281 W/m^2 and kh = 2.7 m^2/s during the passage of dust storms. After passing of dust storm the vertical heat flux varies from -7.3 to 6.0 W/m^2, and the coefficient of turbulent exchange equals 0.7 m^2/s (Fig. 6).

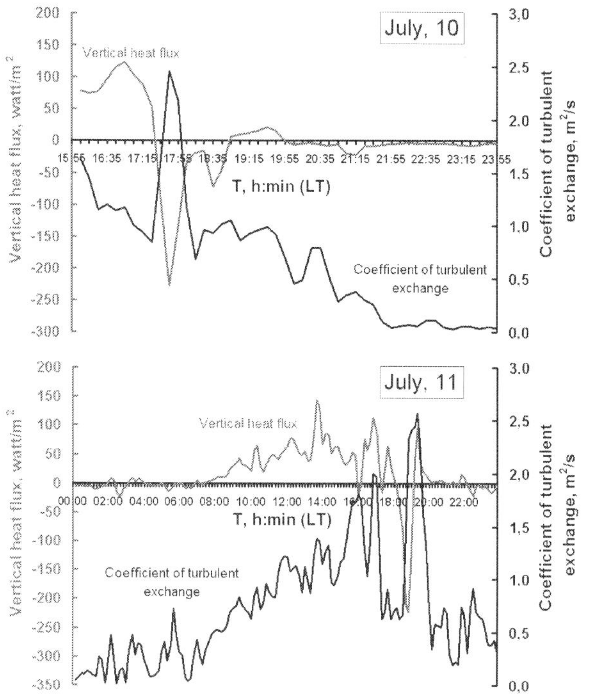

Fig. 6. Temporal variations of vertical heat flux, coefficient of turbulent exchange in atmosphere of Baruun-Urt during dust storm (10, 11 July 2006)

5. Mass concentration of dust aerosol PM10 and PM2.5 in Gobi Desert

In recent decades the problem of global and regional changes of environment and climate are actual due to increasing of air pollution by aerosol or particulate matter (PM). The aerosol changes the radiative balance in the earth-atmosphere system and, consequently, in weather and climate change. Also a harmful effect of particulate matter on human health is noted. The epidemiological studies reveal that under increase $PM_{2.5}$ (particles with an aerodynamic diameter of 2.5 μm or less) and PM_{10} (particles with an aerodynamic diameter of 10 μm or less) concentrations the morbidity, hospitalization and premature mortality of peoples are increased due to particle penetrate into the lower respiratory tract [WHO, 2006, 2007]. Particularly the most attention is devoted to study of $PM_{2.5}$ fine aerosols due to their relatively long lifetime in atmosphere and therefore far transfer. Consequently, the air pollution by suspended particles is not only local but also global problem. Monitoring measurement of mass concentration of PM_{10} and $PM_{2.5}$ at Sainshand station are conducted from 2008 within the framework of the program KOSA Monitor on development of network of dust and sand storms in Central Asia. Samples are collected using high volume samplers Partisol FRM Model 2000 Air Sampler (Japan).

We analyzed daily fund data of 2008 to reveal of daily and annual changes the aerosol mass concentration of PM_{10} and $PM_{2.5}$ at Sainshand and Zamyn - Uud. It is revealed the mean annual course of mass concentration of PM_{10} and $PM_{2.5}$ fine aerosols at Sainshand and Zamyn -Uud. Average annual concentrations of aerosol don't exceed 30 $\mu g/m^3$ of PM_{10} and 8 $\mu g/m^3$ of $PM_{2.5}$. The elevated monthly average PM_{10} and $PM_{2.5}$ concentrations are observed in spring and winter but minimum is autumn. The maximal average concentrations 30 $\mu g/m^3$ of PM_{10} and 20 $\mu g/m^3$ of $PM_{2.5}$ are marked in May 2008. Concentrations of these fractions are minimum value in autumn and up to 5 and 3 $\mu g/m^3$, according. Under the stable weather the daily average concentrations of 5 - 8 $\mu g/m^3$ of PM_{10} and 3 - 5 $\mu g/m^3$ of $PM_{2.5}$ are observed usually. But at dust storms the highest hourly concentrations exceed 1400 $\mu g/m^3$ (PM_{10}) and 380 $\mu g/m^3$ ($PM_{2.5}$). Average annual concentrations of aerosol at Zamyn-Uud located on the border of Mongolia and China considerably exceeds values at Sainshand and is 83 $\mu g/m^3$ of PM_{10} and 55 $\mu g/m^3$ of $PM_{2.5}$. The maximal average concentrations 100 $\mu g/m^3$ of PM_{10} and 59 $\mu g/m^3$ of $PM_{2.5}$ are marked in May 2008. Concentrations of these fractions are minimal in July and up to 11 $\mu g/m^3$ (PM_{10}) and 13 $\mu g/m^3$ ($PM_{2.5}$) in September. The daily average concentrations are varied range 18 - 20 $\mu g/m^3$ (PM_{10}) and 16 - 18 $\mu g/m^3$ ($PM_{2.5}$) under week wind. Maximum values of 1230 $\mu g/m^3$ (PM_{10}) and above, the 700 $\mu g/m^3$ ($PM_{2.5}$) and above under dust storms are observed. Data of mass concentration of PM_{10} at Sainshand and Zamyn - Uud good agree with data Erdene station (44^0 27′N, 111^0 05′E). The Erdene station is located in the about 90 km southeast from the station Sainshand and 100 km northwest from the station Zamyn - Uud. According to [Park et al., 2010] 10-day averaged daily maximum concentration of PM_{10} is 140 $\mu g/m^3$ in early May 2009.

6. Experimental studies of fine and chemical composition of arid aerosol in the atmosphere of the Gobi Desert

6.1 Study of fine arid aerosol

Studies of the daily dynamics of fine arid aerosol were carried out using the diffusion spectrometer DSA range in size from 1.6 to 400 nm at the Sainshand station (Gobi Desert, Mongolia) in August 2009.

The presence of night minima in the diurnal variation of total number concentration of fine aerosol is founded (Fig. 7). In addition, unlike the diurnal cycle of fine aerosol concentrations with peaks during the daytime in other regions, the maximum of the nuclei mode particles (d <0.01 μm) in the daily dynamics is often observed in the morning due to the rise of the particles of fine fraction due to more rapid heating of the earth's surface in arid territories and intensification of the turbulent processes in these hours [Ayurzhanaev et al., 2009].

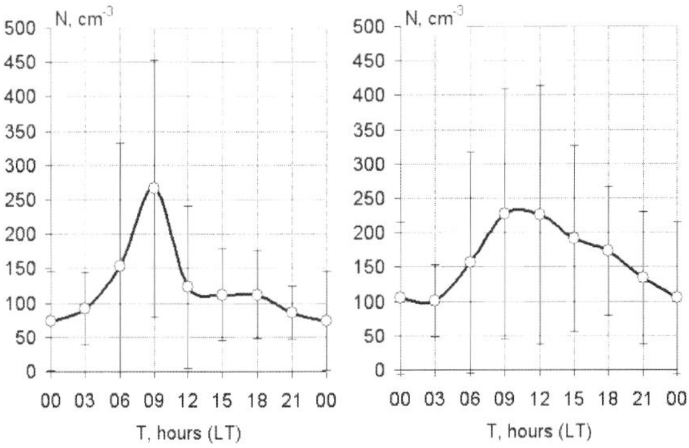

a) b)

Fig. 7. Daily variations of the total number concentration of aerosol (August 2009, Sainshand station, Mongolia): a) nuclei mode; b) the Aitken mode

Reduction of number concentration of nuclei mode particles during the daily hours apparently associate with increased of particle coagulation of this fraction and their transition to the range of Aitken particles fraction (0.01 <d <0.08 μm). The diurnal variations of Aitken particle concentration is confirmed by this conclusion (Fig. 7 b).

Diurnal variation of Aitken particle concentrations is similar to the behavior of the total concentration of fine faction. Number concentration maximum of particles of this fraction is often observed during daily hours, which may be the result of the increased coagulation of nuclei mode particles and the equalization of the generation and destruction rates of this aerosol fraction in this period [Ayurzhanaev et al., 2009; Zhamsueva et al., 2009]. Figure 8 shows a comparison of diurnal variations of soil and air temperatures during the experiments.

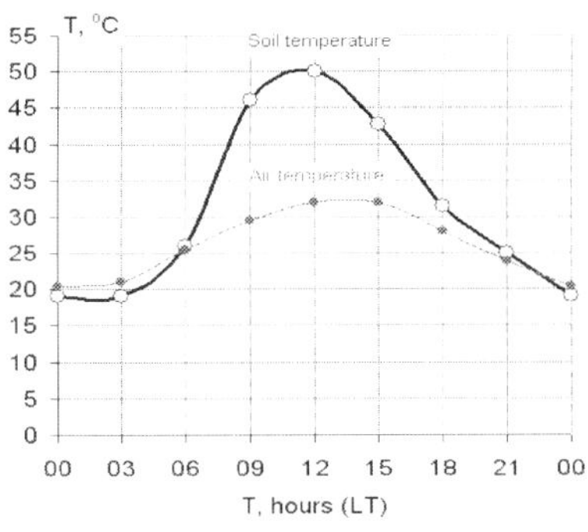

Fig. 8. Daily variations of soil and air temperatures (11 August 2009, Sainshand station, Mongolia).

It is revealed that the number concentration of particles of nuclei mode and Aitken mode differ depend on meteorological and weather conditions of the experiment. Figure 6 shows diurnal variation of the fine particles distribution for days with different weather conditions: under clear weather, with weak winds (10 and 12 August) and in cloudy weather, with gusty winds (13 and 14 August). As can be seen from the figure, the Aitken mode in the distribution of size spectrum particle presents for all days and is a major fraction the fine aerosol.

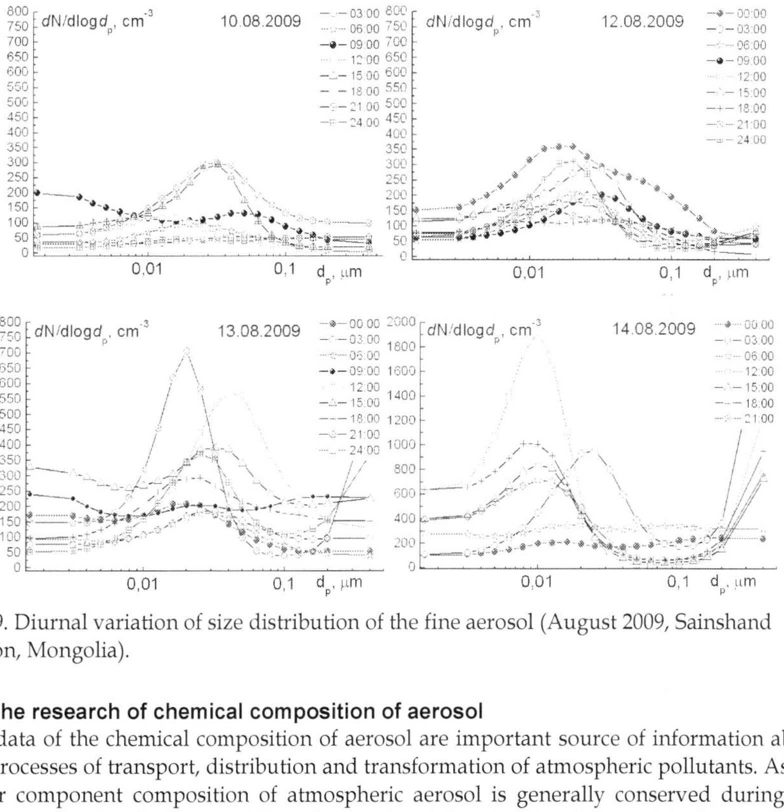

Fig. 9. Diurnal variation of size distribution of the fine aerosol (August 2009, Sainshand station, Mongolia).

6.2 The research of chemical composition of aerosol

The data of the chemical composition of aerosol are important source of information about the processes of transport, distribution and transformation of atmospheric pollutants. As the tracer component composition of atmospheric aerosol is generally conserved during the transport, the chemical composition of aerosol and its microstructure contain the important information about sources and ways of long-range transport of aerosol, including trace gases. Sampling of aerosol particulates with an size less than or equal to 10 μm was carried out on «Whatman-41" filters by high volume sampler PM-10 of General Metal Works Inc. with the volumetric rate of 0.4-1.7 m^3/min and accuracy of ± 0.03 m^3/min in the temperature range 0 ° C - 45 ° C.

The analysis of ionic composition of aerosols was carried out by liquid chromatography Milichrom A-02 on anions and atomic absorption method AAS-30 on cations in the Limnological Institute of SB RAS. The dates of sampling and their duration are presented in Table 2.

Sample number	Date	Time	
		Sampling start	Sampling end
30	11.08.09 - 12.08.09	09:34	08:20
31	12.08.09 - 13.08.09	09:13	08:52
32	13.08.09 - 14.08.09	09:56	07:00
33	15.08.09 - 16.08.09	20:05	17:45
34	16.08.09 - 17.08.09	19:24	16:28
35	17.08.09 - 18.08.09	17:05	15:40
36	18.08.09 - 19.08.09	16:51	16:13

Table 2. Periods of aerosol sampling

Figure 10 shows the results of the analysis of ionic composition of aerosols in the atmosphere of the arid area of Mongolia (Sainshand, August 2009).

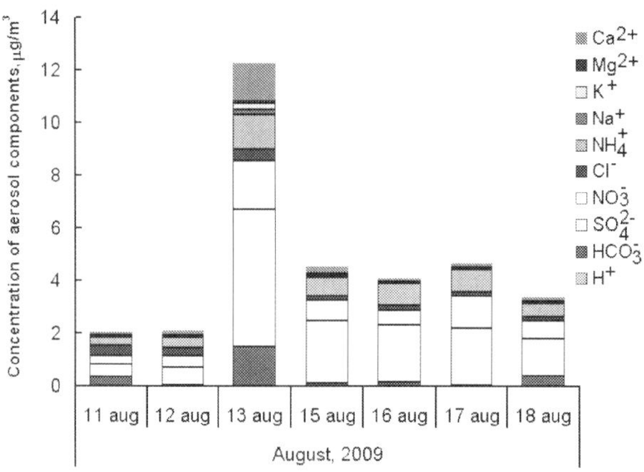

Fig. 10. The chemical composition of aerosols in the atmosphere of Gobi Desert (August 2009)

Apparently from Fig. 7 the anions SO_4^{2-}, NO_3^- and cations NH_4^+ are major components of aerosols in the atmosphere. These anions and cations are the most typical components of industrial emissions. Also the anions HCO_3^+, the cations Ca, Mg, K are contained in significant amount. Samples pH is basically slightly acidic due to the predominance strong anions SO_4^{2-}, NO_3^- in aerosol.

The proportion of these ions is high and is range 15-18%-eq. for nitrate-ions and is range 31-73% eq for sulphate-ions. Also content of chloride is heightened (7-34% eq) (Fig. 11).

Observation site was located in a relatively clean area at a distance of 6-7 km from the Sainshand and 500 km or more from the nearest major industrial centers of China and Mongolia. Influence of local anthropogenic emission sources is minimal in the summer.

Possible mechanism of transport of anthropogenic contaminants to the region of study could be a long-range atmospheric pollutants transfer. The results of calculations of the trajectories of air masses by reanalysis trajectory model HYSPLIT is confirmed this fact (Fig. 12 a, b).

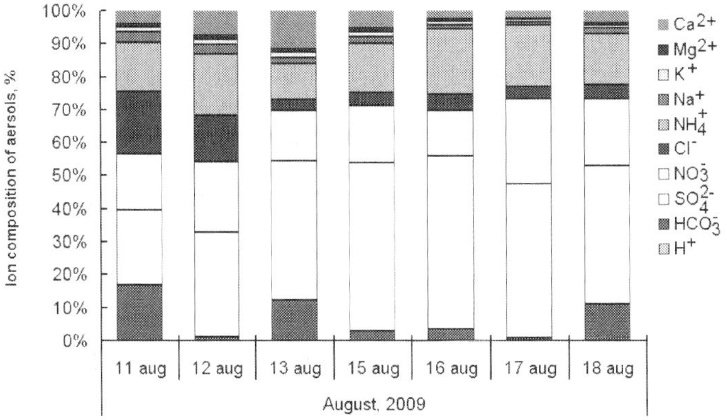

Fig. 11. Contents of the main components in aerosols in the atmosphere of Gobi Desert in August 2009

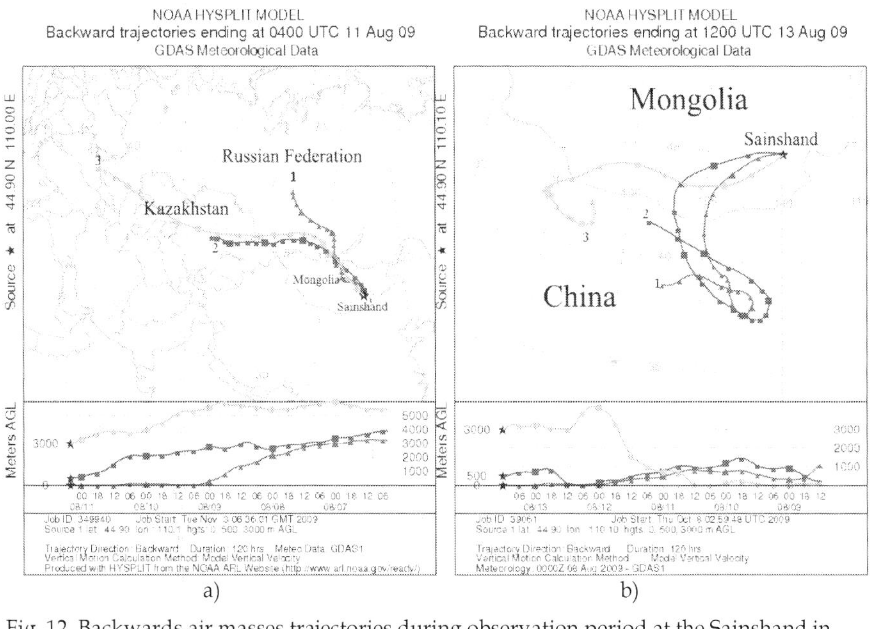

Fig. 12. Backwards air masses trajectories during observation period at the Sainshand in August 2009 on the model HYSPLIT: a) 11August, b) 13 August

The total content of ions at Sainshand are less than 2 μg/m³ under the western and north-western direction of air masses movement from the Eastern Siberia, Kazakhstan and clear stable weather 11-13 August 2009 (Fig. 12). The sulphate-ion (0.46-0.67 μg/m³), nitrate-ion (0.34-0.45 μg/m³), ammonium-ion (0.30-0.40 μg/m³), chloride-ion (0.29-0.39 μg/m³) are predominant ions in aerosols in this time. The total ion content is increased up to 12.3 μg/m³ with the passage of cold atmospheric front and the change of synoptic situation from 13 August 2009. The content of main aerosol components is increased, for example, calcium ions up to 1.4 μg/m³, nitrate-ion up to 1.8 μg/m³, sulphate-ion up to 5.2 μg/m³, hydrocarbonate-ion content up to 1.5 μg/m³. Calculations of air masses movement by model HYSPLIT indicate the transport of atmospheric pollutants from the industrial regions of China during this period. Under the development of local circulation processes from 15 August to 18 August 2009 the concentration of suspended particles is decreased to 3.4-4.6 μg/m³. In these days we observed the high proportion of some ions such as sulphate-ions (73%-eq.), nitrate-ions (28%-eq.) and ammonium ions (80%-eq.) (Fig. 11). These ions are typical components of industrial emissions. The obtained data testify to strong influence of anthropogenic sources on the air composition of arid areas of Eastern Gobi due to long-range transport.

7. Conclusion

In this chapter the investigations of pathways and basic directions of the atmospheric impurities transport in arid and semi-arid territories of Mongolia using of the reanalysis model NCEP/NCAR HYSPLIT (http://www.arl.noaa.gov/ready/hysplit4.html) and archival meteorological data (archive FNL) are conducted. It is established that east, northeast and southeast carrying out of air mass prevail in this region.

As a whole the wind regime within a year in Gobi Desert repeats a direction of the general northwest, characteristic for free atmosphere. During the summer the influence of local circulation is significant. It is revealed that the duration and number of dusty days increase in Gobi Desert. The repeatability of dust storms from 1991 to 2006 has increased in 3 times. It is established that dust storms duration from 2003 to 2007 has increased in 40 times. It is noted that dust storms observe in autumn and winter since 2004 that it was not marked in previous years. The most repeatability of dust storms in atmosphere of arid territories of Mongolia is observed in April-May in day and evening with the maximum concentration of fine aerosols exceeding 1400 μg/m³ (PM_{10}) and 380 μg/m³ ($PM_{2.5}$) at Sainshand and 1230 μg/m³ (PM_{10}) and 700 μg/m³ ($PM_{2.5}$) at Zamyn-Uud station. Under the days with stable, settled weather the mass mean concentration of aerosols PM_{10} and $PM_{2.5}$ changes 5-8 μg/m³ (PM_{10}) and 3-5 μg/m³ ($PM_{2.5}$) at Sainshand. In similar conditions daily mean concentration at Zamyn-Uud change within 18-20 μg/m³ (PM_{10}) and 16-18 μg/m³ ($PM_{2.5}$).

For research of dust storm formation in the Central Asia in May, 2008 we analyzed the surface and high altitude (500 hPa) meteorological maps (http://www.aari.nw.ru/odata/_d0010.php) of Northern hemisphere for 2008. It is established that Lake Baikal is the special climatic area which lies on frontier of Arctic and moderate air mass and influences on trajectories of air mass moving in arid and semi-arid areas of Mongolia and China. In this time the cold Arctic air masses passed across Lake Baikal mixing with air mass of moderate latitudes. Then they moved to the territories of Mongolia and China creating the significant gradients of the temperature and pressure. These meteorological and synoptical conditions were cause to intensive dust storms in arid and semi-arid territories of Mongolia and China.

The chemical composition of aerosol, number concentration and size distribution of submicron fraction of aerosol are analyzed. It is established that the SO_4^{2+}, NO_3^- and NH_4^+ are major components of aerosol particles in atmosphere of Sainshand. These anions and cations are the most typical components of industrial emissions. Also the anions HCO_3^+, the cations Ca, Mg, K are contained in significant amount. The revealed high ion concentration SO_4^{2+}, NO_3^- and NH_4^+ in aerosol at station Sainshand located far from industrial centers and results of modeling by HYSPLIT confirms the strong influence of anthropogenic sources of China on the air composition of arid areas of Eastern Gobi due to long-range transport.

The investigation of fine aerosol content revealed that the Aitken mode in the distribution of size spectrum particle is a major fraction and depends on meteorological and weather conditions.

8. Acknowledgment

This work was carried out within the framework of the Complex Integrate Projects № 4.13 and № 75 supported by Russian Academy of Science. The authors gratefully acknowledge the Hydrometeorology and Environment Monitoring Center of Dornogobi and Sukhbaatar aimags for allowing work to be carried out at the study sites. Thanks also to N. Enkhmaa, K. Enkhbayar for help with the fieldwork.

9. References

Arimoto R. (2001). Eolian dust and climate: relationships to sources, tropospheric chemistry, transport and deposition. Earth-Science Reviews, 54, 29-42

Ayurzhanaev A.A., Zhamsueva G.S., Zayakhanov A.S. & Tsydypov V.V. (2009). Microstructure of submicron aerosols in the Gobi Desert. XVI Workshop "Aerosols of Siberia". Tomsk, 24-27 November 2009, 7.

Azbukin A.A., Bogushevich A.Ya., Ilichevskii V.S., Korolkov V.A., Tikhomirov A.A. & Shelevoi V.D. (2006). Automated ultrasonic meteorological complex AMK-03. Meteorology and hydrology, 11, 89-97

Bezuglaya E.Yu., Berlyand M.E., 1983 Climatic characteristics of distribution of impurities conditions of atmosphere. Hydrometeoizdat, pp. 328

Chan Y.C., McTainsh G., Leys J., McGowan H. & Tews K. (2005). Influence of the 23 October 2002 dust storm on the air quality of four Australian cities. Water Air and Soil Pollution, 164, 329-348

Draxler R.R. & Rolph G.D. (2003). HYSPLIT (HYbrid Single-Particle Lagrangian Integrated Trajectory) Model access via NOAA ARL READY Website (http://ready.arl.noaa.gov/HYSPLIT.php). NOAA Air Resources Laboratory, Silver Spring, MD.

Gillette D.A. (1986). Dust production by wind erosion: necessary conditions and estimates of vertical fluxes of dust and visibility reduction by dust. In: El-Baz F., Hassan M.H.A. (Eds.). Physics of Desertification. Martinus Nijhoff Publishers, Dordrecht. 361-371

Kim H. & Chung Y. (2008). Satellite and ground observations for large-scale air pollution transport in the Yellow Sea region. Atmos. Chem. DOI: 10.1007/s10874 008 9111-4.

Lee B.K., Lee H.K. & Jun N.Y. (2006). Analysis of regional and temporal characteristics of PM_{10} during an Asian dust episode in Korea. Chemosphere, 63, 1106-1115

McTainsh G., Chan Y.C., McGowan H., Leys J. & Tews K. (2005). The 23rd October 2002 dust storm in eastern Australia: characteristics and meteorological conditions. Atmospheric Environment, 39, 1227-1236

Monin A.S. & Yaglom A.M. (1965). Statistical hydrodynamics. M.: Publishers "Nauka", 640 p.

Natsagdorj L., Jugder D. & Chung Y.S. (2003). Analysis of dust storms observed in Mongolia during 1937-1999. Atmospheric Environment, 37, 1401-1411.

Obukhov A.M. (1988). Turbulence and atmospheric dynamics. L.: Gidrometeoizdat, 1988, 413 p.

Park S.-U., Park M.-S. & Chun Y. (2010). Asian dust events observed by a 20-m monitoring tower in Mongolia during 2009. Atmospheric Environment, 44, 4964-4972

Sun J., Zhang M., & Liu T. (2001). Spatial and temporal characteristics of dust storms in China and its surrounding regions, 1901–1999: relations to source area and climate. J. Geophys. Res. 106, 10325–10333.

Tverskoy P.N., 1951. Meteorological course. Hydrometeoizdat. - L., pp. 888.

Wang S., Wang J., Zhou Z. & Shang K. (2005). Regional characteristics of three kinds of dust storm events in China. Atmospheric Environment, 39, 509-520

Wang T. & Zhu Z. (2001). Studies on the sandy desertification in China. Chinese Journal of Eco-Agriculture, 9 (2), 7-12

Wei A.L. (2006). Evaluation of micronucleus induction of sand dust storm fine particles (PM2.5) in human blood lymphocytes. Environmental Toxicology and Pharmacology, 22, 292-297

WHO (2006). Health risk of particulate matter from long-range transboundary air pollution- Joint WHO/UNECE Convention Task Force on the Health Aspects of Air Pollution. World Health Organisation, European Centre for Environment and Health, Bonn Office.

WHO (2007). Health relevance of particulate matter from various sources, Report on a WHO Workshop, Bonn, Germany, 26-27 March 2007.

Xuan J., Sokolik I.N., Hao J., Guo F., Mao H. & Yang G. (2004). Identification and characterization of sources of atmospheric mineral dust in East Asia. Atmospheric Environment, 38, 6239-6252

Zhadambaa Sh., Neushkin A.I. & Tuvdendorzh D. (1967a). Circulation factors of Mongolia climate.

Zhadambaa Sh., Neushkin A.I. & Tuvdendorzh D. (1967b). Seasonal particularities of atmospheric circulation above Mongolia. Meteorology and hydrology, 5, 29-32

Zhamsueva G.S. Zayakhanov A.S, Tsydypov V.V, Ayurjanaev A.A Azzaya D. & Oyunchimeg D. (2008). Assessment of small gaseous impurities in atmosphere of arid and semi-arid territories of Mongolia. Atmospheric Environment, 42(3), 582-587.

Zhamsueva G.S., Zayakhanov A.S, Tsydypov V.V., Ayurjanaev A.A., Golobokova L.P., Khodzher T.V., Azzaya D. & Oyunchimeg D. (2009). Temporal variability of small gaseous impurities, chemical and dispersive composition of aerosol in arid territories of Mongolia. XVI Workshop "Aerosols of Siberia". Tomsk, 24-27 November 2009, 70.

5

Assessment of Areal Average Air Quality Level over Irregular Areas: a Case Study of PM10 Exposure Estimation in Taipei (Taiwan)

Hwa-Lung Yu[1], Shang-Chen Ku[1], Chiang-Hsing Yang[2], Tsun-Jen Cheng[3] and Likwang Chen[4]
[1]*Dept of Bioenvironmental Systems Engineering, National Taiwan University, Taipei,*
[2]*Dept of Health Care Management, National Taipei University of Nursing and Health Sciences*
[3]*Institute of Occupational Medicine and Industrial Hygiene, National Taiwan University*
[4]*Center for Health Policy Research and Development, National Health Research Institutes, Miaoli, Taiwan*

1. Introduction

Assessment of public health risk from the exposure of ambient pollutants has been heavily based upon the analyses of the associations between the pollutant exposure and its potential consequence, e.g. disease occurrences (Abbey et al., 1995; AckermannLiebrich et al., 1997; Beeson et al., 1998; Jerrett et al., 2005; Pope et al., 2002). In many air pollution epidemiologic investigations, individual health datasets at nationwide or regional scales are available to assess the subtle risks of pollution exposure. Among them, the understanding of the spatial or spatiotemporal distribution of ambient pollutants is essential due to their prevalent heterogeneity across space and/or time. In these cases, governmental agencies has established ambient air-quality monitoring networks, such as the Air Quality System (AQS) operated by the U.S. Environmental Protection Agency (EPA), regularly recording important and useful environmental data sources concerning the acute and chronic effects of ambient pollutants (TWEPA, 2006; USEPA, 1992). Based upon these databases, an ideal exposure assessment is performed by applying techniques of spatial or/and spatiotemporal analysis for the estimation of the pollutant level at the locations of individuals in health dataset. However, due to the raising concerns of privacy and confidentiality of personal information, the sensitive personal information, such as residential addresses, of health dataset is usually not allowed to be accessed. As a result, individual information of health dataset obtained from institutes or governmental agencies is often removed or degraded. Among them, the spatial locations of individuals of health dataset are usually aggregated into a larger spatial unit, e.g. higher administrative level, in which no personal identities can be identified. These aggregations raise challenges for environmental epidemiologists to assess the exposure levels and to investigate the potential health risks.

Spatial interpolation techniques have been increasingly used by environmental epidemiologists to estimate the spatiotemporal distribution of air quality levels. Among them, the deterministic approaches such as nearest-neighbor method (NN) and inverse distance weighted method (IDW) are the most widely used methods (Hendryx et al., 2010; Hoek et al., 2001; Michelozzi et al., 2002; Sohel et al., 2010), in which the estimation results are derived from certain assumed functional relationship of geographical distances between the observation and estimation locations without considering the stochastic associations among the air quality measurements. The stochastic techniques, in particular kriging method, have been applied with increasing frequency in exposure assessment studies to address the spatial heterogeneity among the air quality data as well as the estimation uncertainty of predictions (Brauer et al., 2003; Brimicombe, 2000; Buzzelli and Jerrett, 2004; Hoek et al., 2001). Most spatial interpolators are originally developed for the estimations with geographical support of estimation and observation locations at point scale. As a result, most studies of exposure assessment apply spatial techniques for point estimation at the centroid of the geographical unit of interest to characterize its average air pollution level (Chen and Schwartz, 2008; Chen and Schwartz, 2009; Lertxundi-Manterola and Saez, 2009; Maheswaran and Elliott, 2003; Miller et al., 2007). The inconsistency of geographical support between the exposure estimations and aggregated health dataset can potentially distort the results of their associated epidemiological studies (Young et al., 2009; Young et al., 2008).

This study investigates and compares the estimation results of areal-averaged air quality level by several popular spatial mapping techniques, i.e. NN, IDW, ordinary kriging (OK) and block kriging method (BK). Among them, BK is a kriging-based upscaling method which can perform spatiotemporal estimation with the consideration of the stochastic dependence among the locations with irregular sizes and shapes. This comparison is performed on the spatiotemporal PM10 estimation of the townships over Taipei area (Taiwan) during 2004-2006 on the basis of data during 1997-2007.

2. Materials and methods

2.1 Study area

Taipei is the largest metropolitan area in Taiwan including Taipei city and Taipei county with the vehicle density as high as over 6000 vehicles per km^2. Except for traffic emissions, the three incineration plants are the major stationary emission sources in the area (Chang and Lee, 2007). The Taipei area is bounded by mountains, i.e. Yangming mountains to the north, Linkou mesa to the west, and ridge of Snow mountains to the southeast which forms the second largest basin of the island (see Figure 1). Because of the significant variation of the basin topography, the air convection and circulation are generally degraded in this area. In addition, the alluvial plains at basin floor materialized the highly urbanization area of Taipei. As a result, the ambient pollutant concentration across the basin floor is generally higher than that over its surrounding mountain areas. Taiwan Environmental Protection Agency (TWEPA) has established the air quality monitoring network which regularly records the ambient pollutants, including PM10 and other criteria pollutants (TWEPA, 2006), and meteorological covariates throughout the island since September, 1993. Figure 2 shows the eighteen TWEPA monitoring stations in Taipei area. As it can be noticed, the distribution of monitoring locations is spatially imbalanced that most of the stations in Taipei are located at the high urbanized area. In this study, all the PM10 observations from these stations are aggregated into monthly data following the procedure suggested by USEPA (USEPA, 2004). The areal concentration estimations are performed at each of the townships shown in Figure 2 by various spatial interpolation techniques.

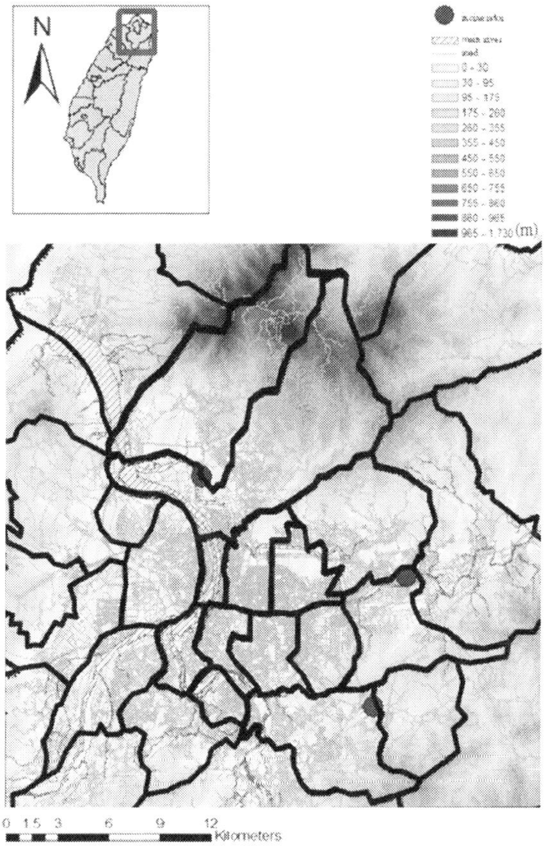

Fig. 1. The highways, rivers and topography in Taipei metropolitan area

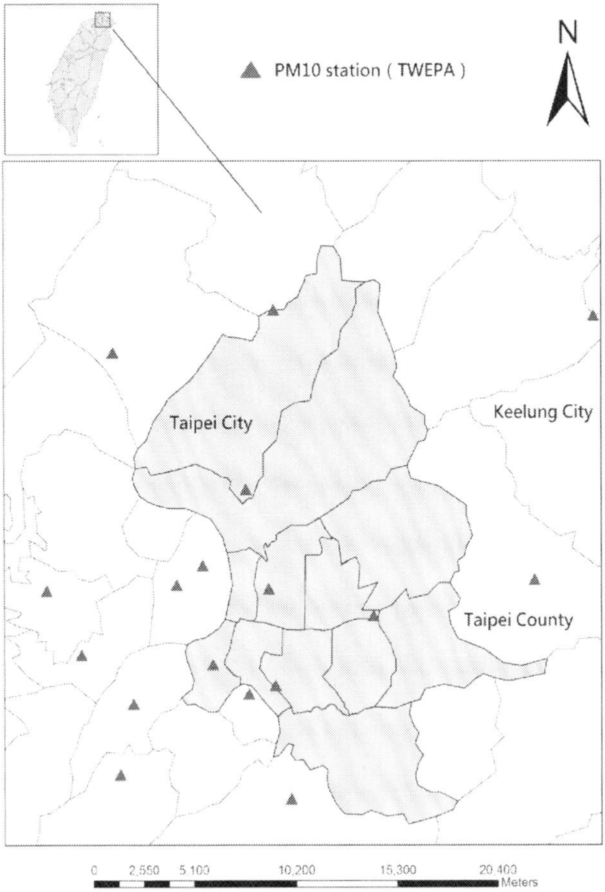

Fig. 2. Spatial distribution of PM10 monitoring stations operated by TWEPA

2.2 Methods

This study investigates several common used spatial interpolation methods for ambient pollutant exposure estimation. Among them, the nearest neighbor method (NN), also called polygon method and Theissen method, is the simplest method for spatial estimation. It assumes the air quality level at an ungauged location is completely determined by its closest observations so that spatial distribution of pollutants is composed of a set of polygon across space. The inverse distance weighted method (IDW) is a weighted average of neighboring observation values. The weights given to each observation is a function of distances between observation and estimation locations (Waller and Gotway, 2004) as below

$$\hat{z}(s_k,t_k)=\sum_{i=1}^{N} z(s_i,t_i)d_{ki}^{-p} \Big/ \sum_{i=1}^{N} d_{ki}^{-p} \quad (1)$$

where d_{ki} is the space-time distance between the estimation location (s_k,t_k) to the observation locations $z(s_i,t_i)$. p is the modeler-specified degree. The increase of p can decreases the weights of distant observations. The space-time distance is assumed to be the function of the geographical distance and time interval (Christakos, 2000; Christakos et al., 2000). It can be expressed into the form of $d_{ki} = |s_k - s_i| + m_{st}|t_k - t_i|$, where m_{st} is the space-time metric to account for the relationship between spatial and temporal metric (Christakos et al., 2002). Both NN and IDW methods are deterministic methods which provide the estimations without the information of estimation uncertainty.

Stochastic techniques, on the other hand, can account for the uncertainty of the space-time data under the framework of random field theory which forms a multivariate joint distribution among the space-time attribute of interest. Among them, kriging method is the most popular technique in space-time mapping which is so-called best linear unbiased estimator (BLUE) in the sense of providing the minimum estimation uncertainty (Olea, 1999). Various types of kriging method have been developed in terms of different assumptions and analytical goals. Among them, ordinary kriging (OK) is the most widely used one which assumes the unknown mean among the space-time dependent data. The basic equations of ordinary kriging are shown below (Goovaerts, 1997)

$$\begin{aligned} &\sum_{i=1}^{n(p_k)} \lambda_{i,p_k} C(p_i,p_j) + \mu_{OK,p_k} = C(p_k,p_j) \\ &\sum_{i=1}^{n(p_k)} \lambda_{i,p_k} = 1 \quad i,j = 1,2...,n(p_k) \end{aligned} \tag{2}$$

where λ_{i,p_k} are the OK weights for the observation at $p_i = (s_i,t_i)$; $C(p_i,p_j)$ denotes the covariance of the attribute between p_i and p_j, $n(p_k)$ is the number of observations used for the kriging estimation at p_k, and μ_{OK,p_k} is the Lagrange multiplier. The OK estimation at p_k can be the linear combination of observations expressed as form of $\hat{z}(p_k) = \sum_{i=1}^{n(p_k)} \lambda_{i,p_k} z(p_i)$. Block kriging (BK) inherits the framework of kriging method and considers the geographical support of observation and estimation locations, i.e. their size and shape. The average value of attribute of concern over a block V can be represented as $\hat{z}_V(p_k) = \int_{V(p_k)} du' z(u') / |V|$, where $|V|$ is the areal size of block V and u' is the point locations within the block V. The block kriging system is very similar to Eq. (2) with the covariance $C(p_k,p_j)$ replaced by the covariance between point observations and estimation block $\bar{C}(V(p_k),p_j)$ (Goovaerts, 1997), where $\bar{C}(p_k,p_j) = \int_{V(p_k)} du' C(u',p_j) / |V|$.

This study compares the township-based estimations of PM10 level by the methods above, i.e. NN, IDW, OK and BK. Among them, three different spatial resolutions are used in BK method, i.e. dividing the irregular areas into 5x5, 10x10 and 40x40 grids. The estimations are performed at all townships in Taipei during 2006 with the support of space-time PM10 observations during 1997-2007. Among them, the township-level PM10 estimations by NN, IDW and OK methods follow the conventional approach that uses the estimations at geographical centroid to characterize the areal average level of pollutant concentration (Chen and Schwartz, 2009; Lertxundi-Manterola and Saez, 2009; Maheswaran and Elliott, 2003). In addition, to assure the comparison is performed at the same basis, the nonstationarity of PM10 process in space and time is removed by locally weighted smoothing regression method (LOESS) (Cleveland, 1979; Cleveland and Devlin, 1988) in advance to the applications of the spatial-time interpolation techniques.

3. Results

The observed PM10 levels across the stations in Taipei vary significantly as shown in Figure 3. The average and variance of monthly PM10 observations at the stations located in highly urbanized areas are generally higher than those in city's surrounding areas. Temporal variation of averaged PM10 across stations is shown in Figure 4 in which the increased PM10 variability happens at the time of higher average PM10 value, i.e. proportional effect. In general, the average of observed PM10 is higher during the seasons of spring and winter. In this study, the spatiotemporal processes of monthly PM10 is decomposed into an nonstationary trend in space and time to account for the general pattern of PM10 observations, e.g. PM10 variation resulting from the changes of seasons and distribution of emissions, and stationary residuals which account for the spatiotemporal dependence among the PM10 transport process. The spatiotemporal trend is estimated by using LOESS method which applies low-order polynomials to obtain the general pattern of PM10 variation at each space-time locality (Cleveland, 1979; Cleveland and Devlin, 1988). The estimation of spatiotemporal trend is performed by R software package. To characterize the spatiotemporal dependence, a space-time seperable function is used as shown below (see Figure 5)

$$c(h,\tau) = c_0 \exp(-\frac{3h}{a_r})\exp(-\frac{3\tau}{a_t}) \tag{3}$$

where $c_0 = 150$ and $[a_r, a_t] = [2000\text{m}, 3 \text{ month}]$. Eq. (3) is used in kriging methods, i.e. OK and BK, to account for the space-time covariance at point scale in space and time. In addition, the covariance function identifies the influential ranges of PM10 data in space and time, and therefore characterizes the space-time metric which is used to estimate the space-time distance for IDW and kriging methods as discussed above.

The estimations of the areal PM10 level during the period of 2004-2006 at all townships of irregular sizes and shapes are compared in this study and their differences are shown in Figure 6 in which BK1, BK2, and BK denote the results of block kriging with spatial resolution of 5 by 5, 10 by 10, and 40 by 40 elements of each township. Results show that the spatiotemporal dependence is important to the estimations, i.e. significant differences between the results from deterministic methods, i.e. NN and IDW, and stochastic methods, OK and BK. The spatial distribution of the mean squared difference of estimations between the results from BK and other methods are shown in Figure 7. Figures 7 (a) and (b) shows the comparison between BK and the deterministic methods, i.e., NN and IDW, in which the mean squared differences (MSD) are generally high and increase as the estimation location further apart from the cluster of monitoring stations located at city central. The comparisons among spatial patterns of kriging results in Figures 7 (c) and (d), for OK-BK and BK1-BK

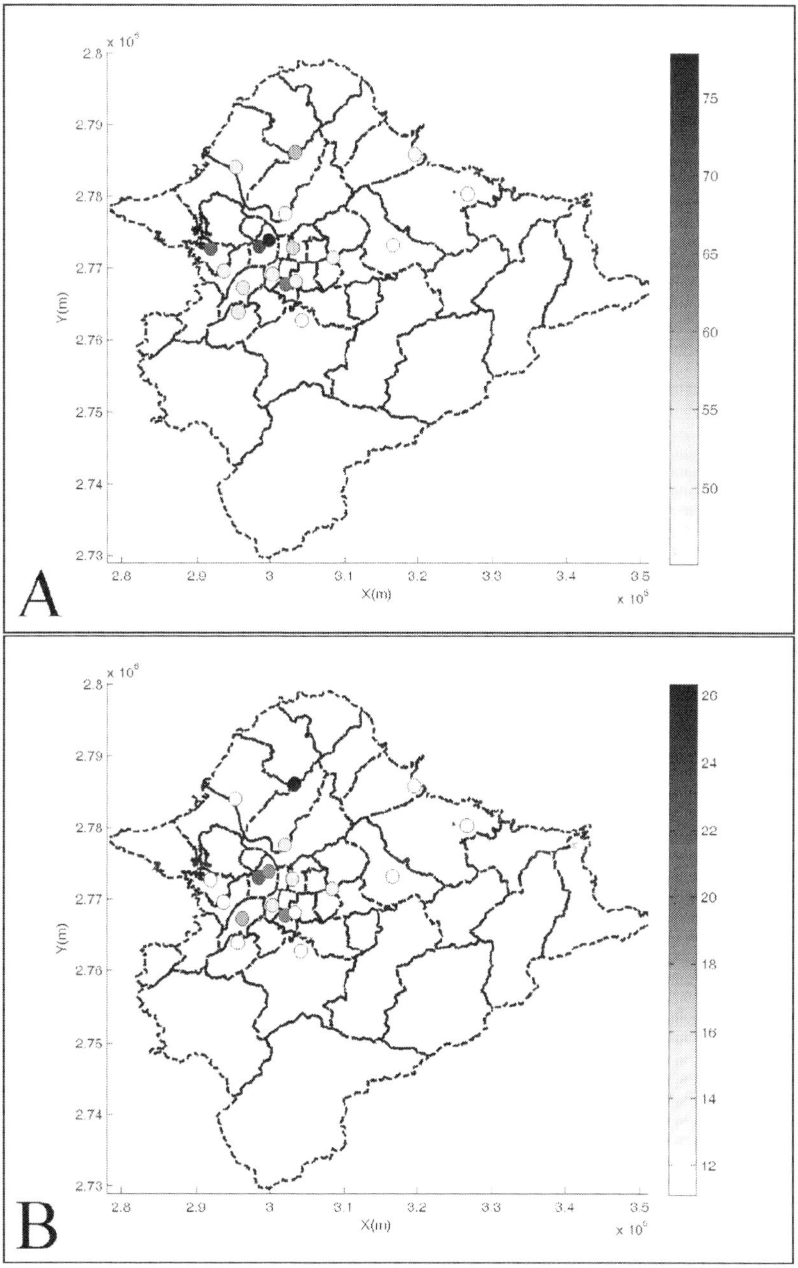

Fig. 3. Spatial distribution of (a) mean and (b) standard deviation of PM10 observations

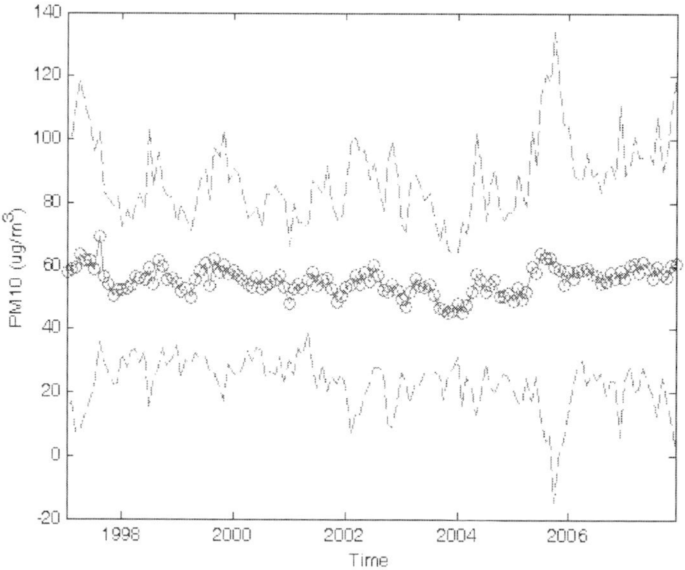

Fig. 4. Temporal variation of averaged PM10 observations and its associated 95% confidence interval.

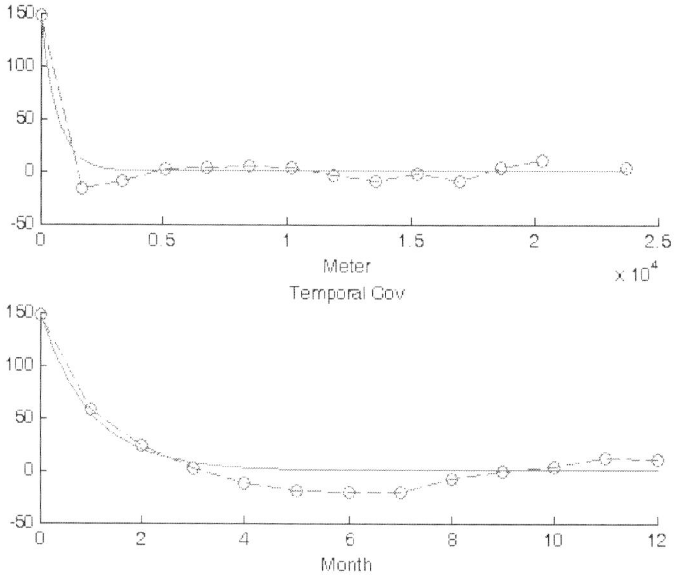

Fig. 5. Spatiotemporal covariance of PM10 observations across (upper) space and (bottom) time.

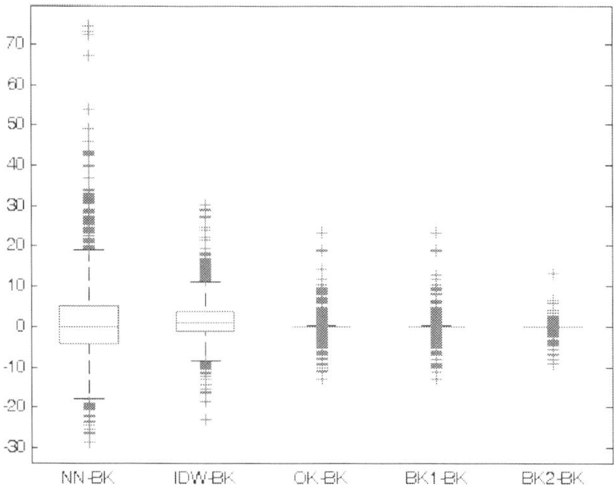

Fig. 6. Boxplot of the estimation differences between the results of block kriging with highest resolution and other methods.

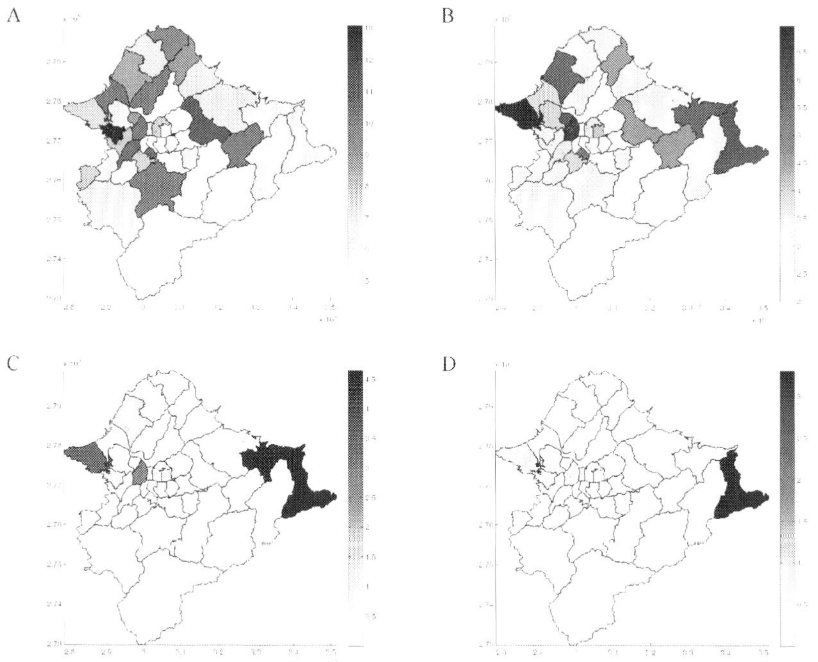

Fig. 7. Spatial distribution of the mean squared differences between the methods of (a) NN-BK, (b) IDW-BK, (c) OK-BK, and (d) BK1-BK

respectively, show the relatively higher MSDs are located at the boundary of study area and townships surrounding by multiple stations with relatively drastic variations of their observations. It should be noted that estimation differences between deterministic methods and BK are generally higher than those in the comparisons between kriging methods. Similar results are shown in the comparisons of over time in Figure 8 in which the seasonality of MSDs is shown in all the comparisons that elevated estimation variability is elevated in spring and winter.

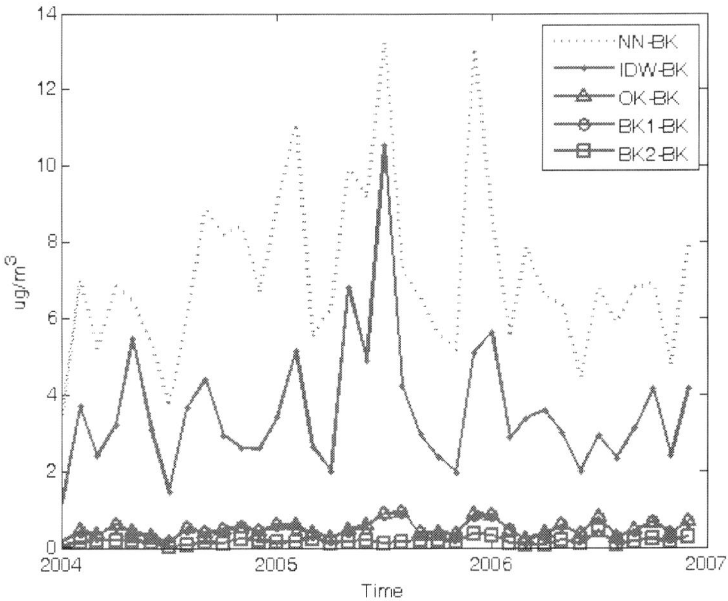

Fig. 8. Temporal distribution of mean squared differences among the methods of NN, IDW, OK, and BK with various spatial resolutions.

4. Discussions

Various methods of different level of complexity have been proposed and used in spatiotemporal estimation of air pollution exposure. Due to the privacy issue of personal data access, the spatiotemporal exposure estimation is often performed on the aggregated dataset with various geographical supports. As a result, certain approximations or assumptions of common techniques of point scale are required to be applied on the estimation of the average level over an area of concern, e.g. administrative division. Among them, the most common approximation is to use the air pollution estimation at the centroid to represent the average pollution level of the entire area of concern (Chen and Schwartz, 2008; Chen and Schwartz, 2009; Lertxundi-Manterola and Saez, 2009; Maheswaran and Elliott, 2003; Miller et al., 2007). The impact of the size and shape of geographical units to the estimation results is seldom considered (Goovaerts, 2008). The assumption of the similar spatial resolution among observation and Estimation areas can be dubious. In addition, the estimation ignores the variability of pollution levels within each of the studied units. Such approximation can be inadequate, especially, in the exposure estimation in metropolitan area where, in addition to its complex topography, various emission sources spreading across space elevates the spatiotemporal variability of air pollution. Though it is theoretical defective, exposure estimation at centroids usually provides a quick and convenient way to assess the air quality level in the corresponding areas of interest. The impact of this assumption of spatial prediction is worthwhile to be examined because the potential biased estimation results of air quality can distort their associated environmental epidemiology studies (Son et al., 2010).

As shown in Figure 6, comparing with the results of BK, the inconsistency is clearly shown between the estimations obtained from the two groups of methods, i.e. deterministic and stochastic, that the variability of the between-group differences are significantly larger than the within group differences. Among them, the results from NN method, i.e. a very common used method for ambient exposure estimation in environmental health studies (Basu et al., 2004; Miller et al., 2007), are significantly variable compared to those from other methods, i.e. its standard deviation is about 20% of common level of PM10. Figure 7 (a) shows that the discrepancy levels of NN results can vary across space and its generally high variability can be reduced at the areas of the higher density of monitoring stations. The periodic feature of the variation in Figure 8 shows that the temporal characteristics of NN results are still distinct from those of the results by other methods. It may resort to the fact that NN method is the only method disregarding the space-time dependence among the PM10 dataset in this study. Despite of its deterministic assumption, IDW accounts for the spatiotemporal dependence by considering the space-time metric among the observation and estimation locations. As a result, IDW reduces the discrepancy in both space and time across Taipei area between its results and those by kriging methods which considering the spatiotemporal dependence by not only space-time metric but the similarity among the observations. The standard deviation of the estimation variability compared to BK is about 10% of common PM10 level in Taipei. In addition, the results of deterministic approach are overestimated, especially at the townships with scarce monitors around, due to the preferential sampling at the highly polluted areas of Taipei central area (see Figure 3), i.e. the observations are mostly sampled at high concentration areas.

Figures 6-8 show the results from kriging methods, i.e. OK and BK with two spatial resolutions, are relatively similar. Among them, the standard deviation of estimation variability is lower than 5% of PM10 level. The distributions of MSDs among kriging methods in both space and time are relatively close to each other shown in Figure 7 (c) and (d). Among them, the areal estimation variability is further reduced by using BK instead of OK that their 95% confidence ranges are 2.8 and 6.8 $\mu g / m^3$, respectively. Because of the short spatial influential range of local PM10 transport, i.e. 2000m (see Eq. (3)), it implies the high spatial variability of PM10 distribution within the townships, i.e. point estimation results can vary significantly from location to location within townships. As a result, the consideration of geographical characteristics can effectively improve the understanding of the areal average level of PM10 concentration. This effect can be especially obvious at the townships containing and surrounding by multiple stations, e.g. San-chung township (the township with two stations), because the nature of kriging methods completely appreciates the observations and therefore result in the higher variability of estimation results among these townships. It should be noted that all the estimations by the present spatial interpolation techniques only depend upon the PM10 observations. The general characteristics of these methods can only provide reasonable estimations within the convex of data locations (Olea, 1999).

5. Conclusions

This study applies several popular spatial techniques to assess the average areal PM10 level of townships in Taipei area. The comparison shows that the importance of the inclusion of space-time metric among the observation and estimation locations as well as the consideration of spatiotemporal dependence among the observations for the estimations. This study shows the consideration of the shape and size of the townships is important to the performance of the estimations of areal average concentration. The MSDs between the estimations by kriging methods with and without considering the geographical characteristics of townships are up to 6% and 2% of the common PM10 level across space and time, respectively. This study provides insights for the impact of geographical support to the exposure estimation in Taipei for environmental epidemiological studies.

6. Acknowledgement

This research was supported by the National Science Council of Taiwan (NSC 99-2625-M-002-007-).

7. References

Abbey DE, Lebowitz MD, Mills PK, Petersen FF, Beeson WL, Burchette RJ. Long-Term Ambient Concentrations of Particulates and Oxidants and Development of Chronic Disease in a Cohort of Nonsmoking California Residents. Inhalation Toxicology 1995; 7: 19-34.

AckermannLiebrich U, Leuenberger P, Schwartz J, Schindler C, Monn C, Bolognini C, et al. Lung function and long term exposure to air pollutants in Switzerland. American Journal of Respiratory and Critical Care Medicine 1997; 155: 122-129.

Basu R, Woodruff TJ, Parker JD, Saulnier L, Schoendorf KC. Comparing exposure metrics in the relationship between PM2.5 and birth weight in California. Journal of Exposure Analysis and Environmental Epidemiology 2004; 14: 391-396.

Beeson WL, Abbey DE, Knutsen SF. Long-term concentrations of ambient air pollutants and incident lung cancer in California adults: Results from the AHSMOG study. Environmental Health Perspectives 1998; 106: 813-822.

Brauer M, Hoek G, van Vliet P, Meliefste K, Fischer P, Gehring U, et al. Estimating long-term average particulate air pollution concentrations: Application of traffic indicators and geographic information systems. Epidemiology 2003; 14: 228-239.

Brimicombe A. Constructing and evaluating contextual indices using GIS: a case of primary school performance tables. Environment and Planning A 2000; 32: 1909-1933.

Buzzelli M, Jerrett M. Racial gradients of ambient air pollution exposure in Hamilton, Canada. Environment and Planning A 2004; 36: 1855-1876.

Chang SC, Lee CT. Evaluation of the trend of air quality in Taipei, Taiwan from 1994 to 2003. Environmental Monitoring and Assessment 2007; 127: 87-96.

Chen JC, Schwartz J. Metabolic syndrome and inflammatory responses to long-term particulate air pollutants. Environmental Health Perspectives 2008; 116: 612-617.

Chen JC, Schwartz J. Neurobehavioral effects of ambient air pollution on cognitive performance in US adults. Neurotoxicology 2009; 30: 231-239.

Christakos G. Modern Spatiotemporal Geostatistics. New York, NY: Oxford Univ. Press, 2000.

Christakos G, Bogaert P, Serre ML. Temporal GIS: Advanced Functions for Field-Based Applications. New York, NY: Springer-Verlag, 2002.

Christakos G, Hristopulos DT, Bogaert P. On the physical geometry concept at the basis of space/time geostatistical hydrology. Advances in Water Resources 2000; 23: 799-810.

Cleveland WS. Robust Locally Weighted Regression and Smoothing Scatterplots. Journal of the American Statistical Association 1979; 74: 829-836.

Cleveland WS, Devlin SJ. Locally Weighted Regression - an Approach to Regression-Analysis by Local Fitting. Journal of the American Statistical Association 1988; 83: 596-610.

Goovaerts P. Geostatistics for natural resources evaluation. New York: Oxford University Press, 1997.

Goovaerts P. Kriging and semivariogram deconvolution in the presence of irregular geographical units. Mathematical Geosciences 2008; 40: 101-128.

Hendryx M, Fedorko E, Anesetti-Rothermel A. A geographical information system-based analysis of cancer mortality and population exposure to coal mining activities in West Virginia, United States of America. Geospatial Health 2010; 4: 243-256.

Hoek G, Fischer P, Van den Brandt P, Goldbohm S, Brunekreef B. Estimation of long-term average exposure to outdoor air pollution for a cohort study on mortality. Journal of Exposure Analysis and Environmental Epidemiology 2001; 11: 459-469.

Jerrett M, Burnett RT, Ma RJ, Pope CA, Krewski D, Newbold KB, et al. Spatial analysis of air pollution and mortality in Los Angeles. Epidemiology 2005; 16: 727-736.

Lertxundi-Manterola A, Saez M. Modelling of nitrogen dioxide (NO2) and fine particulate matter (PM10) air pollution in the metropolitan areas of Barcelona and Bilbao, Spain. Environmetrics 2009; 20: 477-493.

Maheswaran R, Elliott P. Stroke mortality associated with living near main roads in England and Wales - A geographical study. Stroke 2003; 34: 2776-2780.

Michelozzi P, Capon A, Kirchmayer U, Forastiere F, Biggeri A, Barca A, et al. Adult and childhood leukemia near a high-power radio station in Rome, Italy. American Journal of Epidemiology 2002; 155: 1096-1103.

Miller KA, Siscovick DS, Sheppard L, Shepherd K, Sullivan JH, Anderson GL, et al. Long-term exposure to air pollution and incidence of cardiovascular events in women. New England Journal of Medicine 2007; 356: 447-458.

Olea RA. Geostatistics for engineers and earth scientists. Boston: Kluwer Academic Publishers, 1999.

Pope CA, Burnett RT, Thun MJ, Calle EE, Krewski D, Ito K, et al. Lung cancer, cardiopulmonary mortality, and long-term exposure to fine particulate air pollution. Jama-Journal of the American Medical Association 2002; 287: 1132-1141.

Sohel N, Kanaroglou PS, Persson LA, Haq MZ, Rahman M, Vahter M. Spatial modelling of individual arsenic exposure via well water: evaluation of arsenic in urine, main water source and influence of neighbourhood water sources in rural Bangladesh. Journal of Environmental Monitoring 2010; 12: 1341-1348.

Son JY, Bell ML, Lee JT. Individual exposure to air pollution and lung function in Korea Spatial analysis using multiple exposure approaches. Environmental Research 2010; 110: 739-749.

TWEPA. Air quality in Taiwan annual report. Taiwan Environmental Protection Agency, Taipei, 2006.

USEPA. Technology Transfer Network:Air Quality System(AQS). U.S. Environmental Protection Agency, 1992.

USEPA. *AQS Raw Data Summary Formulas Draft*. In: Standards OoAQPa, editor. U.S. Environmental Protection Agency, Washington, DC, 2004.

Waller LA, Gotway CA. Applied spatial statistics for public health data. Hoboken, N.J.: John Wiley & Sons, 2004.

Young LJ, Gotway CA, Kearney G, Duclos C. Assessing Uncertainty in Support-Adjusted Spatial Misalignment Problems. Communications in Statistics-Theory and Methods 2009; 38: 3249-3264.

Young LJ, Gotway CA, Yang J, Kearney G, DuClos C. Assessing the association between environmental impacts and health outcomes: A case study from Florida. Statistics in Medicine 2008; 27: 3998-4015.

6

Evaluation of Hexamethylene Diisocyanate as an Indoor Air Pollutant and Biological Assessment of Hexamethylene Diamine in the Polyurethane Factories

Seyedtaghi Mirmohammadi[1], M. Hakimi. Ibrahim[2] and G. N. Saraji[3]

[1]Executive Deputy Dean of University Vice- Chancellor (MAZUMS),
[1]University Chancellor (Toxicants Safety), Department of Occupational Health,
[1]Faculty of Health, Mazandaran University of Medical Sciences, Mazandaran,
[2]Environmental Technology Division, School of Industrial Technology,
Universiti Sains Malaysia, Pulau Pinang,
[3]Tehran University of Medical Sciences ,
School of Public Health, Occupational Health Dept.
[1,3]Iran
[2]Malaysia

1. Introduction

Isocyanates are widely used in surface coatings, polyurethane foams, adhesives, resins, elastomers, binders, and sealants. In general, the types of exposures (inhale and dermal) encountered during the use of isocyanates (i.e., monomers, prepolymers, polyisocyanates, and oligomers) in workplace are related to vapor pressure of the individual chemical compounds (Bello *et al.*, 2002; Tury *et al.*, 2003). Isocyanates exist in many different physical forms in the workplace. The workers are potentially exposed to unreacted monomer, prepolymer, polyisocyanate, and/or oligomer species found in a given product formulation (Rosa *et al.*, 1999). They can also be exposed to partially reacted isocyanate containing intermediates formed in the course of polyurethane production (Woskie *et al.*, 2004; Tinnerberg and Sennbro, 2005). The second type of exposure might be more hazardous, as a number of isocyanates reactions are exothermic in nature and also evolve sufficient heat for the volatilization isocyanate compounds (NIOSH, 1994).

Isocyanate exposure is irritative to the skin, mucous membranes, eyes, and respiratory tract. The most common adverse health outcome associated with isocyanate exposure is asthma due to sensitization; less prevalent are contact dermatitis (both irritant and allergic forms) and hypersensitivity pneumonitis (Vandenplas *et al.*, 1992). Hexamethylene diisocyanate (HDI) is a colorless compound or may be slightly yellow liquid and is not much heavier than water (Adam *et al.*, 2002). This substance forms oily droplets in water and hydrolyses rapidly.

If isocyanates are inhaled, they are metabolised or broken down in body, eliminated and discharged through the urine. The metabolized form of Hexamethylene diisocyanate (HDI) is hexamethylene diamine (HDA) which is an organic compound and isocyanate-derived diamines from protein which is conjugated with urine (Dalene *et al.*, 1994b, 1996; Williams *et al.*, 1999; Dalene, 2004). HDA has chemical formula of $H_2N(CH_2)_6NH_2$. The molecule is a diamine, consisted of a hexamethylene hydrocarbon chain terminated with amine.

The level of isocyanate metabolites in urine is an indicator of how much isocyanate has been absorbed and how well the pollution controllers and the prevention units are working. The levels of HDA are reported as "μmol/mol creatinine". The guidance standard value for HDA is at a level of 1 μmol/mol creatinine and each sample above the guidance standard value is an indication of exposure to contaminated environment (Williams *et al.*, 1999 and 2004).

The specific goals of this chapter are to answer the following questions:

- What are the determinants of HDI pollution level in the polyurethane factories?
- Does any correlation exist between HDI concentration from air sampling and HDA concentration from biological sampling (worker's urine sample) in the polyurethane factories?
- What is the pollution condition in terms of HDI and HDA concentration in the different selected factories?

2. Materials and methods

2.1 Polyurethane factories

In this study, five HDI polyurethane factories were selected for air sampling and biological monitoring. These factories are situated in three provinces namely T (factories coded as H_1 and H_2), K (H_3 and H_4) and M (H_5).

The main uses of HDI in the HDI factories (H_1, H_2, H_3, H_4 and H_5) are for the production of adhesives, process regulators and paints and lacquers and varnishes, as well as for the production of polishing paints and adhesives. 5 HDI polyurethane factories were selected for this study; the number of workers working in factories H_1, H_2, H_3, H_4 and H_5 were 265, 130, 320, 120 and 130, respectively.

The workers were exposed to HDI in the indoor environment. There were some workers who did not work full-time in the workplace; they worked as officers and sometimes came and went into workplaces and are identified as unexposed workers. In the HDI polyurethane factories all the exposed workers wear simple gloves and simple paper respirable mask (Model: 2600 Half-mask with elastic) when working at their workplaces in the factories.

2.2 Air sampling and analysis

There were two group workers in the polyurethane factories. The first group as office workers who had least exposure (n= 100). The second one were workers working inside factories (n= 400) and they were sufficiently exposed to HDI. Only five samples were collected from office spaces as blank samples for least exposed environment.

Sample handling and preparation include those steps taken to stabilize the sample or make the sample more compatible with the analytical procedure. Sample handling considerations actually began before sample collection. For air sampling in the field, sample pumps [low flow (SIBATA, MP302 Model) Japan] were attached to individual adsorbent tubes using 30 cm lengths of clear, inert polythene tubing suitable for low-level organic compounds. Each pump and tube assembly was calibrated using a bubble flow meter (SKC UK, UK) in order to measure and adjust the sampling flow-rate precisely. The pumps were adjusted to provide a flow-rate of 2 L/min ~ Samples (static and personal) were taken adjacent to an operation where HDI was being handled in the factory for the periods of 2 h. On completion of field sampling, the tubes were sealed with vials end-caps and returned to the analytical laboratory. Samples were chilled to 2°C in a laboratory refrigerator and analysed within 24 h.

The first step in the analysis of a solution is derivatization of isocyanates for the separation through high-performance liquid chromatography (HPLC) for their qualitative as well as quantitative analysis. All the air samples were analyzed in the laboratory for HDI using HPLC through standard method of analysis (NIOSH, 1994)

2.3 Biological sampling and analysis

The biological sampling was carried out by collecting the workers' urine at the end of working shift; it was collected into polystyrene containers with citric acid and transferred to laboratory for GC Mass spectrometry analysis (Wu *et al.*, 1990; Williams *et al.*, 1999).

Due to the short half life (about 1.5 - 3 h) of HDA in urine, samples were collected at the end of the shift to detect any short term exposure as well as an estimation of the 8 h time weighted average exposure (Williams *et al.*, 2004).

In this study the urine samples were collected and dispatched in a similar way, frozen and sent blind to the laboratory for analysis. Based on Williams's method, subjects provided urine samples on the working day at the end of the working shift. The time of exposure and the personal protective equipment worn were recorded. For the factories securities, the samples obtained were labeled with code numbers and either the time or by sample number with other details like whether they were exposed worker, unexposed, or bystander.

2.4 Determinants of HDI pollution level

During air sampling for HDI concentration inside the factories, data for indoor relative humidity, indoor dry bulb temperature, altitude and dimension of factory were recorded. HDI was considered as dependent variable and the rest were independent variables. These independent indoor air variables were divided into two groups. Regression analysis procedure was used to state the statistical relationship between the variables and identify any meaningful relationship between those variables. Due to the fact that the number of independent variables was more than one, multiple linear regression analysis was used in the present analysis (Bahatin and Ibrahim, 2007). A multiple regression equation, which has four independent variables was used, and it can be expressed as follows:

$$Y = B_0 + B_1\,Rh + B_2\,Td + B_3\,D + B_4\,Alt + e_i \tag{1}$$

where:

Y is the dependent variable (HDI concentration)

Rh, Td, D and Alt are independent variables (relative humidity, dry bulb temperature factory dimension and altitude,) as the predictors in this model.

B_0, B_1, B_2, B_3 and B_4 are the model coefficients.

e_i is the residual error

The values for the constant and the coefficients are determined using the least-squares method which minimizes the error as 'e' in the above regression equation. The significance level of the constant and coefficients are statistically tested using t-distribution. The R^2

(coefficient of determination) determines the direction and significance level of relation between the variables in the mathematical model and shows how much the dependent variable is affected by the independent variables. All statistical analysis was performed using SPSS software Ver.16. In this study the alpha level is 0.05 ($P \leq 0.05$) similar to other indoor air pollution studies (Marek *et al.*, 1999; Molander *et al.*, 2002; Dalene, 2004).

3. Results and discussion

3.1 Air sampling and indoor air variables

The sampling protocols described in NIOSH Method 5522 were used throughout the experiments. The psychrometric parameters (relative humidity and dry bulb temperature) and factory parameters (dimension of factory and altitude) were measured for all HDI polyurethane factories and work stations. A total of 300 air samples were collected randomly from working places inside the 5 factories for the exposed workers (20 samples per each HDI factory and for three times per working shift).

Table 1. shows the maximum, minimum and mean values of indoor air independent variables with respect to HDI concentration in the polyurethane factories. The lowest minimum HDI concentration in all the factories was 61 μg/m³ and also the highest of the maximum HDI concentration was 96 μg/m³. These values can be considered as high, when compared to NIOSH exposure limit of 35 μg/m³. The mean HDI concentration was 78.8 μg/m³, the mean indoor relative humidity was 37.7%, and the mean dry bulb temperature was 28.3°C. The size of workplace for the five polyurethane factories ranged from 5,000 to 9,800 m³ and the altitude of factories were from as low as 22 to as high as 1200 m.

Average	H_5	H_4	H_3	H_2	H_1	Factories code Variables
						HDI concentration
91.2	88	90	90	92	96	(μg/m³), Max
64.8	61	64	66	66	67	Min
78.8	76.7	77	78.7	79.8	82.2	Mean
						Relative Humidity (%),
46.8	40	45	45	52	52	Max
31.2	31	31	31	31	32	Min
37.7	34	37	37	40	40.5	Mean
						Dry bulb temperature (°C),
31.8	30	32	32	32	33	Max
24	23	24	24	24	25	Min
28.3	27.1	28.7	28.7	27.8	29.3	Mean
7460	9800	9000	7400	6100	5000	Dimension of factory (m³)
882	22	890	1100	1200	1200	Altitude (m)

Table 1. Values of indoor air variables in the HDI polyurethane factories

It has been summarized and stated in Table 1. that several factories show variable concentration of diisocyanates, with respect to relative humidities of each factory. The highest mean value HDI concentration was 82.2 μg/m³ which corresponds to the highest mean relative humidity of 40.5% inside of factory H_1. Factories H_1 and H_2 were located in T province. Factory H_2 had the second highest mean HDI concentration and also the second highest mean relative humidity (79.8 μg/m³ and 40%, respectively). Factory H_5 had the lowest mean value HDI concentration which was 76.7 μg/m³. Factory H_5 was situated in M province and this factory had the lowest mean relative humidity of 34%.

3.2 Relationship of HDI and indoor air variables

In this study, the data (n = 100) of psychrometric parameters (indoor relative humidity and dry bulb temperature) and factory parameters (factory dimension and altitude) were collected and used to predict HDI pollution level. Correlation analysis was carried out after checking the normality assumptions for both variables, all parameters are strongly correlated with HDI concentration where $P < 0.05$. The calculated R^2 values for RH is 0.5461 ($R^2 = 0.739^2$), Td is 0.77 ($R^2=0.88^2$), D is 0.8767 ($R^2= -0.887^2$) and Alt is 0.7225 ($R^2= 0.85^2$). Since all parameters are strongly correlated, all of them are put in the regression model. Table 2. showed the regression model summary where it can be seen that 83.7% of the HDI concentration can be attributed to any or all the independent variables (relative humidity, dry bulb temperature, dimension and altitude) ($R^2 = 0.837$).

Model	r	R^2	Adjusted r^2
	0.915	0.837	0.83

Predictors: (Constant), Altitude (m), Dimension of factory (m³), Relative humidity (%), Dry bulb temperature (°C)

Table 2. Regression model summary of HDI

The results of the summary imply that all or some of parameters (altitude, dimension of factory, relative humidity and dry bulb temperature) can be significant predictors of HDI concentration in the polyurethane workplaces.
Table 3. below reports an analysis of variance for HDI concentration in the polyurethane factories. From the table, it can be seen that *F* is 121.9 which is significant at $P < 0.05$. We can conclude that the regression model predicts the concentration level of HDI significantly well.

Model		Mean Square	F	P value
	Regression	1155.895	121.934	0.0001
	Residual	9.48		
	Total			

Predictors: (Constant), Altitude (m), Dimension of factory (m³), Relative humidity (%), Dry bulb temperature (°C), Dependent Variable: HDI concentration (μg/m³)

Table 3. Regression model for HDI polyurethane factories factors

Since the results of regression model test in Table 1.3 illustrate that the independent variables are significant predictors of HDI concentration, we can employ equation 1.1 to stand for the different psychrometric and factory parameters in order to measure the predictive regression correlation between the parameters and HDI concentration.
Table 4. shows the results of regression analysis between HDI concentration and polyurethane indoor air parameters. Both indoor relative humidity and dry bulb temperature can be seen to be significant predictors of HDI pollution ($P< 0.05$). Both of these parameters fall under the

psychrometric parameters group. The other two parameters (factory dimension and altitude) are the factory parameters and both are not significant (P>0.05).

Model		Coefficients		T	P-value.
		B	SE		
	(Constant)	43.267	17.837	2.426	0.017
	Relative humidity (%)	0.367	0.071	5.182	0.0001
	Dry bulb temperature (°C)	1.112	0.399	2.785	0.006
	Dimension of factory (m^3)	-0.001	0.001	-1.479	0.142
	Altitude (m)	0.003	0.002	1.475	0.144

Dependent Variable: HDI concentration ($\mu g/m^3$)

Table 4. Result of regression analysis between HDI concentration and polyurethane indoor air parameters

The independent variables (relative humidity and dry bulb temperature) were reproduced for the model to find the regression coefficients for HDI pollution in the polyurethane factories. The coefficients with respect to the constant, relative humidity and the dry bulb temperature as well as the collinearity statistics are shown in Table 5.

Model		Coefficients		t	P value	Collinearity Statistics	
		B	SE			Tolerance	VIF
	(Constant)	27.771	2.638	10.528	0.0001		
	Relative humidity (%)	0.368	0.067	5.454	0.0001	0.592	1.689
	Dry bulb temperature (°C)	1.492	0.119	12.588	0.0001	0.592	1.689

Dependent Variable: HDI concentration ($\mu g/m^3$)

Table 5. Collinearity statistical model coefficients

The Variance Inflation Factor (VIF) measures the impact of collinearity among the variables in a regression model. Myers (1990) suggests that a value of more than 10 means there is a concern to worry about collinearity. Menard (1995) suggests tolerance (or 1/VIF) below 0.2 indicates a potential collinearity problem. For the current work, the suggestion by Myers (1990) is used and it can be seen in Table 1.5 the VIF values are well below 10 and the tolerance is above 0.2 indeed. Therefore, it can be safely said that there is no collinearity within the current data.

The two factory predictor variables (dimension of factory and altitude) were found to be not significant in the model. They were eliminated for making a new regression model based on relative humidity and dry bulb temperature. By replacing Y, X_1 and X_2 with HDI, Rh (relative humidity) and Td (dry bulb temperature), respectively and eliminating X_3 (factory dimension) and X_4 (altitude) and also substituting the relevant coefficients from Table 5., Equation 2, can be written as follows:

$$HDI = 27.77 + 0.368Rh + 1.492Td \quad (2)$$

Equation 2 implies that relative humidity and dry bulb temperature affect diisocyanates pollutant concentration in the work places. The background HDI concentration was about 27.77µg/m³ as indicated by the value of the constant in the regression equation.

3.3 HDI concentration determinants

Figure 1. shows the relationship between HDI concentration and psychrometric parameters (relative humidity and dry bulb temperature) in five polyurethane factories based on equation 1.2.

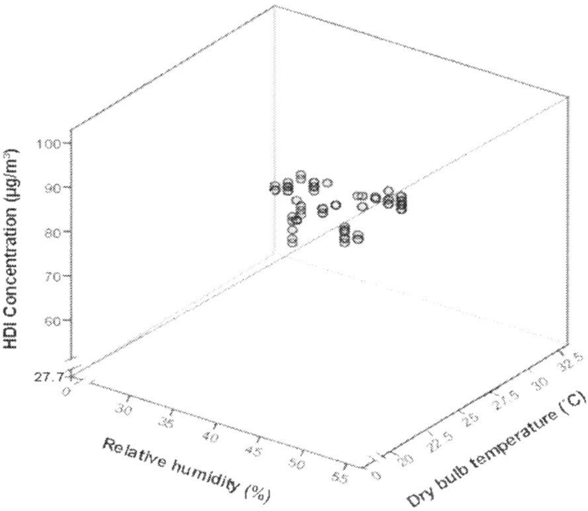

Fig. 1. Relationship between HDI concentration and psychrometric parameters (relative humidity and dry bulb temperature)

Both indoor air relative humidity and dry bulb temperature are significantly contributing to the variability of the HDI concentration ($R^2 = 0.837$) and both factors also show a straight positive relationship with the HDI concentration. This means that as the indoor relative humidity or the dry bulb temperature increases, the HDI concentration also increases.

As for the effects of relative humidity on isocyanate concentration, Ludwig and Urban (1996) observed that reactions of isocyanate groups with OH groups during cross-linking are inversely proportional to the relative humidity of the environment. They explained that the presence of competing reactions between water and isocyanate, hinder the degree of cross-linking between them. Abram and Bowler (2005) studied the effect of relative humidity on the curing and dielectric properties of the polyurethane-based composites. They have found that a polyurethane factory at 87% relative humidity (RH) gave more noticeable effect as compared to one at 37.7% RH. The electromagnetic properties of the polyurea/polyurethane-based composites studied were found to be strongly influenced by the presence of water vapor during the curing process, as evidenced by the significant difference in the real relative permittivity of samples cured in different RH environments.

This difference was caused primarily by water uptake into the polymer matrix. The RH alters the dielectric properties of the composite material due to its strong polar nature and high value of real relative permittivity. In this study, the RH was in the range of 31 to 52%. The working range for RH in this study is lower than that reported by Abram and Bowler (2005) for the noticeable effects (87% RH). Hence the effect of RH is expected to be not as significant as to the effects of temperature.

The mean isocyanate concentration in this study was 78.87μg/m³ (0.079mg/m³). This is comparable to an investigation conducted in southern Australian auto body shops which reported a geometric mean concentration of 0.07 mg/m³ isocyanates (range <0.01-3.5 mg/m³ NCO) (Pisaniello and Muriale, 1989). Swedish auto body shops exposures during spray painting showed higher values between 0.26 and 1.1 mg/m³ (Torling *et al.*, 1990). A survey of Oregon auto body shops measured a geometric mean concentration of 0.35 mg/m³ isocyanate with a maximum of 4 mg/m³ isocyanate (Janko *et al.*, 1992). Maître *et al.* (1996) have reported an arithmetic mean level of 0.33 mg/m³ isocyanate (range 0.05-0.65 mg/m³) in a French isocyanates spray factory. These findings suggest that the measurements in this research are comparable to other related literatures. However comparison of such results is quite difficult because of the sampling time, sampling method and manufacturing process were totally different (Maître *et al.*, 1993).

Previous studies often involved isocyanate exposure levels in auto body shops (Pisaniello and Muriale, 1989; Torling *et al.*, 1990; Woskie *et al.*, 2004) and up to date little work has been done to evaluate the determinants of isocyanate concentration levels in factories producing polyurethane foams with respect to HDI and indoor air quality variables. The current work points to psychrometric factors (indoor air temperature and relative humidity) as important predictors of HDI within the factories. Woskie *et al.* (2004) have pointed out that for small, low volume auto body shops as a consistent predictor of higher exposures especially during the colder months when buildings were closed up and general ventilation was reduced. However in this research dimension or size of the factories along with the altitude were seen to be not significant as HDI pollution determinants. Large size factory has difficulty in interpretation. For instance, Woskie *et al.* (2004) in a work based on auto body shops pointed that 'large shop' might mean more activities or jobs per day or a larger less cramped work area with greater general dilution ventilation.

The current work admittedly has limitations but may be a useful initiative in estimating possible HDI pollution situation in the polyurethane workplaces based on indoor air temperature and relative humidity. It can hopefully help to provide a basis to prioratise future exposure evaluation and intervention efforts and reduce the workers to exposure to isocyanates which are well known as asthmagens and respiratory irritants.

3.4 Biological monitoring

The aims of the present investigation regarding biological monitoring are to: (i) show the relationship between air pollution and exposed worker health; (ii) characterize worker diisocyanate exposure and examine the relative negative impact of pollution and inhalation routes in HDI polyurethane factories.

3.4.1 Subject individual characteristics

Individual exposure data as well as information regarding each individual were extracted, pooled, and entered into SPSS, V. 16 software for statistical analysis. The sample size for biological monitoring was 50 workers. At the end of the shift, workers took a shower and changed clothes before going to the factory's medical service in order to give urine samples. All samples were collected in polystyrene bottles containing 10 g citric acid, and stored at 4 °C until analysis. A negative ion chemical ionization mass spectrometry was used to determine the urinary HDA.

Table 6. describes the mean age of workers, who work in the HDI polyurethane factories was 34.5 years. The mean duration of work history was 4.9 years and the mean weight of the workers was 69.6 kg.

	N	Mean	SD	Min	Max
Age (year)	50	34.48	7.465	24	47
Weight (kg)	50	69.64	12.348	50	88
Work history (year)	50	4.74	2.94	2	13

Table 6. Characteristics of subjects

Table 7. shows that 54% of total workers, working in the HDI factories were categorized as smokers and also the same percentage of HDI workers had some symptoms [sore eyes, running nose, sore throat, coughing, wheezing (asthma) and chest tightness] relevant to diisocyanates exposure.

Variables	N	Percent
Smoking	27 - Smoker	54
	23 - None	46
Symptoms of disease	27- with	54
	23- without	46

Table 7. Frequency and percentage of smokers and symptoms of disease

Table 1.8 shows the age, weight and years of the services of the workers in the HDI factories. The information is presented in terms of the ranges, frequency and percentage.

Variables	Range	Frequency	Percent
Age (year)	20-30	19	38
	31-40	16	32
	41-50	15	30
Weight (kg)	40-60	20	40
	61-80	11	22
	81-100	19	38
Years of services (year)	0-5	35	70
	6-10	10	20
	11-15	5	10

Table 8. Statistical data for the age, weight and years of services of the workers

3.4.2 Urinary concentration of HDA in the factories

Table 9 shows the descriptive statistics of urinary hexamethylene diamine (HDA) in different factories. The maximum concentration of HDA measured from worker's urine in the HDI polyurethane factories was 4 μmol/mol creatinine and the mean value was in the range of 3.01 to 3.58 μmol/mol creatinine for all factories. The urine results for all of the workers indicated high exposure with respect to HDI, because of the lowest concentration of HDA in their urine was 2.33 μmol/mol creatinine.

Factory code	**Mean**	**SD**	**Minimum**	**Maximum**
H_1	3.52	0.630151	2.67	4
H_2	3.58	0.540577	2.83	4
H_3	3.30	0.604230	2.67	4
H_4	3.01	0.547881	2.33	4
H_5	3.01	0.553021	2.33	4

Table 9. Descriptive statistics of hexamethylene diamine (HDA) in different factories (n= 10)

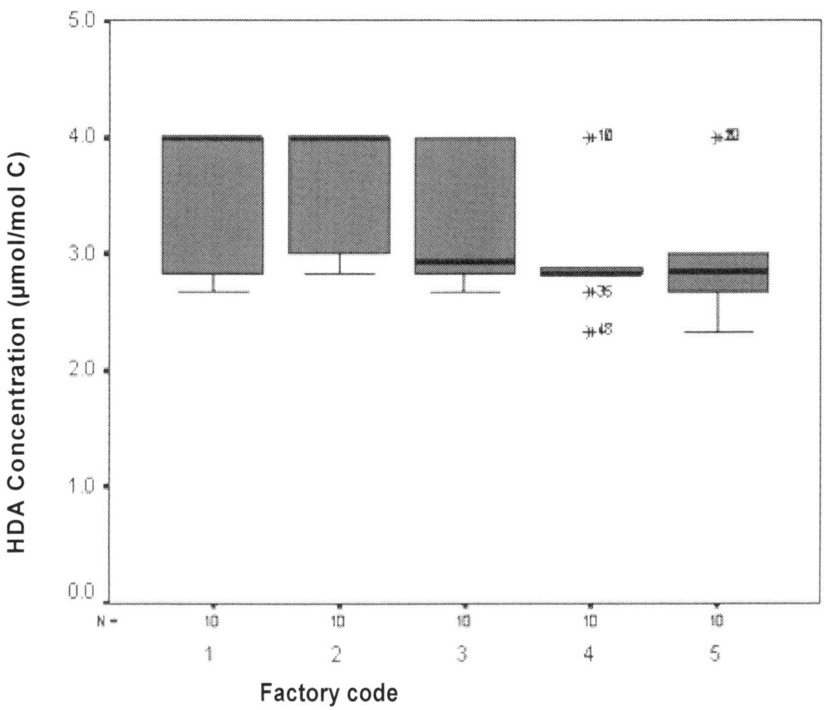

Note: Guideline value: 1 μmol/mol creatinine (Williams et al., 1999)

Fig. 2. HDA concentrations in different factories

Figure 2. indicates different HDA concentration compare to different factories because there are some factors that affected on workers exposure in the different work area, for example working process, space of work station, etc.

3.4.3 HDI and HDA relationship in the factories

Fifty male workers in the five polyurethane factories, in an age range of 20-50 years, participated in the present investigation. They were engaged in injection glue device, painting sprayer and varnishing. Based on the diisocyanates biomonitoring method (Rosenberg *et al.*, 1986; Skarping *et al.*, 1995; HSE, 2005), the urine samples were collected at the end of exposure. This is due to the urinary half-life is about 2 hours and results reflected after 2 - 4 h exposure. Table 1.10 shows the HDA values in the urine associated with HDI concentration for the 50 workers. Air samples taken near the gluing operation contained high concentration of HDI (83.1 to 92.7 µg/m^3) as compared to the other operations. The workers at the gluing operations also showed high urinary HDA concentration of between 2.88 to 4 µmol/mol creatinine. Workers involved in other operations had lower HDA concentration in their urine and also they had lower values of HDI exposure. The lowest HDI exposure was in varnishing operation (62.24 µg/m^3) where the HDA value was found to be also the lowest at 2.33 µmol/mol creatinine but the value is also above the guideline value of 1 µmol/mol creatinine (Williams *et al.*, 1999).

Subject Frequency	Operation	HDI (µg/m^3)	HDA (µmol/molC)
4	Gluing	92.7	4
4	Gluing	90.1	2.88
3	Gluing	87.62	2.88
3	Gluing	85.67	2.88
3	Gluing	83.1	4
4	Paint sprayer	80.57	2.88
3	Paint sprayer	79.29	2.88
3	Varnishing	78.38	2.88
3	Paint sprayer	74.67	2.88
3	Paint sprayer	68.5	3
4	Paint sprayer	67.72	2.83
3	Varnishing	66.94	2.83
3	Paint sprayer	65.33	2.67
3	Paint sprayer	64.11	2.67
4	Varnishing	62.24	2.33

Table 10. Exposure to HDI and excretion of HDA in workers urine (n=50)

Table 11. shows the correlated results between HDI concentration in the air and HDA in the urine sample taken from workers. It can be seen that HDI concentration in the air is directly related to HDA concentration in the urine of workers with a Pearson correlation coefficient at $r = 0.857$ and the significance value is less than 0.001. This means that, as the HDI concentration in the air in factories increase, the HDA level in the urine of workers also increase.

The regression model summary for HDI and HDA is shown in Table 12. It indicates that the value of R^2 is 0.735. This means that the exposure to HDI pollution in the factories account for 73.5% of the level of HDA in the workers' urine.

Table 12. shows the F-test in the regression analysis; it is a relationship test between HDI in air samples and HDA in urine samples. For these data F is 133.247, which is significant at $P< 0.05$. Thus the regression model predicts the exposure to HDI pollutant significantly well.

		HDA Concentration	HDI Concentration
Pearson Correlation	HDA Concentration	.	0.857
	HDI Concentration	0.857	1
Sig. (2-tailed)	HDA Concentration	.	0.0001
	HDI Concentration	0.0001	.
N	HDA Concentration	50	50
	HDI Concentration	50	50

Correlation is significant at the 0.05 level.

Table 11. Generalized correlation HDI concentration and HDA

Model		**Sum of Squares**	**F**	**P value**
	Regression	13.161	133.247	< 0.0001
	Residual	4.741		
	Total	17.903		

(Constant), HDA Concentration, HDI Concentration

Table 12. Generalized regression model for HDI and HDA

The regression model used in this study is a simple linear model in the form of

$$Y = B_0 + B_1 X_1. \quad (3)$$

From Table 13, B_0 is -0.62 and B_1 is 0.051.
Since B_0 is rather small in magnitude, the B_0 value can be ignored but B_1 is significant in model ($P< 0.05$). The predictive relationship for HDI and HDA can be suggested in a linear regression equation ($R^2= 0.735$) as shown in Equation 4:

$$HDA = 0.051\ HDI \quad (4)$$

where
HDA is dependent variable
HDI is independent variable

Model		**Coefficients**		**T**	**P value**
		B	SE		
	(Constant)	-0.62	0.341	-1.818	0.075
	HDI Pollution	0.051	0.004	11.543	0.0001

a Dependent Variable: HDA Concentration

Table 13. Generalized regression model coefficients for HDI and HDA

Figure 3. shows the graph for the relationship between factory air HDI concentration and urine HDA concentration of the workers. The graph shows a linear relationship between HDI and HDA. No background HDA was detected in the workers and the graph validates HDA as an initial indicator of a preceding exposure of workers to HDI with $R^2 = 0.735$.
In a related research conducted by Maitre *et al.* (1993) focusing on biological monitoring of occupational exposure among nine (9) workers; they were exposed to toluene diisocyanate (TDI) in TDI-based polyurethane production. The study has validated the use of urinary

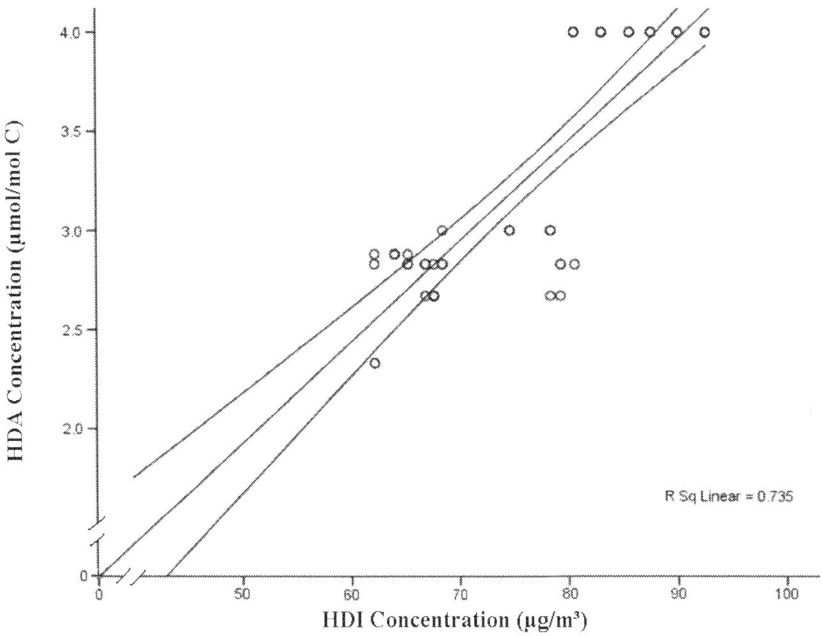

Fig. 3. Relationship between air HDI concentration and urinary HDA concentration

toluene diamine (TDA) as an indicator of preceding exposure to TDI. Their TDI exposure level was 9.5 to 94 µg/m^3, and TDA concentrations varied from 6.5 to 31.7 µmol/mol creatinine. TDI and TDA were found to be linearly related and no background TDA was detected. It was noted that the workers in the research carried out by Maitre *et al.* (1993) did not wear personal protective equipment and there was also no ventilation system. That possibly explains the higher values of TDA and TDI. The workers were also engaged in different jobs (injection molding, mold stripping, cutting and gluing foam). Some were directly exposed and some were indirectly exposed to isocyanates.

In glue injection, the concentration of exposure to HDI in the air was higher than the working area near the paint spraying area or varnishing. In the other study, a high levels exposure has also been found by Lesage *et al.* (2001) at isocyanate injection workplace. The concentrations of HDA detected in this study are less than those found in the other studies (Brorson *et al.*, 1990; Maitre *et al.*, 1993). Brorson *et al.* (1990) have reported a mean value of HDA by the end of the shift was 20 µmol/mol creatinine after 7.5 h exposure to 25 µg/m^3 HDI. Whereas Maitre *et al.* (1993) have measured a value of 12 µmol/mol creatinine for HDI exposure at the same concentration from a separate survey of workers.

In a developing country, the above situations reinforced the reasons why Kakooei *et al.* (2006) had expressed the importance of studying indoor air pollution in such industries. For the imported technology and the machinery used without suitable application of adequate engineering controls and proper safe work practice together with the levels of training and awareness can cause great exposure to air pollutants and may result in more occupational health problems than in the developed countries.

In a similar study that has been conducted by Cocker (2007) the urine samples which were also analysed for the diamine metabolites of hexamethylene diisocyanate had detectable levels with 22 μmol/mol creatinine. This value was much higher than that the maximum urinary concentration of the current study (4 μmol/mol creatinine).
Comparison of obtained results with previous work was difficult; because of the sampling time, sampling method, and manufacturing process were totally different. The present study detected a straight correlation between the isocyanate concentration in the air and its biological results in the workers' urine. The obtained biological results from the present conducted research was lesser than that of the results reported by Lesage *et al.* (2001) that was related to manufacturing of foam plants.
The obtained results from the analysis of worker's urine showed that all of the workers were exposed via isocyanates in their workplaces based on guidance value for HDA. This implies that the high level of metabolites concentration in the urine of the workers can be attributed to the polluted situation of the workplaces. The concentrations of diisocyanates metabolite detected in the polyurethane workers were higher than the guidance value (>1 μmol/mol creatinine). However, this result is low compared to work performed by Brorson *et al.* (1990) where it was reported that the mean value of isocyanates metabolite at the end of working shift was 20 μmol/mol creatinine. Similarly, Maitre *et al.* (1996) have measured a value of 12 μmol/mol creatinine for diisocyanates exposure at the same concentration from a survey of workers.
The result of present investigation showed that the HDA concentration was also less than the research conducted by Tiljander *et al.* (1989) in US polyurethane factories for methylene diamine (5-30 μg/l in urine) and it was similar to results that have been reported by Selden *et al.* (1992).
The results of present study showed that biological monitoring can be a useful tool to assess isocyanate exposure in this group of exposed workers in the polyurethane factories. This work is comparable to the biological monitoring work done by Williams *et al.* (1999) when assessing exposure from use of isocyanates in motor vehicle repair-shop. They have noted that urinary biological monitoring has the potential to measure a worker's body burden received by all routes of exposure (oral, inhalation and dermal). However biological monitoring does not determine the route of exposure which means exposure could have been through any of the three methods although inhalation is the most likely route given the lack of published evidence for the absorption of isocyanates through the skin (Williams *et al.*, 1999).
Tinnerberg and Sennbro (2005) have cautioned that both air and biological monitoring methods have limitations. They have discussed that the exposures are complex and different in various environments. In their studies, biological monitoring has been used as a standard method to survey the exposure to aromatic diisocyanates and air monitoring to assess peak exposures and to find emission sources. They have also found that by using biomarkers it was easier to collect and analyse more samples as the sampling was not as time-consuming as air monitoring. Analyzing urine may be simple however in this study, it was also noted that the workers were quite reluctant to give their urine possibly due to urinary analysis is often associated to drug screening tests.

3.4.4 HDI work-related disease risk assessment

Diisocyanate work-related disease risk assessment is defined as an estimate of daily human exposure to diisocyanates pollution (HSE, 2005). Risk assessment is derived whilst adequate data exist to identify the high risk factors in the polyurethane factories. Eventually, it is important to comprehend how HDI interact with humans, with other species and physical

environment. The risk assessment regression model summary where about 89.5% of the HDI risk can be attributed by any or all the independent variables (age and weight of workers, smoking, years of services and HDI pollution level) ($R^2 = 0.895$, Adjusted $R^2 = 0.883$).
A multiple linear regression analysis was performed to evaluate individual factors for developing asthma induced by diisocyanates (Table 14.). Two factors were statistically related to risk of work-related disease of diisocyanate at the 0.05 significance level. These were: (*a*) weight of exposed workers, (*b*) HDI pollution. None of them had experienced previous incidents of exposure to diisocyanates.

Model		**Coefficients**		t	**P value.**
		B	SE		
	(Constant)	0.201	0.171	1.178	0.245
	HDI Pollution	0.631	0.112	5.612	0.0001
	Age	-0.030	0.031	-0.972	0.336
	Smoking	-0.049	0.095	-0.512	0.611
	Weight	0.218	0.087	2.509	0.016
	Years of services	0.046	0.040	1.134	0.263

Dependent Variables: Symptoms of HDI work-related disease

Table 14. Regression summary model for HDI work-related disease risk assessment

Risk assessment of exposure to diisocyanates among 50 workers increased with HDI pollution with potentially higher exposure with trends in TWA concentrations with regard to individual parameters (workers' weight and HDI pollution) (Redlich and Meryl, 2002).
For HDI work-related disease risk assessment were assessed for possible risk factors experiencing an exposure to diisocyanate through inhalation. The assessment was based on determining the total risk set at the time of each event and accumulating evidence across events in favor or against specific associations. The likelihood of HDI work related disease risk assessment was not statistically related to age, history of work, or current level of cigarette smoking and years of services in the factory. In this study significant associations were found for diisocyanates concentration (expressed as a TWA) and workers' weight and HDI pollution. There was a strong correlation between symptoms of HDI work-related disease and the two independent variables (workers' weight and HDI pollution), as summarized in Table 15.

Model		**Sum of Squares**	F	**P value**
	Regression	1412.599	117.938	< 0.0001
	Residual	105.401		
	Total	1518.000		

Predictors: (Constant), HDI pollution, workers' weight. Dependent Variable: Symptoms of HDI work-related disease

Table 15. Summarized model for isocyanates risk assessment

Under ascertainment is less likely among long term workers at the factories because of the asthmatic people who continue to work in exposed areas often develop increasingly severe symptoms, but surprisingly, there is no significance between years of services and incidence of work-related disease for diisocyanates workers. The result of this study is similar to the other research conducted by Diem *et al.* (1982) and Weill *et al.* (1981). The present study detected the effects of weight of the workers on HDI work-related disease and the results were similar to the study conducted by Lange *et al.* (1998).

4. Conclusion

Both indoor air relative humidity and temperature significantly contribute to the variability of the HDI concentration ($R^2 = 0.837$) and both factors also showed a linear relationship with the HDI concentration. The relationship between HDI and relative humidity and indoor air temperature was found through multiple linear regression according to equation 1.2.

$$HDI = 27.77 + 0.37\ Rh + 1.49\ Td \tag{2}$$

It was also shown that urinary HDA is detectable following HDI exposure. Urinary HDA concentration of the workers was found to be linearly related to HDI concentration of the air in the factories as correlated by following relation:

$$HDA = 0.051\ HDI \text{ with } R^2 = 0.735. \tag{4}$$

The HDI work-related disease risk assessment of the present study showed that two personal and work condition variables were correlated with Symptoms of HDI work-related disease in the polyurethane factories (workers' weight and HDI pollution).

Finally, this study has shown that air monitoring together with biological monitoring can be used to estimate the HDI pollution situation as well as the exposure to workers of polyurethane factories. Such a study at polyurethane factories having more than 100 workers in a factory is not known till to date. Future research should focus among others on HDI exposure in relation to outdoor psychrometric factors, ventilation or application of adequate engineering controls, respiratory protective equipment, behavioral issues, the levels of training and awareness, occupational health background as well as the usage of other possible biomarkers of HDI.

5. References

Abram, E R. and Bowler, N. (2005) Effect of relative humidity on the curing and dielectric properties of polyurethane-based composites. This paper appears in: *Electrical Insulation and Dielectric Phenomena, 2005. CEIDP '05. 2005 Annual Report Conference on Publication* Date: 16-19 Oct. 2005

Adam, V. Wisnewski., Qing, L., Jing-Jing, M., Nadine, M. and Carrie, A R. (2002) Effects of hexamethylene diisocyanate exposure on human airway epithelial cells: in vitro cellular and molecular studies. *Environ Health Perspect.* 110 (9): 901–907.

Bahattin, M., and İbrahim, K. (2007) The Relation Between Meteorological Factors and Pollutants Concentrations in Karabük City. G.U. *J Sc.* 20 (4): 87-95

Bello, D., Streicher, R P. and Woskie, S R. (2002) Evaluation of the NIOSH draft method 5525 for determination of the total reactive isocyanate group (Trig) for Aliphatic Isocyanates in auto body repair shops. *J Environ Med.* 4:351–360.

Brorson, T., Skarping, G. and Nielsen, J. (1990a) Biological monitoring of isocyanates and related amines. II. Test chamber exposure of humans exposed to 1,6-hexamethylene diamine as the trifluoroethyl chloroformate derivative. *Int Arch Occup Environ Health.* 62: 385-389.

Cocker, J. (2007) Biological monitoring for isocyanates. *Occupational Medicine.* 57: 391–396.

Dalene, M., Skarping, G. and Brorson, T. (1996) Chromatographic determination of amines in biological fluids with special reference to the biological monitoring of isocyanates and amines. IV. Determination of 1, 6-hexamethylenediamine in human urine using capillary gas chromatography and selective ion monitoring. *J Chromatogra A*. 516: 405- 413.

Dalene, M., Skarping, G. and Tinnerberg, H. (1994b) Biological monitoring of hexamethylene and isophorone isocyanate by the determination of hexamethylene and isophorone diamine in hydrolysed urine using liquid chromatography and mass spectrometry. *Analyst*. 119: 2051-2055.

Dalene, M. (2004) Determination of airborne isocyanates as di-*n*-butylamine derivatives using liquid chromatography and tandem mass spectrometry. *Anal Chim Acta*. 534, 263-269.

Diem, J E., Jones, R N. and Hendrick, D J. (1982) Five year longitudinal study of workersemployed in a new toluene diisocyanate manufacturing plant. *Am Rev Respir Dis*. 126: 420-428.

HSE. Health and Safety Executive. (2005) Methods for the determination of hazardous substances; MDHS 25/3. Organic isocyanates in air. *Sudbury UK*: Health and Safety Laboratory.

Janko, M., McCarthy, K., Fajer, M. and van, Raalte, J. (1992) Occupational exposure to 1, 6-hexamethylene diisocyanate- based polyisocyanates in the State of Oregon, 1980-1990. *Am Ind Hyg Assoc J*. 53:331-338.

Kakooei, H., Shahtaheri, S J. and Karbasi, H A. (2006) Evaluation of workers' exposure to methylene diphenyl diisocyanate (MDI) in an automobile manufacturing company, Iran. *Int J Occup Saf Ergon*. 12(4): 443-449.

Lange, P., Parner, J. and Vestbo, J. (1998) A 15-year follow-up study of ventilatory function in adults and asthma. *N Engl J Med*. 339:1194-1200.

Lesage, J., Carlton, G., Streicher, R. and Song, R. (2001) Erratum to "comparison of sampling methods for monomer and polyisocyanates of 1,6-hexamethylene diisocyanate during spray finishing operations. *Appl Occup Env Hyg* Jan; 16(1):11- 17.

Ludwig, B W. and Urban, M W. (1996) Quantitative determination of isocyanate concentration in crosslinked polyurethane coatings. J Coating Techn. *Available from Word Wide Web: www.superiortire.com*

Maitre, A., Berode, M., Perdrix, A., Romazini, S. and Savolainen, H. (1993a) Biological monitoring of occupational exposure to toluene diisocyanate. *Int Arch Occup Environ Health*. 65:97-100.

Maitre, A., Berode, M., Perdrix, A., Stoklov, M., Mallion, JM. and Savolainen, H.(1996) Urinary hexane diamine as an indicator of occupational exposure to hexamethylene diisocyanate. *Int Arch Occup Environ Health*. 69:65-68.

Marek, W., Potthast, J., Marczynski, B., Mensing, T. and Baur, X. (1999) Subchronic Exposure to Diisocyanates Increases Guinea Pig Tracheal Smooth Muscle Responses to Acetylcholine. *Respiration*. 66:156-161.

Menard, S. (1995). Applied logistic regression analysis. Sage university paper series on quantitative applications in the social sciences. 07-106, Thousand Oaks, CA: Sage.

Molander, P., Levin, J O. and Ostin, A. (2002) Harmonized Nordic strategies for isocyanate monitoring in workroom atmospheres. *J Environ Monit*. 4: 685-687.

Myers, R. (1990). Classical and modern regression with applications (2nd ed.). Boston, MA: Duxbury.

NIOSH. (1994) Determination of airborne isocyanate exposure. NIOSH manual of analytical methods (Chapter K, 4th ed). Cincinnati, OH: US Department of Health and

Human Services, Public Health Service, Centers for Disease Control and Prevention, National institute for occupational safety and health, *DHHS* (NIOSH) publication no. 94-113.

Pisaniello, D.L. and Muriale, L. (1989) The use of isocyanate paints in auto refinishing—a survey of isocyanate exposures and related work practices in South Australia. *Ann Occup Hyg*. 33: 563–572.

Redlich, C and Meryl, K. (2002) Diisocyanate asthma: clinical aspects and immunopathogenesis. *International Immunopharmacology*. 2: 213- 224.

Rosa, J K, and Samuel, P T. (1999) An Approach to Area Sampling and Analysis for Total Isocyanates in Workplace Air. *AIHA*. 60. 200-207.

Rosenberg, C. and Savolainen, H. (1986) Determination of occupational exposure to toluene diisocyanate by biological monitoring. *J Chromatogr*. 367:385–392.

Selden, A., Berg, P., Jakobsson, R. and Laval, J. (1992) Methylene dianiline: assessment of exposure and cancer morbibity in power generator workers. *Int Arch Occup Environ Health*. 63:403–408.

Skarping, G., Dalene, M. and Littorin, M. (1995) 4, 4'-methylenedianiline in hydrolysed serum and urine from a worker exposed to thermal degradation products of methylene diphenyl diisocyanate elastomers. *Int Arch Occup Environ Health*. 67: 73–77.

Tiljander, A., Skarping, G. and Dalene, M. (1989) Chromatographic determinations of amines in biological fluids with special reference to the biological monitoring of isocyanates and amines. III. Determination of 4, 4-methylenedianiline in hydrolysed human urine using derivatization and capillary gas chromatography with selected ion monitoring. *J Chromatogra A*. 479:145–152.

Tinnerberg, H. and Sennbro, C J. (2005) Assessment of exposure to aromatic diisocyanates air or biological monitoring. In *IOHA* Pilanesberg, Page 61-64.

Torling, G., Alexandersson, R., Hedenstierna, G. and Plato, N. (1990) Decreased lung function and exposure to diisocyanates (HDI and HDI-BT) in car repair painters: observations on re-examination 6 years after initial study. *Am J Ind Med*. 17: 299–310.

Tury, B., Pemberton, D. and Bailey, R E. (2003) Fate and potential environmental effects of methylenediphenyl diisocyanate and toluene diisocyanate released into the atmosphere. *J Air Waste Manage Assoc*. 53: 61-66.

Vandenplas, O., Malo, J C., Cartier, A., Perrault, G. and Cloutier, Y. (1992) Closed-circuit methodology for inhalation challenge tests with isocyanates. *Am Rev Respir Dis*; 145:582-7.

Weill, H., Butcher, B. and Dharmarajan, V. (1981) Respiratory and immunologic evaluation of isocyanate exposure in a new manufacturing plant. Cincinnati, OH: Department of Health and Human Services (NIOSH). (Publ No 81–125.)

Williams, E., Brown, Allen. H., Green, Mery., Karol, H. and Yves, C E. (2004) Immobilized cholinesterase to detect airborne concentrations of hexamethylene diisocyanate (HDI). *Toxicol App Pharmacol*. 73(1), 105-109.

Williams, N R., Jones, K. and Cocker, J. (1999) Biological monitoring to assess exposures from use of isocyanates in motor vehicle repair. *Occup Environ Med*. 56, 598–601.

Woskie, S R., Sparer, R J. Gore, M., Stowe, D., Bello, Y., Liu, F., Youngs, C., Redilich, E. and Cullen. (2004) Determinants of Isocyanate Exposures in Auto Body Repair and Refinishing Shops. *Ann Occup Hygin*. 48(5), 393-403.

Wu, W S., Stoyanoff, R E., Szklar, R S., and Gaind, V S. (1990) Application of tryptamine as a derivatizing agent for airborne isocyanate determination: Part 3. Evaluation of total isocyanates analysis by high performance liquid chromatography with fluorescence and amperometric detection. *Analyst*. 115, 801–807.

7

Assessment on the Ozone Air Pollution in a Medium Metropolitan Area: Seville (Spain)

Jose A. Adame[1], Antonio Lozano[2], Juan Contreras[3] and Benito A. de la Morena[1]
[1]*National Institute for Aerospace Technology (INTA), Atmospheric Research and Instrumentation Branch, Atmospheric Sounding Station "El Arenosillo", Mazagón - Huelva*
[2]*The Environmental Management Company (EGMASA), Seville*
[3]*Environmental Council of the Junta de Andalucía, Seville*
Spain

1. Introduction

Problems caused by the air pollution which are generated mainly in metropolitan areas and industrial complexes, become in the major challenges for the current society. The air pollution has its origin in the chemical species emission into the atmosphere which may be harmful for human health (Lin, et al., 2008; Doherty et al., 2009), vegetation and ecosystems (Cape, 2008). The pollutants in the atmosphere can be classified in primary, directly emitted to the atmosphere, or secondary, formed in the atmosphere from chemical or photochemical reactions.

Nowadays, the main air pollution problems are originated by secondary chemical species, as particles and surface ozone (Kumar et al., 2010; Lefohn et al., 2010). In Europe, there are several Directives to protect air quality, as the Directive 2008 which defines the new threshold for different pollutants.

Generally, the highest concentrations of secondary species are registered in areas far away from the emissions sources, as suburban and rural areas. Nevertheless, there are urban areas that due to both meteorological and emissions conditions can suffer high levels of secondary substances (Papanastasiou and Melas, 2009; Alvim-Ferraz et al., 2006).

In many countries there are air quality networks which measure concentrations both primary (NO, CO, SO_2) and secondary pollutants and provide air quality information in real-time. However, in spite of the elevated economic cost of these networks, it is usual that the collected data are only used to determinate if a pollutant exceed the legal threshold according to the current Directive. Nevertheless, with the information obtained in the air quality networks is possible to carry out studies in order to improve the understanding of the atmospheric pollution and to help in the definition of plans and control strategies to the air quality.

The work presented in this chapter is linked with a secondary pollutant, surface ozone, using data from an air quality network in a sensible region suitable to photochemical pollution problems. The study has been carried out in the Seville metropolitan area (low

Guadalquivir valley) and it is focused on the identification and analysis of ozone events. The events or episodes are defined such as situations with elevated ozone concentrations, exceeding the European Directive threshold, which are produced frequently in this region and affect to one millions inhabitants.

The aim of this paper can be double. To provide information of the ozone events in the metropolitan area with highest population in the southwest of Europe. In addition, to present a methodology which could be applied in other regions with air quality networks.

The first section of the chapter has been devoted to present a general vision of the photochemical air pollution. The second, Material and methods, shows the main features of the studied area, monitoring stations and dataset used in this work. The section 3 exposes the results obtained. The results section starts with a meteorological overview of this area. Following, seasonal and daily variation on ozone and NO_x are presented. In order to know the ozone events, criteria to extract of episodes has been applied and from the days selected as episodes have been analysed its features. In addition, a comparative analysis of days with and without events has been carried out. The phases of an event have been defined; and a specific episode has been studied with detail. Finally, conclusions have been included in the last section.

2. Materials and methods

In this section are presented the main characteristics of the study area as well as the monitoring stations and database used in this work

2.1 Description area

The Seville metropolitan area is located in the low Guadalquivir valley (south of Spain), about 100 km away from the gulf of Cadiz (Fig. 1). This metropolitan area has a population

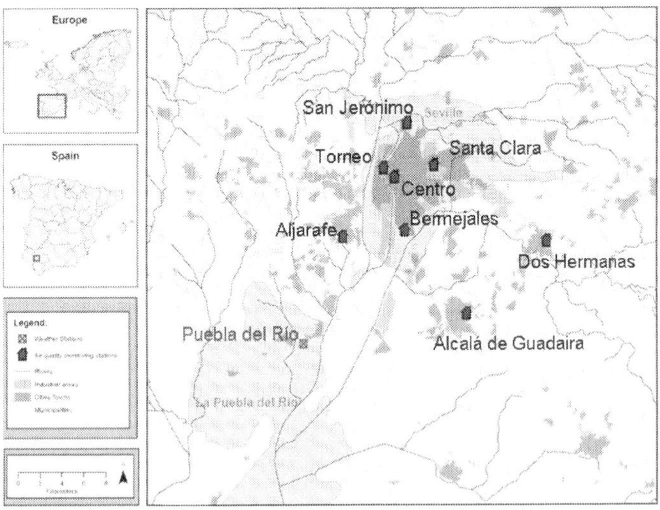

Fig. 1. Seville metropolitan area location in Europe. Air quality stations and meteorological tower used in this work.

of more than a million inhabitants, being the biggest urban zone of Andalusia. This region enjoys a Mediterranean climate with temperature between 4 and 15 °C in winter and autumn, while summer temperature is ranging from 18 to 35 °C.
According to the Andalusia government annual emission report 2003 (Junta de Andalusia, 2003), based on data taken in the year 2002 in Seville, the NO_x emission was about 26 500 tons per year, whose 59 % is attributed to traffic emissions. Meanwhile, the emission of non-methane volatile organic compounds (NMVOC) was 63 403 tons per year, with main contributions coming from the use of solvents 25% and traffic 13%. Seville metropolitan area shows the busiest road of the south of the Iberian Peninsula.

2.2 Monitoring stations and dataset

This work has been performed with surface ozone concentrations and its precursors (NO and NO_2) collected in eight air quality stations sited in the Seville metropolitan area and belong to the Survey and Air Quality Control Network of the Environmental Department, Regional Government of Andalusia. Five stations are in the urban hull of Seville town: Santa Clara, San Jerónimo, Bermejales, Centro and Torneo, while the other three are in the metropolitan area: Aljarafe (west of the city), Dos Hermanas (southwest) and Alcalá de Guadaira (east), between 20-25 km from city centre (Fig. 1). The period used to perform this work has been from 2003 to 2006.
Moreover, meteorological parameters recorded in a meteorological station located in the south of Seville area has also been used, Puebla del Río. This meteorological station belongs to the Network of Climate Information of Andalusia.
Ozone and NO_x data have been taken each 10 or 15 minutes from which the average hourly values have been calculated. Ozone measurements were taken with ultraviolet absorption-based instruments while NO_x data have been measured by chemiluminescence. Both types of instruments have been tested and calibrated periodically. They undergo weekly maintenance of operational parameters and manifold; changing air filters and span calibration every two weeks. Meanwhile, monthly and annual maintenance are done according to the manufacture guidance (e.g. tubing and electrovalves).

3. Results

3.1 Meteorological overview

Photochemical pollution is conditioned by several factors: emissions, orograpy and meteorology, among others. The meteorology is an aspect to be considered in all the air quality studies. The levels and behaviour of the air pollutants will be defined by the meteorological scenarios and the values of the meteorological parameters. Several meteorological variables affect to the photochemical air pollution but perhaps with major intensity are necessary to study two variables: temperature and wind. Temperature is a parameter that can influence aspects such as the photochemical reaction velocity or the atmospheric stability. On the other hand, wind (speed and direction) point out the atmospheric ability to disperse or accumulate chemical species (depending of wind speed) as well as the origin the air masses (wind direction). In order to investigate the wind regime and the temperature evolution in this area, in the Fig. 2 are presented the annual wind rose and the daily and monthly temperature evolution measured in the mentioned meteorological station of Puebla del Río.

During the nights of November to February the temperature can vary between 5 and 10 °C. Nevertheless, in spring and summer months the nocturnal values can reach values ranging between 15 and 25 °C. In the diurnal period and in the autumn and winter months the temperature in Seville area vary from 15 to 20° C. In spring and summer these values are increase from 25 ° C to levels higher than 30 °C.

Wind regime in this area is characterised by two main flows from SW and NE which coincide with the Guadalquivir valley axis. Therefore, the valley acts as a natural channel for the travel of air masses. In winter months winds usually come from the first quadrant, with preferential synoptic origin, due to air masses originated in the north of Europe and the north Atlantic. In spring and autumn there is a meteorological scenario characterized by both the synoptic and mesoscale, with wind blowing from the first and third quadrant along the valley axis. The air masses coming from north cause the NE-ENE wind. Meanwhile, Atlantic air masses are quite frequent during these months; these air masses are channelling towards the Guadalquivir valley originating wind mainly from the SW. Moreover, these months show anticyclonic situation with clear sky, high solar radiation and high mean temperatures, and also an increase in earth's surface temperature. All these follow a daily evolution of a typical mesoscale pattern. These local scenarios show diurnal and nocturnal regimes with winds from the third and first quadrants respectively, according to the valley axis and perpendicular to the gulf of Cadiz coast line.

In summer months, the daily wind evolution reveals that mesoscale processes predominate over most of the days. During the day the main directions are SSW-SW, both directions coming from the third quadrant towards the valley and perpendicular to the cost line, which is coincident with the sea breeze. For the nocturnal period most frequent directions are from NE, wind speed records show lower values than for diurnal period since these directions mainly correspond to land breeze, which are weaker than sea breeze.

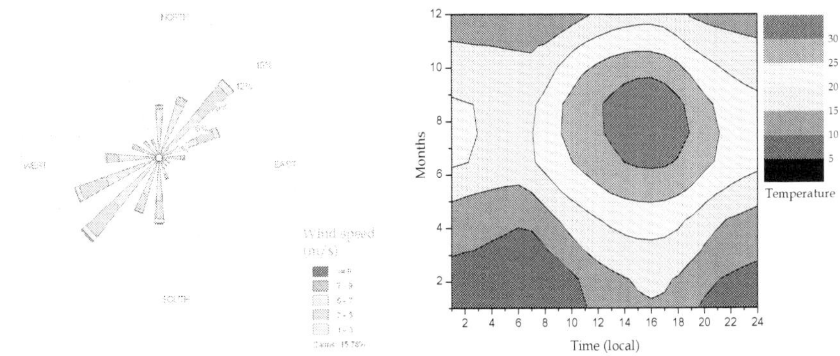

a) b)

Fig. 2. a) Annual wind rose. b) Daily and monthly temperature evolution. Puebla del Río (2003-2006).

3.2 Seasonal and daily variation on ozone and NO_x

In this section are presented both seasonal and daily cycles of surface ozone and NO_x. These pollutants show a marked seasonal evolution in the study area. As expected at mid-latitudes, NO_x present the maximum monthly values in winter and autumn and lower in summer months. Opposite seasonal behaviour is found for surface ozone. Fig. 3 shows the monthly evolution of the three species in Torneo and Aljarafe stations as representative urban and suburban environments. The photochemical activity is more intense in the warm season, hence, ozone formation and transport processes are more effectives. In addition, there is a higher elimination of NO by these photochemical mechanisms and low levels of NO_x are found in the lower layers of the troposphere in these months.

The NO_2 monthly values are higher than NO, with the exception of January and December months in some stations such as Torneo, Bermejales and San Jerónimo. These facts could be attributed the high NO emissions and the unfavourable conditions to photochemical activity (low solar radiation and temperature levels) and favourable conditions (vertical atmospheric stability) to accumulate chemical species such as NO in the lower layers.

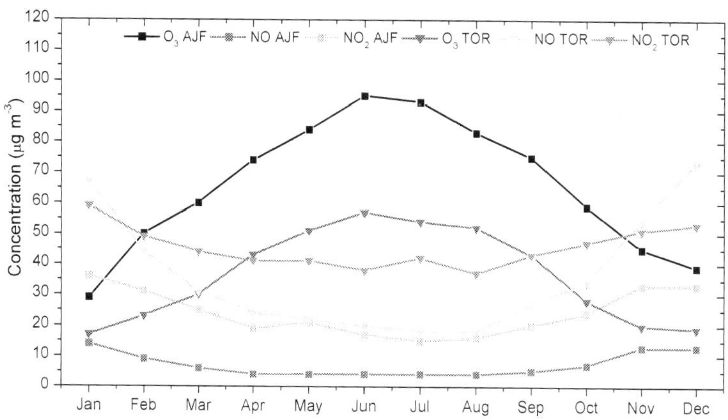

Fig. 3. Monthly evolution of O_3, NO and NO_2 in Aljarafe (AJF) and Torneo (TOR). 2003-2006.

The monthly cycle of NO and NO_2 is different according to the monitoring stations. In Bermejales and Tornero stations is obtained annual amplitude (difference between monthly maximum and minimum) for NO of 55 $\mu g/m^3$. In these stations NO_2 cycle presents a lower range with values of 21 and 32 $\mu g/m^3$ in Bermejales and Torneo respectively. These sites are strongly affected by the traffic emissions with elevated NO emissions which are accumulated in the surface layers under stability conditions. The formation of NO_2 from NO is similar in these stations in summer and winter time, i.e. elevated NO concentrations does not produce high NO_2 levels.

On the other hand, in stations located in the surrounding, the NO monthly cycle is weak with amplitudes between 10 and 14 $\mu g/m^3$, values measured in Alcalá de Guadaira and Aljarafe. In these same stations of NO_2 cycle amplitude is higher with values of 21 and 32 $\mu g/m^3$. These results could be associated to these stations are less affected by the traffic emissions that the sited in the city centre.

Fig. 4 shows the average daily evolution of O_3, NO and NO_2 registered during the study period in the eight air quality stations. A daily cycle is obtained in the three atmospheric species considered. This behaviour has its origin mainly in the traffic emissions, solar radiation levels and human behaviour patterns. These cycles are similar to find in other urban and suburban areas (Minoura, 1999; Mazzeo et al., 2005).

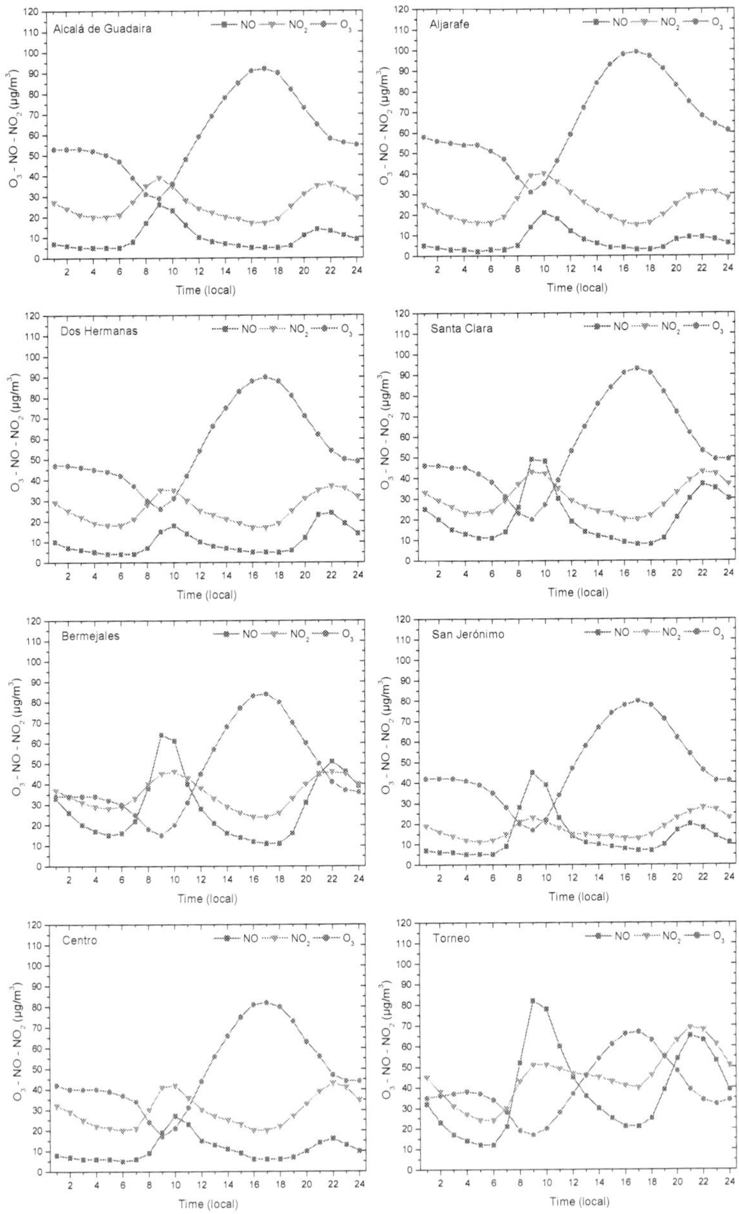

Fig. 4. Annual daily average evolution of O_3, NO and NO_2 in the eight air quality stations of Seville metropolitan area.

In the night time the concentrations remain with similar levels or with a slight increase trend with NO concentrations are always lower than NO_2. At 6:00 has been observed an increase in the NO_x levels due to the start of human activity and therefore the emissions from the traffic. In addition, the ground begins to heat up by the solar radiation and as consequence the mixing processes are produced in the lower layers. Due to fresh emissions of NO in rush hours and to the mixing mechanism, the residual ozone down from upper layers and is eliminated by the reaction with NO to produce NO_2, observing a decrease on ozone levels. At 9:00 is reached a NO and NO_2 maximum and the daily ozone minimum.

The NO_x morning maximum shows two behaviours according to the station type, urban or suburban. In some stations the NO maximum is higher than NO_2 (urban monitoring sites) while in others the behaviour is opposite (suburban). In stations such as Torneo, Bermejales or Santa Clara, NO maximum is higher than NO_2 due to are stations strongly affected by traffic emissions and the NO emissions are more elevated that the NO_2 production by the oxidation of NO with O_3. However, in sites located in the periphery such as Aljarafe, Alcalá de Guadaira and Dos Hermanas the NO_2 are higher than NO. In these cases of traffic emissions are less intense and the conversion of NO to NO_2 is more relevant that the emissions.

Maximum concentrations of NO show clear differences between the stations located in the Seville city and the sites located in the periphery less affected by emissions. Nevertheless, NO_2 maximum are very similar, not showing elevated differences between stations sited in the city and in the periphery.

During the morning, increase the solar activity, mixed ability and photochemical mechanism; have as consequence that NO_x present a decrease trend and surface ozone an increase trend. At 17:00 occurs the ozone maximum in all the air quality stations. The maximum ozone level is different, with higher values in stations away from city centre, such as Aljarafe and Alcalá de Guadaira and low values in stations as Torneo.

In the evening, with the cease of solar activity and therefore of photochemical reactions, ozone concentrations start to decrease. At the same time, it is observed an increase of the NO_x levels caused by several factors: there is an increase of the human activity in the city and their surroundings, hence fresh NO_x are emitted by traffic; decrease of mixing processes and starts of atmospheric stability, therefore accumulation of species in the lower atmospheric layers.

At 22:00 is recorded a secondary NO_x maximum. In all the stations, to exception of Bermejales, NO_2 level registered to this time is higher than the NO level. However, if compared the morning and nocturnal NO_x maximums, are found differences according to the station and the chemical specie (NO or NO_2).

In the NO case, the nocturnal maximum always is lower than the morning maximum in all the stations. The percentage differences vary, thus are found differences of 20-25 % in stations as Torneo, Bermejales and San Jerónimo (sited in the city). Whereas, in stations such Aljarafe or Alcalá de Guadaira (in the periphery) these differences can reach values between 30 to 50%. Due to the NO levels recorded are a clear demonstration of the direct traffic emissions, it is possible to conclude that the traffic activity in this area is more intense during the morning that during the last hours of the day.

For NO_2 in urban stations, Torneo, Bermejales or San Jerónimo, the NO_2 maximum in the first hours of the night is higher than the registered during the morning, with percentage differences between 5 and 33 %. Nevertheless, in suburban stations (Aljarafe and Alcalá) the NO_2 nocturnal maximum is lower than the morning with differences between 9 and 20 %.

3.3 Identification, ocurrence and ozone events characteristics

To the identification of ozone events it is necessary to apply criteria to extract from ozone data series of days considered as events. Using previous works which looking for to know episode situations (Cristofanelli et al., 2006; Oh et al., 2006), have been used two parameters. The daily ozone maximum and the threshold established to protect the human health (Directive, 2008). This threshold has been defined with a value of 120 μg/m³ as the maximum daily 8h, using this criteria ensure that a day with ozone event has been an exceedance of this legal limit. The other parameter has been taken as a value of 160 μg/m³ as daily maximum, this concentration has been defined using the ozone data series levels. Therefore, a day defines as ozone event should satisfy simultaneously of two parameters.

Appling the mentioned criteria to the hourly ozone series of the eight stations, the days with ozone events have been extracted. The total of days with events has been classified in function of the month. The results are shown in the Fig. 5. Ozone events are registered between the period from May to October with a high frequency in June and July; both months sum the 60% of all the events.

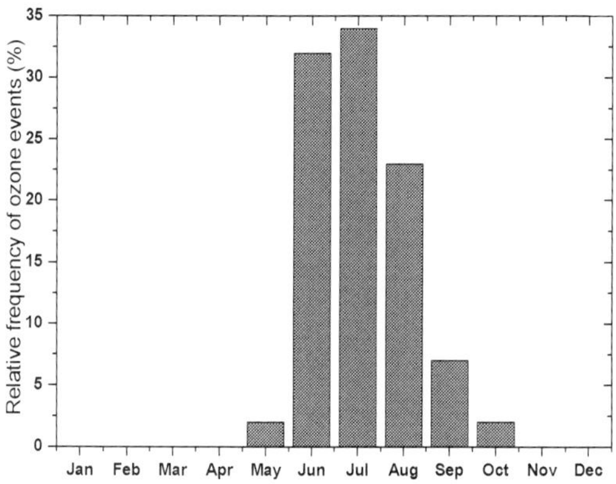

Fig. 5. Relative frequency of ozone events in Seville metropolitan area (2003-2006).

The highest ozone concentrations are recorded between May to September since in this time have the favour meteorological conditions to the formation and transport of surface ozone. In the four-year period analysed, 2003 to 2006, have been identified 373 days with ozone events. Table 1 exposes these events days in function of the year and monitoring station. An annual variability has been obtained, thus, 2003 and 2004 are the years with highest number of ozone episodes. However, in 2005 and 2006 this number has decreased.

The causes to explain the existence of an ozone event number in a specific year are mainly determined by the meteorological conditions. In function of the meteorological scenarios that characterise the study area between May to October, will condition the atmospheric features in the lower atmosphere: temperature, solar radiation and humidity levels, mesoscale processes, mixing mechanism, Saharan intrusions, etc. According to the intensity and occurrence of these processes an ozone events number will occur this year.

Under episodic situations the maximum daily 8h ranging between 125 μg/m³ of Torneo and 156 μg/m³ of Aljarafe. The maximum daily average is close to 180 μg/m³ (information threshold defined in the Directive 2008) in several stations and even is exceeded in Aljarafe

station. This result point out that in days with ozone events the daily maximum ozone is elevated and the absolute maximum exceed 200 μg/m^3 in all the stations to exception of San Jeronimo.
On the other hand, the days with ozone events present a lower percentage of occurrence if is considered of whole period. Aljarafe has been the station with the highest number of events follows by Santa Clara, Bermejales, Dos Hermanas and Alcalá de Guadaira. These five stations sum the 75% of the events identified.

	2003	2004	2005	2006	Total
AJF	47	46	17	12	122
ALG	--	20	8	8	36
BER	14	20	5	8	47
CEN	4	11	9	6	30
DHN	9	25	2	3	39
SCL	17	26	18	18	79
SJE	9	1	7	0	17
TOR	0	3	0	0	3
Total	100	152	66	55	373

Table 1. Number of days with ozone events (exceedance of the threshold to protect of human health and daily maximum of 160 μg/m^3) in the air quality stations for the period from 2003 to 2006.

3.4 Comparative analysis of days with and without events: daily evolution

Surface ozone series could be divided in days with ozone events and days without events. Since more than 75% of the ozone events have been identified in five stations, it has been considered to perform a comparative analysis using the ozone data from these stations.
The average daily ozone evolution, NO and NO_2, recorded in days with events has been compared with the days without events, during the period from May to October since is the period which have been identified the events. On the other hand, it is interesting to know the levels and variation that present the meteorological variables most influential in the ozone concentration such as temperature, relative humidity and wind speed. For this reason, the average daily evolution of these parameters has been studied both in days with and without events. The meteorological data has been obtained from Puebla del Río station. As a representative case in Fig. 6 is shown the results obtained for Santa Clara station.
In all the stations analysed the ozone concentrations are higher in days with events that in days without events. In the nocturnal time has been observed an ozone increase which varies between 5 to 30 μg/m^3. In the diurnal period the differences found are not constant but an elevated photochemical activity produce higher differences. From the first hours in the morning, the differences show an increase trend, reaching maximum values in the noon. In episodic days the daily maximum ozone reaches values of 160 to 170 μg/m^3 while in non

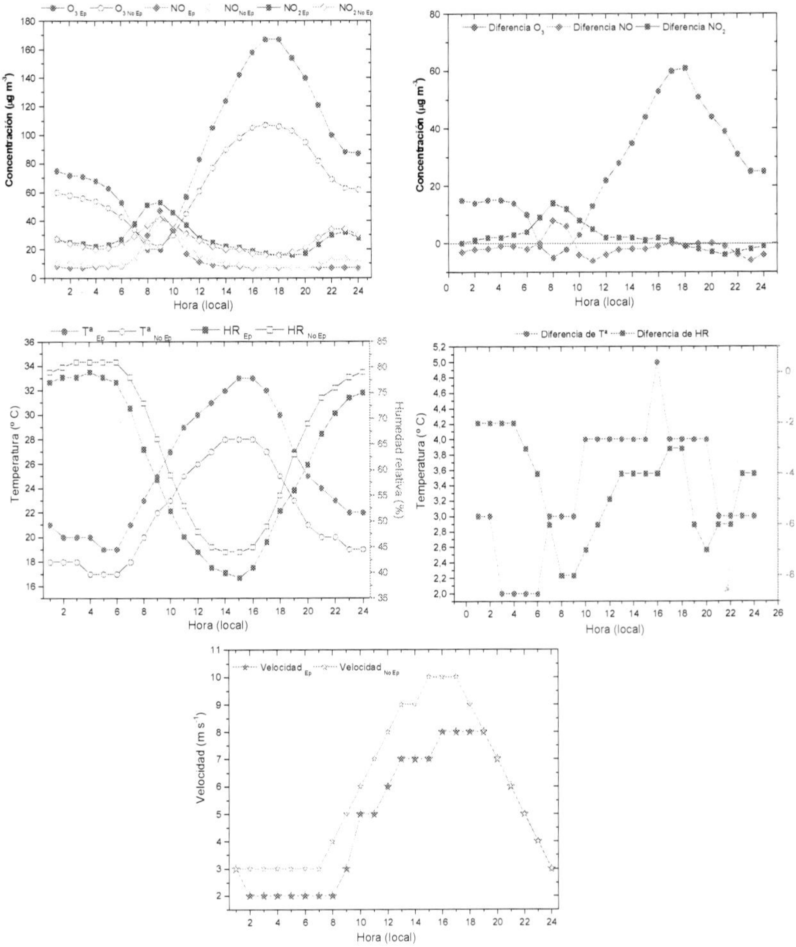

Fig. 6. Ozone, NO and NO_2 daily average evolution recorded in episodic and non episodic days as well as the difference between them in Santa Clara station. Temperature, relative humidity and wind speed daily average evolution collected in episodic and non episodic days and difference between them in Puebla del Río station. 2003-2006.

episodes days the daily maximum varies between 50 to 60 μg/m^3. Therefore, in days with ozone events there are an increase of 50 to 60 μg/m^3.

In Aljarafe and Alcalá de Guadaira stations of NO concentrations are very similar both episodic and non episodic days. This result could point out that the traffic emissions in these suburban stations are equally affected in both types of days. However, in stations affected by the traffic emissions and sited in the Seville urban area such as Bermejales and Santa Clara, have been observed an increase in the NO concentrations in days with events respect to days without events, in the period from 7:00 to 10:00. Accordingly, in these stations and under event conditions the NO maximum is of 70 μg/m^3 while in days without events reaches 50 μg/m^3. These results could be attributed to a low wind speed in days with events, therefore, an atmosphere less dispersive and favouring the accumulation of NO close to the ground.

The daily evolution of NO_2 states two different behaviours. In the first, NO_2 concentrations always are higher in events days. This pattern is obtained, for instance, in Bermejales station. The second pattern identified is characterised by NO_2 levels higher in events days but only in the time interval between 7:00 to 17:00 approximately. In this case, the highest differences are found in the early morning between 9:00 to 11:00. While in events days are detected NO_2 values of 50 to 55 μg/m^3 in not episodic days the NO_2 concentration was of 38 to 45 μg/m^3. NO_2 levels are similar from 11:00 to 17:00 both episodic and non episodic days.

The daily temperature evolution shows higher levels in episodic days. The events days present average nocturnal temperature between 19 and 21 °C, with approximately 2 °C higher that the registered in nights with non episodic days. In the diurnal time the difference found are higher, between 3 and 4 °C. In some situations have been measured temperatures in days defined as events up 30 °C in the period from 11:00 to 17:00 and daily maximums of 33 °C. On the contrary, the relative humidity, in the nocturnal time, presents lower values than in non episodic days, approximately between 2 to 4 %. The values registered during the night in episode days are of 77 to 80 %. During the day the relative humidity decreases to 39 to 41 %, values 4 to 8 % lower than in days without ozone events.

It is well known that the photochemical activity is favour by high temperatures and low relative humidity. These conditions are generally associated with an atmosphere not dispersive, hence, with low level of wind speed. Analysing this variable, during the night of events days the wind speed is 1m/s lower than in days without events. These differences are higher during the day, for example, from 8:00 to 19:00 the differences found are of 2 to 3 m/s. From 20:00 the wind speed measurements are very similar in both cases.

3.5 Ozone events: study case

This section has been devoted to analysis of ozone events using O_3, NO and NO_2 as well as meteorological parameters in several episodic situations. The ozone event duration oscillates between 2 to 10 days with an average value of 3.5 days. The two longest episodes occurred from 29 July to 12 August 2003 and from 9 to 18 June 2004. During 2005 and 2006 years have not occurred long episodes. To present the main characteristics of the ozone events the episode of 2004 will be used.

In the occurrence of an ozone event have been identify three phases. In the first, the ozone concentrations are low. The meteorological conditions that govern this stage is a synoptic flow from S-SW with origin in an high centre pressure system located in the Atlantic Ocean and a low pressure system in the British isles. The SW of Iberian Peninsula under these synoptic conditions could register wind coming from W and NW. However, in summer months under these situations a thermal low is developed in the centre of Iberian Peninsula which produces a flow in the diurnal time from coastal area to inland using as natural channels of valleys (Millán et al., 2002). For these reasons, in Seville area (100 km far from coastal line) is measured wind coming from S-SW.

A similar condition to expose above was registered during 6 and 7 June 2004 which was the starts one long ozone events. In the initial stage the Iberian Peninsula was affected by anticyclonic conditions and a thermal low was developed in the centre of Iberian Peninsula, due to this configuration the wind blows from SW in the study area.

The second phase (central stage) of the event starts thank to the movement of the Atlantic high pressure system toward Europe continent. This new isobaric configuration provokes weak isobaric gradient over the Iberian Peninsula. These conditions favoured the development of mesoscale processes in coastal areas.

From the analysis of many ozone events occurred in Seville metropolitan area have been identified sea-land breeze conditions or a weak (low wind speed) synoptic flow from SW as the main meteorological scenarios in the central phase of the event. Due to the higher ozone concentrations are recorder under mesoscale processes, these will be analysed with detail. In this area, the nocturnal breeze coming from the first quadrant, higher frequency from NE sector, starts approximately between 3:00 to 5:00 and remain during the next 8 hours. At 11:00 to 12:00 cease the land breeze and begin the marine breeze with wind blowing from SW. The marine breeze present higher values of wind speed and remain during the next 12 to 14 hours.

In order to illustrate with detail the behaviour both chemical species and meteorological parameters of an ozone event, Figs. 7 and 8 show the specific case of June 2004. Under episodic conditions, from 00:00 to 3:00 the ozone measured in the air quality stations of Seville metropolitan area presents a decrease trend. Nevertheless, NO and NO_2 shows similar concentrations during this period. At this time, the temperature ranges between 15 to 17 °C and the relative humidity is close to 80%.

From 3:00 to 5:00, the ground is sufficiently cold and starts the transition between the regimes of the breeze, ceases the marine regime and starts to blow from land. This transition is identified by a change in the wind direction and a decrease in the wind speed. Under these conditions has been observed that NO levels remain similar both urban and suburban station. However in some urban stations the NO_2 could present slight increase while in the suburban stations this increase has not been observed. On the other hand, surface ozone can state two behaviours, to remain with similar values or well increase with a rise of 20 to 40 $\mu g/m^3$. In the Fig. 7, this variation is observed in the nights 16 and 17 June 2004.

This increase in the nocturnal ozone can have two origins. The transition between the two breeze regimes is associated to a decrease in the wind speed, as has been mentioned before, therefore atmospheric stagnation conditions can lead to an increase in the ozone levels. On the other hand, the ozone has been transported during the previous hours from third to first quadrant, with the change in the breeze regime, the air masses follow the same pathway but in opposite sense, recirculation processes. This behaviour has been also measured and documented in others areas (Eliasson et al., 2003).

From 6:00, under nocturnal regime of breeze (blowing from NE), there is an increase of NO levels due to the fresh emissions from the traffic. In breeze periods the increase in the NO levels at this time is relevant, and more pronounced in urban stations close to traffic sources, such as Torneo and Bermejales. Under land breeze the wind speed is low or even calms, then an atmosphere non dispersive which favours the processes of NO accumulation close

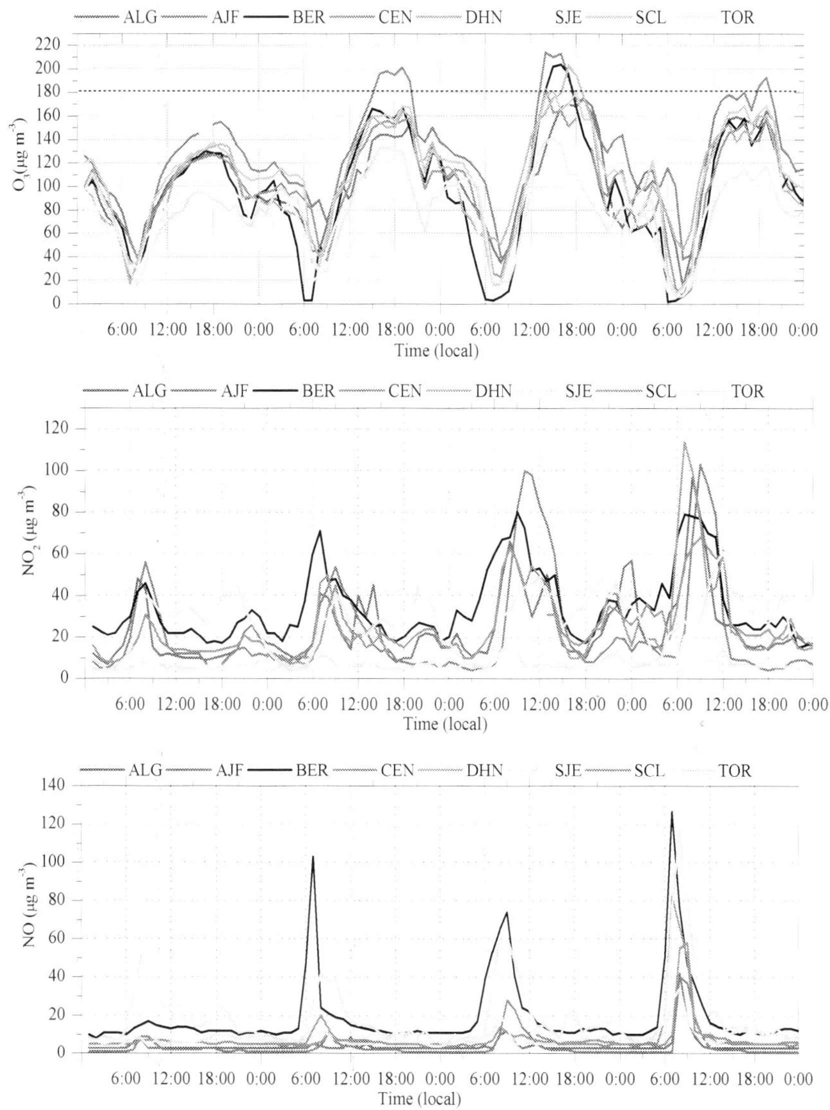

Fig. 7. Hourly ozone, NO and NO_2 evolution in Seville stations from 15 to 18 June 2004.

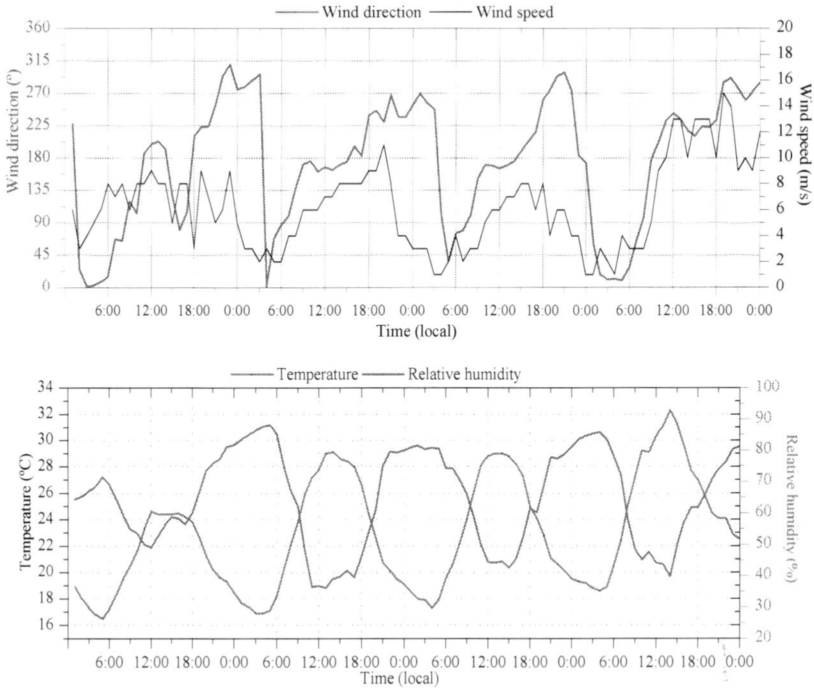

Fig. 8. Hourly wind (direction and speed), temperature and relative humidity evolution at Puebla del Río from 15 to 18 June 2004.

to the ground. NO_2 in these early hours also increase due to the oxidation processes with NO. The suburban stations are the sites with the higher NO_2 increases.

At this time, ozone levels starts to increase which could be associated to several processes. The onset of the solar activity helps to begin the photochemical reactions; hence, ozone can have its origin in this first local photochemical formation. Solar radiation leads to the broken of stable nocturnal boundary layer and the formation of the mixing layer. Mixing mechanisms can be brought from uppers layers both residual ozone and NO_2.

From 10:00 to 12:00 starts to blow wind from S-SW. In this period the increase of the ozone could be caused by horizontal (air masses with ozone from SW) and vertical transport (mixing processes). In addition, NO_2 also starts to increase. In suburban stations as Aljarafe and Alcalá NO values exceed the 40 $\mu g/m^3$. Fig. 7 shows the NO_2 levels reached in these stations 17 and 18 June 2004, probably caused by the transports processes before commented.

In the afternoon the ozone continues to increase and NO_2 values to decrease. Thank to fresh emissions from the traffic in the first hours of the night, ozone starts to decrease and NO_2 to increase. However, NO concentrations remain constant due to NO emitted is converted to NO_2 thanks to the high ozone levels available to react with NO.

A change in the synoptic configuration, movement of the high pressure system and arrival of a low pressure system, provokes and higher isobaric gradient over the Iberian Peninsula.

Hence, unfavourable conditions to mesoscale developments. In the studied area is frequent that the events finish, third stage, with wind blowing from W-SW which has an effect of ventilation and cleaning of the atmosphere.

4. Conclusions

In this work has been evaluated the ozone, NO and NO_2 concentrations measured in eight air quality stations of Seville metropolitan area during the period from 2003 to 2006. Seasonal and daily variation of this three air pollutants have been studied. While higher values of NO and NO_2 have been measured in winter and autumn months, in summer months are recorder the highest ozone levels. The monthly cycles are different according to the stations considered. Urban stations, more influenced by traffic emissions, show higher NO and NO_2 levels and a higher seasonal range (difference between maximum and minimum monthly). NO_2 monthly values are, in general, higher than NO monthly levels. On the other hand, suburban station presents higher ozone concentration and a higher pronounced seasonal cycle.

A well define daily cycle has been found for the three species analysed. NO_x show two peaks in the morning and in the first hours of the night. In general the NO_2 concentrations are higher than the NO. Morning peak of NO is higher than the nocturnal peak. Nevertheless, in NO_2 the opposite behaviour is found. Ozone shows the daily peak in the afternoon with elevated levels in suburban stations.

A criterion has been applied to identify and extract of the days with ozone events from the historical series. A day with ozone event was considered if there is an exceedance of the threshold to protect the human health and exceeds its daily maximum in 160 $\mu g/m^3$. Ozone events occur between May to October with a high frequency in June and July.

A comparative analysis of days with and without events has been performed. The maximum of NO is higher in episode days in urban stations but not in suburban. NO_2 can be higher in events days in urban stations while in suburban in an specific time period. In days with events the ozone peaks could be increased between 50 to 60 $\mu g/m^3$. In the afternoon of days with events the temperature can be 3-4 °C higher than in non episodic days. The relative humidity 4 to 8% lower and the wind speed also 2-3 m/s lower.

Several periods with ozone events have been investigated with detail. During the studied period, 2003 to 2006, have been found long episodes (up to the week) but the average was of 3.5 days. In an event have been identified three phases. The meteorological conditions of each stage have been also defined. In the central phase, where the ozone concentrations are episodic, the surface atmospheric dynamic in this region is governed by the mesoscale processes or wind coming from SW with slow wind speed.

Once analysed and indentified the ozone events in this area, in order to improve the understanding of these processes, modelling tools are being applied. Specific ozone events are being simulated using meteorological mesoscale models. The aim is to know not only the surface behaviour both chemical and meteorological parameters but the profile of these substances. Thanks to use of meteorological model will be possible to carry out the atmospheric boundary layer characterisation of this area which will be essential to understanding the vertical transport and the formation of ozone residual layers.

5. Acknowledgment

The authors want to thank to the Network of Climate Information of Andalusia by the supply of surface meteorological information.

6. References

Alvim-Ferraz, M.C.M., Sousa, S.I.V., Pereira, M.C. & Martins, F.G. (2006). Contribution of anthropogenic pollutants to the increase of tropospheric ozone levels in the Oporto Metropolitan Area, Portugal since the 19th century. *Environmental Pollution*, Vol. 140, 3, Pp. 516-524.

Cape, J.N. (2008). Surface ozone concentrations and ecosystem health: Past trends and a guide to future projections. *Science of the Total Environment*, Vol. 400, 1-3, pp. 257-269.

Cristofanelli, P., Bonasoni, P., Tositti, U., Bonafé, U., Calzolari, F., Evangelisti, F., Sandrini, S. & Stohl, A. (2006). A 6-year análisis of stratospheric instrusions and their influence on ozone at Mt. Cimone (2165 m above sea level). *Journal Of Geophysical Research*, Vol. 111, D03306, doi:10.1029/2005JD006553.

Directive 2008/50/EC. On ambient air quality and cleaner air for Europe. The European Parliament and of the Council.

Doherty, R.M., Heal, M.R., Wilkinson, P., Pattenden, S., Vieno, M., Armstrong, B., Atkinson, R., Chalabi, Z., Kovats, S., Milojevic, A. & Stevenson, D.S. (2009). Current and future climate- and air pollution-mediated impacts on human health. *Environmental Health*, Vol. 8, S8.

Eliasson, I., Thorsson, S. & Andersson-Sköld, Y. (2003). Summer nocturnal ozone maxima in Göteborg, Sweden. *Atmospheric Environment*, Vol. 37, pp. 2165-2627.

Kumar, P., Robins, A., Vardoulakis, S. & Britter, R. (2010). A review of the characteristics of nanoparticles in the urban atmosphere and the prospects for developing regulatory controls. *Atmospheric Environment*, Vol. 44, 39, pp. 5035-5052.

Lefohn, A.S., Shadwick, D. & Oltmans S.J. (2010). Characterizing changes in surface ozone levels in metropolitan and rural areas in the United States for 1980-2008 and 1994-2008. *Atmospheric Environment*, Vol. 44, 39, pp. 5199-5210.

Lin, S., Liu, X., Le, L.H. & Hwang SA. (2008). Chronic Exposure to Ambient Ozone and Asthma Hospital Admissions among Children. *Environmental Health Perspectives*, Vol. 116 , 12, pp. 1725-1730.

Mazzeo, N.A., Venegas, L.E. & Choren, H. (2005). Analysis of NO, NO2, O3 and NOx concentrations measured at a green area of Buenos Aires City during wintertime. *Atmospheric Environment*, Vol. 39, pp. 3055-3068.

Millán, M., Sanz, M.J., Salvador, R. & Mantilla, E. (2002). Atmospheric dynamics and ozone cycles related to nitrogen deposition in the western Mediterranean. *Environmental Pollution*, Vol. 118, pp. 167-186.

Minoura, H. (1999). Some characteristics of surface ozone concentration observed in an urban atmosphere. *Atmospheric Research*, Vol. 51, pp. 153-169.

Oh, I.B., Kim, Y.K., Lee, H. & Kim, C. (2006). An observational and numerical study of the effects of the late sea breeze on ozone distributions in the Busan metropolitan area, Korea. *Atmospheric Environment*, Vol. 40, pp. 1284-1298.

Papanastasiou, D.K. & Melas, D. (2009). Statistical characteristics of ozone and PM10 levels in a medium-sized Mediterranean city. *International Journal of Environment and Pollution*, Vol. 36, 1-3, pp. 127-138.

8

Investigation on the Carbon Monoxide Pollution over Peninsular Malaysia Caused by Indonesia Forest Fires from AIRS Daily Measurement

Jasim M. Rajab, K. C. Tan, H. S. Lim and M. Z. MatJafri
School of Physics, Universiti Sains Malaysia, Penang, Malaysia

1. Introduction

Carbon monoxide (CO) is an important pervasive atmospheric trace gas affecting climate and more than 50% of air pollution nationwide and worldwide, which also plays as a significant indirect greenhouse gases due to its influences on the budgets of hydroxyl radicals (OH) and ozone (O_3). We present a study on Atmosphere Infrared Sounder (AIRS), onboard NASA's Aqua Satellite, detection of CO emission from large forest fire in the year 2005 in the Sumatra, Indonesia. AIRS daily coverage of 70% of the planet symbolizes an important evolutionary advance in satellite trace gas remote sensing. AIRS is one of several instruments onboard the Earth Observing System (EOS) launched on May 4, 2002, with its two companions microwave instruments the Advanced Microwave Sounding Unit (AMSU) and the Humidity Sounder for Brazil (HSB) form the integrated atmospheric sounding system. AIRS providing new insights into weather and climate for the 21st century, as well as AIRS' channels include spectral features of the key carbon trace gases CO_2, methane(CH_4), and CO. AIRS is an infrared spectrometer/radiometer that covers the 3.7-15.4 m spectral range with 2378 spectral channels. Troposphere CO abundances are retrieved from AIRS 4.55 mm spectral region, and measure CO total column by 52 channels with the uncertainty, which is estimated approximately 15-20 % at 500mb. Results from the analysis of the retrieved CO daily Level 3 standard (AIR×3STD) and Monthly product (AIRX3STM) were utilized in order to study the impact of Indonesia forest fire on CO distribution, and the monthly CO distributions in Peninsular Malaysia. AIRS daily CO maps from 12 - 25 August 2005 for study area show large-scale, long-range transport of CO from anthropogenic and natural sources, most notably from forest fire biomass burning. The sequence of daily maps shows the CO advection from central Sumatra to Malaysia. AIRS can also capture the temporal variation in CO emission from forest fires through 6-day composites so it may offer a chance to enhance our knowledge of temporal fire emission over large areas. The result was compared with daily CO emission (13-24) August 2007. The daily measurements of CO concentration on August 2005 are higher than August 2007. The northern region (uppers the latitude 4) was more affected by forest fires than the rest area. Substantial seasonal variations demonstrate season-to-season changes in rainfall and drought patterns in different seasons. We see such seasonal variations in the biomass burning emissions in the late dry season, while industrial contributions are evident at smaller magnitudes on monthly distributions. The study shows that AIRS can reliably detect CO plumes from forest fires in $1^{o}*1^{o}$ spatial resolutions. The CO maps were generated using Kriging Interpolation technique.

CO is an important atmospheric constituent affecting climate and air quality, which also acts as an important indirect greenhouse gases as it significantly impacts the OH budget, and thus indirectly affects the CH_4 and O_3 concentrations. This is due to the characteristic of CO by the indirect radiative forcing effect and increases the concentrations of troposphere O_3 and CH_4 through chemical reactions with other atmospheric constituents. It is approximately 75 % of OH sinks, which is capable to influence the concentration of greenhouse gases, the oxidizing agent in troposphere and finally contributes to climate change (Daniel and Solomon, 1998). CO is not a significant greenhouse gas because of it weak absorption of infrared radiation from the Earth. In addition, CO is useful in many modern technology applications, such as in bulk chemicals manufacturing.

It is very important to observe and document changes in the forcing terms such as gases, in order to understand and assess their influences of climate change, and to achieve more dependable longer range projections. Over the past three decades, the abundances of the atmosphere parameters (gases) were obtained from a lot of sources such as balloons, airplane and sparsely distributed measurement sites. While the observations were mostly limited to the surface and give important insights on flux variability the estimates have large uncertainties because there is not much atmospheric concentration data and more sensitive to sources and sinks. In situ measurements normally have the best accuracy from ground and aircraft, but the major shortfall is not being able to make daily global variation evaluations as well as cost a lot of money and strenuous efforts. There is a lack of data both in the lower - particularly over land - and upper troposphere (Tiwari et al., 2005).

A potentially favourable measurement method for closing some of these data gaps are the retrieval of gases from space. The satellite remote sensing has very good global coverage increase our capability to access the influence of human activities on the chemical composition of the atmosphere and on the climate changes (Clerbaux et al., 2003). Furthermore, can provide the quantitatively data with high spatial or temporal resolution (Dousset & Gourmelon, 2003). In addition, the free download satellite data provided by infrared measurements from the Atmospheric InfraRed Sounder (AIRS) on the NASA AQUA satellite to establish and produce the high quality spectrally resolved radiance climatologic for observing and detecting climate change, to understand the distribution of trace gases, sources and sinks, and to validate weather and climate models.

From 705 km above the Earth's surface, the AIRS measure the integrated impact of numerous atmospheric molecules emitting and absorbing radiation at various temperatures throughout the atmospheric path from the surface to the instrument. The AIRS, included on the EOS Aqua satellite launched on May 4, 2002, is the first of the new generation of meteorological advanced sounders for operational and research use (Aumann et al., 2003).

In Malaysia, one of Southeast Asia country, industrialization, urbanization and rapid traffic growth have contributed significantly to economic growth. Pockets of heavy pollution are being created by emissions from major industrial zones, a dramatic increase in the number of residences, office buildings, manufacturing facilities, increases in the number of motor vehicles and trans-boundary pollution. Besides that, Malaysia is situated in a humid tropical zone with heavy rainfall and high temperatures (Mahmud & Kumar, 2008; Tangang et al.,

2007); the cloudy conditions cover the study area becomes the obstacle to acquire a high quality and resolution for satellite data.

The AIRS ability to provide simultaneous observations of the Earth's atmospheric temperature, water vapour, land surface temperature and the distribution of greenhouse gases (GHG's) with cloud clearing system, makes AIRS/AMSU a very functional space instrument to observe and study the atmosphere reaction to increased several gases such as CO. AIRS with its two companion microwave instruments, (AMSU) and (HSB), form the integrated atmospheric sounding system (Lambrigtsen, 2003). AIRS is an infrared spectrometer/radiometer that covers the 3.7–15.4 m spectral range with 2378 spectral channels and used to measure the upwelling radiance from Earth (Strow et al., 2003).

In addition, AIRS is the first hyperspectral infrared radiometer designed, in order to provide a valuable data for National Ocean and Atmospheric Administration's National centers for Environmental Prediction (NCEP) and other weather forecasting centers, for the operational requirements in terms of weather forecasting in medium-range (Aumann, 2003). AMSU is a 15-channel microwave radiometer operating between 23 and 89 GHz. HSB is a four-channel microwave radiometer that makes measurements between 150 and 190 GHz (Lambrigtsen, 2003). It also provides information for several greenhouse gases, CO_2, CH_4, CO and O_3 as well as to improve weather prediction and study the water and energy cycle. AIRS measure CO total column by 52 channels with the uncertainty, which is estimated approximately 15-20 % at 500mb.

Between 1 August 2005 and 15 August 2005, the central, northern and eastern parts of Peninsular Malaysia experienced severe haze (DOE, 2005). Immense plumes of the gas emitted from forest and grassland burning in Indonesia forest fires in 2005 caused serious air pollution in Malaysia, and the air pollution reached extremely hazardous levels and forced schools and an airport to close. The fires affected 10,000 hectares of peat forest in Sumatra between Riau and north Sumatra. NOAA recorded 5420 hotspots from satellite images over the area between mid-July and mid-August (Jasim et al., 2009). The land and forest fires in the Riau Province of central Sumatra, Indonesia were the primary cause of transboundary pollution which was aggravated by the stable atmosphere conditions during the period. The hazy conditions reached its peak on 11-12 August 2005 when a haze emergency was declared in two areas in the Klang Valley and Kuala Selangor (DOE, 2005).

This study is based on the results from the analysis of CO total column retrievals from AIRS, Standard Level-3 Daily gridded product (AIRX3STD) and Monthly product (AIRX3STM) $1\times1°$ Spatial resolution, Version 5 ascending data, using AIRS IR and AMSU, without-HSB, to investigate the daily distribution map of CO and the impact of Indonesia's forest fire, and the monthly CO distributions in the Peninsular Malaysia.

The presented AIRS daily Peninsular Malaysia CO maps from 12 - 25 August 2005 for study area show large-scale, long-range transport of CO from Indonesia forest fires. The sequence of daily maps shows the CO advection from central Sumatra to Malaysia. The daily measurements of CO concentration on August 2005 are higher than August 2007. The northern region (uppers the latitude 4) was more affected by forest fires than the rest area. Substantial seasonal variations demonstrate season-to-season changes in rainfall and drought patterns in different seasons. We see such seasonal variations in the biomass burning emissions in the late dry season, while industrial contributions are evident at smaller magnitudes on monthly distributions. The CO maps were generated using Kriging Interpolation technique. In this study, the CO concentration was accurately and precisely mapped from AIRS data over Peninsular Malaysia.

2. Southeast Asia, climatology, and air pollution

Southern Asia is one of the most heavily populated regions of the world with a vibrant mixture of cultures, which comprises over a quarter of the world's population. One activity which is prevalent to all people amidst this massive diversity is energy consumption, from biomass burning in cook stoves to fossil fuel usage in trucks and rickshaws (Lawrence, 2004). Southeast Asia is experiencing a similar rapid economic growth to that in Northeast Asia. Furthermore, a large source of several air pollutants may make an important contribution to regional and global pollution because of increasing anthropogenic emissions associated with biogenic emissions from large tropical forests. The greater oxidizing capacity in the tropical regions is due to a higher UV intensity, humidity, rapid development and industrialization (Streets et al., 2001).

In addition, Southeast Asia has many social, economic, and environmental impacts caused by forest and land fires. Fires considered one of the largest anthropogenic influences on terrestrial ecosystems after agricultural activities and urban, and its indeed critical elements in the Earth system, vegetation, linking climate, and land use (Lavorel et al., 2007). Tropical haze from peatland has serious negative impacts on the human health and regional economy, and peatland fires affect global carbon dynamics (De Groote et al., 2007).

The strong monsoon and the associated movement of the inter-tropical convergence zone (ITCZ) were dominated the climatology of continental Southeast Asia which mostly located in the equatorial; seasons are not as discrete as in more temperate zones. When the (ITCZ) relocates southern with the sun across Southeast Asia into the Southern Hemisphere, the wintertime winds over considerable of southern Asia usually come from the NE to SW. This region experiences a dry season for about six months (November - April) before the (ITCZ) moves back to the Northern Hemisphere and long-range transport of continental air masses from the Indian Ocean in the summer monsoon prevails during the subsequent wet season (May - October). In the Southeast Asia the dry and wet seasons occur at opposite times of the year due to the difference circulations in the southern and northern hemisphere (Pochanart, 2004).

In the onset of the northeast monsoon, cold surges originating from Siberia and northeast Asia brings significant amount of pollution to southeast Asia while crossing through the heavily-polluted regions of east Asia (Pochanart et al., 2003) and also caused heavy rains which lead to grave flooding in December of some years along east coast and Johor (Tangang et al., 2007). At the same time with strong subsidence in the early dry season, there are important increases in air pollutant levels in continental Southeast Asia (Pochanart, 2004). Maximum air pollutants levels occurs over continental Southeast Asia in the late months of dry season due to strong regional biomass burning and long-range transport of air masses from the western Asia and Middle East. The tropical biomass burning in Southeast Asia is a major source of atmospheric pollutants and strongly influenced by anthropogenic post-agricultural waste burning (Pochanart et al., 2003). In some years the high pollutant levels can be observed in late dry season, were results of La Niña influences combined with the convection in the ITCZ, was considerably weaker than normal, the resulted in incompetent ventilation of the pollutants out of the continental outflow and accumulation of aerosol levels and trace gas in the lower troposphere region (Lawrence, 2004).

3. Carbon monoxide pollution

Carbon monoxide is a colourless, odourless and at high concentration, a poisonous gas. The resulting from fossil fuel combustion has become a major issue, especially in air pollution. CO is emitted mainly from motor vehicle exhaust, industrial processes and open burning activities (Buchwitz et al., 2007). The situation become worsens for the countries with the instinctive fuel consumption and the increasing energy demand. Carbon monoxide is an important atmospheric constituent affecting climate and a major troposphere air pollutant (Buchwitz et al., 2006). Therefore, it is very important to acquire information regarding the distribution of carbon monoxide. The investigation of carbon monoxide globally has gained attention from researchers recently. It also affects the concentration of greenhouse gases such as methane and ozone. Besides that, carbon monoxide also being a secondary pollutant regarding the respiratory problem and affect the crop yields (Buchwitz et al., 2006).

World-wide, the anthropogenic sources produce about 50% of CO emissions with the remainder coming from biomass burning and oxidation of naturally occurring volatile hydrocarbons. CO is a product of incomplete produced by combustion of fossil fuel and biomass, and having an average lifetime of 2-4 months in the atmosphere (McMillan et al., 2005).

It is generally agreed that biomass burning accounts for about one quarter of CO emission to the atmosphere and its concentrations in the northern Hemisphere are much higher, where human population and industry are much greater than in the southern Hemisphere. CO emissions are generally 5-15% of CO_2 emissions from burning, depending on the intensity of the burn (Liu et al., 2005). Andreae & Merlet (2001) estimated the annual emission of CO from vegetation fires in tropical forests and savannas' is 342 Mt (1 Mt=109 kg) CO per year, while the total CO emission for all non-tropical forest fires is 68 Mt CO per year. A concentration of as little as 400ppm (0.04%) CO in the air can be fatal. The levels of normal carboxyhemoglobin in an average person are less than 5%, whereas cigarette smokers (two packs/day) may have to levels up to 9% (Delaney et al., 2001).

CO can direct observe from space due to its strong absorption properties in the thermal infrared (4.7 μm) and in the solar shortwave infrared (2.3 μm). Since past decade, there are a number of satellite instruments available, in order to measure troposheric CO globally, including Measurements Of Pollution In The Troposphere (MOPITT) [Emmons et al., 2007, 2009], SCanning Imaging Absorption spectroMeter for Atmospheric CartograpHY (SCIAMACHY) [Bovensmann et al., 1999; Burrows et al., 1995], AIRS (McMillan et al., 2005, 2008), Atmospheric Chemistry Experiment-Fourier Transform Spectrometer (ACE-FTS) [Clerbaux et al., 2005, 2008], Technology Experiment Satellite (TES) [Lopez et al., 2008; Luo et al., 2007], and Infrared Atmospheric Sounding Interferometer (IASI) [Fortems-Cheiney et al., 2009; Turquety et al., 2009]. These satellite data overcome the problem faced by in-situ measurements and commercial aircraft (Nedelec et al., 2003).

The AIRS CO retrievals shown here were produced as operational products by the NASA Goddard Earth Sciences (GES) Data Information and Service Center (DISC) have been employed in many studies. The first observation of tropospheric CO was presented by McMillan et al. (2005) during track the impact of major south American fire 22-29 September 2002, and the daily global maps showed the advection of a large CO plume with forward trajectories conforming to long-rang transport, from biomass burning, as far east as the southern Indian Ocean. The comparisons to in-situ aircraft profiles indicate AIRS CO retrievals at 500 mbar are accurate to at least 10% in the northern hemisphere and approaching the 15% accuracy target set by per-launch simulations. McMillan et al. (2007) investigated the wealth of AIRS CO retrievals information contained more than five years, and improved decisively both the characterization of the vertical distribution of CO information in the AIRS retrievals and the sensitivity to the lower tropospheric CO. The analysis showed the capabilities of AIRS see Megacities and large-scale interannual variations in emissions linked to El Nino.

Warner et al. (2007), Kopacz et al. (2010) compared AIRS retrievals against MOPITT for AIRS validation of global CO mixing ratios at 500 mbar. Furthermore, presented are the comparisons of the satellite profiles with in-situ profiles. Because of the simultaneous measurements of troposphere CO from instruments, the comparison and a cross reference are necessary to understand these two data sets and their effects to the scientific conclusions developed from them. Yurganov et al. (2010) compared troposphere CO profiles from AIRS against Atmospheric Emitted Radiance Interferometer (AERI) over North America, the Northern Hemisphere (NH) mid-latitudes, and over the tropics for the period 2002-2009, and there is a good agreement between AIRS and AERI in the surface and total column measurements. They showed that the three months temporal of CO burden minima between the NH mid-latitudes and tropics could effect from transport of lower CO burdens from south to north. In addition, tropics contain about half of the global air mass and CO burden anomaly are higher there than in the NH.

4. Study area

The study area is Peninsular Malaysia, which is located between 1° to 7° latitudes north and 99° to 105° longitudes east, south of Thailand, north of Singapore and east of the Indonesian island of Sumatra. An area (Fig. 1), covering 3.575×105 km^2, with a center at Pahang (102° E and 4° N) was selected for this study. The central dimensions of the study domain are 550 km E-W and 650 km N-S. The Titiwangsa Mountain is a range from the Malaysia–Thai border in the north running approximately south-southeast over a distance of 480km forms the backbone of the Peninsular and separating the western part from the western part. Surrounding the central high regions are the coastal lowlands (Suhaila & Jemain, 2007).

As a characteristic for a humid tropical climate, in Peninsular Malaysia the weather is warm and humid throughout the year with temperatures ranging from 21° C to 32° C. In Malaysia never experienced the excessive day temperatures which are found in continental tropical areas, very rarely been recorded air temperature of 38°C, the days are oftentimes hot and the nights are reasonably cool everywhere (Dasimah, 2009). The highest average temperatures are at April-May and July-August in most places, and the lowest average monthly temperatures are at November-January. Although the variations of the spatial and seasonal temperature are relatively small, they are, nevertheless, equitably definite in some respects and are deserving mention (Shahruddin & Mohamed, 2005). There is a definite variation of monthly mean temperature coincided with the monsoons, the annual fluctuations of roughly 1.5-2° C. The monsoons significantly affect the climate of Peninsular Malaysia. It experiences two rainy seasons throughout the year associated with the Northeast Monsoon (NEM) from November to February and the Southwest Monsoon (SWM) from May to August (Wong at el., 2009).

During the inter-monsoon months (usually occur in April and October), the wind light and variable, and thunderstorms develop causing Substantial rainfall in each of the two transition periods, especially in the west coast states. Monsoon changes as well as the effects of topography are the main factors that affect the rainfall distributions. The monsoon rainfalls form 81% of annual rain that falls in the entire Peninsular Malaysia, which was estimated approximately 2300mm (Suhaila & Jemain, 2009). Maximum rainfall is occurred near the end of the year during the early NEM in most of the areas, while second maximum rainfall during the intermonsoon months (April or May). The high intensity rainfall is absolutely absent during the SWM period except in the west coastal stations. The lowest monthly rainfall occurs in February, and the highest monthly rainfall in December (Varikoden et al., 2010).

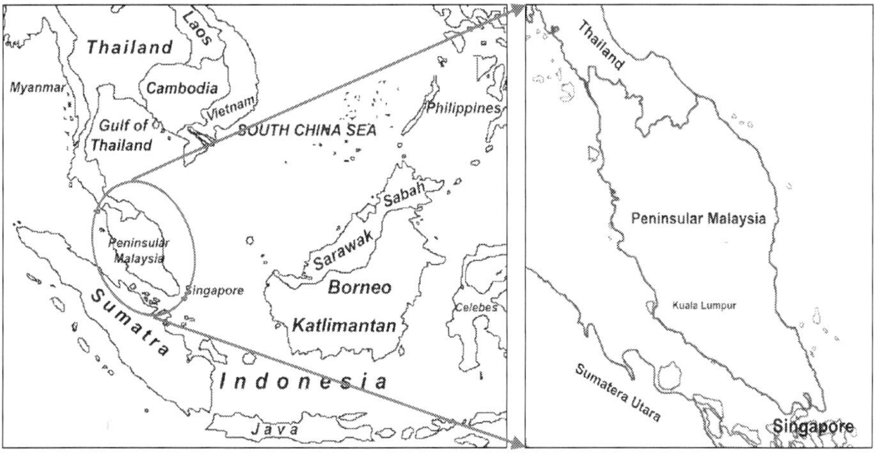

Fig. 1. The geographical feature of study area.

5. Acquisition and specification

The first new generation of meteorological advanced sounders for operational and research use was AIRS, one of several instruments onboard the EOS Aqua spacecraft launched May 4, 2002. AIRS instrument and its two companions microwave instruments, the Advanced Microwave Sounding Unit (AMSU) and the Humidity Sounder for Brazil (HSB), form the integrated atmospheric sounding system are characterizing and observing the entire atmospheric column from the surface to the top of the atmosphere in terms of temperature and surface emissivity, cloud amount and height, atmospheric temperature and humidity profiles and the spectral outgoing infrared radiation (Fishbein et al., 2007). The Aqua spacecraft coverage is pole-to-pole, and covers the globe two times a day, at orbit is polar sun-synchronous with a nominal altitude of 705 km (438 miles) and an orbital period of 98.8 minutes. The platform equatorial crossing local times are 1:30 in the afternoon (ascending) and 1:30 in the morning (descending). The respect cycle period is 233 orbits (16 days) with a ground track repeatability of +/- 20 km (Aumann et al., 2003).

The AIRS is a "facility" instrument developed by NASA as an experimental demonstration of the benefits of high resolution infrared spectra to science investigations and advanced technology for remote sensing (Pagano et al., 2006). The capability of AIRS/AMSU/HSB to supply simultaneous observations of the Earth's atmospheric temperature, land surface temperature, and ocean surface temperature, water vapor, cloud amount and cloud height, albedo, as well as new research products of greenhouse gas and aerosols, makes AIRS the most important EOS instrument for investigating several interdisciplinary issues to be addressed in Earth science (Pagano et al., 2003). The instruments suite was designed to measure the Earth atmospheric water vapor to an accuracy of 10% in 2 km layers in the lower troposphere and an accuracy of 50% in the upper troposphere. It also provides atmosphere temperature profiles with 1K/km accuracy in the troposphere and 1K/4 km layer in the stratosphere up to an altitude of 40 km. Also, provides integrated column burden for several trace gases (Chahine et al., 2006). Figure 2.1 shows the scanning geometry of AIRS and AMSU-A.

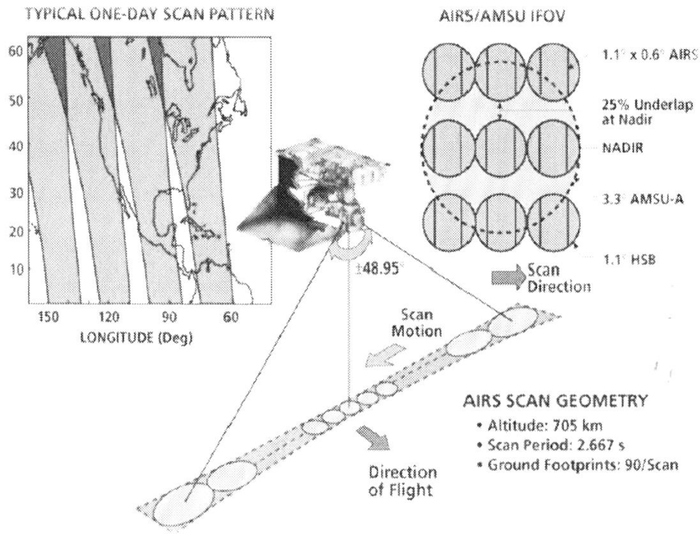

Fig. 2. Schematic showing the AIRS and AMSU scan geometrics [Jason, 2008].

AIRS is a continuously operating cross-track scanning sounder, consisting of a telescope that feeds a scale spectrometer. The AIRS instrument views the atmospheric infrared spectrum in 2378 channels with a nominal spectral resolving power $\lambda/\Delta\lambda$ ranging from 1086 to 1570 covering more than 95% of the earth surface and returning about three million spectra daily, in the 3.74-4.61 μm, 6.20-8.22 μm and 8.8-15.4 μm infrared wavebands at a nominal spectral resolution, also includes four visible/near-IR (Vis/NIR) channels between 0.40 and 0.94 μm, with a 2.3-km FOV (Strow et al., 2003).

The AIRS Science Processing System (SPS) is a set of programs, or Product Generation Executives (PGEs), utilize to process AIRS Science Data. These PGEs process raw, AIRS Visible (VIS), low level AIRS Infrared (AIRS), AMSU, and HSB instrument data to acquire temperature and humidity profiles. There are four distinct processing phases for processing the AIRS PGEs: Level 1A, Level 1B, Level 2 and Level 3 (Fishbein et al., 2007). Level 1A, Level 1B and Level 2 produce 240 granules of different products every day. Each product granule contains six minutes of data (Ye et al., 2007).

The L3 data are created from the L2 data product by binning them in 1°x1° grids. Level 3 products are statistical summaries of geophysical parameters that have been temporally aggregated and spatially re-sampled from lower level data products (e.g., Level 2 data) (Pagano et al., 2006). There are three AIRS Level 3 data products separately derived from Microwave-Only (MW-Only) retrievals and combined Infrared/Microwave (IR/MW) retrievals: daily, weekly and monthly as summarized in Table 1 Each product provides separate ascending (daytime) and descending (nighttime) binned data sets. When there is no coverage for that day, the daily Level 3 products will have gores between the satellite paths. The weekly Level 3 products may have missing data because of data dropouts. Monthly Level 3 products will probably contain complete global coverage with little missing data and without gores (Aumann et al., 2001).

Data Set	Short Name	Granule Size
L1B AMSU-A radiances	AIRABRAD	0.5 MB
L1B HSB radiances	AIRHBRAD	1.7 MB
L1B AIRS radiances	AIRIBRAD	56 MB
L1B VIS radiances	AIRVBRAD	21 MB
L1B AIRS QA	AIRIBQAP	5.6 MB
L1B VIS QA	AIRVBQAP	1.1 MB
L2 Cloud-cleared radiances	AIRI2CCF	10 MB
L2 Standard Product	AIRX2RET	5.4 MB
L2 Support Product	AIRX2SUP	20 MB
L3 standard daily product	AIRX3STD	~70 MB
L3 8-day standard product	AIRX3ST8	~103 MB
L3 monthly standard product	AIRX3STM	~105 MB

Table 1. Level 1, Level 2 and Level 3 data set products. (Jason, 2008).

6. Materials and methods

The study has been carried out for daily [six-day (12-23) August 2005 and six-day (13-24) August 2007], and monthly 2009 data. In order to evaluate and analysis the impact of Indonesia forest fire on CO distribution, and the monthly CO distributions in the study area. Five dispersed stations were selected across Peninsular Malaysia: Subang, Penang, Kuantan, Johor, and Kota Bharu to compare the daily data from August 2005 and August 2007. Results from the analysis of the retrieved for the CO obtained from AIRS ascending Level 3 daily standard (AIR×3STD) and Monthly standard product (AIRX3STM) data. Generally, twelve daily and twelve monthly L3 ascending granules were downloading to obtain the desired output. Extract the (AIR×3STD) and (AIRX3STM) product's files from the Satellite using the AIRS website, and saves in HDF-EOS4 files; this is a convenient file extension that can be easily extracted data from it and arrange in table using MS Excel.

Data including the corresponding location and time along the satellite track in a HDF (Hierarchical Data Format) format on monthly basis. Map of the study area was conducted by using Photoshop CS and SigmaPlot 11.0 software to analyze CO data distribution along the study period. To better assess the impacts and distribution of CO above Peninsular

Malaysia, the maps of CO were generated by using Kriging interpolation technique for the daily (August 2005 and 2007), and monthly 2009. The CO data were obtained from $1^{o}*1^{o}$ degree (latitude × longitude) spatial resolution ascending orbits.

7. Data analysis and results

7.1 Daily peninsular Malaysia CO maps and comparisons

The skies over Peninsular Malaysia were noticeably hazier than normal on mid-August 2005. In this section, we examine the impact of one smoke transport event on CO levels in Peninsular Malaysia. The elevated levels of CO found in smoke plumes from incomplete combustion of forest fires can lead to important further CO production downwind of the fires. To better assess the influences of forest fires and the transport smoke filled air mass on the local air quality in Peninsular Malaysia. We examined the data from AIRS for two different periods daily (12-23) August 2005 and (13-24) August 2007.

The six maps in Fig. 3 (12-23) August 2005 illustrate the extent of AIRS Daily coverage of Total column CO, the nominal peak of AIRS vertical sensitivity and the magnitude of the variations in atmospheric CO over Peninsular Malaysia. It observed an elevation in the CO values higher than normal rates during the period from mid-July to mid-August 2005 in Malaysia (Rajab et al., 2009).

The AIRS data in Fig. 3 indicate numerous fires in the regions of enhanced CO over Malaysia. In Indonesia, the fire counts peaked on 13-15 August in Sumatra (DOE, 2009). From Fig. 3 (12, 14 and 16 August), can see the advection of the CO plume from Sumatra, Indonesia, to Peninsular Malaysia. The CO Pollution for this event was characterized the increase in CO by 60% in the northern region (uppers the latitude 4, e.g. Penang and Kota Bahru), 35% in the central and east coast district (e.g. Kuala Lumpur and Kuantan), and 20% in the southern area (e.g. Johor). Plainly evident the CO total column values are high in 12 August, increase to be highest in 14 August, and a gradual decrease of CO values after 16 with a reduction impact of the forest fire, compared to previous days (DOE, 2005). The highest value was when the hazy conditions reached its peak on 12 August over Selangor (2.68×10^{18} molecules/cm^2).

Fig. 3 (18, 21 and 23), while the CO value is still high in the Sumatra, shows the gradual decrease of the concentration of CO in most areas in coinciding with the declaration of the end of the hazy conditions. The lowest value of CO was on 23 August over Terengganu (1.75×10^{18} molecules/cm^2).

From Fig. 4, illustrate the CO values from 13 to 24 August 2007, were normal circumstances in the absence of any event can observe the highest CO total column over central areas (e.g. Selangor and Negeri Sembilan), Penang and Johor. These regions represent biomass burning sources, long-range transport and industrial/domestic fuel sources. Enhanced CO values in Selangor on 13 August (1.95×10^{18} molecules/cm^2). The greater draw down of CO total column occurs over pristine marine environment in the north east coast of Terengganu and in the Inland region of north Kelantan (1.56×10^{18} molecules/cm^2).

To illustrate the biases between the two data sets, which are caused by biomass burning, the daily August 2005 and 2007 zonal averaged CO total column is showed in Fig. 5. In Fig. 5, a daily series graph of CO August 2005 (solid line) compared to daily CO August 2007 (dots line) from observed AIRS for five stations; kota Bahru, Penang, Kuantan, Subang, and Johor, where the day is on the x-axis and CO total column (molecules/cm^2) on the y-axis. It can be seen clearly, in all the stations the daily CO concentration on August 2005 is higher than August 2007. There is a clear difference between two lines, especially in 14 August when the impacts of forest fire up to a maximum. The discrepancies become less after 17 August with a lower impact of fires. In general, south to north the CO values differences are larger than West to East between two lines over study areas. The largest differences are in the north regions (uppers the latitude 4).

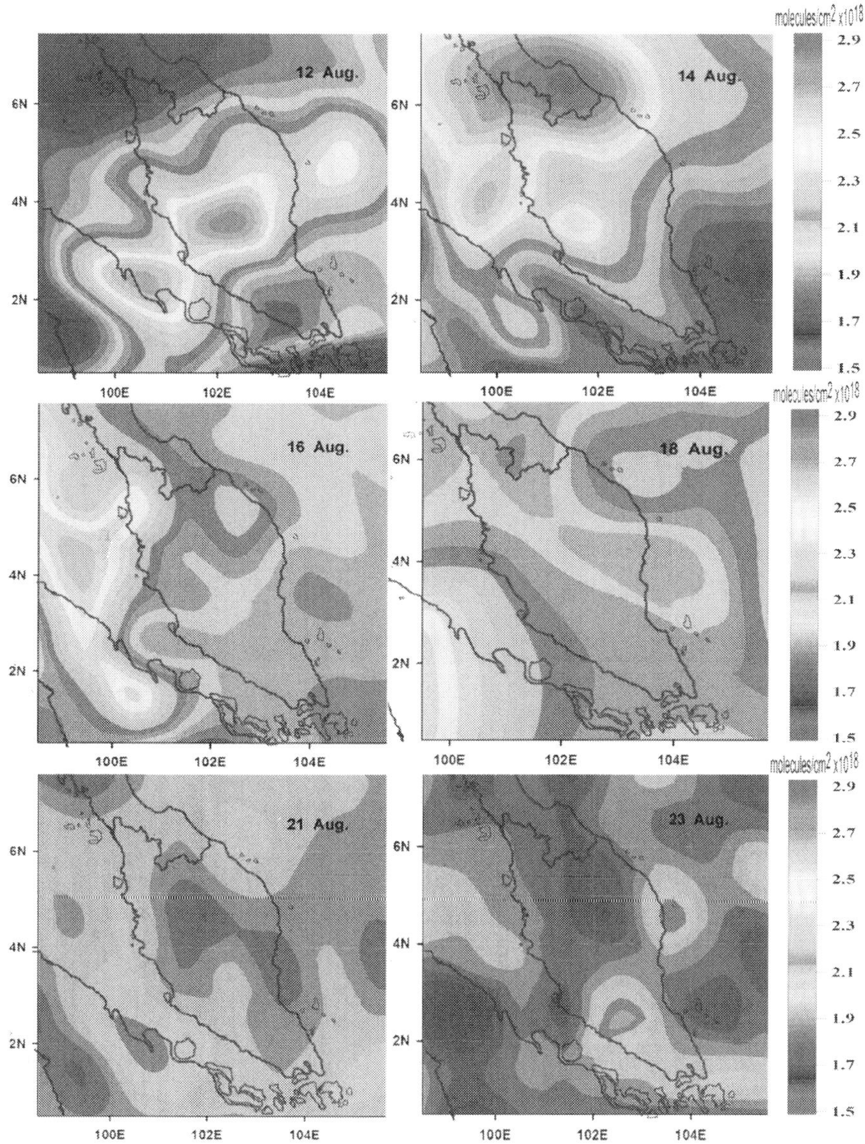

Fig. 3. CO total column (molecules/cm^2) for daily (12-23) August 2005 in Peninsular Malaysia.

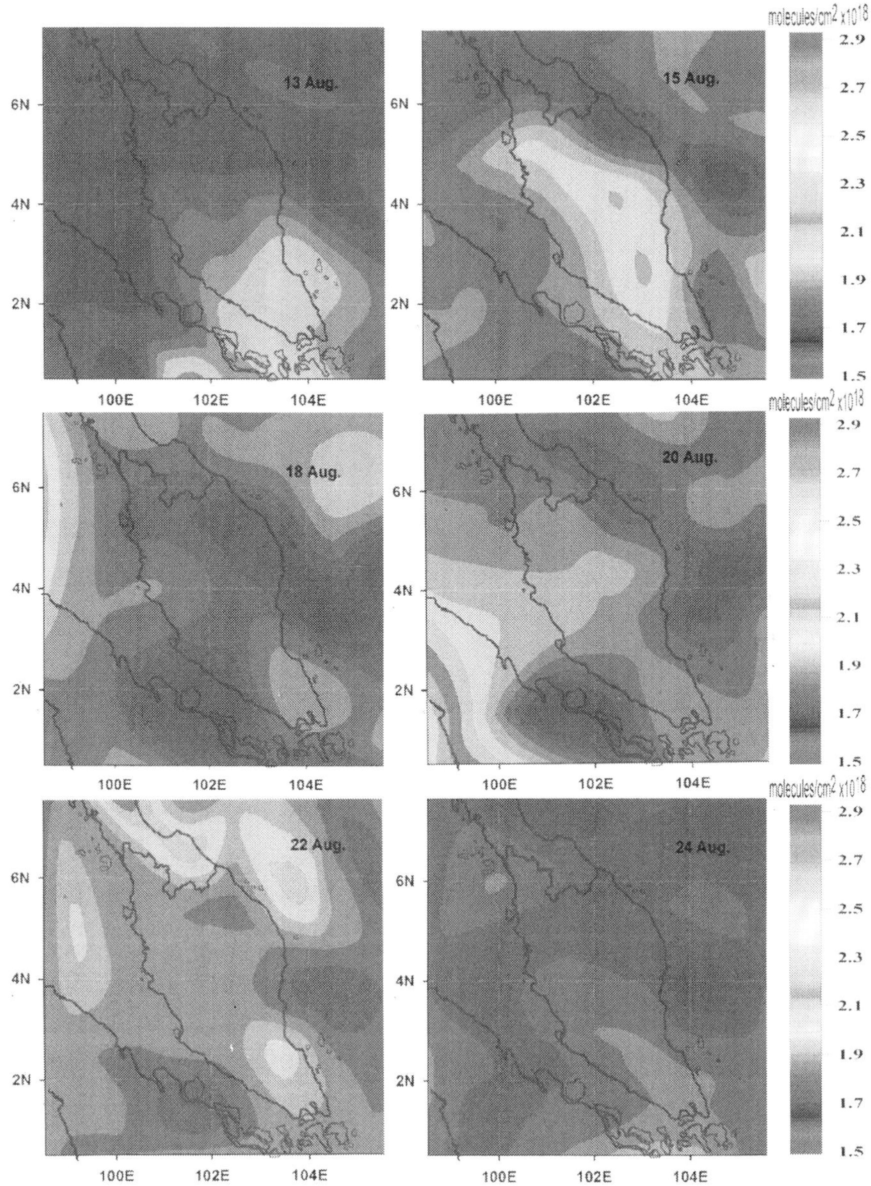

Fig. 4. CO total column (molecules/cm²) for daily (13-24) August 2007 in Peninsular Malaysia.

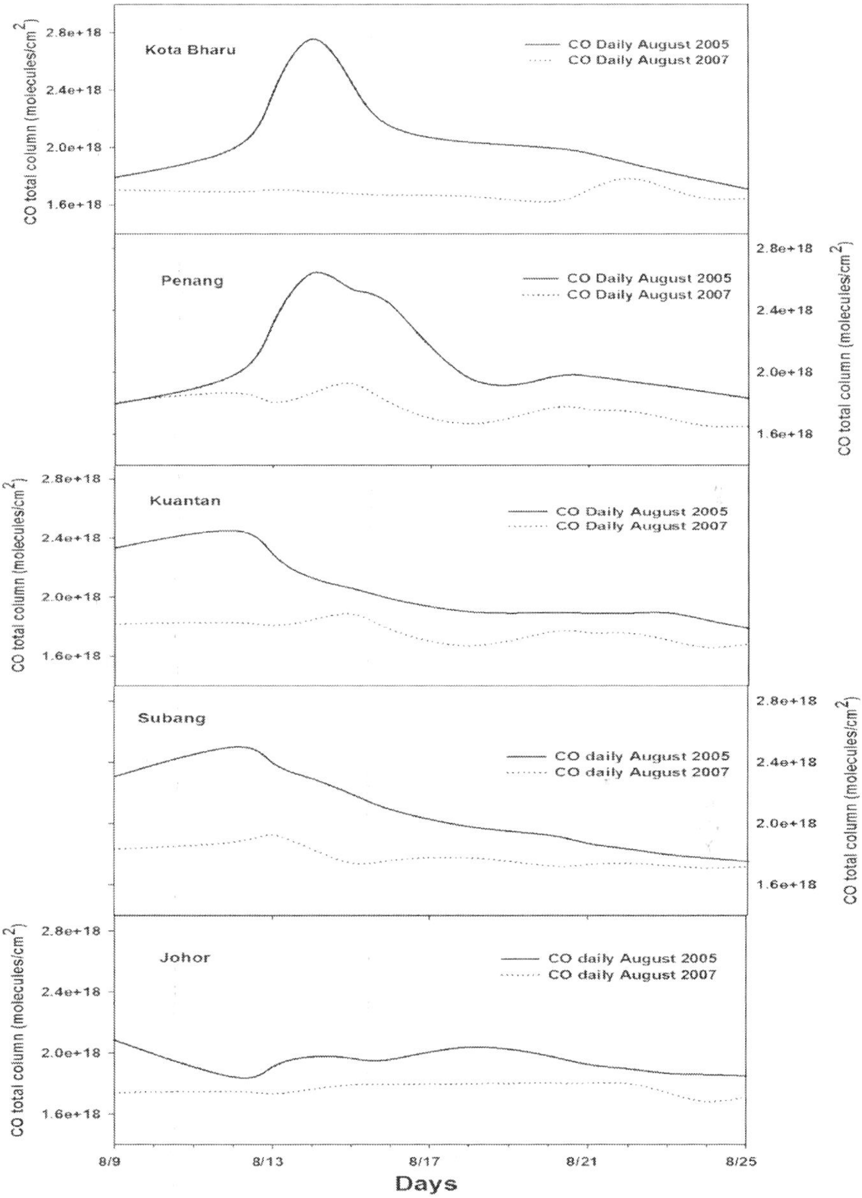

Fig. 5. The CO total column August 2005 (solid line) compared to daily CO total column August 2007 (dots line) from observed AIRS for five stations; Kota Bahru, Penang, Kuantan, Subang, and Johor.

7.2 Monthly peninsular Malaysia CO maps for 2009

The retrieved CO total column amount (CO_total_column_A) Level-3 monthly (AIRX3STM) 1°×1° spatial resolution used for mapping CO for 2009. Based on the combination of rich local sources of CO along with the transport of additional significant amount of pollution from Siberia and northeast Asia brings in the beginning of the northeast monsoon led to increase the values of CO at early dry season (Pochanart et al., 2004), that are shown in Fig. 6 for dry season.

Plainly evident the impact of peatland fires in several areas in Sumatra, Indonesia, on the values of CO in the north and central of Peninsular Malaysia during January, with continuing its impact on the central and southern regions during February. Unusually, there was a decrease in the values of CO at the end of the dry season during March-April due to unexpected reasonable rainfall during this period throughout the Peninsular Malaysia (Malaysian Meteorological Department, 2009). The highest value of CO in dry season was (2.19×10^{18} molecules/cm^2) over Selangor on January {at 101.5°×3.5°, red pixels}, while the less value was (1.75×10^{18} molecules/cm^2) over Perak on April {at 101.5°×5.5°, violet pixels}. The strong effects of regional biomass burning during the dry season, a regular annual occurrence in January to April, makes it extremely difficult to detect any evidence of the long-range transport of air pollution to Southeast Asia from other regions (Pochanart et al., 2004).

Normally, in the late dry season, large regional biomasses burning occur and the impact of air masses pollution transport has significant from the western Asia, Middle East and East Asia (Pochanart et al., 2003). The El-Niño episodes can lead to immense displacement of rainfall regions in the tropics, bringing torrential rain to otherwise arid regions and drought to vast areas. Uniform warming prevail all the regions in Malaysia during an El-Niño event, especially during the November to March (Cheang, 1993). When El-Niño has a drought effect, there will be large biomass burning and increase CO emission (McMillan et al., 2007).

Fig. 7 shows that the CO have low value in the north regions, upper the latitude 5, and low to moderate in the rest of regions during the early wet season (May-July), due to the high levels of moisture and incident solar irradiation causes the production of the OH, the primary oxidizer for CO. When OH abundance is near a maximum; CO is near minimum (McMillan et al., 2007). Considerable values of OH are sufficient to provide significant spoilage rates of trace gases in the haze outflow, outputting for example in a CO lifetime of only ~15 days near the surface. The subtle differences peak of biomass burning indicate differences in transport patterns as well as differences in rainfall patterns across the region, the enhanced CO emission correlates with occasions of less rainfall (McMillan et al., 2007). Additionally, the lack of rain results in long carbonaceous lifetimes, this lengthens the time during which oxidation reactions can take place (Rasch et al., 2001).

In the late wet season (August-October), observed moderate value in the north regions, upper the latitude 5, while slightly high to high values in the rest of regions due to the impact of peatland fires from Sumatra and transboundary pollution which was aggravated by hot weather conditions (DOE, 2009). The highest value was (2.153×10^{18} molecules/cm^2) over Johor on October {at 103.5°×1.5°, red pixels}, while the less value was (1.702×10^{18} molecules/cm^2) over Perak on June {at 101.5°×5.5°, violet pixels} during the wet season.

During the summer monsoon (wet season) May - October, the marine air masses from the middle and low latitudes of the Southern Hemispheric Indian Ocean are dominated continental Southeast Asia. These marine air masses carry small to moderate amounts of air pollution to southwest Asia. In addition, less regional biomass burning occurred and the impact of air masses is less significant from the Pacific and Indian Oceans (Pochanart et al., 2003).

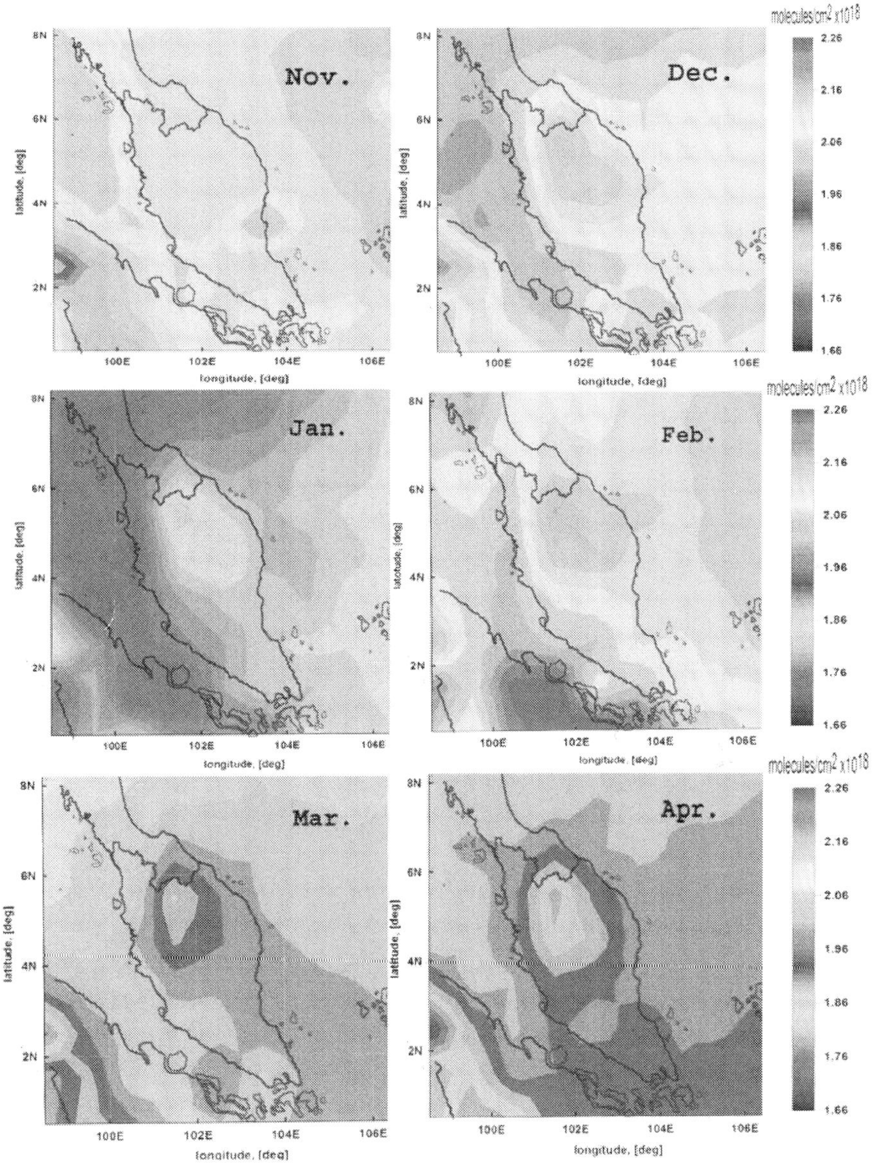

Fig. 6. AIRS monthly coverage retrieved total column carbon monoxide (CO) for dry season [November to April] 2009.

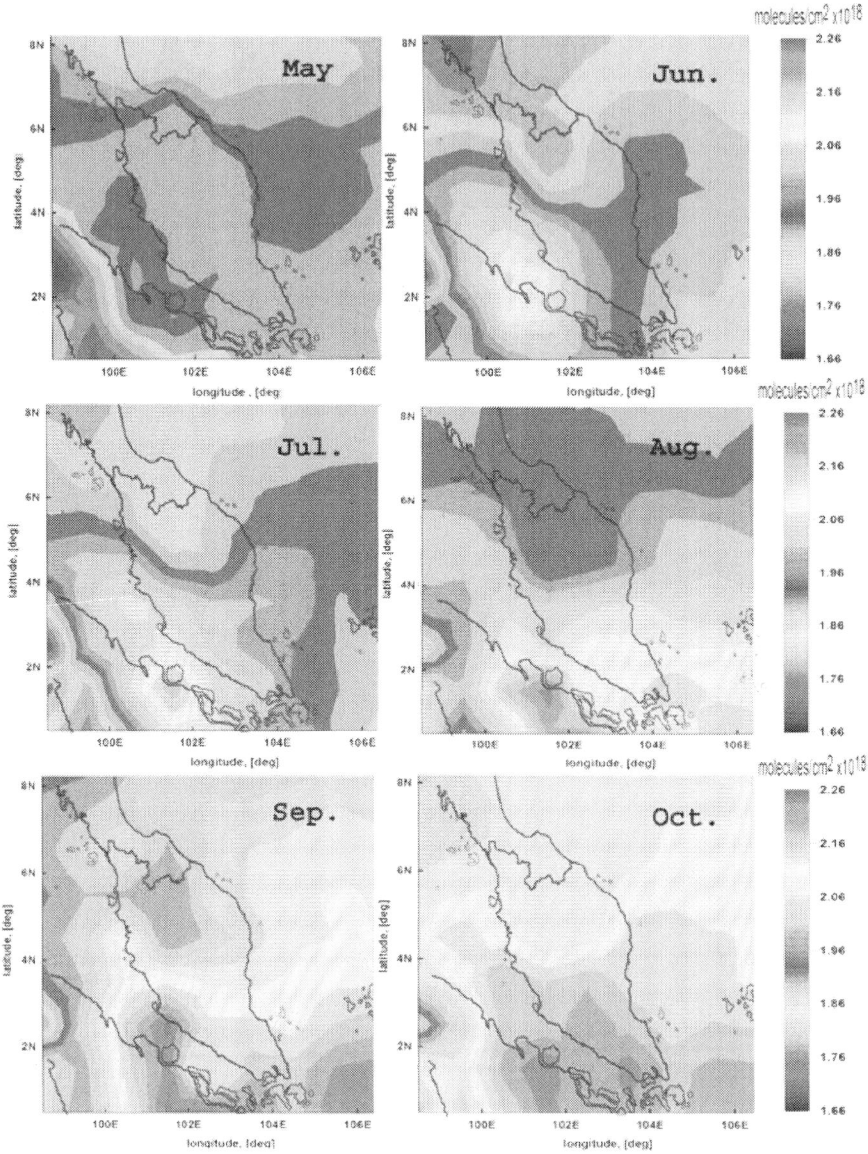

Fig. 7. AIRS monthly coverage retrieved total column carbon monoxide (CO) for wet season [May to October] 2009.

In short, observed the highest values of CO occurred when biomass burning during dry season, and also over the industrial and congested urban zones (it is the most abundant pollutant in urban atmosphere and very stable, having an average lifetime of about two months) where the main source of emission was vehicles which contributed to 95 percent of CO emission load in Malaysia during 2009 (DOE, 2009; Kopacz, et al., 2010). A greater draw down of the CO occurs over the pristine continental environment in the northeast region regions on June at Perak (101.5°×5°) during wet season. This was due to lack of sources CO as well as the direct influence of south westerly wind, which remove polluting gases continuously (CO slightly lighter than air) (Jasim et al., 2010). Furthermore, the rain is a great cleanser of the atmosphere so the (Hoskins, 2001).

In addition, the CO values were higher in the central and southern regions than the rest of regions throughout the year because there are many sources, crowded cities, and more affected by the peatland fires in Sumatra. Can be seen also the impact of El-Nino on CO values during June-August from the unusually high values of CO in this period, especially in southern region.

8. Conclusion

This chapter has reviewed the impact of the CO pollutants by long-range transport from Sumatra, Indonesia, forest fires during August 2005, and monthly distributions 2009 in Peninsular Malaysia. While an immense amount about these issues has been analyzed over the forest fires, there are still many regional sources making it extremely difficult to detect it. The monsoon regimes provide the main climatological controls on the air pollution in Southeast Asia. During the winter monsoon, cold surges originating from Siberia and northeast Asia brings a significant amount of pollution to Southeast Asia while crossing through the heavily-polluted regions of East Asia and caused the heavy rain falls. During the summer monsoon, the impact of air pollution transport is less significant due to the less regional biomass burning occurred, and the marine air masses carry small to moderate amounts of air pollution to southwest Asia from the Pacific and Indian Oceans.

As demonstrated here, AIRS' daily views of atmosphere CO total column across the study area enables detailed analyses of both the spatial and temporal variations in emissions and the visualization of subsequent transport. Focusing on a major Sumatra, Indonesia, fire event, the AIRS daily Peninsular Malaysia maps show the advection of a large CO plume with forward trajectories confirming long-range transport as far as north east as Peninsular Malaysia. The daily maps also distinctly identified the plainly evidence with high values of CO occurred when biomass burning from Indonesia's forest fires reached its peak.

The local CO maximum was in a region experienced extensive the intense fires. The daily maps shows characterized elevated CO values by 60% in the northern region (uppers the latitude 4), 35% in the central and east coast district, and 20% in the southern area. The highest value was when the hazy conditions reached its peak on 12 August over Selangor (2.68×1018 molecules/cm2). And the lowest value of CO was on 23 August over Terengganu (1.75×10^{18} molecules/cm^2).

From the daily maps in August 2007 for normal circumstances, the highest CO values occurred above the industrial and congested urban zones (1.95×10^{18} molecules/cm^2). The greater draw down of CO total column occurs over pristine marine environment in the north east coast of Terengganu and in the Inland region of north Kelantan (1.56×10^{18} molecules/cm^2).

It can be seen clearly from the comparison between the daily August 2005 and 2007; in all the stations, the daily CO concentration on August 2005 is higher than August 2007. There is a clear large difference between two lines, especially in 14 August when the impacts of forest fire up to a maximum. The discrepancies become less after 17 August coincides with a reduction effect of the forest fires.

The AIRS/Aqua Level 3 Monthly gridded product (AIRX3STD) 1×1° spatial resolution, Version 5 data using AIRS IR and AMSU, without-HSB, employed to investigate a monthly distribution map of CO total column over study area for 2009. The high CO values observed during the dry season, due to the strong influences of regional biomass burning, a regular annual occurrence in January to April, and coupled with the significant impacts of air mass pollution transport from the East Asia to by winter monsoon. The highest value was (2.19×10^{18} molecules/cm^2) over Selangor, while the less value was (1.75×10^{18} molecules/cm^2) over Perak.

The low CO values prevail on the wet season due to the less biomass burning occurred, and the marine air masses from the middle and low latitudes of the Southern Hemispheric Indian Ocean carried small to moderate amounts of air pollution. Furthermore, the high levels of moisture and incident solar irradiation lead to the production of the OH, the primary oxidizer for CO. The OH abundance is near a maximum; CO is near minimum. The highest value was (2.153×10^{18} molecules/cm^2) over Johor.

In short, the highest CO values was during biomass burning occurred in dry season, and over industrial and congested urban zones. The cleaner areas and the less CO value areas throughout the year in Peninsular Malaysia was Perak at (long. 101.5°× lat. 3.5°), where the green lands, vast forest and lack sources of pollution.

This study has provided evidence for the impact of remote CO total column emissions and forest fire on CO pollution levels above study area using satellite data. Furthermore, it enhanced our knowledge on AIRS detection of CO emission from forest fire, the utility and accuracy of remotely sensed atmospheric CO total column and abundances from AIRS. Satellite measurements are able to measure the increase of atmospheric CO values over different areas. CO maps from AIRS will lead to a more understanding of the CO budgets.

9. Acknowledgement

The authors gratefully acknowledge the financial support from the Relationship between Heavy Rain, Flash Floods and Central Pressure in Malaysia Grant, account number: 1001/PFIZIK/811152 and USM-RU-PRGS Grant, account number:1001/PFIZIK/841029. We would like to thank the technical staff who participated in this study. Thanks are also extended to USM for support and encouragement.

10. References

Andreae, M. O. & Merlet, P. (2001). Emission of Trace Gases and Aerosols From Biomass Burning. *Global Biogeochemical Cycles*, Vol.15, pp.955–966.

Aumann, H. H., Goldberg, M., Mcmillin, L., Rosenkranz, P., Staelin, D., Strow, L. & Susskind, J. (2001). AIRS-Team Retrieval For Core Products and Geophysical Parameters. In AIRS Team - NASA (Ed., Jet Propulsion Laboratory.

Aumann, H. H., Chahine, M. T., Gautier, C., Goldberg, M. D., Kalnay, E., McMillin, L. M., Revercomb, H., Rosenkranz, P. W., Smith, W. L., Staelin, D. H., Strow, L. L. & Susskind, J. (2003). AIRS/AMSU/HSB on the Aqua mission: Design, Science Objectives, Data Products, and Processing Systems. *IEEE Transaction On Geoscience And Remote Sensing*, Vol.41, No.2, pp.253-264.

Bovensmann, H., Burrows, J. P., Buchwitz, M., Frerick, J., Noël, S., Rozanov, V. V., Chance, K. V. & Goede, A. P. H. (1999). SCIAMACHY - Mission Objectives and Measurement Modes. *Atmospheric Chemistry and Physics*, Vol.56, pp.127-150.

Buchwitz, M., Khlystova, I., Bovensmann, H. & Burrows, J. P. (2007). Three Years of Global Carbon Monoxide From SCIAMACHY: Comparison With MOPITT and First Results Related To The Detection of Enhanced CO Over Cities. *Atmospheric Chemistry and Physics*, Vol.7, pp.2399-2411.

Buchwitz, M., de Beek, R., Noël, S., Burrows, J. P., Bovensmann, H., Schneising, O., Khlystova, I., Bruns, M., Bremer, H., Bergamaschi, P., Körner, S. & Heimann, M. (2006). Atmospheric Carbon Gases Retrieved From SCIAMACHY By WFM-DOAS: Version 0.5 CO and CH_4 and Impact of Calibration Improvements on CO_2 Retrieval. *Atmospheric Chemistry and Physics*, Vol.6, pp.2727-2751.

Burrows, J. P., Hölze, E., Goede, A. P. H., Visser, H. & Fricke, W. (1995). SCIAMACHY - Scanning Imaging Absorption Spectrometer for Atmospheric Chartography. *Acta Astronaut*, Vol.35, No.7, pp.445-451.

Chahine, M. T., Thomas, S. P., Hartmut, H. A., Robert, A., Christopher, B., John, B., Luke, C., Murty, D., Eric, J. F., Mitch, G., Catherine, G., Stephanie, G., Scott, H., Fredrick, W. I., Ramesh, K., Eugenia, K., Bjorn, H. L., Sung-Yung, L., John, L. M., McMillan, W. W., Larry, M., Edward, O. T., Henry, R., Philip, R., William, S. L.,David, S., Strow, L. L., Joel, S., David, T., Walter, W. & Lihang, Z. (2006). AIRS Improving Weather Forecasting and Providing New Data on Greenhouse Gases. *American Meteorological Society*, Vol.87, pp.911 -926.

Cheang, B. K. (1993). Interannual Variability of Monsoons In Malaysia and Its Relationship With ENSO. *Proc. Indian Acad. Sci. (Earth Planet. Sci.)*, Vol 102, pp. 219-239.

Clerbaux, C., Hadji-Lazaro, J., Turquety, S., Mégie, G. & Coheur, P. -F. (2003). Trace Gas Measurements From Infrared Satellite For Chemistry and Climate Applications. *Atmospheric Chemistry and Physics*, Vol.3, pp.1495-1508.

Clerbaux, C., Coheur, P. F., Hurtmans, D., Barret, B., Carleer, M., Colin, R., Semeniuk, K., McConnell, J. C., Boone, C. & Bernath, P. (2005). Carbon Monoxide Distribution From The ACE-FTS Solar Occulation Measurements. *Geophysical Research Letters*, Vol.32, pp.1-4, DOI: 10.1029/2005GL022394.

Clerbaux, C., George, M., Turquety, S., Walker, K. A., Barret, B., Bernath, P., Boone, C., Borsdorff, T., Cammas, J. P., Catoire, V., Coffey, M., Coheur, P. -F., Deeter, M., De Mazière, M., Drummond, J., Duchatelet, P., Dupuy, E., de Zafra, R., Eddounia, F., Edwards, D. P., Emmons, L., Funke, B., Gille, J., Griffith, D. W. T., Hannigan, J., Hase, F., Höpfner, M., Jones, N., Kagawa, A., Kasai, Y., Kramer, I., Le Flochmoën, E., Livesey, N. J., López-Puertas, M., Luo, M., Mahieu, E., Murtagh, D., Nédélec, P., Pazmino, A., Pumphrey, H., Ricaud, P., Rinsland, C. P., Robert, C., Schneider, M., Senten, C., Stiller, G., Strandberg, A., Strong, K., Sussmann, R., Thouret, V., Urban, J. & Wiacek, A. (2008). CO Measurements from The ACE-FTS Satellite Instrument: Data Analysis and Validation Using Ground-Based, Airborne and Spaceborne Observations. *Atmospheric Chemistry and Physics*, Vol.8, pp.2569-2594.

Daniel, J. S. & Solomon, S. (1998). On The Climate Forcing of Carbon Monoxide. *Journal of Geophysical Research*, Vol.103, pp.13249-13260.

Dasimah, O. (2009). Urban Form and Sustainability of a Hot Humid City of Kuala Lumpur. European Journal of Social Sciences vol.8, pp.353-359.

De Groot, W. J., Field, R. D., Brady, M. A., Roswintiarti, O. & Mohamad, M. (2007). Development of The Indonesian and Malaysian Fire Danger Rating Systems. *Mitig Adapt Strat Glob Change*, Vol.12, pp.165-180.

Delaney, K., Ling, L., & Erickson, T. (2001). In Ford Md, Clinical Toxicology, WB Saunders Company, ISBN 0-7216-5485-1.

Department of Environment (DOE), M. (2005). Malaysia Environmental Quality Report. Petaling Jaya.

Department Of Environment (DOE), M. (2009). Malaysia Environmental Quality Report. Petaling Jaya.

Dousset, B., & Gourmelon, F. (2003). Satellite Multi-sensor Data Analysis of Urban Surface Temperatures and Landcover. *ISPRS Journal of Photogrammetry and Remote Sensing*, Vol 58, pp. 43-54.

Emmons, L. K., Pfister, G. G., Edwards, D. P., Gille, J. C., Sachse, G., Blake, D., Wofsy, S., Gerbig, C., Matross, D. & Nedelec, P. (2007). Measurements of Pollution in the Troposphere (MOPITT) Validation Exercises During Summer 2004 Field Campaigns Over North America. *Journal of Geophysical Research*, Vol.112, D12S02, DOI: 10.1029/2006JD007833, 2007.

Emmons, L. K., Edwards, D. P., Deeter, M. N., Gille, J. C., Campos, T., Nédélec, P., Novelli, P. & Sachse, G. (2009). Measurements of Pollution In The Troposhere (MOPITT) validation through 2006. *Atmospheric Chemistry and Physics*, Vol.9, pp.1795-1803.

Fishbein, E., Granger, S., Lee, S. Y., Manning, E., Weiler, M., Blaisdell, J., & Susskind, J. (2007). AIRS/AMSU/HSB Version 5 Data Release User Guide. In Atmospheric Infrared Sounder, A., EOS, Ed., California Institute of Technology.

Fortems-Cheiney, A., Chevallier, F., Pison, I., Bousquet, P., Carouge, C., Clerbauz, C., Coheur, P. -F, George, M., Hurtmans, D. & Szopa, S. (2009). On The Capability of IASI Measurements to Inform about CO Surface Emissions. *Atmospheric Chemistry and Physics*, Vol.9, pp.8735-8743.

Hoskins, A. J. (2001). Ozone Matters. *Indoor and Built Environment*, Vol.10, pp.1-2.

Jasim, M. R., MatJafri, M. Z., Lim, H. S. & Abdullah, K. (2009). Indonesia Forest Fires Exacerbate Carbon Monoxide Pollution Over Peninsular Malaysia During July to September 2005. *Proceedings of Sixth International Conference on Computer Graphics, Imaging and Visualization*, DOI: 10.1109/CGIV.2009.96, Tianjin University, Tianjin, China, August, 2009.

Jasim, M. R., Lim, H. S., MatJafri, M. Z. & Abdullah, K. (2010). Daily Carbon Monoxide (CO) Abundance from AIRS Over Peninsular Malaysia. *Journal of Materials Science and Engineering*, Vol.4, pp.93-99.

Jason, L. (2008). README Document for AIRS Level-2 Version 005 Standard Products. In Goddard Earth Sciences Data And Information Services Center (Ed., National Aeronautics and Space Administration (NASA).

Kopacz, M., Jacob, D. J., Fisher, J. A., Logan, J. A., Zhang, L., Megretskaia, I. A., Yantosca, R. M., Singh, K., Henze, D. K., Burrows, J. P., Buchwitz, M., Khlystova, I., McMillan, W. W., Gille, J. C., Edwards, D. P., Eldering, A., Thouret, V. & Nedelec, P. (2010). Global Estimates of CO Sources with High Resolution by Adjoint Inversion of Multiple Satellite Datasets (MOPITT, AIRS, SCIAMACHY, TES). *Atmospheric Chemistry and Physics*, Vol.10, pp.855-876.

Lambrigtsen, B. H. (2003). Calibration of The AIRS Microwave Instruments. *IEEE Transaction On Geoscience And Remote Sensing*, Vol.41, No.2, pp.369-378.

Lavorel, S., Flannigan, M. D., Lambin, E. F. & Scholes, M. C. (2007). Vulnerability of Land Systems to Fire: Interactions Among Humans, Climate, the Atmosphere, and Ecosystems. Mitig Adapt Strat Glob Change, Vol. 12, pp. 33–53.

Lawrence, M. G. (2004). Export of Air Pollution From Southern Asia and Its Large-Scale Effects. IN STOHL, A. (Ed.) The Handbook of Environmental Chemistry. Berlin, Springer.

Liu, J., Drummond, J. R., Li, Q., Gille, J. C. & Ziskin, D. C. (2005). Satellite Mapping of CO Emission from Forest Fires in Northwest America Using MOPITT Measurements. *Remote Sensing of Environment*, Vol.95, pp. 502–516.

Lopez, J. P., Luo, M., Christensen, L. E., Loewenstein, M., Jost, H., Webster, C. R. & Osterman, G. (2008). TES Carbon Monoxide Validation during Two AVE Campaigns Using the Argus and ALIAS Instruments on NASA's WB-57F. *Journal of Geophysical Research Atmospheres*, Vol.113, D16S47, DOI: 10.1029/2007JD008811.

Luo, M., Rinsland, C. P., Fisher, B. M., Sachse, G., Diskin, G., Logan, J. A., Worden, H. M., Kulawik, S. S., Osterman, G., Eldering, A., Herman, R. & Shephard, M. W. (2007). TES Carbon Monoxide Validation with DACOM Aircraft Measurement During INTEX-B 2006. *Journal of Geophysical Research*, Vol.112, D24S48, DOI: 10.1029/2007JD008803.

Mahmud, M. & Kumar, T. S. V. V. (2008). Forecasting Severe Rainfall in the Equatorial Southeast Asia. *GEOFIZIKA*, Vol. 25, pp. 109-127.

Malaysian Meteorological Department (2009). General Climate of Malaysia. Selangor, Ministry of Science, Technology and Innovation (MOSTI).

McMillan, W. W., Barnet, C., Strow, L., Chahine, M. T., McCourt, M. L., Warner, J. X., Novelli, P. C., Korontzi, S., Maddy, E. S. & Datta, S. (2005). Daily Global Maps of Carbon Monoxide From NASA's Atmospheric Infrared Sounder. *Geophysical Research Letters*, Vol.32, L11801, DOI: 10.1029/2004GL021821.

McMillan, W. W., Yurganov, L., Evans, K., & Barnet, C. (2007). Global Climatology of Tropospheric CO from the Atmospheric InfraRed Sounder (AIRS). *20th Conference on Climate Varibility and Change*, Vol. 5B.3, pp. 217 - 228.

McMillan, W. W., Warner, J. X., McCourt Comer, M., Maddy, E., Chu, A., Sparling, L., Eloranta, E., Hoff, R., Sachse, G., Barnet, C., Razenkov, I. & Wolf, W. (2008). AIRS Views Transport from 12 to 22 July 2004 Alaskan/Canadian fires: Correlation of AIRS CO and MODIS AOD with Forward Trajectories and Comparison of AIRS CO Retrievals with DC-8 in Situ Measurements During INTEX-A/ICARTT. *Journal of Geophysical Research*, Vol.113, D20301, DOI: 10.1029/2007JD009711.

Nedelec, P., Cammas, J. -P, Thouret, V., Athier, G., Cousins, J. -M., Legrand, C., Abonnel, C., Lecoeur, F., Cayez, G. & Marizy, C. (2003). An Improved Infrared Carbon Monoxide Analyzer for Routine Measurements aboard Commercial Airbus Aircraft: Technical Validation and First Scientific Results of The MOZAIC III Programme. *Atmospheric Chemistry and Physics*, Vol.3, pp.1551-1564.

Pagano, T. S., Aumann, H. H., Hagan, D. E. & Overoye, K. (2003). Prelaunch and In-Flight Radiometric Calibration of The Atmospheric Infrared Sounder (AIRS). *IEEE Transactions on Geoscience and Remote Sensing*, Vol.41, pp.265 - 273.

Pagano, T. S., Chahine, M. T., Aumann, H. H., Tian, B., Lee, S. Y., Olsen, E., Lambrigtsen, B., Fetzer, E., Irion, F. W., McMillan, W., Strow, L., Fu, X., Barnet, C., Goldberg, M., Susskind, J. & Blaisdell, J. (2006). Remote Sensing of Atmospheric Climate Parameters from the Atmospheric Infrared Sounder. IEEE International Geosience and Remote Sensing Symposium (IGARSS), pp.2386-2389, ISBN: 0780395107.

Pochanart, P., Akimoto, H., Kajii, Y. & Sukasem, P. (2003). Carbon Monoxide, Regional-Scale Transport, and Biomass Burning in Tropical Continental Southeast Asia: Observations in Rural Thailand. *Journal of Geophysical Research*, Vol.108, pp. 4552, 15.

Pochanart, P., Wild, O. & Akimoto, H. (2004). Air Pollution Import To and Export From East Asia. IN STOHL, A. (Ed.) The Handbook of Environmental Chemistry. Berlin, Springer.

Rasch, P. J., Collins, W. D. & Eaton, B. E. (2001). Understanding the Indian Ocean Experiment (INDOEX) Aerosol Distributions with an Aerosol Assimilation. *Journal of Geophysical Research*, Vol.106, pp.7337-7355.

Shaharuddin, A. & Mohamed, E. Y. (2005). Urban climate research in Malaysia. IAUC Newsletter, 5-10.

Streets, D., Tsai, N., Akimoto, H. & Oka, K. (2001). Trends in Emissions of Acidifying Species in Asia, 1987-1997. *Water, Air, and Soil Pollution*, Vol.130, pp. 187-192.

Strow, L. L, Hannon, S. E., Souza-Machado, S. D., Motteler, H. E. & Tobin, D. (2003). An Overview of the AIRS Radiative Transfer Model. *IEEE Transaction on Geoscience and Remote Sensing*, Vol.41, No.2, pp.303-313.

Suhaila, J. & Jemain, A. A. (2007). Fitting Daily Rainfall Amount in Malaysia using the Normal Transform Distribution. *Journal of Applied Sciences*, Vol.7, pp.1880 - 1886.

Suhaila, J. & Jemain, A. A. (2009). Investigating the Impacts of Adjoining Wet Days on the Distribution of Daily Rainfall Amounts in Peninsular Malaysia. *Journal of Hydrology*, Vol.368, pp.17-25.

Tangang, F. T., Juneng, L. & Ahmad, S. (2007). Trend and Interannual Variability of Temperature in Malaysia: 1961-2002. *Theoretical and Applied Climatology*, Vol.89, pp.127-141.

Tiwari, Y. K., Gloor, M., Engelen, R., Rodenbeck, C. & Heimann, M. (2005). Comparing Model Predicted Atmospheric CO_2 with Satellite Retrievals and In-situ Observations - Implications for the Use of Upcoming Satellite Data in Atmospheric Inversions. *Geophysical Research Abstracts*, Vol. 7, 09823.

Turquety, S., Hurtmans, D., Hadji-Lazaro, J., Coheur, P. -F, Clerbauz, C., Josset, D. & Tsamalis, C. (2009). Tracking the Emission and Transport of Pollution from Wildfires Using the IASI CO Retrievals: Analysis of the Summer 2007 Greek fires. *Atmospheric Chemistry and Physics*, Vol.9, pp.4897-4913.

Varikoden, H., Samah, A. A. & Babu, C. A. (2010). Spatial and Temporal Characteristics of Rain Intensity in The Peninsular Malaysia Using TRMM Rain Rate. *Journal of Hydrology*, Vol.387, pp.312-319.

Warner, J., Comer, M. M., Barnet, C. D., McMillan, W. W., Wolf, W., Maddy, E. & Sachse, G. (2007). A Comparison of Satellite Tropospheric Carbon Monoxide Measurements from AIRS and MOPITT during INTEX-A. *Journal of Geophysical Research*, Vol.112, D12S17.

Wong, C. L., Venneker, R., Uhlenbrook, S., Jamil, A. B. M. & Zhou, Y. (2009). Variability of Rainfall in Peninsular Malaysia. *Journal Hydrology and Earth System Sciences Discussions*, Vol. 6, pp.5471-5503.

Ye, H., Fetzer, E. J., Bromwich, D. H., Fishbein, E. F., Olsen, E. T., Granger, S. L., Lee, S. Y., Chen, L. & Lambrigtsen, B. H. (2007). Atmospheric Total Precipitable Water from AIRS and ECMWF during Antarctic Summer. *Geophysical Research Letters*, Vol.34, L19701.

Yurganov, L., McMillan, W., Wilson, C., Fischer, M. & Biraud, S. (2010). Carbon Monoxide Mixing Ratios over Oklahoma Between 2002 and 2009 Retrieved from Atmospheric Emitted Radiance Interferometer Spectra. *Atmospheric Measurement Techniques*, Vol. 3, pp. 1263-1301.

9

Air Quality Plans for the Northern Region of Portugal: Improving Particulate Matter and Coping with Legislation

Borrego, C., Carvalho, A., Sá, E., Sousa, S., Coelho, D., Lopes, M., Monteiro, A. and Miranda, A.I

CESAM & Department of Environment and Planning, University of Aveiro, Portugal

1. Introduction

Air quality is one of the environmental areas in which the European Union (EU) has been most active, in particular designing and implementing legislation on air quality and on the restriction of pollutant emissions to the atmosphere. The air quality Framework Directive (FD) (Directive 96/62/CE of 27 September 1996) established the obligations of the EU member states (MS) and redefined the guidelines for the assessment and management of air quality, namely the creation of an air quality management system comprising measuring networks, pollutant emission inventories and air quality modelling. Also, the FD states the obligation to elaborate and implement Plans and Programmes to improve air quality when the air quality standards are not fulfilled. The FD was transposed into the Portuguese legislation by the Decree-Law (DL) nr 276/99 of 23 July 1999. In order to state limits for specific pollutants, four "daughter" directives emerged later, which were also transposed to national legislation by three decree-laws (DL nr 111/2002, DL nr 320/2003, DL nr 351/2007). In May 2008, a new Directive on Ambient Air Quality and Cleaner Air for Europe (Directive 2008/50/EC) was published, integrating the former FD, three of the four previous "daughter" directives and one Council decision into a single document. The new Directive was transposed to national legislation by the Decree-Law nr 102/2010 of 23 September 2010. The European and national legislation regarding air quality management is illustrated in Figure 1.

The new directive (Directive 2008/50/EC) introduces new objectives for fine particles, but does not change existing air quality standards. Additionally, the application of numerical air quality models is highlighted as a fundamental tool to better assess and manage air quality. One of the main MS obligations is the implementation of Air Quality Plans (previously known as Plans and Programmes) when the air pollutant concentrations exceed the air quality standards in zones or agglomerations. The implementation of Air Quality Plans should be based on the design of measures to reduce the pollutant atmospheric concentrations and meet the legal requirements.

Exceedances of the thresholds of particulate matter with a 50 % efficiency cut-off at 10 µm aerodynamic diameter (PM10) have been reported by the majority of the EU member states, mainly in urban agglomerations, where human exposure is also higher (EEA, 2005). The

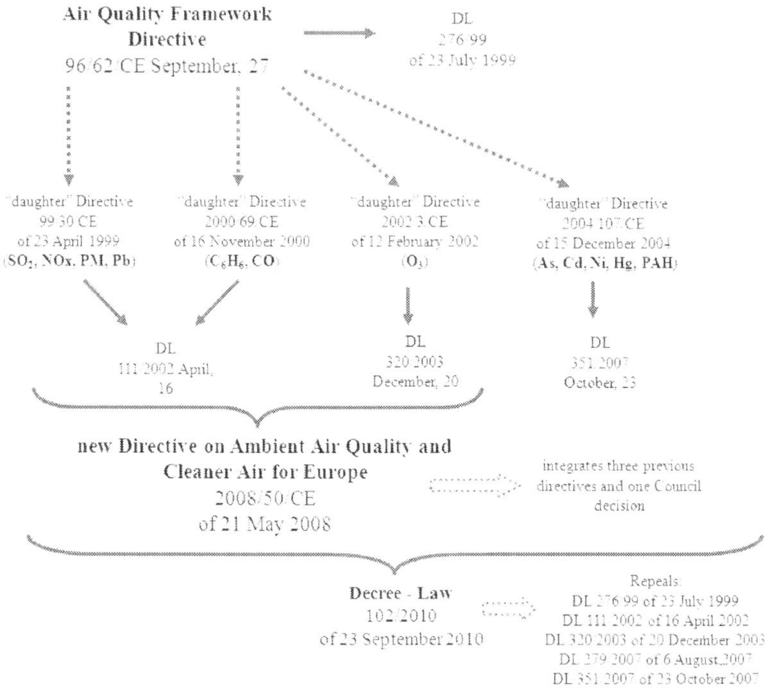

Fig. 1. European and Portuguese air quality legislation framework.

daily limit value for PM10 is 50 μg.m^{-3} plus the margin of tolerance (LV+MT), and cannot be exceeded more than 35 times in a calendar year. Like what is being registered in the EU, over the last years Portugal has been surpassing the legislated limit values for PM10 concentrations in ambient air at several air quality monitoring stations (Monteiro et al., 2007), namely in the northern region.

The North Regional Coordination and Development Commission (CCDR-N) is the entity in charge of controlling and improving the air quality in the region. This is achieved through the development and implementation of Air Quality Plans to improve the PM10 levels, taking into account environmental benefits and economic costs of the proposed mitigation measures. In this chapter, data from the air quality monitoring stations between 2000 and 2007 (Air Quality Plans reference period) and the proposed PM10 emissions reduction measures are analysed. The environmental benefits of these measures for the improvement of the PM10 levels in the atmosphere are assessed through the application of a numerical modelling system.

2. Air quality assessment over the northern region of Portugal, 2000-2007

The northern region is divided into 2 zones (*Norte Interior* and *Norte Litoral*) and 4 agglomerations, namely *Porto Litoral, Vale do Ave, Vale do Sousa,* and *Braga,* defined according to the population density criteria settled by legislation (EU Directive 2008/50/EC). The air quality assessment for the period 2000-2007 imposed the development of Air Quality Plans in order to improve the PM10 levels in the agglomerations of *Porto Litoral, Vale do Ave* and *Vale do Sousa*.

2.1 The air quality monitoring network

The air quality monitoring network managed by the CCDR-N includes several background, traffic and industrial stations, with a total of 21 sites where PM10, among other air pollutants, is continuously measured (Borrego et al., 2008a). Figure 2 presents the northern region's zones and agglomerations with the respective location of the monitoring stations, together with the type of influence (traffic, industrial and background).

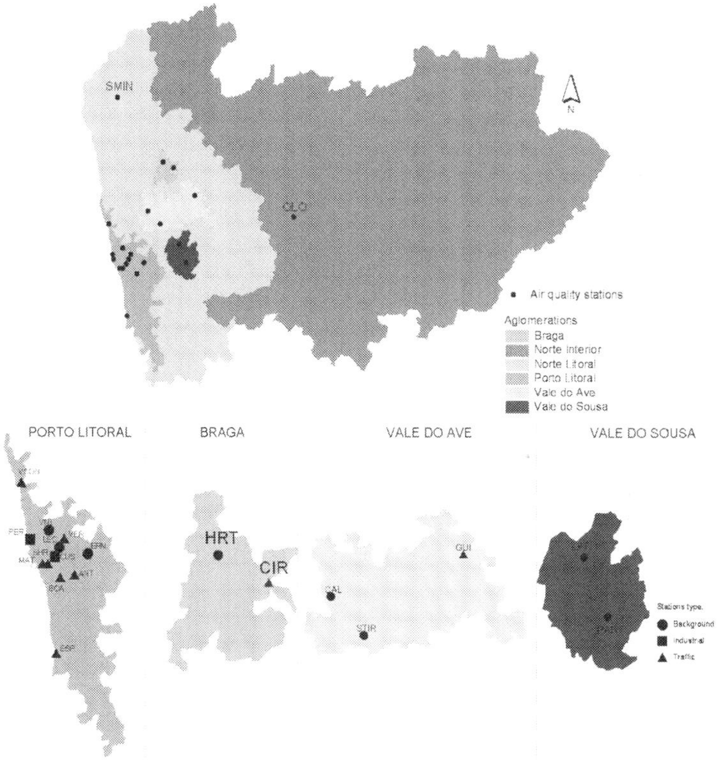

ANT-Antas; BOA-Boavista; MAT-Matosinhos; SHOR-Senhora da Hora; ESP-Espinho; VER-Vermoim;PAR-Paredes; GUI-Guimarães; CIR-Circular Sul; ERM-Ermesinde; LAT-Centro Lacticínios; STIR-Santo Tirso; PER-Perafita; CUS-Custóias; VCON-Vila do Conde; LEC-Leça do Balio; VNT-Vila Nova da Telha;CAL-Calendário; HRT-Horto; SMIN - Senhora do Minho; OLO- Lamas D'Olo

Fig. 2. Representation of the CCDR-N PM10 air quality monitoring network and respective zones and agglomerations in 2007

The agglomeration of *Porto Litoral* has the most extensive PM10 monitoring network, with 2 background, 7 traffic and 2 industrial sites. The other agglomerations are only covered by 2, at maximum 3, stations. This network is, however, very recent. Before 2003, only 1/3 of the stations were in operation. Due to the EU commitments and obligations the number of monitoring stations increased and between 2004 and 2007 the PM10 network improved substantially and reached the coverage exhibited in Figure 2.

Nevertheless, for the air quality diagnosis of the northern region the data collection efficiency has to be taken into account, but since not all of these stations fulfill the annual minimum PM10 data collection criteria required by legislation (85%, considering around 5% of data loss due to calibration and maintenance procedures), they can not all be considered for the present data analysis. Figure 3 shows the number of monitoring stations with the required annual data collection efficiency from 2000 till 2007.

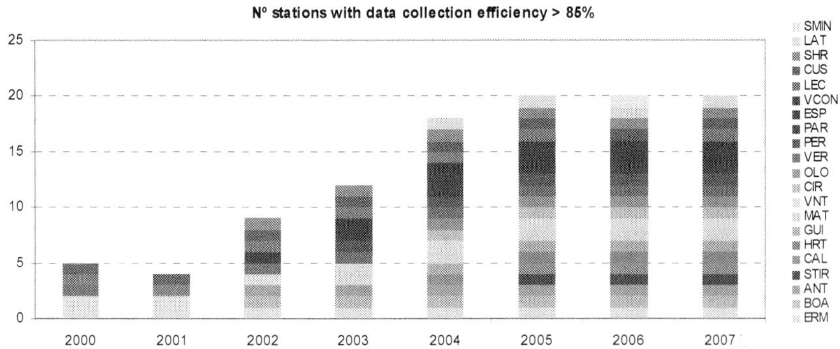

Fig. 3. Number of PM10 monitoring stations with annual data collection efficiency above 85%.

After 2004 the number of monitoring stations with adequate data collection efficiency achieved an almost constant value, around 20 monitoring stations available for the analysis. In this sense, the following data analysis is based on data from these air quality stations, namely regarding the PM10 exceedances and general trends.

2.2 PM10 data analysis

A long-term evaluation of the PM10 concentration values registered in the northern region of Portugal is presented based on the hourly values collected over the study period of 2000-2007. This analysis was focused on the legislation accomplishment, as concerns the daily (50 $\mu g.m^{-3}$) and annual (40 $\mu g.m^{-3}$) limit values for human health protection (EU Directive 2008/50/EC).

The fulfilment of the daily limit value is shown in Figure 4 over all the monitoring stations (Figure 4a) and considering only the background sites (Figure 4b).

Exceedances of the daily limit value (red line in the plots) are observed in the majority of the monitoring sites (Figure 4a), including background sites (Figure 4b) and over the entire analyzed period (2000-2007). Besides that, several values higher than 200 $\mu g.m^{-3}$ were registered in different monitoring stations, and at specific locations values above 400 $\mu g.m^{-3}$ and 800 $\mu g.m^{-3}$ can also be detected (in this case, due to a special episode of construction works close to the station). These data highlight that pollution due to particulate matter is a severe problem over this region, with systematic exceedances and particular high peaks/episodes.

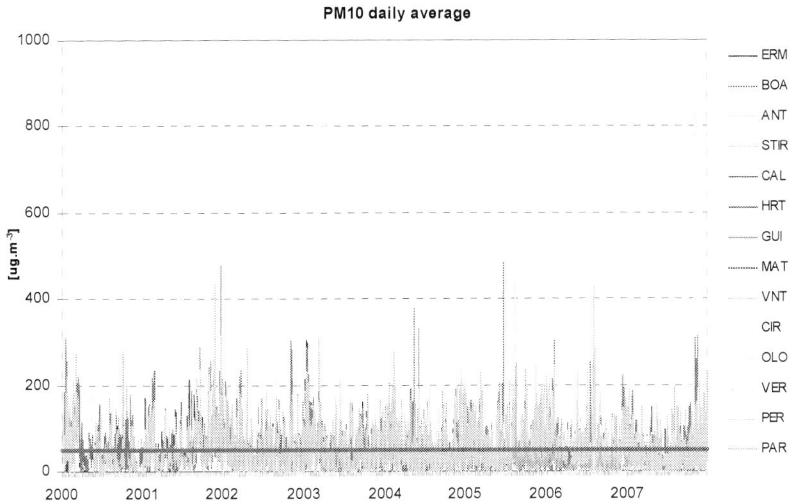

(a)

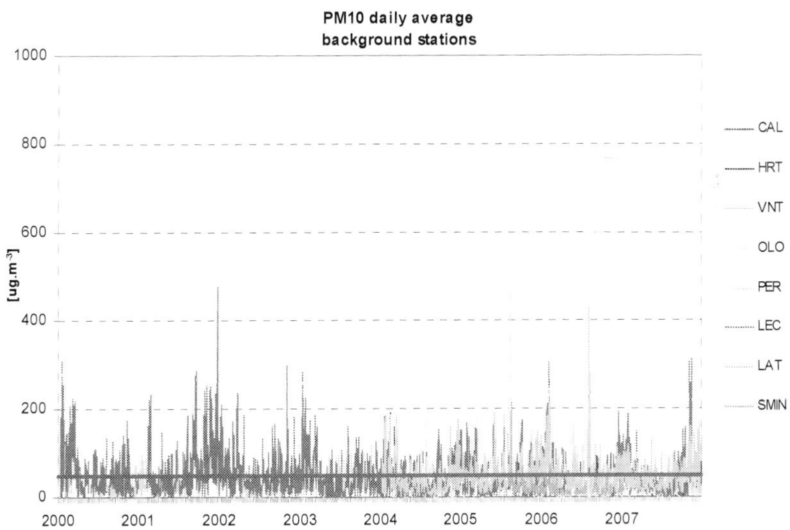

(b)

Fig. 4. PM10 daily average concentrations between 2000 and 2007 for all stations (a) and only considering the background stations (b).

The number of days with daily limit exceedances, as well as the number of stations reporting these exceedances are shown in Figure 5, over the study period 2000-2007.

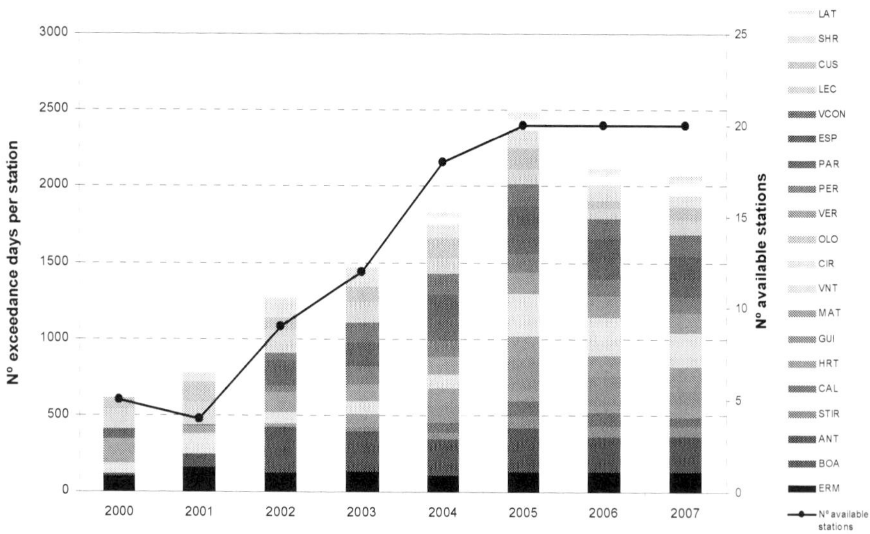

Fig. 5. Number of PM10 daily value exceedances over the 2000-2007 period for all available stations. The right axis presents the total number of available stations for each analyzed year in 2000-2007.

The number of stations with exceeding days increased over the 2000-2007 period, but this was mainly due to the increase in the number of available stations. The year 2005 was the most critical in terms of the number of stations that recorded concentrations above the limit value and of exceeding days.

Beyond the short-term impact, PM10 should also be assessed for the long-term. Legislation settles a limit value of 40 $\mu g.m^{-3}$ for PM10 annual average. Figure 6 shows the annual average of PM10 registered in each monitoring station during the analyzed period.

As well as with the daily average values, there are several stations that exceed the annual average limit value (Figure 6a). Nevertheless, the majority of these are traffic sites (Figure 6b), which explains the high levels detected (40-60 $\mu g.m^{-3}$ on average). The unique exceptions are the background stations of LEC and CUS, whose high concentrations can be explained based on their suburban influence.

The total number of stations with exceedances of the annual limit value is summarized in Figure 7 for the study period (2000-2007).

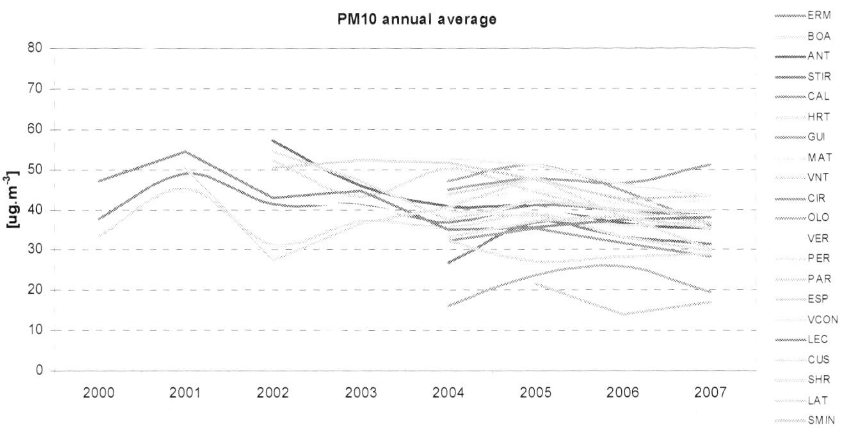

(a)

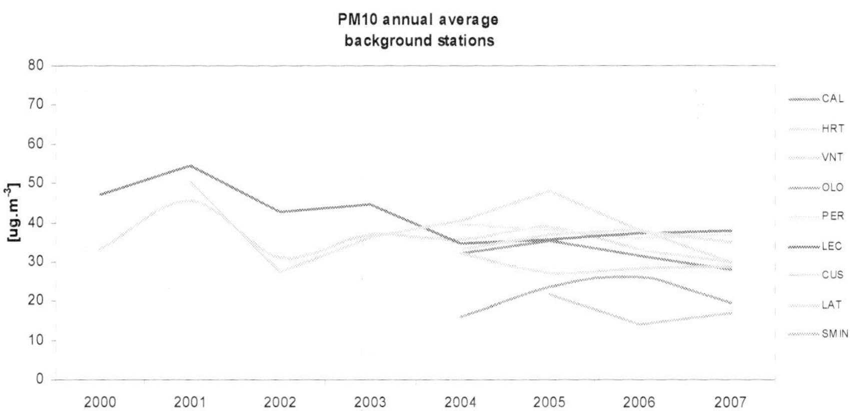

(b)

Fig. 6. PM10 annual average over the 2000-2007 period, for all available stations (a) and only for the background stations (b).

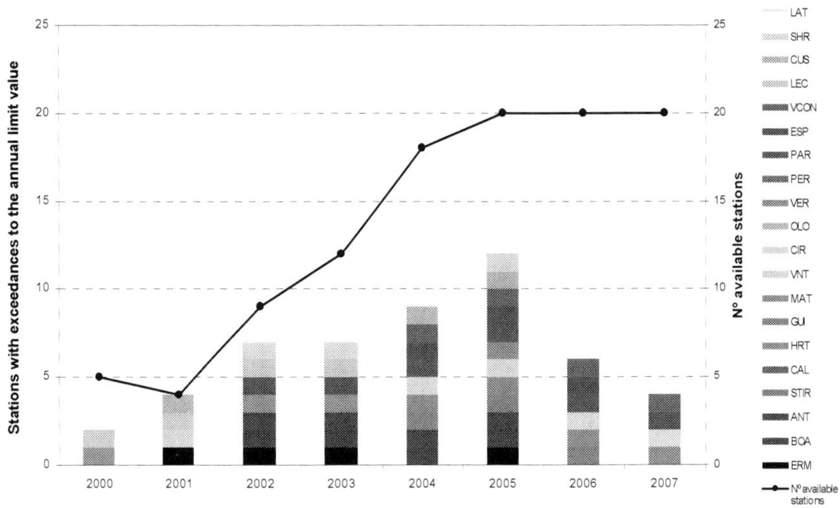

Fig. 7. Number of PM10 annual value exceedances over the 2000-2007 period for all the monitoring stations. The right axis indicates the total number of available stations for each analyzed year.

There is a clear trend to a reduction of PM10 annual values during the last two analyzed years (2006 and 2007). Until 2006, the majority of the stations registered exceedances every year, in particular between 2001 and 2003. 2005 was a critical year in terms of the number of stations with exceedances of the annual limit value.

There is a group of stations that have exhibited exceedances since they became active, namely, MAT, ESP and VCON. These three stations are located in the agglomeration of *Porto Litoral*.

In order to identify the more critical areas and respective agglomerations, the analysis of exceedances of the limit values was repeated grouping the stations by agglomeration (Figure 8).

The agglomeration of *Porto Litoral* presents the highest number of exceedances, for both daily and annual limit values, and during the entire study period (2000-2007). The remaining agglomerations (*Braga, Vale do Ave* and *Vale do Sousa*) registered exceedances of both limit values only after 2004, although, in fact, the monitoring stations were only active after this date.

The exceedances of the PM10 limit values occurred in traffic, industrial, urban and suburban background stations with an estimated exposed population of around 500,000 inhabitants. This situation highlights the need for the design and implementation of Air Quality Plans to improve the PM10 levels in the region. The next section presents the measures that were selected and studied in order to reduce the PM10 emissions to the atmosphere in the northern region of Portugal.

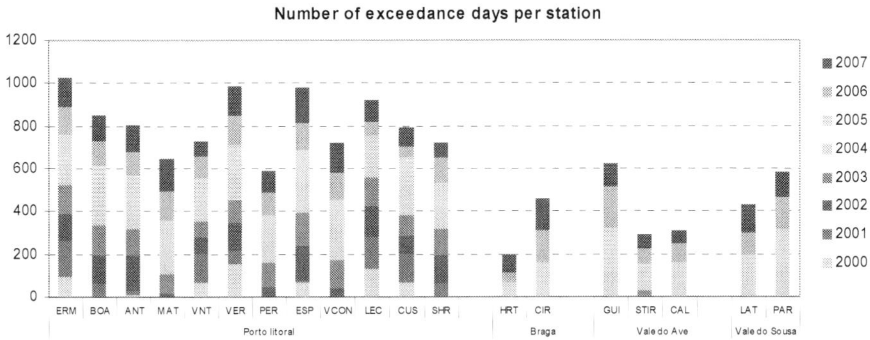

(a)

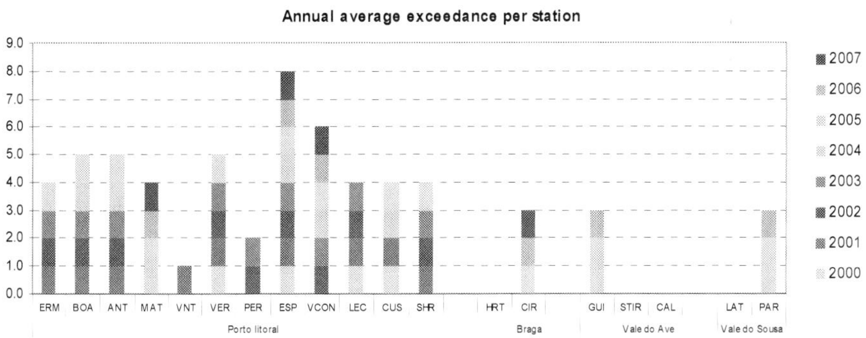

(b)

Fig. 8. PM10 exceedances for the daily (a) and annual (b) target values for each station within the analyzed agglomerations.

3. Measures to reduce the PM10 emissions over the northern region of Portugal

As previously mentioned, the development of Air Quality Plans for the agglomerations of *Porto Litoral*, *Vale do Ave* and *Vale do Sousa* is mandatory. The design of efficient Air Quality Plans implies the comprehension of the sources of PM10 affecting these agglomerations and the identification of the main sources of the measured levels. The PM10 concentrations can derive from natural or anthropogenic emission sources. Studies performed over the last decade consider that natural emissions could make a significant contribution to the PM10 levels. The anthropogenic sources of PM10 are usually located in urban and industrial areas. Over the urban areas, the emission of PM10 is closely related to residential combustion (Borrego et al., 2010a), construction and road traffic. Traffic emissions are due not only to the

direct exhaust of the vehicles, but also to brakes wear, road abrasion, tyre wear and road dust re-suspension. According to Ketzel et al. (2005), dust re-suspension accounts for about 50 to 85% of the total emissions from traffic-related emissions. The PM10 industrial sources include ceramics, clinker and cement production or mining processes (Rodriguez, 2002).
Based on the emission inventory for the northern region (Borrego et al., 2008b) traffic, industrial and commercial activity (restaurants), residential combustion and construction sites were identified as the main anthropogenic sources of PM10 in those agglomerations. Figure 9 presents the main PM10 emission sources at national level and in the northern region.

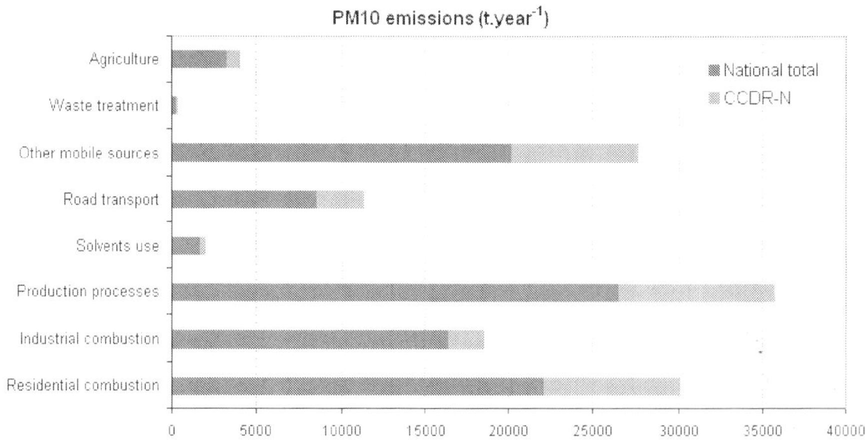

Fig. 9. PM10 emissions for different sectors at national level and over the northern region (CCDR-N) (Borrego et al., 2008b).

Taking into account the main emission sources, several mitigation measures to reduce PM10 levels in the three agglomerations were defined according to their environmental benefits and economic costs (Borrego et al., 2008a). These measures were discussed with the entities involved in their implementation and finally a list of selected measures was designed. The Execution Programme (EP), which is the Portuguese legal document that promotes and enforces the application of the emission reduction measures, describes these selected measures in detail. Figure 10 summarizes the entire process from the elaboration of the Air Quality Plan to the Execution Programme.
The measures included in the EP result from multiple contacts and meetings established between the CCDR-N and the entities identified as responsible for their implementation. The number of local, regional and national entities participating in the selection and implementation of the measures to reduce PM10 emissions in the northern region rose to 35. Several protocols were signed with the local and regional entities to formalize the implementation of the selected measures within the EP framework. These measures were grouped in three main sectors: (i) traffic; (ii) industry; and (iii) residential combustion. A brief description of the measures considered in the EP is presented below.

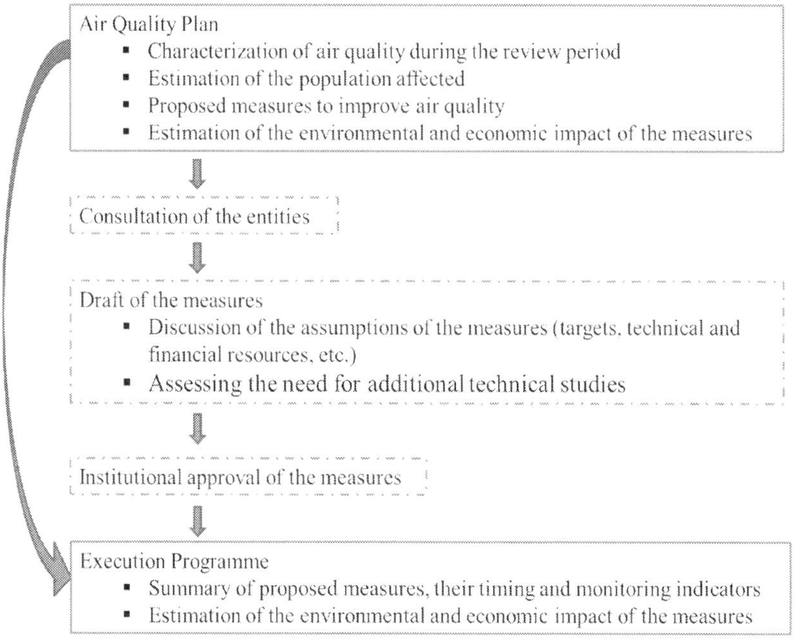

Fig. 10. Process related to the design of the Execution Programme.

3.1 Traffic sector

- Improving the public transport system
  The improvement of the public transport system aims to reduce the number of kilometers travelled, as being economically attractive and at the same time reducing PM10 emissions.
- Low Emission Zones (LEZ)
  LEZ are urban spaces where circulation of the most polluting heavy diesel vehicles (with emissions over a certain level or not meeting a certain European emission standard - Euro standards) is restricted or, in some cases, charged according to their emission levels.
- Installation of particle filters on heavy vehicles
  The percentage of heavy vehicles in comparison with the total number of vehicles in circulation is small, although, the associated PM10 emissions are significant, being one of the categories that most contribute to the PM10 traffic emissions. In diesel engines, the introduction of filters reduces by 90% the emissions of particulate matter (Tente et al., 2011). The Portuguese government launched legislation to support, among others things, the retrofitting of vehicles with particulate filters. Financial incentives to support the renewal and update of the public transportation fleet were also launched.
- Renewal of the municipal solid waste management fleet
  According to an inquiry held in the municipalities, over 80% of the fleet of municipal solid waste collection vehicles was old (prior to 2000). Thus, the renewal of that fleet

was proposed through the introduction of emission control systems, including particulate filter systems with exhaust gas recirculation or the replacement of old vehicles by modern ones that meet most recent Euro standards.

- **Introduction of public fuel stations for natural gas**
  According to the Portuguese Association of Natural Gas Vehicles in Portugal, in 2007 there were 377 natural gas vehicles, most of which were owned by public transport companies and municipalities. At national level, there was only one fuel station for natural gas for the general public, and the remaining four were inactive or were for private use. An increase in supply will tend to encourage the public to purchase more such vehicles, thereby reducing the emission of particulate matter. It is intended to build a public natural gas station in the northern region.
- Street sweeping and washing
  In urban areas, the PM10 emissions from resuspension caused by road traffic are comparable with or even larger than those from exhaust gases (Amato et al., 2009). Consequently, street sweeping and washing contributes to the removal of the particles deposited on the pavement and reduces the resuspension of such material. When water adheres to deposited particles, it increases their mass and surface tension forces, decreasing the likelihood of suspension and transport, especially as cohesion of wetted particles often persists after the water has evaporated due to the formation of aggregates (Amato et al., 2009).

3.2 Industrial sector

- Publication of a legal diploma with new emission limit values (ELV) for stationary sources
  The reduction of the current emission limit value (ELV) for particulate matter coming from stationary sources associated with industrial and commercial activities represents a large contribution to PM10 emission decrease in the northern region of Portugal.
  Moreover, enhanced surveillance of point sources will have a major role in fulfilling the legal obligations in this regard. The obligation to comply with more stringent ELV will lead, in some cases, to the need for installation of emission reduction systems or their replacement with more efficient ones.

3.3 Residential combustion sector

- Certification of residential combustion equipment
  Residential combustion is an important source of particulate matter emissions. Borrego et al. (2010a) estimated an 18% contribution of this residential combustion to the total PM10 emissions in Portugal, and that 98% of these originate from the burning of biomass.
  Consequently, the regulation of this sector, particularly the certification of equipment with lower PM10 emission rates, will contribute to air quality improvement (both outdoors and indoors). The implementation of this measure follows a complex process that needs the involvement of several entities and stakeholders.

3.4 Summary

The PM10 emission reduction was estimated for each described measure based on different approaches and methodologies. For instance, for the traffic sector measures the emission

reduction was estimated using national emission factors for different age and circulation speeds of each vehicle. At the end, the foreseen emission reduction values are: for the residential and commercial combustion sector 92% for all the northern region; for the industry sector 50% for all the northern region, excluding Vila Nova de Famalicão (45%) and Matosinhos (18%) municipalities; for the traffic sector, the mean reduction value is 14%, with a maximum reduction of 28% and the lowest reduction of 1%.
Table 1 summarizes the overall PM10 emission reduction, including the base scenario (no measures) and the reduction scenario (concerning traffic, industrial processes and residential combustion).

Sector	PM10 Emissions (Base scenario) ($ton.year^{-1}$)	PM10 Emissions (Reduction scenario) ($ton.year^{-1}$)	% reduction
Road traffic	32899	31254	5%
Industry	1476	1314	11%
Residential	382	378	1%
Total	**34747**	**32946**	**5.1%**

Table 1. Total PM10 emissions for each sector for the base and the reduction scenarios.

The implementation of these measures accounts for a 5.1% overall reduction of the total northern region PM10 emissions.
In addition to these main measures, additional measures were also considered and implemented, namely regarding the reduction of dust emissions from construction sites and specific measures related to environmental education.
The construction works represent a significant source of particulate matter in the atmosphere, leading to air quality problems at local level. Different approaches can be implemented in order to reduce the PM10 emissions generated by this activity (e.g., washing roads and work fronts, especially on dry and windy days and in unpaved areas). Several actions have been compiled in a manual that was distributed among the construction companies to proceed to actions of training and awareness. The associations of industrial construction also have an important role in dissemination of best practice among their members. Municipalities should also implement these practices under the Municipal Regulation of Building and Urbanization.
Many measures to reduce PM10 are related to changes in habits and behaviours. Therefore, education and recommendations to the population, entrepreneurs and municipalities are the keys to acquiring habits with reduced environmental impact. Environmental education and awareness also allows a greater social acceptance of the measures to reduce PM10 emissions and, consequently, greater public support. The execution of these initiatives is primarily a responsibility of the CCDR-N and the municipalities involved, in cooperation with other agencies and associations that are closer to specific target groups (e.g., industry associations, schools), and is being conducted through the media, forums, workshops, etc.
Several authorities, institutions and 21 municipalities were involved in the implementation of the selected measures, which will achieve an overall cost of almost €45 million.

Engagement in the application of all the measures constitutes an important step forward to achieve the main goal of PM10 emission reduction and air quality improvement over the northern region of Portugal.

4. Air quality modelling

Estimation of the PM10 emissions reduction is not, however, enough to assess their effects on the air quality of the region. In order to better evaluate the impact on the northern region's air quality resulting from the selected PM10 emission reduction measures, a numerical modelling system has been applied.

The Air Pollution Model (TAPM) was selected since it is suitable for long term simulations (e.g., one year), has a user friendly interface and runs fast. Additionally, TAPM has already been applied and validated in several studies over Portugal (Ribeiro, 2005; Bandeira et al., 2011) and specifically over the northern region (Borrego et al., 2010b).

4.1 The air quality model

The Air Pollution Model (TAPM) is a 3-D Eulerian model which predicts meteorology and air pollution concentrations, based on fundamental fluid dynamics and scalar transport equations. Technical details of the model equations, parameterisations and numerical methods are described by Hurley et al. (2005). The model integrates two modules: the meteorological and the air pollution component. The first predicts the meteorological parameters that are one of the inputs for the second, which predicts the concentration and deposition of the air pollutants. In the TAPM meteorological component, global databases of terrain and land use from the Earth Resources Observation Systems (EROS), surface temperature from the US National Centre for Atmospheric Research (NCAR), and synoptic conditions from the Limited Area Prediction System (LAPS) and Global Analysis and Prediction (GASP) models from the Bureau of Meteorology (BOM) were used.

Besides the meteorological outputs, the air pollution component considers the air pollutants emissions from several sources, such as: point sources, line sources, gridded surface emissions, biogenic surface emissions, among others. TAPM has some options regarding pollution calculations, including chemistry mode with sulphur and fine particle chemistry. In the chemistry mode, gas-phase is based on a semi-empirical mechanism entitled the Generic Reaction Set (GRS), having 10 reactions for 13 species.

TAPM includes a nested approach regarding meteorology and air pollution, for computational efficiency, with the pollution grids optionally being able to be configured for a sub-region and/or at finer grid spacing than the meteorological grid, which allows the user to zoom in to an urban region of interest quite rapidly. This option is particularly important for this study. Meteorological and pollution outputs are in various formats and include vertical and temporal profiles for several parameters (Hurley et al., 2005).

4.2 Air quality modelling application

In order to investigate the impact of the designed PM10 reduction measures on the air quality of the northern region, the TAPM was applied over the study region, which includes the agglomerations of *Porto Litoral, Vale do Ave* and *Vale do Sousa* where the PM10 concentrations exceeded the legislated limit values. The application considered three simulation domains through the use of the nesting approach: the outer domain covers an area of 1080×1080 km^2 with a spatial resolution of 43.2×43.2 km^2, the second domain covers an area of 360×360 km^2 with a spatial resolution of 14.4×14.4 km^2 and the inner domain with an area of 120×120 km^2 has a spatial resolution of 4.8×4.8 km^2. TAPM was applied along the year 2004, which was chosen as the base year for the analysis of the impact of the selected measures, for two different emissions scenarios: base scenario and reduction scenario. For the base scenario, annual emissions were obtained from the Portuguese national inventory and spatially downscaled to the sub-municipality level for each pollutant and for each activity sector. The national emissions inventory takes into account annual emissions from line sources (streets and highways), area sources (industrial and residential combustion, solvents and others) and large point sources. The annual emission data for each pollutant and activity sector were spatially and temporally disaggregated in order to obtain the resolution required for the selected simulation domains. For the reduction scenario, the PM10 emission values were estimated based on the implementation of the designed reduction measures, which were previously described. The background concentrations required by the model were obtained estimating the average values of the background air quality stations of the study area for 2004.

The meteorological and air quality simulations were performed for the defined domains over the study region. The validation of the base scenario simulation was performed for the innermost domain, which includes all the air quality stations. Three quality indicators have been used to evaluate the TAPM performance over the study region, namely: Index of Agreement (IOA) (Eq.1), Measures of Skill ($SKILL_R$) (Eq.2) and Bias (BIAS) (Eq.3).

Index of Agreement:

$$IOA = 1 - \frac{\sum_{i=1}^{N}(P_i - O_i)^2}{\sum_{i=1}^{N}\left(|P_i - O_{mean}| + |O_i - O_{mean}|\right)^2}, \text{ where } O_{mean} = \sum_{i=1}^{N} O_i \quad (1)$$

Measures of Skill:

$$SKILL_R = \frac{RMSE}{O_{std}}, \text{ where } O_{std} = \sqrt{\frac{1}{N-1}\sum_{i=1}^{N}(O_i - O_{mean})^2} \text{ and } RMSE = \sqrt{\frac{1}{N}\sum_{i=1}^{N}(P_i - O_i)^2} \quad (2)$$

Bias:

$$BIAS = \frac{\sum_{i=1}^{N}(P_i - O_i)}{N} \quad (3)$$

$O_i \rightarrow$ observed values
$P_i \rightarrow$ predicted values
$RMSE \rightarrow$ Root Mean Square Error

The Index of Agreement (IOA) gives the difference between the predicted values and the observed ones. If this value is higher than 0.5 the model presents a good performance. RMSE values lower than the Standard Deviation of the observed values (O_{std}) also indicate a good performance of the model.

The BIAS indicates if the model follows the variations of the observed values: if positive, the model overestimates the results, and if negative the model underestimates the results.

4.3 Air quality modelling system performance

A statistical analysis of the model's performance was made. The meteorological module was evaluated against the measured values of temperature and horizontal wind components (U and V) at two different meteorological stations of the northern region (*Pedras Rubras* and *Viana do Castelo* meteorological stations). Table 2 summarizes the model's performance for temperature and wind at the two weather stations.

	Pedras Rubras			Viana do Castelo		
	TEMP	U	V	TEMP	U	V
IOA	0.94	0.79	0.86	0.91	0.66	0.74
$SKILL_R$	0.46	1.03	0.83	0.63	1.39	1.00
BIAS	0.46	-0.83	-0.21	-0.94	-0.76	-0.48

Table 2. TAPM performance statistics for temperature (°C) and horizontal wind components ($m.s^{-1}$).

TAPM shows a good performance simulating temperature and both wind components, with no significant biases, low $SKILL_R$ and high IOA for both meteorological stations. The *Pedras Rubras* simulation presents a slightly better performance than *Viana do Castelo*. These results highlight that the TAPM model is capable of simulating the main meteorological conditions over the study region.
The same statistical indicators have been applied for the PM10 simulated values at the air quality stations that presented exceedances of the limit values. Figure 11 presents the IOA, BIAS and $SKILL_R$ for each air quality station, for the base scenario simulation.
According to Figure 11, only some air quality stations have an IOA higher than 0.5, highlighting the model's difficulty in simulating the PM10 values over the study region. The negative BIAS over all the analyzed air quality stations reveals a clear underestimation of the measured PM10 concentrations. The $SKILL_R$ also presents a very poor score. These results show that the TAPM model has some difficulties in simulating the PM10 levels over the study region. This fact can be related to the complex topography of the study domain and uncertainty related to the PM10 emissions estimation. PM10 is one of the most difficult pollutants to simulate by air quality modelling systems. In addition, the TAPM uses a very simple chemical mechanism that to some extent may explain the poor results for the PM10 simulation.
Notwithstanding this evaluation, TAPM can be an interesting tool for decision-makers to assess the efficiency of the measures selected and included in the Air Quality Plans. Since the model can rapidly simulate different emission scenarios, the decision-making can be facilitated through the study of the differences between scenario results, allowing their relative comparison, instead of the analysis of the absolute predicted values.

4.4 Air quality results for the emission reduction scenario

In order to study the environmental effects of the application of the selected PM10 mitigation measures on the air quality of the northern region of Portugal, the reduction scenario was also simulated with TAPM. Figure 12 shows the spatial differences between the reduction and the base scenarios in terms of the annual mean PM10 concentrations (absolute values (a) and percentages of change (b)).

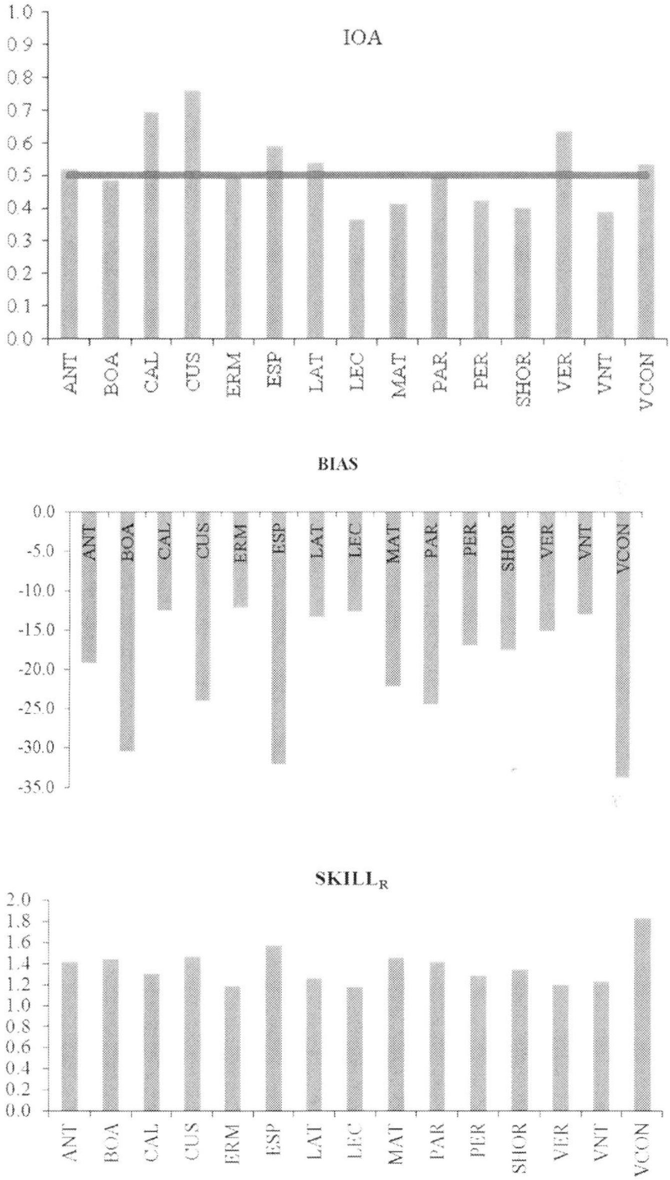

Fig. 11. IOA, BIAS and SKILLR statistical indicators for each air quality station for the TAPM base scenario simulation for the year 2004 regarding PM10 daily concentrations.

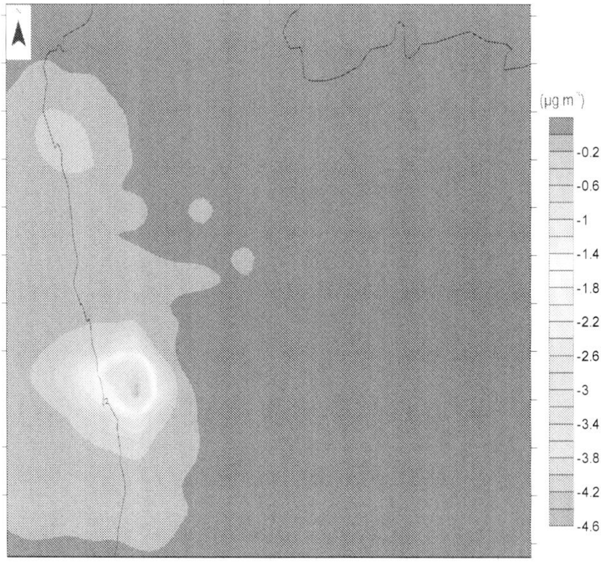

(a)

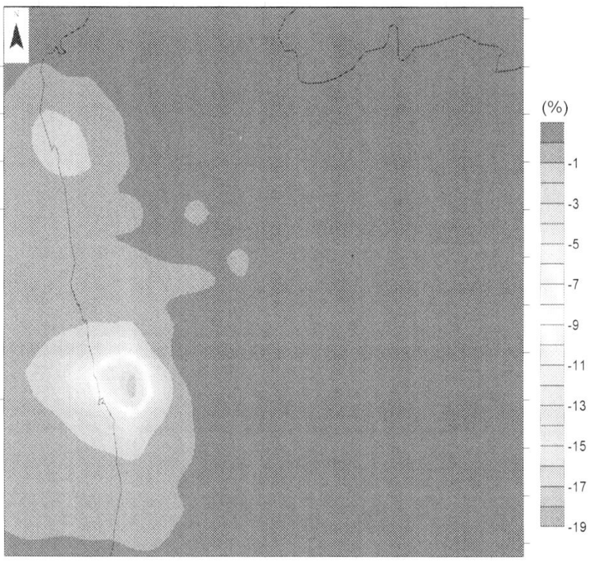

(b)

Fig. 12. PM10 annual mean concentration differences between the reduction scenario and the base scenario in terms of absolute (a) and relative changes (in %) (b).

According to Figure 12, the maximum absolute reduction is achieved over the Porto region and reaches almost 5 μg.m-3, which corresponds to a 19% reduction of the PM10 average concentrations.

Another way to study the effects of the implementation of the reduction measures consists in estimating the reduction obtained by comparing the results from each scenario simulation (base and reduction) for each day, and then applying this reduction profile to the daily measured values at the monitoring stations. This approach has only been applied to the air quality stations with IOA higher than 0.5.

Figure 13(a) illustrates the measured annual mean and the annual mean after the application of the reduction profile, and Figure 13(b) represents the number of observed exceedances of the daily limit value (plus the margin of tolerance for 2004) and the number of exceedances after the application of the reduction profile.

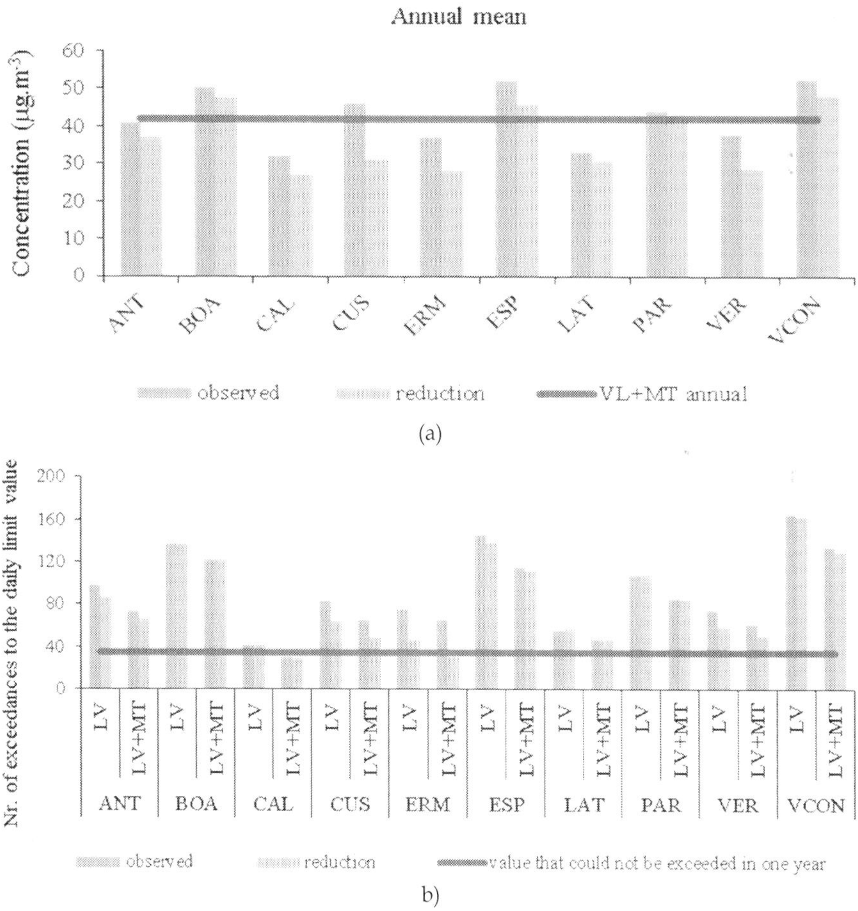

Fig. 13. PM10 annual mean (a) and number of exceedances (b) for the observed values and for the observed values after the application of the reduction profile.

Figure 13 shows that although the PM10 annual mean and the number of exceedances decrease in the selected air quality stations, the legislated limit values continue to be surpassed. Despite the efforts needed for the application of the selected measures, they lead to a slight decrease of the PM10 levels in the atmosphere. It should also be stated that air quality improvements can be achieved, although these are not sufficient to eliminate all the exceedances of the PM10 limit values that can be detected over the northern region of Portugal. However, it is important to reinforce the fact that not all the measures considered in the Air Quality Plan were considered in the reduction scenario simulation, since it was not possible to estimate the PM10 emissions' reduction associated with them (e.g., environmental education, construction sites). Therefore, considering all the measures contained in the Air Quality Plan, it is possible to achieve a higher decrease in the overall PM10 levels.

5. Final comments

In the northern region of Portugal high PM10 concentrations were measured at several agglomerations and, according to the European and national legislation, action plans must be designed and implemented to reduce these PM10 levels. For this purpose reduction measures have been defined at the local level for three main sectors: traffic, industry, and residential combustion. Also, measures related to construction activities and environmental education were developed. The impact of the measures on the air quality of Northern Portugal was the main objective of this study.

It was possible to conclude that the application of the selected reduction measures is important to improve the PM10 concentrations in the northern region. A decrease of the annual mean and of the number of exceedances was estimated at the air quality monitoring stations location. The modelling results for the reduction scenario show a PM10 annual averaged concentration decrease of almost 5 $\mu g.m^{-3}$ over the Porto urban area, which corresponds to a 19% maximum reduction. However, even after the application of the reduction measures, most of the air quality stations continue to surpass the legislated limit values. Thus, additional measures must be considered in order to fulfill the air quality requirements. The involvement of local authorities and stakeholders is very important, but some potential reduction measures, probably the most effective and efficient, should be taken at national level.

Finally, it should also be pointed out that under climate change conditions the northern region is one of the areas that will be most affected by changes in weather patterns, and predicted changes in the boundary layer height, relative humidity, temperature, solar radiation, wind speed and precipitation may be responsible for a significant impact in the PM10 levels in the atmosphere (Carvalho et al., 2010). The design and execution of additional measures is therefore even more important due to the current conditions, but also to the future climate change driving force.

6. Acknowledgements

The authors wish to thank the North Regional Coordination and Development Commission (CCDR-N) for their support. Also, the authors thank the Portuguese Foundation for Science and Technology for the PhD grant of E. Sá (SFRH/BD/60474/2009).

7. References

Amato, F., Querol, X., Alastuey, A., Pandolfi, M., Moreno, T., Gracia, J., Rodriguez, P. (2009) - Evaluating urban PM10 pollution benefit induced by street cleaning activities, Atmospheric Environment 43, 4472-4480.

Bandeira, J.M., Coelho, M.C., Sá, M.E., Tavares, R., Borrego, C. (2011) - Impact of land use on urban mobility patterns, emissions and air quality in a Portuguese medium-sized city, The Science of Total Environment 409, 6, 1154-1163. doi:10.1016/j.scitotenv.2010.12.008.

Borrego, C., Miranda, A.I., Sousa, S., Carvalho, A., Sá, E., Martins, H., Valente, J., Varum, C., Jorge, S. (2008a) - Planos e Programas para a Melhoria da Qualidade do Ar na Região Norte - Uma visão para o período 2001-2006 (Plans and Programmes for the improvement of the air quality over the norther region - an overview over the perid 2001-2006), University of Aveiro, Portugal. AMB-QA-7/2008.

Borrego, C., Miranda, A.I., Monteiro, A., Tchepel, O., Martins, H., Tavares, R., Gonçalves, L. and Barbedo, P. (2008b) - Inventário das emissões de poluentes atmosféricos da Região Norte, Relatório R3 (Atmospheric polluttant emissions report for the northern region - report R3), Departamento de Ambiente e Ordenamento da Universidade de Aveiro.

Borrego, C., Valente, J., Carvalho, A., Sá, E., Lopes, M., Miranda, A.I. (2010a) - Contribution of residential wood combustion to PM10 levels in Portugal, Atmospheric Environment. 44, 642-651.

Borrego, C., Sá, E., Carvalho, A., Sousa, S., Miranda, A.I. (2010b) – Plans and Programmes to improve air quality over Portugal: a numerical modelling approach. Proceedings of the 13th International Conference on Harmonisation within Atmospheric Dispersion Modelling for Regulatory Purposes, 1-4 June, Paris, France.

Carvalho, A., Monteiro, A., Solman, S., Miranda, A.I. and Borrego, C. (2010) - Climate-driven changes in air quality over Europe by the end of the 21st century, with special reference to Portugal, Environment Science & Policy I3, 445-458.

EEA (European Environment Agency), 2005: The European environment - State and outlook 2005. State of Environment report No 1/2005. Copenhagen.

Hurley, P.J., Physick, W.L., Luhar, A.K. (2005) - TAPM: a practical approach to prognostic meteorological and air pollution modelling, Environment Modelling and Software 20, 737-752.

Ketzel, M., Omstedt, G., Johanson, C., During, I., Gidhagen, L., Lohmeyer, A., Berkowicz, R., Wahlin, P. (2005) - Estimation and validation of PM2.5/PM10 exhaust and non-exhaust emission factors for street pollution modelling, Proceedings of the 5th International Conference on Urban Air Quality, eds. Ranjeet Sokhi, Millán M. Millán and Nicolas Moussiopoulos, Proceedings on CD-Rom. Valencia, Spain.

Monteiro, A., Miranda, A.I., Borrego, C., Vautard, R., Ferreira, J. and Perez, A.T. (2007) - Long-term assessment of particulate matter using CHIMERE model, Atmospheric Environment 41, 7726-7738.

Ribeiro, C. (2005) - Aplicação de um Modelo Meteorológico e de Qualidade do Ar a Portugal (Application of a meteorological and air quality model over Portugal). Masters Science Thesis, University of Aveiro, Portugal.

Rodriguéz, S.G. (2002) - Sources and processes affecting levels and composition of atmospheric particulate matter in the Western Mediterranean. PhD Thesis, Universidade Politécnica da Cataluna, Barcelona, Spain.

Tente, H., Gomes, P., Ferreira, F., Amorim, J.H., Cascão, P., Miranda, A.I., Nogueira, L., Sousa, S. (2011) - Evaluating the efficiency of Diesel Particle Filters in high-duty vehicles: field operational testing in Portugal. Atmospheric Environment (Accepted for publication)

10

Changeability of Air Pollution in Katowice Region (Central Europe, Southern Poland)

Mieczysław Leśniok
University of Silesia
Poland

1. Introduction

The air, as an elementary component of the biosphere, is a conveyor of many pollutants in the form of dusts, gases and aerosols. Among others the basic factors shaping the quality of the air are: the size of the emission and ambient concentration of the pollutants and local conditions- the lie of the land, ground cover and meteorological conditions. Among the meteorological factors the biggest significance has the direction and the velocity of the wind, the height and the intensity of the rainfall and the thermic stratification of the atmosphere. There have been created many classifications of the pollutants, according to the basic one the pollutants fall into two categories: the natural and anthropogenic ones. Anthropogenic sources of pollutants' emission basically include: industrial plants and service centres, boiler plants and the housing estates' boiler rooms, domestic fires of the individual households and means of transport, which emit the pollutions in the organized or random (not organized) way. Regarding the quality we can distinguish the particulate pollutants (suspended dust and falling dust) and the gas pollutants (SO_2, NO, NO_2, O_3, CO, CO_2, etc.), which ambient concentration can occur because of dry and wet deposition.

2. Monitoring system

Environment monitoring is the research, analysis and the evaluation of the natural environment in order to observe the happening changes. The environment monitoring can be conducted on the local, nationwide, continental or global scale. The source of the systematic information concerning the state of the environment are the stable networks of observation and measurement or the moving measurement points. Generally, environment monitoring in Poland involves: air monitoring, monitoring of the surface and underground waters and monitoring of the soils, forests and noise.

In Silesian province air monitoring has been functioning for 30 years and it is the most developed monitoring system in the country. In 2006 it consisted of the network of 200 collecting points for taking the falling dust along with the contained heavy metals (manual measurements), and 30 points for measurement of the suspended dust concentration and the gas pollutants (automatic measurements).

Moreover, due to the European Union funds, since 1993 a regional air pollution monitoring has been functioning within the Upper Silesian Industrial Region (it was reorganized in

2005). It consists of eleven air pollution measurement stations, one meteorological conditions measurement station, one forest monitoring station and one station for the atmosphere stratification studies (Figure 1). The characteristics of the aforementioned measurement points is presented in Table 1. This monitoring system cooperates with the network of posts belonging to the Institute of Meteorology and Water Management in Katowice.

3. Air pollution

3.1 Spatial distribution and the dynamics of the dust fall changes

Air is considered to be polluted when it contains additional elements which are not included in its original composition and when the share of the stable elements exceeds their average contents in clean and dry air.

Within the area of Katowice Region the aerosanitary conditions are mostly shaped by the emission of local dust-gas pollutants, the source of which are domestic fires of the individual households, industrial plants and service centres, boiler plants and the housing estates' boiler rooms, means of transport and the pollution emitted from the adjoining areas.

The analysis of the spatial distribution of the dust fall during the industrial period (at the turn of the 70s and 80s of the XX century) clearly indicates its increased values in the central part of the Katowice Region, nevertheless this distribution is not characterized by one, uniform area but by a few centres of increased (highest) values. These centres are: the areas of the central part of Gliwice, Ruda Śląska, Chorzów, Dąbrowa Górnicza, border areas

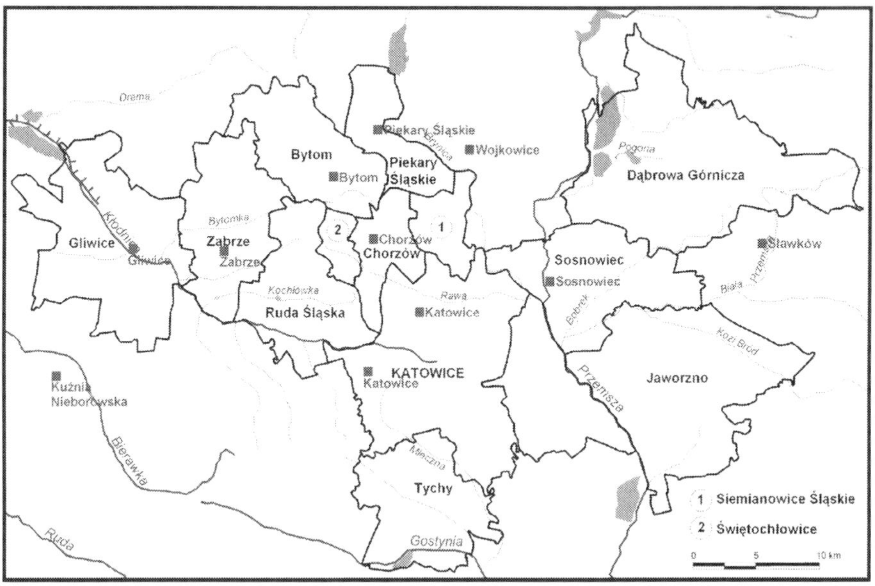

Fig. 1. Distribution of the research points of the Regional Air Pollution Monitoring within the area of Katowice Region

Name of the station	Coordinates: latitude longitude altitude	Type of the station	Measure-ment period	Polutants recorded	Sources of pollution around the station
A-1 Katowice	E 19°00′17″ N 50°15′18″ 275 m above MSL	Urban station	1993-2004	PM10, SO_2, NO, NO_2, O_3, CO, meteorological conditions	Industrial works, vehicle emission, domestic emission
A-2 Chorzów	E 18°58′19″ N 50°18′17″ 308 m above MSL	Urban station	1993-2004	PM10, SO_2, NO, NO_2, O_3, CO, meteorological conditions	Industrial works, vehicle emission domestic emission
A-3 Bytom	E 18°53′58″ N 50°20′00″ 285 m above MSL	Urban station	1993-2004	PM10, SO_2, NO, NO_2, O_3, CO, meteorological conditions	Industrial works, vehicle emission domestic emission
A-4 Zabrze	E 18°47′52″ N 50°18′07″ 257 m above MSL	Urban station	1993-2004	PM10, SO_2, NO, NO_2, O_3, CO, meteorological conditions	Industrial works, vehicle emission domestic emission
A-5 Gliwice	E 18°41′03″ N 50°17′07″ 232 m above MSL	Urban station	1993-2004	PM10, SO_2, NO, NO_2, O_3, CO, meteorological conditions	Industrial works, vehicle emission domestic emission
A-6 Piekary Śląskie	E 18°57′08″ N 50°23′42″ 285 m above MSL	Urban station	1993-2004	PM10, SO_2, NO, NO_2, O_3, CO, meteorological conditions	Industrial works, vehicle emission domestic emission
A-7 Wojkowice	E 19°03′40″ N 50°20′58″ 263 m above MSL	Urban station	1993-2004	PM10, SO_2, NO, NO_2, O_3, CO, meteorological conditions	Industrial works, vehicle emission domestic emission
A-8 Sosnowiec	E 19°09′00″ N 50°16′27″ 255 m above MSL	Urban station	1993-2004	PM10, SO_2, NO, NO_2, O_3, CO, meteorological conditions	Industrial works, vehicle emission domestic emission
A-9 Kuźnia Nieborowicka	E 18°37′00″ N 50°12′29″ 235 m above MSL	Suburban station	1993-2004	PM10, SO_2, NO, NO_2, O_3, CO, meteorological conditions	Small industrial, vehicle and domestic emission
A-10 Sławków-Okradzionów	E 19°22′48″ N 50°19′56″ 321 m above MSL	Suburban station	1993-2004	PM10, SO_2, NO, NO_2, O_3, CO, meteorological conditions	Mean industrial, vehicle and domestic emission

Table 1. Characteristic of the measurement points

between Zabrze and Bytom and Bytom and Piekary Śląskie. Within these regions yearly values of the dust fall in places exceeded 1000 g/m^2.

In this distribution the maximum values, accounting for over 2 000 $g \times m^{-2}$ ($t \times km^{-2}$), were registered in the central part of the Silesian province (Zabrze, Chorzów, Bytom, Katowice, Sosnowiec). In 1980 over the whole area the dust fall decreased and its maximum value was over 1 500 $g \times m^{-2}$ (permissible norm is 200 $g \times m^{-2}$). The following ten years resulted in further, more visible improvement- maximum value of the dust fall registered at the beginning of the 90s was 760 $g \times m^{-2}$. However, the fundamental change was observed at the end of the 90s when the highest level of the dust fall was only 200 g m^{-2} and corresponded to the binding permissible norm (Leśniok & Radomski, 1999). The example given here can be the city of Sosnowiec where in the years 1995-2001 the doubled decrease of the dust fall was observed (Puszczewicz, 2003). The comparison of the dust fall within the area of Katowice Region in 1997 and 2002 as well as the dynamics of its changes is presented in the figures 2, 3 and 4.

Situation when the level of the dust fall decreases or stays on the same level takes place until these days, although within the present borders of the Silesian province there are some points where the yearly dust fall norm is exceeded (e.g. Zabrze Makoszowy). The decrease of the dusts emission was mainly caused by the general economic recession which took place in Poland at the turn of the 80s and 90s, as well as by the industry restructuring within the Upper Silesian Industrial Region (changes of the production profiles, moving the productions to the other regions, closing down the plants and factories) and by the pro-ecological activities aiming at the reduction of pollutants emission, mainly the dust ones which are emitted by the large and medium-size sources.

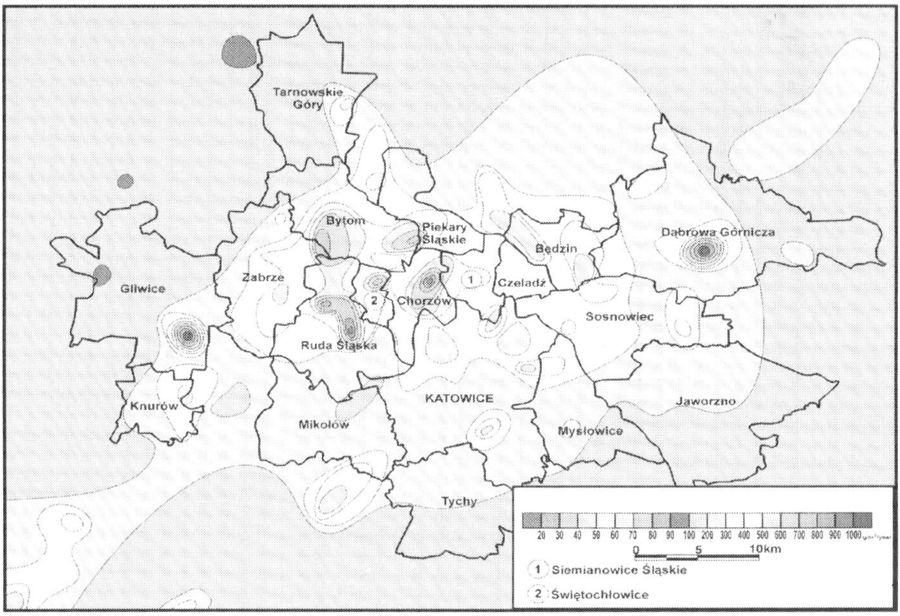

Fig. 2. Distribution of the dust fall within the area of Katowice Region in 1977 [in g/m^2/year]

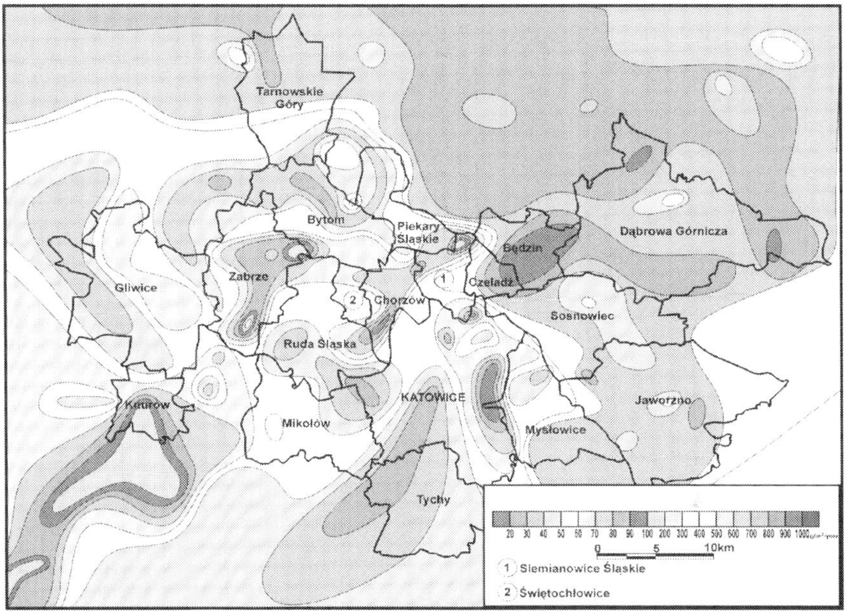

Fig. 3. Distribution of the dust fall within the area of Katowice Region in 2002 [in g/m2/year]

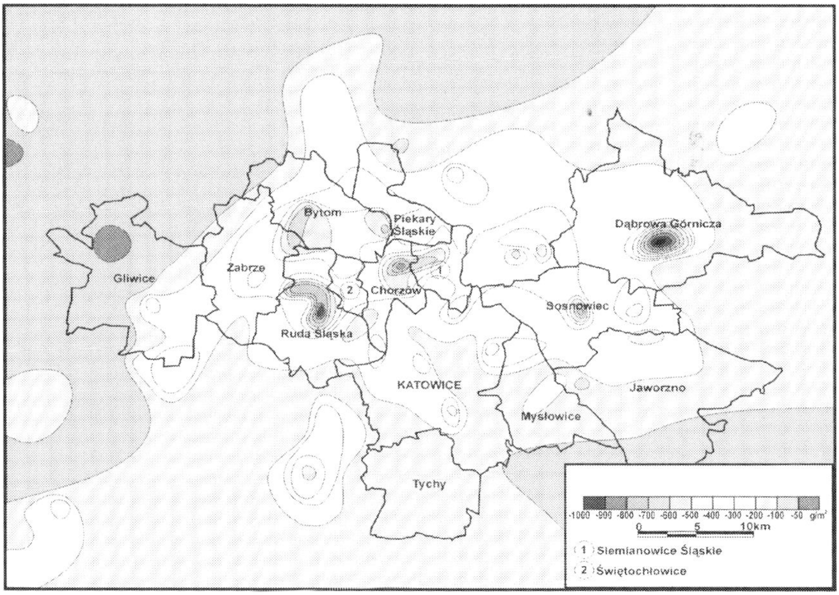

Fig. 4. Dynamics of the dust fall change within the area of Katowice Region in 1977–2002 [g/m2]

3.2 Differentiation of the distribution of the pollutants' concentrations

Apart from the dust fall the visible decrease of the suspended dust concentration can also be observed (PM10 – dust of the fraction up to 10 μm), nevertheless its average value for the whole area of Silesian Upland still exceeds the permissible norm of 40 $\mu g \times m^{-3}$ (fig. 5).

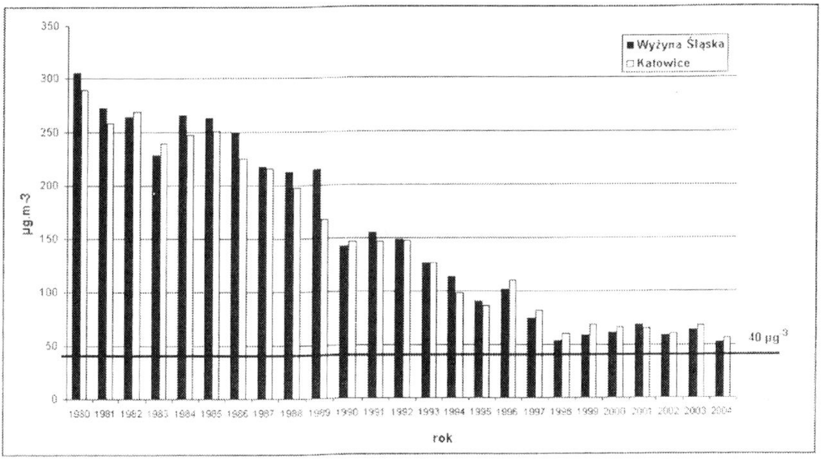

Fig. 5. Mean yearly distribution of the suspended dust concentration (PM 10) for Katowice Region and Katowice city in 1980–2004

As a result of the industry restructuring within the area of the Upper Silesian Industrial Region the significant improvement of the air quality is visible. The amount of the dust emitted into the atmosphere has decreased, what can be mainly observable during the measurement of the mean yearly concentrations of the suspended dust (Hubicki, 2007).

Complete analysis of the differentiation of the dust pollution covering the present-day period and the years before the industry restructuration is presented in the pictures 6-10.

The pictures illustrate the formation of the mean yearly suspended dust concentrations for five selected cities of Katowice Region. For each of these cities the decreasing tendency in contents of dust suspended in the air is visible (Hubicki, 2007). It can be clearly observed in the years when the industry restructuring was in progress.

In case of gas pollutants the decreasing tendency is also visible, however it is not so clear. Until now within the area of Katowice Region high concentrations of SO_2, NO, NO_2 and O_3 are registered, moreover in case of these pollutants we can observe considerable daily differences (rush hours, heating peaks) and seasonal changes in concentrations' values (summer and winter season).

Among the gas pollutants the biggest amount of studies has been devoted to the sulphur dioxide. Sulphur is an element which occurs in all environment components, and is rated on the 16th place in the sequence of elements which are widely common in nature. It is included in the contents of the living organisms, occurring in many compounds which fulfil the metabolic, structural and functional functions. In the atmosphere it mainly appears in the form of SO_2, which excessive concentration has bad influence on chemical and physical properties of the soils, the condition of the plants and health of human beings and animals. It is also the cause of accelerated corrosion, damages of constructions, devices and cultural monuments (Degórska, 2008).

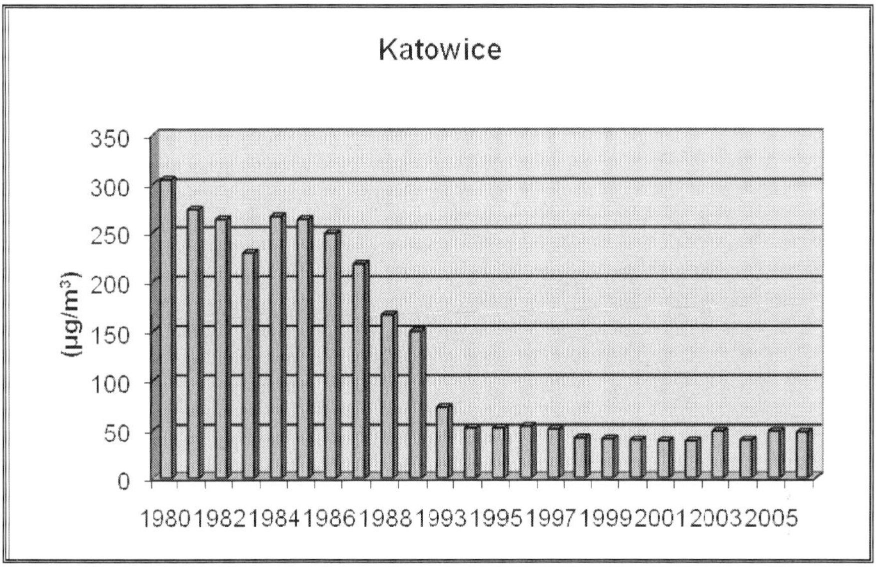

Fig. 6. Distribution of mean yearly suspended dust concentrations on the area of Katowice in 1980–2006

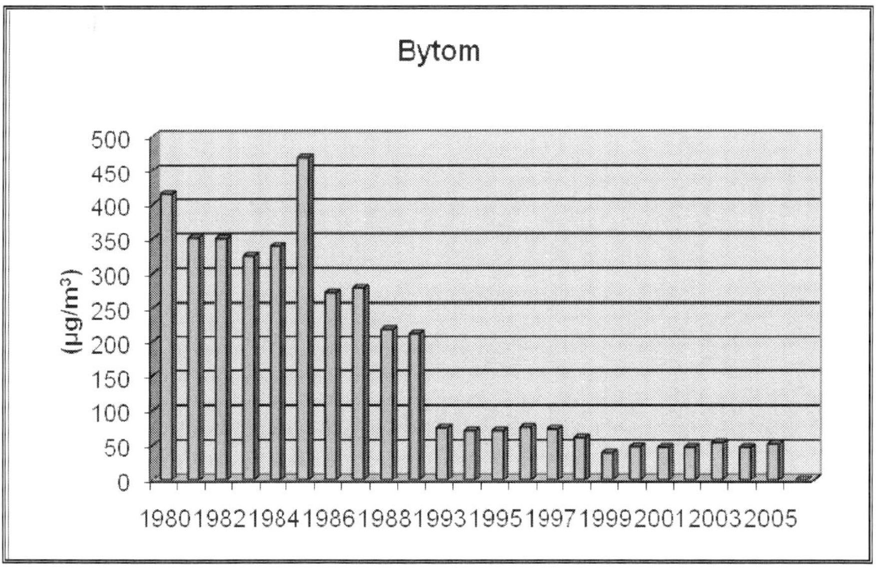

Fig. 7. Distribution of mean yearly suspended dust concentrations on the area of Bytom in 1980-2006

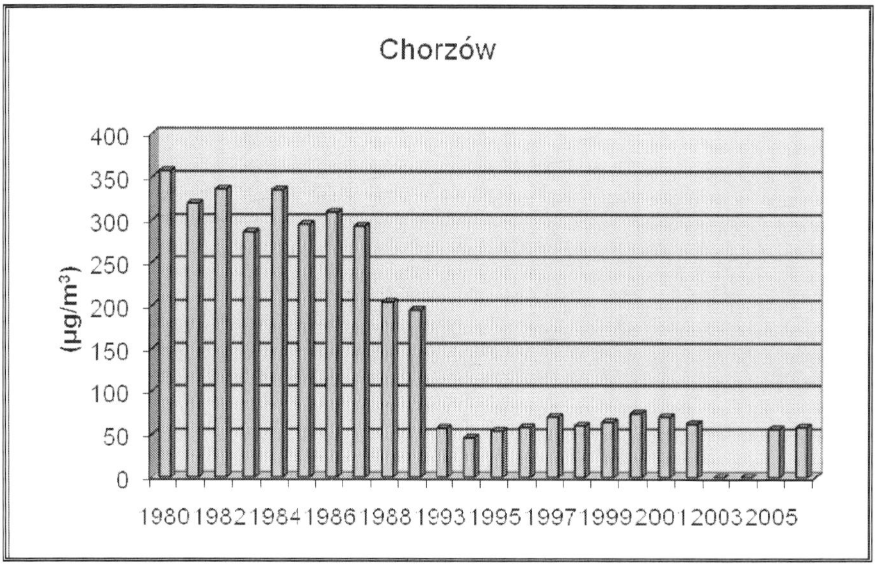

Fig. 8. Distribution of mean yearly suspended dust concentrations on the area of Chorzów in 1980-2006

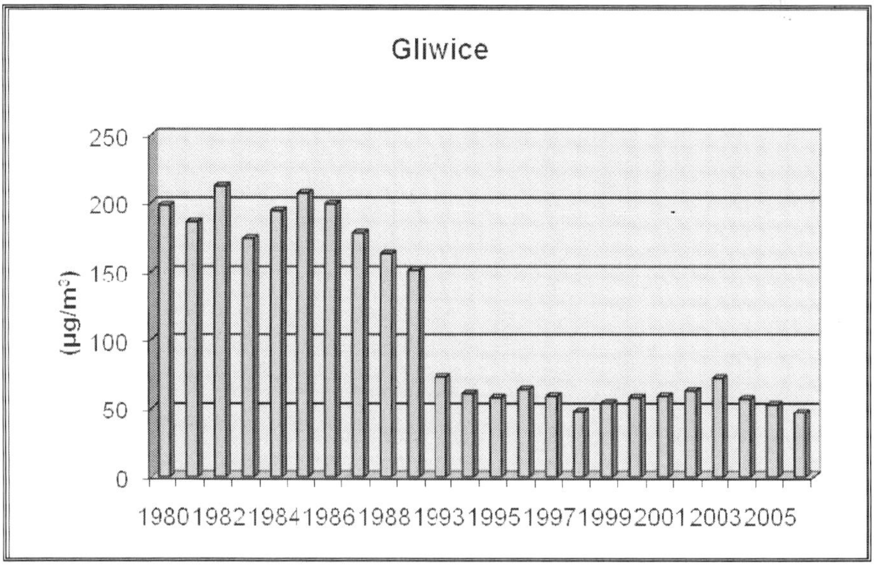

Fig. 9. Distribution of mean yearly suspended dust concentrations on the area of Gliwice in 1980-2006

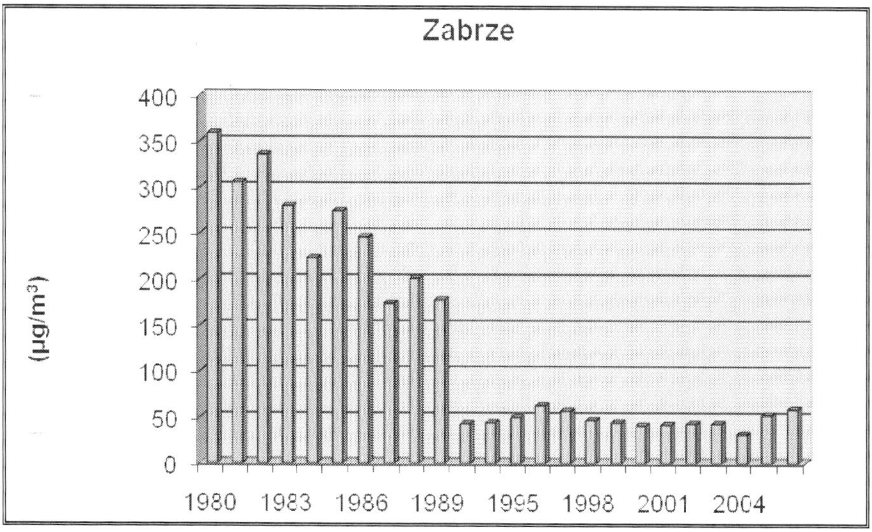

Fig. 10. Distribution of mean yearly suspended dust concentrations on the area of Zabrze in 1980-2006

Data juxtaposition presented in the form of the graphs, which refer to the mean yearly concentrations of the sulphur dioxide from the years before the industry restructuration and the present-day period, allows to analyse the formation process of the aforementioned pollution in the area of Katowice Region (fig. 11-15).

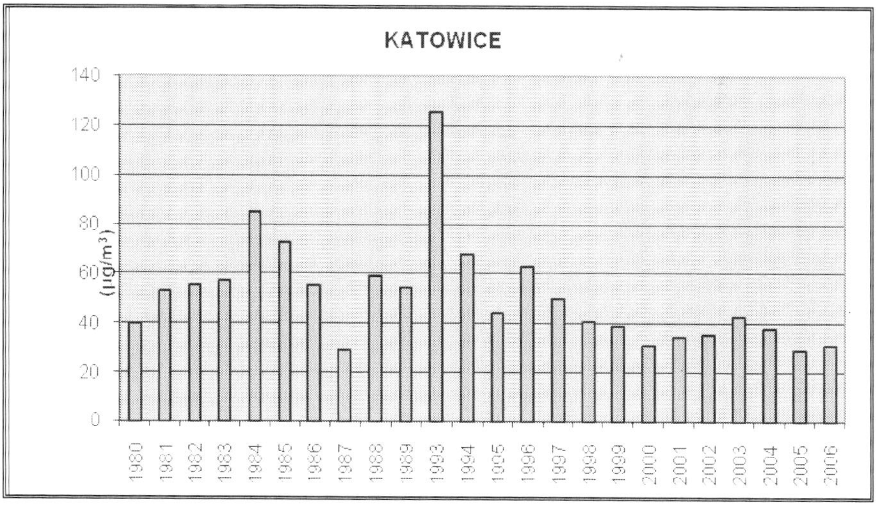

Fig. 11. Distribution of mean yearly sulphur dioxide (SO_2) concentrations on the area of Katowice in 1980-2006

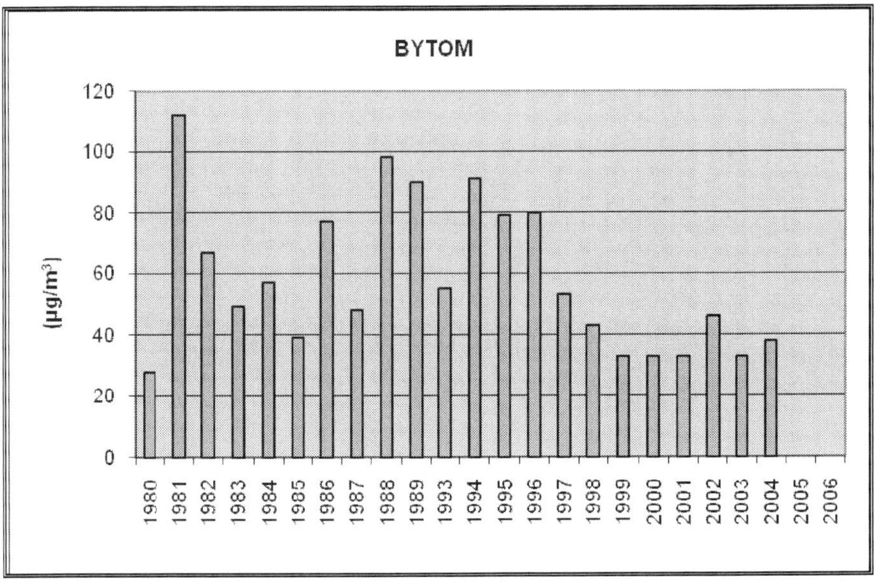

Fig. 12. Distribution of mean yearly sulphur dioxide (SO_2) concentrations on the area of Bytom in 1980-2006

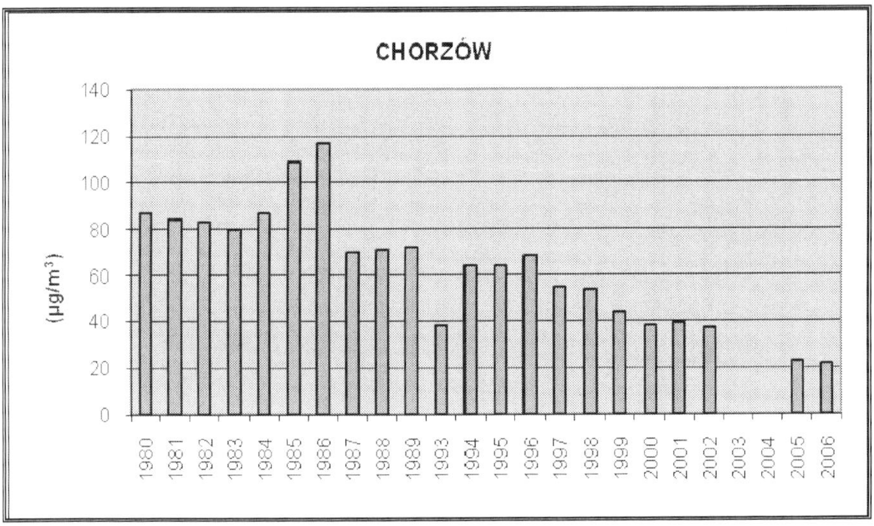

Fig. 13. Distribution of mean yearly sulphur dioxide (SO_2) concentrations on the area of Chorzów in 1980-2006

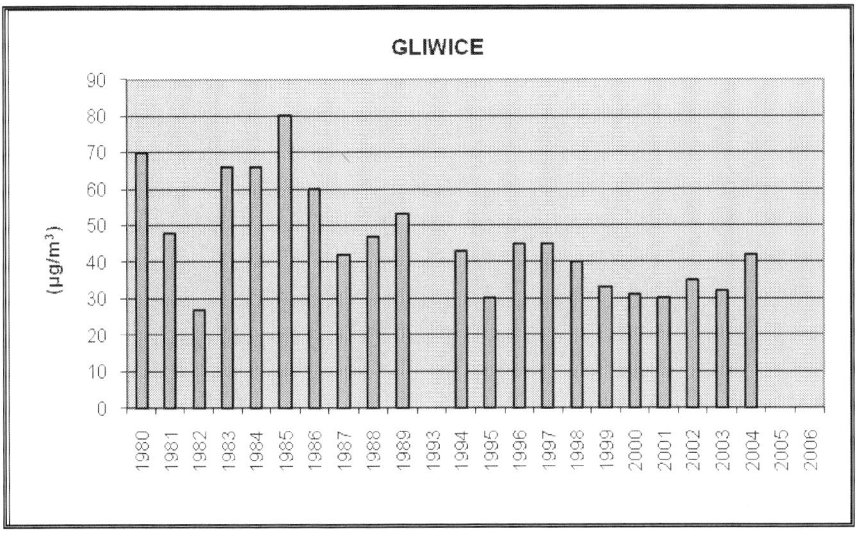

Fig. 14. Distribution of mean yearly sulphur dioxide (SO_2) concentrations on the area of Gliwice in 1980-2006

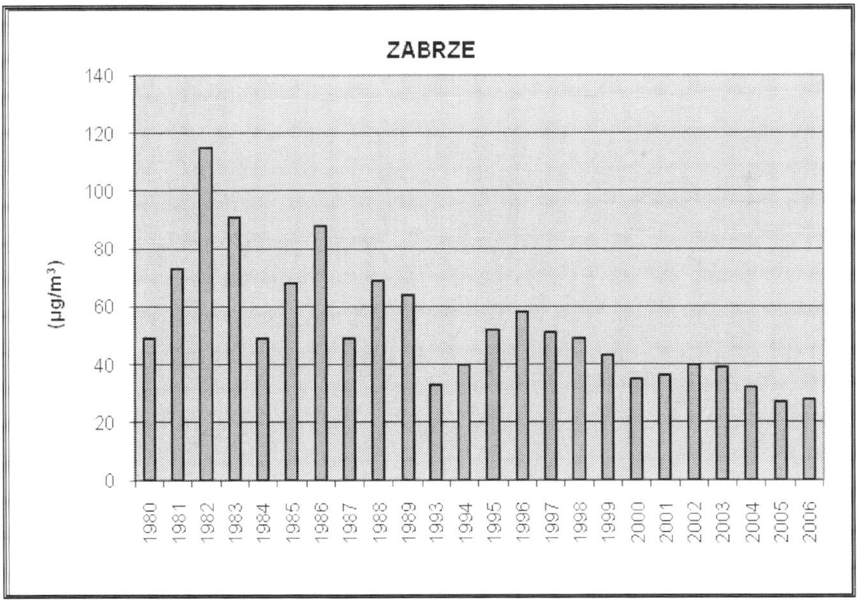

Fig. 15. Distribution of mean yearly sulphur dioxide (SO_2) concentrations on the area of Zabrze in 1980-2006

The above figures present formation of the mean yearly concentrations of sulphur dioxide for the particular cities in the years before the industrial restructuring and present-day period. For the majority of the cities period before 1989 is characterized by the chaotic account of the mean yearly concentrations of SO_2. Frequent increases and decreases of the yearly concentrations of the sulphur dioxide are noticeable. Only after 1996 there occurs a gradual decreasing tendency of the mean yearly concentrations of the sulphur dioxide (Hubicki, 2007).

Nitric compounds, just after the sulphur compounds, are the most frequent cause of air contamination. They are less toxic than sulphur dioxides or ozone, but they play a role of the forerunners for the other pollutants, which regarding fitotoxicity exceed the initial products. Like in the case of sulphur nitrogen is an important compound (main building material of the protein) for plants, therefore the influence of NO_x should be examined regarding both its positive and negative effect, and the result of their absorption depends on the taken dose and physiological condition of the plant (Degórska, 2008).

Complete analysis of the differentiation of the nitric oxides atmospheric pollution within the area of Katowice Region, and particularly with nitric dioxide, is presented in the figures 16-20. The pictures show the mean yearly concentration of the nitric oxide in the years 1980-1987, as well as the values of the nitric dioxide concentration from 1988 and at intervals until the year 2006. This division is related to the fact that during the years 1980-1987 the Voivodship Sanitary and Epidemiological Station gave only the general information concerning the mean yearly concentrations for the nitric oxides, regardless the separate division into NO i NO_2. Only since 1988 the results regarding the mean yearly concentrations of the nitric dioxide have been revealed. Therefore the following analysis

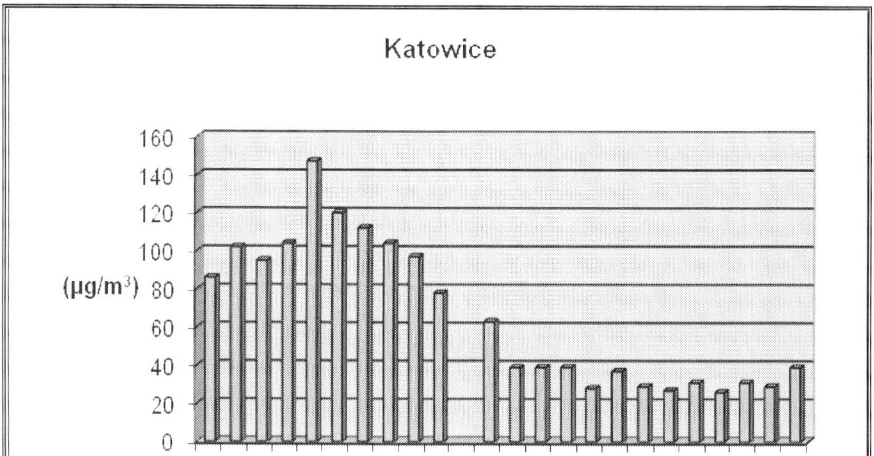

Fig. 16. Distribution of mean yearly nitric oxides concentrations on the area of Katowice in 1980-1987 and nitric dioxide concentrations on the area of Katowice in 1988-2006

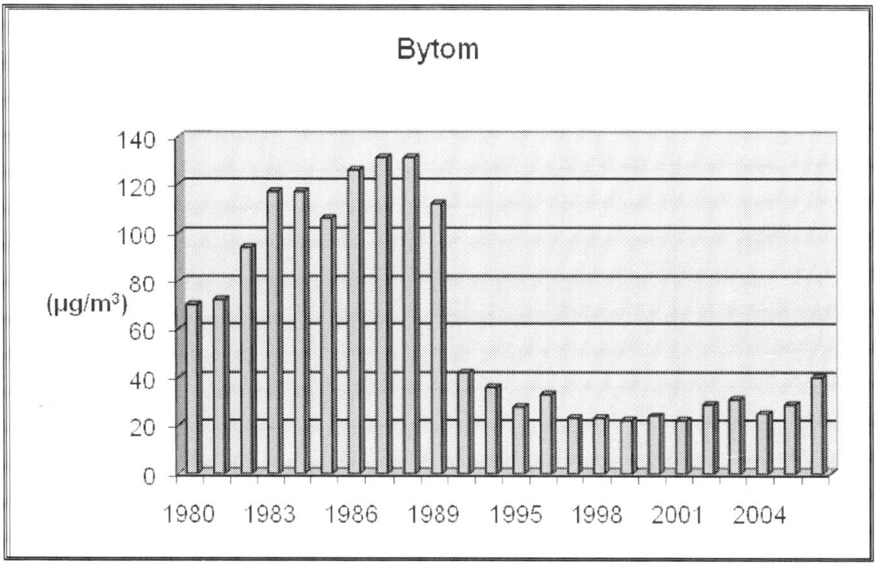

Fig. 17. Distribution of mean yearly nitric oxides concentrations on the area of Bytom in 1980-1987 and nitric dioxide concentrations on the area of Bytom in 1988-2006

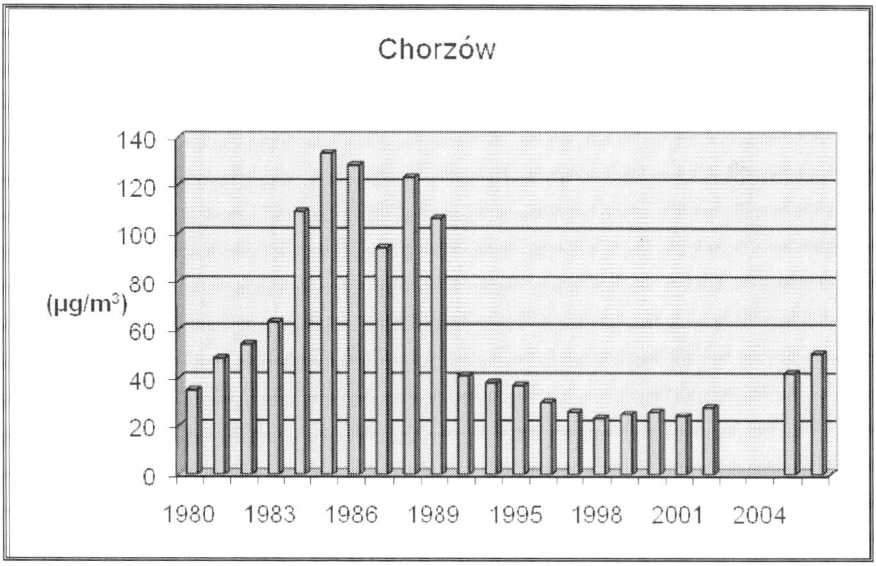

Fig. 18. Distribution of mean yearly nitric oxides concentrations on the area of Chorzów in 1980-1987 and nitric dioxide concentrations on the area of Chorzów in 1988-2006

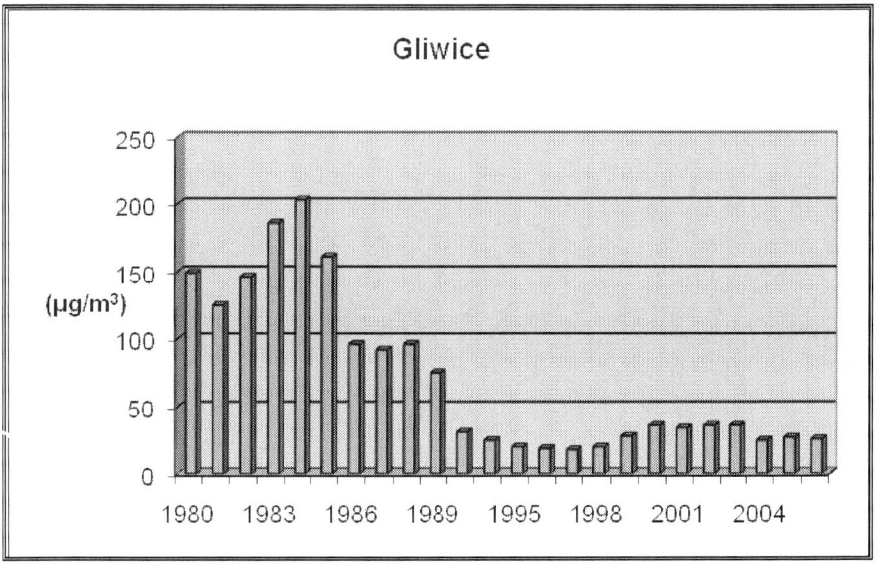

Fig. 19. Distribution of mean yearly nitric oxides concentrations on the area of Gliwice in 1980-1987 and nitric dioxide concentrations on the area of Gliwice in 1988-2006

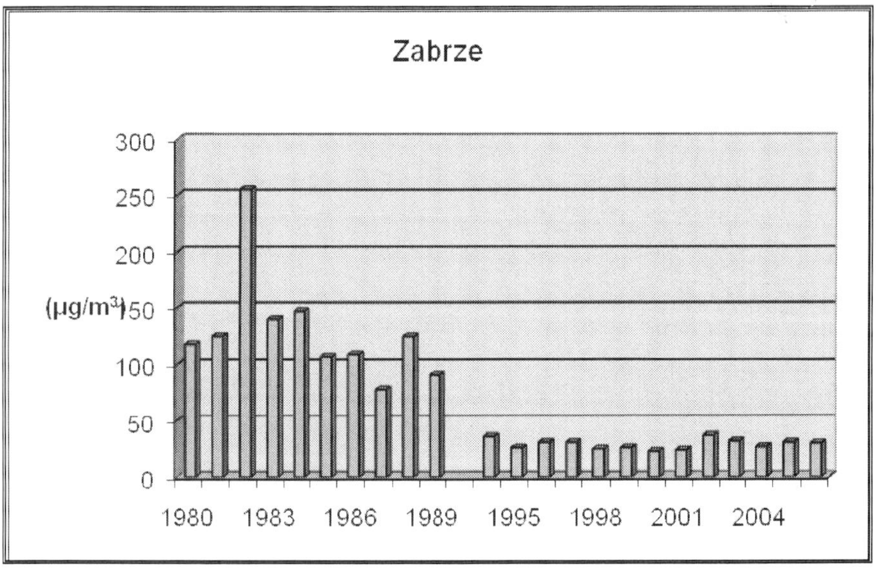

Fig. 20. Distribution of mean yearly nitric oxides concentrations on the area of Zabrze in 1980-1987 and nitric dioxide concentrations on the area of Zabrze in 1988-2006

should be treated symbolically. The values given for the nitric oxide in the years 1980-1987, roughly show the contents of the nitric dioxide for this period (Leśniok, 1996). Such perspective is possible due to the fact that majority of the pollutants are emitted in the form of NO which in the atmosphere, as the result of dilution of the smoke trail and the presence of the oxygen, is quickly transformed into NO_2, and therefore becomes the dominant component (Degórska, 2008).

As the result of the industry restructuring which took place at the beginning of the 90s, within the area of Katowice Region mean yearly concentrations of the nitric dioxide were decreased in a significant way. For the majority of the analysed cities, after 1997 a rapid decrease in the level of nitric dioxide contents can be noticed. In the majority of the measurement stations the decrease of the nitric dioxide contents in the air lasts until 1998. While in the years 1998-2006 within the area of Katowice Region we can observe a slight increase in mean yearly concentrations of NO_2 (figures 16-20) which is mainly related to the dynamic development of transport industry as well as to the clearer tendency towards replacing the compact cars with the ones having a bigger cylinder capacity.

4. Pollution of the precipitation waters

Rainfalls are one of the main elements shaping the quality of the air. The pollutants which are removed along with the rainfalls, in the diluted or non-diluted form, by getting into the ground quickly enter the particular stages of the hydro-geo-chemical circulation. Regarding the quantity and quality of the transported substances and the scope of their effect, rainfalls are good indicators for the evaluation of the degree of atmospheric pollution. Simultaneously, however, they contribute to degradation of the remaining elements of the environment.

As the result of soil acidification we can observe the increase in the mobility of the heavy metals, suppression of the mineralization processes and humification of the organic matter as the result of dying of bacteria and protozoans which are replaced with the selected fungi species. Similarly, in the rivers and lakes the increase in the water acidity caused by the atmospheric acid rainfall leads to the increase in contents of diluted, potentially toxic metals, what results in significant limitation of the species diversity (Degórska, 2008).

The quality of the precipitation waters is mainly influenced by the acid-forming and alkalescent compounds. The biggest role in the process of precipitation acidification is attributed to the gas compounds, which mostly include: sulphur dioxide (SO_2), nitric dioxides (NO_x), carbon dioxide (CO_2) and ozone (O_3). Alkalization, and thus weakening of the process of precipitation acidification, is caused by the calcium compounds (Ca) which are elements in the composition of the emitted dusts and ammonia(NH_3). All the aforementioned substances get into the atmosphere as the result of energetic and kinetic fuel combustion and photochemical and natural processes.

So far in Poland there has not been created one, uniform system for analysing the precipitation pollution. Many institutions are involved in this type of research (universities, research institutes, The State Forests National Forest Holding, national parks, etc.). But only within the area of the Lower Silesian province the research are conducted by the Institute of Meteorology and Water Management in Wrocław. In the years 1986-1995 in the Department of Climatology of the University of Silesia the detailed research on the chemical composition of the rainfall in connection with meteorological conditions were being carried out at the 12 points located within the area of Silesian-Cracow Upland (Leśniok,1996). Currently these research are being carried on at three points located in different areas (in Sosnowiec- the industrialized area, in Cieszyn- the area located in the zone of transboundary air pollution transfer, in Ojców National Park- the protected area).

The dusts of the upper Silesian region contain many elements (quickly or slowly soluble in water), which are acidificating and alkalizing, and thus neutralizing "acid rains", these are mostly calcium compounds (e.g. $CaCO_3$).

Due to the significant air pollution in the 70s and 80s, especially with the dust compounds containing alkalizing calcium carbonate, the phenomena od the "acid rain" (pH < 5) did not occur at all or it happened very episodically. During these years in the central part of the Upper Silesian Industrial Region the average value of pH was fluctuating from 6,5 to 7,4 (Jankowska, 1982). After the decrease of pollination and slight fall in the contents of the acid-forming compounds, especially in the central part of the Upper Silesian Industrial Region occurred the increase in the frequency of the acid rains, for example in Sosnowiec the frequency was risen from 60% (1990–1992) to 91% (1996–1998), and the average pH value decreased from 6,1 in 1987 to 4,0 in 2004 (fig. 21, Table 2).

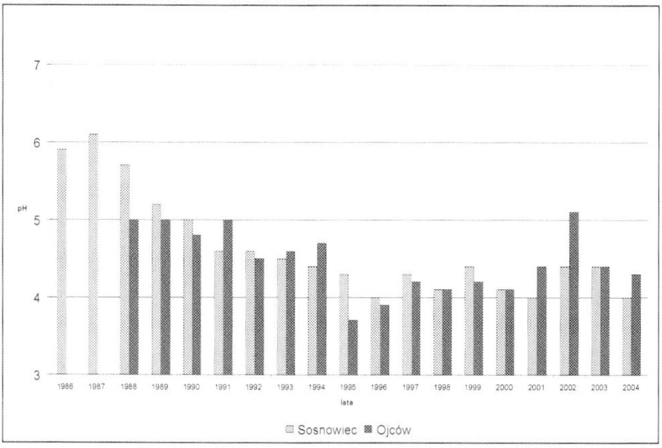

Fig. 21. Distribution of mean yearly values of the precipitation pH reaction in Sosnowiec and Ojców

pH scope	Sosnowiec				Ojców				Cieszyn			
	1990 1992	1993 1995	1996 1998	2000 2003	1990 1992	1993 1995	1996 1998	2000 2003	1990 1992	1993 1995	1996 1998	2000 2003
< 5,0	60	78	91	84	72	80	90	76	71	87	88	71
5,0–6,0	32	19	9	14	16	17	9	15	25	11	12	23
> 6,0	8	3	0	2	12	3	1	9	4	2	0	6

Table 2. Percentage of the precipitation pH reaction in the particular scopes in Sosnowiec, Ojców and Cieszyn, in the years 1990-2003

Similar changes can be observed in the remaining points, although they are not as dynamic. Currently the pH reaction remains on the 4,2-4,4 level. The amount of the dissolved compounds has also decreased, especially of calcium ions (Ca^{2+}), however the sulphur ions are still dominating in the composition of the participation waters and are mainly responsible for their acidification (Leśniok el al., 2005).

5. Conclusions

- The results of the research on the air pollution and precipitation revealed significant spatial differentiation of the degree of air pollution and precipitation which is the result of the change in proportion between the quality and quantity of the pollutants emitted into the atmosphere.
- As the result of the industry restructuring after the year 1989 the quantity of the emitted dust and gas pollutants within the area of Katowice Region decreased, however, in comparison with the other regions in Poland, it is still very big.
- On the whole area of Katowice Region the significant decrease in emission of dust pollutants was observed. This decrease contributed to the increase in the occurrence of "acid rains" phenomenon. "Acid rains" currently are present in 85% of the rainfall and their frequency is still increasing. Due to this fact central part of Katowice Region becomes similar to the peripheral areas.
- Gradual increase in the acidification of the precipitation during the last 10 years can be the signal for the initiation and further progress of the acidification processes of the remaining environment elements such as soil, surface waters and underground waters, what can have negative effect on flora and fauna. This problem is especially crucial for Katowice Region area which has been polluted for over one hundred years. The pollution accumulated on this area (mostly the heavy metals), due to the increase of acidification, can become active and contribute to the irreversible changes in the sphere of the remaining environment elements.
- Precipitation can be an important element in the evaluation of the air pollution. Knowing the quantity and quality of the pollutants, which are transported along with the rainfalls, is significant information which in a serious way can contribute to prevent the phenomenon of the natural environment acidification.
- In spite of the noticeable interference of the human and strong transformation of the natural environment, part of the areas within Katowice Region has been considered valuable and are under the different types of environment protection e.g. natural reserves ("Murckowski Forest", "Segiet", „Ochojec"), ecological sites (swamps in Antoniów), Natura 2000 areas (Tarnogórsko-Bytomskie Undergrounds and Błędowska Desert under the Habitat Directive).

6. Acknowledgements

The author thanks the Voivodeship Inspector of Environmental Protection in Katowice Mr. Jerzy Jamrocha for providing data from Regional Air Monitoring Network and Prof. Jacek Jania, the dean of the Faculty of Earth Sciences at the University of Silesia, for financial support for the purchase of the aforementioned data.

7. References

Degórska, A. (2008) *Zmiany zanieczyszczenia powietrza na terenie konurbacji* katowickiej w latach 1995–2004 na przykładzie dwutlenku siarki i tlenków azotu. Katedra Ekologii, WBiOŚ UŚ, Katowice, Poland, pp.90-101

Hubicki, T. (2007) *Analiza porównawcza współczesnego zróżnicowania zanieczyszczenia atmosfery nad GOP, z okresem przed restrukturyzacją przemysłu.* Katedra Klimatologii, WNoZ UŚ, Sosnowiec, Poland

Jankowska, K. (1982) *Wstępne wyniki badań składu chemicznego wód opadowych.* Geographia. Studia et Dissertationes, Vol. 6, UŚ, Katowice, Poland, pp. 114–124

Leśniok, M; Radomski, J. (1999) *Zróżnicowanie zapylenia atmosfery i zakwaszenia wód opadowych na obszarze Wyżyny Śląskiej. [W:] Materiały sesji naukowej nt. „Zanieczyszczenia pyłowe atmosfery".* Prace Specjalne Polskiego Towarzystwa Mineralogicznego, ISSN 0867-7360, Kraków, Poland, Vol. 15, pp. 73–82.

Leśniok, M. (1996) *Zanieczyszczenie wód opadowych w obrębie Wyżyny Śląsko-Krakowskiej.* Wydawnictwo Uniwersytetu Śląskiego, ISBN 83-226-0709-1, Katowice, Poland

Leśniok, M.; Barczyk, M; Hibszer, B: Krawczyk,W.E; Węgierek, M (2005) *Zanieczyszczenia pyłowe atmosfery a kwaśne deszcze.* [W:] *Priorytety środowiskowe aglomeracji wielkoprzemysłowych.* Rekultywacja-Powietrze, no. 2, InfoMax, Katowice:, Poland, pp. 27–32.

Puszczewicz, Z. (2003) *Zanieczyszczenie powietrza na tle wybranych elementów meteorologicznych w Sosnowcu.* Katedra Klimatologii WNoZ UŚ, Sosnowiec, Poland

11

Catalytic Converters and PAH Pollution in Urban Areas

Federico Valerio, Anna Stella, Mauro Pala, Daniele Balducci, Maria Teresa Piccardo and Massimo Cipolla
National Cancer Research Institute, Environmental Chemistry Service, Genoa, Italy

1. Introduction

Every organic substance containing carbon and hydrogen yields a large number of polycyclic aromatic hydrocarbons (PAH) during incomplete combustion or pyrolysis or during the formation of petroleum and coal. PAHs are included in a class of chemical compounds characterised by two or more condensed aromatic rings. Several PAHs have been classified by the International Agency for Research on Cancer (IARC) as probable or possible human carcinogens (IARC, 1987).

Due to their carcinogenic and mutagenic effects, sixteen PAHs were included in the US EPA priority pollutant list (Yan, Wang et al., 2004); between them, benzo(a)pyrene (BaP), with five condensed aromatic rings, is recognised as the PAH with the highest carcinogenic potency and is considered to be an indicator of the presence of the PAH group in environmental matrices: air, water, soil, food. In fact in every analysed mixture, the BaP concentration fits linearly with the concentrations of other semi volatile PAHs, prevalently adsorbed to airborne particulate matter.

2. PAH sources

Major sources of PAHs include transportation vehicles, coke ovens, steel and iron furnaces, metal smelter, manufactured gas plants, petroleum cracking, coal tar pitch, power generation using fossil and bio fuels, incineration of wastes and domestic heating. Traffic and coal related sources, have been considered to be the two most important source categories in many metropolitan areas in 20th and 21th century (Menichini, 1992; Harrison, Smith et al., 1996).

Even if contribution of motor vehicles to global PAH emissions may be less than biomass burning and wildfire, as estimated in China (Zhang, Tao et al., 2009) and in UK (DEFRA, 2001), traffic emissions occur mostly in urban areas where population densities are much higher than the global average and drivers (Piccardo, Stella et al., 2004), passengers of surface public transport (Kaur, Nieuwenhuijsen et al., 2007) and people working near heavy traffic streets (Piccardo, Stella et al., 2003) may be heavily exposed to traffic pollutants and particularly to PAHs. It was estimated that the most important source of environmental BaP, inhaled by Chinese population, is traffic emission (Zhang, Tao et al., 2009).

Occurrence of PAHs in vehicle emissions is mainly due to incomplete fuel combustion, which also explains another important traffic pollutant: carbon monoxide (CO) (WHO, 1999). For this reason, contemporary measures of CO and PAH concentrations along heavy traffic streets are frequently linearly correlated (Valerio, Stella et al., 2000; Lodovici, Venturini et al., 2003).

3. Trends of PAH concentrations

Several studies carried out in Europe during the last few decades demonstrated a generalized temporal decrease in the concentration of polycyclic aromatic hydrocarbons (PAH) in air (Fertmann, Tesseraux et al., 2002), in biological samples (Jacob, Grimmer et al., 1997) and in urban areas (Fertmann, Tesseraux et al., 2002; Schauer, Niessner et al., 2003; Menichini, Belladonna et al., 2006). Possible explanations for the short time observed decline of PAH concentrations in European countries were seasonal factors as photo degradation (Schauer, Niessner et al., 2003), different contamination of air masses impacting monitored sites, specific local emissions, i.e. domestic heating (Prevedouros, Brorstrom-Lunden et al., 2004) and industrial activities (Fertmann, Tesseraux et al., 2002). Regulations designed to reduce emissions from mobile sources were identified as possible explanation for long time PAH pollution trends (Menichini, Belladonna et al., 2006), but no clear demonstration was offered by authors for these hypotheses.

Temporal declines of PAHs were observed also in USA and Canada. A study on PAH concentration in precipitation at rural sites was conducted by Brun and co-workers who measured the concentrations of 14 PAHs at five sites close to the Gulf of St. Lawrence (Brun, Vaidya et al., 2004). Over the period 1980-2001, these authors reported an apparent long-term decreasing trend at one of their sampling sites (Kejimkujik National Park, in Nova Scotia), where PAH concentrations in rain and snow decreased with a half-life of 6.4 years. The PAH decrease was found to be correlated ($P<0.05$) with decreasing sulphate ion concentrations in precipitations and this suggested to authors that the implementation of air pollution abatement programs in Canada and the Unites States, switching from coal to cleaner sources of energy and improved technology for emission abatement was, most likely, responsible for the observed decline.

Similar measurements in precipitations collected near the Great Lakes, from 1997 to 2003, observed a long-term decreasing trend for PAHs collected at Chicago (Sun, Backus et al., 2006).

In a site near Lake Michigan, BaP concentrations, measured from 1991 to 1997, decreased with an apparent half-life of 4.9 years (Cortes, Basu et al., 2000).

In New York City, PAH levels (Σ 8 PAHs) from personal monitoring of pregnant mothers, collected from 1998 to 2006, showed an important decrease (from 3.3 ng m^{-3} in 1998 to 0.4 ng m^{-3} in 2006). After controlling for environmental tobacco smoke, indoor heat and cooking, years of personal monitoring remained a predictor of this decline (Narvaez, Hoepner et al., 2008) but specific explanation of this trend was not found.

3.1 Half-lives of PAH concentrations in atmosphere

To better describe the progressive decline of PAH concentrations in the atmosphere, Prevedouros proposed the calculation of apparent half-life, i.e. time (years) expected for the reduction to half of the mean annual PAH concentrations (Prevedouros, Brorstrom-Lunden et al., 2004).

According to this method, firstly, data are logarithmically transformed to reduce their skewness, a common practice when examining environmental datasets. In such a way the standard deviation/variability in the data was reduced, making it less likely that outliers drive the observed trends. Then, the natural logarithm of daily PAH concentrations (Y) are regressed with the progressive number of days (X) during which the trend was observed.
If a significant linear trend line is obtained, the half-life of this trend ($t_{1/2}$) may be calculated by the slope (A) of the regression line, applying the following equation:

$$t_{1/2} = -\log_n 2/A \quad (1)$$

The obtained half-lives allow comparing different long time PAH trends available from studies carried out in different situations. The only condition for a correct comparison is that available data cover a long observation period with constant sampling frequency during each year, to avoid seasonal artefacts.

4. PAH regulations

Because of their toxic and persistent properties, in 1998, PAHs were included in the UNECE Convention on Long-Range Transboundary Air Pollution (LRTAP) Protocol on Persistent Organic Pollutants (POP) which entered into force in 2003 (UN-ECE, 1998).
This protocol obliges parties to reduce emissions of PAHs and other persistent compounds, below their levels occurring in 1990.
According to estimated PAH emissions in European countries, the implementation of this protocol induced an important reduction of the quantity of PAHs annually emitted into the atmosphere (Denier van der Gon, van het Bolscher et al., 2007). In the EEA-32 region, estimated emissions of PAHs have fallen by 60% between 1990 and 2008. A combination of targeted legislation coupled with improved controls and abatement techniques has led to significant progress been made in most European countries to reduce PAH emissions. While the majority of individual countries report PAH emissions which have decreased since 1990, there are eight countries (Denmark, Estonia, Portugal, Latvia, Italy, Iceland, Norway, and Sweden) in which increased PAH emissions have occurred (European_Environment_Agency, 2010). In particular, total PAHs estimated to be emitted in Italy, during 2008, were 156.2 tons; their main sources (40%) were domestic heating plants. During the same year, Italian passenger vehicles emitted 1.67 tons of PAHs (1%) (SINANET, 2009).
The BaP indicator approach was used for regulatory purposes in the PAH risk assessment, and gave rise to air quality objectives for BaP in several countries: in Italy, since 1999, the BaP air quality was set at 1 ng m^{-3} (Menichini, 1992); the same target value (1 ng m^{-3}) was introduced in the European Community's fourth Air Quality Daughter Directive (2005/107/EC), while in UK, the proposed air quality standard for BaP is 0.25 ng m^{-3}, as annual mean concentration (DEFRA, 2001).

4.1 European Regulations to reduce vehicle emissions

In 1993, all European Union countries adopted Directive 91/441/CEE 06/26/1991, which introduced acceptable limits for exhaust emissions of new vehicles sold in EU member states, and defined the first set, referred to as Euro 1, of acceptable emission limits. Increasingly stringent standards for the regulation of air pollutants (CO, hydrocarbons, nitrogen oxide, particulates) were introduced in 1996 (Euro 2), in 2000 (Euro 3), in 2005 (Euro 4) and in 2008 (Euro 5). Accordingly, since 1993, catalytic converters have been obligatory on new gasoline vehicles sold in European countries, and new diesel motors and motorcycle engines have been improved progressively to reduce their emissions.

In 1994, the Italian Health and Environment Ministries established technical regulations to monitor BaP in Italian towns with more than 150,000 residents. To reduce high pollutant levels found in Italian urban areas, in 1997, the Italian Environment Ministry passed a decree containing economic incentives favouring replacement of old cars (EURO 0) with less polluting cars.

To reduce air pollution in urban areas, local measures to limit the number of circulating vehicles were also taken, such as payment of a ticket for cars entering the city (Beevers and Carslaw, 2005) or restriction of the passage of EURO 0 cars, during rush hours.

PAHs are classified as unregulated automobile exhaust constituents, which means that no legal limit exists for their concentrations in vehicle emissions. Thus, less information is available on the effects of catalyst systems on reducing PAH and BaP concentrations in vehicle emissions. A thorough review on PAH emission factors for motor vehicles was conducted by Shen (Shen, Tao et al., 2011) who revealed that Emission Factors reported by different laboratories varied more than 4 orders of magnitude, primarily due to the influence of many factors, including country were the test was performed, vehicle model year, vehicle and fuel type, operation mode, ambient temperature (Baek, Field et al., 1991).

Ambient temperature is a critical variable for amount of pollutants emitted by catalysed vehicles. Catalytic converters must reach high temperatures to significantly reduce the concentration of pollutants emitted and this condition is reached after a few minutes after the ignition time and this time, in turn, depends on air temperature (Weilenmann, Soltic et al., 2005). For example, a three-way catalyst car, run in accordance with the European driving Cycle, at different ambient temperature, emitted, each kilometre run, 0.16 μg of BaP at 22 °C, 26 μg at -7°C, 116 μg at -20°C (Ludykar, Westerholm et al., 1999).

Notwithstanding these uncertainties, the introduction of catalytic converters decreased PAH emissions of diesel and gasoline vehicles and Table 1 shows emission factors of CO (Wikipedia) and BaP (DEFRA, 2001) of European vehicles, according to different EURO classifications.

		Diesel cars		Gasoline cars	
	date	CO	BaP	CO	BaP
EURO 0	Pre-1992		2.85	13.9	2.38
Euro 1	July 1992	2.72	0.63	2.72	0.26
Euro 2	Jan 1996	1.0	0.38	2.2	0.26
Euro 3	Jan 2000	0.64		2.3	
Euro 4	Jan 2005	0.50		1.0	
Euro 5	Sept 2009	0.500		1.000	

Table 1. Emission factors for CO (g/km) and BaP (μg/km) of European vehicles, according to EURO classifications and date of their becoming effective

5. Trends of BaP daily concentrations in two Italian streets

Genoa (610,307 residents according to 2001 census) is one of the most important Italian ports of the Mediterranean Sea. In this city, since 1994, the Genoa Province, in collaboration with the National Cancer Institute and the Regional Environmental Protection Agency, has implemented a network of six monitoring stations to measure regulated air pollutants and PAHs adsorbed to airborne particulate (Valerio, Pala et al., 1996; Menichini, Belladonna et al., 2006). Two traffic oriented sampling stations (Site A and Site B), placed along two canyon streets, were selected to study long term trends of BaP concentrations with prevalent origin from vehicular traffic.

Both monitoring sites are within the urban area where limitations on the driving of EURO 0 vehicles have been applied by the Genoa Council, since September 1998.

Site A was located along a three-carriageway street with a total width of 15 m and a mean building height of about 20 m along both sides. The mean daily flow was about 27,000 vehicles, mainly passenger cars. This is the only monitoring site in Genoa where the local traffic flow of four wheels auto vehicles was measured automatically, by electromagnetic coils embedded in the asphalt layer, till 2002.

Site B was near a five-carriageway street, with an estimated daily traffic flow of about 35,000 vehicles, with a total width of 21 m and a mean building height of about 20 m along both sides. In both sites, sampling started in September 1994 and the last samples evaluated in this study were collected in 2009. Altogether, 292 daily samples from site A and 261 from site B were analysed.

Two different periods were studied separately. The first period, from September 1994 to September 2002, assessed the possible effect of the gradual reduction of EURO 0 cars, which began in Italy in July 1992, replaced by EURO 1 cars (since July 1992), EURO 2 (January 1996) and EURO 3 (January 2000).

The second period evaluated trend of BaP daily concentrations from October 2002 to September 2009. During this second period, all new cars registered since January 2005 were classified EURO 4 and around 2006, according to our estimates, the majority of EURO 0 cars were no longer in circulation along the streets of Genoa. Details on sampling, analysis and quality control were previously published (Valerio, Stella et al., 2000; Valerio, Stella et al., 2009). In brief, samples of airborne particulate matter (PM_{10}) were collected on glass-fibre filters using high-volume samplers. Samples for 24 hours were collected from midnight to midnight. Sampling days were selected randomly and, according to the sampling program, four daily samples per month were collected simultaneously at each site. Occasional sampler failures, other technical problems during analyses and interruptions during important maintenance works along the monitored streets, reduced the planned frequency of samples collection.

The cyclohesane soluble compounds present on the filters were extracted by sonication; the PAHs were separated by SPE technique and analyzed by GLC with a capillary column and flame ionization detector.

5.1 Results

At each site, during the first observation period, the slopes of the natural logarithm of daily concentrations of BaP versus Julian days (progressive day counter from the first sampling day, September 1, 1994 until September 15, 2002) were calculated and results are shown in Figure 1. Apparent half-lives of BaP were evaluated (Prevedouros, Brorstrom-Lunden et al., 2004), dividing each calculated slope by ln 2.

In Figure 1 the typical seasonal BaP trend (Larsen and Baker, 2003; Schauer, Niessner et al., 2003) may be observed, with the higher daily concentrations in winter and the lowest in summer.

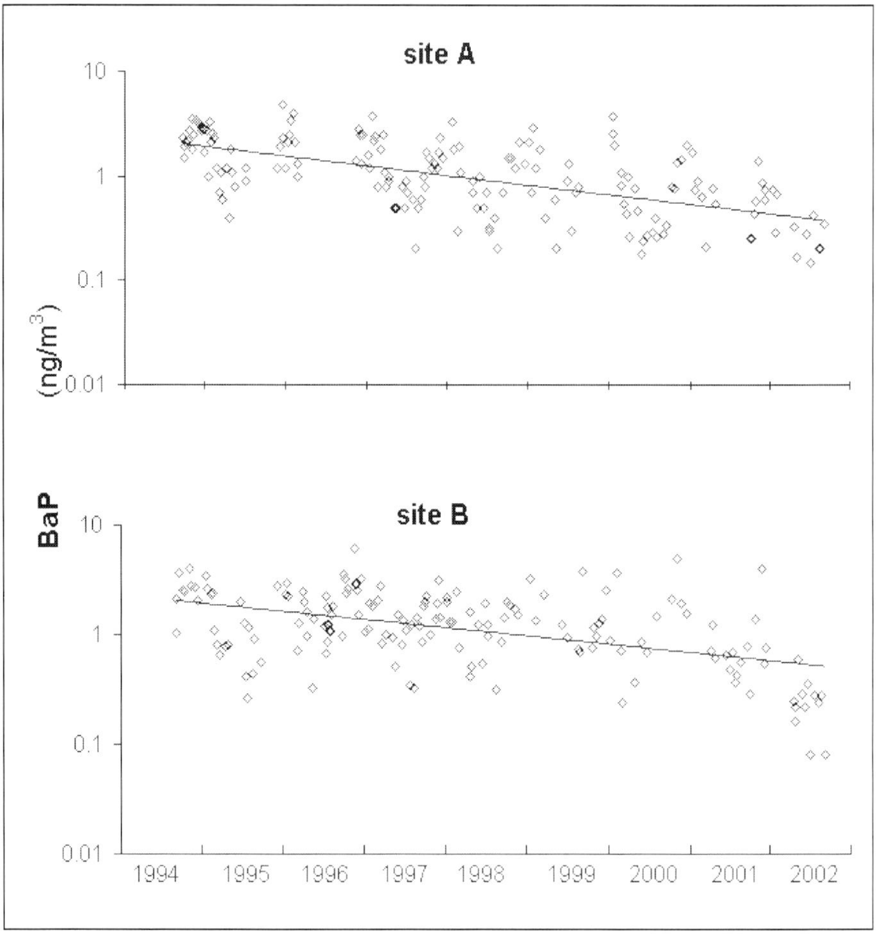

Fig. 1. Temporal trends for BaP (ng m^{-3}) concentrations at two oriented-traffic sites in Genoa (site A and B) from 1994 to 2002. Logarithmic daily BaP concentrations with regression lines

Figure 1 shows that, in both sampling sites, during the first nine years of monitoring (1994-2002) there was a constant decrease in the daily BaP concentrations and the calculated regression lines were both statistically significant ($p < 0.0001$).

Table 2 shows the mean apparent half-lives and confidence intervals (95%) of BaP at Site A and at Site B found in this study, compared with other half lives found during similar periods in Europe, in United States and Canada.

Mean half-lives for BaP found in the two traffic oriented sites in Genoa are between 3.3 and 4.0 years, respectively, values no statistically different ($p<0.01$).

Figure 2 shows the BaP trends during the second period, Oct 2002- Sept 2009.

During the second period, 125 samples were collected in Site A and 99 in Site B.

The daily BaP concentrations, during the eight monitored years, showed no statistically significant trend (Figure 2).

Therefore BaP annual means may be considered constant and their values, in the second period, were 0.39 ng m^{-3} (95% Cl = 0.08) in Site A and 0.54 ng m^{-3} (95% Cl = 0.08).in Site B, well below Italian air quality standards for BaP (1.0 ng/m^3, daily annual mean).

More precisely, BaP air quality standard was constantly met at Site A, starting from 1998 and at Site B from 2001, till the last available annual samplings, carried out during 2010 in Site A and during 2009, in Site B.

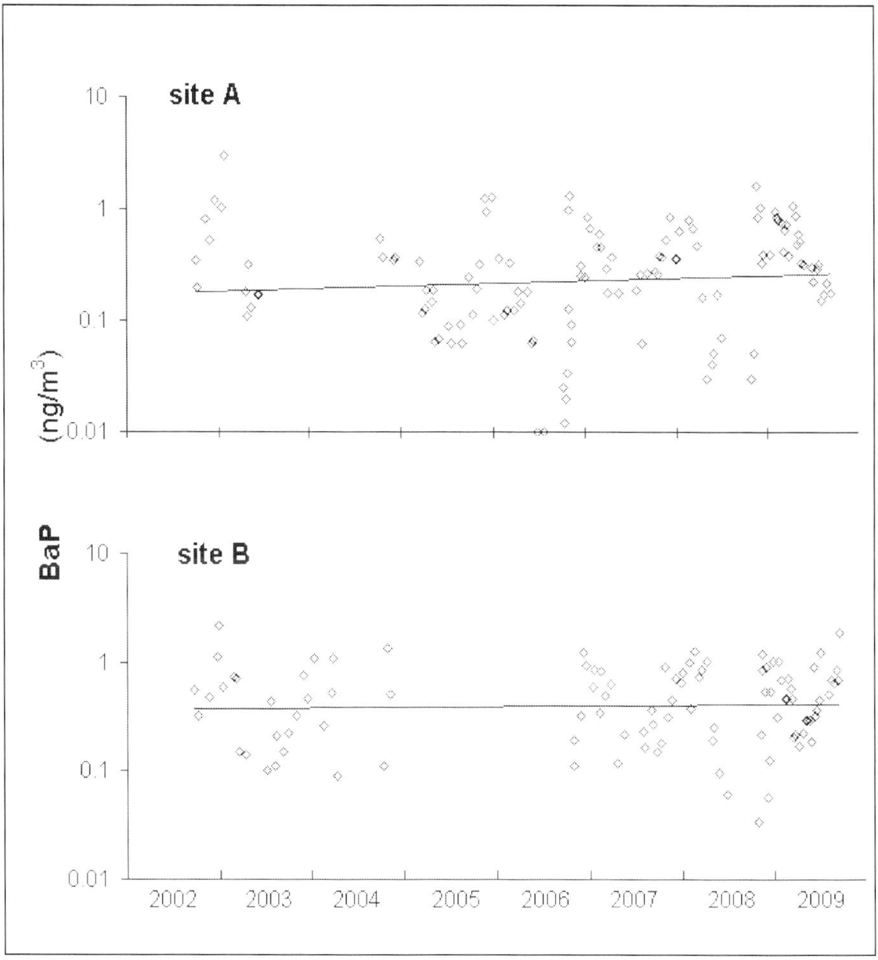

Fig. 2. Temporal trends for BaP (ng m^{-3}) concentrations at two oriented-traffic sites in Genoa (site A and B) from October 2002 to September 2009. Logarithmic daily BaP concentrations with regression lines

References	Period	site	half life year	lower Cl year	Upper Cl year
Cortes,2000	1991-1997	Lake Michigan	4.9		
Sun, 2006	1997-2003	Chicago	2.4±0.6		
Meijer, 2008	1991-2005	London	4.6	3.1	7.6
Meijer, 2008	1991-2005	Manchester	4.8	3.2	9.4
Meijer,2008	1991-2005	Middlesbrough	7.4	5.0	14.6
Menichini, 2006	1993-2004	Rome	9		
Menichini, 2006	1996-2004	Florence	5		
Valerio	1994-2002	Genoa A	3.3	2.9	3.8
Valerio	1994-2002	Genoa B	4.0	3.2	5.4

Table 2. Half lives (years) and lower and upper confidence limits at 95%, for BaP calculated in this study compared with other long term samplings carried out in USA, Canada, in UK and in Italy.

5.1.1 Multivariate analyses on BaP concentration

At site A the daily traffic flow, measured from 1994 to 2002, was constant with a mean value of 26,128 vehicles day^{-1} (range: 14,360 - 30,946). According to percentage of catalysed vehicles estimated to circulate in Genoa, daily passages of EURO 0 vehicles (NCcars) were calculated; NCcars, daily mean temperature (T; °C), daily ozone concentration (O_3; µg m^{-3}) and relative humidity (RH; %) were tested in a multivariate analyses as possible variables of daily BaP and CO concentrations, measured in site A.

Significant multiple regression equations found in this study (Valerio, Stella et al., 2009) are the followings:

$$\ln[CO] = 0.85 - 0.005\,[O_3] + 52\,10^{-6}\,NCcars \qquad R^2=0.62 \tag{2}$$

$$\ln[BaP] = -0.07 - 0.013\,[O^3] - 0.007\,RH + 101\,10^{-6}\,NCcars \qquad R^2=0.67 \tag{3}$$

Both CO and BaP mean daily concentrations correlated positively with the estimated number of EURO 0 vehicles passing along this street, during the sampling. The high standardised β coefficients for NCcars variable (0.57 for CO and 0.51 for BaP) confirm the relevance of EURO 0 vehicle emissions in air pollution measured along this canyon street.

Daily BaP concentration was negatively correlated with O_3 concentration and relative humidity, confirming similar previous studies (Valerio and Pala, 1991; Valerio, Stella et al., 2000; Prevedouros, Brorstrom-Lunden et al., 2004).

The only ambient variable found significant for CO concentrations measured in Site A was the O_3 concentration

According to the mean annual substitution of EURO 0 vehicles, estimated to occur in Genoa in 2006, no EURO 0 cars were in use in this town.

The BaP and CO concentrations, regressed using equations 2) and 3), for zero EURO 0 transits (NCcars = 0) and using mean annual temperature, humidity and ozone concentrations measured in Genoa in 2006, are 0.2 ng m^{-3} and 1.7 mg m^{-3}, respectively .
These values are comparable to BaP and CO annual means, actually measured at site A in 2006 (BaP: 0.18 ± 0.06 ng m^{-3}, N = 27; CO: 0.9 ± 0.7 mg m^{-3}).

5.2 Discussion on BaP trends found in Genoa

Traffic is certainly the main source of BaP measured in the two Genoa streets. All results agree with this statement.
Indeno (cd) pyrene and benzo (ghi) perylene are two PAHs analysed for this study. Their mean ratio, in 553 samples collected during the whole sampling period (1994-2009), was 0.44±0.08, in agreement with values found in emissions of gasoline vehicles (Caricchia, Chiavarini et al., 1999) and gasoline is the fuel that in 1994 was prevalently used by 293.514 motor vehicles circulating in Genoa. In the same year, light duty diesel cars were about the 7 % of used motor vehicles and their number increased progressively (30% in 2008). Also circulating motor vehicles increased during the studied period: 66,000 in 1994, 112,000 in 2003, and 132,148 in 2008. However, diesel cars and motorcycles prevalently substituted gasoline cars as the total numbers of vehicles circulating in Genoa changed from 382,565 in 1994 to 412,456 in 2008.
Multivariate analysis carried out in this study (equations 2 and 3) estimated the role of EURO 0 cars on air pollution and confirmed that the predominant source of BaP found along the two streets of Genoa was the number of EURO 0 vehicles passing along these streets.
Therefore, the progressive substitution of EURO 0 vehicles with EURO 1, EURO 2 and EURO 3, is the main explanation of the important reduction of mean BaP concentrations occurred between 1994 and 2002: 86% in site A and 89% in site B.
The importance of replacing EURO 0 vehicles to reduce air pollution in urban areas, was confirmed by the constant low BAP concentrations in the entire second period (Figure 2), during which only a small percentage of Euro 0 vehicle was estimated to be still in circulation. Statistics on the age of vehicles used in Genoa were available only since 2008 (ACI, 2009); during this year, the EURO 0 cars circulating in Genoa were only 10% of the total vehicles fleet (gasoline and diesel powered cars and motorcycles).
The BaP decrease found in the atmosphere in two streets in Genoa from 1994 to 2002 (Figure 1) was similar to that found in other European towns during the same period.
In a suburban sampling location in Munich, the air mass concentration of PAHs with 3 to 6 aromatic rings was 8 ng m^{-3} in 1991 and 2 ng m^{-3} in 2001/02. The decrease was 76-94% for the 3 and 4 ring PAHs and 92-96% for 5 and 6 ring PAHs. In particular, the abundance of BaP was reduced by 95%.
A significant decrease of BaP was observed in air samples collected in the North Rhine Westphalia area: the median concentration was 1.6 ng m^{-3} in 1991 and 0.7 ngm^{-3} in 1997 (Fertmann, Tesseraux et al., 2002)
Airborne PAHs, monitored regularly at rooftop height from 1991 to 1998 in London and Manchester (UK), exhibited a clear decreasing trend at both sites: 4.1 years was the half-life calculated for BaP (Prevedouros, Brorstrom-Lunden et al., 2004).
More recently, another study updated to 2005 the PAH concentrations found in six UK locations (Meijer, Sweetman et al., 2008). Particularly, in the three urban sites (London, Manchester and Middlesbrough) BaP concentrations, from 1991 to 2005, decreased

dramatically and since 2001, in these three urban sites, the BaP air quality standard proposed by UK (0.25 ng m^{-3}) exceeded only in winter samples. BaP half-lives, calculated in UK towns are shown in Table 2.

Concentrations of BaP measured near traffic-oriented sites in nine Italian towns between 1993 and 2004 (Menichini, Belladonna et al., 2006) showed a generalized decreasing trend in the last few years. In six of these towns (Florence, Genoa, Padua, Trieste, Bolzano, Bologna, Rome, and Verona) concentrations of BaP roughly halved as compared to initial concentrations. In 2004, in seven of these towns, BaP annual mean values ranged from 0.5 ng m^{-3} to 1.0 ng m^{-3}. BaP concentrations measured in Florence (1996-2004) and Rome (1993 .2004) near traffic oriented sites (Menichini, Belladonna et al., 2006) decreased with half lives of 5 and 9 years, respectively.

In European towns (Table 2) the longest half-life (9 years) was found in Rome and the lowest (3.3 years) in the site A of Genoa. To explain these differences it is necessary to have information about any change in the number of circulating cars, occurred during the studied period. Indeed, it is possible that a sharp increase of circulating cars may neutralize the lower emissions of new catalyzed vehicles. Therefore, according to the experience shown in Genoa, it is important to monitor traffic intensity near traffic oriented sampling stations.

The decrease of BaP found in Genoa and in other European towns may share a common explanation: the European Decree that gave to all EU countries, the mandatory requirements for all new gasoline and diesel vehicles sold from 1/1/1993, to respect stringent emission limits. In all European countries, this law caused a progressive substitution of non-catalyser gasoline vehicles with new, less polluting cars and the relevant contribution of traffic emissions to PAH pollution found in European urban areas in the last two decades was confirmed by several studies carried out in Italy (Valerio, Brescianini et al., 1992), the United Kingdom (Harrison, Smith et al., 1996) and Denmark (Nielsen, Jorgensen et al., 1996).

In this case, according to Emission Factors in Table 1, the substitution of EURO 0 vehicles with EURO 1 vehicles may reduce BaP emissions of gasoline cars by 89% and of diesel cars by 78%.

When, in 1996, all new vehicles sold in European countries were complying with EURO 2 emissions, based on emission factors provided in Table 1, the purchase of a new diesel car, replacirg a EURO 0 diesel car, should have further reduced BaP emissions.

For the following years, emission factors for BaP in EURO 3, 4, 5 cars are not available. Therefore predictions are not possible. However, according to the aforementioned correlation between the concentrations of CO and BaP, since 2005, further significant reductions of CO and probably of BaP would not have been expected.

6. Conclusions

Before this study, it was not taken for granted the drastic air pollution reductions occurred in Genoa could be linked to vehicles fleet renewal. The effectiveness of catalytic converters in urban courses was doubtful (Pornet, Beaubestre et al., 1995) due to the prevalent cold starts and frequent traffic congestions occurring in urban areas.

Notwithstanding mean daily course of only 14 kms, it is possible that, in Genoa, the rapid reaching of optimal motor temperatures may have been helped by the mild Mediterranean winters when daily temperatures rarely fall below 0 °C.

BaP half-lives found in London and Manchester suggest that catalytic converters and new diesel motors can be effective to reduce air pollution also in climatic conditions of these towns.

The European experience confirms that, in urban areas, the stringent air quality standards proposed for BaP and related PAHs may be respected.

7. Acknowledgements

The authors are indebted to the Genoa Province Administration, the Air Quality Department and, in particular, to Dr E. Daminelli for the financial and technical support that made this study possible.

We thank the Statistic Office of Genoa municipality and Liguria Region, Economic Development Department for their useful collaboration and doctor Marcello Ceppi for his support to statistical analysis of our data.

8. References

ACI, (2009).Analisi dell'anzianità del parco veicoli in alcuni comuni capoluogo di provincia, Available from http://www.aci.it/fileadmin/documenti/studi_e_ricerche/dati_statistiche/Anzianita/Capitolo_4.pdf

Baek, S.O., R.A. Field, M.E. Goldstone, P.W. Kirk, J.N. Lester R. Perry (1991). A review of atmosheric polycyclic aromatic hydrocarbons: sources, fate and behavior. *Water Air and Soil Pollution,* Vol. 60,No. 3-4, pp. 279-300.

Beevers, S.D. D.C. Carslaw (2005). The impact of congestion charging on vehicle emissions in London. *Atmospheric Environment,* Vol.39, No. 1, pp. 1-5.

Brun, G.L., O.C. Vaidya M.G. Leger (2004). Atmospheric deposition of polycyclic aromatic hydrocarbons to Atlantic Canada: geographic and temporal distributions and trends 1980-2001. *Environ Sci Technol,* Vol. 38,No. 7, pp. 1941-1948.

Caricchia, A.M., S. Chiavarini M. Pezza (1999). Polycyclic aromatic hydrocarbons in the urban atmospheric particulate matter in the city of Naples (Italy). *Atmos Environ,* Vol.3 3, pp. 3731-3738.

Cortes, D.R., I. Basu, C.D. Sweet R.A. Hites (2000). Temporal Trends in and Influence of Wind on PAH Concentrations Measured near the Great Lakes. *Environmental science & technology,* Vol. 34, pp. 356-360.

DEFRA, (2001).Assessment of benzo(a)pyrene atmospheric concentrations in the UK to support the establishment of a national PAH objective, Available from http://uk-air.defra.gov.uk/reports/empire/aeat-env-r-0620.pdf

Denier van der Gon, H., M. van het Bolscher, A. Visschedijk P. Zandveld (2007). Emissions of persistent organic pollutants and eight candidate POPs from UNECEEurope in 2000, 2010 and 2020 and the emission reduction resulting from the implementation of the UNECE POP protocol. *Atmospheric Environment,* Vol. 41, No. 40, pp. 9245-9261.

European_Environment_Agency, (2010).Persistent organic pollutants (POP) emissions (APE 006), Available from http://www.eea.europa.eu/data-and-maps/indicators/eea32-persistent-organic-pollutant-pop-emissions-1

Fertmann, R., I. Tesseraux, M. Schumann H. Neus (2002). Evaluation of ambient air concentrations of polycyclic aromatic hydrocarbons in Germany from 1990 to 1998. *J Expo Anal Environ Epidemiol,* Vol. 12,No. 2, pp. 115-123.

Harrison, R.M., D.J.T. Smith L. Luhana (1996). Source apportionment of atmospheric polycyclic aromatic hydrocarbons collected from an urban location in Birmingham, U.K. *Environ Sci Technol,* Vol. 30,No., pp. 825-832.

IARC (1987). Monographs on the evaluation of the carcinogenic risk of chemicals to humans. Supplement 7. Lyon, France.

Jacob, J., G. Grimmer A. Hildebrandt (1997). Long-term decline of atmospheric and marine pollution by polycyclic aromatic hydrocarbons (PAHS) in Germany. *Chemosphere,* Vol. 33, No. 9-10, pp. 2099-2108.

Kaur, S., M.J. Nieuwenhuijsen R.N. Colvile, (2007). Fine particulate matter and carbon monoxide exposure concentrations in urban street transport microenvironments, Available from http://www.sciencedirect.com/science?_ob=GatewayURL&_origin=ScienceSearch&_method=citationSearch&_piikey=S1352231007001343&_version=1&_returnURL=http%3A%2F%2Fwww.scirus.com%2Fsrsapp%2F&md5=4f2a07d19537f6d08f3f1c00b5a61ca9

Larsen, R.K., 3rd J.E. Baker (2003). Source apportionment of polycyclic aromatic hydrocarbons in the urban atmosphere: a comparison of three methods. *Environ Sci Technol,* Vol. 37,No. 9, pp. 1873-1881.

Lodovici, M., M. Venturini, E. Marini, D. Grechi P. Dolara (2003). Polycyclic aromatic hydrocarbons air levels in Florence, Italy, and their correlation with other air pollutants. *Chemosphere,* Vol. 50, No. 3, pp. 377-382.

Ludykar, D., R. Westerholm J. Almen (1999). Cold start emissions at +22, -7 and -20°C ambient temperatures from a three-way catalyst (TWC) car: regulated and unregulated exhaust components. *The Science of The Total Environment,* Vol. 235, No. 1, pp. 65-69.

Meijer, S.N., A.J. Sweetman, C.J. Halsall K.C. Jones (2008). Temporal trends of polycyclic aromatic hydrocarbons in the U.K. atmosphere: 1991-2005. *Environ Sci Technol,* Vol. 42, No. 9, pp. 3213-3218.

Menichini, E. (1992). Opinion adopted by the Italian National Advisory Toxicological Committee on polycyclic aromatic hydrocarbons. Rome.

Menichini, E. (1992). Urban air pollution by polycyclic aromatic hydrocarbons: levels and sources of variability. *Sci Total Environ,* Vol. 116, No. 1-2, pp. 109-135.

Menichini, E., V. Belladonna, F. Bergoglio, C. Gabrieli, M. Ceccanti, I. Rossi, L. Cellini, E. Corradetti, D. Grechi, V. Tricarico, M. Rosa, C. Zemello, A. Spiazzi, A. Stella, F. Valerio G.R. Trevisani (2006). Trend of atmospheric benzo(a)pyrene in Italy before the adoption of the European directive on PAHs. *Polycyclic Aromatic Compound,* Vol. 26, pp. 79-92.

Narvaez, R.F., L. Hoepner, S.N. Chillrud, B. Yan, R. Garfinkel, R. Whyatt, D. Camann, F.P. Perera, P.L. Kinney R.L. Miller (2008). Spatial and temporal trends of polycyclic aromatic hydrocarbons and other traffic-related airborne pollutants in New York City. Environ Sci Tech Vol 42 N° 19 pp 7330-7335

Nielsen, T., H.E. Jorgensen, J.C. Larsen M. Poulsen (1996). City air pollution of polycyclic aromatic hydrocarbons and other mutagens: occurrence, sources and health effects. *Sci Total Environ,* Vol. 30, No. 20, pp. 41-49.

Piccardo, M.T., A. Stella, A. Redaelli, D. Balducci, R. Coradeghini, C. Minoia F. Valerio (2004). Personal daily exposures to benzo(a)pyrene of taxi drivers in Genoa, Italy. *Sci Total Environ,* Vol. 330, No. 1-3, pp. 39-45.

Piccardo, M.T., A. Stella, A. Redaelli, C. Minoia F. Valerio (2003). Newsagents' daily personal exposures to benzo(a)pyrene in Genoa, Italy. *Atmospheric Environment,* Vol. 37, No. 5, pp. 603-613.

Pornet, P., C. Beaubestre, Y. Courtois, H. Ing, B. Lopez J.L. Marduel (1995). Impact des conditions de conduite sur l'efficacite des pots catalytiques de vehicules a essence et diesel. *Sci Total Environ,* Vol. 169, pp. 321-329.

Prevedouros, K., E. Brorstrom-Lunden, J.H. C, K.C. Jones, R.G. Lee A.J. Sweetman (2004). Seasonal and long-term trends in atmospheric PAH concentrations: evidence and implications. *Environ Pollut,* Vol. 128, No. 1-2, pp. 17-27.

Schauer, C., R. Niessner U. Poschl (2003). Polycyclic aromatic hydrocarbons in urban air particulate matter: decadal and seasonal trends, chemical degradation, and sampling artifacts. *Environ Sci Technol,* Vol. 37, No. 13, pp. 2861-2868.

Shen, H., S. Tao, R. Wang, B. Wang, G. Shen, W. Li, S. Su, Y. Huang, X. Wang, W. Liu, B. Li K. Sun (2011). Global time trends in PAH emissions from motor vehicles. *Atmospheric Environment,* Vol. 45, No. 12, pp. 2067-2073.

SINANET, (2009).Emissioni nazionali di idrocarburi policiclici aromatici (IPA) (PAH) in Italia per l'anno 2008, Available from http://www.sinanet.isprambiente.it/it/sinanet/sstoriche/visualizza

Sun, P., S. Backus, P. Blanchard R.A. Hites (2006). Annual variation of polycyclic aromatic hydrocarbon concentrations in precipitation collected near the Great Lakes. Environ Sci Technol Vol 40 N° 3 pp. 696-701

UN-ECE, (1998).UN/ECE Convention on Long Range Transboundary Air Pollution. Protocol on Persistent Organic Pollutants, Available from http://www.unece.org/env/lrtap/welcome.html

Valerio, F., C. Brescianini, M. Pala, A. Lazzarotto, D. Balducci V. Fontana (1992). Sources and atmospheric concentrations of polycyclic aromatic hydrocarbons and heavy metals in two Italian towns (Genoa and La Spezia). *Sci Total Environ,* Vol. 114, pp. 47-57.

Valerio, F. M. Pala (1991). Effects of temperature on the concentration of polycyclic aromatic hydrocarbons (PAHs) adsorbed onto airborne particulates. *Fresenius J Anal Chem,* Vol. 339, pp. 777-779.

Valerio, F., M. Pala, A. Lazzarotto, A. Stella, F. Ciccarelli D. Balducci (1996). Air quality standard for benzo(a)pyrene in Genoa (1994-1995). *Polycyclic Aromatic Compound,* Vol. 9, pp. 61-66.

Valerio, F., A. Stella A. Munizzi (2000). Correlations between PAHs and CO, NO, NO2, O3, along an urban street. *Polycyclic Aromatic Compounds,* Vol. 20, No., pp. 235-244.

Valerio, F., A. Stella, M. Pala, D. Balducci, M.T. Piccardo M. Cipolla (2009). The effect of euro-0 vehicle substitution on polycyclic aromatic hydrocarbon and carbon monoxide concentrations in an urban area. *Atmospheric Environment,* Vol. 43, pp. 1520-1526.

Weilenmann, M., P. Soltic, C. Saxer, A.-M. Forss N. Heeb (2005). Regulated and nonregulated diesel and gasoline cold start emissions at different temperatures. *Atmospheric Environment,* Vol. 39, No. 13, pp. 2433-2441.

WHO (1999). Carbon monoxide. Environmental Health criteria. Geneva, Worl Health Organization.

Wikipedia, Available from http://en.wikipedia.org/wiki/European_emission_standards#Emission_standards_for_passenger_cars

Yan, J., L. Wang, P.P. Fu H. Yu (2004). Photomutagenicity of 16 polycyclic aromatic hydrocarbons from the US EPA priority pollutant list. *Mutation Research/Genetic Toxicology and Environmental Mutagenesis,* Vol. 557, No. 1, pp. 99-108.

Zhang, Y., S. Tao, H. Shen J. Ma (2009). Inhalation exposure to ambient polycyclic aromatic hydrocarbons and lung cancer risk of Chinese population. *Proceedings of the National Academy of Science of the United States of America,* Vol. 106, No. 50, pp. 21063-21067.

Part 2

Air Quality Modelling

12

Imprecise Uncertainty Modelling of Air Pollutant PM_{10}

Danni Guo[1], Renkuan Guo[2], Christien Thiart[2] and Yanhong Cui[2]
[1]*Climate Change and Bioadapation Division, South African National Biodiversity Institute, Cape Town*
[2]*Department of Statistical Sciences, University of Cape Town, Cape Town, South Africa*

1. Introduction

Particulate matter (PM) refers to solid particles and liquid droplets found in air. Many manmade and natural sources produce PM directly, or produce pollutants that react in the atmosphere to form PM. The resultant solid and liquid particles come in a wide range of sizes, and particles that are 10 micrometers or less in diameter (PM_{10}) can be inhaled into and accumulate in the respiratory system and are believed to pose health risks (Environmental Protection Agency, 2010). Particulate matter is one of the six primary air pollutants the Environmental Protection Agency (EPA) regulates, due to exposure to high outdoor PM_{10} concentrations causes increased disease and death (Environmental Protection Agency, 2010). Therefore, PM_{10} concentrations, amongst many other air pollutants, are sampled and measured in various places in California, United States.

The general trend of PM air pollutant concentrations in the air in California are on the decrease, but it continues to be monitored and observed. The California standards for annual PM_{10} concentrations is that the annual arithmetic mean is 20 $\mu g/m^3$, and the national standard is 50 $\mu g/m^3$ before 2006 (California Environmental Protection Agency Air Resources Board, 2010, Environmental Protection Agency, 2010). The State of California sets very high standards for their air quality, and air pollutants are carefully monitored.

However, in reality, it is too costly in terms of time, finance, and manpower to keep all the 213 sites to be monitoring and recording. In Fig. 1, a complete map of all 213 sample locations for PM_{10} are shown. However, one must note that these sample sites are never all used at any given year, PM_{10} samples are taken at different locations each year. At best, a maximum of 102 PM_{10} samples are collected during some years, and at worst, 61 PM_{10} samples are collected at that year. Therefore comparisons of PM_{10} between years are difficult, due to missing data at sample sites. It is difficult to construct kriging maps in terms of actual observations annually since the air pollutants were measured in different locations each year although the site design originally planned was quite delicate statistically.

Each year, approximately 40% of the 213 sites were actually observed. We call a site that does not have a recorded PM_{10} value as "missing value", and since there are no patterns so that serious problems would twist the kriging map constructions. In Fig. 2, this is clearly demonstrated. In 1989, there are 61 PM_{10} samples collected (29% of 213 locations), and in 2000, there are 94 PM_{10} samples collected (44%).

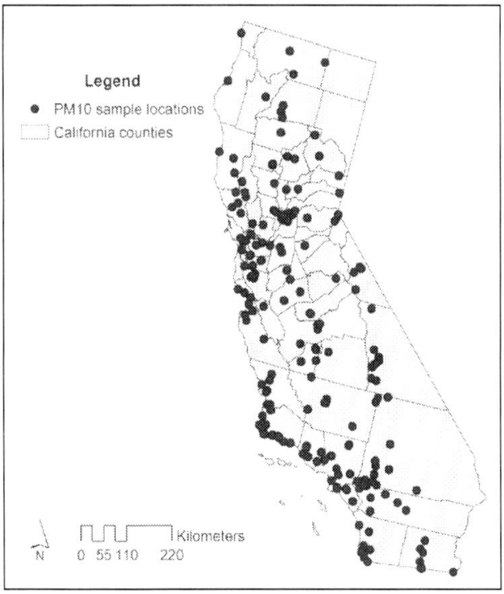

Fig. 1. Complete 213 Observational Sites in the California State

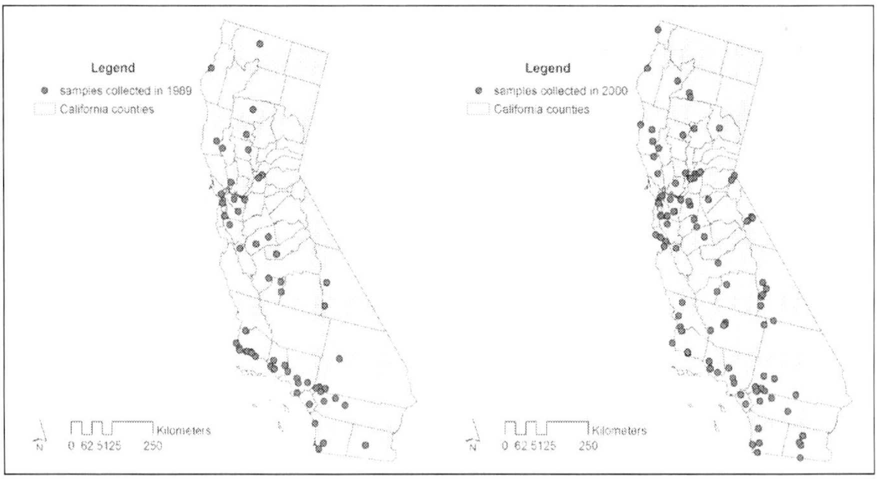

Fig. 2. PM_{10} Samples Collected in California in 1989 (61 sites) and 2000 (94 sites)

The data scarcity brings in a series of (five) fundamental issues into the spatial-temporal modelling and prediction practices for California PM_{10} data, namely:

1. The necessity to recognize the impreciseness in analyzing the spatial-temporal pattern in terms of California PM_{10} records, which inevitably acts the solidness of a geo-statistical analysis;

2. Which theoretical foundations are appropriate for modelling impreciseness uncertainty;
3. How to fill up the "missing value" sites so that the "complete" records are available, which is either an original annual average from the original observations (40%) recorded on the site or a or predicted value by "neighbourhood sites" (60%), i.e., to facilitate spatial-temporal imprecise PM_{10} value by interpolations and extrapolations;
4. How to estimate the parameters of uncertain processes (temporal patterns), particularly the rate of change parameter α_i, $i = 1,2,\cdots,213$;
5. Create annual kiging maps (19 maps) under spatially isotropy and stationarity assumptions so that the changes between annual maps can be analyzed by kriging map difference between 2007 and 1989 and kriging map of location rate of change.

These issues will be addressed in the remaining sections sequentially.

2. The necessity of modelling impreciseness in California PM_{10} spatial-temporal analysis

Impreciseness is a fundamental and intrinsic feature in the PM_{10} spatial-temporal modelling, due to the observational data shortage and incompleteness. Spatially, there are 213 sites involved, and temporally, PM_{10} observations were collected from 1989 to 2007, over a 19-year period. During the 19-year period, there are only two sites (Site 2125 and Site 2804) having complete 19 year records. There are 16 sites having only 1 record (8%) and 70 sites having 10 or above records. To have a statistically significant time-series analysis, 50 data points are minimal requirement for each site, so classical time-series analysis (probabilistic analysis) cannot be performed. In order to have a quick overall evaluation of PM_{10} records on each site, we borrow the statistical quality control idea here (Electric, 1956, Montgomery, 2001). But we do not carry on traditional 6-sigma rule, rather, classify the PM_{10} records into four groups: $1\text{-}(5,20]$, $2\text{-}(20,35]$, $3\text{-}(35,50]$, $4\text{-}(50,160]$. These four-group limits in Table 1 reflect the national standard, (50 $\mu g/m^3$) and California state standard (20 $\mu g/m^3$) respectively. For example, $1\text{-}(5,20]$ is for a location whose PM_{10} fall in 5 to 20 ($\mu g/m^3$).

County name	$1\text{-}(5,20]$	$2\text{-}(20,35]$	$3\text{-}(35,50]$	$4\text{-}(50,160]$	No. of Sites
Los Angeles			3	7	10
Kern	1	2	1	5	9
Riverside	3		3	3	9
San Diego			5	3	8
Imperial	3			4	7
Lake		3			3
Inyo	2	2	4		8

Table 1. PM_{10} Hazard level evaluation over selected 7 counties

One must be aware that the classification is not in absolute sense, rather, additional rules are adding (similar to quality control chart pattern analysis (Electric, 1956):
(1) if a single point, then, classify the site hazard level according to which group it falls in;
(2) if a sequence of records, some of them, particularly early points may fall in higher (or lower) hazard level, but if last three points fall in a lower (or higher) hazard level, the later level would be chosen for the site.

The additional rule 1 can attribute to expert's knowledge confirmation, while the additional rule 2 can be regarded as an expert's decision based on trend pattern.

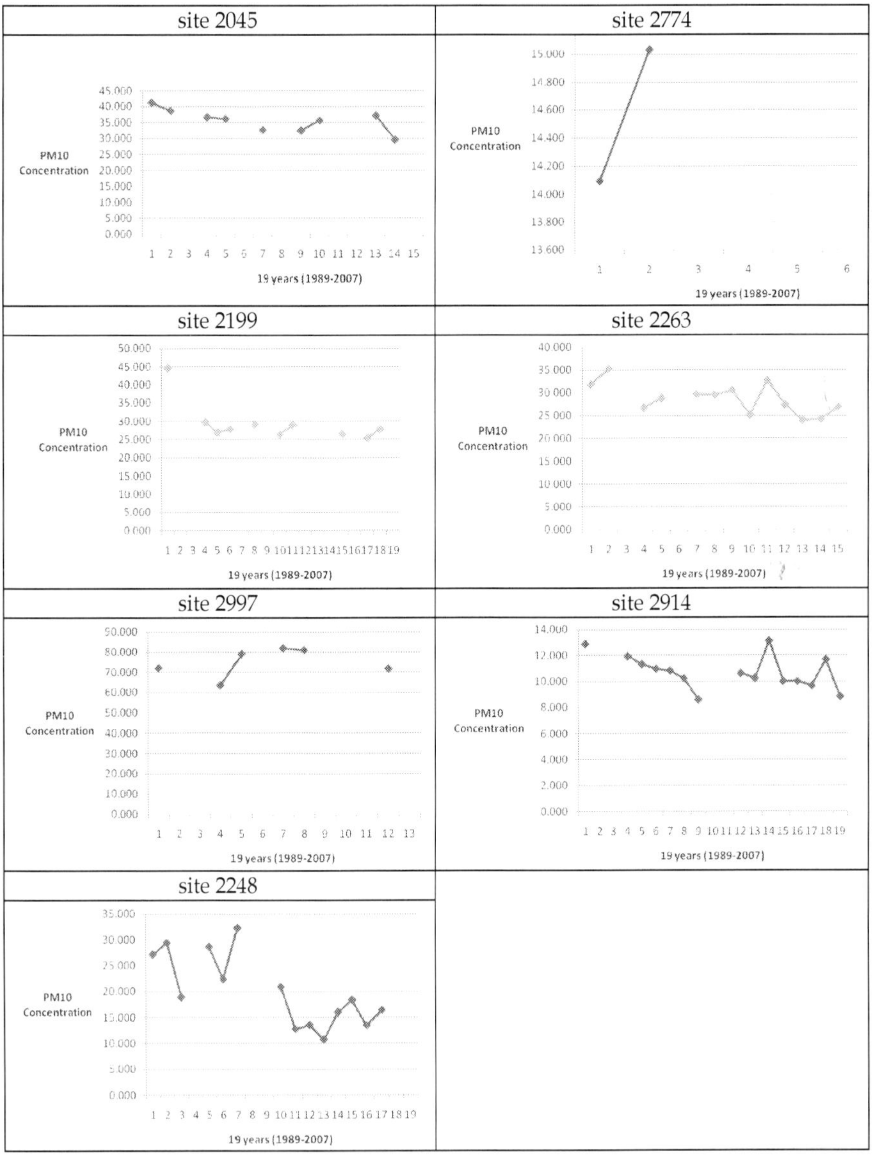

Fig. 3. The 7 sites from the selected 7 counties with original PM_{10} data plots and the hazard level classifications

Fig. 3 shows the classifications of a seven sites from the selected 7 counties in Table 1, each county one site is picked up for illustration purpose. The red coloured plot means the hazard level 1 $(5,20]$; the green coloured plot means the hazard level 2$(20,35]$; the purple coloured plot means the hazard level 3 $(35,50]$; and the black coloured plot means the hazard level 4 $(50,160]$.

It is evident that facing the impreciseness caused by incomplete data recording, one has to rely on expert's knowledge to compensate the inadequacy and accuracy in collected observational data. Impreciseness is referred to a term with a connotation specified by an uncertain measure or an uncertainty distribution for each of the actual or hypothetical members of an uncertainty population (i.e., collection of expert's knowledge). An uncertain process is a repeating process whose outcomes follow no describable deterministic pattern, but follow an uncertainty distribution, such that the uncertain measure of the occurrence of each outcome can be only approximated or calculated.

The uncertainty modelling without a measure specification will not have an rigorous mathematical foundations and consequently the modelling exercise is baseless and blindness. In other words, measure specification is the prerequisite to spatial-temporal data collection and analysis. For example, without Kolmogrov's (1950) three axioms of probability measure, randomness is not defined and thus statistical data analysis and inference has no foundation at all.

Definition 2.1: Impreciseness is an intrinsic property of a variable or an expert's knowledge being specified by an uncertain measure.

It is therefore inevitably to seek appropriate form of uncertainty theory to meet the impreciseness challenges. In the theoretical basket, interval uncertainty theory (Moore, 1966), fuzzy theory (Zadeh, 1965, 1978), grey theory (Deng, 1984), rough set theory (1982), upper and lower provisions (or expectations) (Walley, 1991), or Liu's uncertainty theory (2007, 2010) may be chosen.

While imprecise probability theory (Utikin and Gurov, 1998) may be a typical answer to address the observational data inaccuracy and inadequacy. However the imprecise probability based spatial modelling requires too heavy assumptions. Just as Utikin and Gurov (2000) commented, "the probabilistic uncertainty model makes sense if the following three premises are satisfied: (i) an event is defined precisely; (ii) a large amount of statistical samples is available; (iii) probabilistic repetitiveness is embedded in the collected samples. This implies that the probabilistic assumption may be unreasonable in a wide scope of cases." Guo et al. (2007) and Guo (2010) did attempt to address the spatial uncertainty from the fuzzy logic and later Liu's (2007) credibility theory view of point.

Nevertheless, Liu's (2007, 2010) uncertainty theory is the only one built on an axiomatic uncertain measure foundation and fully justified with mathematical rigor. Therefore it is logical to engage Liu's (2007, 2010, 2011) uncertainty theory for guiding us to understand the intrinsic character of imprecise uncertainty and facilitate an accurate mathematical definition of impreciseness in order to establish the foundations for uncertainty spatial modelling under imprecise uncertainty environments.

3. Uncertain measure and uncertain calculus foundations

Uncertainty theory was founded by Liu in 2007 and refined in 2010, 2011. Nowadays uncertainty theory has become a branch of mathematics.

A key concept in uncertainty theory is the uncertain measure, which is a set function defined on a sigma-algebra generated from a non-empty set. Formally, let Ξ be a nonempty set (space), and $\mathfrak{A}(\Xi)$ the σ-algebra on Ξ. Each element, let us say, $A \subset \Xi$, $A \in \mathfrak{A}(\Xi)$ is called an uncertain event. A number denoted as $\lambda\{A\}$, $0 \le \lambda\{A\} \le 1$, is assigned to event $A \in \mathfrak{A}(\Xi)$, which indicates the uncertain measuring grade with which event $A \in \mathfrak{A}(\Xi)$ occurs. The normal set function $\lambda\{A\}$ satisfies following axioms given by Liu (2011):

Axiom 1: (Normality) $\lambda\{\Xi\} = 1$.

Axiom 2: (Self-Duality) $\lambda\{\cdot\}$ is self-dual, i.e., for any $A \in \mathfrak{A}(\Xi)$, $\lambda\{A\} + \lambda\{A^c\} = 1$.

Axiom 3: (σ- Subadditivity) $\lambda\left\{\bigcup_{i=1}^{\infty} A_i\right\} \le \sum_{i=1}^{\infty} \lambda\{A_i\}$ for any countable event sequence $\{A_i\}$.

Axiom 4: (Product Measure) Let $\left(\Xi_k, \mathfrak{A}_{\Xi_k}, \lambda_k\right)$ be the kth uncertain space, $k = 1,2,\cdots,n$. Then product uncertain measure λ on the product measurable space $(\Xi, \mathfrak{A}_{\Xi})$ is defined by

$$\lambda = \lambda_1 \wedge \lambda_2 \wedge \cdots \wedge \lambda_n = \min_{1 \le k \le n} \{\lambda_k\} \tag{1}$$

where

$$\Xi = \Xi_1 \times \Xi_2 \times \cdots \times \Xi_n = \prod_{k=1}^{n} \Xi_k \tag{2}$$

and

$$\mathfrak{A}_{\Xi} = \mathfrak{A}_{\Xi_1} \times \mathfrak{A}_{\Xi_2} \times \cdots \times \mathfrak{A}_{\Xi_n} = \prod_{k=1}^{n} \mathfrak{A}_{\Xi_k} \tag{3}$$

That is, for each product uncertain event $\Lambda \in \mathfrak{A}_{\Xi}$ (i.e, $\Lambda = \Lambda_1 \times \Lambda_2 \times \cdots \times \Lambda_n \in \mathfrak{A}_{\Xi_1} \times \mathfrak{A}_{\Xi_2} \times \cdots \times \mathfrak{A}_{\Xi_n} = \mathfrak{A}_{\Xi}$), the uncertain measure of the event Λ is

$$\lambda\{\Lambda\} = \begin{cases} \sup\limits_{A_1 \times \cdots \times A_n \subset \Lambda} \min\limits_{1 \le k \le n} \lambda\{\Lambda_k\} & \text{if} \sup\limits_{A_1 \times \cdots \times A_n \subset \Lambda} \min\limits_{1 \le k \le n} \lambda\{\Lambda_k\} > 0.5 \\ 1 - \sup\limits_{A_1 \times \cdots \times A_n \subset \Lambda^c} \min\limits_{1 \le k \le n} \lambda\{\Lambda_k\} & \text{if} \sup\limits_{A_1 \times \cdots \times A_n \subset \Lambda^c} \min\limits_{1 \le k \le n} \lambda\{\Lambda_k\} > 0.5 \\ 0.5 & \text{otherwise} \end{cases} \tag{4}$$

Definition 3.1: (Liu, 2007, 2010, 2011) A set function $\lambda : \mathfrak{A}(\Xi) \to [0,1]$ satisfies *Axioms 1-3* is called an uncertain measure. The triple $(\Xi, \mathfrak{A}(\Xi), \lambda)$ is called an uncertainty space.

Definition 3.2: (Liu, 2007, 2010, 2011) An uncertainty variable is a measurable function ξ from an uncertainty space $(\Xi, \mathfrak{A}(\Xi), \lambda)$ to the set of real numbers, i.e., for any Borel set B of real numbers, the set $\{\tau \in \Xi : \xi(\tau) \subset B \in \mathfrak{B}(\mathbb{R})\} \in \mathfrak{A}(\Xi)$, i.e., the pre-image of B is an event.

Remark 3.3: Parallel to revelation of the connotation of randomness in geostatistics, impreciseness occupies an fundamental position in geospatial-temporal uncertainty statistical analysis. In California PM_{10} spatial-temporal study, nearly 60% sites do not have "complete" temporal sequences so that in order to fill the "missing" observations, we have to engage expert's knowledge to pursue "complete sequences" (i.e., to have 19 PM_{10} values at each individual site), which is inevitably imprecise and incomplete. Impreciseness is referred to a term here with an intrinsic property governed by an uncertainty measure or an uncertainty distribution for each of the actual or hypothetical members of an uncertainty population (i.e., collection of expert's knowledge). An uncertainty process is a repeating process whose outcomes follow no describable deterministic pattern, but follow an uncertainty distribution, such that the uncertain measure of the occurrence of each outcome can be only approximated or calculated.

Remark 3.4: Impreciseness exists in engineering, business and research practices due to measurement imperfections, or due to more fundamental reasons, such as insufficient available information, ... , or due to a linguistic nature, because it is an unarguable fact that impreciseness exists intrinsically in expert's knowledge on the real world.

Definition 3.5: Let ξ be a uncertainty quantity of impreciseness on an uncertainty measure space $(\Xi, \mathfrak{A}(\Xi), \lambda)$. The uncertainty distribution of ξ is

$$\Psi_{\xi}(x) = \lambda\{\tau \in \Xi \mid \xi(\tau) \le x\} \tag{5}$$

An imprecise variable ξ is an uncertainty variable and thus is a measurable mapping, i.e., $\xi : \mathbb{D} \to \mathbb{R}$, $\mathbb{D} \subseteq \mathbb{R}$. An observation of an imprecise variable is a real number, (or more broadly, a symbol, or an interval, or a real-valued vector, a statement, etc), which is a representative of the population or equivalently of an uncertainty distribution $\Psi_{\xi}(\cdot)$ under a given scheme comprising set and σ-algebra. The single value of a variable with impreciseness should not be understood as an isolated real number rather a representative or a realization from the uncertain population.

Definition 3.6: (Lipschitz condition) Let $f(x)$ be a real-valued function, $f : \mathbb{R} \to \mathbb{R}$. If for any $x, y \in \mathbb{R}^n$, there exists a positive constant $M > 0$, such that

$$|f(x) - f(y)| < M|x - y| \tag{6}$$

Definition 3.7: (Lipschitz continuity) Let $f : \mathbb{R}^m \to \mathbb{R}^m$

1. for $\forall B \subset \mathbb{R}^m$, B to open set, f is Lipschitz continuous on B if $\exists M > 0$ such

$$\|f(x) - f(y)\| < M\|x - y\|, \ \forall x, y \in B \tag{7}$$

where $\|\cdot\|$ is some metric (for example, Euclidean distance in $\mathbb{R}^m$), such

$$d(f(x), f(y)) < Md(x, y), \ \forall x, y \in B \tag{8}$$

2. for each $z \in \mathbb{R}^m$, f is Lipschitz continuous locally on the open ball B of center z radius $M > 0$ such

$$B_M(z)=\{y\in\mathbb{R}^m \mid d(y,z)<M\} \tag{9}$$

3. if f is Lipschitz continuous on the whole space $\mathbb{R}^m$, then the function is called globally Lipschitz continuous.

Remark 3.8: For continuity requirements, Lipschitz continuous function is stronger than that of the continuous function in Newton calculus but it is weaker than the differentiable function in Newton differentiability sense. In other words, Lipschitz-continuity does not warrant the first -order differentiability everywhere but it does mean nowhere differentiability. Lipschitz-continuity does not guarantee the existence of the first-order derivative everywhere, however, if exists somewhere, the value of the derivative is bounded since

$$\frac{|f(x)-f(y)|}{|x-y|}<M \tag{10}$$

by recalling the definition of the Newton derivative

$$\lim_{y\to x}\frac{f(x)-f(y)}{x-y}=f'(x) \tag{11}$$

Similar to the concept of stochastic process in probability theory, an uncertain process $\{\xi_t, t\geq 0\}$ is a family of uncertainty variables indexed by t and taking values in the state space $\mathbb{S}\subseteq\mathbb{R}$.

Definition 3.9: (Liu 2010, 2011) Let $\{C_t, t\geq 0\}$ be an uncertain process.

1. $C_0=0$ and all the trajectories of realizations are Lipschitz-continous;
2. $\{C_t, t\geq 0\}$ has stationary and independent increments;
3. every increment $C_{t+s}-C_s$ is a normal uncertainty variableb with expected value 0 and variance t^2, i.e., the uncertainty distribution of $C_{t+s}-C_s$ is

$$\Psi_{C_{t-s}-C_s}(z)=\left(1+\exp\left(-\frac{xz}{\sqrt{3}t}\right)\right)^{-1} \tag{12}$$

Then $\{C_t, t\geq 0\}$ is called a canonical process.

Remark 3.10: Comparing to Brownian motion process $\{B_t, t\geq 0\}$ in probability theory, which is continuous almost everywhere and nowhere is differentiable, while Liu's canonical process $\{C_t, t\geq 0\}$ is Lipschitz-continuous and if $\{C_t, t\geq 0\}$ is differentiable somewhere, the derivative is bounded. Therefore $\{C_t, t\geq 0\}$ is smoother than $\{B_t, t\geq 0\}$.

Definition 3.11: (Liu, 2010, 2011) Suppose $\{C_t, t\geq 0\}$ is a canonical process, and f and g are some given functions, then

$$d\xi_t=f(t,\xi_t)dt+g(t,\xi_t)dC_t \tag{13}$$

is called an uncertain differential equation. A solution to the uncertain differential equation is the uncertain process $\{\xi_t, t\geq 0\}$ satisfying it for any $t>0$.

Remark 3.12: Since dC_t and $d\xi_t$ are only meaningful under the umbrella of uncertain integral, i.e., the an uncertain differential equation is an alternative representation of

$$\xi_t = \xi_0 + \int_0^t f(s,\xi_s)ds + \int_0^t g(s,\xi_s)dC_s \tag{14}$$

Definition 3.13: The geometric canonical process $\{G_t, t \geq 0\}$ satisfies the uncertain differential equation

$$dG_t = \alpha G_t dt + \sigma G_t dC_t \tag{15}$$

has a solution

$$G_t = \exp(\alpha t + \sigma C_t) \tag{16}$$

where α can be called the drift coefficient and $\sigma > 0$ can be called the diffusion coefficient of the geometric canonical process $\{G_t, t \geq 0\}$ due to the roles played respectively.

4. Spatial interpolation and extrapolation via inverse distance approach

Statistically, spatial interpolation and extrapolation modeling is actually a kind of linear regression modeling exercises, say, kriging methodology. Considering the shortage of California PM_{10} data records, we will utilize a weighted linear combination approach, which was first proposed by Shepard (1968). The weights are the inverse distances between the missing value cell to the actual observed PM_{10} value cells. The weight construction is a deterministic method, which is neutral and does not link to any specific measure theory. It is widely used in spatial predictions and map constructions in geostatistics, but is not probability oriented, rather, molecular mechanics stimulated. A unique aspect of geostatistics is the use of regionalized variables which are variables that fall between random variables and completely deterministic variables. The weight of an observed PM_{10} value is inversely proportional to its distance from the estimated value. Let:

C_{ij} — The jth cell on the Site i, (i represent the actual site number), $i = 1, 2, \cdots, 213$. Note index j points to the cell where no PM_{10} value is recorded, i.e., missing value cell.

x_i — Longitude of site i

y_i — Latitude of site i

d_{ij} — The distance between site i and site j (where missing value j^{th} cell is located)

$$d_{ij} = \sqrt{(x_j - x_i)^2 + (y_j - y_i)^2}$$

w_{ij} — Weight, $w_{ij} = \begin{cases} 0 & \text{Cell } (i,j) \text{ has no } PM_{10} \text{ obs.} \\ 1/d_{ij} & \text{Cell } (i,j) \text{ has } PM_{10} \text{ obs.} \end{cases}$

Then the inverse distance formula is,

$$C_{ij} = \frac{w_{ij}}{\sum_{i=1}^{213} w_{ij}} z_{ij} \tag{17}$$

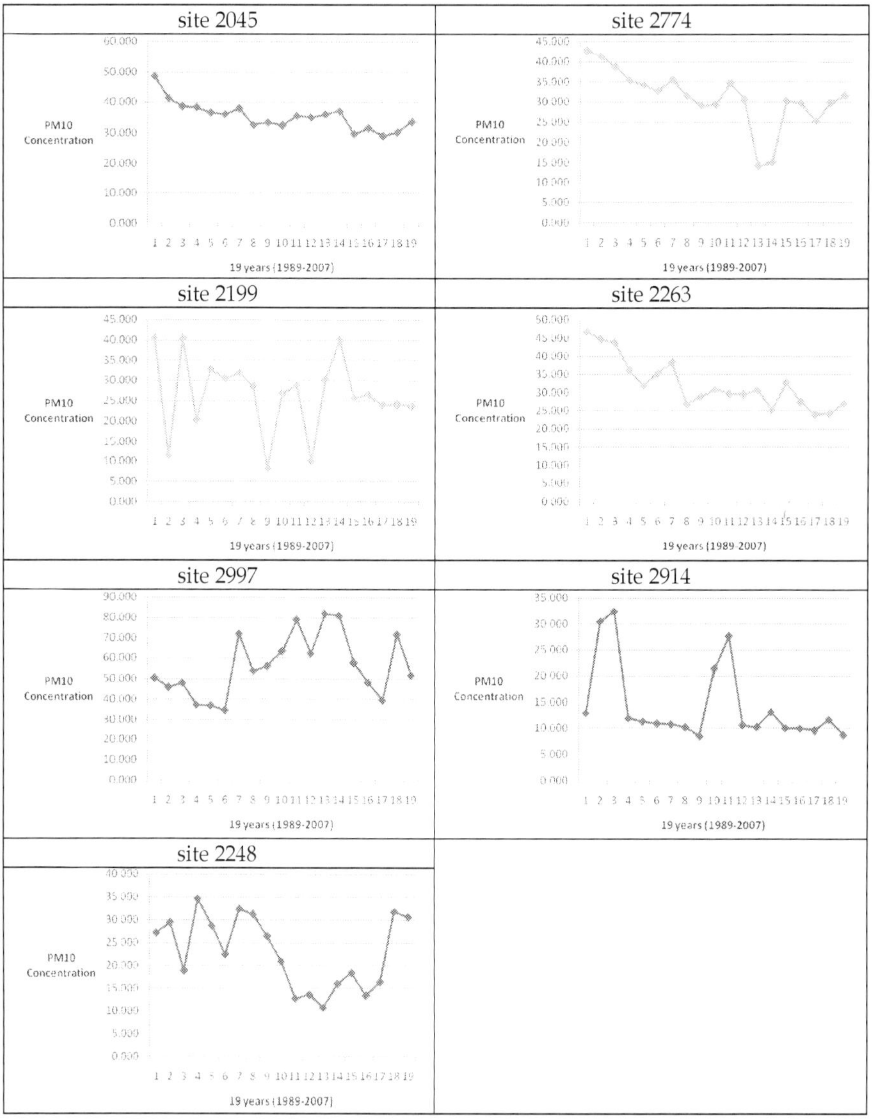

Fig. 4. The 7 sites from the selected 7 counties with completed 19 year observations of PM_{10}

We wrote a VBA Macro to facilitate the interpolations and the extrapolations to "fill" up the 2048 missing value cells in terms of the 1639 cells with PM_{10} values. With the interpolations and the extrapolations, every site has 19 PM_{10} values now. As to whether the inverse distance approach can facilitate highly accurate predictions for each cell without a observed PM_{10} value, we performed a re-interpolation and re- extrapolation scheme (by deleting a true PM_{10} record, then fill it by the remaining records one by one) to evaluate the mean square value for error evaluation, the calculated mean of sum of error squares is 59.885, which is statistically significant (asymptotically).

We plotted sites 2045, 2744, 2199, 2263, 2297, 2914, and 2248 (appeared in Fig. 3) respectively in Fig. 4. By comparing Fig. 3 and 4, it is obvious that only Site 2744 the hazard level changed (moving up to next higher hazard level), while the hazard level of other six sites are unchanged. This may give an justification of the inverse distance approach. Keep in mind, the aim of this article is investigate whether the PM_{10} level is changed over 1989 to 2007 19-year period. The change is not necessarily be accurate but reasonably calculated because of the impreciseness features of PM_{10} complete records.

5. Uncertain analysis of site temporal pattern

Once the interpolations and the extrapolations in terms of the inverse distance approach is completed, a "complete" data set is available, containing 4047 data records of 213 sites over 19 years. The next task is for a given site, how to model the uncertain temporal pattern. It is obvious that the "complete" data set contains impreciseness uncertainty due to the interpolations and the extrapolations. We are unsure that the impreciseness uncertainty is of random uncertainty, so that we still use uncertain measure theory to pursue the temporal uncertainty modelling.

Recall that the **Definition 3.13** in Section 3 facilitates a uncertain geometric canonical process, $\{G_t, t \geq 0\}$. Notice that $G_0 = 0$ may not fit the data reality so that we propose a modified uncertain geometric canonical process, $\{G_t^*, t \geq 0\}$ with $G_0 > 0$:

$$G_t^* = G_0 G_t = G_0 \exp(\alpha t + \sigma C_t) \tag{18}$$

Note that

$$\ln G_t^* = \ln G_0 + \alpha t + \sigma C_t \tag{19}$$

Let $y_t = \ln G_t^*$, $\alpha_0 = \ln G_0$, then we have

$$y_t = \alpha_0 + \alpha t + \sigma C_t, \; t = 1, 2, \cdots, 18 \tag{20}$$

Recall the relevant definitions in Section 3, we have

$$\mathrm{E}[C_t] = 0, \text{ and } V[C_t] = t^2 \tag{21}$$

But note that for $\forall s < t$,

$$\begin{aligned} \mathrm{E}[C_t C_s] &= \mathrm{E}\left[\left(C_s + (C_t - C_s)\right)C_s\right] \\ &= \mathrm{E}\left[C_s^2\right] + \mathrm{E}\left[C_s(C_t - C_s)\right] \\ &= s^2 + \mathrm{E}\left[C_s(C_t - C_s)\right] \end{aligned} \tag{22}$$

Notice that the increment $C_t - C_s$ is independent of C_s, i.e., C_{t-s} is independent of C_s. Therefore,

$$
\begin{aligned}
E[C_{t-s}C_s] &= \int_{-\infty}^{\infty}\int_{-\infty}^{\infty} z_1 z_2 d\Phi_{C_{t-s},C_s}(z_1,z_2) \\
&= \int_{-\infty}^{\infty}\int_{-\infty}^{\infty} z_1 z_2 d\min\left\{\Psi_{C_{t-s}}(z_1),\Upsilon_{C_s}(z_2)\right\} \\
&= \int_{-\infty}^{\infty}\int_{-\infty}^{\infty} z_1 z_2 d\min\left\{\left(1+\exp\left(-\frac{\pi z_1^2}{\sqrt{3}(t-s)}\right)\right)^{-1},\left(1+\exp\left(-\frac{\pi z_2^2}{\sqrt{3}s}\right)\right)^{-1}\right\}
\end{aligned}
\tag{23}
$$

since if ξ_1 and ξ_2 are independent uncertain variables with uncertainty distributions Ψ_{ξ_1} and Υ_{ξ_2} respectively, then the joint uncertainty distribution of (ξ_1,ξ_2) is $\Phi_{\xi_1,\xi_2}(z_1,z_2)=\min\left\{\Psi_{\xi_1}(z_1),\Upsilon_{\xi_2}(z_2)\right\}$. Hence we obtain the expression of $\sigma_{s,t}$:

$$
\sigma_{s,t} = E[C_sC_t] = s^2 + \int_{-\infty}^{\infty}\int_{-\infty}^{\infty} z_1 z_2 d\min\left\{\left(1+\exp\left(-\frac{\pi z_1^2}{\sqrt{3}(t-s)}\right)\right)^{-1},\left(1+\exp\left(-\frac{\pi z_2^2}{\sqrt{3}s}\right)\right)^{-1}\right\} \tag{24}
$$

Then the ith "variance-covariance" matrix for uncertain vector $(y_{i1},y_{i2},\cdots,y_{i19})'$

$$
\Gamma_i = \sigma^2\left(\sigma^i_{j,k}\right)_{19\times 19} \tag{25}
$$

where i is the site index, $j,k=1,2,\cdots,19$ are the entry pair in Γ_i matrix. Hence we have a regression model (Draper and Smith, 1981, Guo et al., 2010, Guo, 2010).

For the ith site, the regression model is

$$
y_{it} = \alpha_{i0} + \alpha_i t + \sigma_i C_{i,t} \tag{26}
$$

Then in terms of the weighted least square criterion we can define an objective function as

$$
J_i(\alpha_{i0},\alpha_i) = \left(\underline{Y}_i - X\underline{\beta}_i\right)'\Gamma^{-1}\left(\underline{Y}_i - X\underline{\beta}_i\right) \tag{27}
$$

where

$$
\underline{Y}_i = \begin{bmatrix} y_{i1} \\ y_{i2} \\ \vdots \\ y_{i19} \end{bmatrix} \quad \underline{\beta} = \begin{bmatrix} \alpha_{i0} \\ \alpha_i \end{bmatrix} \quad X = \begin{bmatrix} 1 & 1 \\ 1 & 2 \\ \vdots & \vdots \\ 1 & 19 \end{bmatrix} \tag{28}
$$

We further notice that

$$
\begin{aligned}
r_{ij} &= y_{i,j} - y_{i,j-1} \\
&= \alpha_i + \sigma_i\left(C_{ij} - C_{i,j-1}\right) = \alpha_i + \sigma_i \Delta C_{ij} \\
j &= 2,3,\cdots,19
\end{aligned}
\tag{29}
$$

Then it is reasonable to estimate α_i by

$$\hat{\alpha}_i = \frac{1}{18}\sum_{j=2}^{19} r_{ij} \tag{30}$$

Furthermore, we notice that

$$\hat{\sigma}_i = \sqrt{\frac{1}{18}\sum_{j=2}^{19}\left(r_{ij} - \hat{\alpha}_i\right)^2} \tag{31}$$

Also, we can evaluate

$$\sigma_{j,k}^{i} = j^2 + \int_{-\infty}^{\infty}\int_{-\infty}^{\infty} z_1 z_2 d\min\left\{\left(1+\exp\left(-\frac{\pi z_1}{\sqrt{3}(k-j)}\right)\right)^{-1}, \left(1+\exp\left(-\frac{\pi z_2}{\sqrt{3}j}\right)\right)^{-1}\right\} \tag{32}$$

in terms of numerical integration, Then an estimate for Γ_i matrix is obtained:

$$\hat{\Gamma}_i = \hat{\sigma}_i^2\left(\sigma_{j,k}^{i}\right)_{19\times 19} \tag{33}$$

Finally, we use the approximated objective function

$$\hat{J}_i(\alpha_{i0}, \alpha_i) = \left(\underline{Y}_i - X\underline{\beta}_i\right)'\hat{\Gamma}_i^{-1}\left(\underline{Y}_i - X\underline{\beta}_i\right) \tag{34}$$

to obtain a pair of estimates $(\tilde{\alpha}_{i0}, \tilde{\alpha}_i)$. Repeat this estimation process until all the 213 weighted least square estimate $(\tilde{\alpha}_{i0}, \tilde{\alpha}_i)$ are obtained.

Recall the definition of coefficient α_i so that the sign and the absolute value of α_i indicates the geometric change over the 19 years. Since the estimation procedure of α_i involves all the spatial-temporal information, it is reasonable to have them plotted in a kriging map to reveal the overall changes over 19-year period.

6. Kriging maps and time-change maps based on completed PM_{10} data

Kriging map presentation is vital for a geostatistian's visualization, and maps reveal hidden information or the whole picture. A sample statistic is typically condensing the wide-spread information into a numerical point. While, a kringing map is actually a map statistic (or a statistical map) which contains infinitely many information aggregated from limited "sample" information (i.e., observations). Kriging itself is not specifically probability oriented, it is another weighted linear combination prediction, but requires more mathematical assumptions. In fuzzy geostatistics, the fuzzy kriging scheme has also been developed (Bardossy et al., 1990).

Ordinary kriging (abbreviated as OK) is a linear predictor, see Cressie (1991) and Mase (2011). The formula is

$$Z(s_0) = \sum_{j=1}^{N}\lambda_j Z(s_j) \tag{35}$$

where s_j are spatial locations with observation $Z(s_j)$ available and the coefficients λ_j satisfy the OK linear equation system

$$\begin{cases} \sum_{j=1}^{N} \lambda_j \gamma\left(\varepsilon(s_j)-\varepsilon(s_i)\right)-\psi=\gamma\left(\varepsilon(s_0)-\varepsilon(s_i)\right),\ i=1,2,\cdots,N \\ \sum_{j=1}^{N} \lambda_j = 1 \end{cases} \tag{36}$$

The OK system is generated under the assumptions of an additive spatial model

$$Z(s)=\mu(s)+\varepsilon(s) \tag{37}$$

where $\mu(s)$ is the basic (expected) spatial trend and $\varepsilon(s)$ is a Gaussian error $N\left(0,\sigma^2(s)\right)$, i.e., Gaussian variable with mean and variance

$$\mathrm{E}\left[\varepsilon(s)\right]=0,\ V\left[\varepsilon(s)\right]=\sigma^2(s) \tag{38}$$

respectively. Accordingly, the variogram 2γ of the random error function $\varepsilon(\cdot)$ is just defined by

$$2\gamma(h)=\mathrm{E}\left[\left(\varepsilon(s+h)-\varepsilon(h)\right)^2\right] \tag{39}$$

where h is the separate vector between two spatial point $s+h$ and s under the isotropy assumption.

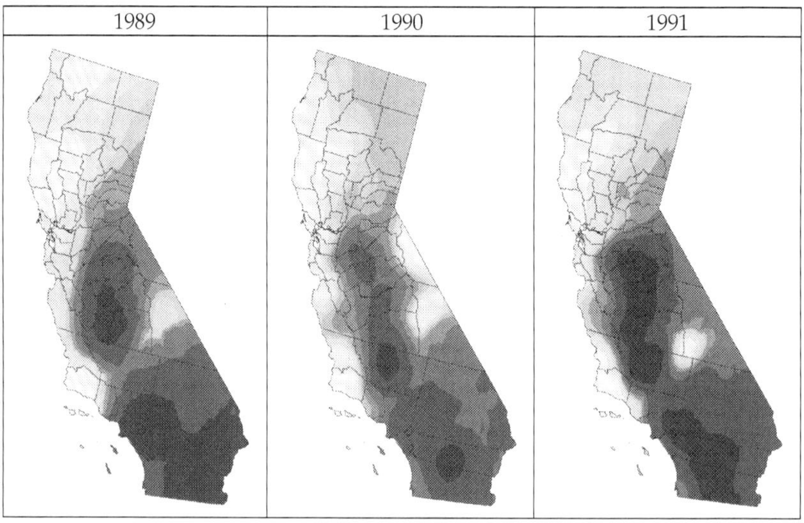

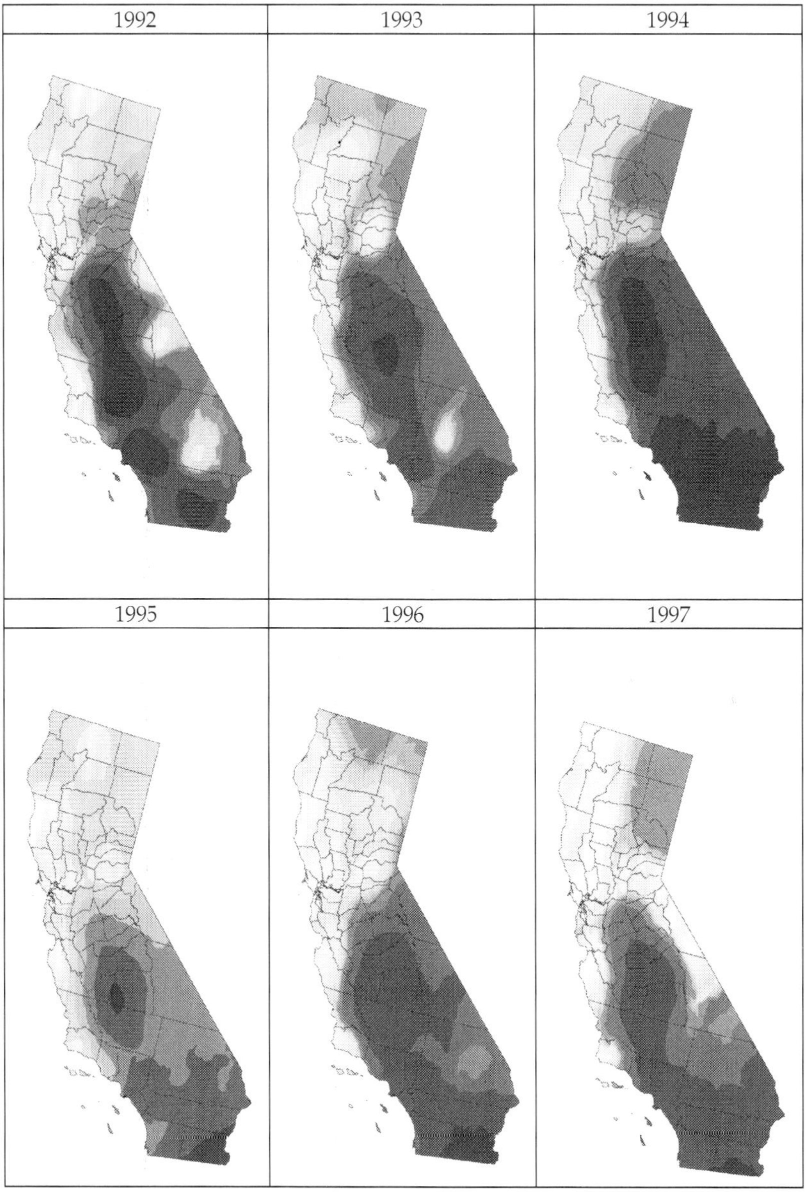
1992
1993
1994
1995
1996
1997

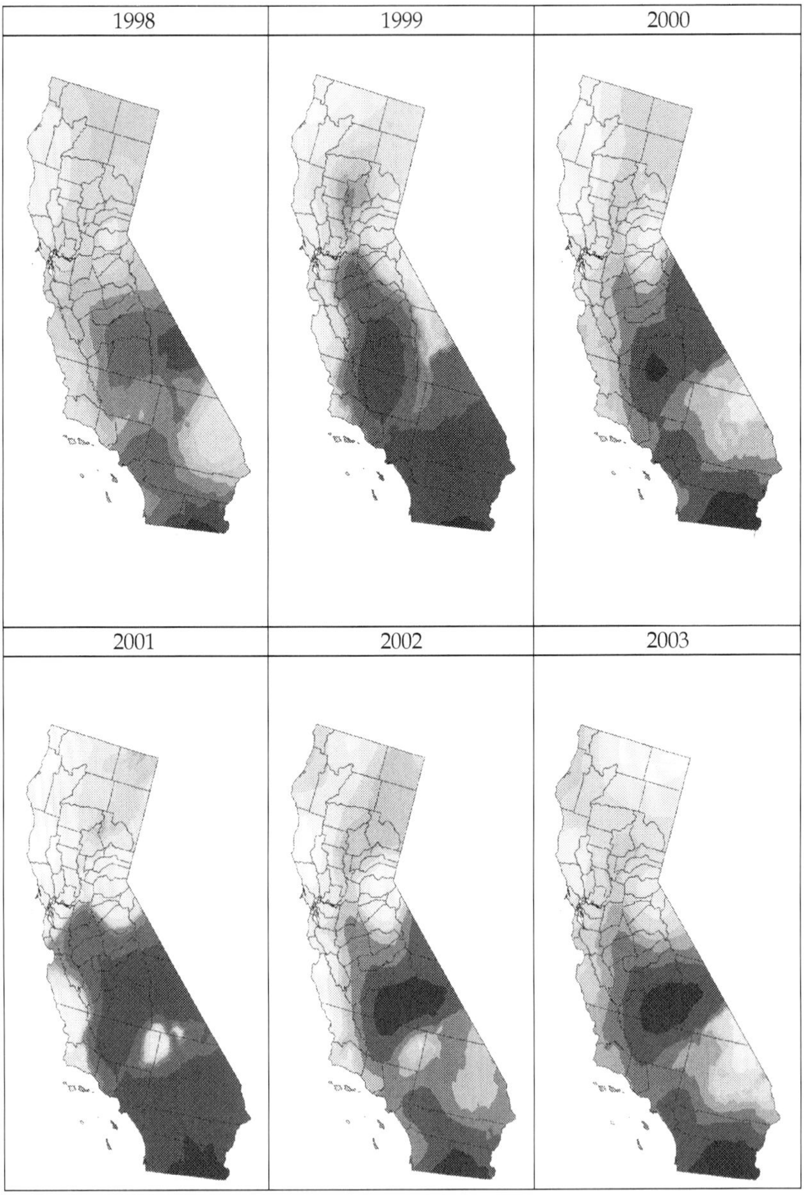
1998
1999
2000
2001
2002
2003

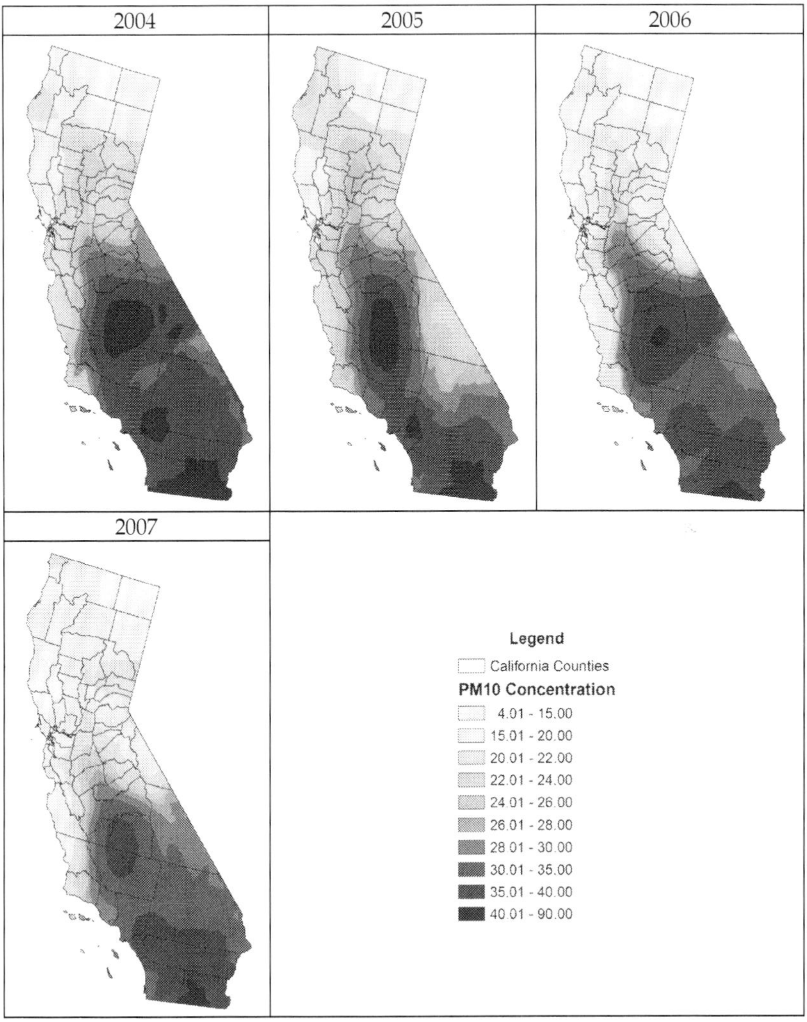

Fig. 5. Kriging Prediction Maps for PM_{10} in California 1989-2007.

The 213 observation sites now have 19-year PM_{10} values, a "complete" data set is now available, containing 4047 data records of 213 sites over 19 years, and then the 19 ordinary kriging pred4iction maps are generated for comparisons. In Fig. 5, all 19 years of PM_{10} concentration in California State are shown. It is very interesting to examine the change in PM_{10} concentrations through the 19 years, based upon the modelled complete 213 site data. In particular, 1998 shows to have an extremely low PM_{10} concentration. Although air quality is varied over the years, but in general, the PM_{10} concentration is decreasing, showing an improvement of air quality trend.

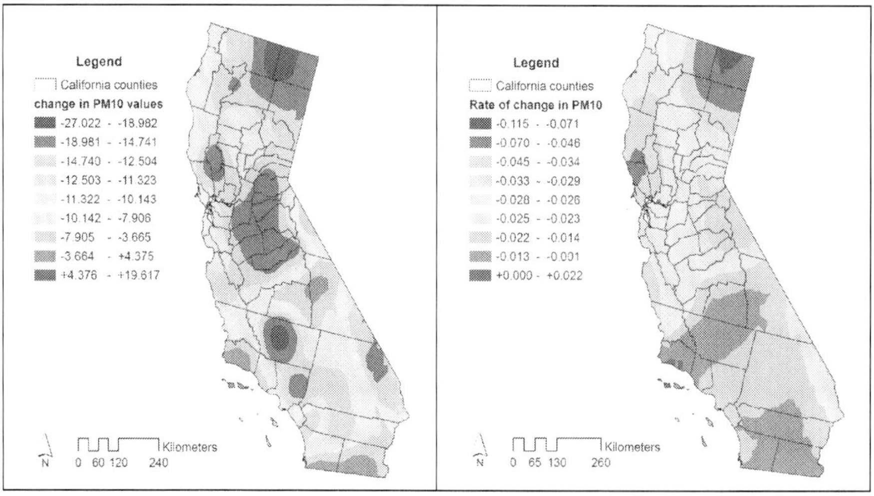

Fig. 6. Changes in PM_{10} values and the rate of change of PM_{10} in California between 1989 and 2007.

As one can clearly see from Fig. 6, that PM_{10} concentration has clearly decreased over the 19 years, and air quality has improved remarkably over the years. The blue and green colours show negative changes, and red shows positive changes or near positive changes. Counties such as San Diego, Inyo, Santa Barbara, Imperial, still show an increase in PM_{10} concentration in the air, and indicate bad air quality. While Kern, Modoc, Siskiyou counties show the most improvement in air quality. The left map in Fig. 6 is PM_{10} record difference between 2007 and 1989 at each location, in total 231 values, and then a difference map is constructed. It is obvious that the difference map only utilizes 1989 and 2007 two-year PM_{10} records, 1990, 1991, ..., 2006 seventeen years' information do not participate the change map construction. The right map in Fig. 6 show completed $\tilde{\alpha}_i$, $i = 1,2,\cdots,213$, the rate of change over 1989 to 2007 19-year period.

Note that the calculations of $\tilde{\alpha}_i$, $i = 1,2,\cdots,213$ involve all nineteen years by temporal regression, the dependent variable y are estimated form the actual PM_{10} observations cross over all the available locations. Therefore, the rate of change parameter α_i at each individual location contains all spatial-temporal information. It is reasonable to say the rate of change parameter $\tilde{\alpha}_i$ is an aggregate statistic for revealing the 19-year changes over 213 locations. $\tilde{\alpha}_i$ kriging map is thus different from 2007-1989 kriging maps. The positive sign of $\tilde{\alpha}_i$ indicates the increasing trend in PM_{10} concentration, while the negative sign f $\tilde{\alpha}_i$ indicates the decreasing trend in PM_{10} concentration. The absolute value of $\tilde{\alpha}_i$ reveals the magnitude of change of PM_{10} concentration. It is worth to report, among 213 locations, 193 locations have negative $\tilde{\alpha}_i$, while the negative $\tilde{\alpha}_i$ locations are 20 (9% approximately).

7. Discussion

Air quality and health is always a central issue to public concern on the quality of life. In this chapter, we examined PM_{10} levels over 19 years, from 1989 to 2007, in the California State.

Facing the difficult task of a lack of "complete" PM_{10} observational data, we utilised the inverse distance weight methodology to "fill" in the locations with missing values. By doing so, the impreciseness uncertainty is introduced, which is not necessarily explained by probability measure foundation. We noted the character of a regionalized variable in geostatistics and therefore engage Liu's (2010, 2011) uncertainty theory to address the impreciseness uncertainty. In this case, we developed a series of uncertain measure theory founded spatial-temporal methodology, including the inverse distance scheme, the kriging scheme, and the geometric canonical process based weighted regression analysis in order to extract the change information from the incomplete 1989-2007 PM_{10} records. The use of the rate of change parameter alpha is a new idea and it is an aggregate change index utilized all spatial-temporal data information available. It is far better than classical change treatments. However, due to the limitations of our ability, we are unable to demonstrate the detailed uncertain measure based spatial analysis model. In the future research, we plan to develop a more solid uncertain spatial prediction methodology.

8. Acknowledgements

I would like to thank the California Air Resources Board for providing the air quality data used in this paper. This study is supported financially by the National Research Foundation of South Africa (Ref. No. IFR2009090800013) and (Ref. No. IFR2011040400096).

9. References

Bardossy, A.; Bogardi, I. & Kelly, E. (1990). Kriging with imprecise (fuzzy) variograms, I: Theory. *Mathematical Geology*, Vol. 22, pp. 63–79.

California Environmental Protection Agency Air Resources Board. (2010). Ambient Air Quality Standards (AAQS) for Particulate Matter. (www.arb.ca.gov)

Cressie, N. (1991). *Statistics for Spatial Data*. Wiley-Interscience, John-Wiley & Sons Inc. New York.

Deng, J. L. (1984). *Grey dynamic modeling and its application in long-term prediction of food productions*. Exploration of Nature, Vol. 3, No. 3, pp. 7-43.

Draper, N. & Smith, H. (1981). *Applied Regression Analysis*. 2nd Edition. John Wiley & Sons, Inc. New York.

Electric, W. (1956). *Statistical Quality Control Handbook*. Western Elctric Corporation, Indianapolis.

Environmental Protection Agency (EPA). 2010. National Ambient Air Quality Standards (NAAQS). U.S. Environmental Protection Agency. (www.epa.gov)

Guo, D.; Guo, R. & Thiart, C. (2007). Predicting Air Pollution Using Fuzzy Membership Grade Kriging. *Journal of Computers, Environment and Urban Systems*. Editors: Andy P Jones and Iain Lake. Elsevier, Vol. 31, No. 1, pp. 33-51. ISSN: 0198-9715

Guo, D.; Guo, R. & Thiart, C. (2007). Credibility Measure Based Membership Grade Kriging. *International Journal of Uncertainty, Fuzziness and Knowledge-Based Systems*. Vol. 15, No. Supp. 2, (April 2007), pp. 53-66. B.D. Liu (Editor). ISSN 0218-4885

Guo, D. (2010). *Contributions to Spatial Uncertainty Modelling in GIS*. Lambert Academic Publishing (online.lap-publishing.com). ISBN 978-3-8433-7388-3

Guo, R.; Cui, Y.H. & Guo, D. (2010). Uncertainty Statistics. (Submitted to Journal of Uncertainty Systems, under review)

Guo, R.; Cui, Y.H. & Guo, D. (2010). Uncertainty Linear Regression Models. (Submitted to Journal of Uncertainty Systems, under review)

Guo, R.; Guo, D. & Thiart, C. (2010). Liu's Uncertainty Normal Distribution. *Proceedings of the First International Conference on Uncertainty Theory*, August 11-19, 2010, Urumchi & Kashi, China, pp 191-207, Editors: Dan A. Ralescu, Jin Peng, and Renkuan Guo. International Consortium for Uncertainty Theory. ISSN 2079-5238

Kolmogorov, A.N. (1950) *Foundations of the Theory of Probability*. Translated by Nathan Morrison. Chelsea, New York.

Liu, B.D. (2007). *Uncertainty Theory: An Introduction to Its Axiomatic Foundations*. 2nd Edition. Springer-Verlag Heidelberg, Berlin.

Liu, B.D. (2010). *Uncertainty Theory: A Branch of Mathematics of Modelling Human Uncertainty*. Springer-Verlag, Berlin.

Liu, B.D. (2011). *Uncertainty Theory,* 4th Edition, 17 February, 2011 drafted version.

Liu, S.F. & Lin, Y. (2006). *Grey Information*. Springer-Verlag, London.

Mase, S. (2011). GeoStatistics and Kriging Predictors, In: *International Encyclopedia of Statistical Science*. Editor: Miodrag Lovric, 1st Edition, 2011, LVIII, pp. 609-612, Springer.

Montgomery, D.C. (2001). *Introduction to Statistical Quality Control*. 4th Edition. John Wiley & Sons, Now York.

Moore, R.E. (1966). *Interval Analysis*. Prentice-Hall, Englewood Cliff, NJ. ISBN 0-13-476853-1

Pawlak, Z. (1982). Rough Sets. *International Journal of Computer and Information Sciences*, Vol. 11, pp. 341-356.

Shepard, D. (1968). A two-dimensional interpolation function for irregularly-spaced data. *Proceedings of the 1968 ACM National Conference*, pp. 517–524.

Utkin, L.V. & Gurov, S.V. (1998). New reliability models on the basis of the theory of imprecise probabilities. *Proceedings of the 5th International Conference on Soft Computing and Information Intelligent Systems*, Vol. 2, pp. 656-659.

Utkin, L.V. & Gurov, S.V. (2000). New Reliability Models Based on Imprecise Probabilities. *Advanced Signal Processing Technology, Soft Computing*. Fuzzy Logic Systems Institute (FLSI) Soft Computing Series - Vol. 1, pp. 110-139, Charles Hsu (editor). Publisher, World Scientific. November 2000. ISBN 9789812792105

Walley, P. (1991). *Statistical Reasoning with Imprecise Probabilities*. London: Chapman and Hall. ISBN 0412286602

Zadeh, L. A. (1965). Fuzzy sets. *Information and Control*, Vol. 8, pp. 338-353.

Zadeh, L. A. (1978). Fuzzy sets as a basis for a theory of possibility. *Fuzzy Sets and Systems*, Vol. 1, pp. 3-28.

13

Modeling the Dynamics of Air Pollutants: Trans-Boundary Impacts in the Mexicali-Imperial Valley Border Region

Alberto Mendoza[1], Santosh Chandru[2], Yongtao Hu[2],
Ana Y. Vanoye[1] and Armistead G. Russell[2]
[1]*Tecnológico de Monterrey, Campus Monterrey,*
[2]*Georgia Institute of Technology,*
[1]*Mexico*
[2]*United States of America*

1. Introduction

Air pollution continues to be an increasing problem in the largest metropolitan areas and regional industrial and commercial corridors in the world. This is also the case in Mexico. Current air quality trends in Mexico indicate that major urban centers continue to exceed the Mexican Ambient Air Quality Standards (MAAQS) for ozone (O_3) and particulate matter with less than 10 microns of aerodynamic diameter (PM_{10}), while other cities are starting to show warning signs of future air quality problems (Zuk et al., 2007). $PM_{2.5}$ monitors are just starting to be deployed around the country, thus no extensive historical data is available on this pollutant.

Some of the urban centers of concern share a common airshed with twin cities across the international border with the United States of America (USA), bringing additional complexity to the study of air pollution dynamics in the region. In this sense, trans-boundary air pollution across USA and Mexico has become a rising problem due to increased commercial and industrial activities in the border region. Trans-boundary air pollution has been studied at different levels in different areas of the border region (Mukerjee, 2001). Two main areas can be identified as the ones that have drawn most of the attention. The first one is the Lower California Area: Tijuana/San Diego, Mexicali/Calexico-Imperial Valley (Figure 1). Here, most of the attention has been on primary PM (e.g., Osornio-Vargas et al., 1991; Chow et al., 2000; Sheya et al., 2000; Kelly et al., 2010), with some studies addressing secondary pollutants (e.g., Zielinska et al., 2001). The second area is the airshed formed by Ciudad Juarez-El Paso-Sunland Park. Perhaps, this area is the one that has received most of the attention regarding trans-boundary air pollution and in a more comprehensive fashion (Currey et al., 2005).

Two of the key steps to improving air quality in a region are identifying, quantitatively, the emissions from sources that affect the area, and assessing how those emissions evolve in the atmosphere to impact pollutant concentrations. Both are difficult, and both can be subject to uncertainties. Air quality modeling is key to both steps because it provides a means to do

Fig. 1. Location of the twin cities of Mexicali-Calexico and Tijuana-San Diego in the Mexico-US border region.

both in a consistent, supportable fashion (Russell & Dennis, 2000). Armed with such information, policy makers can then identify environmentally and economically effective strategies to improve air quality (McMurry et al., 2004).

As indicated, trans-boundary air pollution has been studied at different levels in different areas of the US-Mexico border region. However, limited modeling studies exist where comprehensive chemistry-transport air quality models (CTMs) have been applied at a regional level to understand trans-boundary air pollution in the Mexicali-Imperial Valley border region. In the present study we used a CTM to describe pollutant formation and transport around the Mexicali-Imperial Valley border area, as well as to estimate source contributions to O_3 and $PM_{2.5}$. Even though the principal attention in this study was on the Mexicali-Imperial Valley area, we also expanded our attention outside this area to track down pollutant transport from major urban centers and point sources outside it, but close enough to affect the air quality of the valley (e.g., Tijuana in Mexico, and San Diego and Los Angeles [LA] in the USA).

2. Past air pollution studies in the Mexicali-Imperial Valley Border Area

Tijuana-San Diego has been a border economical belt for a long time. However, over the last 15 years, Mexicali has been one of the fastest-growing cities in Mexico in terms of industrial development, job creation, and energy demand (Quintero-Núñez et al., 2006). This has resulted in that Mexicali on the Mexican side of the border is non-compliant with respect to CO, O_3 and PM_{10} MAAQS, as Calexico is in non-attainment for PM_{10}, $PM_{2.5}$, and O_3 USA standards.

Harmful contaminants in the Mexicali-Imperial Valley border region originate from a number of sources (Sweedler et al., 2003; Quintero-Núñez et al., 2006), including motor vehicles, farms, power plants (natural gas fired and geothermal), and factories. Light manufacturing operations, waste disposal sites, mining, and aggregate handling are also located near the border. In particular, poorly maintained vehicles contribute heavily to the levels of CO, NOx ($NO+NO_2$), and hydrocarbons (HCs) in the air; driving on unpaved roads contributes heavily to PM emissions. Burning of trash, tires, and other materials are also sources of PM, SO_2, and CO.

Several studies have been conducted to understand the composition, spatial variability, and sources of air pollution in the Mexicali-Imperial Valley region. Cerro Prieto, the largest geothermal plant in Latin America (720 MW) is located ~30 km to the south of downtown Mexicali. Since it started operations in the 1970's, H_2S emissions and transport from this facility to Mexicali and Imperial Valley has been a concern (Gudiksen et al., 1980; Deane, 1984). However, atmospheric conversion of H_2S to SO_2 was estimated as not significant. A major effort to understand PM_{10} pollution in Mexicali and Imperial Valley was undertaken in the early 1990's (Chow et al., 2000; Chow & Watson, 2001; Watson & Chow, 2001). This study demonstrated that PM_{10} in the region is mainly composed of crustal material (50% to 62% of the mass) and organic matter (over 25% of the mass). Receptor modeling gave evidence that pollution transport from LA to Mexicali and Calexico could be a concern. In addition, preliminary pollutant flux estimates indicated that the total PM_{10} flux from Mexico to the USA was about 1.5 times the total flux from the USA to Mexico. PM_{10} levels in Mexicali are consistently higher than in Imperial Valley, however wind patterns tend to be in a higher percentage from the north. Other studies have also given evidence of the potential transport of emissions originating in Mexicali and Imperial Valley to areas to the north like the Grand Canyon National Park (Eatough et al., 2001). However, these results have relied on the use of receptor models rather than comprehensive CTMs.

More recently, the fact that the Mexicali Valley and Imperial Valley continue to experience high air pollutants levels made it relevant to conduct an integrated study of the air quality problem in the region. Partial results of this integrated study have been published elsewhere, particularly on levels and chemical composition of fine PM (Mendoza et al., 2010), chemical speciation and source apportionment of VOCs (Mendoza et al., 2009), mobile source emissions characterization using a mobile laboratory (Zavala et al., 2009), and numerical experiments to address the meteorological patterns that foster air pollution episodes (Vanoye & Mendoza, 2009). Here we present our findings on the application of a regional three-dimensional comprehensive CTM to the Mexicali-Calexico border region to follow the dynamics of gas-phase and particulate-phase air pollutants. Particular emphasis is placed on the relevance of understanding trans-boundary air pollutants transport and the implications on emission control strategies on both sides of the border.

3. Description of the modeling system and its application

3.1 Modeling platform

Three-dimensional CTMs continue to be the most scientifically sound tool to assess how emissions from multiple sources impact air quality (Russell & Dennis, 2000). The modeling effort in this study consisted in the application of an advanced air quality and emissions modeling system to the border region to assess how particular sources impact O_3 and $PM_{2.5}$ levels. Specifically, we applied an extended version of the Models-3 suite, including the

Community Multiscale Air Quality model -CMAQ- (Byun & Ching, 1999) for air quality modeling, the PSU/NCAR 5th generation Mesoscale Meteorological model -MM5- (Grell et al., 1995) for meteorological modeling, and the Sparse Matrix Operator Kernel for Emissions model -SMOKE- (Houyoux & Vukovich, 1999) was used to process emissions.

3.2 Modeling domains

The modeling system was applied using nested grids (Figure 2). At the coarser scale, horizontally, 36 km × 36 km grid cells were used. This mother domain is the same as the one defined by the Regional Planning Organizations (RPOs) that oversee other major modeling efforts in North America (i.e., VISTAS, WRAP, CENRAP, LADCO, MANE-VU). At the horizontal mid-resolution level, we included a 12 km × 12 km grid (Figure 3). The coarse grid system allows relatively rapid simulation to set appropriate boundary conditions for the finer grid, serves to stabilize the meteorological and air quality model solutions, and to consider long-range transport from very particular sources. A 4 km × 4 km grid was specified in the Mexicali area for simulations that suggested that fine scale features existed and could not be accurately represented using the coarser grids (i.e., capture the dynamics at the urban scale in Mexicali-Calexico). In this work, we only present the results obtained with the 12 km x 12 km grid, which are ones than provide details on the mid-range pollutants transport in the border region of interest. Details on the extent of each modeling domain are presented in Table 1. The horizontal resolution (including grid nesting) was kept consistent between MM5, SMOKE and CMAQ.

3.3 Episodes selection

The adequate selection of modeling episodes constitutes a fundamental part of the modeling process. If representative episodes are not selected adequately, the modeling results might not characterize effectively the meteorological features that foster high pollution level episodes. Here we used Classification and Regression Tree (CART) Analysis (Breiman et al., 1998) as the formal statistical tool to select the modeling episodes of interest. In essence, CART is a recursive binary partition technique. It divides a set of observations in subgroups taking as reference the value of a particular variable defined by the user (e.g., maximum daily ozone concentration). Each partition in the decision tree is conducted to minimize the classification error of the decision variable. This technique has demonstrated its capacity to help in the selection of days with similar meteorological conditions that give rise to similar pollution levels, using a formal procedure and eliminating the effects of meteorological variability (Kenski, 2004).

CART was applied to obtain decision trees to classify daily maximum O_3, CO and PM_{10} concentrations (separately). The database used was composed of observations (chemical and meteorological) taken by air quality stations in the border region for the years 2001 and 2002. The purpose was to group days with similar O_3, CO, and PM_{10} levels and influenced by similar meteorological condition. The results obtained from CART application were compared against time series plots to corroborate that the episodes selected in fact represented a continuum of days with relatively high pollutant concentrations levels. One of the parameters that can be manipulated while applying the CART technique is the number of final bins that the decision tree will have. Typically, as the number of bin increases the error is reduced; however, if the number of bin increases the probability of having consecutive days in a bin decreases and thus it is harder to construct episodes for air quality

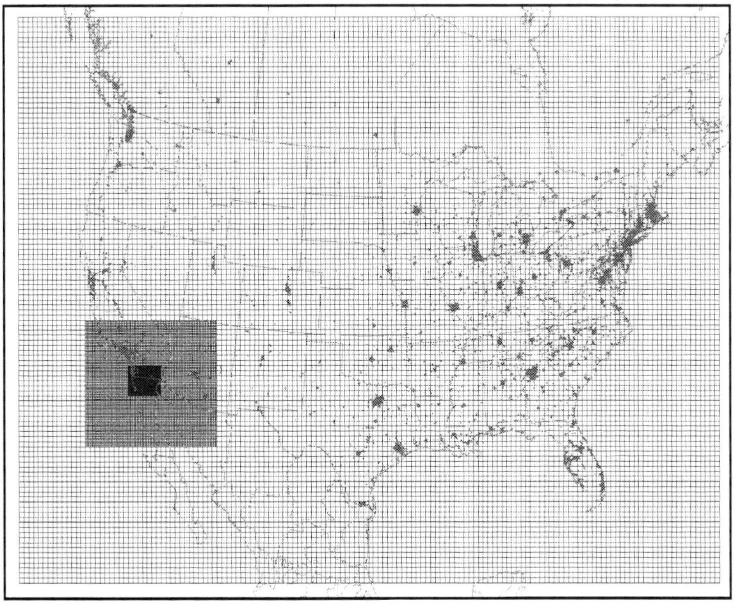

Fig. 2. Horizontal resolution of the nested modeling domain.

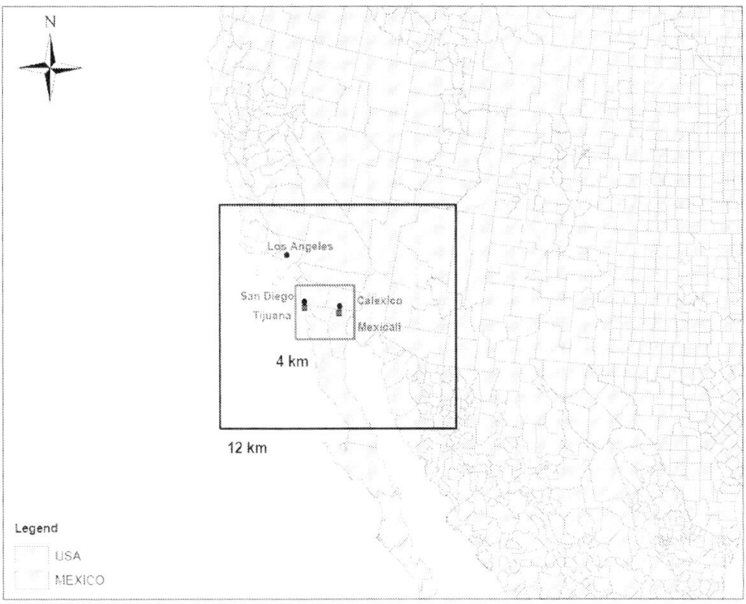

Fig. 3. Location of the 12 km and 4 km horizontal modeling domains.

Grid ID	Origin (*x*, *y*) in km	Horizontal domain (# columns, # rows)
USMEX36	(−2736.0, −2088.0)	(148,112)
USMEX12	(−2232.0, −1160.0)	(84,75)
USMEX4	(−1908.0, −756.0)	(63,54)

Table 1. Modeling domains specifications. Origin coordinates are based on a Lambert Conformal Conic projection with centre lat. and long. as 40 and −97 degrees, respectively

modeling purposes. A convenient number of consecutive days for a modeling episode is between 10 and 15, so with this in mind the number of bins was varied until decision trees with low classification errors and high number of consecutive days in the bins were obtained.

Based on the CART Analysis application, the following modeling episodes were defined: August 18-27, 2001 and July 17-25, 2001 for high O_3 events, and January 6-16, 2002 for high CO and PM events that are typical during autumn and winter times. Additional details on the episode selection process can be found elsewhere (Vanoye & Mendoza, 2009).

3.4 Emissions modeling

SMOKE is a computational engine used to generate the gridded emissions inventory, and its main purpose is to speciate and allocate spatially and temporally area and point emissions and to couple emission estimation tools for mobile and biogenic emissions to spatial and temporal allocation routines.

Base emissions inventory data for the USA side of the border were obtained from the 2001 US National Emissions Inventory (NEI) prepared for the Clean Air Interstate Rule (CAIR). Emissions for the Mexican side came from combining the 1999 BRAVO Mexican inventories (Pitchford et al., 2004) with the 1999 Six Border States Mexican inventory (MNEI) (ERG et al., 2004). Biogenic emissions for both sides of the border were prepared using BEIS3 (Vukovich & Pierce, 2002), and USA mobile emissions were prepared using MOBILE6 (US EPA, 2003). Mobile emissions for Mexico were directly obtained from BRAVO and MNEI.

The emissions inventory generated considers O_3 and PM precursors, as well as primary PM emissions and some toxics (particularly VOC species). The modeling episodes selected were not the same ones as the base years used to derive the raw emissions inventories used for the Mexican side of the border; thus, scaling was needed to update the emissions (e.i., MNEI base year is 1999 and modeling years for our applications were 2001 and 2002). This scaling was based primarily on population growth. VOCs speciation was conducted based on the chemical mechanism selected for the CTM application: SAPRC-99 chemistry (Carter, 2000). Spatial surrogate ratios used to allocate emissions on both sides of the border considered population, highways, total railroads, airport points, and marine ports.

As an example of the results obtained from the application of SMOKE, Figure 4 illustrates CO and biogenic isoprene emission inventories for the 12 km resolution domain. It can be seen, for example, that the CO emissions inventory contains the expected spatial structure (main roads are clearly shown and emissions follow general population patterns). Overall, mobile sources contribute to ~65% of the NOx and ~30% of the VOCs emitted in the LA area; area sources represent ~15% of the NOx and 25% of the VOCs emitted in LA. Values for San Diego are similar as the ones for LA. In contrast, 34% of the NOx and 13% of the VOCs emitted in Mexicali come from mobile sources; area sources (including non-road mobile sources) represent 37% of the NOx and 51% of the VOCs emitted in Mexicali. In Tijuana, mobile sources emit 61% of the NOx and 23 of the VOCs, whilst area sources emit 29% of the NOx and 60% of the VOCs. PM in the Mexicali-Calexico region comes from area sources that are dominated by wood-fuel combustion, agricultural burning, and paved and unpaved road dust.

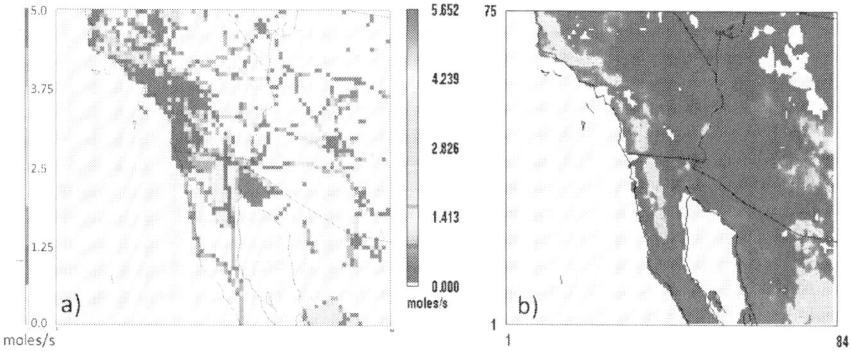

Fig. 4. Examples of (a) CO and (b) biogenic isoprene emissions allocated in the 12 km resolution modeling domain.

3.5 Meteorological modeling

MM5 (Grell et al., 1995) version 3.7 was the meteorological model used here to develop the fields needed to drive the CTM simulations and to provide meteorological information needed to estimate meteorological-variable emissions (e.g., biogenic emissions depend on solar radiation and temperature, while mobile emissions depend on temperature). MM5 is a non-hydrostatic mesoscale meteorological model with grid nesting and four-dimensional data assimilation capabilities. Here we briefly describe the model setup and the input data used to run the model. Additional details, including model performance statistics on the MM5 application, can be found elsewhere (Vanoye & Mendoza, 2009).

MM5 was run with 34 vertical layers with the top of the domain set at 70 mb; horizontal resolution was described earlier. Following a set of sensitivity tests, the MM5 parameterization configuration that gave the best statistical performance of the model for the July and August episodes is presented in Table 2. Of note, the Pleim-Xiu Land Surface model is the recommended scheme by the US Environmental Protection Agency (US EPA) and is the one that has demonstrated to give the best meteorological fields for CMAQ (Olerud & Sims, 2003; Morris et al, 2004). Another advantage of the Pleim-Xiu scheme is that it allows using CMAQ's dry deposition scheme which is technically superior to the conventional Wesley scheme.

MM5 was executed enabling its four-dimensional data assimilation capabilities for the 36 km and 12 km domains. One-way nesting was selected as the way MM5 transferred information from the outer grids to the inner grids. Finally, a relaxation scheme was chosen for the manipulation of the boundary conditions, i.e. the five outermost points are used to damp the information flowing from the boundaries to the inner domain.

Initial and boundary condition were prepared using the National Center for Atmospheric Research (NCAR)/National Centers for Environmental Prediction (NCEP) Eta analyses data. The data consist of regional meteorological analyses for North America based on the output of the Eta model, which generates data every 12 hours from observations of over 600 stations in the region. To complement this information and increase the effectiveness of the data assimilation step, additional observations with a temporal resolution of 6 hours were extracted from NCAR archives, through its Data Support Section of the Scientific Computing Division. This included observations from surface and marine stations, as well as from aerial soundings. Basic landuse, vegetation cover and topography was also obtained from NCAR. Landuse information was based on the USGS 24 categories, and topographic resolutions of 10 min, 5 min, 2 min, and 30 sec were used.

Parameter ID	Description	Selected option
IMPHYS	Explicity Moisture Scheme	Mix Phase
MPHYSTBL	Intrinsic Exponent for Calculating IMPHYS	Use Look-up table for moist physics
ICUPA	Cumulus Schemes	Grell
IBLTYP	Planetary Boundary Layer	Pleim-Xiu
FRAD	Radiation Cooling of Atmosphere	Rapid Radiative Transfer Model
ISOIL	Multilayer Soil Temperature Model	Pleim-Xiu Land Surface Model
ISHALLO	Shallow Convection Option	No Shallow Convection

Table 2. MM5 parameterization options that gave the best model performance for the simulation of meteorological conditions in the Mexicali-Imperial Valley border area.

3.6 CMAQ application

3.6.1 Base case simulations

CMAQ is an Eulerian photochemical model that simulates the emissions, transport, and chemical transformations of gases and PM in the troposphere (Byun & Ching, 1999). Similar to other photochemical models, CMAQ solves the species conservation equation:

$$\frac{\partial C_i}{\partial t} = -\nabla \cdot (\mathbf{u}C_i) + \nabla \cdot (\mathbf{K}\nabla C_i) + R_i + E_i \tag{1}$$

where, C_i is the concentration of species i, $\mathbf{u}$ is the wind field, $\mathbf{K}$ is the eddy diffusivity tensor, R_i is the net rate of generation of specie i, and E_i is the emission rate of species i. Meteorological parameters such as $\mathbf{u}$ and $\mathbf{K}$ in eq. 1, as well as temperature and humidity fields come from the MM5 application, while emission rates from SMOKE. CMAQ contains state-of-the-science descriptions of atmospheric processes and has a "one-atmosphere" approach for following the dynamics of gas-phase and particulate matter pollutants. The latter is an important characteristic to assess simultaneously O_3 and aerosols.

CMAQ, as MM5, allows for grid nesting. The horizontal grid structure used was described earlier. The vertical structure of all domains has 13 layers with its top at about 15.9 km above ground. Seven layers are below 1 km and the first layer thickness is set at 18 meters.

Initial and boundary conditions used for the mother domain were the same as the ones suggested by other RPO applications (Russell, 2008). Results from the simulations were compared to data from ground-based monitors for NO_2, O_3, CO, SO_2, PM_{10}, and $PM_{2.5}$. Observational data was obtained from the California Air Resources Board (CARB) for monitoring stations in the State of California, and from Mexican border municipalities of interest (i.e., Tijuana and Mexicali). Observational data were also obtained from US EPA's Air Quality Data system. In each episode, the first two days were considered ramp-up days and were not further used for additional analysis.

CMAQ has been used extensively to study air pollutant dynamics in the continental USA (e.g., Tagaris et al., 2007; Liao et al., 2007) and in particular regions of that country (e.g., Dennis et al., 2010; Ying & Krishnan, 2010), as well as in other countries around the world (e.g., Che et al., 2011; Im et al., 2011). Only one additional application using CMAQ as the CTM has looked at trans-boundary air dynamics in the USA-Mexico border using fine scale grid resolutions. Choi et al. (2006) looked at high PM events over the sister cities of Douglas, Arizona (USA) and Agua Prieta, Sonora (Mexico). In that application, model performance was acceptable, and it was concluded that secondary processes contributed marginally to the modeled PM events. Primary local sources dominated high PM events.

3.6.2 Sensitivity analysis

Sensitivity analysis is an important tool that can be used to understand the impacts of emissions from various sources on ambient air concentrations of specific pollutants. The ability to conduct sensitivity analyses in an efficient fashion is critical to obtain robust descriptions of the response fields of pollutant concentrations to changes in model inputs (particularly emissions), which then are used in source attribution analyses and control strategy design (e.g., Bergin et al., 2007). Among the different choices to estimate the sensitivity fields, the direct decoupled method for three-dimensional models (DDM-3D) has proven to be superior to other techniques (Yang et al., 1997; Hakami et al., 2003). DDM-3D is an implementation of the Decoupled, Direct Method (Dunker, 1984; Dunker et al., 2002) for sensitivity analysis. The version of CMAQ used in our applications was extended with DDM-3D (Cohan et al., 2005). The method directly calculates the response of model outputs (concentrations) to parameters and inputs, i.e., the semi-normalized sensitivities S_{ij}:

$$S_{ij} = \frac{\partial c_i}{\partial e_j} \tag{2}$$

where c_i is the concentration of species i and e_j is the relative perturbation on parameter j $-p_j-$ (e.g., NOx emissions) from its nominal value $p_j°$ (i.e., $e_i = p_j/p_j°$). This is an efficient approach for directly assessing the sensitivity of model results to various inputs and parameters, and replaces the need to use the traditional brute force approach of re-running a model after modifying a parameter. More importantly, it does not suffer from numerical noise problems that can overwhelm brute force approaches. In addition, it is a linear method. In prior studies, the atmospheric chemistry has been found to respond relatively linearly for emissions changes on the order of 25% or more (Dunker et al., 2002; Hakami et al., 2004).

Of particular interest was to explore the sensitivity due to variations on the emissions inventory. The implementation of DDM to CMAQ allows defining spatial- and source-specific emissions categories as the input being perturbed, and in a single model run the sensitivities of all species tracked by the model to changes in a set of emissions sources can be calculated. To accomplish this, the region of interest was divided into three different areas: Mexicali-Calexico (abbreviated as MXC), Tijuana-Tecate-San Diego (abbreviated as

TSD), and Los Angeles-Riverside-Orange-Ventura (abbreviated as LAR). O_3 sensitivities to area-, mobile-, and point-source emissions of NOx and VOC were calculated for each of the regions defined for both summer episodes. Additionally, $PM_{2.5}$ sensitivities to changes in the same source categories were calculated.

4. Results

4.1 Air quality model performance

Domain-wide episode performance statistics were determined to ascertain the confidence of the simulation results. Table 3 presents the average model performance for the 12 km domain in terms of Mean Bias Error (MBE), Root Mean Squared Error (RMSE), Mean Normalized Bias (MNB), and Mean Normalized Error (MNE). Established performance guidelines indicate that the model should have a MNB of ±5–15%, and a MNE 30–35% for O_3 (Tesche et al., 1990). Based on these guidelines, CMAQ performed well in predicting the observed O_3 concentrations. No guidelines exist for the rest of the gas-phase species, though the results are comparable to results obtained by others using different CTMs (e.g., Mendoza-Dominguez & Russell, 2001). Overall, the gas-phase species results indicate a tendency of the model to underestimate the pollutant concentrations. This is in line with the results obtained from mobile laboratory measurements that indicate an underestimation of the official emissions inventory for Mexicali (Zavala et al., 2009). PM proved to be more difficult to simulate correctly, which is a known setback of current CTMs (Russell, 2008).

		MBE	RMSE	MNB (%)	MNE (%)
August-01	O_3	−1.64E-03	1.60E-02	−0.21	19.7
	CO	−3.52E-01	6.56E-01	−18.6	63.2
	NOx	−1.25E-02	2.52E-02	−34.9	74.2
	SO_2	−1.40E-03	5.40E-03	−19.4	87.7
	$PM_{2.5}$	−6.76E+00	9.14E+00	−36.9	39.2
	PM_{10}	−3.00E+01	3.57E+01	−76.2	76.2
July-01	O_3	−9.19E-02	1.76E-01	−18.3	61.2
	CO	−1.43E+00	1.97E+00	−22.0	56.8
	NOx	−6.18E+00	7.58E+00	−31.0	60.0
	SO_2	−5.31E+00	6.53E+00	−29.2	60.2
	$PM_{2.5}$	−7.10E+00	8.72E+00	−33.3	63.0
	PM_{10}	−8.12E+00	9.97E+00	−35.3	59.5
January-02	O_3	2.86E-03	7.76E-03	7.0	14.1
	CO	−7.33E-01	1.33E+00	−26.5	73.1
	NOx	−3.99E-02	7.98E-02	−34.6	76.0
	SO_2	−1.10E-03	5.40E-03	−29.6	74.4
	$PM_{2.5}$	−2.54E+00	1.01E+01	-4.6	37.2
	PM_{10}	−2.53E+01	3.21E+01	−56.1	61.7

Table 3. Average performance metrics for the 12 km domain during the diferent episodes modeled. MBE and RMSE are in ppmv for gas-phase species and μg/m^3 for PM species.

Peak O_3 concentrations in Calexico during the August episode were observed at Ethel Street (MBE -4.0 ppbv, MNB -3.0%), and East Calexico (MBE 5.0 ppbv, MNB 12.0%) sites. Calexico and Mexicali, being adjacent to each other in the border region, experience similar O_3 concentrations. The inability to capture the minimums in Ethel Street and Calexico East sites can be attributed to the fact that both these locations are located close to roadways, hence experience strong O_3 sinks in the night time due to its reaction with NO which the model is unable to capture using the 12 km grid structure (Figure 6).

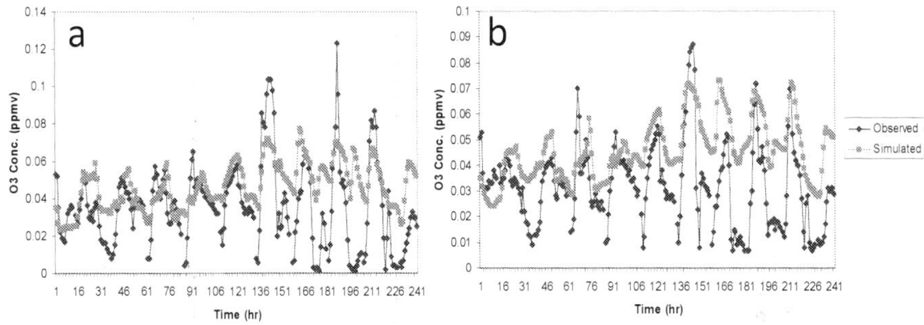

Fig. 6. Observed vs. simulated O_3 concentrations at representative sites in Calexico during August 2001 at (a) Ethel Street and (b) Calexico East site. Time scale represents hours of simulation.

Also, the LAR area was tested for model performance with respect to spatial as well as temporal variability (Figure 7). This region was chosen because of the high density of monitoring stations located in it, which can give a more realistic comparison with the modeled average hourly concentration values in the region. During the August episode, the peak O_3 concentrations occurred on August 26th, 2001: 189 ppbv at the Azusa site and 190 ppbv at the Glendora Laurel station, respectively. These sites are located in the San Gabriel Valley and come under the same 12 km grid cell. Simulated concentrations correlated well with the observed concentrations at Azusa (MBE 5.0 ppbv, MNB 12.2%) and Glendora Laurel (MBE 0.0 ppbv, MNB 3.5%) on most days. For the January episode, January 12 and 13, 2001 were the high concentration days with peaks of ~60 ppbv in the Los Angeles area, and ~75 ppbv in the Mexicali area.

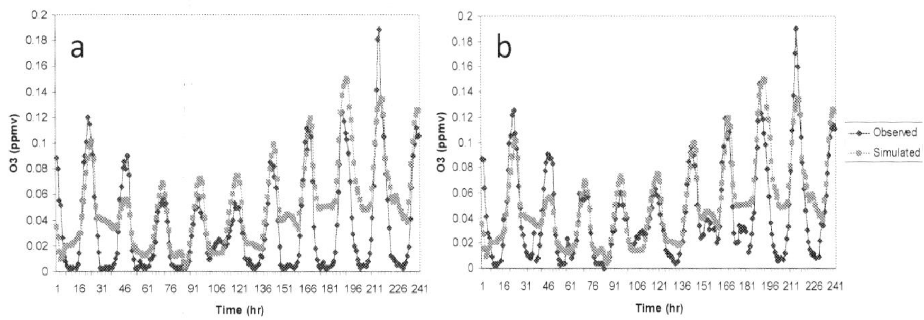

Fig. 7. Observed vs. simulated O_3 concentrations at representative sites in LA, during August 2001 at (a) Azusa, (b) Glendora Laurel. Time scale represents hours of simulation.

4.2 Modeled pollutant concentration fields

Resulting O_3 fields for the July and August episodes illustrate the influence of regional transport across the domain. During the July 2001 episode, a peak of 125 ppbv O_3 was simulated in the LA area on July 23, 23:00 hrs UTC (Figure 8d). The plume from LA can be seen transported towards the east (Figure 8 a-c). Plumes of up to 78 ppbv O_3 emerge from San Diego-Tijuana and travel eastwards and reach the Mexicali-Calexico region (Figure 8 a-c). Peak $PM_{2.5}$ concentrations of over 50 $\mu g/m^3$ were simulated in the LA area on July 15th, while $PM_{2.5}$ concentrations did not exceed 15 $\mu g/m^3$ in Tijuana-San Diego and Mexicali-Calexico during the July episode.

On August 24th (20:00 hrs UTC), strong O_3 plumes started to develop, and plumes from Mexicali-Calexico and Los Angeles almost converged (Figure 9a). At the same time plumes from Tijuana-San Diego build up as well and move eastwards towards Mexicali-Calexico (Figure 9b). Similar patterns start to emerge on August 25th (19:00 hrs UTC) (Figure 9c); peaks reach 162 ppbv in the LA area on August 26th (00:00 hrs UTC) (Figure 9d). A peak concentration of 162 ppbv is reached about 30 km northwestfrom the Glendora Laurel site where a simulated peak of 144 ppbv is reported. Similar to the July episode, O_3 plumes from San Diego-Tijuana border area are transported eastward towards Mexicali-Calexico during the August episode as well. Peak $PM_{2.5}$ concentration of 100 $\mu g/m^3$ are seen close to the LA area on August 25th. In the Mexicali-Calexico region, Mexicali showed a peak of 42 $\mu g/m^3$ on August 25th (14:00 hrs UTC).

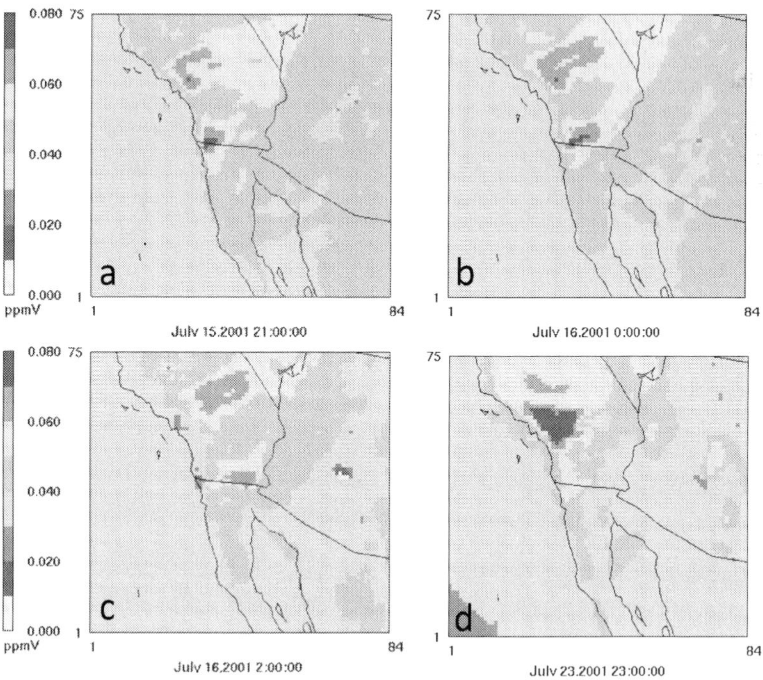

Fig. 8. Regional dynamics of O_3 plumes during the July 2001 episode (see text for details).

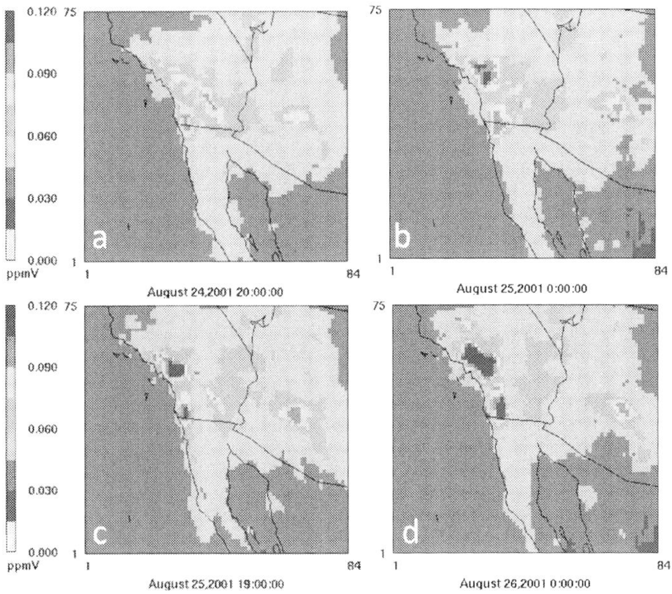

Fig. 9. Regional dynamics of O_3 plumes during the August 2001 episode (see text for details).

In addition to O_3, the dynamics of other primary (e.g., CO, NO_2, and SO_2) and secondary PM species (e.g., sulfate in fine PM) are also of interest when analyzing the output data obtained from CMAQ (Figure 10). For the August episode, CO concentrations peak around 7 PM (PDT), with LA showing the highest concentration, followed by San Diego-Tijuana. CO levels in Mexicali-Calexico are lower and more localized. In general, NO_2 distribution is very similar to that of CO, highlighting the importance of mobile source emissions. SO_2 emissions are highest in the Tijuana region. Thus, with the wind blowing in the northeast direction during the morning hours, much of the SO_2 is transported inland into the San Diego region. Consequently, the sulfate aerosols have a high regional effect encompassing the whole of San Diego region, and also showing its effect on Imperial Valley and Mexicali during late evening hours.

As PM concentrations are a major concern during the winter season, we limit our discussion of the January 2002 episode to $PM_{2.5}$. A peak of 188 $\mu g/m^3$ was simulated on January 12, 2002 (18:00 hrs UTC) near LA. The movement of regional $PM_{2.5}$ plumes is represented in Figure 11. Plumes from San Diego-Tijuana, LA and Las Vegas move towards the Mexicali-Calexico region with impacts of 10 to 35 $\mu g/m^3$. $PM_{2.5}$ originated in the USA and transported to Mexicali-Calexico, along with local fresh emissions, is carried further southeast inside Mexico. Mexicali-Calexico shows peak $PM_{2.5}$ concentration of 50 $\mu g/m^3$. Primary organic mass was the main contributor to fine PM in LA (98 $\mu g/m^3$). The maximum contribution from primary organic matter to the fine PM in Mexicali-Calexico was 10 $\mu g/m^3$. Peak soil dust concentrations of 40 $\mu g/m^3$ were found in Pheonix and Las Vegas areas. The soil dust contributions from LAR, TSD and MXC range between 5-25 $\mu g/m^3$ (Figure 11).

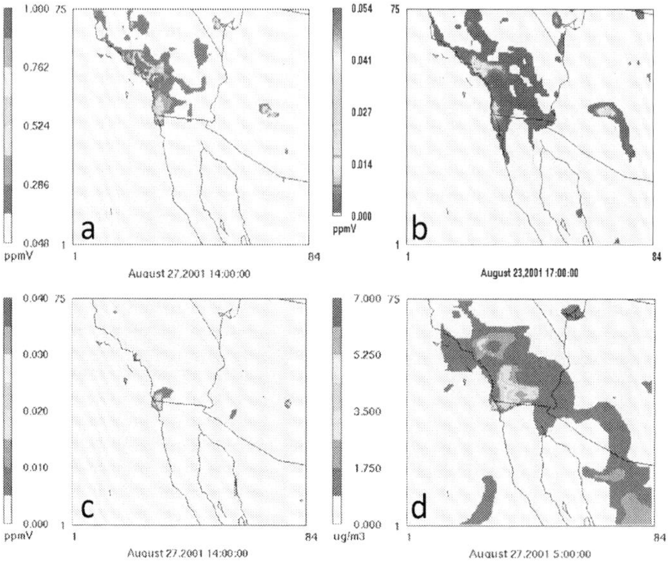

Fig. 10. Concentration fields for gas-phase and aerosol species (August 27, 2001): a) CO, b) NO_2, c) SO_2, d) sulfate $PM_{2.5}$.

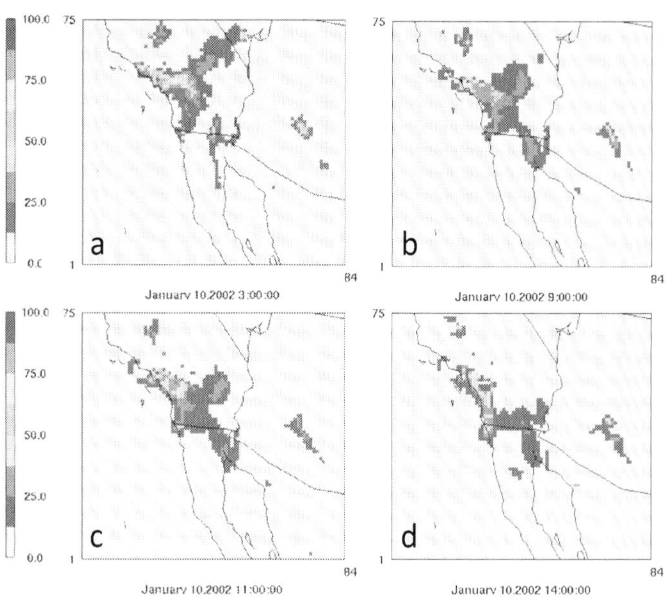

Fig. 11. Dynamics of $PM_{2.5}$ plumes during the January 2002 episode (see text for details).

4.3 Source contribution

4.3.1 Source contribution to specific NOx and VOC emission sources: August episode

To understand source contributions in the modeling domain, sensitivity fields were estimated using CMAQ/DDM. Results are presented only for the August 2001 episode; values for the July 2001 episode were similar. First, the sensitivity of the regional O_3 field to changes in NOx or VOC emissions from specific sources is presented. The figures presented are "response surfaces" and are interpreted as the amount of increment in pollutant concentration per 10% increase in emissions (or amount of reduction per 10% decrease in emissions) from certain source. The sensitivity coefficients are linear (first order) in nature and thus can be used in the manner described. In general, is reasonable to imply a linear response over a range of emissions perturbations (±30%) even for species that it is well known their non-linear response in the atmosphere (e.g., O_3; Hakami et al., 2003).
The impact of NOx emissions from MXC was the highest on August 26, with sensitivity response reaching 9 ppbv of O_3 per 10% change in the emissions (Figure 12). The area of influence of NOx emissions extends northwards into Imperial Valley, and partially into Arizona. The results indicate that down-wind the atmosphere is not NOx-inhibited; that is, an increase in NOx does not give as a response a decrease in O_3 down-wind as has been the case in other areas of the Mexico-US border (Mendoza-Dominguez & Russell, 2001). Change in VOC emissions from sources in MXC produce a smaller change -localized- in O_3 concentrations (maximum of 3 ppbv per 10% change), indicating the benefits of NOx control over VOC control in the region.

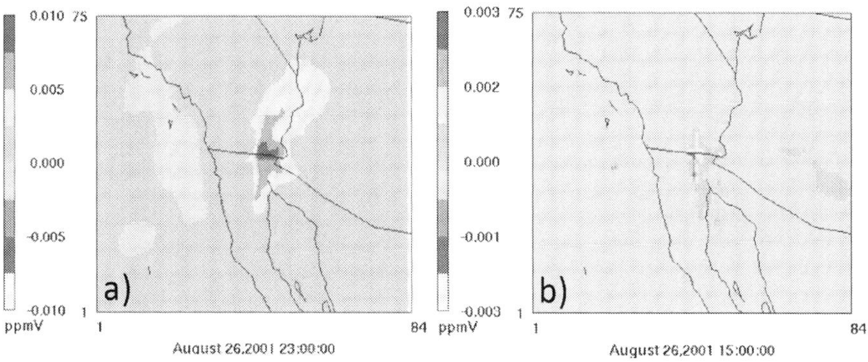

Fig. 12. Maximum sensitivity of O_3 to (a) NOx emissions and to (b) VOC emissions from the MXC region during the August 2001 episode.

The influence of emissions from other geographic locations was also tested. Figure 13a illustrates the response of O_3 to changes in NOx emissions from mobile sources located in the TSD area. The highest impact is almost 17 ppbv, occurring near San Diego, with a strong influence in the MXC region as well. This result indicates that emission controls implemented in San Diego (or increment in emissions) will impact the MXC area accordingly. Of interest is also the small NOx-inhibited region located down-wind of San Diego, toward the Tijuana border, that implies that a decrease in mobile emissions will result in an increase in O_3 concentrations. In contrast, and as expected, impact from VOCs emitted by mobile sources located in the TSD area is limited to less than 2 ppbv per 10% change in emissions (Figure 13b). The spatial extent of influence is also more limited than the sensitivity to NOx emissions, influencing the northern region of Imperial Valley County.

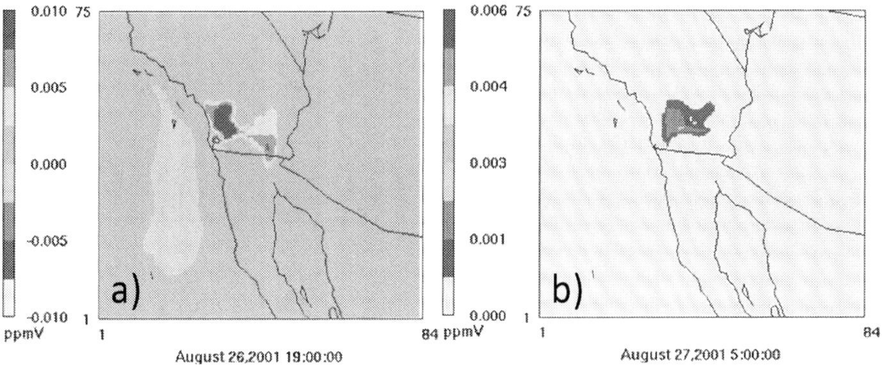

Fig. 13. Maximum sensitivity of O_3 to (a) mobile NOx emissions and to (b) mobile VOC emissions from the TSD region during the August 2001 episode.

Finally, example sensitivity values due to changes in NOx emissions from mobile and area sources from the LAR region are presented (Figure 14). For the case of mobile sources, the increment of NOx emissions results in a decrease in ozone (~7.5 ppbv per 10% increase in emissions) in downtown LA, with a corresponding increase (~30 ppbv) in neighboring counties of Ventura, Orange, and Riverside. From the extent of the sensitivity field, it is possible that under the right meteorological conditions, the influence can reach the Imperial Valley area. On the other hand, the sensitivity to area source NOx is smaller in value and extent because area emissions are smaller than the mobile emissions in Southern California.

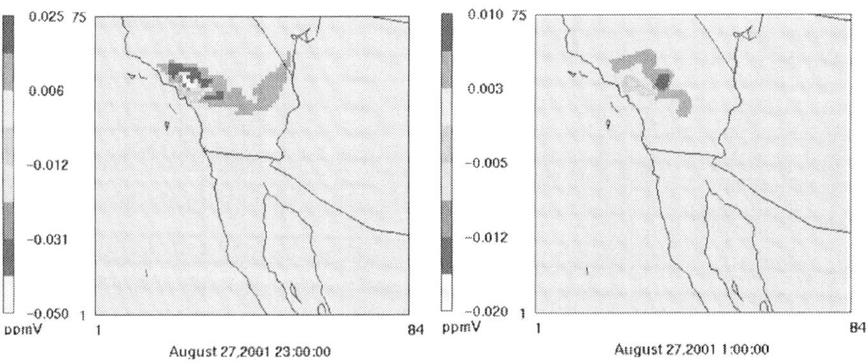

Fig. 14. Maximum sensitivity of O_3 to (a) mobile NOx emissions and (b) area NOx emissions located in the LAR area during the August 2001 episode.

4.3.2 Source contribution to overall emission sources: August episode

When considering the overall emissions from mobile sources, the LAR area made an overall contribution of 44 ppbv on the surrounding region i.e., east of the city of LA towards Glendora Laurel and Azuza on August 26 (00:00 hrs UTC) (Figure 15c). Presence of high concentrations of NOx results in negative sensitivities up to -46 ppbv in urban LA (Figure 15a). As seen in the base case simulations where O_3 plumes from LAR, MXC and TSD formed a triangle over southern California, the O_3 sensitivity fields extends towards MXC with increments of up to 10 ppbv (Figure 15a). Due to the northeasterly direction of the winds, plumes also reach the Grand Canyon National Park area, again with increments of about 10 ppbv (Figure 15d). LAR area sources contribute up to 8 ppbv of O_3 in the Riverside area.

Mobile traffic passing through Mexicali's border crossings is of concern. However, the overall mobile contribution to O_3 is found to be small in the simulation results. The impact from Mexicali vehicles alone is very small, with a peak impact of only 1.3 ppbv over Calexico and Mexicali (Figure 16a). Possible emission inventory underestimates can be a potential reason for low simulated impacts, and should be further explored.

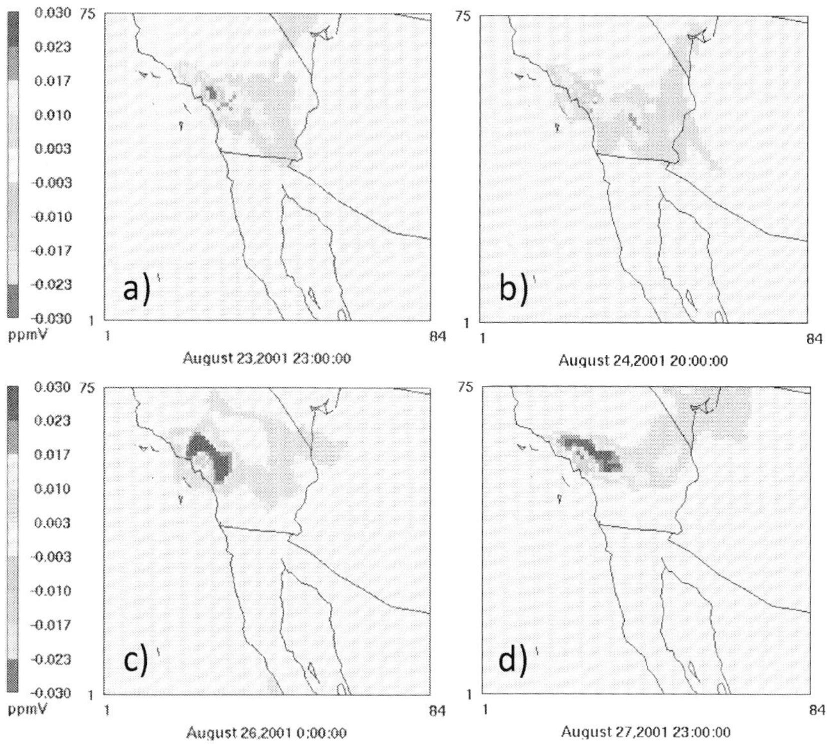

Fig. 15. O_3 sensitivity to LA mobile source emissions (see text for details).

The maximum impact from Calexico mobile sources is 2 ppbv of O_3 seen over the Calexico region itself, and the border between California and Arizona (Figure 16b). The primary areas of mobile emissions are the two border crossing areas (seen in blue as negative sensitivities). Area sources in MXC contribute a simulated maximum of 8 ppbv O_3 during the summer episode (Figure 17a). The area of influence can be seen encompassing California, and the border regions of California-Arizona. O_3 impacts up to 4 ppbv in the Grand Canyon area can be attributed to area sources in the Mexicali-Calexico region (Figure 17b).

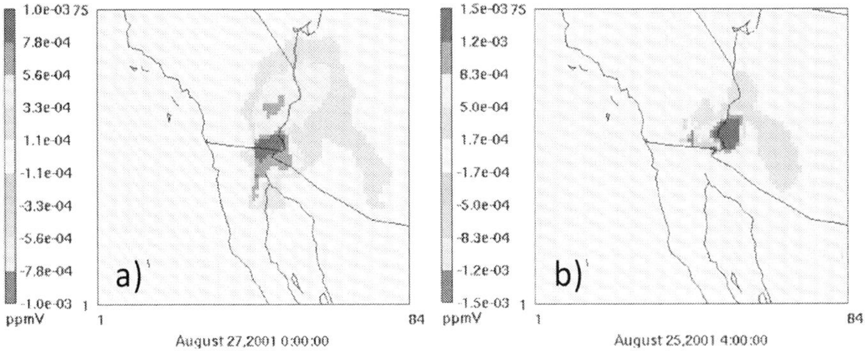

Fig. 16. Contribution to O_3 concentrations from (a) Mexicali and (b) Calexico mobile sources.

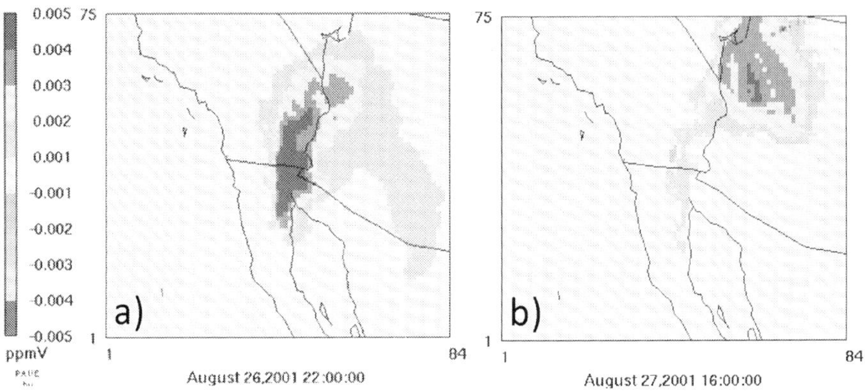

Fig. 17. Contribution to O_3 concentrations from MXC area sources (see text for details).

Area sources from TSD have a peak impact of 40 ppbv of O_3 over the San Diego area and this plume is carried eastwards into the USA close to the border region. The contribution of Tijuana emissions extends to the southeast into inner Baja California and impacting up to 20 ppbv of O_3 (Figure 18a,b). Also, on August 26th, the sensitivity field of O_3 from the TSD region extends eastwards towards Calexico, thus adding O_3 to the already polluted air in Calexico and Mexicali (Figure 18c,d).

Tijuana mobile source impacts reach up to 6 ppbv on both sides of the border depending on the wind direction (Figure 19b,c). Since the dominant wind pattern is towards the northeast,

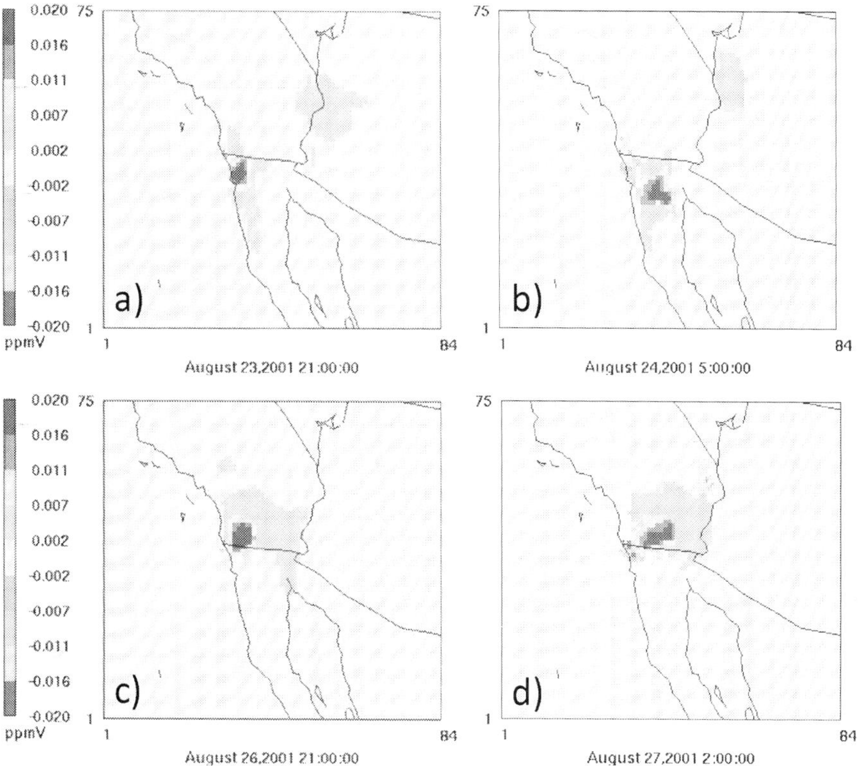

Fig. 18. O_3 sensitivity to TSD area sources during the August 2001 episode (see text for details).

O_3 is transported through the California-Baja California border towards Calexico (Figure 19a,c). Tijuana mobile sources impacts up to 3 ppbv of O_3 in Mexicali-Calexico (Figure 19a). It is also of interest the areas of negative sensitivity observed in downtown Tijuana of more than 3.0 ppbv.

Mobile sources from San Diego contribute up to 26 ppbv of O_3 in the region itself, and also over the park areas such as Anza Borrego Desert State Park located southeast of San Diego (Fig. 20a). The base case scenario showed O_3 plumes from TSD area transported to Mexicali-Calexico. A contribution of up to 11 ppbv of O_3 in Mexicali-Calexico can be attributed to the high density of vehicles in and around the San Diego region (Fig. 20b). This contribution is higher than the contribution from MXC mobile sources.

The peak $PM_{2.5}$ concentration simulated over the MXC region was 42 $\mu g/m^3$. Of this, MXC area sources contributed to 21 $\mu g/m^3$ of primary $PM_{2.5}$. Thus, 50% of the $PM_{2.5}$ levels in MXC can be attributed directly to MXC area sources during August 2001. $PM_{2.5}$ contribution from MXC mobile sources was very small, with peak contributions less than 1 $\mu g/m^3$. MXC point sources contributed the remaining share of up to 7 $\mu g/m^3$. Simulations found similar results for TSD with contributions of up to 33 $\mu g/m^3$ of $PM_{2.5}$ from TSD area sources, less than 2 $\mu g/m^3$ from mobile sources, while the point sources in the region contributed up to 13 $\mu g/m^3$.

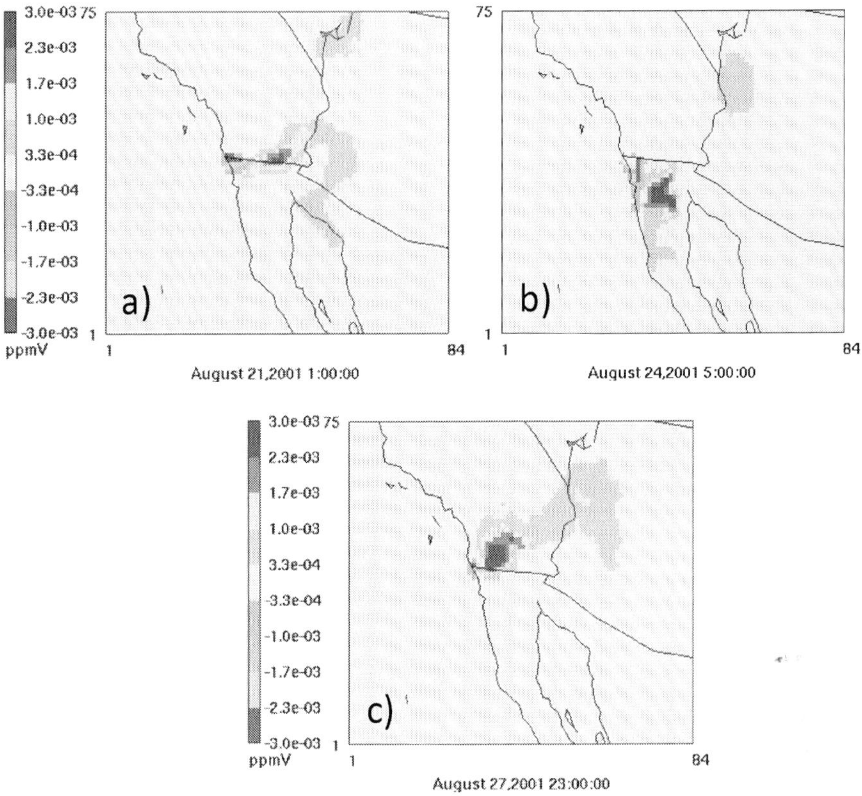

Fig. 19. O_3 sensitivity to Tijuana mobile sources during the August 2001 episode (see text for details).

4.3.3 Source contribution during the winter episode

Contributions to O_3 from various sources in the region were simulated for the winter episode. Impact of LAR mobile sources of up to 14 ppbv was seen over the Pacific Ocean. High contributions were also observed along the coast from Los Angeles to San Diego, which represents a major travel road route (Figure 21a). MXC area sources had simulated impacts of up to 6 ppbv over the southern regions of Baja California.

However, much of the time fresh NOx emissions led to decreases (negative sensitivities) over the local urban areas and positive impacts downwind. Peak impacts of 11 ppbv of O_3 were simulated over the Los Angeles area during the winter episode which originate from TSD area sources (Figure 21b). This same changes originated O_3 reductions of more than 7 ppbv over TSD.

Similar values as that in summer episode were simulated with peak impacts of up 2 ppbv O_3 on the Baja California region from Mexicali mobile emissions. Tijuana, San Diego and Calexico mobile sources contribute to less than 6 ppbv O_3 during the winter episode.

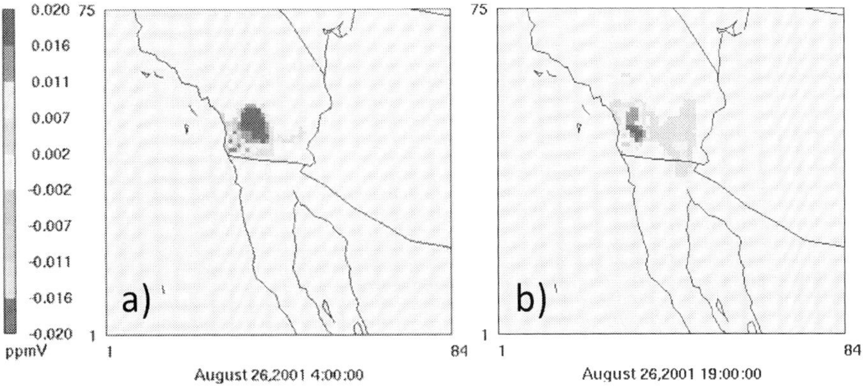

Fig. 20. O_3 sensitivity to San Diego mobile sources during the August 2001 episode (see text for details).

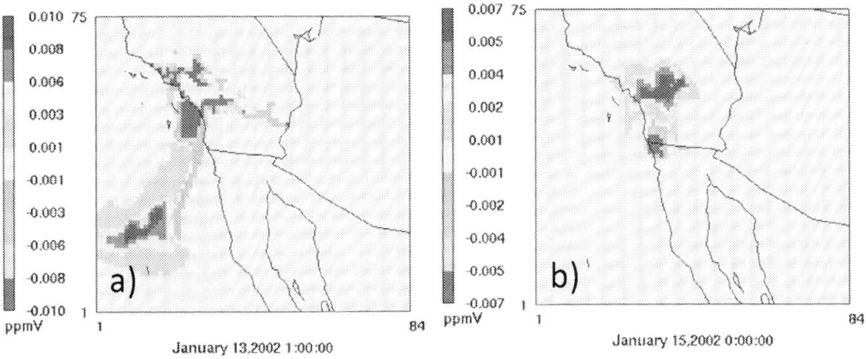

Fig. 21. Peak O_3 sensitivity to (a) LAR mobile sources and (b) to TSD area sources, during the January 2002 episode.

MXC area sources contribute to a simulated $PM_{2.5}$ maximum of 34 μg/m³ (Figure 22a). The pattern is very localized. Primary $PM_{2.5}$ emissions from MXC mobile sources contribute negligibly with peak contributions of 0.5 μg/m³. MXC point sources, primarily present in Mexicali contributed to a maximum of 12 μg/m³ over the border region. Area sources in TSD had very large contributions, ranging up to 52 μg/m³ (Figure 22b). However, once again the extent of the sensitivity field is constrained to the vecinitiy of the cities of Tijuana and San Diego. TSD mobile sources contributed to less than 3 μg/m³ of primary $PM_{2.5}$. Point sources in San Diego contributed to a maximum 13 μg/m³ of primary $PM_{2.5}$ in the region.

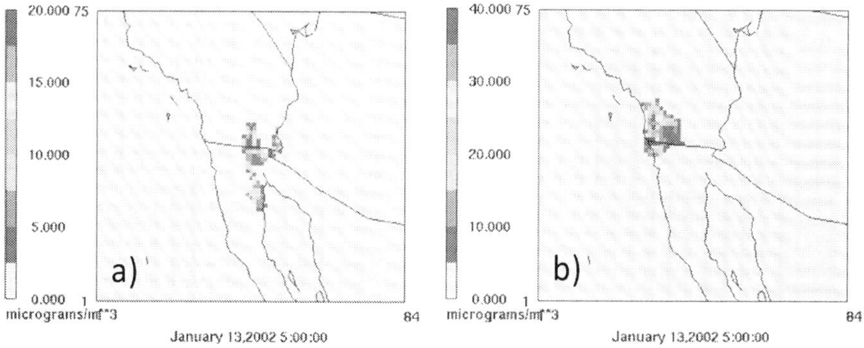

Fig. 22. Peak contribution to $PM_{2.5}$ from (a) MXC primary $PM_{2.5}$ area sources and (b) TSD primary $PM_{2.5}$ area sources, during the January 2002 episode.

5. Conclusion

Results suggest relevant information on trans-boundary impacts of air pollutants in the Mexicali-Imperial Valley border area. Simulated O_3 and $PM_{2.5}$ concentrations in the domain were the highest in the LA area, as expected. However, limited contribution of sources in the LAR area to O_3 and $PM_{2.5}$ levels in the border region was observed. Mobile sources, the most abundant sources in the LAR area contributed up to 10 ppbv of O_3 in MXC, but meteorological events that favored the transport of pollutants from LAR to MXC were few compared to prevailing conditions that favored transport to the east and northeast of LAR during the summer episodes or the southwest during the winter episode. Emissions from the TSD region play a much more important role in the air quality of the MXC area, particularly on the levels of O_3 during the summer episodes. Again, mobile sources contributed the most to the observed impacts from TSD to MXC. Even more, MXC O_3 levels were more sensitive to NOx changes in TSD mobile emissions than VOC changes in that same source. Even though, mobile sources are of concern in the MXC area, O_3 impacts from precursors emitted within the region were small. Area sources in MXC contributed the most: up to a maximum of 8 ppbv of O_3 during the summer episodes. O_3 plumes reached the border regions of California-Arizona and O_3 concentrations up to 4 ppbv in the Grand Canyon area can be attributed to area sources in the MXC region. The MXC region is more sensitive to NOx controls than to VOCs controls. In regards to $PM_{2.5}$, about 50% of the $PM_{2.5}$ in MXC during the summer episode can be attributed directly to area sources. During the winter episode, plumes from TSD, LAR and Las Vegas unite and move towards the MXC region with impacts of 10-35 $\mu g/m^3$. Soil dust contribution from LAR, TSD and MXC ranges between 5-25 $\mu g/m^3$. MXC area sources contribute a maximum of 34 $\mu g/m^3$ $PM_{2.5}$.

6. Acknowledgments

This study was supported by LASPAU: Academic and Professional Programs for the Americas, under its Border Ozone Reduction and Air Quality Improvement Program. Additional support was obtained from Tecnológico de Monterrey through grant CAT-186.

7. References

Bergin, M.S.; Shih, J.S.; Krupnick, A.J.; Boylan, J.W.; Wilkinson, J.G.; Odman, M.T. & Russell, A.G. (2007). Regional air quality: Local and interstate impacts of NOx and SO_2 emissions on ozone and fine particulate matter in the Eastern United States. *Environ. Sci. Technol.*, 41(13), 4677-4689.

Breiman, L., Friedman, J.; Olshen, R. & Stone, C. (1998) Classification and regression trees. Chapman & Hall/CRC, ISBN 0-412-04841-8, Boca Raton, FL.

Byun, D.W. & Ching, J.K.S. (1999). *Science Algorithms of the EPA Models 3 Community Multiscale Air Quality (CMAQ) Modeling System.* EPA/600/R-99/030, US EPA, Washington DC.

Carter, W.P.L. (2000). *Documentation of the SAPRC99 chemical mechanism for VOC reactivity assessment.* Center for Environmental Research and Technology, University of California, Riverside, CA.

Che, W.; Zheng, J.; Wang, S.; Zhong, L. & Lau, A. (2011). Assessment of motor vehicle emission control policies using Model-3/CMAQ model for the Pearl River Delta region, China. *Atmos. Environ.*, 45, 1740-1751.

Choi, Y.-J.; Hyde, P. & Fernando, H.S.J. (2006). Modeling of episodic particulate matter events using a 3-D air quality model with fine grid: Applications to a pair of cities in the US/Mexico border. *Atmos. Environ.*, 40, 5181-5201.

Chow, J. C.; Watson, J. G.; Green, M. C.; Lowenthal, D. H.; Bates, B.; Oslund, W. & Torres, G. (2000). Cross-border transport and spatial variability of suspended particles in Mexicali and California's Imperial Valley. *Atmos. Environ.*, 34, 1833-1843.

Chow, J.C. & Watson, J.G. (2001). Zones of representation for PM_{10} measurements along the US-Mexico border. *The Sci. of the Total Environ.*, 276, 49-68.

Cohan, D. S.; Hakami, A.; Hu, Y. T. & Russell, A. G. (2005). Nonlinear response of ozone to emissions: Source apportionment and sensitivity analysis. *Environ. Sci. Technol.*, 39, 6739-6748.

Currey, R.C.; Kelly, K.E.; Meuzelaar, H.L.C. & Sarofim, A.F. Eds. (2005). *The U.S.-Mexican Border Environment: Integrated Approach to Defining Particulate Matter in the Paso del Norte Region.* SCERP Monograph Series, No. 12, San Diego State University Press, ISBN 0-925613-47-9, San Diego, CA.

Deane, M. (1984). Epidemiological monitoring plan for geothermal developments. *Sci. Total Environ.*, 32(3), 303-320.

Dennis, R.L.; Mathur, R.; Pleim, J.E. & Walker, J.T. (2010). Fate of ammonia emissions at the local to regional scale as simulated by the Community Multiscale Air Quality model. *Atmos. Pollution Res.*, 1, 207-214.

Dunker, A.M. (1984). The Decoupled Direct Method for Calculating Sensitivity Coefficients in Chemical-Kinetics. *Journal of Chemical Physics*, 81, 2385-2393.

Dunker, A.M.; Yarwood, G.; Ortmann, J.P. & Wilson, G.M. (2002). Comparison of source apportionment and source sensitivity of ozone in a three-dimensional air quality model. *Environ. Sci. Technol.*, 36, 2953-2964.

Eatough, D. J.; Green, M.; Moran, W. & Farber, R. (2001). Potential particulate impacts at the Grand Canyon from northwestern Mexico. *Sci. Total Environ.*, 276, 69-82.

ERG (Eastern Research Group), Acosta y Asociados & TransEngineering. (2004). Mexico National Emissions Inventory, 1999: Six Northern States, Final Report prepared for SEMARNAT, INE, US EPA, WRAP and CEC; Report #3393-00-011-002.

Grell, G.; Dudhia, J. & Stauffer, D.R. (1994). *A description of the Fifth-Generation Penn State/NCAR Mesoscale Model (MM5)*. NCAR Technical Note, NCAR/TN-398+STR.

Gudiksen, P.H.; Ermak, D.L.; Lamson, K.C.; Axelrod,M. C. & Nyholm, R.A. (1980). Potential air quality impact of geothermal power production in the Imperial Valley. *Atmos. Environ.*, 14(11), 1321-1330.

Hakami, A.; Odman, M. T. & Russell, A. G. (2003). High-order, direct sensitivity analysis of multidimensional air quality models. *Environ. Sci. Technol.*, 37, 2442-2452.

Hakami, A.; Odman, M.T. & Russell A.G. (2004). Nonlinearity in atmospheric response: A direct sensitivity analysis approach. *J. Geophys. Res.*, 109 (D15), D15303.

Houyoux, M.R. & Vukovich, J.M. (1999) Updates to the Sparse Matrix Operator Kernel Emissions (SMOKE) Modeling System and Integration with Models-3. *Conference The Emission Inventory: Regional Strategies for the Future,* Air & Waste Manage. Assoc., Raleigh, NC.

Im, U.; Poupkou, A.; Incecik, S.; Markakis, K.; Kindap, T.; Unal, A.; Melas, D.; Yenigun, O.; Topcu, S.; Odman, M.T.; Tayanc, M. & Guler, M. (2011). The impact of anthropogenic and biogenic emissions on surface ozone concentrations in Istanbul. *Sci. Total Environ.*, 409, 1255-1265.

Kelly, K.E.; Jaramillo, I.C.; Quintero-Núñez, M.; Wagner, D.A.; Collins, K.; Meuzelaar, H.L.C. & Lighty, J.S. (2010). Low-Wind/High Particulate Matter Episodes in the Calexico/Mexicali Region. *J. Air & Waste Manage. Assoc.*, 60, 1476-1486.

Kenski, D.M. (2004). CART Analysis of Historic Ozone Episodes. In *Proceedings of the AWMA 97th Annual Conference and Exhibition*, Indianapolis, USA.

Liao, K.-J.; Tagaris, E.; Manomaiphiboon, K.; Napelenok, S.L.; Woo, J.-H.; He, S.; Amar, P. & Russell, A.G. (2007). Sensitivities of Ozone and Fine Particulate Matter Formation to Emissions under the Impact of Potential Future Climate Change. *Environ. Sci. Technol.*, 41, 8355-8361.

McMurry, P.; Sheperd, M. & Vickery, J. (Eds.) (2004) *Particulate Matter Science for Policy Makers: A NARSTO Assessment*, Cambridge University Press, ISBN 0-521-84287-5, New York, USA.

Mendoza-Dominguez, A.; Wilkinson, J. G.; Yang, Y.-J. & Russell, A. G. (2000). Modeling and Direct Sensitivity Analysis of Biogenic Emissions Impacts on Regional Ozone Formation in the Mexico-U.S. Border Area. *J. Air & Waste Manage. Assoc.*, 50, 21-31.

Mendoza, A.; Gutiérrez, A.A. & Pardo, E.I. (2009). Volatile organic compounds in the downtown area of Mexicali, Mexico during the spring of 2005: analysis of ambient data and source-receptor modeling. *Atmósfera*, 22, 195-217.

Mendoza, A.; Pardo, E.I. & Gutierrez, A.A. (2010). Chemical Characterization and Preliminary Source Contribution of Fine Particulate Matter in the Mexicali/Imperial Valley Border Area. *J. Air & Waste Manage. Assoc.*, 60, 258-270.

Morris, R.E.; Koo, B.; Lau, S.; Tesche, T.W.; McNally, D.; Loomis, C.; Stella, G.; Tonnesen, G. & Wang, Z. (2004). *VISTAS Emissions and Air Quality Modeling - Task 4: Model Performance Evaluation and Model Sensitivity Tests for Three Phase I Episodes*. Report prepared by ENVIRON International Corporation, Novato, California.

Mukerjee, S. (2001). Selected air quality trends and recent air pollution investigations in the US-Mexico border region. *Sci. Total Environ.*, 276, 1-18.

Olerud, D. & Sims, A. (2003). *MM5 sensitivity modeling in support of VISTAS (Visibility Improvement – State and Tribal Association*. Report prepared for the VISTAS Technical Analysis Workgroup by Baron Advanced Meteorological Systems, LLC.

Osornio-Vargas, A.R.; Hernandez-Rodriguez, N.A.; Yanez-Buruel, A.G.; Ussler, W.; Overby, L.H.; Brody, A.R. (1991) Lung cell toxicity experimentally induced by a mixed dust from Mexicali, Baja California, Mexico. *Environ. Res.*, 56(1), 31-47.

Pitchford, M.L.; Tombach, I., Barna, M.; Gebhart, K.A.; Green, M.C.; Knipping, E.; Kumar, N.; Malm, W.C.; Pun, B.; Schichtel, B.A.; Seigneur, C. (2004). *Big Bend Regional Aesorol and Visibility Observational Study (BRAVO) Final Report*, September 2004.

Quintero-Núñez, M.; Reyna, M.A.; Collins, K.; Guzmán, S.; Powers, B. & Mendoza, A. (2006). Issues Related to Air Quality and Health in the California-Baja California Border Region, In *The U.S. Mexican border environment: Binational Air Quality Management*, SCERP Monograph Series No. 14, R. Pumfrey (Ed.), 1-46, San Diego State University Press, ISBN: 0-925613-50-9, San Diego, California, USA.

Russell, A. & Dennis, R. (2000). NARSTO critical review of photochemical models and modeling. *Atmos. Environ.*, 34, 2283-2324.

Russell, A.G. (2008). EPA Supersites Program-Related Emissions-Based Particulate Matter Modeling: Initial Applications and Advances. *J. Air & Waste Manage. Assoc.*, 58, 289-302.

Sheya, S.A.N.; Meuzelaar, H.L.C.; Jeon, S.J.; Dworzanski, J.P.; Jarman, W.; Kasteler, C.; Lighty J.; Sarofim, A.; Li, W.W.; Valenzuela, V.; Anderson, J.; Banerji, S.; Perry, D.; Mejia, G.; Zavala, M. & Simoneit, B. (2000) Novel Analytical Dimensions in Exploratory Field Studies of Air Particulate Matter, *Proceedings of the 93rd Air & Waste Management Association's Annual Conference & Exhibition*, pp. 1104-1130, Salt Lake City, UT, USA, June 18-22, 2000.

Sweedler, A.; Fertig, M.; Collins, K. & Quintero-Núñez, M. (2003). Air Quality in the California-Baja California Border Region, In *The U.S. Mexican border environment: Air Quality Issues along the US-Mexico Border*, SCERP Monograph Series No. 6, A. Sweedler (Ed.), 15-58, San Diego State University Press, ISBN: 0-925613-38-X, San Diego, California, USA.

Tagaris, E.; Manomaiphiboon, K.; Liao, K.-J.; Leung, L.R.; Woo, J.-H.; He, S.; Amar, P. & Russell, A.G. (2007). Impacts of global climate change and emissions on regional ozone and fine particulate matter concentrations over the United States. *J. Geophys. Res.*, 112(D14), D14312/1-D14312/11.

Tesche, T.W.; Georgopoulos, P.; Seinfeld, J.H.; Cass, G.; Lurmann, F.L. & Roth, P.M. (1990). *Improvement of procedures for evaluating photochemical models*, Report prepared by Radian Corporation for the State of California Air Resources Board, Sacramento, CA.

US EPA. (2003) *User's Guide to Mobile6.1 and Mobile6.2 Mobile Source Emission Factor Model*. Office of Transportation and Air Quality, EPA420-R-03-010, Ann Arbor, MI.

Vanoye, A.Y. & Mendoza, A. (2009). Mesoscale Meteorological Simulations of Summer Ozone Episodes in Mexicali and Monterrey, Mexico: Analysis of Model Sensitivity to Grid Resolution and Parameterization Schemes. *Water, Air, & Soil Pollution: Focus*, 9, 185-202.

Vukovich, J. & Pierce, T. (2002). The Implementation of BEIS3 within the SMOKE Modeling Framework. *United States Environmental Protection Agency Emissions Inventory Conference*, Atlanta, GA.

Yang, Y. J.; Wilkinson, J.G. & Russell, A.G. (1997). Fast, direct sensitivity analysis of multidimensional photochemical models. *Environ. Sci. Technol.*, 31, 2859-2868.

Ying, Q. & Krishnan, A. (2010). Source contributions of volatile organic compounds to ozone formation in southeast Texas. *J. Geophys. Res.*, 115(D17), D17306/1-D17306/14.

Watson, J.G. & Chow, J.C. (2001). Source characterization of major emission sources in the Imperial and Mexicali Valleys along the US/Mexico border. *Sci. Total Environ.*, 276, 33-47.

Zavala, M.; Herndon, S.C.; Wood, E.C.; Jayne, J.T.; Nelson, D.D.; Trimborn, A.M.; Dunlea, E.; Knighton, W.B.; Mendoza, A.; Allen, D.T.; Kolb, C.E.; Molina, M.J. & Molina, L.T. (2009). Comparison of emissions from on-road sources using a mobile laboratory under various driving and operational sampling modes. *Atmos. Chem. Phys.*, 9, 1-14.

Zielinska, B.; Sagebiel, J.; Harshfield, G. & Pasek, R. (2001). Volatile organic compound measurements in the California-Mexico border region during SCOS97. *The Sci. of the Total Environ.*, 276, 19-31.

Zuk M.; Rojas Bracho, L. & Tzintzun Cervantes, M.G. (2007). *Tercer almanaque de datos y tendencias de la calidad del aire en nueve ciudades mexicanas*. Instituto Nacional de Ecología, ISBN 978-968-817-840-9, México, D.F.

14

Uncertainty in Integrated Modelling of Air Quality

Piotr Holnicki
Systems Research Institute/Polish Academy of Sciences
Poland

1. Introduction

Forecasting models of air quality and the more complex Integrated Assessment Systems (IAM) are recently used for supporting decisions concerning air quality management and emission control policy (ApSimon et al., 2002; Warren & ApSimon, 1999). The natural application of environmental models is predicting dispersion of pollutants, analysis of ecological results of some specific meteorological conditions or evaluation of environmental influence of emission sources. In most of deterministic models of air quality, the process of pollution transport is considered as distributed parameter system and is mathematically described by the set of advection-diffusion equations, along with respective boundary and initial conditions. To quantify possible ecological, economic or health benefits of emission abatement, there is a need to estimate an incremental contribution of the respective group of emission sources to ambient concentrations with a reasonable accuracy. However, due to a very complex, multidisciplinary structure of such systems, there exist many sources of imprecision and uncertainty in the modelling of environmental effects of atmospheric pollution and also in resulting regulatory decisions.

To assess the accuracy of modelling results and a connected decision support process, performance and uncertainty of the model should be evaluated. The straightforward method of such an assessment is usually based on examining a relative agreement between measurements and results of computer simulations. However, the accuracy of such a comparison is usually insufficient because of different spatial scales of these two quantities, where the point measurements are compared with volume-averaged results of simulation. To better characterize the problem, the main sources of variability (temporal, spatial, or inter-individual differences of input data) and uncertainty (imprecise information or lack of information about unknown quantity) should be identified and assessed (Park et al., 2006; Sax & Isakov, 2003). In addition, implementations of operational models of air pollution usually involve some specific simplifications or parameterizations and cannot completely characterize complex physical processes. This is the source of conceptual uncertainty which is also reflected in the final results. In particular, it relates to uncertainty in deriving trajectories in Lagrangian approach or sub-grid effects in Eulerian models. Moreover, due to performance requirements, in operational models some atmospheric processes are parameterized or described in a simplified way. For example, the height of the mixing layer and atmospheric stability are usually evaluated in course of an imprecise heuristic procedure, which is another source of the final uncertainty. However, numerous previous studies have revealed that major uncertainties (measurement or estimation error) are due to input datasets, e.g. emission inventory and meteorological data, involving more than the model itself (Russel & Dennis, 2000). The uncertainty analysis of emission data is especially significant and challenging in the case of urban or industrial areas. The main problem follows from a high spatial concentration of the large number of emission sources with different technological characteristics, used fuel type and related fuel parameters, composition of polluting compounds, emission intensities, and as a consequence - different range of emission uncertainty.

This chapter addresses the problem of uncertainty in computer modelling of air pollution dispersion, but it also takes into account the impact of this uncertainty on an environment-oriented decision support process. The following sections include an analysis. Section 2 shortly characterizes basic implementations of air pollutions transport models, their applications in supporting strategic decisions, and also the general idea of Integrated Assessment Models (IAM). Sections 3 and 4 discuss the main sources of uncertainty (emission intensity, basic meteorological data, model parameters) in air quality modelling and present results of the uncertainty analysis for a regional case study. In Section 5 a regional-scale decision problem is formulated as an emission abatement task, with solution depending on the outcomes of the previous forecasting model. It is mathematically formulated as an integer-type optimization problem. The selected results of the optimization task quantitatively illustrate how uncertainty of air pollution predictions influences decisions generated in the optimization procedure. Section 6 describes general conclusions related to the applicability of air quality forecasting models in environmental decision support systems. Due to the mostly qualitative character of such strategic decisions, the obtained modelling accuracy (including overall uncertainty) is sufficient in the majority of practical applications. Moreover, suggestions concerning the accuracy of optimization algorithms are formulated.

2. Air quality models in decision support systems

In basic applications of air quality models, the processes of air pollution transport are considered as a distributed parameter system, which is governed by the set of transport equations, along with respective boundary and initial conditions. The exact form and structure of a model usually depends on its practical application, type of the polluting compounds considered and the scale of modelling. A model usually takes into account the input data (emission field and meteorological data) as well as the main physical and chemical processes which decide on the transport in the atmosphere and transformations of air pollution components. The mathematical description of these processes within time interval $(0,T)$, related to one polluting compound and a single vertical layer of the atmosphere, usually takes a general form of an advection-diffusion equation

$$\frac{\partial c}{\partial t} + \vec{v}\nabla c - K_h \Delta c + \gamma c = Q \tag{1}$$

along with the respective boundary conditions

$$c = c_b \quad \text{on} \quad S^- = \{ \partial\Omega \times (0,T) \mid \vec{v} \cdot \vec{n} < 0 \}, \tag{2}$$

$$K_h \frac{\partial c}{\partial \vec{n}} = 0 \quad \text{on} \quad S^+ = \{ \partial\Omega \times (0,T) \mid \vec{v} \cdot \vec{n} \geq 0 \}$$

and the initial condition

$$c(0) = c_0 \quad \text{in} \quad \Omega . \tag{3}$$

The following notation is used: Ω - domain with the boundary $\partial\Omega = S^+ \cup S^-$, c - concentration, $\vec{v}$ - wind field vector, K_h - horizontal diffusion coefficient, γ - transformation coefficient, Q - emission field, $\vec{n}$ - outward normal vector. In the case of multi-species or multi-layer models this is the respective set of transport equations which are coupled by the chemical transformation rates and/or by vertical diffusion coefficient.
Implementation of the used model relates to the requirements of an application and to the pollutants which are considered. But the basic types of air quality models can differ significantly in approach to the analysis of equations (1)-(3) and also to the scale of modelling. As shown in Fig. 1, spatial and temporal scales of the environmental impact of air pollution are correlated with and, moreover, they directly depend on the *lifetime* of pollutant. This parameter can differ significantly between compounds. Thus, depending on the analysis scale, there are respective categories of modelling: local, regional and global.

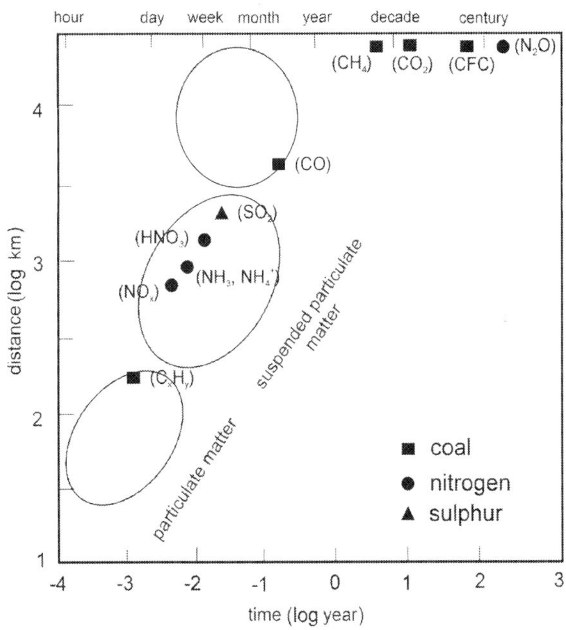

Fig. 1. Correlation of spatial and temporal scale of modelling

Regarding the practical application as well as the scale of modelling, the most common types (implementations) of air pollution models are (Markiewicz, 2004; Sportisse, 2007):

- Gaussian model - based on a simplified, analytical solution of transport equations, originally used mainly for local scale analysis. However, the new generation of

Gaussian models is now available, where variability of main meteorological fields is taken into account, with models also used on regional scale (Scire et al., 2000).

- Lagrangian model - where the related trajectory of an air polluting parcel is observed and analyzed, according to the wind field and other meteorological parameters. Mathematical description takes a form of the respective set of ordinary differential equations (source-oriented approach). The advantage of this method is a natural ability to assess individually the environmental impact of selected emission sources via *transfer matrices*. This approach is utilized in the analysis of emission abatement strategy (compare, e.g. Cofała et al., 2010; Holnicki, 2010b; Kelly, 2006).
- Eulerian model - mathematically governed by finite-dimensional approximation of equations (1)-(3), where a modelling region is horizontally and vertically discretized into the respective number of cells. Parameters of numerical scheme (temporal and spatial discretization steps) must be accordingly set, to satisfy stability and monotonicity conditions (Jacobson, 2005). Implementations usually include evolution of pollutant concentrations, including advection, diffusion, chemistry, sedimentation and deposition. This category of models, characterized by high computing requirements, is utilized in the most complex regional and multi-scale implementations (receptor-oriented approach).

Models of air pollution dispersion are recently used in numerous projects regarding air quality analysis and management. Such a model, based on the input dataset (emission inventory and meteorological forecast), can quantitatively evaluate environmental quality and suggest the ways of improvement. For example, the models can be directly used for:

- forecasting of air pollution distribution in the domain,
- assessment of air quality standard violation (critical levels for concentrations or critical loads for depositions),
- assessment of environmental impact of some specified sources,
- selection of optimal locations for new investments,
- simulation and analysis of emission abatement strategies.

Apart from such a direct use of air pollution models, many countries have recently been trying to develop so called *Integrated Assessment Models* (IAM) (ApSimon et al., 2002) which are designed for complex analysis of environmental quality and are to be used for supporting decisions in environmental quality management. The integrated assessment models (compare Fig. 2, according to (Russel & Dennis, 2000)), recently developed, aim to combine a classical pollution transport model with some economic, technological and other constraints and standards. Such a system, aside from the natural scenario analysis, gives also the possibility to formulate and solve optimization problems, which take into consideration certain specified environmental standards. Some general optimization methods give the possibility of implementation of complex air quality control strategies.

A traditional air pollution transport model is one of the components of such an integrated system. Complementary modules allow us to take into consideration some additional relations and constraints, for example technological, economic, demographic, ecological, and others. This system is a tool for a complex analysis of environment-oriented development strategies as well as for solving the respective optimization problems (compare Holnicki, 2006, 2010a). However, irrespective of how complex such a system is, its main component is usually an air pollution dispersion model, with other modules including respective constraints and limits. An important stage of this complex analysis consists in the

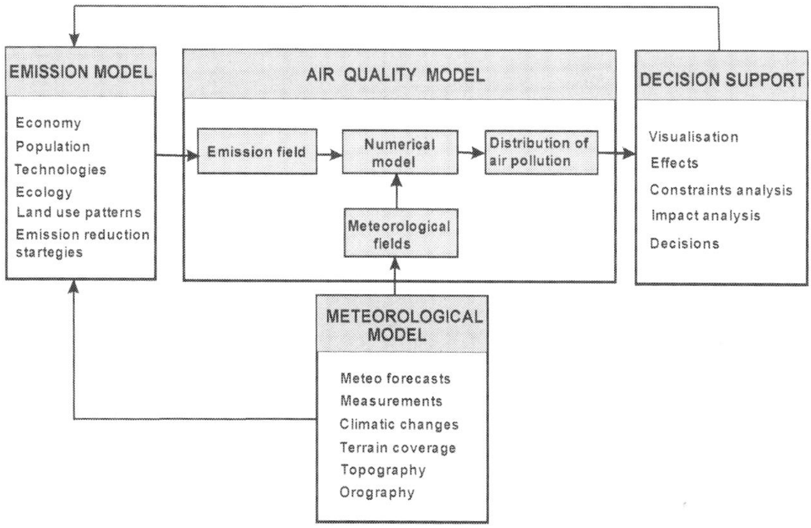

Fig. 2. Exemplary block-diagram of Integrated Assessment Model

assessment of environmental impact of individual emission sources. Such an evaluation is more natural by use of Lagrangian models, where the total pollution is usually calculated as the superposition of individual sources contribution. The task is more challenging in the case of Eulerian models, where the entire emission field, composed of many individual sources, is taken into account in one forecasting run of the model. A possible approach in this case was presented in (Holnicki, 2006).

To implement each strategy of air quality management, an air quality damage (air quality cost) function must be defined. Definition of such an index usually relates to the main polluting factors, such as concentration of pollutants (temporary or long-term averaged), cumulated deposition or exceedance of critical loads (Cofała et al., 2010; Holnicki, 2006; Mill & Schlama, 2010). Another important index that is considered in formulation of the optimal emission reduction strategy is the cost of strategy implementation. Denoting, respectively:

$\Phi_1(c\,(\vec{u}))$ - environmental damage index related to air pollution,

$\Phi_2(\vec{u})$ - cost of each emission abatement action,

($\vec{u}$ is the emission vector of controlled sources) the following two basic formulations of the optimization problems related to air quality control can be considered:

a. Minimization of the environmental damage subject to the total cost constraint

$$\begin{cases} \Phi_1(c\,(\vec{u})) \rightarrow \min, \\ \Phi_2(\vec{u}) \leq \Phi_{2,MAX}, \end{cases}$$

b. Obtaining the assumed air quality standard at the minimum total cost.

$$\begin{cases} \Phi_2(\vec{u}) \rightarrow \min, \\ \Phi_1(c\,(\vec{u})) \leq \Phi_{1,MAX}. \end{cases}$$

In (Holnicki, 2010a) an emission control problem is considered, where the objective function is formulated in a more general form as compared to the weighted sum of two components, representing environmental damage (related to the concentration of polluting factor) and emission reduction cost in the controlled (or modernized) emission sources. This general index is as follows:

$$J(c(\vec{q})) = \alpha_1 \int_0^T \int_\Omega \varphi_1(c(\vec{q}))\, d\Omega\, dt + \alpha_2 \int_0^T \varphi_2(\vec{q}) dt \,. \tag{4}$$

The respective regularity of the sub-integral functions φ_1 and φ_2 is assumed. The time interval $(0,T)$ depends on the temporal scale of an analysis, and can vary from several hours (short-term forecasts, emission control) to one year (long-term strategy analysis). In all optimization algorithms, it is necessary to assess sensitivity of the quality index to emission of individual sources. An exemplary implementation, based on adjoint transport equations, is discussed in (Holnicki, 2006, 2010a).

In the sequel (compare Section 5) another implementation of an environment-oriented decision task is considered. The problem consists in optimal selection of emission reduction technologies in controlled emission sources in order to obtain the predefined air quality standard at the minimum cost. In this case the main aim is not to solve the optimization problem itself, but to assess how uncertainty of the input data influences the accuracy of a solution to this final decision problem.

3. Sources of uncertainty in air quality modelling

Depending on the model considered, there are numerous potential sources of imprecision and uncertainty in general forecasts of air pollution. The main sources of such uncertainty can be (compare Hanna et al., 1998; Sax & Isakov, 2003; Sportisse, 2007):

a. input data (emission inventory, meteorological forecast),
b. model structure, simplifications and parameterizations,
c. numerical implementation (temporal and spatial domain discretization, applied numerical approximation scheme),
d. other specific model parameters.

In the case of model input uncertainty, for instance, meteorological data or emissions are uncertain themselves due to measurement errors, estimation errors and inherent variability. Model conceptual uncertainty appears because a single model parameter or mathematical description can never precisely characterize the considered process. Another conceptual uncertainty occurs because a discrete representation of numerical code is used to represent continuous physical processes and their natural variability.

In the case of emission sources, uncertainty is mainly related to the source category, namely emission intensity, composition of polluting compounds, technological characteristics, etc. For example, relatively low uncertainty characterizes high point sources (representing the energy sector, power plants and heating plants), mainly due to well defined and stable combustion processes and known fuel parameters. Uncertainty is more significant in the case of small and intermediate industrial point sources, where all the combustion parameters are less precise and can change in time. On the other hand, high uncertainty must be assigned to area sources (residential sector of urban agglomerations, distributed industrial sources) and to linear sources (usually representing the transportation system).

Meteorological forecast consists of a set of input data which are the result of a multi-step process of data assimilation (Jacobson, 2005). The consecutive steps of this procedure encompass: measurements, numerical weather prediction (by the specialized meteorological model) and analysis, in order to merge the results and generate the final meteorological dataset for a given time interval. Each of these steps generates certain error which contributes to the final accuracy of the forecast. An additional source of uncertainty is connected with the fact that air pollution models usually apply different space discretization grids than meteorological predictor, and data must be additionally interpolated.
Model induced uncertainty relates directly to the type of the mathematical structure. As an example, one can observe that errors generated by relatively simple Lagrangian models are related to uncertainty in deriving wind trajectories, especially in multi-layer implementations. Moreover, some additional errors are due to neglecting a turbulent diffusion process in most of simplified trajectory approaches. Due to these problems and less precise mathematical representation of the Lagrangian approach, Eulerian models are becoming the dominating forecasting tools in air quality analysis. On the other hand, solving the finite-dimensional approximation of transport equations in this category of models leads to the known numerical problems, such as numerical diffusion effect, spurious oscillations and shape modifications. Some special shape-preserving numerical schemes must be applied to eliminate or reduce these negative effects (compare e.g. Holnicki, 2006; Jacobson, 2005).

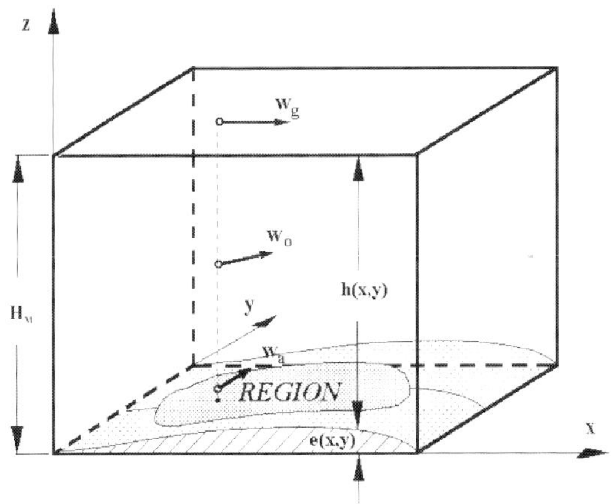

Fig. 3. Mixing height, H_m and vertical wind profile

Commonly used operational models (applied for decision support) of air pollution, to improve performance utilize a simplified mathematical representation which requires some additional meteorological data, usually not included in the standard dataset of weather prediction. These can include, for example, *mixing height* or *atmospheric stability* parameters, which are evaluated in a rather imprecise (heuristic) process, based on some standard meteorological data (Markiewicz, 2004). Mixing height is a key parameter, particularly in models with a rough vertical resolution or in single-layer implementations, where the air pollution concentration is calculated as averaged over the mixing layer (compare Fig. 3).

Approximation of a vertical wind profile is another important feature which characterizes the model accuracy. Wind velocity and direction change vs. elevation, as shown in Fig. 3, due to a wind shear effect. Geostrophic wind vector, $\vec{w}_g = [u_g, \upsilon_g]$, at the upper bound of the mixing layer results from the pressure distribution over the domain. The shear of this vector in lower layers is the result of the Coriolis force and the friction between air masses and the Earth surface. In some implementations, the vertical wind profile is assessed based on geostrophic and anemometric wind vectors: $\vec{w}_g$, $\vec{w}_a$ (Holnicki, 2006; Markiewicz, 2004). This procedure also contributes to the final uncertainty of model predictions.

Atmospheric stability is another key meteorological parameter which strongly contributes to the overall uncertainty. It depends on the vertical gradient of temperature in the mixing layer and has a significant impact on air pollution transport. The reference here is a so called *adiabatic gradient* - when temperature decreases about 10 $^{\circ}$C for each 1000 m of height (Jacobson, 2005). This situation refers to neutral conditions of atmospheric stability. If the gradient is greater than adiabatic (temperature decreases more rapidly), there are unstable conditions and it means an intensive mixing process of air pollution. On the other hand, for stable conditions when temperature gradient is less than adiabatic (or reversed), the air pollution mixing process is minimal and pollution plume is narrow. In many practical applications of an operational model, the Pasquill classification of atmospheric stability is applied, as shown in Table 1.

Category	Atmospheric stability class
A	strongly unstable
B	unstable
C	slightly unstable
D	neutral
E	slightly stable
F	stable

Table 1. Atmospheric stability classes

The above mentioned and other model parameters depend on the implementation which is applied. In the next section air pollution model uncertainty is determined based on some specific Eulerian model implementation on regional scale. Emission field is composed of power plants located in the region. To evaluate uncertainty of the model prediction, uncertainty of the main input data as well as the key conceptual model parameters are taken into consideration in a Monte Carlo analysis. Results of the model uncertainty assessment are then used in Section 5 in an analysis of the impact on decision process.

4. Uncertainty analysis – regional case study

The uncertainty analysis has been performed for a regional-scale 3-layer Eulerian model REGFOR3 (Holnicki, 2006, 2010a) applied for Upper Silesia rectangle region shown in Figure 4. Dispersion of SO_2 pollution emitted by 20 dominating power plants has been considered. Space discretization step applied in a numerical process is h=2 km. Resulting concentration values were recorded in 5 fictitious receptor points. Location of receptors and emission sources are shown in Fig. 4.

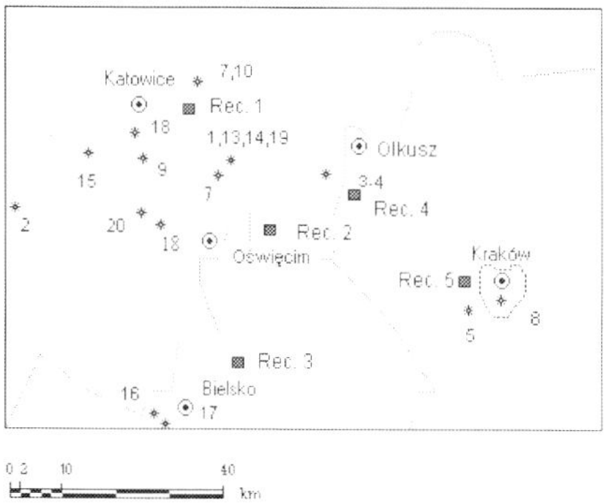

Discretization parameters:
- rectangle domain 110 km x 76 km
- homogeneous discretization, h = 2 km
- discrete problem dimensions: 55 x 38

Fig. 4. Computational domain and location of emission sources

The Monte-Carlo algorithm (Hanna et al., 1998; Moore & Londergan, 2001) was applied to carry out the calculations concerning the level of forecast uncertainty. The calculations were performed for the winter season 2005 (meteorological data and emission inventory). The nominal emission intensities within the specified period are shown in Table 2. A numerical experiment consists of three consecutive steps, whose aim is to evaluate the impact of: a) uncertainty of the source emission intensity, b) uncertainty due to the technological

No.	Source	Coordinates	He [m]	Emission SO_2 [t/d] (Winter)	Emission SO_2 [t/d] (Summer)
1	Jaworzno III	(21,24)	250	303.2	227.2
2	Rybnik	(1,20)	200	225.2	167.6
3	Siersza A	(30,23)	150	104.0	88.0
4	Siersza B	(30,23)	260	91.8	68.0
5	Skawina	(43,11)	120	90.1	58.6
6	Łaziska I	(8,20)	200	78.0	55.6
7	Będzin B	(18,31)	200	65.0	15.2
8	Łęg	(46,12)	250	52.0	37.2
9	Katowice	(13,25)	250	52.0	37.2
10	Będzin A	(18,31)	160	45.1	30.2
11	Łaziska II	(8,20)	160	34.7	23.1
12	Łaziska III	(8,20)	100	33.8	23.5
13	Jaworzno IIA	(21,24)	120	29.9	19.2
14	Jaworzno IIB	(21,24)	100	25.1	17.7
15	Halemba	(8,25)	110	26.0	17.3
16	Bielsko-Biała	(14,2)	140	18.7	11.2
17	Bielsko-Kom.	(15,1)	250	16.9	7.5
18	Chorzów	(12,27)	100	15.1	7.5
19	Jaworzno I	(20,23)	152	12.3	6.8
20	Tychy	(13,19)	120	11.6	8.6

Table 2. Characteristics of emission sources

parameters of the source, c) uncertainty of the basic meteorological data. The final results represent the total uncertainty regarding these factors.
Table 3 presents the range of uncertainty of the input data, i.e. the emission of sulphur dioxide and technological parameters for 20 analysed sources and the basic meteorological parameters. 1000 sets of test parameters were randomly generated for the specified range of input data uncertainty and the assumed distribution. In order to avoid creating unrealistic meteorological episodes, a correlation between geostrophic and anemometric wind vectors was assumed. Preliminary calculations were carried out for the normal (N) and log-normal distribution (L-N), respectively, for all analysed variables. In the majority of publications (ApSimon et.al., 2002; Hanna et al., 1998), the analysis of uncertainty deals with log-normal distribution. Since, however, the differences in results obtained for both schedules were immaterial (decisive was the width of the uncertainty range), the essential calculations and the results presented in the sequel refer to the normal distribution. The resulting distribution of mean SO_2 concentration is shown in Fig. 5.

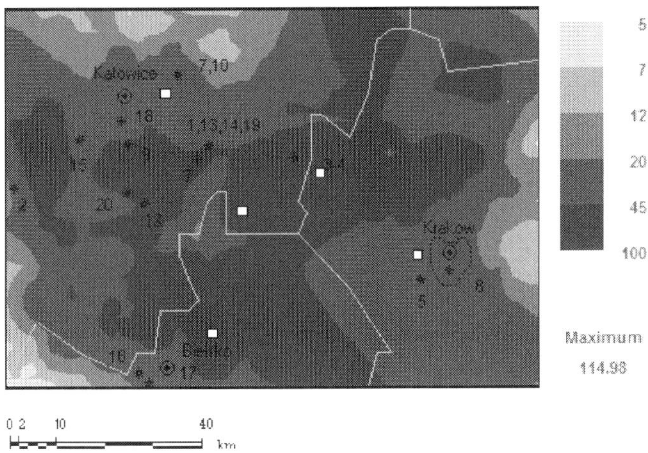

Fig. 5. Mean seasonal SO_2 concentration [μg/m³] in the domain (nominal emission)

Parameter	Uncertainty range (for 95% of data)	Distribution
Emission [g/s]	± 20%	N / L-N
Velocity of outlet gasses [m/s]	± 15%	N / L-N
Temperature of outlet gasses [°K]	± 15%	N / L-N
Mixing height [m]	± 25%	N / L-N
Geostrophic wind components [m/s]	± 25%	N / L-N
Anemometric wind components [m/s]	± 25%	N / L-N
Temperature [°C]	± 25%	N / L-N
Precipitation intensity [mm/h]	± 25%	N / L-N
Atmospheric stability class [-]	± 1	Discrete

Table 3. Uncertainty range of the input parameters

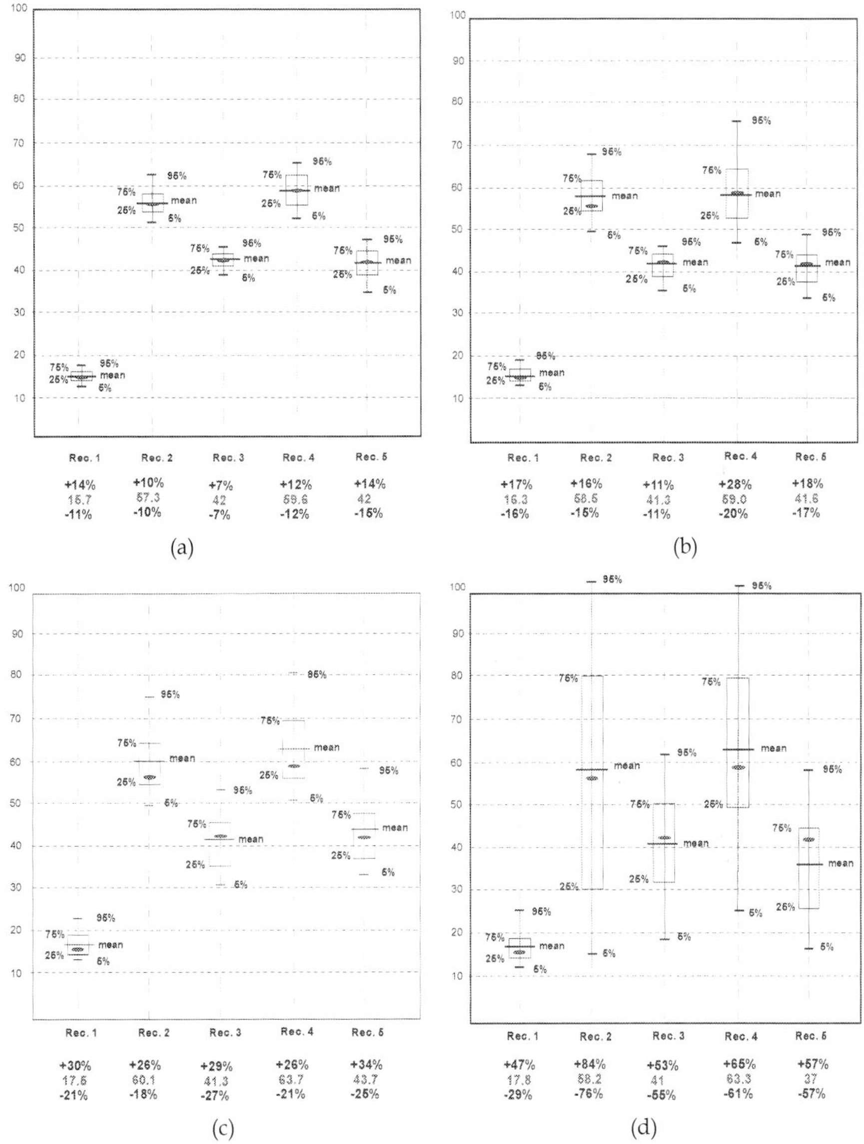

Fig. 6. Impact of input data categories on the resulting forecast uncertainty:
a) uncertainty of the sources' emission intensity,
b) + uncertainty of technological stack parameters,
c) + uncertainty of meteorological forecast (except for atmospheric stability),
d) + impact of atmospheric stability.

The results of calculation were recorded as mean values of SO_2 concentration at receptor points (compare Fig. 4 and Fig. 5 for receptor location), averaged over the simulation period (winter season of the year 2005). Contributions of assumed parameters to the resulting uncertainty of the model forecast are presented in Fig. 6 as standard "box plots". The presented figures concern 50 and 90 percentile uncertainty ranges of the input dataset, respectively. Full results containing empirical distribution functions and regarding the sample values can be found in (Holnicki, 2010b).

Fig. 6a-b demonstrates that the impact of emission intensity and technological parameters of the source cause uncertainty up to ±20% (for 90% confidence interval). Basic meteorological data (see Fig. 6c), in turn, cause an increase of uncertainty range up to approximately ±30%.

What is more, the ranges of uncertainty logged in all receptors are broadly similar. A considerable increase in uncertainty occurs after the impact of atmospheric stability is included. In this case, also the substantial variability of results depends on the receptor location. The largest recorded uncertainty of results is connected with receptors No. 2 and 4 (about ±80% and ±63%, respectively), with the smallest uncertainty regarding receptor No. 1 (approximately ±30-45%).

There are several reasons of visible and, at the same time, diversified impact of atmospheric stability that must be taken into account. Emission sources considered in the framework of this analysis belong to the category of LCP (Large Combustion Plants - high sources of very large emission intensity). In this category of sources, technological parameters exert a significant impact on the pollution plume dispersion. In conjunction with the fluctuations in atmosphere stability, this gives a very large difference in the spatial scale of a given source impact. On the other hand, very important factors which influence the concentrations recorded at receptor sites are: locations of dominating emission sources, location of the receptors and the dominating wind directions in the specified time interval. The analysis should also take into account the impact of atmospheric balance parameterization, used in the pollution transport model. Moreover, high sensitivity of the model to atmospheric stability may result from the applied parameterization. This, however, requires a more thorough and detailed discussion.

5. Uncertainty in decision support problem

The aim of this section is to illustrate and assess how uncertainties in modelling air pollution transport influence results of the more complex decision support process, where a pollution dispersion model is one of the key components. The decision problem considered in the sequel consists in optimal (at minimum costs) selection of emission reduction technologies within a set of power plants. A regional scale pollution transport model is used as the main forecasting tool, whose uncertainty contributes to the overall uncertainty of the final decision.

5.1 Optimization problem statement

A regional scale decision problem presented in this section constitutes an exemplary implementation of the integrated decision support structure, discussed more generally in Section 2. The task defined in the sequel is sulphur-oriented and addresses the optimal strategy of emission control. It concentrates on the selection of emission reduction technologies (desulphurization) within a given set of power plants. We assume that there is a set of controlled emission sources (power plants) located in the region under question. Moreover, a set of emission abatement technologies is available. Each desulphurization technology is characterized by certain effectiveness of emission reduction and the unit cost (both for investment and operational costs). Based on these data, optimization problems concerning environmental quality can be formulated (see e.g. Holnicki, 2006, 2010a). One of the possible questions can deal with the optimal allocation of desulphurization technologies to emission sources. The task presented below is aimed at obtaining the required threshold of environmental quality index at minimal costs.

To formally state the optimization problem, the necessary notation must be introduced. Assume that in domain Ω there are N controlled emission sources and we have M desulphurization technologies available. Moreover, we denote:

$\vec{u} = [u_1, u_2, \ldots, u_N]$ - emission vector of controlled sources,

$\vec{e} = [e_1, e_2, \ldots, e_M]$ - effectiveness vector of desulphurization technologies,

$\mathbf{F} = \{f_{ij}\}, (1 \le i \le N,\ 1 \le j \le M)$ - matrix of technology abatement cost per unit emission,

$\mathbf{X} = \{x_{ij}\}, (1 \le i \le N,\ 1 \le j \le M)$ - "0-1" matrix of abatement technology assignment to the controlled sources (decision variable matrix).

The environmental quality index (used in the formulation of required environmental constraints) depends on SO_2 concentration which is the model output. We use a global environmental cost function of the following form:

$$J(c) = \frac{\alpha}{2} \int_{\Omega} w(x,y) \cdot \max^2(0, c(x,y) - c_{ad})\, d\Omega, \tag{5}$$

where: $w(x,y)$ - area sensitivity function, c_{ad} - admissible level of SO_2 concentration. Due to the scaling factor α, index J is considered in the sequel as a dimensionless quantity. The concentration forecast, considered as the solution to (1)-(3), is calculated as:

$$c(x,y) = c_0(x,y) + \sum_{i=1}^{N} A_i(x,y) \cdot u_i, \qquad (x,y) \in \Omega, \tag{6}$$

where: $c_0(x,y)$ - background concentration (impact of uncontrolled sources), $A_i(x,y)$ - the unit transfer matrix (relation emission $\rightarrow$ concentration) of the *i*-th source, u_i - current emission of *i*-th source.

The unit *transfer matrices*, A_i for the controlled sources are pre-processed off-line by the respective forecasting model. A similar way is used to compute the background pollution field for uncontrolled emissions, including the inflow from the neighbouring regions. The current emission intensity of the *i*-th source depends on the initial emission level - u_i^0 and the efficiency of the applied abatement technology, according to the formula:

$$u_i = u_i^0 \sum_{j=1}^{M} (1 - e_j) \cdot u_i, \quad \sum_{j=1}^{M} x_{ij} = 1, \quad x_{ij} \in \{0,1\}, \quad 1 \le i \le N. \tag{7}$$

Cost of emission reduction in each source consists of two components: investment cost and operational cost. Both components depend on the specific abatement technology and on parameters of an energy installation where this technology is to be applied. In this work a simplified approach is utilized. The investment cost of the *j*-th abatement technology installed in the *i*-th emission source is calculated as an annual cost, averaged over the entire amortization period. Thus, the total emission abatement cost per year, considered as a sum of desulfurization costs in the respective plants, is considered in the following form

$$C_T = \sum_{i=1}^{N} u_i^0 \sum_{j=1}^{M} f_{ij} \cdot x_{ij} = \sum_{i=1}^{N} u_i^0 \sum_{j=1}^{M} (f_{ij}^1 + f_{ij}^2) \cdot x_{ij} , \qquad (8)$$

where the coefficients: f_{ij}, f_{ij}^1, f_{ij}^2 denote the unit, annual investment and operational cost, respectively, of the *j*-th technology applied to the *i*-th emission source. Finally, we can state the following:

ALLOCATION TASK (AT): *For a given set of admissible solutions,* X_{ad}

$$X_{ad} = \{x_{ij} \in \{0,1\}: \sum_{j=1}^{M} x_{ij} = 1, \quad 1 \le i \le N, \quad 1 \le j \le M\}$$

assign abatement technologies to emission sources to minimize the total cost

$$C_T = \sum_{i=1}^{N} u_i^0 \sum_{j=1}^{M} f_{ij} \cdot x_{ij} \Rightarrow \min$$

and obtain the assumed air quality index $J(c^*) \le J_O$.

Due to the character of decision variable, the allocation task (AT) is an example of an integer-type optimization. Its numerical analysis requires special algorithms (e.g. heuristic or evolutionary algorithms). In one of the approaches presented in (Holnicki, 2006) the problem is solved by means of a gradient optimization method. For this purpose, the original, integer-type problem is formulated and solved as continuous one. This formulation means that decision variable is continuous, thus the set of admissible solutions is modified as follows:

$$X_{ad} = \{x_{ij} \ge 0: \sum_{j=1}^{M} x_{ij} = 1, \quad 1 \le i \le N, \quad 1 \le j \le M\} . \qquad (9)$$

Although technologically less realistic (admits technology mixing), the formulation is quite satisfactory from a practical perspective of the decision support and allows application of standard optimization algorithms.

Calculations of the quality index J(c) are based on unit transfer matrices for all of the controlled sources which have been computed off-line by the regional air pollution transport model REGFOR3, presented in more detail in (Holnicki, 2006). The same approach is currently used in uncertainty analysis based on the Monte Carlo procedure. The implementation discussed below is to illustrate how uncertainty of the forecasting model contributes to the accuracy of the decision task.

5.2 Uncertainty in decision support – case study results

Imprecision and uncertainties of the input dataset as well as technological parameters in the combustion process imply uncertainty of air pollution forecasts. Uncertain quantities of concentration are next used as input information in an optimization algorithm, whose aim is to support decisions and to select the best strategy of emission control. The main goal of computations, whose selected results are presented in this section, was to evaluate how uncertainty of air pollution forecasts influences the solution of the optimization problem formulated in Section 5.1 and how it contributes to uncertainty of the final decision.

In the example discussed below we consider 20 controlled sources, as shown in Table 2. Moreover, 8 desulphurization technologies are taken into account (5 basic technologies and 3 combined). The technologies and the respective emission reduction efficiencies are as follows:

1. "do nothing" technology ($e_1 = 0$),
2. low-sulfur fuel ($e_2 \cong 0.30$),
3. dry desulphurization method ($e_3 \cong 0.35$),
4. low-sulfur fuel + dry desulphurization ($e_4 \cong 0.55$),
5. half-dry desulphurization method ($e_5 \cong 0.75$),
6. low-sulfur fuel + half-dry desulphurization ($e_6 \cong 0.83$),
7. wet desulphurization method ($e_7 \cong 0.86$),
8. low-sulfur fuel + wet desulphurization ($e_8 \cong 0.91$).

Given initial emissions of the controlled sources (and the initial value of environmental index J_i), the main task is to obtain an assumed final value, J_0 at the minimum cost by the optimal assignment of respective technology to each source. The optimization algorithm used to solve (AT) is based on transfer matrices and it does not need on-line solution of air pollution transport. The Monte Carlo statistical algorithm was utilized to evaluate uncertainty of the optimization task (AT). Thus, 1000 randomly generated sets of emission data for controlled sources were preprocessed and used as the input for the optimization algorithm. Distribution and range of uncertainty of the input emission data should reflect the final uncertainty related to forecasts of air pollution concentrations. To this end, based on the final results presented in Section 4, uncertainty range applied to emission sources used in the optimization process was assumed as the average of 5 resulting receptor values (compare Fig. 6 for forecast uncertainty). Such an approach allows encompassing the impact of all key elements (input emission end meteorological datasets, technological parameters).

Due to editorial limitations, only selected findings of computational tests are shown below (more complete outcomes can be found in the report (Holnicki, 2010b)). They present results of the optimization process (including the impact of uncertainty) for two sets of initial emission intensities: (a) nominal emissions of all sources, as given in Table 2 (initial quality index, $J_i = 3.24 \cdot 10^7$), and (b) emissions increased by 10% (initial quality index, $J_i = 3.92 \cdot 10^7$). Each set of initial input data contains 1000 randomly generated emission episodes for the controlled sources. The normal distribution is assumed with uncertainty range reflecting results of Section 4 (approximately ±50% uncertainty range for 90%

confidence interval). The target (optimal) value of the air quality index in both cases was assumed as $J_o = 1.90 \cdot 10^6$.

src	abatement technologies								emit.
	1	2	3	4	5	6	7	8	
1	.0	.0	.0	.0	.0	.0	**1**	.0	45.6
2	**.4**	.0	**.6**	.0	.0	.0	.0	.0	175.1
3	.0	.0	.0	**.7**	.0	.0	**.3**	.0	38.0
4	.0	.0	.0	.0	.0	.0	**1**	.0	14.3
5	.0	.0	.0	.0	.0	.0	.0	**1**	9.5
6	.0	.0	.0	**1**	.0	.0	.0	.0	35.8
7	.0	**.6**	.0	**.4**	.0	.0	.0	.0	38.4
8	.0	**1**	.0	.0	.0	.0	.0	.0	35.9
9	.0	**.4**	.0	**.6**	.0	.0	.0	.0	29.2
10	**.1**	.0	**.9**	.0	.0	.0	.0	.0	31.3
11	.0	**.1**	.0	**.9**	.0	.0	.0	.0	16.9
12	.0	.0	.0	.0	.0	.0	**.1**	**.9**	3.7
13	.0	.0	.0	.0	.0	.0	.0	**1**	3.1
14	.0	.0	.0	.0	.0	.0	.0	**1**	2.6
15	.0	.0	.0	**.3**	.0	**.5**	.0	**.2**	6.5
16	.0	.0	.0	**.5**	**.3**	**.2**	.0	.0	6.2
17	.0	.0	**.5**	**.5**	.0	.0	.0	.0	9.5
18	.0	.0	.0	.0	.0	**.2**	.0	**.8**	1.8
19	.0	.0	.0	**1**	.0	.0	.0	.0	5.6
20	.0	.0	.0	**.8**	.0	**.2**	.0	.0	4.4

abatement technologies								emit.
1	2	3	4	5	6	7	8	
.0	.0	.0	.0	.0	.0	**1**	.0	45.5
.0	.0	**1**	.0	.0	.0	.0	.0	146.3
.0	.0	.0	**1**	.0	.0	.0	.0	47.3
.0	.0	.0	.0	.0	.0	**1**	.0	13.7
.0	.0	.0	.0	.0	.0	.0	**1**	9.5
.0	.0	.0	**1**	.0	.0	.0	.0	35.5
.0	**1**	.0	.0	.0	.0	.0	.0	45.5
.0	**1**	.0	.0	.0	.0	.0	.0	36.4
.0	**.1**	.0	**.9**	.0	.0	.0	.0	24.2
.0	.0	**1**	.0	.0	.0	.0	.0	29.3
.0	.0	.0	**1**	.0	.0	.0	.0	15.8
.0	.0	.0	.0	.0	.0	.0	**1**	3.5
.0	.0	.0	.0	.0	.0	.0	**1**	3.1
.0	.0	.0	.0	.0	.0	.0	**1**	2.6
.0	.0	.0	**.2**	.0	**.8**	.0	.0	5.7
.0	.0	.0	**.4**	**.6**	.0	.0	.0	6.2
.0	.0	**.6**	**.4**	.0	.0	.0	.0	9.7
.0	.0	.0	.0	.0	.0	.0	**1**	1.6
.0	.0	.0	**1**	.0	.0	.0	.0	5.6
.0	.0	.0	**1**	.0	.0	.0	.0	5.3

Table 4. Optimal fuzzy solution (left) and reference continuous solution (right) for optimization parameters: $J_i = 3.24 \cdot 10^7$; $J_o = 1.90 \cdot 10^6$; $C_T \approx 190$.

src	abatement technologies								emit.	abatement technologies								emit.
	1	2	3	4	5	6	7	8		1	2	3	4	5	6	7	8	
1	.0	.0	.0	.0	.0	.0	1	.0	50.0	.0	.0	.0	.0	.0	.0	1	.0	50.0
2	.2	.0	.8	.0	.0	.0	.0	.0	174.7	.0	.0	1	.0	.0	.0	.0	.0	161.1
3	.0	.0	.0	.3	.0	.0	.7	.0	27.8	.0	.0	.0	1	.0	.0	.0	.0	28.0
4	.0	.0	.0	.0	.0	.0	1	.0	15.2	.0	.0	.0	.0	.0	.0	1	.0	15.1
5	.0	.0	.0	.0	.0	.0	.0	1	10.4	.0	.0	.0	.0	.0	.0	.0	1	10.4
6	.0	.0	.0	1	.0	.0	.0	.0	39.2	.0	.0	.0	1	.0	.0	.0	.0	39.0
7	.0	.3	.0	.7	.0	.0	.0	.0	37.3	.0	1	.0	.0	.0	.0	.0	.0	32.5
8	.0	.8	.0	.2	.0	.0	.0	.0	37.6	.0	1	.0	.0	.0	.0	.0	.0	40.0
9	.0	.2	.0	.8	.0	.0	.0	.0	28.6	.0	.1	.0	.9	.0	.0	.0	.0	26.0
10	.0	.0	1	.0	.0	.0	.0	.0	32.8	.0	.0	1	.0	.0	.0	.0	.0	32.2
11	.0	.0	.0	1	.0	.0	.0	.0	17.8	.0	.0	.0	1	.0	.0	.0	.0	17.3
12	.0	.0	.0	.0	.0	.0	.0	1	4.0	.0	.0	.0	.0	.0	.0	.0	1	3.9
13	.0	.0	.0	.0	.0	.0	.0	1	3.5	.0	.0	.0	.0	.0	.0	.0	1	3.5
14	.0	.0	.0	.0	.0	.0	.0	1	2.9	.0	.0	.0	.0	.0	.0	.0	1	2.9
15	.0	.0	.0	.1	.0	.4	.0	.5	5.0	.0	.0	.0	.2	.0	.8	.0	.0	5.0
16	.0	.0	.0	.2	.3	.4	.1	.0	5.3	.0	.0	.0	.4	.6	.0	.0	.0	5.0
17	.0	.0	.3	.7	.0	.0	.0	.0	9.5	.0	.0	.6	.4	.0	.0	.0	.0	8.7
18	.0	.0	.0	.0	.0	.1	.0	.9	1.8	.0	.0	.0	.0	.0	.0	.0	1	1.7
19	.0	.0	.0	1	.0	.0	.0	.0	6.2	.0	.0	.0	1	.0	.0	.0	.0	6.2
20	.0	.0	.0	.5	.0	.4	.0	.1	3.8	.0	.0	.0	1	.0	.0	.0	.0	5.8

Table 5. Optimal fuzzy solution (left) and reference continuous solution (right) for optimization parameters: $J_i = 3.92 \cdot 10^7$; $J_O = 1.90 \cdot 10^6$; $C_T \approx 208$.

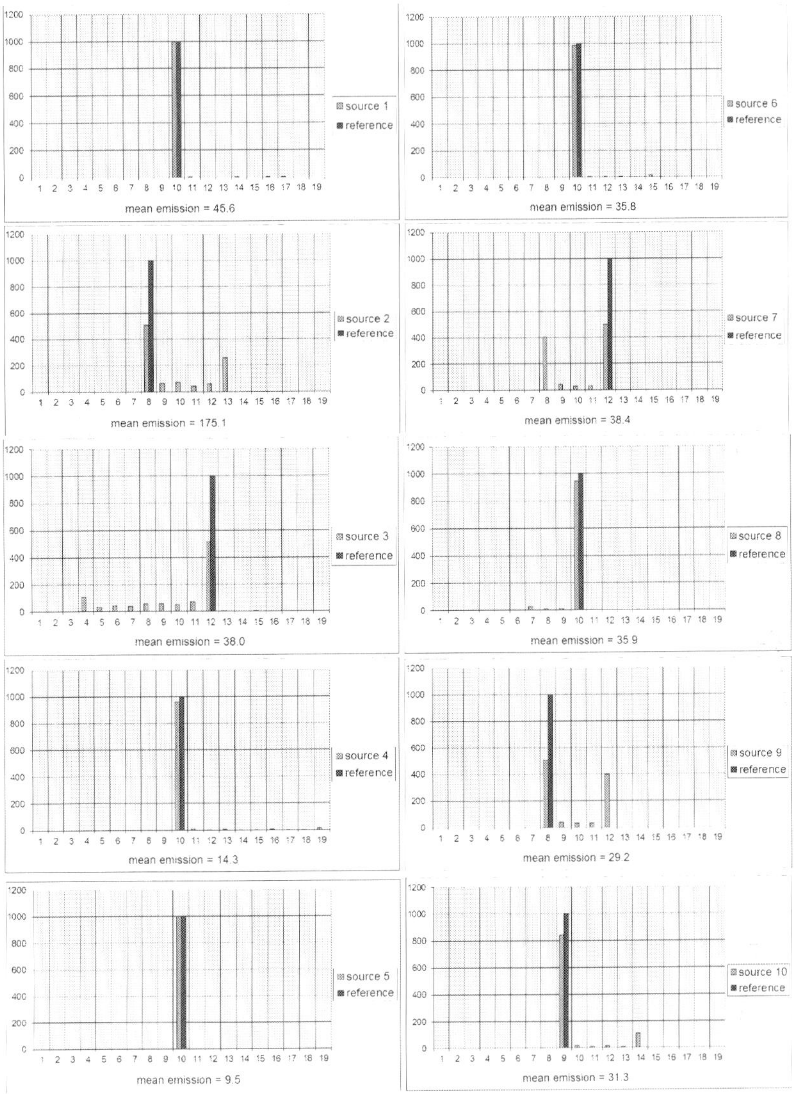

Categories:

1) 0.0 - 0.15
2) 0.15 - 0.25
3) 0.25 - 0.35
4) 0.35 - 0.45
5) 0.45 - 0.55
6) 0.55 - 0.65
7) 0.65 - 0.75
8) 0.75 - 0.85
9) 0.85 - 0.95
10) 0.95 - 1.05
11) 1.05 - 1.15
12) 1.15 - 1.25
13) 1.25 - 1.35
14) 1.35 - 1.45
15) 1.45 - 1.55
16) 1.55 - 1.65
17) 1.65 - 1.75
18) 1.75 - 1.85
19) 1.85 - 1.95

Fig. 7. Histogram of emission [t/d] for optimal solutions (sources 1-10; $J_i = 3.24 \cdot 10^6$)

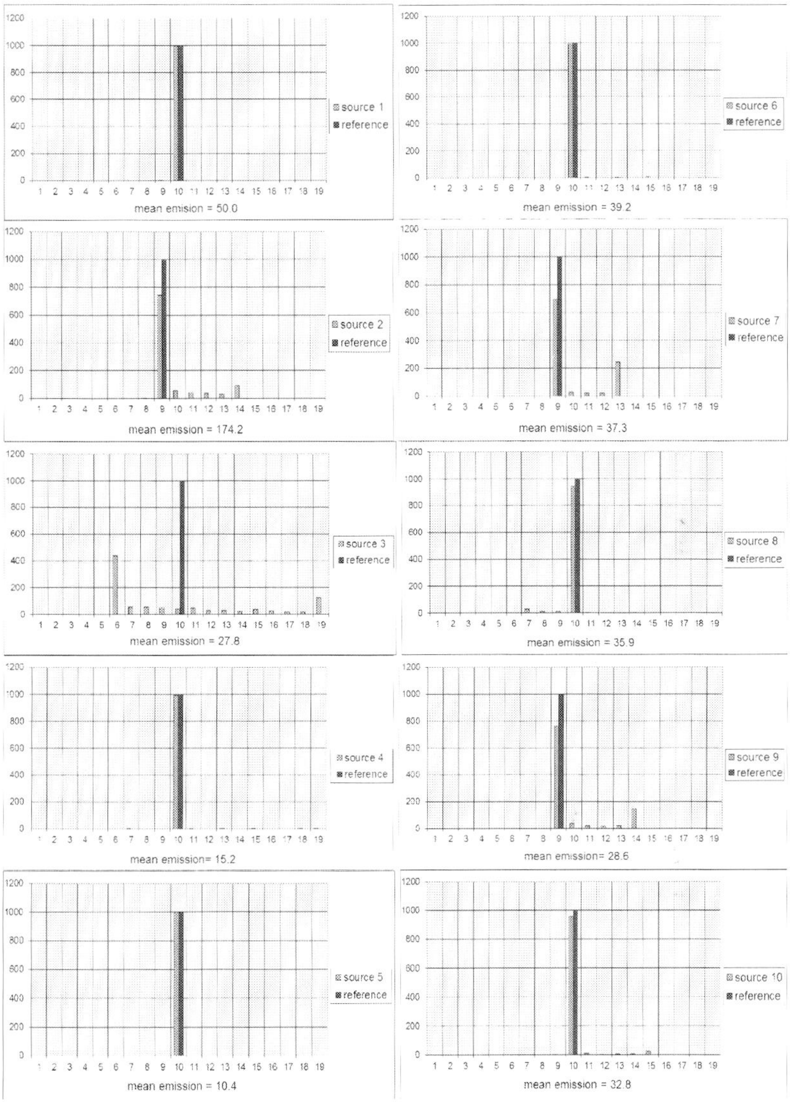

Categories:

1) 0.0 - 0.15	6) 0.55 - 0.65	11) 1.05 - 1.15	16) 1.55 - 1.65
2) 0.15 - 0.25	7) 0.65 - 0.75	12) 1.15 - 1.25	17) 1.65 - 1.75
3) 0.25 - 0.35	8) 0.75 - 0.85	13) 1.25 - 1.35	18) 1.75 - 1.85
4) 0.35 - 0.45	9) 0.85 - 0.95	14) 1.35 - 1.45	19) 1.85 - 1.95
5) 0.45 - 0.55	10) 0.95 - 1.05	15) 1.45 - 1.55	

Fig 8. Histogram of emission [t/d] for optimal solutions (sources 1-10; $J_i = 3.92 \cdot 10^6$)

The basic, discrete-type solutions of the discussed optimization problems are shown in Table 4 and Table 5, respectively. Each set consists of two tables, with the first one relating to the fuzzy (reflecting uncertainty of data) solution and the second one to the reference, basic solution connected with the nominal values of emission in all sources. The original task is defined as an integer optimization problem, but the reference solution may also show some "fuzziness" due to the continuous approach applied in the optimization algorithm. Effect of the solution uncertainty can be observed by comparing both tables: left - for an uncertain solution and right - for a reference solution. It can be observed that in both cases (Table 4 and Table 5, respectively) there are 8—9 sources with an exact solution, with the remaining sources having more fuzzy (which means uncertain) character.
Other type results of the discussed optimization tasks are presented in Figures 7—8. They show the histograms of emission intensities relating to the optimal solution (only 10 dominating emission sources are presented), which include the impact of the input data uncertainty. The fuzzy emission distribution reflecting resulting uncertainty is once again compared with the basic (reference) solution obtained for the nominal initial emission (no uncertainty). It can be seen that the solutions for about 5-6 sources are exact (coincide with the reference solution) while the others demonstrate the impact of the input data uncertainty. One can also observe the correlation between the histograms and discrete-type solutions presented in Tables 4—5.

6. Summary

The aim of the computational tests presented in the previous section was to assess the uncertainty impact of modelling air pollution dispersion, when a pollution transport model is a part of a more complex decision support system. In the discussed implementation, the decision searched is defined as the result of an environment-oriented optimization problem. The utilized implementation addresses the problem of the optimal (minimum cost) allocation of emission abatement technologies, subject to the assumed (required) air quality standard which must be achieved (AT problem). The air quality is measured by means of the criterion function (5), which directly depends on an air pollution concentration, evaluated by a regional scale forecasting model (REGFOR3 Eulerian model (Holnicki, 2006; 2010a) is used). Uncertainties generated by the forecasting model (discussed in Section 4) manifest themselves in the final solution of the decision problem (results in Section 5).
The Monte Carlo analysis algorithm was applied to assess overall uncertainty of the optimization problem (AT). Selected results presented in Section 5 illustrate optimal solutions for two sets of the input data that relate to the mean emission intensity of the sources: (a) nominal emissions as shown in Table 2 (Winter season is considered), and (b) emissions increased by 10%. Randomly generated set of 1000 initial emissions was prepared for both cases, with the normal distribution of the data and uncertainty range corresponding to the results of Section 3.
Two types of results are presented. The first one has the form of a discrete-type solution that shows which emission reduction technologies should be applied in power plants under consideration to satisfy conditions of the (AT) problem. Some general conclusions can be formulated based on the two initial data case studies. Irrespectively of the initial data, identical or very approximate solutions were found for about 10 sources. These are the

sources No. 1, 4, 5, 6, 10-14, and 19. The fuzzy (uncertain) solution is also identical or very approximate to the reference one. Moreover, it can be observed that in the other sources, where fuzziness of the solution is more significant, the suggested choice of abatement technologies corresponds with the reference solution. Generally, more significant uncertainty characterizes the same sources, irrespectively of the assumed initial emissions. These are, first of all, sources with minor emission intensity or those located near the boundary of the computational region. The second group of results presents the histograms of each source's emissions which refer to the optimal solution to (AT). The fuzzy (uncertain) solution is once again compared with the reference one, obtained for the nominal input emission data. Moreover, in the optimal solution to all sources presented in Fig. 7–8, the dominating emission value of the distributed (with uncertainty) emissions coincides with the reference solution. On the other hand, while comparing the correspondence between the respective sources in Figures 7–8 and Tables 4–5, one can also see the correlation between these two types of results.

General conclusion that follows from the above tests relates to the applicability of operational air pollution transport models for supporting decisions in the field of environment protection policy. Most of decisions related to air quality management have a rather qualitative character. Thus, based on the results of the discussed case study, a more general opinion can be stated that, in spite of forecasting models' uncertainty, the accuracy of air quality predictions is sufficient for most of the applications and such predictions can be useful in decision support processes as well as in a strategic policy formulation regarding air quality management. On the other hand, taking both the needs of applications and uncertainty of the modelling process into consideration, utilization of very precise and time-consuming optimization algorithms seems to be unfounded. In this case, a sufficient, accurate and justified decision can be made upon application of simpler and more computationally efficient heuristic methods.

7. References

ApSimon, H.M. ; Warren, R.F. & Kayin, S. (2002). Addressing Uncertainty in Environmental Modeling: a Case Study of Integrated Assessment of Strategies to Combat Long-range Transboundary air Pollution. *Atmospheric Environment*, Vol.36, No.35, pp. 5417- 5426, ISSN 1352-2310.

Cofała, J. ; Amann, M. ; Asman, W. ; Bertok, I. ; Heyes, C. ; Hoeglund, I.L. ; Klimont, Z. ; Schoepp, W. & Wagner, F. (2010). Integrated assessment of air pollution and greenhouse gases mitigation in Europe. *Archives of Environmental Protection*. Vol.36, No.1, pp. 29-39, PL ISSN 0324-8461.

Hanna, S.R. ; Chang, J.C. & Fernau, M.E. (1998). Monte Carlo estimates of uncertainties in predictions by photochemical grid model (UAM-IV) due to uncertainties in input variables. *Atmospheric Environment*, Vol.32, No.21, pp. 3619-3628, ISSN 1352-2310.

Holnicki, P. (2006). *Air Pollution Transport Models in Air Quality Control* (in Polish). EXIT Publishers, ISBN 83-60434-13-1, Warsaw, Poland.

Holnicki, P. (2010a). Some Aspects of Integrated Approach to air Quality Management Based on Optimization Techniques. *Archives of Environmental Protection*. Vol.36, No.1, pp. 145-159, PL ISSN 0324-8461.

Holnicki, P. (2010b). Uncertainty in Integrated Systems of Air Quality Assessment (in Polish). Report RB/6/2010, Systems Research Institute PAS, Warsaw.

Holnicki, P. ; Nahorski, Z. & Tainio, M. (2010). Uncertainty in air quality forecasts caused by emission uncertainty. *Proceedings of HARMO 13th Conference on Harmonisation within Atmospheric Dispersion Modelling*, pp. 119-123, ISBN 2-8681-5062-4, Paris, France, June 1-4, 2010.

Jacobson, M.Z. (2005). *Fundamentals of Atmospheric Modeling*. Cambridge University Press., ISBN-13: 978-0521839709, Cambridge, U.K.

Kelly, J.A. (2006). *An Overview of the RAINS Model*. Environmental Research Centre Report, Environmental Protection Agency, ISBN: 1-84095-208-3, Dublin, Ireland.

Markiewicz, T.M. (2004). *Fundamentals of Air Pollution Modeling* (in Polish). Warsaw University of Technology Publ., ISBN 83-7207-461-5, Warsaw, Poland.

Mill, W. & Schlama A. (2010). Modeling Critical Loads of Airborne Acidity and Eutrophication of Polish Forest Ecosystems - the SONOX Model. *Archives of Environmental Protection*. Vol.36, No.1, pp. 117-127, PL ISSN 0324-8461.

Moore, G.E. & Londergan, R.J. (2001): Sampled Monte Carlo uncertainty analysis for photochemical grid models. *Atmospheric Environment*, Vol.35, No.28, pp. 4863-4876, ISSN 1352-2310.

Park, S.-K. ; Cobb, C.E. ; Wade, K. ; Mulholland, J. ; Hu, Y. & Russel, A.G. (2006). Uncertainty in air quality model evaluation for particulate matter due to spatial variations in pollutant concentrations. *Atmospheric Environment*, Vo.40, No.S2, pp. S563-S573, ISSN 1352-2310.

Russel, A. & Dennis, D. (2000). NASTRO critical review of photochemical models and modeling. *Atmospheric Environment*, Vo.34, No.14, pp. 2283-2324, ISSN 1352-2310.

Sax, T., & Isakov, V. (2003). A case study for assessing uncertainty in local-scale regulatory air quality modeling applications. *Atmospheric Environment*, Vol.37, No.25, pp. 3481-489, ISSN 1352-2310.

Scire, J.S. ; Strimaiti,s D.G. & Yamartino, R.J. (2000). *A User's Guide for the CALPUFF Dispersion Model*. Earth Technology Inc, Concord, MA, USA.

Sportisse B. (2007): A review of current issues in air pollution modeling and simulation. *Computational Geosciences*, Vo.11, No.1, pp. 159-181, ISSN 1420-0597.

Warren, R.F. & ApSimon, H.M. (1999). Uncertainties in integrated assessment modeling of abatement strategies: illustrations with the SAM model. *Environment Science and Policy*, Vol.2, No.6, 439-456, ISSN 1462-9011.

15

Evaluation of Regional Emission Control Based in Photochemical Air Quality Modelling

Ángel Rodríguez et al.*
University of Santiago de Compostela, Spain

1. Introduction

The growing use of natural gas as fuel both for industrial and domestic purposes implies an increment in the emissions of potential photochemical oxidants in the troposphere, particularly, nitrogen oxides (NO_x). This effect can be higher in regions with a significant use of other fossil fuels for electricity production, such as coal fired power plants.

Therefore, even though the air quality levels of typical primary pollutants (SO_2 and coarse particles) can keep to acceptable values, synergies between additional NO_x emissions, biogenic emissions and meteorological conditions can lead to the production of photochemical oxidants, such as O_3. The application of meteorological and air quality modelling techniques to evaluate the impact of new NO_x sources in the O_3 levels over a region can achieve a quantitative estimation of such impact, in terms of the variation of both NO_x and O_3 levels in the region. This approach to the impact of new sources in air quality, now common for primary pollutants, requires the use of comprehensive air quality modelling in the evaluation of photochemical oxidants.

There are several comprehensive air quality models currently available that can be applied to this approach. However, the characteristics of new emission sources, regional conditions (i.e., terrain, sea influence) and pollutants of interest (i.e., photochemical oxidants) result in the selection of the most appropriate model for a particular problem (Chang et al., 1987; Song et al., 2010).

In this paper, comprehensive air quality modelling is applied to evaluate the effectiveness of emission control policy in a European Atlantic coastal region, with complex terrain and mixed land use, and a significant area of forestry. This region, located in the northwest of the Iberian Peninsula, includes both medium-sized cities and large coal fired power plants. Therefore, changes in both industrial and domestic emissions due to an increase in the use of the natural gas can affect the generation of photochemical oxidants, depending on the biogenic emissions (forestry) and typical meteorological conditions.

---

* Santiago Saavedra[1], María Dios[1], Carmen Torres[1,2], José A. Souto[1], Juan Casares[1], Belén Soto[3] and José L. Bermúdez[3]
[1] University of Santiago de Compostela, Spain.
[2] Now at University of Rovira i Virgili, Spain.
[3] Endesa Generación, Spain.

2. A case study

Air pollution on local and regional scales is conditioned by a significant number of parameters grouped into three classes (Seinfeld and Pandis, 1998): topography, meteorology and precursors. Meteorology, which is implicitly linked to topography, plays an important role in air quality. Topography defines air flow paths and pollutant depositions. The influence of emission sources depends on both the amount of pollutants and the location of such sources (Beck et al., 1998). Therefore, the description of the area under study must include these three groups of parameters.

Fig. 1. Regional domain under study, including the Galician region. In addition, the location of the five O_3 ground level concentration stations around As Pontes power plant, to be used identifying air quality episodes, is shown.

2.1 Topography

The area under study (figure 1) is characterized by a complex terrain environment, with large differences. The central part is defined by the As Pontes valley, from west to east, with the river Eume flowing along it. This valley is surrounded by the smooth Atlantic coast in the west and the more complex Cantabrian coast in the north; there are several mountains in the north-east (e.g. Xistral), and the valley boundaries are closed in the east by the Dorsal Gallega mountain range, up to 1000 m high. Along the south, the topography displays smooth, uniform, elevated plains. Therefore, it is a complex terrain, combining granitic mountains and valleys in the same environment.

2.2 Meteorology

This region is characterized (Martínez Cortizas et al, 1999) by rains distributed throughout the year, with an annual precipitation rate between 1000 and 1600 mm, mainly during autumn and spring, and sporadic in summertime (but not unusual, as there are isolated storms in the afternoons). Summers are mild, as the sea breeze refreshes the coastal areas and the height of the mountains regulates the temperature in the inner areas. Summer days are usually sunny, with low moisture and temperatures from 20 to 30 °C. On the other hand, heat waves are not usual in summer, and they only last a few days, as the proximity of the coast keeps the average temperature at 20-25 °C, with higher temperatures (up to 30 °C) in the interior valleys.

Main winds come from the SW and NW during winter and autumn, when pressure is low, whereas high pressure, typical in summertime, usually corresponds to NE winds. The atmosphere-topography interaction is very important in this environment, especially the mountains, which affect the production of rain, and the complex coastal topography, which increases the sea breeze.

2.3 Sources of ozone precursors

The main ozone precursors are taken to be oxidized nitrogen and volatile organic compounds. Sources of these precursors located in the area under study (Casares et al., 2005) can be divided into point sources (usually, industrial sources) and diffuse sources. Industrial sources include the Artabric Arc Industrial Belt (located on the west coast, including the cities of A Coruña and Ferrol), where there are various sectors such as petroleum refining, energy (thermal power plants and cogeneration units), wood, food and metallurgy. Other sparse emission sources, such as industrial waste plants, are also included.

The main diffuse sources are traffic and biogenic emissions from cattle farming and forest cultivation (mainly, eucalyptus and pines). However, as ozone can be transported over long distances, it is necessary to take into account not only the local sources, but also sources in neighbouring regions (Asturias, Castilla-León and Portugal), and even further afield. This was done by analyzing air quality levels in those regions, as a characteristic of air masses that can be carried to the area under study. In order to evaluate the impact of emission changes on air quality, three typical meteorological conditions favourable to increasing tropospheric ozone levels are identified in the NW of the Iberian Peninsula. From these typical conditions, three different periods were selected as reference episodes to apply a comprehensive air quality modelling of different emission scenarios: Episode 1 (14-23 July 2002), episode 2 (16-24 March 2003), and episode 3 (12-22 September 2003). The characterization of these episodes shows a different principal origin of ozone peaks depending on the meteorological conditions: episode 1, North of Portugal; episode 2, Castilian plateau; and episode 3, Galicia.

3. Air pollution episodes

3.1 Air quality data

Hourly ground level concentration registers of O_3, NO_x and NO_2 were collected for the 3-year period 2002-2004. First of all, validated data were obtained from 10 air quality stations included in the Galician Air Quality Network, in order to identify the O_3 episodes, and to obtain a first approach to them. These stations are mainly rural, with only background influence or, sometimes, influence from distant industrial sources. After identifying the main O_3 episodes, the possible influence from surrounding regions was studied by looking at more air quality stations: 19 air quality stations from the National Portuguese Air Quality Network, 18 from the Castilla-León Regional Air Quality Network, and daily average data from the Asturias Regional Air Quality Network (as hourly data were not available).

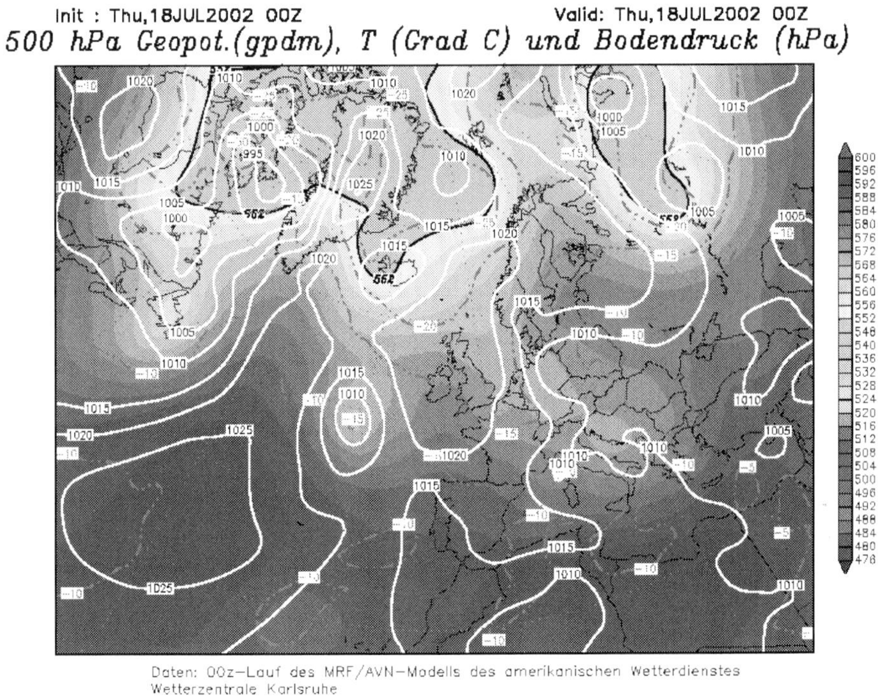

Fig. 2. An example of a synoptic chart provided by Wetterzentrale, applied in the classification of ozone air pollution episodes in the region under study.

3.2 Meteorological dataset

The regional scale meteorological analysis of the episodes was done by means of synoptic maps (Font-Tullot, 1983; Castell et al., 2004), in order to identify the synoptic conditions with more influence in the tropospheric ozone episodes over Galicia.

Synoptic charts (figure 2) were collected from the National Center for Environmental Prediction (NCEP, USA) reanalysis. These charts include synoptic observations which are interpolated and recalculated by diagnostic models, so they are more trustworthy than forecast charts. In these NCEP charts, several variables were included, over geopotentials at 500 mb and sea level pressure; in addition, temperature maps at 500 and 850 mb are included. These charts are routinely collected by the Wetterzentrale (German Meteorological Office) on its web page (www.wetterzentrale.de).

Regional meteorology was completed by observations of surface temperature and wind, precipitations and other typical meteors (clouds, storms, fog, etc.) from Wetterzentrale maps based in European meteorological stations with 6 h observations.
After identifying the ozone episodes, local meteorological observations were collected, in order to obtain a more comprehensive analysis of the meteorological influence over local pollution episodes, taking into account local phenomena, such as katabatic and anabatic winds, sea breeze and thermal inversions, which are not represented on a synoptic scale. This is an important issue in a complex terrain like Galicia.
Local surface meteorological measurements were taken on an hourly basis and included surface temperature (2 m) and surface wind speed and direction (10 m) from five stations of the As Pontes Power Plant air quality network, integrated in the Galician regional air quality network. These data were extended with hourly measurements from A Mourela station (D1 station, close to the Power Plant), providing temperatures at 2, 10, 30 and 80 agl-m (which is useful for stability classification), wind at 10 and 80 agl-m, solar radiation and precipitation.

Episode period	Maximum hourly O_3 glc date/time	Maximum hourly O_3 glc (μg/m^3)
	Year 2002	
14-23 July	18 July - 16 UTC	201
11-17 August	14 August - 16 UTC	142
23-04 September	02 September - 16 UTC	157
09-18 September	14 September - 14 UTC	136
21-30 September	27 September - 17 UTC	145
	Year 2003	
16-24 March	21 March - 16 UTC	148
26-01 June	29 May - 05 UTC	161
18-24 June	20 June - 17 UTC	210
06-13 July	08 July - 15 UTC	139
30-16 August	07 August - 17 UTC	174
12-22 September	16 September - 15 UTC	193
	Year 2004	
14-22 May (12)	19 May - 15 UTC	170
31-11 June (13)	05 June - 15 UTC	156
12-19 June (14)	17 June - 16 UTC	196
28-05 August (15)	01 August - 00 UTC	181

Table 1. O_3 episodes identified in the region under study, from the glc measurements available, taking dual criteria into account: either 150 μg/m^3 threshold or significant rise of glc in a short period.

3.3 Air pollution episodes

A coarse selection of episodes with tropospheric ozone levels higher than usual (risky episodes) was made with reference to the air pollution measurement dataset in the domain for 2002, 2003 and 2004. These measurements correspond to the five air quality stations located mainly in rural areas around As Pontes Power Plant (figure 1), and distributed over

2 to 30 km away from the industrial site. These stations are integrated into the Galician air quality network, controlled by the regional government.
In order to select O_3 risky episodes, two criteria were adopted: (a) tropospheric ozone threshold of 150 $\mu g/m^3$, as an hourly average (20% below first European Union legal threshold); (b) persistent rise of tropospheric ozone levels in several stations, independently on the absolute values achieved.
The period under study for each episode was established as at least 7 days around the maximum concentration, which can be extended depending on the concentration tendency (persistence either in ascent or descent). This period is important because of the time required by ozone production and destruction, taking into account the influence of meteorological conditions in these chemical phenomena. In fact, the duration of an episode was extended until it was clear that clean air masses with low ozone levels were achieved in the domain, ensuring that the episode had ended.
With these criteria, 15 tropospheric ozone air pollution episodes (table 1) were identified in the domain, 5 in the year 2002, 6 in 2003, and 4 in 2004. All the episodes were observed between March and September. Episodes lasted from 7 to 18 days. These episodes can be represented by three typical meteorological conditions which are favourable to the increase in tropospheric ozone levels, and are identified in the NW of the Iberian Peninsula. Therefore, three different periods were selected as reference episodes to apply a comprehensive air quality modelling to different emission scenarios: Episode 1 (14-23 July 2002), episode 2 (16-24 March 2003), and episode 3 (12-22 September 2003). The characterization of these episodes shows a different main origin of ozone peaks depending on the meteorological conditions: episode 1, North of Portugal; episode 2, Castilian plateau; and episode 3, Galicia.

4. Air quality modelling

In this paper, the CAMx model (Environ International Co., 2010; Song et al., 2010) was selected because of its capacity to integrate large aloft emissions from power plant stacks in the model grid with area emissions representing sparse sources (domestic and transport). This approach is essential to obtain a good representation of the impact from large point sources in regional air quality.
The CAMx model is coupled with the PSUN/NCAR MM5 meteorological model (Grell et al., 1995) in order to provide the required meteorological conditions for both dispersion and chemical transformation of pollutants.

4.1 Simulation domains

The area under study covers Galicia, a region in the north-west of Spain, although the simulation domain has to cover a larger area in order to minimize the boundary effect on the model results. However, because of the high resolution required in the complex region under study, applying a high resolution to the larger simulation domain leads to heavy computing work. Therefore, the use of the nesting technique is highly recommended.
Thus, a coarse resolution grid (with a resolution of 27x27 km^2) covering the Iberian Peninsula (see figure 3a) is defined. Simulation results over this larger domain were applied as input to a nested inner grid (with a resolution of 9x9 km^2, see figure 3b), following a one-way nesting approach; i.e, it is thought that the meteorology over the inner grid does not substantially affect the meteorology over the outer grid.

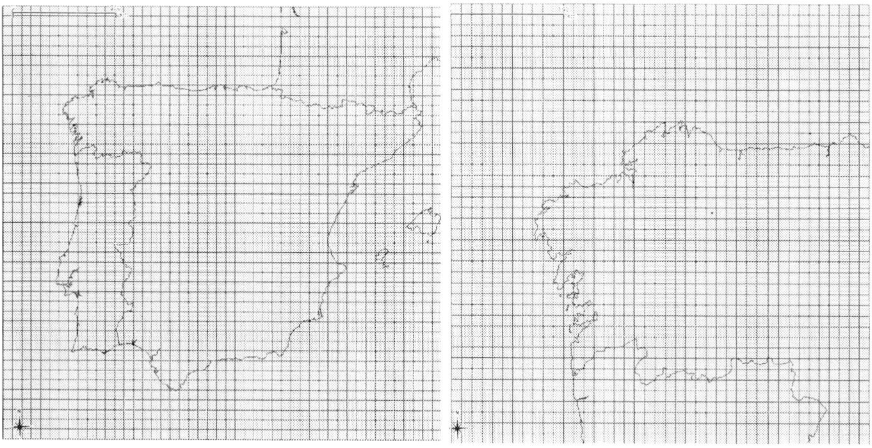

(a) (b)

Fig. 3. Simulation grids covering (a) the Iberian Peninsula, and (b) the northwest of the Iberian Peninsula, including the Galician region.

An additional problem in the application of air quality models relates to geographical projections. These models are usually solved over plane grids, even though the Earth surface is curved. In order to define the plane representation of the Earth surface, the following parameters are defined: (a) geodesic datum, i.e., the shape and dimensions of the ellipsoid to be represented; in our case, it is the Earth. The European Datum 50 (ED50) was adopted; (b) geographical projection, i.e., a sorted relationship between the locations over the curved surface (Earth's surface) and the plane surface over the simulation grid is defined. In this case, UTM ("Universal Transverse Mercator") projection referring to 29 North Time Zone is used.
As a result, the Iberian Peninsula domain is represented by a grid (figure 3a) with 44x47 cells. The inner domain is represented by a grid (figure 3b) with 41x41 cells. However, the computing work required to solve the inner grid is higher, because of the Courant number; i.e., the smaller the size of the grid cells is, the shorter the time step to be applied in the numerical integration of the model equations during the same simulation period. Therefore, in the inner grid, it is necessary to solve the equations more often to simulate the same period. This is a significant limitation in the application of the Eulerian models (such as the CAMx model) in the simulation of the air quality with high resolution grids.

4.2 Emission inventories

As the simulation domains cover the Iberian Peninsula, including the Galician region, a 2001 annual emission inventory was developed by combining the top-down inventory from EMEP (Iberian Peninsula) and the bottom-up industrial emission EPER inventory from REGADE (Casares et al., 2005). SMOKE (The Institute for the Environment, 2003) was applied for speciation, time and spatial segregation of industrial emissions.
In addition, a 2010 annual emission projection (table 2) was carried out using changes in industrial emissions, due to the application of Best Available Techniques (BATs) to reduce emissions from the old power plants, and the installation of new combined cycles in the

region. This new scenario follows the legal restrictions currently approved by the Government, so changes should lead to a significant reduction in primary pollutant levels (SO_2, PM10); however, secondary pollutants, such as ozone, are affected by extremely non-linear physical and chemical processes that should be considered when evaluating the effect of emission changes over their levels. These non-linear processes can be simulated by a comprehensive air quality model.

Source	NO_x (Tm/year) 2001 inventory	2010 projection	Change
Sabón	833 -	738 2764	+320 %
Meirama	9059	1604	-83 %
As Pontes	20035 -	6863 2764	-52 %

Table 2. NO_x emission changes between year 2001 (reference inventory) and 2010 (projected inventory), in the three power plants.

As the main differences in the industrial emissions of ozone precursors between 2001 inventory and 2010 projection, NMVOCs shows an increment of 6.6% and NO_x a reduction of 27%. In addition, a significant spatial redistribution of NO_x industrial sources is observed (table 2) due to: (a) fuel changes in the three Galician high stack power plants, which have reduced their NO_x emissions by 50%; (b) two planned combined cycles, with NO_x emissions at lower height (less than 200 m).

In addition to the anthropogenic contribution, the role of volatile organic compounds (VOCs) in tropospheric ozone production is also taken into account, especially for biogenic organic species (BVOCs), such as isoprene and terpenes (Makar et al., 1999; Geron et al., 2000). These BVOCs emissions were estimated by using MEGAN model (Guenther et al., 1995), as this model uses a high resolution functional plant distribution (1x1 km^2) and the results can be processed as input (such as time-varying emissions) to the CAMx model.

4.3 Emission integration

4.3.1 Simulation grids

For a realistic representation of anthropogenic emissions in the region, different emission inventories were combined: EMEP for the Peninsula domain and EPER-REGADE for the Galician domain. Both inventories were adapted to the corresponding nested grids using SMOKE, in order to provide the emission input to the CAMx model.

a. Peninsula domain

In the Peninsula domain, only the EMEP emission inventory was applied. However, this inventory is aggregated in a 50 km resolution grid with the origin in the North Pole; therefore, it must be adapted to the Peninsula domain simulation grid. This adaptation has been done by applying the GIS extension of the PortgreSQL database, with a distribution weighted to the surface area (figure 4).

b. Galician domain

In the Galician domain, EMEP emissions were applied for CORINAIR sources, other than S1 and S3, following the same process as in the Peninsula domain. With the EMEP inventory

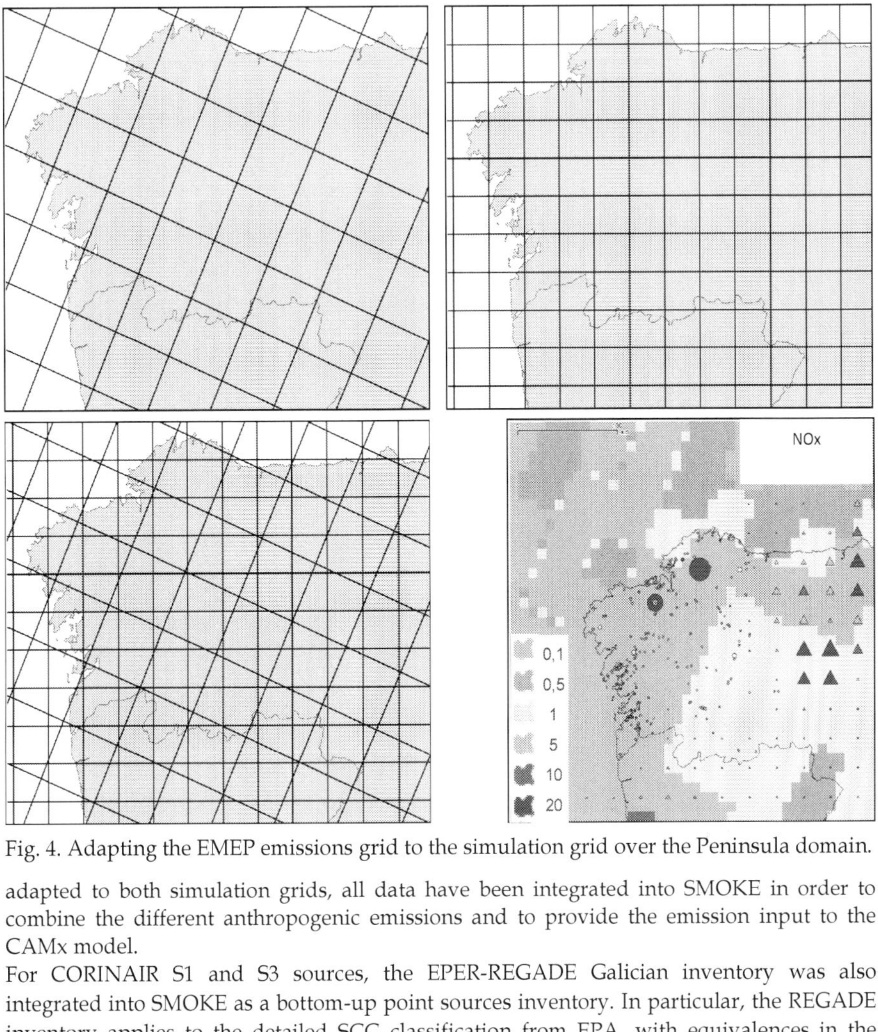

Fig. 4. Adapting the EMEP emissions grid to the simulation grid over the Peninsula domain.

adapted to both simulation grids, all data have been integrated into SMOKE in order to combine the different anthropogenic emissions and to provide the emission input to the CAMx model.

For CORINAIR S1 and S3 sources, the EPER-REGADE Galician inventory was also integrated into SMOKE as a bottom-up point sources inventory. In particular, the REGADE inventory applies to the detailed SCC classification from EPA, with equivalences in the coarser CORINAIR sources classification. In addition, this procedure is important in order to obtain an accurate geographical location of these industrial sources and, particularly, to take into account their different stack heights in the CAMx algorithm for elevated point sources.

4.3.2 Time-varying anthropogenic emissions

Depending on the type of emission sources, the variation in time of their emissions is very different. Industrial emissions are normally quite steady-state. However, non-industrial emissions present varying behaviour depending on their specific activity; therefore, more

complex temporal profiles for the non-industrial emissions have been used, depending on the simulation domain, which are:
Peninsula domain: The industrial emissions from S1 and S3 CORINAIR sectors are applied as steady state annually, weekly and daily. For the remaining emissions (sectors), no annual or weekly variations were used, but a daily profile with daytime emissions three times higher than night-time emissions was applied.
Galician regional domain: As the industrial emissions for this domain are available from the EPER-REGADE inventory with SCC EPA classification, temporal profiles following EPA recommendations were applied. These profiles are included in the SMOKE package for processing SCC classified emissions.

4.3.3 Aloft anthropogenic emission distribution

The height of emissions is a significant factor in the results of air quality models (Kumar and Russell, 1996), especially for large point sources with stacks. In fact, CAMx model considers point sources and surface sources separately, as the model includes a plume rise algorithm to estimate the increment of the emission height for elevated point sources. However, this algorithm requires complex datasets that are not always available. Therefore, depending on the precision required in each of the simulation grids, different approaches to entering the emission height in the model are used, as follows.
Peninsula domain: As the EMEP inventory only includes the geographical location of emissions, their height is assumed from the group of emission sources. Thus, S1 and S3 industrial sectors emissions from every grid cell are converted to one virtual point source by cell, and every virtual point is considered to be emitted at 200 agl-m (as an average value); the rest of the emissions over every cell are taken as surface level emissions.
Galician regional domain: The EPER-REGADE inventory includes all the dataset required to define the stack heights and their plume rises. This dataset is included in the SMOKE package to distinguish surface and elevated sources and to provide the CAMx with the required information for the elevated sources.

4.3.4 Chemical speciation

Emission inventories include a chemical speciation which is usually less detailed than required by the chemical mechanisms applied in the air quality models (Carter, 1990). Therefore, it is necessary to segregate some groups of emissions in order to obtain an approximation to the actual chemical mechanism speciation. In this work, Carbon Bond IV (CBIV) chemical mechanism included in the CAMx model is applied. Therefore, for an initial emission speciation to this mechanism, the Spcmat tool from SMOKE is used.

5. Results

Simulations were performed with CAMx, with two nested grids covering the Iberian Peninsula (27x27 km^2 resolution) and Galicia (9x9 km^2 resolution), using chemical boundary conditions provided by GOCART and MOZART data. Meteorological inputs were provided by PSUN/NCAR MM5 one-way nesting simulations at 27x27 km^2 and 9x9 km^2 resolutions, using NCEP re-analysis as initial and boundary conditions.
Results for the three different reference episodes (figure 7) show that changes in industrial emissions will produce a significant reduction in the nocturnal ozone glc around the new

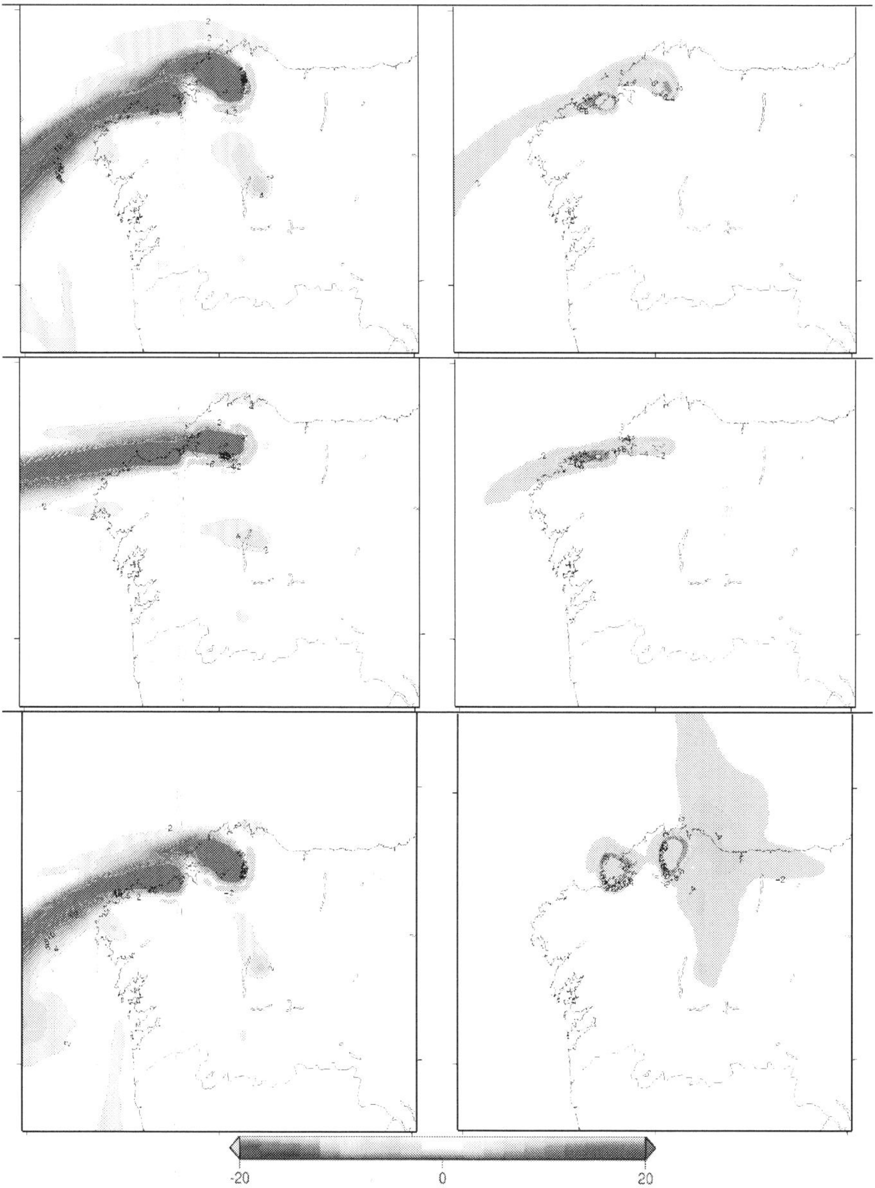

Fig. 7. Relative differences (%) in nocturnal NO_2 (left) and O_3 (right) glc over Galicia between 2001 emission inventory and 2010 emission projection simulations at the central date of the episodes under study. From above to below: 18/July/2002 - 04 UTC, 21/March/2003 - 03 UTC, and 17/September/2003 - 03 UTC, respectively.

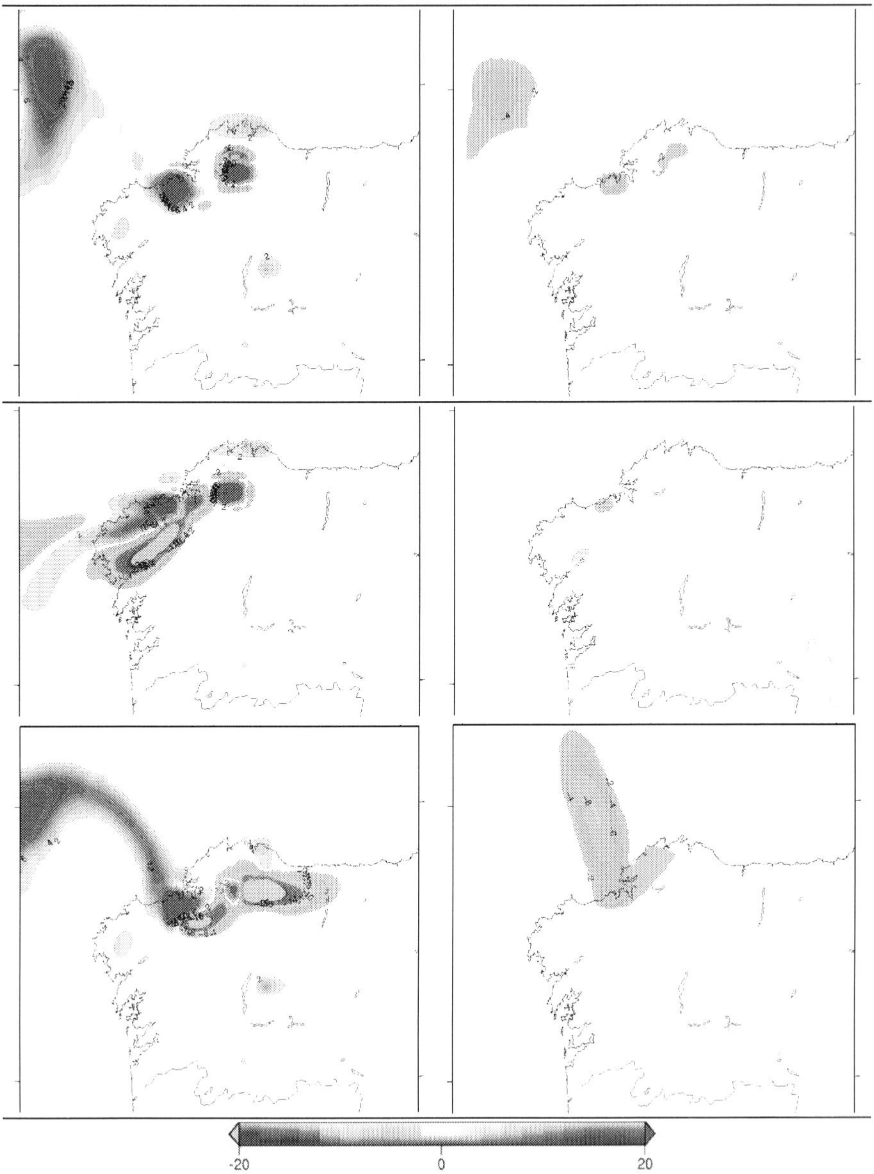

Fig. 8. Relative differences (%) in diurnal NO_2 (left) and O_3 glc (right) over Galicia between 2001 emission inventory and 2010 emission projection simulations at the central date of the episodes under study. From above to below: 18/July/2002 - 16 UTC, 21/March/2003 - 14 UTC, and 17/September/2003 - 14 UTC, respectively.

combined cycles (As Pontes and Sabón) and surrounding areas sporadically affected by their plumes; this reduction is not observed during the diurnal period (figure 8). At least for episodes 1 and 2, nocturnal reduction is due to the significant increment of NO_x glc around both sources (because their stacks are lower than the old coal fired power plant stacks). In episode 3, the ozone reduction is more sparse, even in an opposite direction (east) to the NO_x industrial plumes; this may be due to lower winds and higher dispersion during this episode.
No significant changes (figure 8) are observed in the diurnal ozone glc; only a small reduction over the local areas affected by the plumes from the aloft sources (As Pontes and

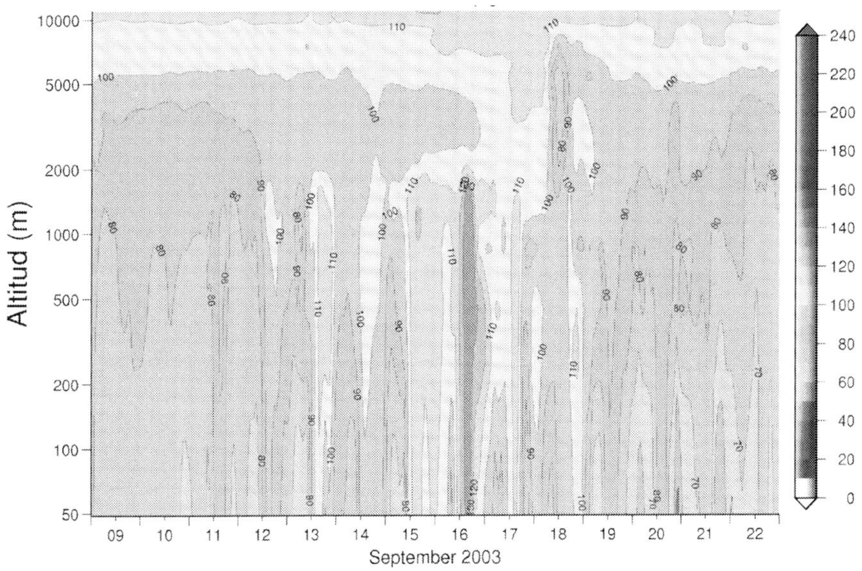

Fig. 9. Relative differences (%) in nocturnal O_3 concentration over the maximum glc location along the episode 9-22 September 2003, between 2001 emission inventory and 2010 emission projection simulations.

Meirama), as the combined effect of the reduction of NO_x emissions aloft (reducing diurnal ozone production) and the increment of NO_x surface emissions (increasing diurnal ozone production and consuming nocturnal ozone).

However, ozone glc reduction is accompanied by an increment in the aloft O_3 levels; as an example, figure 9 shows an increment in aloft O_3 concentration and, simultaneously, a decrease in O_3 concentration at lower levels (including ground level). At the same time, there is no significant changes in nocturnal ozone glc in the rest of the region, even though a significant reduction of NO_x emissions aloft is expected from the old high stack coal fired power plants (see table 2); this is because aloft emissions are dispersed above the nocturnal stable layer.

Changes in air quality are more significant in NO_2 glc: there is an increment of NO_2 glc around Sabón and As Pontes sources (figures 7 and 9), due to the contribution of surface emissions around these areas.

6. Conclusions

From the comparison of 2001 emission inventory and 2010 emission projection at Galician region, simulation results from the CAMx model show a change in ozone distribution. Surprisingly, concentration downwind from the combined cycles is decreased because of their lower stacks (with respect to the old coal and fuel oil power plants), and slightly decreased in remote areas. This effect is especially significant during nocturnal periods.

On the other hand, simulation results show an increment of O_3 aloft concentration, which can be transported to neighbouring regions and increase their glc. This effect is not usually taken into account in the current impact assessment, which is based on either estimated or measured pollutant glc close to the sources. Therefore, although the strategy of emission reduction in this region can achieve better results on a local scale, it could lead to an increase in the contribution of this region to the O_3 levels in the vicinity.

At the same time, an increment in the NO_x ground level concentration is expected both due to the rise in surface NO_x emissions and the lower stack height of the new industrial sources (combined cycles). Therefore, although ozone glc achieves local reductions along the NO_x industrial plumes, the increment of NO_x levels can lead to worse air quality in other areas not affected by these industrial plumes (i.e. urban and rural areas far away from these sources). However, simulation results show no significant effect due to NO_x in the glc values.

7. Acknowledgements

This work has been financially supported by the Spanish Research & Development Programme, Ministry of Science and Technology, under project CTQ15481-PPQ, and Endesa Generación S.A.. Research grants of the "María Barbeito" Programme (Xunta de Galicia) to Angel Rodríguez and María Dios, are acknowledged.

Hourly ground level concentration measurements were provided by the Portuguese Ministry of Environment, the Spanish Ministry of Environment, the Galician Regional Department of Environment, the Castilla-León Regional Department of Environment, and Endesa Generación, S.A..

8. References

Beck, J.P., Krzyzanowski, M., Koffi, B., 1998: Tropospheric Ozone in the European Union - The consolidated report, Topic Report no. 8/1998, European Environment Agency.

Carter, W. P. L., 1990: A detailed mechanism for the gas-phase atmospheric reactions of organic compounds, Atmospheric Environment, 24, pp. 481 - 518.

Casares, J. J., Rodríguez, R., Maceira, P., Souto, J. A., Ramos, S., Costoya, M. A., Sáez, A., Vellón, J.M., 2005: Inventario, Análisis y Proyección de las Emisiones Atmosféricas Industriales de Galicia, Ed. Servizo de Publicacións e Intercambio Científico, Universidade de Santiago de Compostela, Santiago de Compostela.

Castell, N., Mantilla, M., Fernández, F., López, E., 2004: Distribución espacial y temporal de los episodios de contaminación por ozono y su relación con las situaciones sinópticas en la vertiente mediterránea, Proyectos y métodos actuales en Climatología, Asociación Española de Climatología, 2004, Serie B, 1, pp. 600-606.

Chang, J. S., Brost, R. A., Isakson, I. S. A., Madronich, S., Middleton, P., Stockwell, W. R., Walcek, C. J., 1987: A three-dimensional Eulerian acid deposition model: physical concepts and formulation, Journal of Geophysical Research, 92, pp. 14651 - 14700.

Environ International Co., 2010: CAMx: Comprehensive Air Quality Model with Extensions - User's Guide. Novato (Ca), USA.

Font-Tullot, I., 1983: Climatología de España y Portugal, Instituto Nacional de Meteorología, Madrid.

Geron, C., Guenther, A., Sharkey, T., and Arnts, R. R., 2000: Temporal variability in basal isoprene emission factor, Tree Physiology, 20(12), pp. 799-805.

Grell, G.A., Dudhia, J., Stauffer, D.R.,1995: A description of the Fifth-Generation Penn State / NCAR Mesoscale Model (MM5), Technical Report NCAR/TN-398+STR, NCAR, Boulder (Co), USA.

Guenther, A., Hewitt, C. N., Erickson, D., Fall, R., Geron, C., Graedel, T., Harley, P., Klinger, L., Lerdau, M., Mckay, W. A., Pierce, T., Scholes, B., Steinbrecher, R., Tallamraju, R., Taylor, J., and Zimmerman, P. A., 1995: Global-Model of Natural Volatile Organic-Compound Emissions, Journal of Geophysical Research, 100(D5), pp. 8873–8892.

Kumar, N., Russell, A.G., 1996: Development of a computationally efficient, reactive subgrid-scale plume model and the impact in the northeastern United States using increasing levels of chemical detail, Journal of Geophysical Research, 101, pp. 16737 - 16744.

Makar, P., Fuentes, J., Wang, D., Staebler, R., and Wiebe, H. 1999: Chemical processing of biogenic hydrocarbons within and above a temperate deciduous forest, Journal of Geophysical Research, pp. 104(D3), pp. 3581- 3603.

Seinfeld, J. H., Pandis, S.N., 1998: Atmospheric Chemistry and Physics, From Air Pollution to Climate Change, John Wiley and Sons, Inc., New York.

Song, J et al., 2010: Ozone response to emission changes: a modeling study during the MCMA-2006/MILAGRO Campaign, Atmospheric Chemistry and Physics, 10, pp. 3827-3846

Sparse Matrix Operator Kernel Emissions (SMOKE), 2006: SMOKE v2.3 User's Manual, Carolina Environmental Program, University of North Carolina at Chapel Hill

The Institute for the Environment, 2003: SMOKE v2.0 User Manual. CEP (Carolina Environmental Program). The University of North Carolina at Chapel Hill.

Part 3

Advanced GIS Applications in Air Quality Management

16

Development of GIS-aided Emission Inventory of Air Pollutants for an Urban Environment

Sailesh N. Behera[1], Mukesh Sharma[1], Onkar Dikshit[1] and S.P. Shukla[2]
[1]*Department of Civil Engineering, Indian Institute of Technology Kanpur, Kanpur,*
[2]*Department of Civil Engineering, Institute of Engineering and Technology, Lucknow*
India

1. Introduction

Understanding types of pollutants and their emission rates is essential for control of air pollution. The emission rates, along with the rates of various chemical reactions those take place in the atmosphere, the prevailing meteorological conditions and topographical factors determine the extent of pollutions experienced by various environmental components. An emission inventory is a quantitative detailed repository of air pollutants emitted (along with their sources) into the atmosphere from a given area for a specified time period (Behera et al., 2011). Present and future inventories are critical components of air quality planning and modeling. The ultimate goal of the pollution control planning process is to identify and achieve emission patterns that do not result in violations of ambient air quality standards. An emission inventory should be able to provide: (i) a consistent estimate of total emissions of different pollutants, (ii) the spatial and temporal distribution of pollutants, (iii) the time-specific evolution of emissions and their distributions, (iv) the identification and characterization of sources, (v) tracking progress towards attainment of National Ambient Air Quality Standards (NAAQS) and emission reduction plans, (vi) serve as the basis in modeling for prediction of pollutant concentrations in ambient air, and (vii) compliance records with allowable emission rates established by facilities and regulatory agencies.

The broad source categories of air pollution in an urban area include; (i) transport (motor vehicles and railways), (ii) commercial establishments, (iii) industrial, (iv) domestic cooking/heating, (v) fugitive dust and (vi) biomass burning. There could also be some unique or specific sources in a particular area. Procedures and reliability of emission inventory for regular point, area and line sources are well-established. However, identification and quantification of fugitive/non-point emission sources (emissions not released through stacks, vents, ducts or pipes) are quite challenging. For these sources, emission factors (EFs) (EF: a ratio that relates the emission of a pollutant to an activity level) are either not developed or, if developed, are associated with high amounts of uncertainty. The important non-point sources include: traffic related non-exhaust (NE) emissions (tire, road and brake wear, road dust), evaporative emissions, raw material handling, open crushing, and biomass and agricultural residue burning.

The development of a better emission inventory poses a challenge to air quality engineers, as it requires systematic *in situ* surveys to collect activity data and proper management system for handling these data. Therefore, it is important to develop emission inventories based on geographic information system (GIS), which could also help the policy makers for implementation of legislation (Diem and Comrie, 2001). GIS can be defined as a computer-based system for capturing, storing, checking, and manipulating data that are spatially referenced, which can be of immense utility in developing, displaying and querying an information related to emission inventory.

Unlike other criteria air pollutants, particulate matter (PM) in the atmosphere causes several problems to the community in terms of human health effects, reduced visibility and climate change (Brown et al., 2004; Behera et al., 2005; Makar et al., 2003; Zhang et al., 2010). PM can be termed as a complex mixture of organic and inorganic substances, present in the atmosphere. The size of particles is directly linked to the potential for causing health problems. USEPA is concerned about PM_{10} (PM having diameter less than 10 μm) because these particles generally pass through the throat and nose and enter the lungs. Once inhaled, these particles can affect the heart and lungs and cause serious health effects (Suh et al., 2000). The sources of PM_{10} on the basis of its origin may be either from primary sources or/and from secondary transformations (Behera and Sharma 2010a; Behera and Sharma, 2010b). Control of PM_{10} from secondary transformation is not easy, whereas primary sources responsible for PM_{10} can be controlled by regulating legislation on the policies for reductions. It is noteworthy that in the task of reduction of PM_{10} from primary sources, the secondary transformation will be controlled in parallel to some extent. The reason being PM precursor gases (e.g., nitrogen oxides (NO_x), sulfur dioxide (SO_2), ammonia (NH_3) and volatile organic compounds (VOCs)) responsible for secondary formation of PM_{10} will also be controlled from the same sources of primary PM_{10} emission.

The objective of this chapter is to present a systematic set of approaches to prepare a GIS-based emission inventory for an urban environment. We have considered development of PM_{10} emission inventory as an example. The study area is Kanpur city (latitude 26°26′ N and longitude 88°22′ E), which represents typical weather conditions and atmospheric seasonal variability in the Ganga basin (longitudes 73°30′–89°0′ E and latitudes 22°30′–31°30′ N), India (Figure 2, described later). The specific objectives of the present study are: (1) identification of sources responsible for PM_{10} pollution and source categorization, (2) collection of activity level data of identified sources, (3) methods for estimating emissions of PM_{10} from various sources, (4) selection of emission factors of identified sources, (5) development of GIS-based digitized maps, (6) development of spatially resolved GIS-based emission inventory of PM_{10} and (7) assessment of contribution of the identified sources towards the particulate pollution.

2. Materials and methods

The methodology adopted in this study consists of four steps to achieve final goal as 'emission inventory of PM_{10}' (summarized in Figure 1). In the first step, identification of sources and collection of source information including location, population, production, fuel use, height of release, temperature etc. for area, line and point sources are done. In the second step, emission factors and methodology for emission estimation are selected. Next, digitized map of the study area with 2 km × 2 km grid resolutions is evolved. In the third step, an emission inventory of PM_{10} has been developed with ArcGIS after execution of data management of activity levels. In the final step, spatially resolved map of PM_{10} emission loads over the study area is generated and the contributions of identified sources towards PM_{10} pollution are assessed.

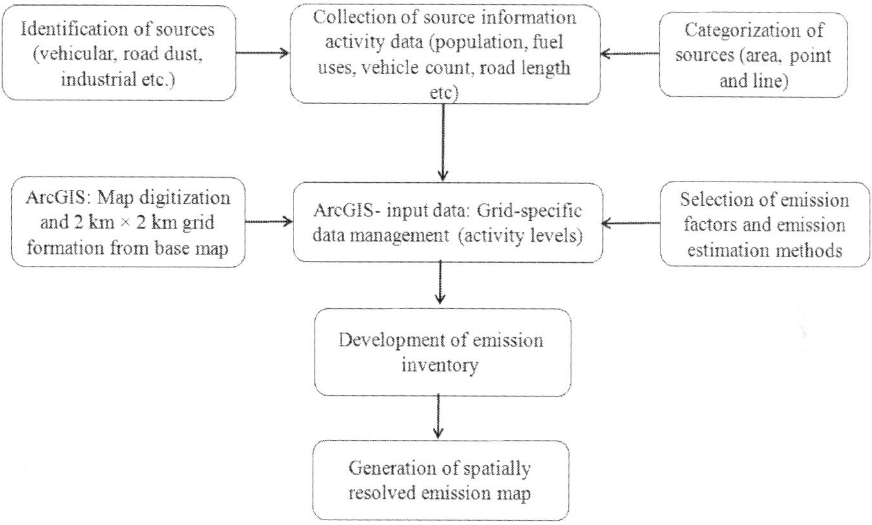

Fig. 1. Summary of the Approaches for Development of Emission Inventory

2.1 The study area

The study area (Kanpur city; Figure 2) represents typical weather conditions and atmospheric seasonal variability in the Ganga basin, India (Behera and Sharma, 2010a). The Ganga Basin is the largest river basin in India, supporting more than 40% of India's population and accounting for 26% of the Indian landmass (Figure 2). As per 2001 census, the basin has 104 urban centers (population > 0.1 million) and 7 large cities (Kanpur, Lucknow, Patna, Agra, Meerut, Varanasi and Allahabad) having a population of more than one million each. Unbridled growth of population and economy in this region has resulted in a wide range of anthropogenic activities including biomass and fossil fuel burning, industry (including coal-based power generation of more than 10000 MW), transport, mining, urbanization and agricultural activities (Behera and Sharma, 2010a). The Ganga basin is characterized to have large uncontrolled emissions of PM, SO_2, NO_x (from combustion sources) and NH_3 (from poor waste management practices). Prevailing westerly winds, dry conditions during winter and low anticyclonic winds provide favorable conditions for advection from the west and confinement of high aerosol loading over the central and eastern parts of the basin (Nair et al., 2007). As a result, consistent hazy conditions prevail from East Pakistan to Bay of Bengal.

Kanpur is a large industrial city (area 270 km^2) with population of about 4 million (census of 2001) having cotton, leather, and wool industries. Temperatures in cold weather drop to freezing, sometimes reaching a minimum of −1°C (lowest −1°C, 2004, 1968) with maximum at almost 12-14°C. In summer (April to June), the maximum temperature can go up to 47.5°C. During the rainy season (July to September), the relative humidity is generally over 70%. Thereafter, the humidity decreases, and by summer, which is the driest part of the year, the relative humidity in the afternoons becomes less than 30%. The average annual rainfall is 792 mm. About 85% of the annual normal rainfall is received during the south-west monsoon months from June to September, with August being the rainiest month. Winds are generally light and are mostly from directions between south-west and north-west (Behera at al., 2011).

The source activities of air pollution in Kanpur can be broadly classified as transport (motor vehicles and railways), commercial, industrial, domestic and fugitive dust sources. Under commercial activities, diesel/ kerosene generators are the most prevailing sources for air pollution in the city. For transport of men, mostly public transport (buses) and private diesel tempos fulfill the transport requirement for the city. The combustion of fuels like coal, kerosene, liquefied petroleum gas (LPG) and wood come under the source for domestic activities. As far as the industrial activities are concerned, the dominant source is the 200 MW coal based thermal power plant (Figure 2). Lots of small and medium scale industries are also responsible for the air pollution. In most of the institutions and offices, the diesel generators (DG) are used at the time of power failure. At several places, garbage and agricultural waste burning is a common practice, which can be an important contributor to air pollution mostly in the evening hour.

2.2 Identification of sources in the study area

The entire study area was divided into 85 grids of 2 km x 2 km. Each grid was assigned an identification number with a land use pattern based on the land use map from Central Pollution Control Board (CPCB) New Delhi. Preliminary surveys have been done at *seven important areas* (as 2 km × 2 km grids), those represent all the source categories of the study area. Based on the preliminary surveys done by the team of experts from Indian Institute of Technology (IIT) Kanpur, the major sectors of sources of PM_{10} have been classified as: (1) industry as point sources (stack height ≥ 25.0 m), (2) industry as area sources (stack height < 25.0 m), (3) vehicles, (4) domestic fuel burning, (5) paved and unpaved road dust, (6) open burning (agricultural residue burning and garbage burning), (7) hotel and restaurant fuel uses, (8) DG sets and (9) rest other sources (e.g., bakery, construction and demolition, medical waste incinerators).

2.3 Categorization of sources and collection of activity level information

The above sectors of the sources of pollution can be broadly classified into three categories: (1) point sources, (2) area sources and (3) line sources. The industries having stack height greater than or equal to 25.0 m are considered as point sources and the industries having stack height less than 25.0 m come under as area sources. Vehicular combustion sources, and paved and unpaved road dust come under line source category. All remaining sources are considered as area sources in this study.

For point source inventories, activity level was obtained through a questionnaire filled by individual plants. The questionnaire sheets collected the detailed information of the industries (i.e., details of stack diameters, stack heights, fuel type, fuel consumption usage and details of the air pollution control devices. In this study, all seven selected primary grids

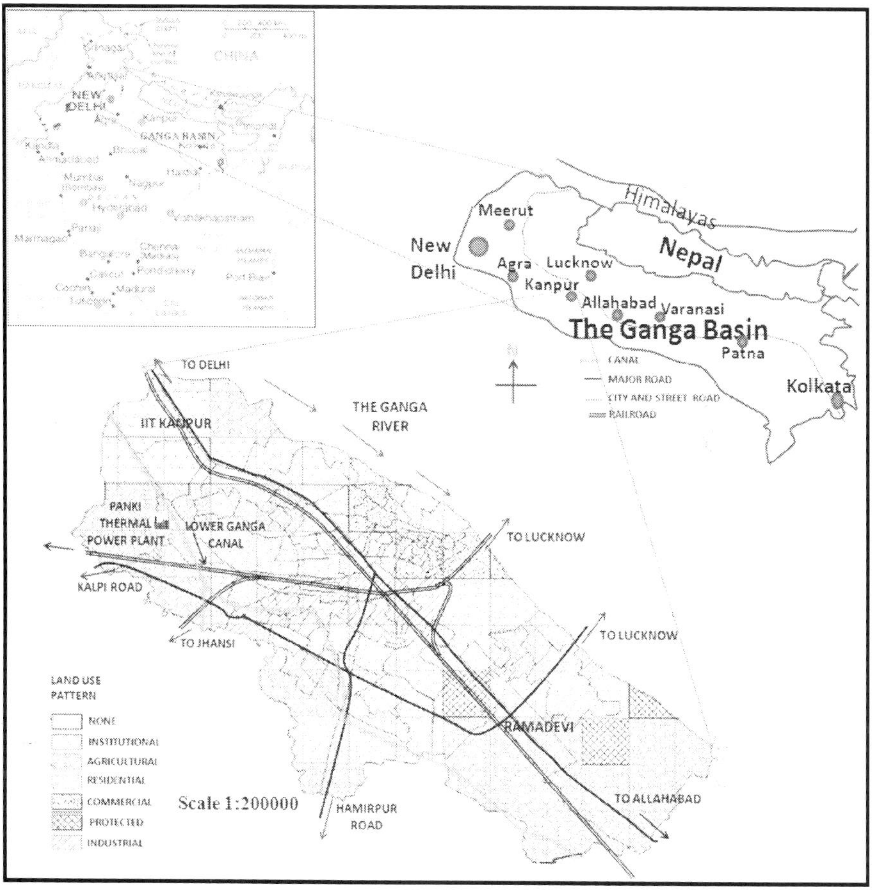

Fig. 2. The Ganga basin, India and gridded map (2×2 km) of Kanpur with land-use patterns

(2 km x 2 km) of Kanpur were surveyed thoroughly for industries as point sources by the team of IIT Kanpur. The activity data for other grids were collected from State Pollution Control Board, Kanpur.

For area source inventories, one can obtain activity level information by the either or by combinations of the two methods as: (i) surveys, and (ii) collection of data from pollution control boards. It is to be noted that the combination of methods can be used to ensure the consistency in data. Surveys can be done to get a representative sample for collection of data in some specific localities and later that can be extrapolated to other localities on the basis of population and/or land-use pattern. State, local, and central agencies such as Central Statistical Organization (CSO), maintain information on population trends, land use, business patterns, agricultural trends, fuel use, chemical production and use. These documents are often valuable resources for determining appropriate activity data or parameters for apportioning data. Several NGOs like, TERI (The Energy Research Institute, New Delhi), CSE (Center for Science and Environment, New Delhi) etc. may also have information on pollution load, inventory procedures and emission factors in Indian context.

In the present study, seven primary grids (2 km × 2 km) were subdivided into small grids (1 km×1 km) with specific grid identification numbers. The activity data were collected from all these 28 (7 x 4) sub-grids with a thorough house-to-house survey. The details of the activity level data for various area sources included in surveys were as follows: (i) production capacity, types of fuel usage, fuel consumption capacity, stack height and details of the air pollution control devices for the pollution from industries as area sources, (ii) capacity of the bakery, types of fuel usage, and daily fuel consumption for pollution from bakeries, (iii) types of fuel usage and daily consumption of each fuel for hotels, restaurants and open eat outs, (iv) type of the family (i.e., high, medium and low, based on economic strata), population of the family, types of fuel usage and daily fuel consumption for the domestic fuel burning, (v) type of the street under survey (i.e., commercial, residential activities etc.), number of garbage collection points in the street, area of the collection points and height of the collection points for pollution from garbage burning, (vi) quantity of agricultural waste generated per unit area and frequency of burning for pollution from agricultural burning, (vii) quantity of area under construction/demolition activities, (viii) capacity of DG set, fuel consumption per hour of each DG set, and (ix) quantity of medical waste.

One example on the activity data survey has been described here for garbage burning. The area of survey was done in the Juhilal Colony of Kanpur. Number of garbage collection points = 1, area of collection point = 1.0 m^2, height = 0.3 m, volume of one point = 1.0 × 0.3 = 0.3 m^3. Dry density of garbage was assumed as 800 kg/m^3. The mass of garbage at that street = 800 × 0.3 = 240 kg. Total population of that street = 200 (all are middle class). So per capita generation of garbage in that street = 240000/200=1200 g/person. This was a four-day garbage generation result. Therefore, garbage generation = 300 g/person/day. Several surveys had been undertaken in the seven primary grids to ascertain the garbage collection. Finally the data for other grids, apart from the seven primary grids, were extrapolated from the data of the seven grids.

For line source activity data, traffic survey of vehicle counts was done at 13 traffic intersections for various vehicle types (two-wheeler motorbikes, three-wheeler auto rickshaw and tempos, four-wheeler passenger cars and jeeps, light commercial vehicles and heavy duty vehicles) covering various road types (e.g., main highway, minor road, service road, etc.). Traffic count on each road in each grid was extrapolated from the data of the traffic survey. The actual road lengths for each grid were estimated from the digitized map in the ArcGIS software. The daily vehicular kilometer traveled (VKT) for all types of vehicles were estimated on the basis of the parking-lot survey done at four important locations in the city. The activity data of PM_{10} pollution from the source of paved and unpaved roads (categorized as line source) were estimated on the basis of mean weight of the vehicle fleet (tons), vehicular kilometer traveled, and silt loading on the road (details given later).

2.4 Digitization of map

GIS has been chosen in the studies related to development of emission inventories because of its usefulness for air quality modeling and it is capable of supporting the development of geospatial air quality models (Puliafito et al. 2003). The Geostatistical Analyst extension of ArcGIS (ArcMap, version 9.2; ESRI Inc., Redlands, WA, USA) was selected for this study because of its relative user friendliness and its frequent use by local authorities and research institutes for air pollution management (Leem et al. 2006; Behera et al., 2011). ArcMap is primarily used to view, edit, create, and analyze geospatial data, and it allows the user to explore data within a data set. The topographical map, issued by the Survey of India (SOI) (prepared in 1977) having scale of 1:50,000, was geo-coded as the base map in the form of polygons for geo-referencing the other maps. This base map was transformed to Universal Transverse Mercator projection with Everest 1956 as the datum. The other three maps, (a) land use, (b) road and railway intersection (from Central Pollution Control Board (CPCB), New Delhi, India), and (3) ward (smallest political unit in a city) boundary (from Kanpur Municipal Corporation) were geo-referenced with respect to the base map.

2.5 Emission estimation methods

Two fundamental approaches have been used by the air quality engineers for the development of emission inventories, i.e., top-down approach and bottom-up approach (Costa and Baldasano, 1996). Top-down approach estimates total emissions for the geographical area under study and then prorates those emissions in the different cells that constitute the spatial ambit of analysis (spatial disintegration), through parameters such as traffic, population and industrial density (Garg et al., 2002). However, the top-down approach is not appropriate at regional scale emission inventory due to many assumptions and inaccuracy present in base data itself and lack of better resolution. Bottom-up approach estimates emissions for all cells comprising selected geographical area. Bottom-up approach is preferred over top-down approach for better resolution and model performance. In this study, a detailed spatially resolved emission inventory has been developed using GIS with bottom-up approach.

The digitized map representing all 85 grids of the city with specific grid identification was embedded into ArcGIS and different files containing Arc attribute tables were created for various source categories. The activity level data for all the sources were stored in respective attribute tables for each grid. It may be noted that the largest source in the study area, the thermal power plant (coal-based), capacity of 200 MW has been accounted under point source. To obtain the total emission in a particular grid, emission from each source (within the grid) were added.

Emission from the vehicles has been calculated using the following expression:

$$E_v = \Sigma(\text{Veh}_i \times D_i) \times \text{EF}_i \tag{1}$$

Where, E_v is PM_{10} mass emission per day, i is the vehicle category, Veh is number of vehicles, D is distance traveled in km in one day by the vehicle i and EF is the mass emission factor for one kilometer (km) travel per vehicle. In this study, we have taken the Indian emission factors (CPCB 2007). The emission factors for CPCB are for different models (year of manufacturing and vehicle type) of all categories of vehicles. The emission factors vary considerably for engine size, fuel uses and age of the vehicles. From the results of parking-lot surveys, a composite emission factors for the broad vehicular categories (two-wheeler motorbikes, three-wheeler auto rickshaw and tempos, four-wheeler passenger cars and jeeps, light commercial vehicles and heavy duty vehicles) were worked out of all individual models of the vehicles indicating fraction of various vehicle categories on the road.

Emissions from fuel burnt in households, hotels and restaurant were estimated using Eq. (3):

$$E_h = \sum(\text{Fuel}_j \times \text{EF}_j) \quad (2)$$

Where E_h is mass emission of PM_{10} per day from households, *j* is the fuel type (e.g. LPG, coal, wood, and kerosene), Fuel is the mass consumption of fuel per day, and EF is mass emission factor. The suitable emission factors were taken from the USEPA AP-42 in the absence of factors developed in India (USEPA 1996).

Emissions from industrial sources were estimated as:

$$E_I = \sum_i \sum_j (A_{i,j} \times \text{EF}_{i,j}) \times \left(1 - \frac{ER_{i,j}}{100}\right) \quad (3)$$

Where E_I is the mass emission of PM_{10} from industrial sources per day, *i* is the type of industry (boiler, cupola, smelter etc.), *j* is the type of fuel used, *A* is the quantity of fuel use, EF is the emission factor and *ER* overall emission reduction efficiency of devices installed (in percentage). The suitable emission factors were taken from the USEPA AP-42 in the absence of factors developed in India (USEPA 1996). These industries under point source categories were located using Leica GS-5 Global position system (GPS) receiver and the information was later utilized in dispersion modeling.

Emission from paved and unpaved road was estimated on the basis of the experimental study conducted in the city (Bhasker and Sharma 2008). The quantity of dust emissions from movement of vehicles on a paved or unpaved road can be estimated using the following empirical expression (Jaecker-Voirol and Pelt 2000):

$$E_R = \sum k(\text{sL}_i / 2)^{0.65} \times (W_i / 3)^{1.5} \quad (4)$$

Where, E_R is emission rate of size specific PM, (units same as *k*), *i* is the road type, sL is silt load (g m^{-2}) on road *i*, *W* is mean weight of the vehicle fleet (tons) on road type. *k* is the base emission factor for particle size range and units of interest constant (function of particle size) in g vkt^{-1} (vehicle kilometer travel) (e.g. $k = 4.6$ g vkt^{-1} for PM_{10}). The values of sL at various locations of the city were taken from Bhasker and Sharma (2008).

The emission from other categories of area sources were estimated as:

$$E_0 = \sum_{\text{source\_cat}} (\text{Activity data}_{\text{source\_cat}}) \times (\text{Emission factor}_{\text{source\_cat}}) \quad (5)$$

Where, E_o is the emission of PM_{10} from all other source categories and source_category is the other area sources (i.e. diesel generator, open burning, construction and demolition, medical waste incinerator). Activity data is the amount of individual source category. Emission factor is the emission factor of PM_{10} by individual source category. The suitable emission factors were taken from the USEPA AP-42 (USEPA 1996).

3. Results and discussions

3.1 Thematic layers for emission inventory

The vector data of digitized maps was converted to different thematic layers to interact and to determine the locations of industries, areas of specific interest and also for editing emission factors. These thematic layers were used for computing emissions and extracting inventories. Different thematic layers were prepared including map of industrial activity, population, transportation layers, area boundaries etc. These are used to determine the surrogates as well as to identity the sources and to visually represent data. The attribute values in the map table can be edited allowing users to tweak values such as a feature's population, number of vehicles on a road in each grid etc.

The entire city was divided into 85 grids of 2 km× 2 km size (as explained earlier) and each grid was assigned a land-use pattern based on the land-use map from CPCB. After obtaining the area of the wards (having irregular boundary), population density in each ward was calculated as follows: population density (person/m^2) = population of ward (person)/ward area (m^2). The area of each ward falling inside a grid was calculated to estimate the population in the grid as:

$$\text{Grid population} = \sum_{i=1}^{N} (\text{intersected ward area}_i \times \text{density of ward}_i) \qquad (6)$$

Where N is the number of wards (fraction of full ward) in that grid, and i represents a ward. Thus, the population for each grid was estimated and stored in GIS. The entire study area was assigned with six types of land-use patterns (Figure 2), i.e., residential, commercial, institutional, industrial, agricultural, and protected (military uses). Figure 3 can be referred as a detailed schematic diagram of the whole process in GIS to develop the emission inventory.

3.2 Activity level data

Figure 4 shows a representative traffic survey for two wheeler motorbikes held at six traffic intersections of the city. It is to be noted that traffic surveys were done at 13 traffic intersection of city. The traffic surveys revealed that two-wheeler motorbikes, three-wheeler auto rickshaw and tempos and four-wheeler passenger cars and jeeps were found to be more at the central part of the city (Rawatpur, Bada Chaurah and Colonelganj). Whereas a larger flow of heavy duty vehicles on the highways, especially the GT Road (IIT Gate and Ramadevi). Overall highest vehicular density was observed between 10:00 AM and 12:00 noon, and between 4:00 PM and 6:00 PM.

Figure 5 shows a representative of parking-lot survey for two wheeler motorbikes done at four important locations of the city in terms of engine size and year of manufacturing. This information is vital in calculating the emission from vehicles on the road. The emission factors vary considerably for engine size, fuel uses and age of the vehicles.

Figure 6 shows location of important industries treated as point sources (stack height ≥ 25.0 m). This figure was the map generated from the GIS and the activity data information was stored in the respective attribute tables to develop the final emission inventory.

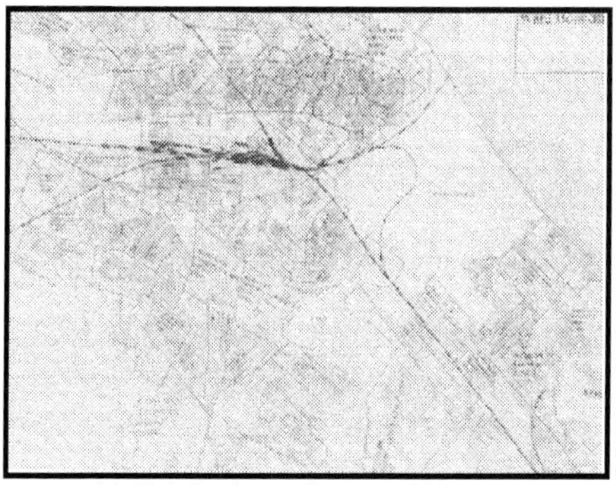

(a) Original ward map of Kanpur

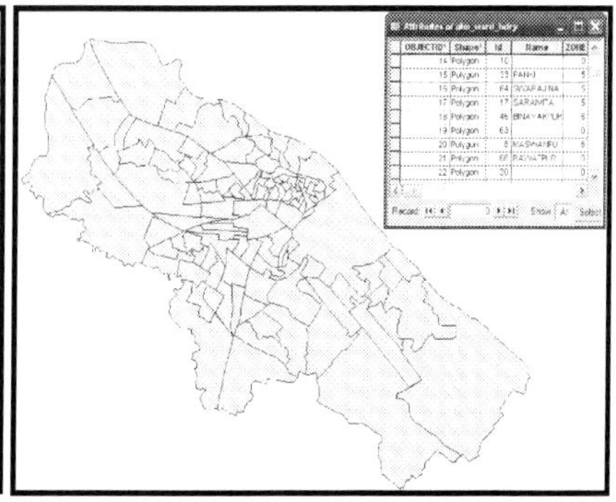

(b) Digitized Ward map with attribute table

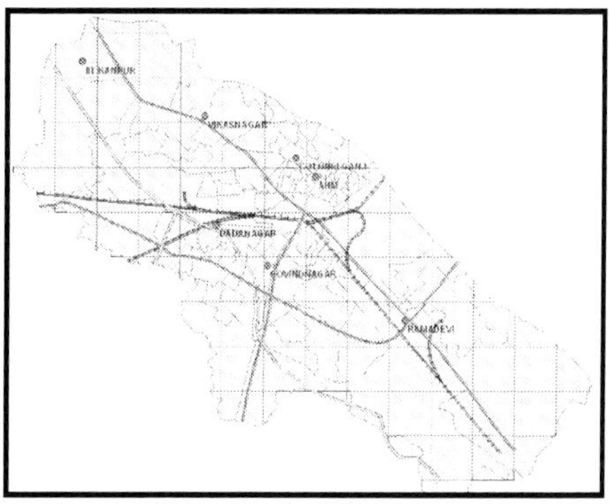

(d) Gridded map of the city (2 km × 2 km grids)

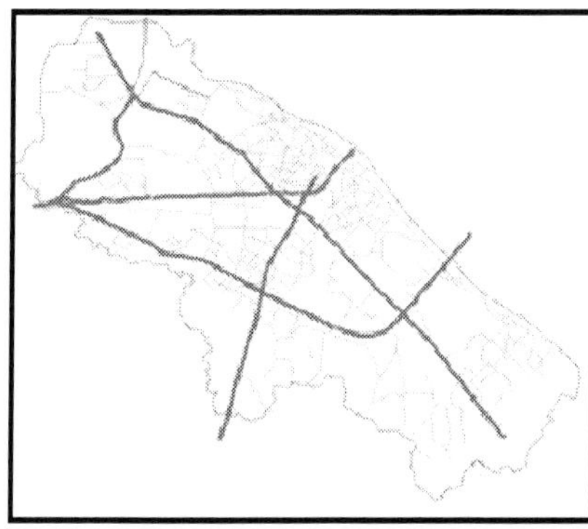

(c) Digitized road network map for all roads

(e) Grid-wise Emission Inventory for PM_{10}

Fig. 3. Schematic Diagram of Thematic layers for GIS based Emission Inventory

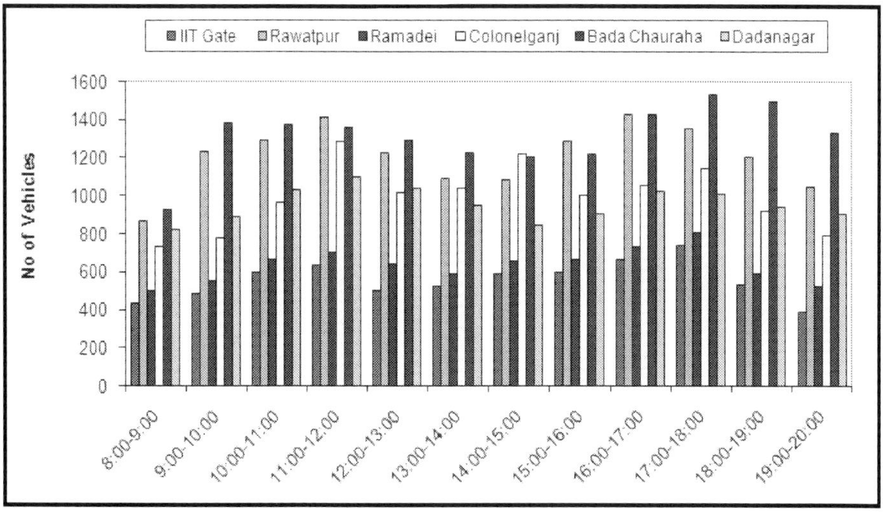

Fig. 4. Patterns of 2W Vehicles at some important traffic junctions

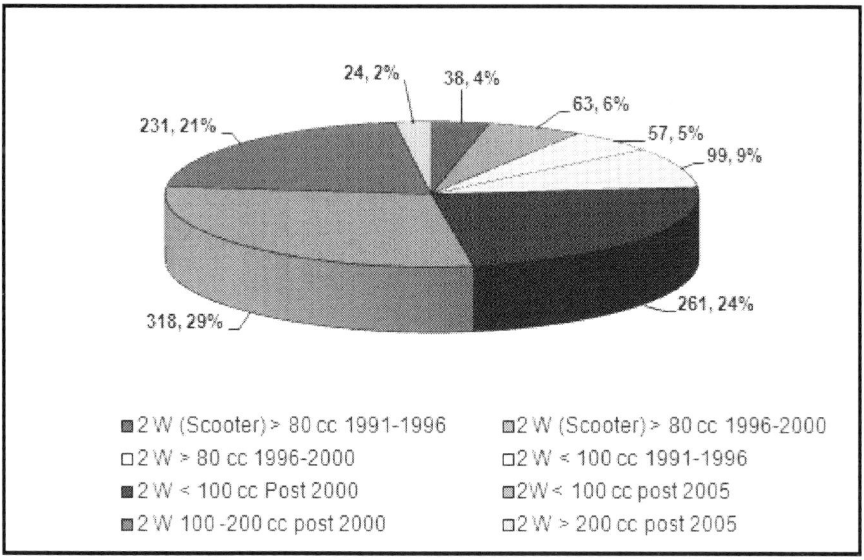

Fig. 5. Results of parking lot survey for two wheeler motorbikes (Nos, %)

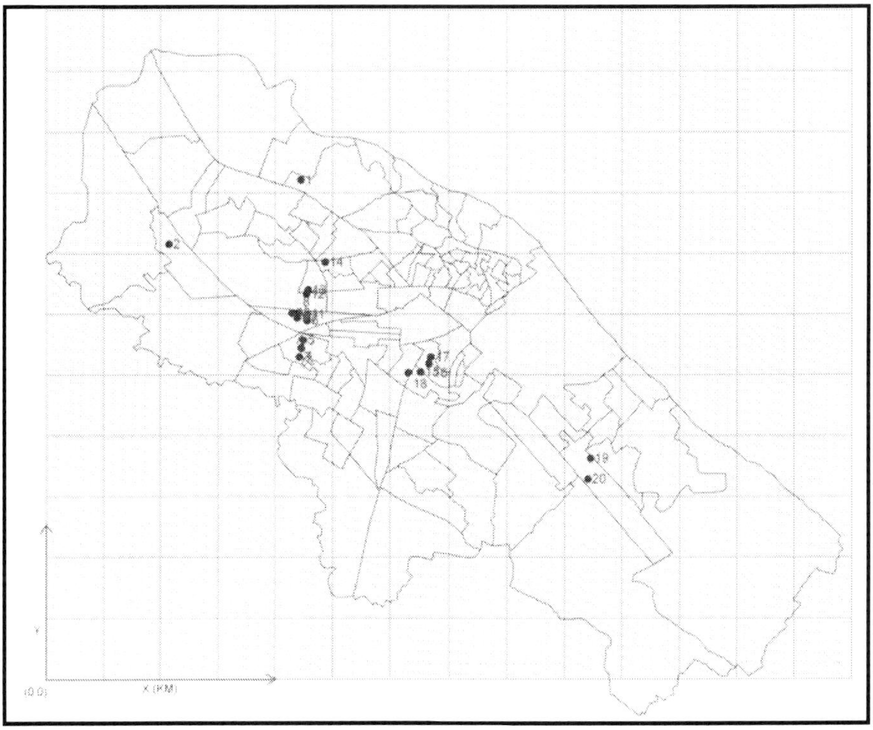

Fig. 6. Locations of point sources in Kanpur

3.3 Emission inventory of PM_{10}

Figure 7 presents emission load of PM_{10} in Kanpur from use of domestic fuel burning. Coal produces a significant pollution load, 985 kg/day, nearly 47 percent of total PM_{10} from domestic sector. Although coal is used by only a small fraction of population, PM emission factor is quite high from coal burning.

Among the vehicular sources, two-wheeler motorbikes contribute 0.14 ton day^{-1}, 3-wheeler auto and tempos as 0.90 ton day^{-1}, 4-wheeler cars and jeeps as 0.16 ton day^{-1}, light commercial vehicles (LCV) as 0.45 ton day^{-1}, and buses and trucks as 0.66 ton day^{-1}. The reason for higher contribution from buses and trucks might be due to use of diesel which has a higher emission factor. Figure 8 shows the pie chart for PM_{10} emissions from various vehicles.

Figure 9 shows the spatial distribution of PM_{10} emission loads over the study area in year 2007. The high pollution areas are identified as the city centre, where most of the pollution activities and congestion problems exist. The emission from domestic sector largely depends on population, type of fuel use in the grid. Emission from vehicles depends on the type of vehicle and type of road. Emissions from industries depend on the capacity of industries, type of fuel for combustion, stack height and particular of the particulate control devices.

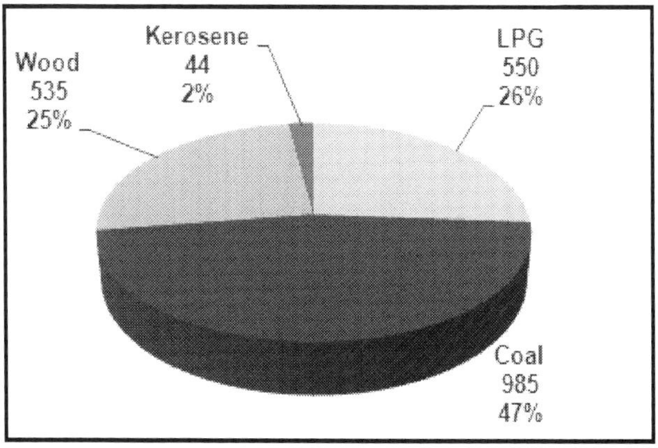

Fig. 7. PM_{10} emission in Kanpur from the domestic fuel burning (kg/d, %)

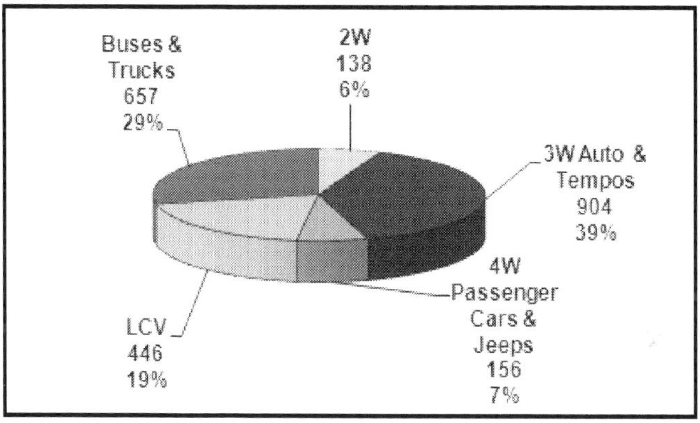

Fig. 8. PM_{10} emission in Kanpur from the vehicles (kg/d, %)

3.4 Assessment for contributions of various sources

The contribution of all identified sources for PM_{10} has been shown in Figure 10. The total emissions of PM_{10} was estimated to be about ~11.2 ton day^{-1} with an overall break-up of: (i) industrial as point sources: 2.9 ton day^{-1} (26%), (ii) industry as area sources: 0.8 ton day^{-1} (7%), (iii) vehicles: 2.3 ton day^{-1} (21%), (iv) domestic fuel burning: 2.1 ton day^{-1} (19%), (v) paved and unpaved road dust: 1.6 ton day^{-1} (15%), (vi) open burning: 1.0 ton day^{-1} (5%), (vii) hotel and restaurant fuel uses: 0.4 ton day^{-1} (4%), (viii) DG sets: 0.1 ton day^{-1} (1%), and (ix) rest other sources 0.2 ton day^{-1} (2%). The highest emission was from the industrial point sources, mainly from a 200 MW coal-based thermal power plant (Panki power plant, Figure 2). The estimated emission suggests that there are many important sources and a composite emission abatement strategy including most of the sources will be required to attain the desired air quality.

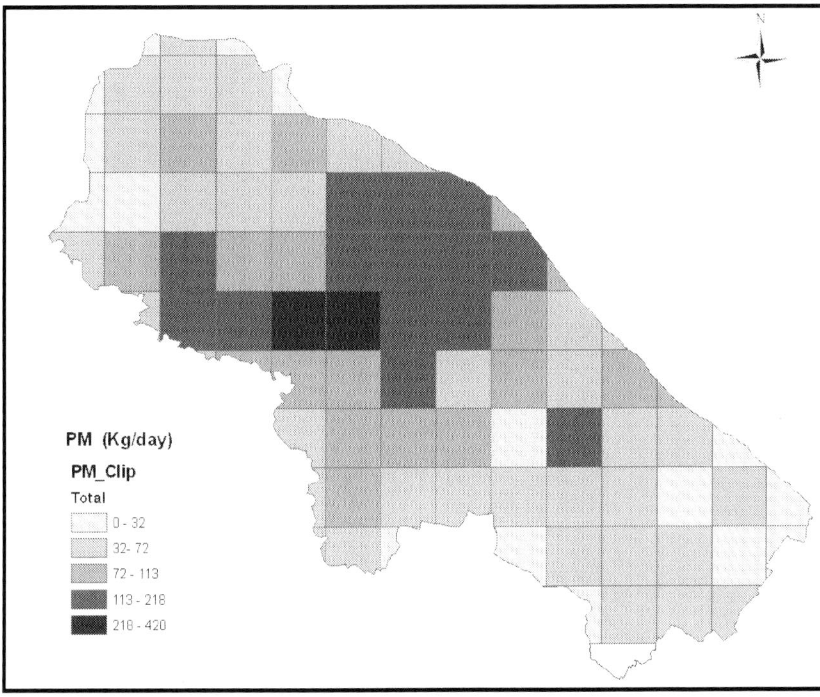

Fig. 9. Spatial Grid-wise emission inventory for PM_{10} in Kanpur

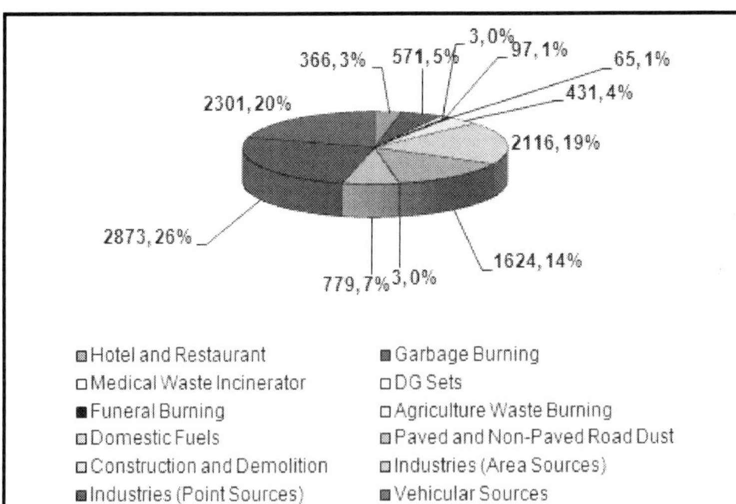

Fig. 9. Overall emission loads of PM_{10} from identified sources in Kanpur (kg/d, %)

4. Conclusions

Particulate pollution in the atmospheres of megacities becomes a major source of concern for public authorities. To understand precisely, the details of the sources responsible for particulate pollution, a PM_{10} emission inventory was carried out. The sources considered were: industries (point and area sources), vehicles, domestic and commercial fuel burning for cooking, paved and unpaved road dust, open burning, diesel generator sets and rest other identified sources. A methodology consisting of four steps was adopted for the desired PM_{10} emission inventory to be generated by using ArcGIS. The total emissions of PM_{10} was estimated to be about 11.2 ton day^{-1} with an overall break-up of: (i) industrial as point sources: 2.9 ton day^{-1} (26%), (ii) industry as area sources: 0.8 ton day^{-1} (7%), (iii) vehicles: 2.3 ton day^{-1} (21%), (iv) domestic fuel burning: 2.1 ton day^{-1} (19%), (v) paved and unpaved road dust: 1.6 ton day^{-1} (15%), (vi) open burning: 1.0 ton day^{-1} (5%), (vii) hotel and restaurant fuel uses: 0.4 ton day^{-1} (4%), (viii) DG sets: 0.1 ton day^{-1} (1%), and (ix) rest other identified sources (bakery, construction and demolition, medical waste incinerators) 0.2 ton day^{-1} (2%). The highest emission was from the industrial point sources, mainly from a 200 MW coal-based thermal power plant. The estimated emission suggests that there are many important sources and a composite emission abatement including most of the sources will be required to attain the desired air quality.

5. References

Behera, S.N., Sharma, M., 2010a. Investigating the potential role of ammonia in ion chemistry of fine particulate matter formation for an urban environment. Science of the Total Environment 408, 3569–3575.

Behera, S.N., Sharma, M., 2010b. Reconstructing primary and secondary components of $PM_{2.5}$ aerosol composition for an urban atmosphere. Aerosol Science and Technology 44,983–992.

Behera, S.N., Sharma, M., Dikshit, O., Shukla, S.P., 2011. GIS-based emission inventory, dispersion modeling, and assessment for source contributions of particulate matter in an urban environment. Water, Air, & Soil Pollution, 218, 423–436.

Behera, S.N., Singh, A.K., Prasad, S.C., 2005. Assessment of health effects due to urban vehicular traffic - a case study. National Academy Science Letters-India, 28, 399-404.

Bhaskar, V. S., Sharma, M., 2008. Assessment of fugitive road dust emissions in Kanpur, India: A note. Transportation Research. Part D, 13, 400–403.

Brown, D.M., Donaldson, K., Stone, V., 2004. Effects of PM_{10} in human peripheral blood monocytes and J774 macrophages. Respiratory Research, 5(9), doi:10.1186/1465-9921-5-29.

Costa, M., M.B., Jose, 1996. Development of a source emission model for Atmospheric pollutants in the Barcelona area. Atmospheric Environment, 30, 309-318.

CPCB (2007). Air Quality Monitoring Project-Indian Clean Air Programme (ICAP). Draft report on"emission factor development for Indian vehicles" as a part of ambient air quality monitoring and emission source apportionment studies. Central Pollution Control Board. Delhi. http://www.cpcb.nic.in/DRAFTREPORT-on-efdiv.pdf.

Diem, J.E., Comrie, A.C., 2001. Allocating anthropogenic pollutant emissions over space: application to ozone pollution management. Journal of Environmental Management, 63, 425–447.

Garg, A., Kapshe, M., Shukla, P.R., Ghosh, D., 2002. Large point source (LPS) emissions from India: regional and sectoral analysis. Atmospheric Environment, 36, 213–224.

Jaecker-Voirol, A., Pelt, P., 2000. PM_{10} emission inventory in Ile de France for transport and industrial sources: PM_{10} re-suspension, a key factor for air quality. Environmental Modeling and Software, 15, 575–581.

Leem, J.-H., Kaplan, B. M., Shim, Y. K., Pohl, H. R., Gotway, C. A., Bullard, S. M., et al. 2006. Exposures to air pollutants during pregnancy and preterm delivery. Environmental Health Perspectives, 114, 905–910.

Makar, P.A., Bouchet, V.S., Nenes, A., 2003. Inorganic chemistry calculations using HETV—a vectorized solver for the SO_4^{2-}-NO_3^--NH_4^+ system based on the ISORROPIA algorithms. Atmospheric Environment, 37, 2279–2294.

Martuzzi, M., Krzyzanowski, M., Bertollini, R., 2003. Health impact assessment of air pollution: providing further evidence for public health action. The European Respiratory Journal, 21(S40), 86–91.

Nair, V.S., Moorthy, K.K., Alappattu, D.P., Kunhikrishnan, P.K., George, S., Nair, P.R., et al. 2007. Wintertime aerosol characteristics over the Indo-Gangetic Plain (IGP): impacts of local boundary layer processes and long-range transport. Journal of Geophysical Research, 112, D13205, doi:10.1029/2006JD008099.

Puliafito, E., Guevara, M., Puliafito, C., 2003. Characterization of urban air quality using GIS as a management system. Environmental Pollution, 122, 105–117.

Suh, H.H., Bahadori, T., Vallarino, J., Spengler, J.D., 2000. Criteria air pollutants and toxic air pollutants. Environmental Health Perspectives, 108(S4), 625-633.

US Environmental Protection Agency (USEPA). (1996). Supplement B to compilation of air pollutant emission factors, vol I: Stationary, point and area sources. Technical Report. Office of Air Quality Planning and Standards, Research Triangle Park.

Zhang, Q.H., Zhang, J.P., Xue, H.W., 2010. The challenge of improving visibility in Beijing. Atmospheric Chemistry and Physics, 10, 7821-7827.

17

Air Pollution, Modeling and GIS based Decision Support Systems for Air Quality Risk Assessment

Anjaneyulu Yerramilli[1], Venkata Bhaskar Rao Dodla[1] and Sudha Yerramilli[2]

[1]*Trent Lott Geospatial & Visualization Research Center @ e-Center, College of Science Engineering & Technology, Jackson State University, Jackson, MS,*

[2]*National Center For Bio-Defense @e-Center, College of Science Engineering & Technology, Jackson State University, Jackson MS, USA*

1. Introduction

Air Pollution is a state of the atmosphere with predominant presence of hazardous substances that are harmful to humans and animals. The air-borne pollutants degrade the air quality and constant exposure to polluted air may lead to several health problems such as cardiopulmonary disease, bronchitis, asthma, wheezing and coughing etc. Average composition of the atmosphere below 25 km indicates that nitrogen, oxygen, water vapor, carbon dioxide, methane, nitrous oxide, and ozone are the major constituents and their balance is important to the maintenance of the Earth's biosphere. An imbalance of these constituents for a considerable long time may lead to serious implications of air quality and weather and climate. The pollutants are categorized as primary and secondary, primary pollutants are directly emitted from a source and secondary pollutants result from reaction of primary pollutants in the atmosphere. The primary pollutants are the ash from volcanoes, carbon monoxide and sulfur dioxide emissions from vehicles and factories and combustion of fossil fuels etc whereas secondary pollutants are tropospheric ozone resulting from photolysis of nitrogen oxides and hydrocarbons in the presence of sunlight and smog from a mixture of smoke and sulfur dioxide. The sources of air pollution are classified as natural and anthropogenic. Volcanoes, forest fires and biological decay providing sulfur dioxide and nitrogen oxides, large barren lands providing dust, vegetation producing volatile organic compounds come under natural sources whereas anthropogenic sources are categorized as mobile and stationary sources. Different forms of transportation such as automobiles, trucks, and airplanes come under mobile sources whereas power plants and industrial facilities are the stationary sources. Stationary sources are further classified as point and area sources, wherein a point source refers to a fixed source such as a smokestack or storage tank that emits the pollutant and an area source refers to several small sources affecting the air quality in a region such as dry cleaners, gas stations, auto body paint shops and a community of homes using woodstoves for heating. Air pollutants are also categorized as 'criteria' and 'hazardous', where the criteria pollutants refer to the commonly and frequently observed six chemicals which are carbon monoxide, lead, nitrogen dioxide, ozone, particulate matter, and sulfur dioxide and hazardous pollutants are toxic pollutants which cause cancer and other serious health problems or lead to adverse environmental effects.

Anthropogenic primary pollutants such as carbon monoxide, particulate matter, nitrogen oxides and lead are detrimental to health as well as environment. Sulfur dioxide and nitrogen oxides get transformed as sulfuric acid and nitric acid in the atmosphere due to chemical reactions and may fall as acid rain. Some details of these pollutants are briefly described as follows:

Carbon monoxide is a colorless, odorless, poisonous gas produced from burning of fuels with carbon and so the major source is road transport vehicles. Due to oxidation process, CO will be transformed as carbon dioxide. The background levels of carbon monoxide are in the range of 10-200 parts per billion (ppb) and urban concentrations generally vary between 10 to 500 parts per million (ppm). Continuous exposure to higher levels (>500 ppm) for longer time periods (> 30 minutes) may lead to headache, dizziness and nausea and also death.

Nitric oxide (NO) is a colorless, odorless gas produced during burning of fuel at high temperatures in cars and other road vehicles, heaters and cookers. Mostly, nitrogen dioxide in the atmosphere is formed from the oxidation of nitric oxide (NO). Nitrogen dioxide reacts to form nitric acid and organic nitrates and plays an important role in the production of surface ozone. Mean concentrations in urban areas are in the range of 10-45 ppb reaching as high as 200 ppb. Continuous exposure to NO_2 leads to respiratory problems and lung damage.

Particulate matter comprises of both organic and inorganic substances, mainly from dust, fly ash, soot, smoke, aerosols, fumes, mists and condensing vapors and is regarded as coarse particulates with a diameter greater than 2.5 micrometers (μm) and fine particles less than 2.5 micrometers. The acid component of particulate matter (PM) generally occurs as fine particles. Primary sources of the particulate matter are from road transport (25%), non-combustion processes (24%), industrial combustion plants and processes (17%), commercial and residential combustion (16%) and public power generation (15%). In urban areas, secondary particulate matter occurs as sulfates and nitrates with mean values in the range 10-40 μg/m3 and may rise up to higher than 100 $\mu g/m^3$. Primary PM sources are derived from both human and natural activities which include agricultural operations, industrial processes, fossil fuel burning etc and secondary pollutants such as SO_2, NOx, and VOCs are considered as precursors as they help form PM. Measures to reduce these precursor emissions will have a controlling impact on PM concentrations. Fine PM will cause asthma, lung cancer, cardiovascular issues, and premature death and estimated to cause 20,000 - 50,000 deaths per year in US.

Sulfur dioxide (SO_2) is a colorless, nonflammable gas with an odor that irritates the eyes and air passages. The most common sources of sulfur dioxide are fossil fuel combustion, smelting, manufacture of sulfuric acid, conversion of wood pulp to paper, incineration of refuse and production of elemental sulfur. Coal burning is the single largest man-made source of sulfur dioxide accounting for about 50% of annual global emissions, with oil burning accounting for a further 25-30%. The most common natural source of sulfur dioxide is volcanoes. The mean concentrations are in the range of 5-15 ppb but hourly peak values may reach to 750 ppb. The health effects include asthma, respiratory illness and cardiovascular disease. Sulfur dioxide pollution can be more harmful when particulate and other pollution concentrations are high which is known as the "cocktail effect".

Secondary pollutants, resulting from the conversion of primary pollutants through complex chemical reactions, are potentially more harmful than their precursors. Since much of the pollutant chemistry is driven by the presence of sunlight, these are commonly referred to as photochemical pollutants. A well-known secondary photochemical pollutant is ozone (O_3) formed due to oxidation of benzene and other volatile organic compounds in the presence of nitrogen oxides. In urban areas with high NOx concentrations, ozone concentrations tend to be lower due to scavenging effect (NOx react with ozone to form NO_2 and O_2), but tend to be higher due to downwind transport of NOx and its reaction with VOCs under sunlight. Ozone is a colorless, pungent, highly reactive gas and is the principal component of smog, which is caused primarily by automobile emissions, predominantly in urban areas. Ozone concentrations in urban areas rise in the morning, peak in the afternoon, and decrease at night. Higher frequencies of higher concentrations are dependent on stable atmospheric conditions such as low level inversions which restrict the pollutants to the boundary layer. Two of the most important factors that affect human health are the concentration of ozone and duration of exposure.

1.1 Air Quality and Air Quality Index

Air Quality depends on the trace gas emissions from the biosphere, from human activities and the chemical reactions which govern the concentrations of trace species in the atmosphere. Air Quality is an assessment of extent of pollutants present in air in a given locality relative to permissible levels while Air Quality Index (AQI) is an index for reporting daily air quality. EPA (Environmental Protection Agency) of USA calculates the AQI for five major air pollutants regulated by the Clean Air Act: ground-level ozone, particle pollution (also known as particulate matter), carbon monoxide, sulfur dioxide, and nitrogen dioxide. For each of these pollutants, EPA has established national air quality standards to protect public health. Ground-level ozone and airborne particles are the two pollutants that pose the greatest threat to human health in USA. AQI as a yardstick ranges from 0 to 500 and higher the AQI value, greater the level of air pollution and greater the health concern. For example, an AQI value of 50 represents good air quality with little potential to affect public health, while an AQI value over 300 represents hazardous air quality. An AQI value of 100 generally corresponds to the national air quality standard for the pollutant, which is the level EPA has set to protect public health. AQI values below 100 are generally thought of as satisfactory. When AQI values are above 100, air quality is considered to be unhealthy at first for certain sensitive groups of people, then for everyone as AQI values get higher.

1.2 Spatial Temporal variability

Air pollution became a widespread problem in the United States (US) and, according to the Environmental Protection Agency (EPA), it is estimated that over 100 million individuals are routinely exposed to levels of air pollution that exceed one or more of their health-based standards [1]. The major air pollutants include oxides of sulfur and nitrogen, suspended particulate matter, oxides of carbon, hydrocarbons, lead. These gases are released into the atmosphere from either stationary sources such as industrial point sources or mobile sources like vehicular pollution. These pollutant gases will also produce secondary pollutant like surface Ozone, which has great oxidative capacity and is the primary component of urban smog. Criteria Pollutants Ozone and particulate matter (PM_{10} and $PM_{2.5}$) are main focus now as they manly drive the AQI (which is based on five criteria pollutants CO, NO_2, SO_2, O_3, PM). These pollutants can affect our health in many ways, with irritation to the eyes, nose and throat and more serious problems such as chronic respiratory disease and cardiovascular diseases. Understanding the long-term (yearly, decadal) trends of these criteria air pollutants is important for assessing chronic exposure to population and the efficacy of control strategies, emissions changes, and the year to- year influence of meteorology.

The air pollutant levels at any location can be explained by three factors: (a) the rate of emissions or production from all sources; (b) the rate of chemical or physical removal via reaction or deposition; and (c) the dispersion and transport of a chemical within, to, or from an area. These three factors are likely to be influenced by meteorological and anthropogenic factors such as temperature, precipitation, proximity to sources, or weekday - weekend emission activity differences. Meteorological conditions strongly influence air quality. These include transport by winds, recirculation of air by local wind patterns, and horizontal dispersion of pollution by wind; variations in sunlight due to clouds and season; vertical mixing and dilution of pollution within the atmospheric boundary layer; temperature; and moisture. The variability of these processes, which affects the variability in pollution, is primarily governed by the movement of large-scale high- and low-pressure systems, the diurnal heating and cooling cycle, and local and regional topography. Changes in climate affect air quality by perturbing ventilation rates (wind speed, mixing depth, convection, and frontal passages), precipitation scavenging, dry deposition, chemical production and loss rates, natural emissions, and background concentrations.

Further researches have shown that $PM_{2.5}$ chemical composition and concentrations vary by location and source [2]. Particulates from different sources also have varying amounts of correlation with adverse health effects [3]. Most epidemiological studies which examine relationships between air pollution levels and human health needs spatial and temporal variation. Effects of changes under different possible emission scenarios are also important in determining pollutant distributions and future air quality risks.

1.3 Air Quality Management

The damages caused by air pollution in many countries are large and it is generally accepted that there is an urgent need for reducing the emissions to the atmosphere. The damages are caused by high ambient air concentrations and depositions of many chemical components. Among the most important components are acidifying constituents (sulfur and both oxidized and reduced nitrogen compounds), photochemical components (including ozone), particulate matter and toxic compounds, such as metals, organic compounds and others. The concentrations and depositions are dependent on (i) the total mass of pollutants emitted to the atmosphere and its spatial and temporal distribution (ii) transport and transformation processes in the atmosphere and (iii) deposition processes.

Assessments of emission reduction strategies must consider all the three factors and the complexity of these problems call for the use of atmospheric models. Average exposure of the ecosystem to concentrations and deposition, emission scenario studies and linkage to economical aspects and cost effectiveness are all examples of areas where the models are needed. Air Pollution is a good example of how models can play an important role in the decision making process toward protocols on emission reductions. Regional scale models quantifying the trans-boundary fluxes of air pollution between the European countries and deposition to ecosystems [4] have successfully been applied together with knowledge on ecosystem critical loads of acidity and the costs involved in emission reduction in order to find optimal solution for the reductions.

2. Air Quality assessment

Pollutants in the atmosphere undergo transportation and dispersion, whose characteristics are dependent on the prevailing atmospheric conditions. Pollutants can be transported across the continents due to globally circulating winds on time scale of months to years, e.g. intercontinental transport of dust from Sahara desert in Northwest Africa and the Gobi desert in East Asia and dispersed over a few hundreds of miles in a few days under the influence of local scale wind circulations. While wind is primarily responsible for transportation and dispersion, topography and atmospheric stability also play an important role. Horizontal and vertical motions arise due to atmospheric pressure gradients and stability conditions. Atmospheric pressure, defined as the weight of the air column above a point, varies due to differential surface heating. Heating and cooling of the earth surface is dependent on the type of surface, for e.g. water bodies have larger specific heat capacity and so takes longer time to get heated or cooled as compared to land surface. Different surfaces have different heating/cooling rates, e.g. asphalt surface gets heated/ cooled faster than vegetation surface; heating/ cooling at the surface causes lower/ higher density vertical columns leading to lower/ higher pressures. As air tends to move from denser to lighter regions, atmospheric wind gets established as movement from high pressure regions to low pressure regions. Similarly vertical motions are dependent on atmospheric stability. Heating at surface causes air to rise mixing with cooler and denser air at higher levels leading to instability and most of the mixing takes place in the lowest part of the atmosphere referred to as atmosphèric boundary layer. This is characterized by turbulent fluctuations of wind velocity, temperature and moisture due to energy exchange with earth surface capped by an infinitesimal transition layer below the free atmosphere with lower/ higher stability conditions associated with stronger/ weaker vertical mixing. Since the pollutants originate near the earth surface, characteristics of the atmospheric boundary layer play an important role in the dispersion of pollutants. This is the reason why winter nights which are cooler with higher stability have stagnation of pollutants causing health hazards. Atmospheric dispersion on the scale of a few days (synoptic time scale) is influenced by the transient surface low and high pressure systems. Moderate to intense low pressure systems characterized by winds > 5m/s, mass convergence, higher instability and upward motions contribute to dispersion over wider horizontal extent ranging up to few thousand kilometers thus reducing the pollution effect through mixing with larger environment volume. Conversely, high pressure systems with calm winds, subsiding downward motion and higher atmospheric stability contain dispersion to smaller volume enhancing the pollution effects. These emphasize the role of atmospheric stability criteria and prevailing atmospheric circulation on the atmospheric dispersion of pollutants. In the absence of synoptic forcing, mesoscale and local circulations developed due to topography and land use variations and with synoptic forcing admixture of synoptic scale and mesoscale circulations control the pollutant dispersion.

Assessment of atmospheric dispersion requires precise information of meteorological parameters such as wind, temperature, humidity and stability at spatial and temporal resolutions as high as possible. Since meteorological observations are made at discrete spatial locations and at synoptic times (twice daily at 0000 and 1200 UTC), these data are to be generated at the required spatial and temporal resolutions. Weather prediction models provide the required quantitative information of the meteorological fields, which are used as input to a pollutant dispersion model to derive the spatial distribution. The integration of the weather prediction and dispersion models is the basis of the air quality models.

3. Meteorological Models - Numerical Weather Prediction

In view of the importance of meteorological variables in estimating atmospheric dispersion, a brief description of weather prediction using numerical models is provided here. Weather is defined as the state of the atmosphere at a given time and place, with respect to variables such as temperature, moisture, wind velocity, and barometric pressure where atmosphere is a gaseous envelop covering the Earth. Weather prediction is the application of science and technology to estimate the future state of the atmosphere (in terms of pressure, temperature, wind, rainfall etc) with reference to a specified region or location. Atmosphere is influenced by transient air movement and predicting that flow provides the status of weather conditions. Although different methods are available, numerical models alone provide quantitative weather forecasting. These models use a closed system of mathematical equations of dynamics developed from fundamental equations based on conservation of mass, momentum and water vapor along with equation of state and thermodynamic energy equation for temperature in differential form and use of numerical methods to solve them to produce future state of atmosphere starting from an observed initial state. To solve these equations numerically, the study region is to be formulated as of rectangular form with equally spaced horizontal intersecting grid in the horizontal direction and enough number of vertical levels as suitable for numerical solution. Due to constraints of mathematical formulation, numerical set up and the computational resources, the horizontal grid resolution is often restricted to a few kilometers. Since atmospheric modeling includes both the dynamical and physical parts, the physical processes of atmospheric radiation, planetary boundary layer, convection, cloud microphysics and surface physics are to be parameterized as these processes occur on scales smaller than the resolvable scales of domain resolution. For the system to predict the weather for a certain area it needs initial conditions and boundary conditions as adjacent areas affect the area of interest. It is known that observations are available at non-uniform locations and are sparser than the model resolution; the atmospheric variables are to be interpolated to the model grid using objective methods. Errors in the observations, interpolation to model grid, limitations in the representation of physical processes and errors due to numerical methods of solution all lead to uncertainties in weather forecasting and increase of errors with the progress of prediction. Rapid advances in computational resources have lead to the design and development of atmospheric models as applicable for real time weather forecasting of various weather phenomena with scales of few kilometers to thousands of kilometers.

There are many meteorological models available for weather prediction and the mainly used are the WRF, MM5, RAMS. Brief description of these models as follows.

3.1 WRF modeling system

Weather Research and Forecasting (WRF) modeling system was developed and sourced from National Center for Atmospheric Research (NCAR), as the next generation model after MM5, incorporating the advances in atmospheric simulation suitable for a broad range of applications. It has two cores, NMM and ARW which are developed independently and have differences in dynamics, methods of solution and physical schemes. Here a description of ARW model is presented as it is widely used for air quality modeling studies. ARW (Advanced Research WRF) model has versatility to choose the domain region of interest; horizontal resolution; interactive nested domains and with various options to choose parameterization schemes for convection, planetary boundary layer (PBL), explicit moisture;

radiation and soil processes. ARW is designed to be a flexible, state-of-the-art atmospheric simulation system that is portable and efficient on available parallel computing platforms and a detailed description was provided by Skamarock et al. (2008) [5]. The model consists of fully compressible non-hydrostatic equations and the prognostic variables include the three-dimensional wind, perturbation quantities of pressure, potential temperature, geo-potential, surface pressure, turbulent kinetic energy and scalars (water vapor mixing ratio, cloud water etc). The model equations are formulated using mass-based terrain following coordinate system, and solved in Arakawa-C grid using Runge-Kutta third order time integration techniques. The model has several options for spatial discretization, diffusion, nesting and lateral boundary conditions. The ARW Solver is the key component of the modeling system, which is composed of several initialization programs for idealized, and real-data simulations, and the numerical integration program. ARW supports horizontal nesting that allows resolution to be focused over a region of interest by introducing an additional grid (or grids) into the simulation with the choice of one-way and two-way nesting procedures. ARW model system was used in this study for its accurate numerics, higher order mass conservation characteristics and advanced physics.

3.2 MM5 modelling system

MM5 is a regional mesoscale model used for creating weather forecasts and climate projections. MM5 modeling system software is mostly written in Fortran and has been developed at Penn State and National Center for Atmospheric Research (NCAR) as a community mesoscale model with contributions from users worldwide. MM5 is a limited-area, non-hydrostatic, terrain-following sigma-coordinate model designed to simulate or predict mesoscale atmospheric circulation [6]. The model is supported by several pre and post-processing programs, which are referred to collectively as the MM5 modeling system [http://www.mmm.ucar.edu/mm5/].

3.3 RAMS modeling system

Regional Atmospheric Modeling System (RAMS) is a highly versatile numerical code developed by scientists at Colorado State University for simulating and forecasting meteorological phenomena, and for depicting the results.
[http://rams.atmos.colostate.edu/rams-description.html].

4. Air Pollution Dispersion modeling

Air pollution models are the only method that quantifies the deterministic relationship between emissions and concentrations/depositions, including the consequences of past and future scenarios and the determination of the effectiveness of abatement strategies. Air pollution measurements give information about ambient concentrations and deposition at specific locations and times, without giving clear guidance on the identification of the causes of the air quality problem. This makes air pollution models indispensable in regulatory, research, and forensic applications. The concentrations of substances in the atmosphere are determined by 1) transport, 2) diffusion, 3) chemical transformation, and 4) ground deposition. Transport phenomena, characterized by the mean velocity of the fluid, have been measured and studied for centuries. For example, the average wind has been studies by man for sailing purposes. The study of diffusion (turbulent motion) is more recent.

Among the first articles that mention turbulence in the atmosphere, are those by Taylor (1915, 1921) [7,8].

Pollutants in the atmosphere get transported longer distances by large scale atmospheric wind flows and dispersed in the atmosphere by small scale turbulent flows and mix with environment. Dispersion is difficult to understand and estimate due to presence of different scales of eddies and their complex interaction. Atmospheric dispersion models shall include the physical and chemical processes of transportation, transformation and dispersion of pollutants in the atmosphere to provide estimates of pollutant concentrations with the information of emission sources and concentrations. Current models are designed to compute the pollutant concentrations using information of characteristics of emission sources, emission rates, terrain variations, meteorological variations and background concentrations. These dispersion models use mathematical based concepts of dispersion and diffusion with proper treatment of complex terrain and land use. The pollutants concentrations in the atmosphere constantly change influenced by weather conditions. Meteorological parameters play an important role in the atmospheric dispersion as transport is controlled by large scale wind flows and dispersion by atmospheric stability, vertical mixing and local mesoscale winds. A proper design and application of the dispersion models can be used for identification of contributors through establishment of source- receptor relationships, assessment of air quality compliance with government norms, planning of new facilities, management of emissions at sources, prediction of high concentration episodes, providing information for risk assessment and management. Above all these, modeling information saves cost of continuous monitoring over large areas and for longer times. All the model estimations are to be evaluated to understand the uncertainties due to inaccuracies in emission strength, meteorological input data, physics of dispersion concentration estimations and analysis before assessment of health effects.

There are two modeling approaches, following Lagrangian and Eulerian mechanics. In the Lagrangian approach, the path of an air parcel (or puff) is followed and the changes tracked along the trajectory path. Lagrangian modeling was mostly used for transport of SO_2 over large distances and longer time-periods [9, 10, 11, 12]. In Eulerian method, the 3-dimensional atmosphere is divided into grid cells, both in horizontal and vertical directions, and the time changes of the properties are identified at each grid cell. Eulerian modeling was adopted for urban area studies on ozone [13]; for SO_2 [14] and for regional scale sulfur [15, 16]. As such Eulerian modeling was used only for specific episodes of a few days. Hybrid approaches, in which particle-in-cell methods were employed, were used by Friedlander and Seinfeld (1969)[17], Eschenroeder and Martinez (1970)[18] and Liu and Seinfeld (1974)[19]. It may be stated that prior to 1980, Lagrangian models were generally used for transport studies of sulfur and other particulate matter whereas Eulerian models were adopted for episodic events of secondary pollutants such as ozone. After 1980, the basic concepts were fine tuned with development and use of 2-D and 3-D global troposphere models. AERMOD and CALPUFF computer packages were developed for simulation of non-reactive chemicals (e.g., SO2). AERMOD is a steady-state Gaussian plume model which uses wind field derived from surface, upper-air, and onsite meteorological observations combined with terrain elevations and land use. CALPUFF is a non-steady state Lagrangian puff dispersion model with advantage of realistically simulating the transport in calm and stagnant conditions and over complex terrain, and coastal regions with sea/land breezes. AEROMOD is suitable for short-range simulations whereas CALPUFF is appropriate for both long-range and short-range simulations.

There are different types of modelling approaches are available and they are photochemical modelling, Plume Rise models, particle models, Deposition Modules, Odor models and Statistical models and a brief description as follows.

4.1 Photochemical modeling

These photochemical models are large-scale air quality models that simulate the changes of pollutant concentrations in the atmosphere using a set of mathematical equations characterizing the chemical and physical processes in the atmosphere. These models are applied at multiple spatial scales from local, regional, national, and global. Some examples of photochemical models are to be CMAQ, CAMX, UAM, WRF/Chem etc.

The Community Multi-scale Air Quality (CMAQ) modeling system has been designed for air quality as a whole by including state-of-the-science capabilities for modeling multiple air quality issues, including tropospheric ozone, fine particles, toxics, acid deposition, and visibility degradation [http://www.epa.gov/asmdnerl/CMAQ/index.html]. CMAQ was also designed to have multi-scale capabilities so that separate models were not needed for urban and regional scale air quality modeling. CMAQ modeling system simulates various chemical and physical processes that are thought to be important for understanding atmospheric trace gas transformations and distributions.

The Comprehensive Air quality Model with extensions (CAMx) is a computer modeling system for the integrated assessment of gaseous and particulate air pollution [http://www.camx.com/]. This model is designed to simulate air quality over many geographic scales; treat a wide variety of inert and chemically active pollutants like Ozone, inorganic and organic $PM_{2.5}/PM_{10}$, Mercury and toxics and to provide source-receptor, sensitivity, and process analyses.

The Urban Airshed Model (UAM) modeling system, developed and maintained by Systems Applications International (SAI), is the most widely used photochemical air quality model [http://uamv.saintl.com/].

4.2 Plume rise models

The simplest way of estimating smoke concentrations is to assume that plumes diffuse in Gaussian pattern along the centerline of a steady wind trajectory. Plume models usually assume steady state conditions during the life of the plume, which means relatively constant emission rates, wind speed, and wind direction. For this reason, they can be used only to estimate concentrations relatively near the source or for a short duration with restrictions to the influence of topography or land use. Most air pollution models include a computational module for computing plume rise i.e., the initial behavior of a hot plume injected vertically into a horizontal wind flow, e.g., PRIME in AERMOD and HYSPLIT.

4.3 Particle models

Particle model simulates the source by the release of many particles over the duration of the burn. The trajectory of each particle is determined as well as random component that mimics the effect of atmospheric turbulence. This allows a cluster of particles to expand in space according to the patterns of atmospheric turbulence rather than following a parameterized spatial distribution pattern, such as common Gaussian approximations. These models tend to be the most accurate way of simulating concentrations at any point in time but restricted to individual point sources with simple chemistry or sources that have critical components such as toxins that must be tracked precisely. These models use Lagrangian coordinates for accurate depiction of place of each time of particle movement e.g. HYSPLIT.

4.4 Deposition modules

Many air pollution models include the computational module for computing the fraction of the plume deposited at the ground as a consequence of dry and wet deposition phenomena.

4.5 Odor modeling

Odor models include algorithms to simulate instantaneous or semi-instantaneous concentrations, since odors are instantaneous to human sensations. The mechanisms of dispersion of odorous chemicals in the atmosphere are the same as the dispersion of other pollutants. In multiple pollutants emission, the relationship between concentrations of individual chemicals and odor is not well defined and odor must be characterized in terms of an odor detection threshold value for the entire mixture of odorous chemicals in the air.

4.6 Statistical models

Statistical models are techniques based on statistical data analysis of measured ambient concentrations. These are non-deterministic, as they do not establish nor simulate a cause-effect, physical relationship between emissions and ambient concentrations. These models are used in air quality forecast and alarm systems and receptor modeling. Statistical techniques have been used to forecast air pollution trends a few hours in advance for the purpose of alerting the population. Receptor models are mathematical or statistical procedures for identifying and quantifying the sources of air pollutants at a receptor location. Receptor models use the chemical and physical characteristics of gases and particles measured at source and receptor to both identify the presence of and to quantify source contributions to receptor concentrations e.g., Chemical Mass Balance (CMB) Model.

Basically there are two approaches for the computation of pollutants dispersion in the atmosphere. One is the offline approach, in which meteorological fields are first simulated and dispersion models use these meteorological data for dispersion and deposition computations with no feedback between pollutant concentrations and meteorological fields. In the online approach, the meteorology and dispersion models are run simultaneously, with exchange of information both ways at each time step. Both these methods have their own advantages, offline method more frequently used for primary pollutants such as mercury, PM2.5 etc and online approach for secondary pollutants such as ozone. We briefly describe here outlines of HYSPLIT offline model and WRF/Chem online model as results from dispersion studies over the Mississippi Gulf coast region with these two models carried out at TLGVRC, Jackson State University under a NOAA supported research program are presented in later sections.

4.7 HYSPLIT model

HYSPLIT (Hybrid Single-Particle Lagrangian Integrated Trajectory) model version 4, developed jointly by National Oceanic and Atmospheric Administration (NOAA) Air Resources Laboratory (ARL) and Australian Bureau of Meteorology, is a computational tool designed to produce air parcel trajectories, and to carry out simulations on different spatial and time scales from local, regional and long-range transport, dispersion, and deposition of air pollutants [20]. The model can take inputs of meteorological data from any of the Mercator, Lambert and Polar map projections but the data is required on grids and at regular time intervals. The model handles the input data from different resolutions, interpolates to an internal terrain following sub-grid and performs the calculations using data from fine to coarse resolutions and also from different map projections, e.g. starting from a Mercator or Lambert conformal regional grid and switching to Polar stereographic global grid. The trajectories and the pollutant dispersion and concentrations are calculated using Lagrangian mechanics in which advection and diffusion are calculated independently and has computational advantages with single point source emissions without constraints on resolution as compared to the application of Eulerian method for complex emission scenarios requiring solutions at all grid points of a region. Hysplit model computes the trajectories based on Lagrangian approach which is the advection of each particle in time following three dimensional velocity fields. In this model, dispersion and deposition can be computed using either puff or particle model or a hybrid in which both puff and particle approaches are adopted. In puff model, pollutant puffs are released from the source at regular intervals with each puff containing an appropriate fraction of the pollutant mass. The puff expands in time according to the atmospheric dispersion and advected following the trajectory. The puffs will split into new smaller puffs when their size exceeds the meteorological grid. The expansion within grid cell represents dispersion due to sub-grid scale turbulence and puff splitting represents grid scale process. Concentrations are calculated on the grid following a defined spatial distribution within the puff. In the particle model, cluster of particles are released from the source which will expand in space and time following the atmospheric dispersion. Advection of each particle will have a component related to in situ atmospheric turbulence. Concentrations are calculated as a sum of all the particles in a grid cell. By default, Hysplit model uses a hybrid approach in which puff and particle dispersions are incorporated in the horizontal and vertical representations respectively, which has the advantages of using puff model for better representation of horizontal distribution limiting the number of particles and better particle representation in the vertical where large discontinuities are possible. Both the horizontal puff and vertical particle dispersions use turbulence velocity components obtained from meteorological data. Regardless of the approach, atmospheric stability, atmospheric mixing and rate of pollutant dispersion are necessary for simulating pollutant dispersion. Atmospheric stability is estimated from wind and temperature parameters, horizontal and vertical mixing coefficients are estimated from velocity deformation and the coefficients of heat respectively, and the dispersion is computed as dependent on the mean and turbulent velocity components. Concentrations are summed up at each time step at all the grid points. This methodology clearly indicates the importance of the input meteorological data towards best representation of the vertical atmospheric structure. This model is simple to use with menu driven operations with the computations requiring only a few minutes operable on both windows and Linux based computer systems. The current version 4.9, suitable for Windows, was released on February 24, 2009 and Beta version suitable for IBM AIX and Linux servers was released on August 25, 2009.

4.8 WRF/Chem model

The Weather Research and Forecasting - Chemistry model (WRF/Chem) is a new generation regional air quality modeling system developed at NOAA (National Oceanic and Atmospheric Administration) [21]. This model uses the Eulerian mechanics, in which meteorology and chemistry variables are predicted at all the grid points at each time step. Although the computations are time consuming, it is most suited for prediction of air quality which requires full and continuous interaction of chemical species with meteorological parameters. This modeling system has two modules for meteorology and chemistry. For meteorology part, ARW mesoscale weather prediction model described in section 3 is used. The chemistry module treats the processes of dry deposition, coupled with the soil/vegetation scheme; aqueous phase chemistry coupled to some of the microphysics and aerosol schemes; biogenic emissions; anthropogenic emissions; gas-phase chemical reaction calculations; photolysis schemes; and aerosol schemes with different choices. The air quality component of the model is fully consistent with the meteorological component; both components use the same transport scheme, the same horizontal and vertical grids, the same physics schemes and the same time step for transport and vertical mixing. The model is consistent with all conservative transport done by the meteorology model. The resolution of the model is flexible, ranging from a few kilometers to hundred kilometers. This modeling system is suitable as coupled weather prediction/ dispersion/ air quality model to simulate release and transport of primary pollutants such as particulate matter and the prediction of secondary pollutants such as ozone. The model can be run on parallel processing computer platforms to save computational time. The current version is 3.2.1 released on August 18, 2010.

5. Air Quality modelling over Mississippi Gulf Coast

Pollutant contamination in the environment is a serious environmental concern for US Gulf coast region as of many other coastal habitats elsewhere. The Mississippi coastal region is environmentally sensitive due to multiple air pollution problems originating as a consequence of several developmental activities such as oil and gas refineries, operation of thermal power plants, rapidly increasing traffic pollution etc. The Mississippi Delta encompasses the largest area of coastal wetlands in the United States and supports one of the most extensive developments of petroleum extraction of any coastal area in the world. This area has been experiencing ecological impacts from energy development related human activities for more than hundred years. Coastal wetlands are vital for protecting developed areas from storm surges, providing wildlife and fish habitat, and improving water quality. Mississippi Gulf Coast area has sensitive ecosystems like National Forests, State parks, wildlife management and refuge areas, conservation areas or wildlife sanctuaries. The growth of industrial and commercial operations has created a need for air pollution dispersion models that can handle complex meteorological conditions of the coastal environment. Differential heating, strong thermal gradients along the land-sea interface and topographic friction cause localized mesoscale phenomena such as land-sea breeze circulations, sea breeze induced convection and formation of thermal internal boundary layer. The horizontal and vertical extents of the land-sea breeze, the internal boundary layer and their spatial heterogeneity under varying synoptic meteorological settings typify the complex dispersion patterns in the coastal region. The Thermal Internal Boundary Layer (TIBL) limits the region of vertical mixing, heating/convection and the low-level circulation characteristics which influence the coastal area dispersion. These spatio-temporal effects are to be included in the dispersion assessment for realistic air quality estimations using appropriate meteorological and dispersion models.

During the recent past years, air quality modeling studies have been carried out under varied meteorological conditions using sophisticated mesoscale models and observational assimilation techniques which have shown significant success [22-31]. At Jackson State University, we have conducted several studies on multi-scale simulation of meteorological fields for wind, temperature, humidity, and planetary boundary layer (PBL) turbulence utilizing ARW mesoscale weather prediction model (described in section 3.1). The simulated high resolution meteorological fields were integrated with HYSPLIT dispersion model (described in section 4.7) for studies on the source- receptor relationships of mercury and $PM_{2.5}$ pollutants. Similarly WRF/Chem model (described in section 4.8) was used to study the evolution of surface ozone over MS Gulf coast region and Jackson city urban metropolitan region. We present here the results from our studies which indicate the importance and usefulness of offline models for study of primary pollutants (Hg and $PM_{2.5}$) and online models for surface ozone.

5.1 Case study of $PM_{2.5}$

The Mississippi Gulf Coast is a typical coastal urban terrain featuring several industries that have been identified as emission sources of $PM_{2.5}$ and its precursor gases. A case study was undertaken to assess the WRF-HYSPLIT modeling approach towards assessment of source-receptor relationships of $PM_{2.5}$ over MS Gulf Coast [32]. For this study, measurements of $PM_{2.5}$ sulfate and nitric acid collected as 6-h samples during 17-20 June 2009 at the two selected locations, Harrison County School (30.5N, 89.1W) and Wiggins Airport (30.8N, 89.13W) were used. During this period, the concentrations of $PM_{2.5}$ sulfate and HNO_3 at Harrison and Wiggins were in the range of 5-8 $\mu g\ m^{-3}$ for sulfate and 1.5-3 $\mu g\ m^{-3}$ for HNO_3. The experimental study constituted two parts, simulation of atmospheric fields using WRF model and deriving back trajectories and forward deposition concentrations using HYSPLIT model. WRF model was used to produce the necessary atmospheric fields for 24-h periods with the outputs were stored at 1-h interval as needed for input to dispersion model. Model-simulated atmospheric flow fields were validated by comparison with the North American Mesoscale (NAM)-12 km regional analysis as they are critically important for computation of back trajectories and forward dispersion. The model simulated wind flow showed anti-clockwise circulation agreeing with NAM analyses. The onset and extent of sea breeze along the coastal regions was well simulated. HYSPLIT model, driven by meteorological inputs from ARW model, was used to produce 24 numbers of back trajectories at 1-h intervals to identify the sources along the paths of trajectories. The back trajectories, drawn from each observation site (Figure 1) shown that the air parcels were mostly confined to heights below 1.0 km within the planetary boundary layer and had paths in the quadrant between south and west, all pointing the origination of parcels from land on west-side and from sea on south-side (i.e.) Gulf of Mexico. Specifically, Watson, Cajun, and Morrow power plants were identified to be possible land sources as located within the path of back trajectories. HYSPLIT model was run in the forward mode, driven by the ARW model-generated atmospheric fields at 1-h interval and the hourly emission rates, starting from each of the three identified coal-fired power plants to produce the 24-h atmospheric dispersion separately for SO_2 and NOx with the hourly emission rates derived from EPA annual emissions data. The forward 24-h atmospheric dispersions from the three sources for SO_2 (Figure 2) showed the dispersion to be towards east/northeast under the prevailing wind flow of southwesterly/westerly relative to the observation sites. The pattern of the pollutant dispersion was a slowly expanding plume due to the low magnitude of the wind speed (2-5 m/s). The contributions from each of the three power plants to the two observation sites have shown that Big Cajun and Morrow plants contributed to a smaller of extent <0.235 μg m^{-3} for Wiggins location while none of three sources contributed for the observations at Harrison although these two sites are only 65 km apart. These results lead to the conclusion that increasing activity in diesel-powered heavy duty vehicles (ship) over the Gulf of Mexico (as 60% of US energy imports are from this area) and ports around, diesel oil emissions from this area may also contribute to precursors SO_2 and NOx of $PM_{2.5}$. The source attribution from the Gulf of Mexico could also be due to the diurnal Gulf breeze; with the night time land breeze carry the land-borne precursor pollutants on to the sea and daytime Gulf breeze would bring back the pollutants from the sea to the land region, thus representing the Gulf of Mexico as a virtual pollutant source. Although the case study does not concern with an episodic event, it brings out the result that the integrated modeling approach does not produce spurious results.

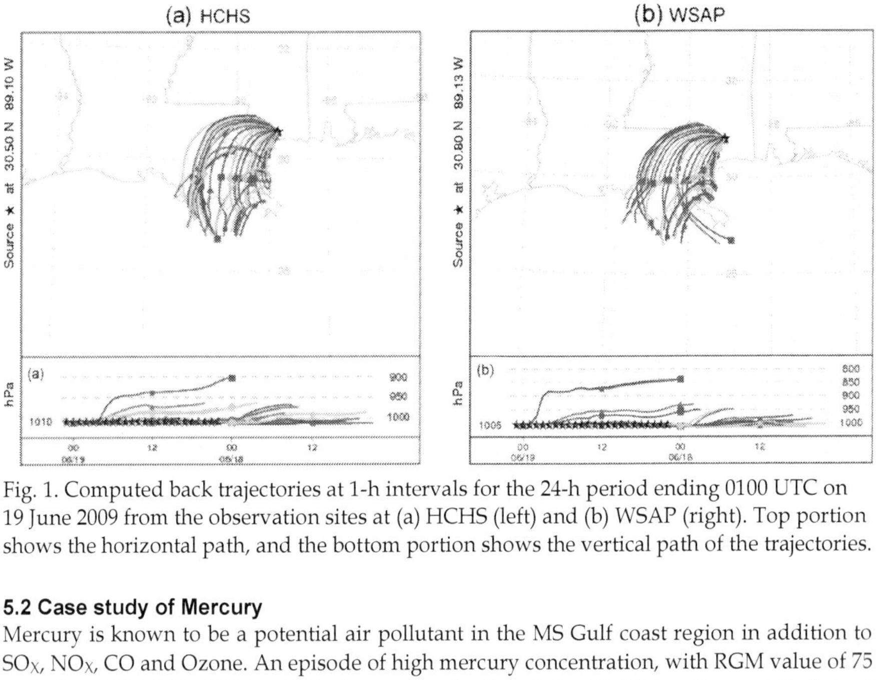

Fig. 1. Computed back trajectories at 1-h intervals for the 24-h period ending 0100 UTC on 19 June 2009 from the observation sites at (a) HCHS (left) and (b) WSAP (right). Top portion shows the horizontal path, and the bottom portion shows the vertical path of the trajectories.

5.2 Case study of Mercury

Mercury is known to be a potential air pollutant in the MS Gulf coast region in addition to SO_X, NO_X, CO and Ozone. An episode of high mercury concentration, with RGM value of 75 pg/m^3 at 1600 UTC of 5 May 2008 observed at the Grand Bay NERR location on Mississippi Gulf Coast (30.41N, 88.4W) was selected as a case study. WRF-ARW model was used to simulate the atmospheric fields at 4 km resolution and with the outputs stored at 1-h interval. The ARW model derived 10m wind flow over the study region was validated with EDAS- 40km analysis. WRF model produced sea breeze with upward vertical motions adjacent to the coast and downward vertical winds in the northeastern and central parts of the land region relatively stronger than EDAS fields. The HYSPLIT model was used to simulate Lagrangian back trajectories from NERR site for the 12-hour period ending at 1600 UTC 5 May 08 with each of inputs from EDAS 40 km, ARW 36 km and ARW 4 km data

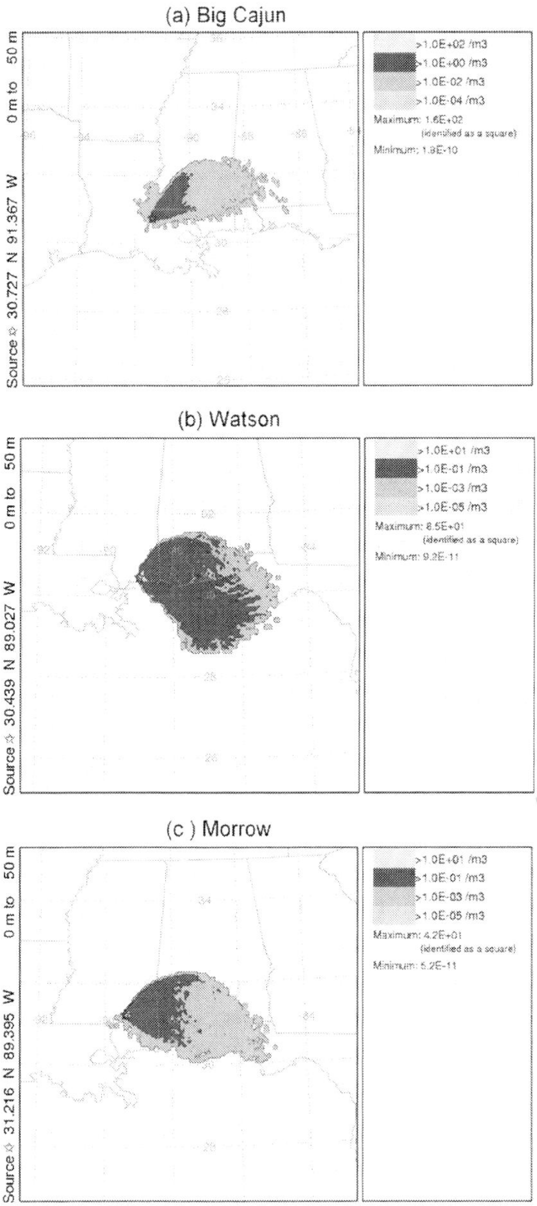

Fig. 2. HYSPLIT-generated SO2 concentration (μg m−3) averaged between 0 and 100 m levels and integrated for 24-h period between 0100 UTC of 18 June 2009 and 0100 UTC of 19 June 2009 sourced from the three identified coal-fired power plants.

(Figure 3). These back trajectories have shown that the air parcels originated towards north-northwest to north and the height of the parcels to be below 100 m level. The trajectories from ARW 4 km data were more towards east and realistic corresponding to mesoscale circulations and associated turbulence and vertical motion fields in the coastal region. The mean trajectory paths indicated four important sources i.e., Charles R Lowman (31.489N, 87.9W) and Barry (30.84N, 88.09W) power plants located in Alabama State and Jack Watson (30.44N, 89.026W) and Daniel (30.64N, 88.59W) power plants located in Mississippi State. The HYSPLIT model produced 24-hour ground level (0 -25 m) RGM concentration patterns with inputs from both the EDAS 40-km analysis (Figure 4) and the high resolution ARW 4km (Figure 5) data were compared. The deposition concentrations with ARW model have shown that Barry plant chiefly contributed with multiple peaks at different times during the day and with the highest concentration occurring 4 hours earlier than the actual measured peak and minute contributions from Daniel and Lowman plants. With EDAS data, Lowman and Daniel plants were identified to contribute chiefly. The concentration values with ARW fields were roughly one order higher than those obtained with EDAS fields. The representation of the wind field in the EDAS and the ARW data lead to major differences in the transport of the RGM from these two nearby sources (Barry and Daniel) in the two simulations. The multiple peaks in the simulated concentrations using ARW data was due to the diurnal circulation at the coast found in the ARW simulations indicating the impact of the sea- land breeze type mesoscale phenomena causing frequent plume transitions. This study helped to establish the sources and their relative contributions to the moderate mercury episode concentrations at NERR observation site using the integrated ARW-HYSPLIT modeling approach.

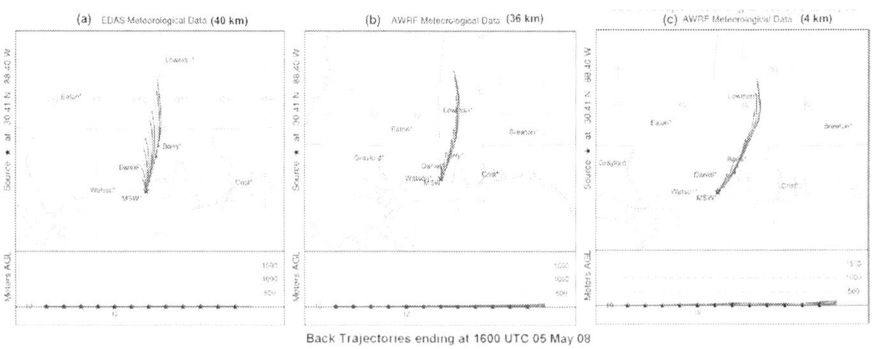

Fig. 3. Back trajectories produced by HYSPLIT model using EDAS-40 km (left); ARW-36km (middle) and ARW-4 km (right).

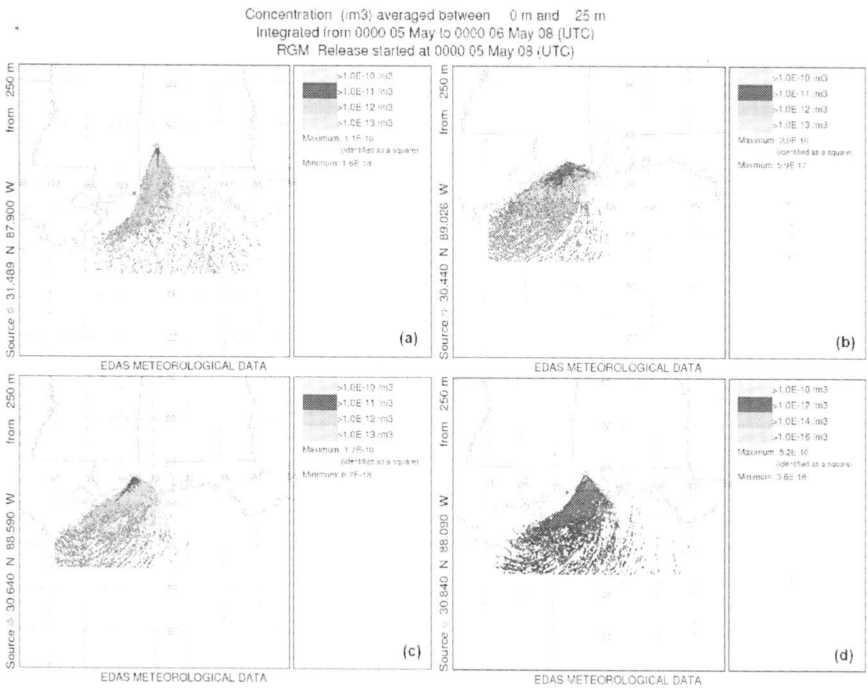

Fig. 4. Atmospheric dispersion as predicted by HYSPLIT model using EDAS-40 km data from the four different sources of Lowman (upper left), Watson (upper right), Daniel (bottom left) and Barry (bottom right).

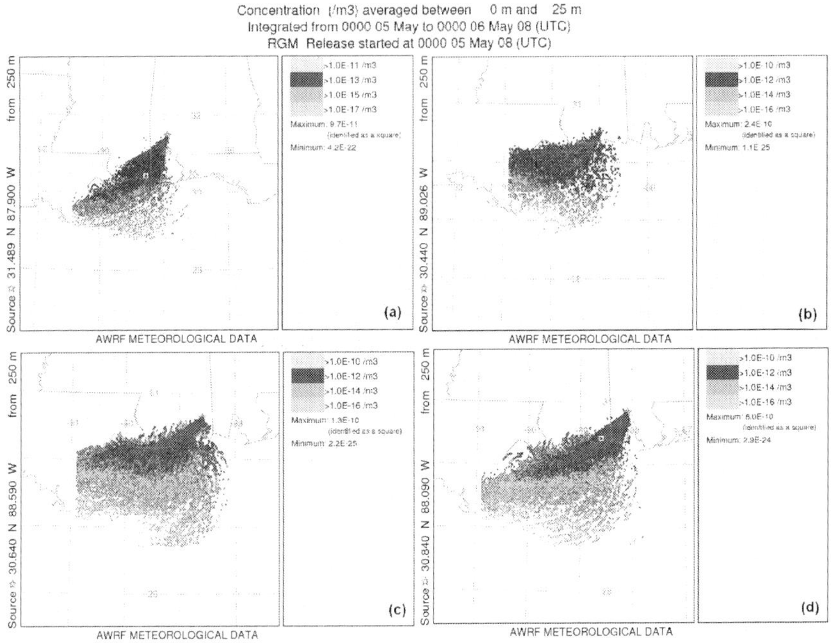

Fig. 5. Atmospheric dispersion as predicted by HYSPLIT model using ARW- 4 km data from the four different sources of Lowman (upper left), Watson (upper right), Daniel (bottom left) and Barry (bottom right)

5.3 Case study of surface ozone over Jackson metropolitan area

As per US EPA (United States Environmental Protection Agency), urban air quality in US with reference to ozone is a growing concern due to its oxidative capacity which will have great impact on the environment. Large number of people are likely to be exposed to the unhealthy ozone concentrations when ground-level ozone gets accumulated in urban metropolitan areas under certain weather conditions [33] and quantitative atmospheric dispersion models will be of great help to provide effective decision support systems for planners and administrators. We made an attempt to study the evolution of surface ozone and other precursor emissions like NOx and CO over southeast parts of US in general and Jackson, MS urban area in particular. We were motivated to take up this study to assess the performance of WRF/Chem in the simulation of a moderate ozone episode over an urban locality in which the atmospheric flow patterns are strongly influenced by local terrain and land cover patterns and the results from this study would provide useful information of urban air pollutants to air quality regulatory agencies and health administrators. WRF/Chem model was adapted to have nested four domains with the horizontal resolution of 36-12-4-1 km (Figure 6), with the innermost 1 km domain covering Hinds County. The chemistry was initialized with idealized profiles of anthropogenic emissions data which consists of area type emissions on a structured 4-km grid and point type emissions at latitude and longitude locations. WRF/Chem model was used to simulate the spatial and

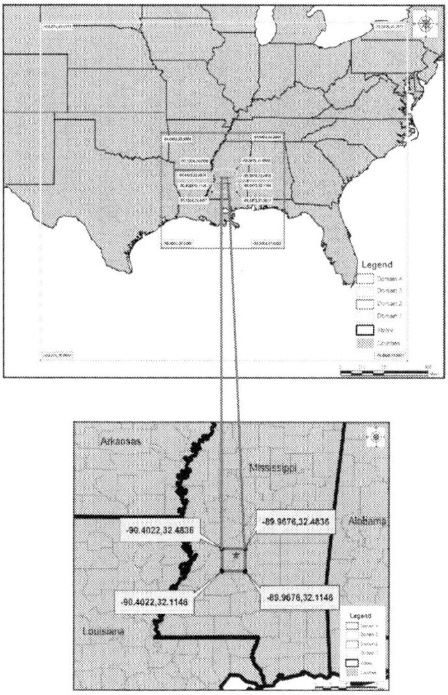

Fig. 6. Model domains. Innermost domain covering the study region of Jackson, MS is shown as red square. Observation site is shown as star.

temporal variations of surface ozone and other precursor pollutants CO (Carbon Monoxide), NO (Nitric oxide), NO_2 (Nitrogen dioxide) and HONO (Nitrous acid). The simulated time series at the specific location (32.38578 N, 90.141006 W) inside Jackson city during 7 PM 6 June to 7 PM 7 June 2006 showed that surface ozone gradually decreased from 7 PM onwards reaching minimum around midnight and stays nearly constant till dawn and gradually increased from dawn to noon reaching a maximum of 50 ppb around 1 PM. The model simulated the time variations and the time of attainment of maximum same as of observations with slight underestimation. The vertical variation of the model simulated ozone at peak time showed that the magnitude of ozone is nearly constant up to 900 hPa level, decreases rapidly up to 800 hPa level and then gradually increases upward. This means that there is good mixing of O_3 in the PBL region extended up to 900 hPa level as common for a summer afternoon. The time series of CO and NO_2 showed a maximum around midnight followed by gradually decrease and with increasing trends at certain times of daytime. The maximum during the first parts of the night could be attributed to these accumulations, decrease during later half to their dispersion and the increasing trends to the reaction between NO with VOC leading to NO_2 and other products. The occurrence of NO maximum at 11 AM, 2 hours prior to ozone maximum indicates the mechanism of NO_2 splitting as NO and O and the formation of O_3 through the amalgamated reaction of O and O_2. Similar explanation holds good for CO, as O_3 is produced in the troposphere by the

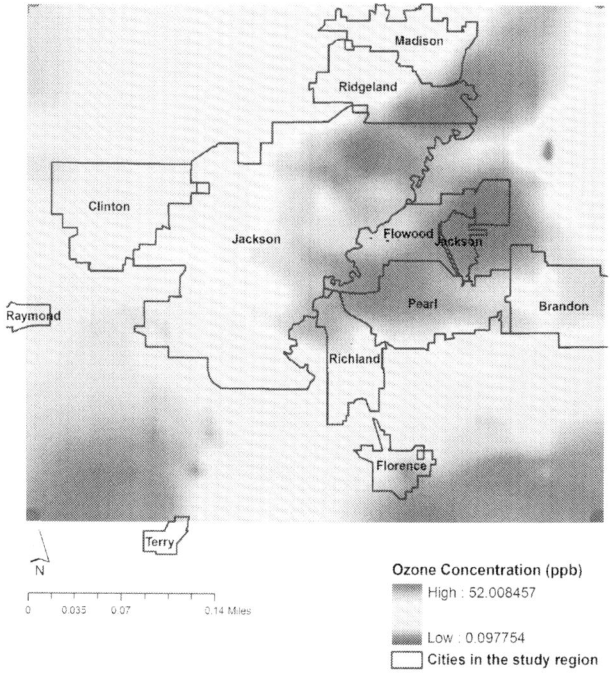

Fig. 7. Spatial distribution of model simulated surface ozone (ppbv) "over domain 4 covering Jackson city and neighbourhood at 1 PM CDT 7 June 2006.

photochemical oxidation of hydrocarbons in the presence of nitrogen oxides (NO and NO_2). The time series of HONO showed a maximum between 10-12 AM indicating contributions from nearby combustion sources and its decrease after 12 noon could be due to its breakdown through reactions with NO and OH in the presence of sunlight. All these indirectly supports formation of ozone as it leads production of NO and OH. The spatial distributions of O_3 corresponding to the peak time at 1 PM (Figure 7) showed that the central, east and northeast parts of model domain, covering the regions of Jackson, Flowood, Pearl and parts of Ridgeland and Madison have the maximum concentrations whereas the western and southern parts covering Clinton, Raymond and Florence were less affected. This spatial pattern indicates the maximum as associated with the identified urban locations due to mobile traffic and combustion sources. The spatial distributions of NO, NO_2, HONO and CO are similar indicating that mobile and combustion sources in Jackson, Ridgeland and Madison were contributing to the production of O_3 through chemical reactions. Back trajectories drawn from the observation site showed the air parcels to originate from west and south, where the road-ways are located which confirm that the pollutants of mobile origin are chiefly responsible for the production of ozone in sunlight hours. This study is important as ozone and other pollutant species were simulated at 1 km resolution over Jackson urban metropolitan region providing important information for assessment, management and mitigation of urban pollutants.

5.4 Case study of surface ozone over MS Gulf Coast

A regional scale study was undertaken to simulate the surface ozone in the central Gulf coast using WRF/Chem model. A moderately severe ozone episode, with ozone values exceeding 80 ppbv that occurred during 8-11 June 2006 was selected for a case study [34]. The model was configured with three two-way interactive nested domains (36-12-4 km resolution) and 31 vertical levels (Figure 8) with the inner finest domain covering the Mississippi coast. The default profiles for chemical species available with the model were used as the initial pollutants. The model simulated meteorological fields and ozone distributions were compared with available observations for validation. Several sensitivity experiments with different planetary boundary layer and land surface model schemes which revealed that YSU PBL scheme in combination with NOAH land surface model provided best simulation of meteorological fields in the lower atmosphere over the MS Gulf Coast region [35]. The patterns of simulated surface ozone and NO_2 concentrations with different PBL and land surface physics have shown variations. While the patterns of ozone were similar with the YSU and MYJ PBL formulations, MYJ PBL produced relatively shallow mixing layers as compared to moderately deep mixed layer development with YSU and ACM schemes. As of soil schemes, 5-layer soil model simulated relatively deeper mixed layers than NOAH LSM. The simulated ozone patterns with ACM PBL scheme were different from other PBL schemes as it produced localized higher concentrations over northern Mississippi, eastern Louisiana and west Florida coast (Figure 9). The peak ozone concentration were $\geq$ 65 ppbv with ACM PBL over the northern parts of Mississippi river and west Florida coast located to the south of Alabama. The convergence in the surface flow along the coast and along Mississippi river in runs with ACM PBL and RUC LSM caused high ozone formation in these areas. The combination of YSU PBL and NOAH land surface schemes gave best simulations for all the meteorological and air quality fields with least BIAS, RMSE and highest correlation values due to better simulation of PBL height, wind speed, temperature and humidity.

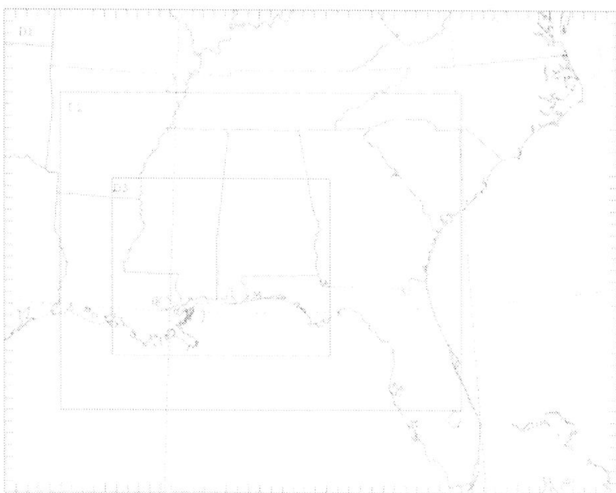

Fig. 8. Modelling domains used in WRF/Chem. Outer domains D01, D02 are coarse with resolutions 36 and 12 km and the inner finer domain (D3) is of resolution 4 km.

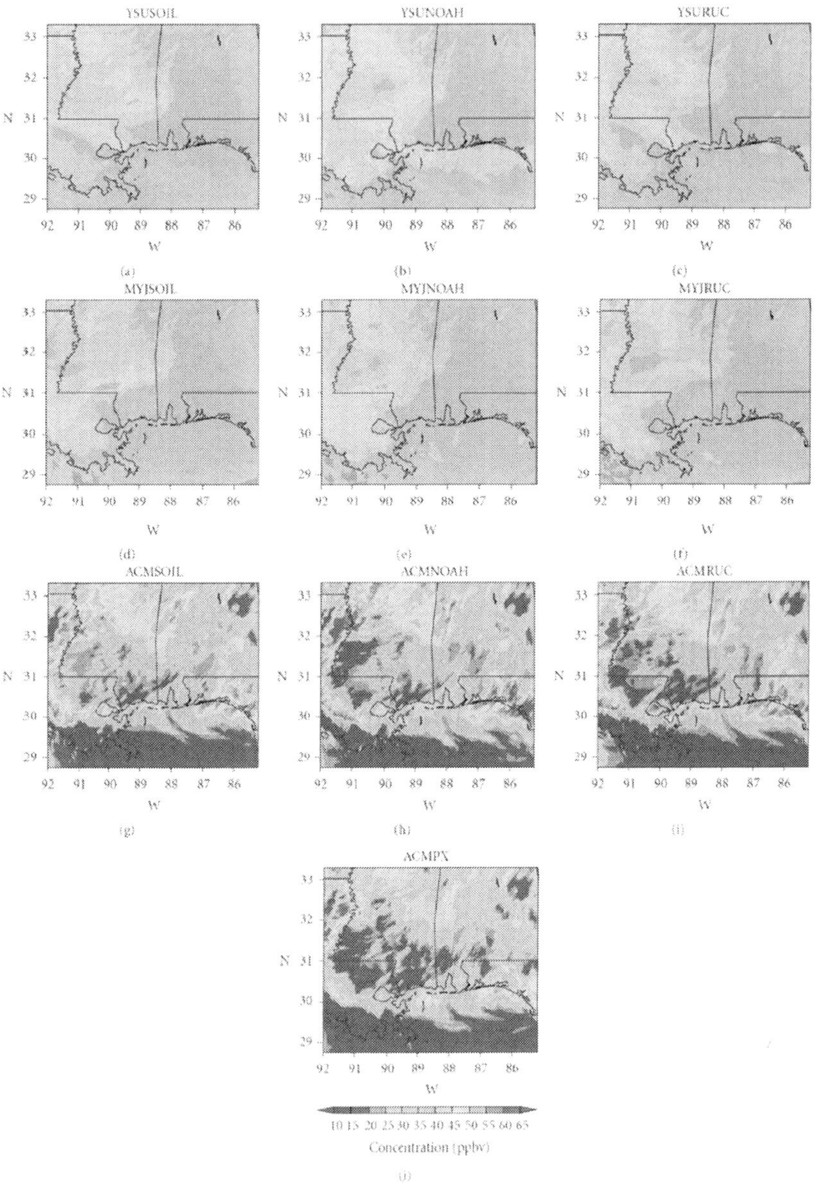

Fig. 9. Simulated ozone concentration (ppbv) in the model fine domain at the lowest level (30m) from experiments with different PBL and LSM options at 10 CST 9 June 2006 for the experiments YSUSOIL, YSUNOAH, YSURUC, MYJSOIL, MYJNOAH, MYJRUC, ACMSOIL, ACMNOAH, ACMRUC and ACMPX.

6. Applications of Geographic Information Systems (GIS)

Geographical Information System (GIS) is a powerful tool that facilitates linking spatial data to non-spatial information [36]. With its embedded relational database component, the system assists in storing, mapping and analyzing geo-referenced data in an organized structure [37]. The database and the geographical base form the two major components of the GIS system that helps in visualizing the data in a map format. Unlike the reports generated by stand alone applications, which summarize the tabular data, these maps illustrate the geographical connections among the spatial variables and visually communicate geo-specific information to a decision maker.

The conceptual approach of GIS not only provides the capability of querying the spatial data but also, with its inbuilt analytical tools, translates the existing spatial patterns into measurable objectives. These analytical capabilities of GIS offer a dynamic dimension to 'Spatial Analysis' factor in determining the principles of behavior or illustrating the inter-relationships among spatial and non-spatial data.

Understanding these relationships or dependencies in a spatial phenomenon is the major crux of any field; health, environment, geology, hydrology, air pollution studies, agronomy and many others. With its framework, GIS facilitates the integration of various field-specific applications into its interface and allows integrated ways to conduct research and develop new analytical approaches to relate their information to the terrestrial activities. These abilities (providing fully functional data processing environment) distinguish GIS from other information systems as it results in enhancing the applications with productive findings that are broadened and deepened in a geographic location.

To build a spatial data model, GIS Systems support three basic types of data (1) Vector Data: a) Events or Points: Pattern expressed as points in space, b) Lines: Patterns expressed as networks and c) Polygons: Patterns expressed as analytical units with defined closed boundaries, (2) Raster data: a) Grid-cell data and (3) Image: a)Satellite Imagery, photographs.

The spatial analysis modeling process involves interpreting and exploring the interactions, associations and relationships among these data types specific to a geographic location. The exploratory process in developing the spatial model is composed of a set of procedures.

- To identify/map the various data layers and describe the attributes associated with the spatial pattern (Representation model)
- To explore/model the relationships, association or interactions between the data layers identified in representation model (Process Model)

While representation model involves deriving necessary input datasets, in most of the cases, the datasets needs to be reclassified by setting a common measurement scale to the attribute variables by giving weight age depending on their influence. Numerous interactions/ relationships between various data components of a region can be captured by the wide range of spatial analytical modeling tools like Suitability analysis, Distance analysis, Hydrological analysis, Surface analysis. Depending on the type of interaction, the spatial model makes use of these analyses in designing the process.

Another dynamic tool GIS offers is the geo statistical analyst tools that include numerous techniques in exploring and modeling the relationships in the spatial data. The inclusion of interpolation methods provides a powerful environment to the users to assess the quality of their analysis. Many disaster mitigation spatial problems are addressed using the Network Analyst tool as required by different research applications. The capabilities of Network

Analyst tool allows to dynamically model realistic conditions and facilitates in examining the flows within and between the natural and man-made networks.
The broad range of many analyst tools, with their modeling and analysis features, not only makes the GIS applications penetrate through number of research fields (from environment, transportation, community *etc.*) but also exploring associations and relations ships between these fields.

6.1 GIS in air pollution studies

The quality of air is a very important factor in projecting or representing the status of environment and health of any region. Air pollution studies that analyze the quality of air provide strategic information to the decision making process and play a significant role in the implementation of the policies that influence the air quality of a region. Most of the air pollution models, in their pollutant distribution simulations, consider the physical characteristics of the pollution such as wind direction, speed, temperature etc., in determining the air pollution trajectory. Integrating these models with GIS presents a geographic dimension to the air quality information by relating the actual pollution concentrations to the plant and human life in that location. With its numerous analyst tools, GIS can demonstrate the relationship of poor air quality and occurrences of deficient human and environmental health [38]. GIS can portray the spatial correspondence between the air quality and the disease statistics in the area that is potentially impacted [39]. In this process, GIS can explore:

- The relative spatial phenomenon of the pollutant with respect to the geographical distribution in terms of location, extent and distance.
- The spatial extent of the pollutant dispersion and its intensity at any geographical location under the impact zone
- The socio-economic characteristics of the populations affected by these pollutants
- The data visualization from various perspectives by classifying and reclassifying the data by determining the class breaks.

Examining the relationships between high pollution concentrations across various demographic thematic layers helps in identifying *hotspots* that are in need of special investigation or monitoring. Data visualization, illustrating such information through a map provides an insight in a more dynamic way that helps the authorities to plan their future strategies.
The output generated from an air pollution dispersant model (Figure 10), illustrates the mercury pollution concentrations in the Gulf Coast region. The intensity and spatial dispersion of the pollutant are presented in a scientific terminology as units and latitude/longitude. The format of this scientific information generated by these models may not be suitable for directly integrating the data into the policy making process. The relevance of this information to a spatial location has to be derived and its implications need to be presented in a format that relates to the socio-economic or health characteristics of the people living in that region. The gap of presenting the scientific format from a geographical dimension can be filled by integrating these outputs with Arc GIS tools. Viewing the same data across various demographic themes projects the vulnerable populations under the impact zone and assists the authorities in making informed choices.
Figure 11 illustrates the HYSPLIT output shown in Figure 10 in a GIS environment. The output of HYSPLIT is obtained in the format of ASCII with relevant data on latitude, longitude and the mercury concentrations at various points in the dispersion trajectory.

Using data management and 3D Analyst tools, the ASCII file is converted into a point data and a TIN (Triangular Irregular Networks) file is created from this vector data. By interpolating the cell Z-values (Mercury concentrations) using natural neighbors method, the TIN is converted to a Raster file. The final mercury dispersion raster file is then mapped over the Gulf Coast states.

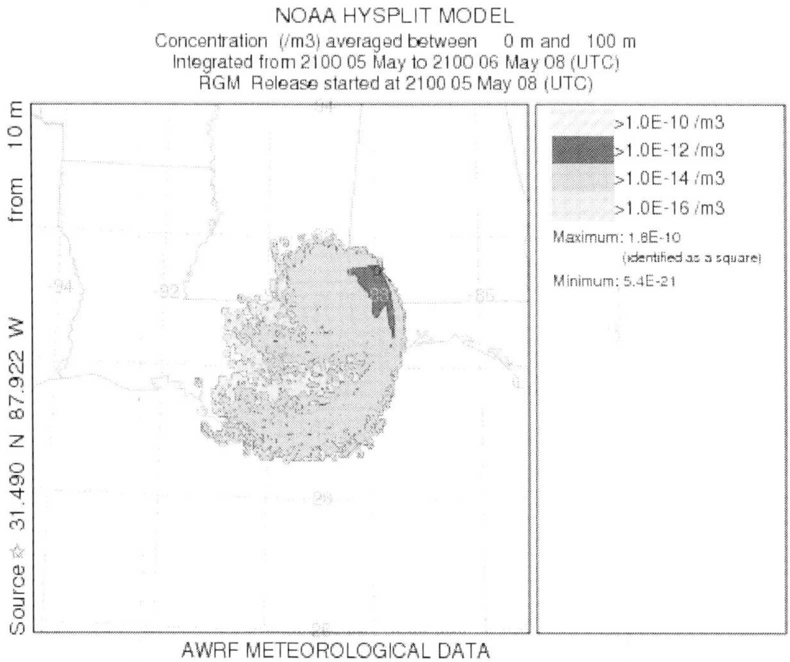

Fig. 10. Output generated from the HYSPLIT model developed by NOAA

Figure 11 describes that parts of Mississippi, Louisiana and Alabama were impacted by the emissions from the Barry Generating Plant in Alabama. The mercury dispersion raster file is overlaid on the classified demographic thematic layer (dark to light color indicates high to low dense population) shows most of the tracts that are under the impact zone are densely populated.

The impact of severe pollution zones: red (very high) and yellow (high) colors, on vulnerable populations (age above 65 years and less than 5 years) are presented in Figure 12 and Figure 13.

Integrating HYSPLIT model output with classified demographic GIS data gives a clear visual representation of the spatial location of highly dense vulnerable populations under impact zone. It assists in projecting the age of the vulnerable population (which in this case ages less than 5 years) under the risk due to the pollution impact.

A spatial phenomenon or a relation can be developed by examining the health statistics of these tracts against the pollution levels they are being exposed. Such spatial phenomenon may reveal hidden facts and can be a significant contribution in designing the policy strategies.

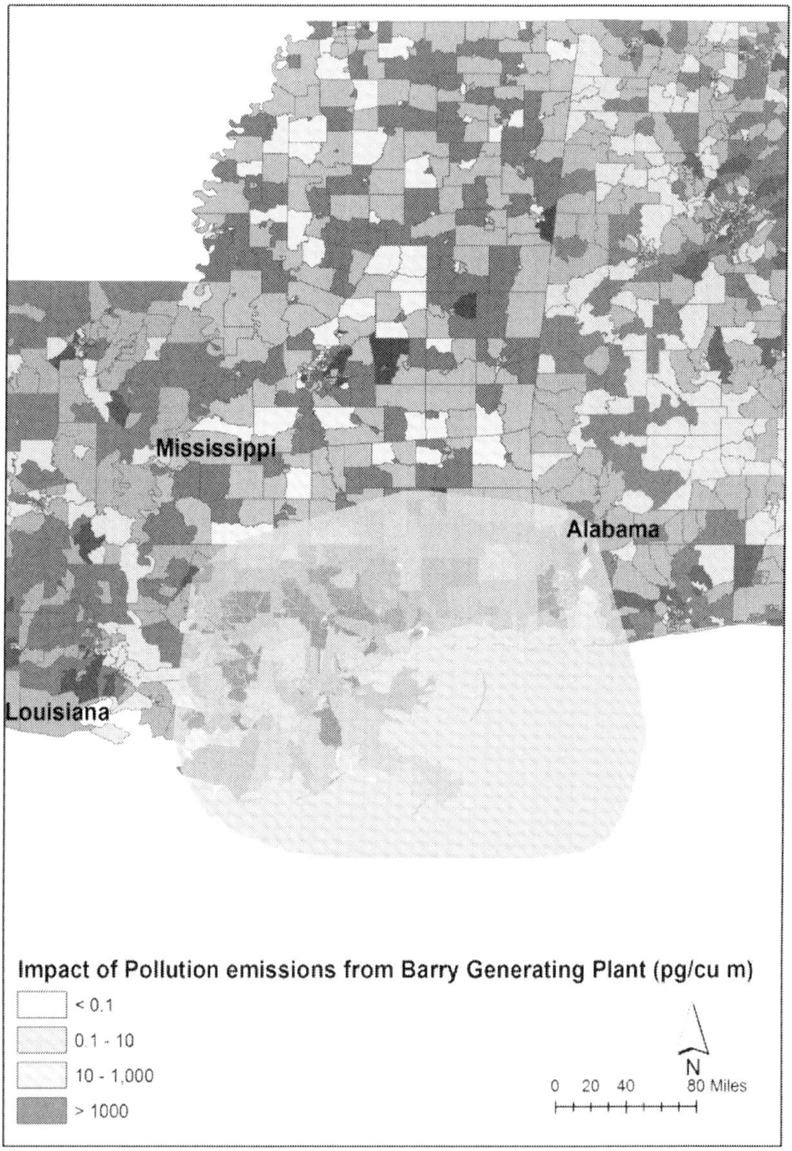

Fig. 11. Representation of HYSPLIT output in GIS environment

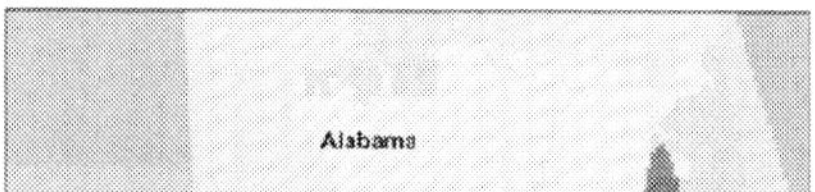

Fig. 12. Population of age above 65 under the impact of Barry Plant-mercury emissions.

Fig. 13. Population of age less than 5 years under the impact of Barry Plant -mercury emissions.

The demographic age classification under the impact zone of 10-1000 pg/m^3 of mercury concentration (yellow color) can be well interpreted from the figures 12 and 13. The GIS functionalities facilitate in estimating the actual numbers: tracts, along with vulnerable population numbers, falling under the impact zone (Table 1). For convenience, four classes of populations were made based on density (i.e.) >1200, 800-1200, 400-800 and <400.

Population / Age	High density population (>1200) (no. of tracts)	Medium density population (800-1200) (no. of tracts)	Low density population (400-800) (no. of tracts)
5 years and below	1797(1)	15761(13)	11665 (28)
65 years and above	1640 (1)	5710 (6)	22652 (38)

Table 1. Numbers of area tracts and population affected from pollutant dispersion

The pollutant dispersion region covered approximately 92 tracts in which about 338,439 habitants live. From the data in Table 1, it may be inferred that elders with 65 years and more are the most affected in the two categories of high and low density population concentrations whereas more children below 5 years were affected in the category of 800-1200. This emphasizes the need to detailed analysis using GIS to clearly understand the relative age groups affected, in this case due to pollution. Presenting such statistical information through GIS mapping leads to risk assessment and provides an idea of not only the areal extent of pollutant influence but also its impact on demographic data lines.

7. References

[1] US EPA, 2003. A Review of the Reference Dose and Reference Concentration Processes. US EPA/630/P-02/002F, December 1, 2002. Risk Assessment Forum, Washington, DC, 192pp. http://cfpub.epa.gov/ncea/raf/

[2] US EPA, 2004. Air quality criteria for particulate matter. Research Triangle Park, NC: Office of Research and Development, Report No. EPA/600/P-99/022aF-bF.

[3] Hannigan, M.P., Busby, W.F., Cass, G.R., 2005. Source contributions to the mutagenicity of urban particulate air pollution. *Journal of Air & Waste Management Association*, 55, 399-410.

[4] Barrett, K. and Seland, Ø. (Eds.), 1995. European Transboundary Acidifying Air Pollution: Ten years calculated fields and budgets to the end of the first Sulphur Protocol. EMEP Research Report No. 17, The Norwegian Meteorological Institute, Oslo.

[5] Skamarock, W.C., Klemp, J.B., Dudhia, J., Gill, D.O., Barker, D.M., Duda, M.G., Huang, X.-Y., Wang, W., and Powers, J.G., 2008. A Description of the Advanced Research WRF Version 3. *NCAR Technical Note, NCAR/TN-475+STR*. Mesoscale and Microscale Meteorology Division, National Center for Atmospheric Research, Boulder, Colorado, USA.

[6] Grell, G. A., Dudhia, J., Staufer, D.R., 1995. A Description of the Fifth-Generation Penn State/ NCAR Mesosscale Model (MM5), NCAR Technical Note, 6-1995.

[7] Taylor, G.I., 1915. Eddy motion in the atmosphere Phil. *Transactions of the Royal Soc. of London.*Series A, 215, 1.

[8] Taylor, G.I., 1921. Diffusion by continuous movements. *Proc. London Math. Soc.* 20, 196.

[9] Rohde, H., 1972. A study of the sulfur budget for the atmosphere over northern Europe. *Tellus*, 24, 128.

[10] Rohde, H., 1974. Some aspects of the use of air trajectories for the computation of large scale dispersion and fallout patterns. *Adv. in Geophysics*, 18B, 95, Academic press.

[11] Eliassen, A., and Saltbones, J., 1975. Decay and transformation rates of SO2 as estimated from emission data, trajectories and measured air concentrations. *Atm. Env*,. 9, 425.

[12] Fisher, B.E.A., 1975. The long-range transport of sulfur dioxide. *Atm.Env.*, 9, 1063.

[13] Reynolds, S., Roth, P., and Seinfeld, J., 1973. Mathematical modeling of photochemical air pollution. *Atm.Env.*, 7.

[14] Shir, C.C. and L.J. Shieh, 1974. A generalized urban air pollution model and its application to the study of SO2-distribution in the St. Louis Metropolitan area. *J. Appl. Met.*, 19, 185-204.

[15] Egan, B.A., Rao, K.S., and Bass, A., 1976. A three dimensional advective-diffusive model for long-range sulfate transport and transformation. 7th ITM, 697, Airlie House.

[16] Carmichael, G.R., and Peters, L.K., 1979. Numerical simulation of the regional transport of SO2 and sulfate in the eastern United States. *Proc. 4th Symp. on turbulence, diffusion and air pollution*, AMS 337.

[17] Friedlander, S.K., and Seinfeld, J.H., 1969. A Dynamic Model of Photochemical Smog. *Environ. Science Technol.*, 3, 1175.

[18] Eschenroeder, A.Q., and Martinez, J.R., 1970. Mathematical Modeling of Photochemical Smog. *Proc. American Institute Aeronautics and Astronautics, Eight Aerospace Sciences Meeting*, New York, Jan 19-21.

[19] Liu, M.K., and Seinfeld, J.H., 1974. On the Validity of Grid and Trajectory Models of Urban Air Pollution. *Atmos. Environ.*, Vol. 9, pp. 555-574.

[20] Draxler, R.R., Rolph, G.D., 2010. HYSPLIT (HYbrid Single-Particle Lagrangian Integrated Trajectory) Model. NOAA Air Resources Laboratory, Silver Spring, MD. Available at NOAA ARL READY Website http://ready.arl.noaa.gov/HYSPLIT.php

[21] Grell, G. A., Peckham, S.E., Schmitz, R., McKeen, S.A., Frost, G., Skamarock, W.C., and Eder, B., 2005. Fully coupled "online" chemistry within the WRF model. *Atmos. Environ.*, 39, 6957–6975.

[22] Indracanti, J., Srinivas, C.V., Hughes, R. L., Baham, J.M., Patrick, C., Rabarison, M., Young, J., Anjaneyulu, Y., and Swanier, S.J., 2007. GIS Assisted Emission Inventory Development for Variable Grid Emission Database for Mississippi Region. 16th Annual International Emission Inventory Conference on Emission Inventories: "Integration, Analysis, and Communications". EPA, NC.

[23] Srinivas, C.V., Jayakumar, I., Baham, J.M., Hughes, R.L., Patrick, C., Young, J., Rabarison, M., Swanier, S.J., Hardy, M.G., and Anjaneyulu, Y., 2008. Sensitivity of Atmospheric dispersion simulations by HYSPLIT to the meteorological predictions from a mesoscale model. *Environ. Fluid Mechanics*, 8, 367-387.

[24] Anjaneyulu, Y., Srinivas, C.V., Jayakumar, I., Hari, P. D., Baham, J.M., Patrick, C., Young, J., Hughes, R.L., White, L. D., Hardy, M.G., and Swanier, S.J., 2008. Some observational and modeling studies of the coastal atmospheric boundary layer at Mississippi Gulf Coast for Air Pollution Dispersion assessment. *Int. J. Environ. Res. Public Health*, 5, 484-497.

[25] Srinivas, C.V., Jayakumar, I., Baham, J.M., Hughes, R.L., Patrick, C., Young, J., Rabarison, M., Swanier, S.J., Hardy, M.G., and Anjaneyulu, Y., 2009. A Simulation Study of Meso-Scale Coastal Circulations in Mississippi Gulf Coast for Atmospheric Dispersion. *J. Atmos. Res.*, 91, 9-25.

[26] Anjaneyulu, Y., Srinivas, C.V., Hari Prasad, D., White, L. D. , Baham, J.M., Young, J.H., Hughes, R.L., Patrick, C., Hardy, M.G., and Swanier, S. J., 2009. Simulation of Atmospheric Dispersion of Air-Borne Effluent Releases from Point Sources in Mississippi Gulf Coast with Different Meteorological Data. *Int. J. Environ. Res. Public Health*, 6, 1055-1074.

[27] Anjaneyulu, Y., Venkata, B.R. D., Hari, P. D., Srinivas, C.V., Francis, T., Julius M. B., John H. Y., Robert, L. H., Chuck, P., Mark G. H., Shelton J. S., Mark, D.C., Winston, L., Paul, K., and Richard, A., 2010. Source-receptor modeling using high resolution WRF meteorological fields and the HYSPLIT model to assess mercury pollution over the Mississippi Gulf Coast region. 90th American Meteorological Society Annual Meeting, Atlanta, USA.

[28] LaToya, M., William, R. P., Christopher, A. V., Anjaneyulu, Y., Venkata, B.R.D., Hari, P. D., Srinivas, C.V., Francis, T., Julius, M. B., Robert, L. H., Chuck, P., John, H. Y., and Shelton J. S., 2010. Evaluation of $PM_{2.5}$ source regions over the Mississippi Gulf Coast using WRF/HYSPLIT modeling approach. 90th American Meteorological Society Annual Meeting, Atlanta, USA.

[29] William, R. P., LaToya, M., Christopher, A. V., Venkata, B.R.D., Hari, P. D., Anjaneyulu, Y., Srinivas, C.V., Francis, T., Julius, M. B., Robert, L. H., Chuck, P., John, H. Y., and Shelton, J. S., 2010. Observation, analysis and modeling of the sea breeze circulation during the NOAA/ARL-JSU Meteorological Field Experiment, Summer-2009. 90th American Meteorological Society Annual Meeting, Atlanta, USA.

[30] William, R. P., LaToya, M., Christopher, A. V., Venkata, B.R.D., Hari, P. D., Anjaneyulu, Y., Srinivas, C.V., Francis, T., Julius, M. B., Robert, L. H., Chuck, P., John, H. Y., and Shelton, J. S., 2010. Analysis and perdition of the atmospheric boundary layer characteristics during the NOAA/ARL-JSU Meteorological Field Experiment, Summer- 2009. 90th American Meteorological Society Annual Meeting, Atlanta, USA.

[31] William, R. P., LaToya, M., Christopher, A. V., Venkata, B.R.D., Hari, P. D., Anjaneyulu, Y., Srinivas, C.V., Francis, T., Julius, M. B., Robert, L. H., Chuck, P., John, H. Y., and Shelton, J. S., 2010. Numerical prediction of atmospheric mixed layer variations over the Gulf coast region during NOAA/ARL JSU Meteorological Field Experiment, Summer-2009 - Sensitivity to vertical resolution and parameterization of surface and boundary layer processes. 90th American Meteorological Society Annual Meeting, Atlanta, USA.

[32] Anjaneyulu, Y., Venkata, B. R.D., Hari, P. D., Srinivas, C.V., Francis, T., Julius, M. B., John, H. Y., Robert, L. H., Chuck, P., Mark, G. H., Shelton, J. S., 2011. An Integrated WRF/HYSPLIT Modeling Approach for the Assessment of $PM_{2.5}$ Source Regions over Mississippi Gulf Coast Region. *Air Quality, Atmosphere & Health,* 1-12, doi 10.1007/s11869-010-0132-1.

[33] Paul, R.A., Biller, W.F., McCurdy, T., 1987. National estimates of population exposure to ozone. Presented at the Air Pollution Control Association 80th Annual Meeting and Exhibition, Pittsburgh, PA, 87-42.7.

[34] Anjaneyulu ,Y., Venkata, B.D., Srinivas, D., Srinivas, C.V., John, H. Y., Chuck, P., Julius, M. B., Robert, L. H., Sudha, Y., Francis, T., Mark, G. H., and Shelton J. S., 2010. Air Quality Modeling for Urban Jackson, MS Region using High Resolution WRF/Chem Model. *International Journal of Environmental Research and Public Health* (Accepted for publication).

[35] Anjaneyulu, Y., Srinivas, C.V., Venkata, B. R.D., Hari, P. D., John, Y., Chuck, P., Julius, M. B., Robert, L. H., Mark, G. H., and Shelton J. S., 2011. Simulation of Surface Ozone Pollution in the Central Gulf Coast Region Using WRF/Chem Model: Sensitivity to PBL and Land Surface Physics. *Advances in Meteorology*, Volume 2010, Article ID 319138, 24 pages, doi:10.1155/2010/319138.

[36] Matejicek, L., 2005. Spatial Modelling of Air Pollution in Urban Areas with GIS: A Case Study on Integrated Database Development. *Advances in Geosciences*, 63–68.

[37] Manjola, B., Elvin, C., Bledar, M., and Albana, Z., 2010. Mapping Air Pollution in Urban Tirana Area Using GIS. *International Conference SDI.*

[38] ESRI. (2007, December). GIS for Air Quality. Retrieved Febraury 2011. [http://www.esri.com/library/bestpractices/air-quality.pdf]

[39] Juliana, M., 2007. The Geography of Asthma and Air Pollution in the Bronx: Using GIS for Environmental Health Justice Research. *Health Care Services in New York: Research and Practice.* Greater New York Hospital Association-United Hospital Fund.

18

Contribution of Geostatistics to the Study of Risks Related to Air Pollution

Jacques Deraisme[1], Michel Bobbia[2]
and Chantal de Fouquet[3]
[1]*Geovariances*
[2]*Air Normand*
[3]*Mines Paristech*
France

1. Introduction

The diffusion of a pollutant in a geographical space, is a phenomenon whose main characteristics are the following: it is driven by complex physical and chemical laws influenced by local environmental parameters, it spreads out in a three dimensional space and, finally it is of dynamic nature. Two main approaches may be proposed to allow a spatial estimation of population exposure: (1) a "deterministic" approach which transforms estimated emissions derived from inventories into atmospheric concentrations (Hauglustaine (1998), and (2) a "probabilistic" approach which consists of spatializing concentrations obtained from an air quality monitoring network. Predicting the pollutant concentration at a given point in space at a given time and assigning to that value an indication of its reliability is out of the scope of entirely deterministic models. By choosing a probabilistic framework geostatistics opens the way to models that can be characterized by actual observations for providing practical answers to a large variety of questions specifically related to risk analysis. Typical questions addressed by this approach are: "what is the probable value of the pollutant concentration at a specific location without measurement?", "what is the probability that the admissible pollution level is reached at the same point?". We must pay attention that unlike deterministic modelling, a sufficient number of measures is required to ensure reliable mapping.

This framework aims at considering the phenomenon under observation as an outcome of a random process (Matheron ,1989). This process is represented by a random function Z(x, y, z, t) where x, y and z are the coordinates in a geographical three dimensional space and t is the time axis. For simplicity we will illustrate the intention of this article by reducing the work space to two dimensions, hence the random function is now noted Z(x,y). By reintroducing the third dimension of the space and especially the temporal component, new difficulties arise for specifying completely the models from actual data. These will be discussed later.

Geostatistics generally proposes that the whole set of possible random functions is restricted to stationary random functions, particularly regarding their mean and variance values.

Even with this important restriction the geostatistical framework is richer than other possible models, like, for instance, the purely statistical models which suppose that

measures taken at two different points are not correlated. Clearly the pollution at two different location is likely to be more similar when these points are close than at a larger distance. Characterizing that correlation is at the centre of the geostatistical approach.
The methods developed under the geostatistical framework are many and depend on the nature of the question. Simple issues can be easily solved by linear geostatistical methods. When the questions are more complex, particularly when dealing with risk analysis, non linear geostatistical methods are applicable.
The outline of the chapter is:

- reminders of the most important concepts underlying the geostatistical methods.
- necessity of non linear methods
- discussion on what is achievable to estimate without resorting to simulations
- example of the estimation of the population exposed to a pollution
- example dealing with a temporal issue

2. Reminders on the models in geostatistics

Characterising a spatial distribution in the general case is complex, because of the numerous parameters that would be necessary to infer from the available measures. In order to get a map of a property measured at some points over a given area, geostatistics proposes a simple method called kriging[1] named after Dr Krige, who first pointed out the drawbacks of traditional estimation methods of gold deposits in South Africa. The method aims to interpolate the value at any point by assigning weights depending on the correlation with the data measures (Chilès, 1999). The correlation is quantified by means of the variogram[2] that can be estimated from bi-point statistics of the data (Cressie, 1998; Goovaerts, 1997). By introducing other measured parameters or factors that may have an influence on the parameter of interest the result of the interpolation can be obtained as a linear combination of all parameters at data point referred to as cokriging (Jeannée, 2006). Like any linear interpolator, kriging fails in providing more information related to confidence levels or risk. In order to go ahead in that direction non linear geostatistics based on Gaussian models has been developed. The particular case of data whose distribution is Gaussian is intensively used in statistics as well as in geostatistics for the simple reason that the density of probability only depends on two parameters, i.e. the mean and the variance. Introducing the idea of Gaussian spatial distribution, the additional knowledge of the variogram will entirely characterise the probabilities at several points simultaneously. When the distribution of values is not Gaussian, which is the case most of the time, we come back to the Gaussian case by means of a transform of the data measures, by a process called normal score transform. The variable of interest Z(x), x designating the location of a point in the space, can be written as the transform by the Gaussian anamorphosis function, noted φ, of a random function Y(x), that has a Gaussian spatial law centred and normalized: $Z(x) = \varphi(Y(x))$.

[1]Kriging is a linear combination of data with weights chosen for minimizing the variance of the estimation error.

[2] The variogram represents the variance of the difference between two measures as a function of the distance between these measures.

When the anamorphosis function φ is bijective, hence strictly increasing on its domain of definition, the knowledge of the measures $Z_\alpha = Z(x_\alpha)$ or of their normal score transforms $Y_\alpha = \varphi^{-1}(Z_\alpha)$ is equivalent. At all locations x, the a priori probability to overcome a threshold « s » can be easily calculated on the gaussian transforms by using the classical cumulated gaussian density function noted G. This can be written :

$$\begin{aligned} P(Z(x) \geq s) &= P(\varphi(Y(x)) \geq s) \\ &= P(Y(x) \geq \varphi^{-1}(s)) \\ &= 1 - G(\varphi^{-1}(s)) \end{aligned}$$

The probabilistic model gives also access to statistical quantities like quantiles. The order ω quantile of a variable is the value of the random variable for the cumulated probability ω. Thus that quantile of the normal variable Y(x), noted q^ω is $P(Y(x) < q^\omega) = \omega$.
Since the anamorphosis φ is increasing, the order ω quantile of Z(x) can be directly deduced from the Gaussian quantile of the same order.
In an analogous way the confidence interval for the statistical risk ε of a random variable $W(x)$ is defined as $P(q^{\varepsilon/2} \leq W(x) < q^{1-\varepsilon/2}) = \varepsilon$. For a normal variable that interval is symmetrical: $q^{\varepsilon/2} = -q^{1-\varepsilon/2}$. For instance the confidence interval at the risk level 5% is defined by difference between the quantiles at the order 97.5% and 2.5%. For a gaussian variable those quantiles are equal to the mean value plus or minus 1.96 standard deviations.
In practice we will be interested to the distribution law of a Random Function in a given point, conditioned by the available measures at the experimental stations. For a Random Function of Gaussian spatial law, that conditional distribution is simply:

$$Y^*(x) + \sigma_K(x).W(x)$$

$Y^*(x)$ is for the result of a kriging with zero mean of Y(x) by the data Y_α, $\sigma_K(x)$ is for the standard deviation of the associated error and W is a Random Function independent of Y^*.
In that case the conditional law of the anamorphosed variable Z of Y is directly deduced of that of the Gaussian transform, i.e. $\varphi(Y^*(x) + \sigma_K(x).W(x))$. The probability of overcoming a threshold is then:

$$\begin{aligned} P(Z(x) \geq s \mid Z_\alpha) &= P\left(W(x) > \frac{\varphi^{-1}(s) - Y^*(x)}{\sigma_K(x)} \right) \\ &= 1 - G\left(\frac{\varphi^{-1}(s) - Y^*(x)}{\sigma_K(x)} \right) \end{aligned} \qquad (1)$$

In absence of any measurement error, at any data location α we have $Y^*(x_\alpha) = Y_\alpha$, and the kriging variance is null. That probability is then either 0 or 1, depending on whether $Y_\alpha = \varphi^{-1}(Z_\alpha)$ is inferior or superior to $\varphi^{-1}(s)$. Far from the experimental data, $Y^*(x)$ tends towards the a priori expectation, equal to 0 and $\sigma_K(x)$ tends towards the a priori variance, equal to 1. We then come back to the a priori probability $1 - G(\varphi^{-1}(s))$. If we replace in the calculations the kriging with known mean (also called simple kriging) by the kriging with unknown mean (also called ordinary kriging) the results will be in line with the local mean of the concentrations. In the case of urban pollution (by NO_2 in the example presented hereafter) it allows to account for the higher pollution at the town centre than at the periphery.

In an analogous way, the confidence interval at the statistical risk level ε of the actual concentration Z(x) is deduced from the confidence interval of the gaussian transform $\left[\varphi\left(Y^*(x)-\sigma_K(x).q^{1-\varepsilon/2}\right),\varphi\left(Y^*(x)+\sigma_K(x).q^{1-\varepsilon/2}\right)\right]$. Because of the non linearity of the anamorphosis φ, that interval is no more symmetrical to the conditional expectation of Z.

3. Pros and cons of linear estimation methods

Let's take the example of the annual mean concentration in NO_2 in the city of Rouen (France). If we imagine we had a perfect knowledge of the actual pollution, the map would resemble that of figure 1 (which is in fact a simulation), showing the high spatial variability of NO_2 from one point to another.

The available data consist of samples measured on passive diffusive tubes (Roth, 2001), that have been located in the Rouen city and suburbs during the year 2000 for 6 fortnights in total. These measures have been made on 89 sites spread over an area of 300 km². On the maps (Fig.1, 2, 5, 6 and 7) only the central area of 9x9 km² with 59 sites equipped with diffusive tubes is represented. Public health regulations require experts to evaluate the probability to overcome the annual threshold of 40µg/m³ as well as the number of inhabitants potentially exposed to excessive pollutant concentrations. We will assume that the threshold applies to the concentration at any point of the space and that the measuring period is large enough to neglect the temporal term of the estimation error.

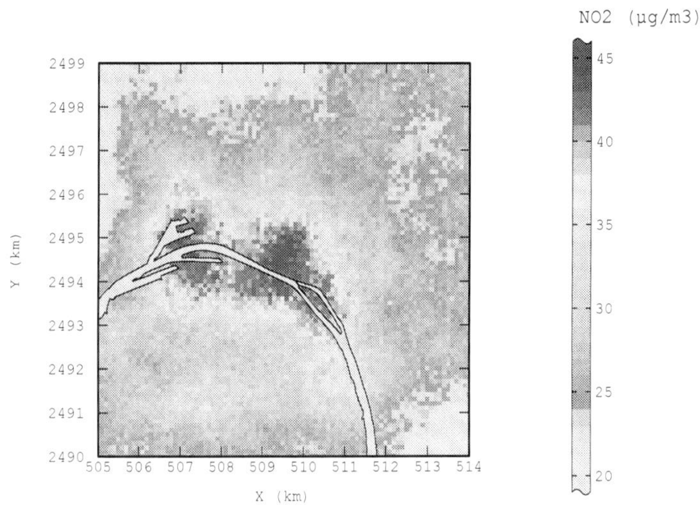

Fig. 1. Map of the true annual mean concentration of NO_2 at the centre of the Rouen city.

A first solution would be to predict the pollution concentrations at unsampled locations. Among different interpolators, the techniques based on kriging provide the best linear

interpolator in the sense that they minimise the variance of the estimation error. To be specific we have used in this example a co-kriging (extension of kriging for processing several correlated variables) with a co-factor combining information related to pollution sources and the spatial distribution of the inhabitants. The resulting map (Fig. 2.) provides a simplified image of the real phenomenon. If it reproduces the main tendencies of the pollution phenomenon, it fails in describing the variations of the NO_2 concentration at a scale of a few hundred metres, i.e. inferior to the spacing between the measures (1km).
By looking at both maps (Fig. 1 and 2) it is immediately visible that applying a threshold (here 40µg/m^3) on the estimated concentrations would lead to an under-estimation of the "contaminated surface" with respect to the right surface obtained from the threshold applied on the real, but unknown, values. The bias cannot be neglected: the contaminated area is estimated as 352 ha whereas its real surface equals 569 ha. The underestimation is observed because the threshold is relatively high with respect to the average level of the measured concentrations. At the contrary for a low threshold, we would have observed an over-estimation. The figure 3 shows that the histogram of the estimated concentrations is more squeezed around the average values, that is known as the smoothing effect of kriging, and that the bimodal distribution is not kept on the estimated values.

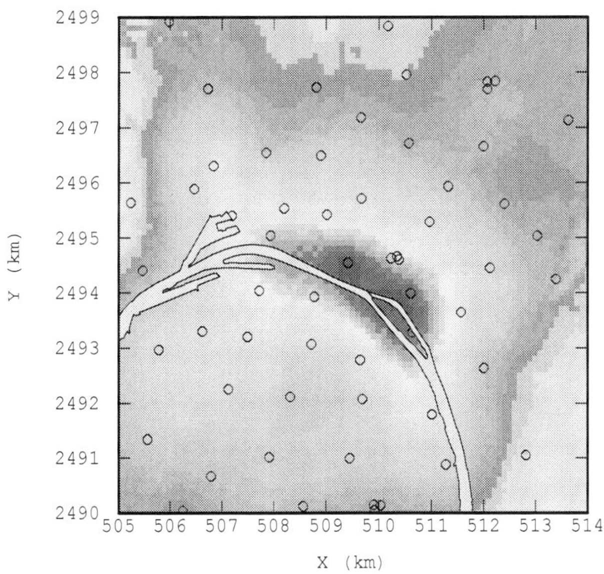

Fig. 2. Map of the co-kriging of annual mean concentration of NO_2 at the centre of Rouen.

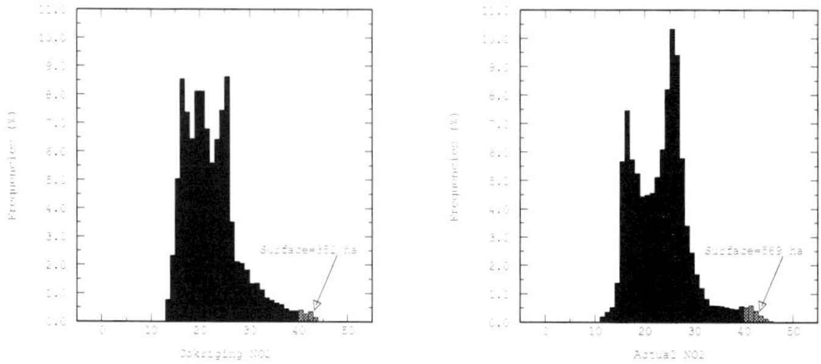

Fig. 3. Histograms of the average annual concentrations of NO_2 at the centre of Rouen for the estimated and actual concentrations and thresholding at 40µg/m^3.

The scatter diagram (Fig. 4) between the "reality" and its estimation illustrates the presence of an error due to the interpolation procedure. This error can only be diminished by adding information, i.e. other measures. The uncertainty resulting from that error can be represented by the notion of statistical risk.

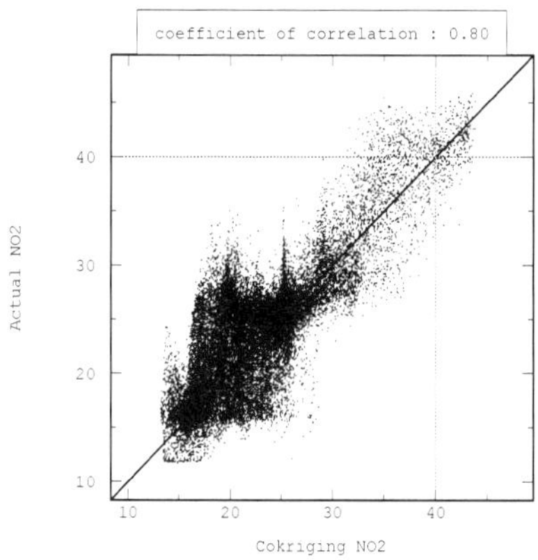

Fig. 4. Scatter diagram of the actual NO_2 concentration and its estimate at the centre of Rouen.

The methods derived from kriging are by nature inadequate for reproducing the actual variability, because the kriging methods aim to minimise the difference between the actual value and its interpolation: the interpolated values by kriging include necessarily less extreme values and more intermediate values, conveying the so-called smoothing effect of kriging. What is at stake is the confusion between the unknown map of the actual concentration and the available maps that are only estimates of the reality. Contrary to a topographical map (which can be considered as "exact") every map of a pollutant calculated from a sampling will never be the reality, is it resulting from geostatistical methods or from physic-chemical models. Those maps will only provide an image more or less close to the reality.

4. Non linear estimation or simulations?

In real situations we will never have access to the actual phenomenon which is unique and largely unknown. The geostatistical framework proposes to consider that phenomenon as a realisation - among other equally possible - of a random phenomenon characterised by its spatial structure. In our example it amounts to say that the real phenomenon could resemble one of the representations of figure 5, analogous to the initial representation from a general point of view while differing notably on detailed local features.

Reality being unknown at most locations, we use a model of a random function, from which it is possible to derive statements correct in probability. We will favour the models that, after non linear operators (like applying a threshold) produce non biased results. The maps of figure 5 have been obtained by simulating outcomes of a Gaussian random function back-transformed by an anamorphosis function. The principle of geostatistical simulations consist in drawing outcomes of the random function, keeping the statistical distribution and the spatial structure. The conditional simulations will recreate the measured values at the experimental data points, i.e. is an exact interpolator. When many simulations are available, the probability of exceeding a threshold can be empirically calculated at every point by computing the proportion of the outcomes that exceed the threshold.

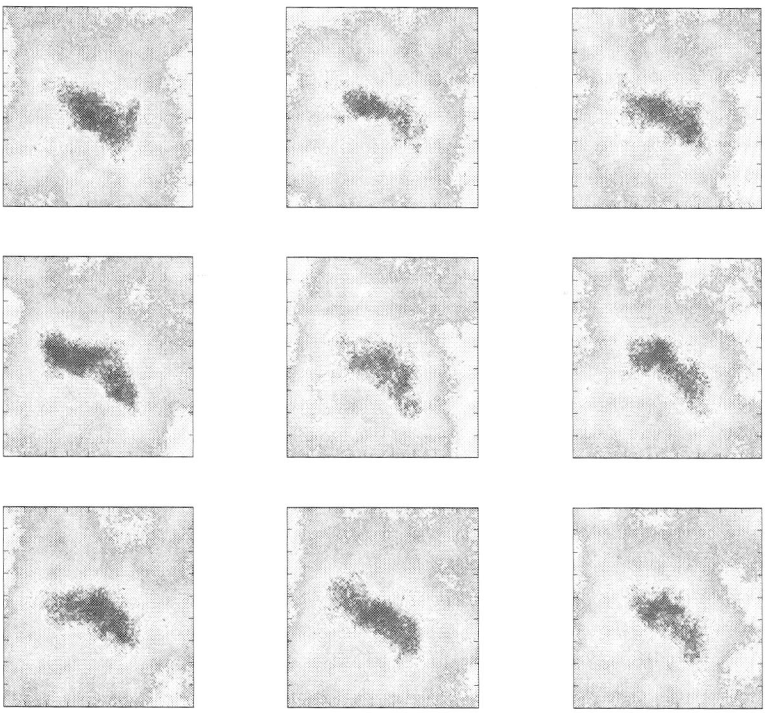

Fig. 5. Possible maps of the average annual concentration in NO2 at the centre of Rouen (the legend is the same as that of Fig. 1).

Nevertheless in the framework of the model, the probability of exceeding any threshold can be directly calculated without resorting to simulations, by just applying the distribution law of the annual concentration Z conditional to the data according to equation (1). The advantage of this approach is that the calculation is analytical and straightforward.
In the example of the pollution by NO_2 in Rouen, the Gaussian model provides at any location the probability of exceeding the threshold of 40µg/m³. The studied area has a surface S of 81km² (i.e. 8100 ha). By multiplying the spatial mean of the probability by S we get an unbiased estimate of the surface probably exposed to the pollution. We get then a surface of 583 ha, close to the true surface of 569 ha (Fig.3.), instead of 352 ha calculated from the kriged map.
If the estimated concentration is less than the threshold, it does not mean that the risk of exceeding the threshold is nil, because of the estimation error. Conversely it may happen that the estimated concentration is above the threshold while the actual concentration is less. In practice the estimation error is unknown but the probabilistic model gives access to its mean (zero because of the unbiasedness condition applied in kriging) and its variance. The calculation of the probability of exceeding a threshold takes into account that variance or more precisely the estimation variance $\sigma_K^2(x)$ of the gaussian transform. Particularly this variance is increasing with the distance of the current location x to the measured points. On the figure 6 the iso-value lines of the probability of exceeding threshold are reported. The area where the estimated concentration is above the threshold corresponds to high values of the probability, even if it is not equal to 1, particularly in locations where the estimation is of poor quality in terms of precision. Conversely if the estimated concentration is less than the

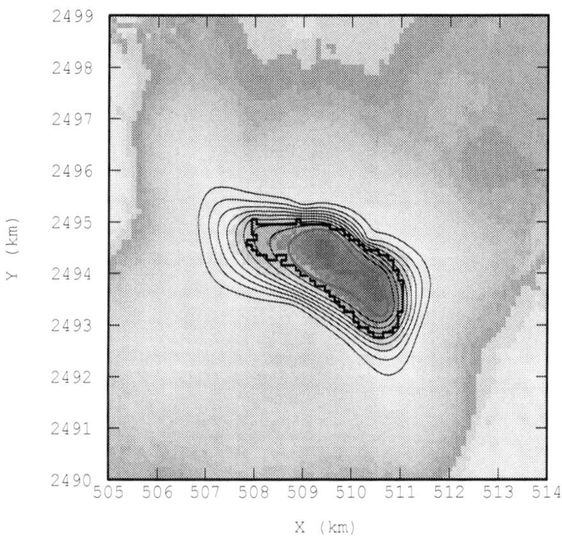

Fig. 6. Map of the cokriging of the annual mean concentration of NO_2 at the centre of Rouen. The curves represent the iso probability by steps of 0.1. Each point inside the red curve has a probability above 0.9 to exceed the threshold of 40µg/m³. The bold black outline surrounds the zone where the cokriging has given an estimate above the threshold.

threshold and the estimation is not precise the probability of exceeding threshold tends to increase. An intermediate probability reveals the uncertainty due to the estimation error, particularly when the estimated value is close to the threshold and the estimation is not precise. The map of the probability can be used to classify the different zones of the urban area with regard to the levels of risks and taking into account the uncertainty in the knowledge of the actual pollution. Such maps are a useful and can help in making decisions related to the protection of the population from pollution.

5. Evaluation of the population exposed to the pollution

More complex questions, like evaluating functions depending on several variables at several points simultaneously, require a simulations approach. Let's take as an example the case of the computation of statistics on the population exposed to a threshold. For sake of simplification we will not consider the movement of the population during the year.
The number of inhabitants b_i is supposed to be known on a fixed grid, the numbers concerned by the exceeding of threshold s is $B(s) = \sum_i b_i I_{z(x_i)\geq s}$.
For the risk δ related to the local exceeding of the threshold, the exposed population is estimated by $B^*_\delta(s) = \sum_i b_i \, I_{P[Z(x_i)\geq s|Z_\alpha]\geq\delta}$. It can be calculated from the map of the population (Fig.7) and the map of the conditional probabilities (Fig.6) cut at δ: it is the sum of the inhabitants living in the cells for which the probability of exceeding threshold is above δ.

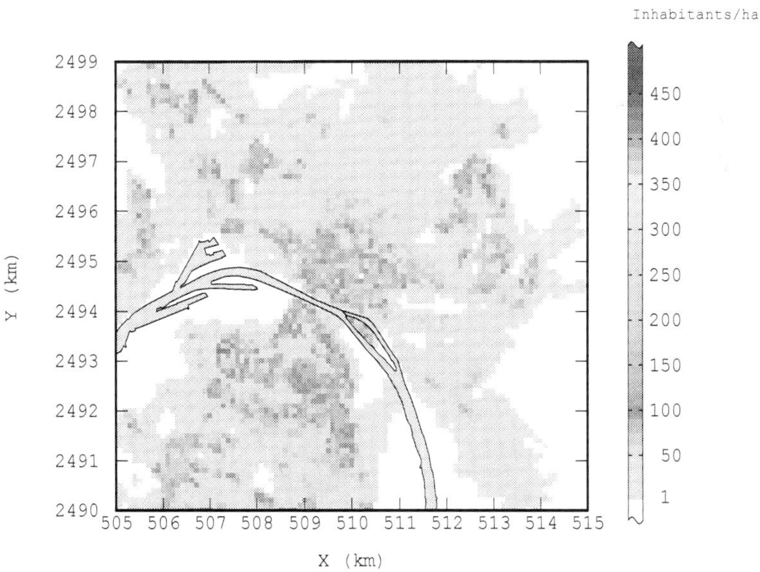

Fig. 7. Map of the density of the population living in Rouen urban area.

This calculation is based on adding up the population of each cell independently. This is a naïve way to proceed.
In fact the quantiles of the exposed population *B(s)* require the spatial law of the concentration. They must be calculated by means of simulations (Deraisme, 2002) because it is mandatory to take into account the probability to exceed the threshold simultaneously in different locations. The spatial distribution of the exposed population is characterized by the simulation of the concentration on a regular grid. For a given simulation the exposed population is simply obtained by adding the inhabitants living in the cells where the simulated concentration exceeds the threshold of 40µg/m³. By repeating this computation on each of the 100 simulations we get a reliable representation of the distribution of the exposed population (Fig. 8). It should be noted that carrying out a large number of simulations is leading to computationally heavy.
For the statistical risk of 10%, the exposed population is estimated as 33 480 inhabitants: the probability that the exposed population is more than 33 480 inhabitants is 0.1. In other words, there is a 90% chance that 33 480 inhabitants are exposed to the excess of the threshold. Another way to say it is that this value represents the quantile at 90% of the exposed population.

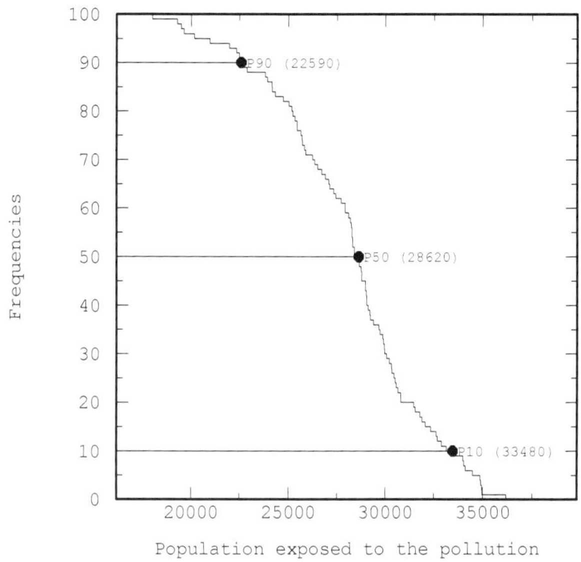

Fig. 8. Distribution of the population of Rouen urban area exposed to a pollution by NO_2 above the threshold of 40µg/m³ obtained from 100 simulations. The quantiles at 10%, 50% and 90% (noted respectively P90, P50 and P10) are reported.

The first "naïve" calculation, multiplying the population map with the probability map truncated at the risk of 10%, would have given 47 880 inhabitants, all living in zones where the probability of exceeding the threshold is above 0.9. The significant difference is due to the fact that, in the first case, the risk is applied to the quantity we want to evaluate, i.e. the

exposed population, while in the "naïve" way the statistical risk is applied on the NO_2 concentration independently of the population. The quantile of the exposed population depends on the probability to exceed the threshold together at different locations, which is ignored by the "naïve" calculation.
For more simple calculations, such as the average population exposed to the pollution, simulations are not necessary. We get the same result, i.e. 28150 inhabitants, by multiplying the population by the probability map (without any cut) and summing up the contribution of each cell. Averaging simulations and calculating directly the conditional expectation are equivalent.

6. Introducing the temporal factor

In the previous example only the spatial variability has been taken into account in the risk evaluation. Yet the variability in time is an important factor inherent to the definition of some official regulations.
In the following example we will examine the calculation of the quantile of daily mean values of SO_2 concentrations on the urban area of Le Havre. For that study, twenty four analysers gave daily records of the SO_2 concentration for a year. The aim is to identify the zones where the daily concentration exceeds the level of 250 µg/m³, prescribed by the regulations, for more than 4 days per year, i.e. 1% of the year. In other words we will look at the spatial distribution of the quantile at 99% of the daily SO_2 concentration.
The evaluation of a quantile of high order requires that the measures recorded every day are complete for all data locations and all day of the year. At least we must be certain that the missing values are inferior to the threshold.
A simple method consists in calculating the quantile at 99% on each analyser and in modelling its spatial distribution, by means of geostatistical simulations of that quantile. (Fig.9).

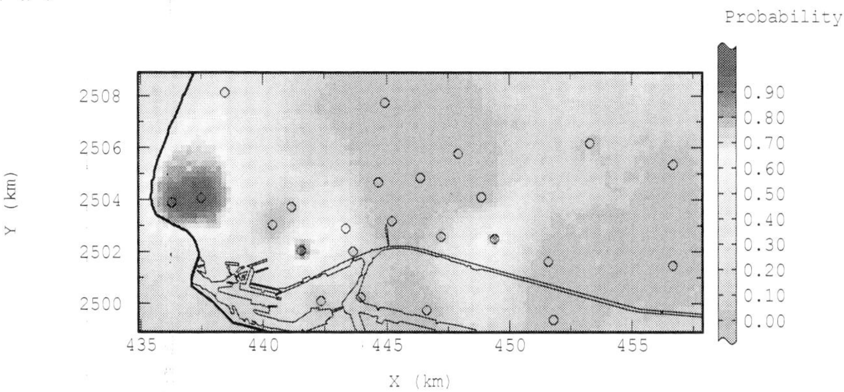

Fig. 9. Map of the probability of exceeding the threshold of SO_2 concentration of 250 µg/m³ for more than 4 days in the urban area of Le Havre.

This simplified method ignores the variability in time: for a given quantile we will get the same results whether the pollution peaks are synchronous between the different analysers or not. Without fully characterising all complex correlations in space and time, an

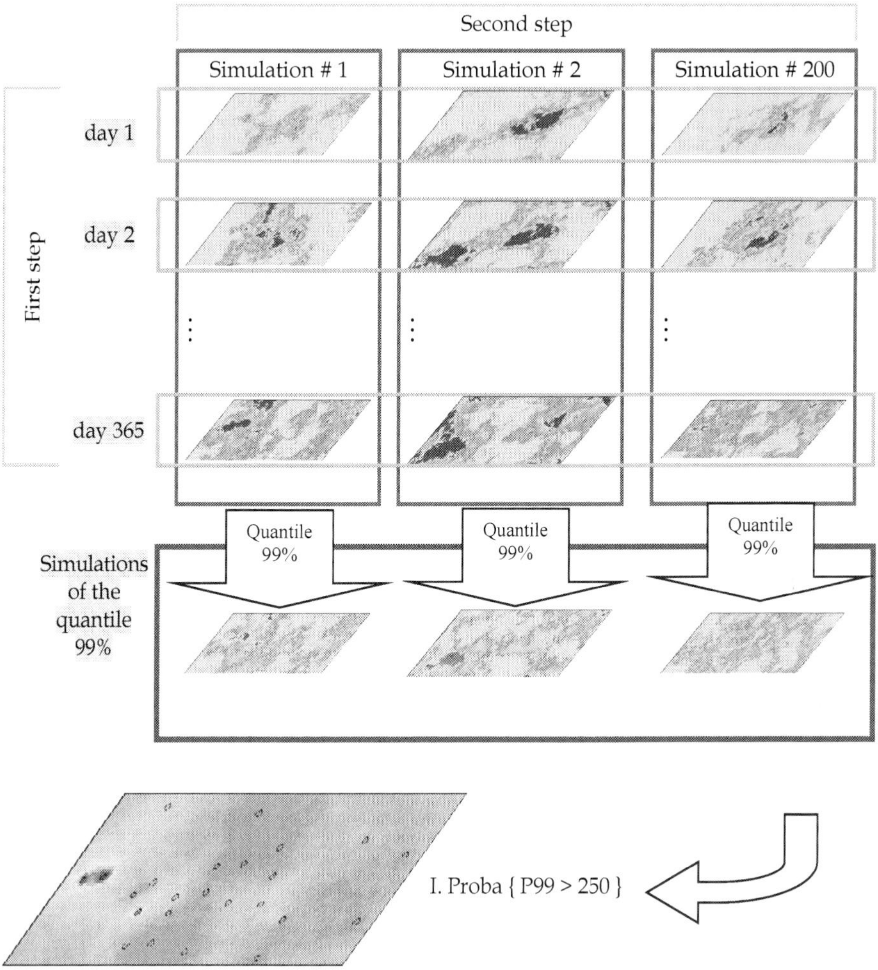

Fig. 10. Flow chart of the procedure used for simulating the daily concentrations in SO_2 in Le Havre for a year.

intermediate solution consists of simulating the daily concentrations for all days of a year, independently for each day (see first step of the figure 10). An automatic procedure providing 200 simulations has been used. We have obtained in total 365x200= 73000 draws of the daily concentration at regular locations of a grid covering the urban area. The results have been validated by analysing the distributions and spatial correlations for several specific situations.

A map of the quantiles at the 99% frequency has been calculated for each simulation of the 365 days as shown in the second step of the scheme of figure 10. By comparing the 200 quantiles to the threshold we can then get the probability of exceeding the threshold (Fig.11): at each location it is the number of occurrences where the simulated quantile is above the threshold divided by the total number of simulations.

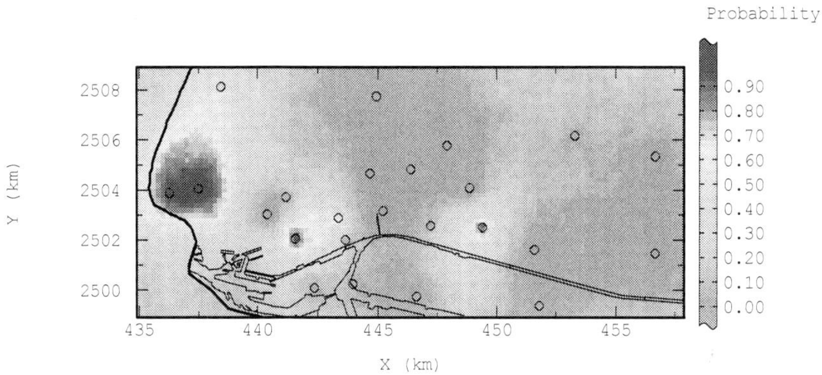

Fig. 11. Probability map of exceeding a threshold of 250µg/m³ of SO_2 for more than 4 days in a year in the Le Havre urban area.

Both maps of the figures 9 and 11 look, at first glance, rather close, however by computing the difference between both calculations, important discrepancies can be observed (see Fig. 12).

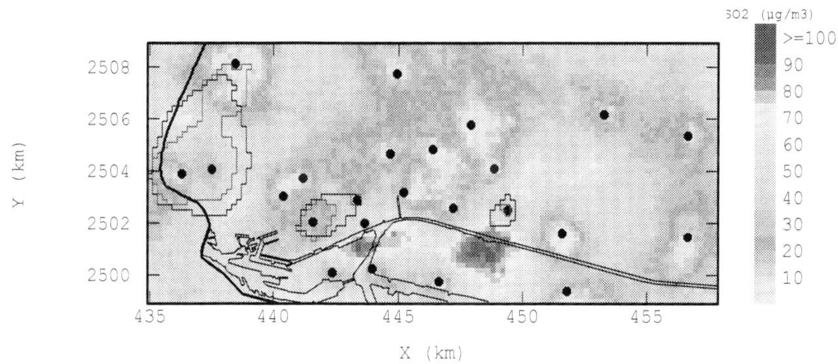

Fig. 12. Map of the differences of SO_2 quantiles at 99% obtained "directly" and from daily simulations. The outlines delineate the zones where the quantile at 99% is above 250µg/m³, from the direct method (black) and from the simulations (red).

The correlation between the mean quantile calculated from 200 simulations and the average of 20 direct simulations of the quantile (Fig.13) shows clearly the bias due to the implicit hypothesis that the pollution events occur simultaneously: the direct simulation of the quantile gives almost systematically values above the quantile calculated from the daily simulations of the SO_2 concentration.

This schematic example illustrates the risk of important bias generated by excessive simplifications. It still remains that the correlation in time should also be introduced.

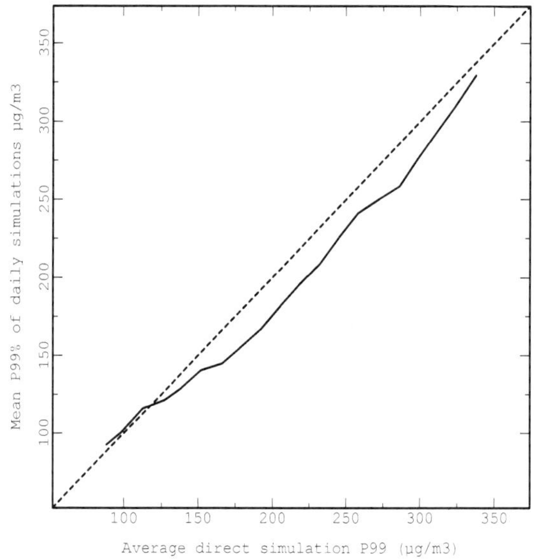

Fig. 13. Conditional expectation of the quantile at 99% calculated from 200 simulations of the daily concentrations of SO_2 versus the average of 200 direct simulation of the quantile at 99%. The first bisector is shown as a dashed line.

7. Conclusion

Geostatistics provides a rigorous conceptual frame work for modelling phenomena spreading in space and time. The application of the geostatistical methods requires some care because, if they are not adapted for answering appropriately the problem, the consequences on the results may be very significant. An effective limitation lies in the use of kriging techniques which provide an optimal solution to estimate quantities depending linearly on the variable of interest. However for evaluating the risk of atmospheric pollution, applying a threshold on the pollutant concentration is typically a non linear operation. If the threshold is applied on the kriged map the calculations will result in a potentially importantl bias. More complex models are then required, they are enhanced by non linear geostatistical methods.

The calculation of the conditional probability using the Gaussian model makes the solution of simple problems quite straightforward: delineation of the geographical area concerned by the pollution or evaluation of the average of the number of inhabitants exposed to it.

The conditional simulations provide a consistent solution and the most adequate results for the evaluation of the population exposed to pollution. These techniques are used to evaluate health risks. When dealing with simulations a sufficient number of outcomes must be generated, particularly to derive high order quantiles. The problems become still more complex when the time dimension of the pollution events has to be taken into account, such as the control of the number of days when the pollutant concentration exceeds a threshold.

The simulations provide a correct solution but are heavy in computations. The example on SO_2 in Le Havre is favourable because the data are available at the same locations all over the year. By performing the simulations independently for each day, the most important bias can be avoided. Nevertheless a complete spatial and temporal model would have the theoretical and practical advantage of handling globally the whole set of measures. Further work is required before being able to apply such models. The present regulations in European countries often refer to complex criteria combining different time scales. The power of simulation techniques is not, in theory, limited regarding this complexity, as soon as the model correctly represents the pollutant concentrations at the origin of the pollution events.

The applications illustrated here have been successfully carried out because the criteria were simple (a threshold applied on one pollutant for a given period of time). When the criteria are multiple (several pollutants, several time scales) the simulations will definitely be the only possible method, but computationally heavy. The need for better models, particularly dealing with the temporal aspect, is still important.

8. Acknowledgement

The data used for illustrating applications of geostatistical methods have been kindly provided by Air Normand, an organization in charge of the air quality control in the Normandy region of France. The computations have been performed using the Isatis software marketed by Geovariances.

9. References

Chilès J.-P., Delfiner P. (1999). *Geostatistics Modeling Spatial Uncertainty*. New York: Wiley & Sons

Cressie N, Kaiser MS, Daniels MJ, Aldworth J, Lee J, Lahiri SN, Cox LH (1998) Spatial Analysis of particulate matter in an urban environment. In: *Second European Conference on Geostatistics for Environmental Application*, eds JJ Gomez-Hernandez, A Soares & R Froidevaux, Kluwer Academic Publishers, pp. 41-51

Deraisme J, Jaquet O, Jeannée N (2002) Uncertainty management for environmental risk assessment using geostatistical simulations. In: *Fourth European Conference on Geostatistics for Environmental Application*, eds X Sanchez-Villa, J Carrera & JJ Gomez-Hernandez, Kluwer Academic Publishers, pp. 139-150

Geovariances (2009) *Isatis technical References* Version 9, 222 p.

Goovaerts P. (1997). *Geostatistics for Natural Resources Evaluation*, Oxford University Press, New York

Hauglustaine, D.A., Brasseur, G.P., Walters, S. *et al.* (1998). MOZART: a global chemical transport model for ozone and related chemical tracers, Part 2. Model results and evaluation. J Geophys Res 1998; 103: 28291-28335.

Jeannée N., L. Mosqueron, V. Nedellec, Ch. Elichegaray, S. Bouallala, H. Desqueyroux, B. Guillaume, C. Liousse, R. Lagache. Assessment of urban population exposure to atmospheric pollution: existing methods and application to PM10 in France. *Pollution Atmosphérique, 190, Avril-Juin 2006*, 197-209.

Matheron G. (1989). *Estimating and Choosing - An Essay on Probability in Practice*, Springer, Berlin

Roth C, Bournel-Bosson C (2001) Mapping diffusive sampling results: including uncertainty and indirect information. In: *Int. conf. Measuring Air Pollutants by Diffusive Sampling*, Montpellier, Sept. 26-28 2

19

Spatial Interpolation Methodologies in Urban Air Pollution Modeling: Application for the Greater Area of Metropolitan Athens, Greece

Despina Deligiorgi and Kostas Philippopoulos
National and Kapodistrian University of Athens
Greece

1. Introduction

Air pollution in urban environments has serious health and quality of life implications. A wide variety of anthropogenic air pollution sources increase the levels of background air pollutant concentrations, leading to the deterioration of the ambient air quality. Principal sources of urban air pollution are vehicular traffic, industrial activity and in general fossil fuel combustion, introducing a mixture of chemical components, particulate matter and biological material into the atmosphere. The deterioration of urban air quality is considered worldwide one of the primary environmental issues and current scientific evidence associate the exposure to ambient air pollution with a wide spectrum of health effects like cardiopulmonary diseases, respiratory related hospital admissions and premature mortality (Analitis et al. 2006; Ito et al., 2005; Samet et al., 2000). Direct measurements of sensitive population groups' exposure to air pollution are scarce and therefore methods of accurate point and areal air quality estimations are prerequisite. This fact highlights the importance of generating accurate fields of air pollution for quantifying present and future health related risks.

In the field of air pollution modeling, two different approaches have been adopted by the scientific community, differentiated by their applied fundamental principles. The first approach involves the numerical simulation of atmospheric dispersion based on the current understanding of physics and chemistry that govern the transport, dispersion and transformation of pollutants in the atmosphere. The modeling process typically requires a set of parameters such as meteorological fields, terrain information along with a comprehensive description of pollution sources. An alternative approach is based on statistical analysis of pollutant concentrations collected from air quality monitoring networks commonly deployed in urban areas. The reasoning of the statistical approach is that physical processes are likely to induce correlations in air quality data collected over space and time. Statistical models generate predictions by exploiting these spatio-temporal patterns, enabling the estimation of pollutant concentrations in unmonitored locations.

The chapter's main objective is to present and review the statistical spatial interpolation methodologies which are commonly employed in the field of air pollution modeling. An additional scope of the chapter is to compare and evaluate the accuracy of the interpolation methods for point estimations, using data from a real urban air quality monitoring network located at the greater area of metropolitan Athens in Greece.

2. Spatial interpolation methodologies

Spatial interpolation is the procedure of estimating the values of the variable under study at unsampled locations, using point observations within the same region. Statistical interpolation methodologies are applied in air pollution modeling for estimating the spatial distribution of pollutants, based on data provided from an existing air quality monitoring network. Such networks consist of a number of irregularly distributed stations and therefore the issue of air pollution spatial interpolation is essentially a problem in the field of scattered data approximation. The monitoring sites are typically located to detect high concentrations and although such a configuration is appropriate for identifying the maximum potential exposure and risk, in many cases it fails to describe the spatial variability of air pollution. Recent developments in the design and the modification of existing air quality monitoring networks address this issue (Lozano et al., 2009; Mofarrah & Husain, 2010; Nejadkoorki, 2011) and propose methodologies for an optimum configuration of monitoring sites used for more reliable and cost-effective spatial predictions.

The spatial interpolation methodologies addressed in this review consist of some of the most widely used interpolation schemes and can be classified according to their inherent general features. The following classifications and characteristics have been adopted in a number of spatial interpolation reviews (Eberly et al., 2004; Li & Heap 2008) as they serve the purpose of illustrating the suitability of each methodology for the spatial process of interest.

- *Point - Areal.* Point based interpolation methodologies produce estimates in discrete unsampled locations based on point observations of the variable under study, whereas the areal based schemes estimations refer to a whole predefined area or zone.
- *Global - Local.* Global methodologies utilize all available data of the variable under study and develop a global estimation function for the study area, while local schemes segment the area and operate independently within a window, using only a fraction of the available data.
- *Exact - Approximate.* The methodologies which produce an estimation surface that essentially passes from the observed data are called exact interpolators, while approximate interpolators make use of the available data to produce the estimation surface, which is not required to predict exactly the observed values.
- *Stochastic - Deterministic.* Stochastic methodologies incorporate the concept of randomness providing therefore additionally the uncertainty associated with the prediction, while deterministic methods make no use of the probability theory.
- *Gradual - Abrupt.* Depending on the degree of smoothness of the resulting estimation surface the methodologies are classified as gradual or abrupt interpolators.

The reviewed methodologies include the nearest neighbor, the triangulated irregular network interpolators, the natural neighbor, which utilize the tessellation of the spatial field of interest, the inverse distance weighted methodologies, the radial basis function and thin plate spline schemes and the geostatistical Kriging set of interpolation methodologies. Additionally the artificial neural network approach to spatial interpolation is presented. The majority of the interpolation schemes are based on the weighted average approach, differentiating in the procedure used to calculate the interpolation weights (w_i).

2.1 Nearest neighbor

The nearest neighbor scheme is the simplest methodology of multivariate interpolation and it is based on the Voronoi diagram (Dirichlet or Thiessen tessellation). Given a set of distinct points in the Euclidean plane, for constructing the Voronoi diagram, all locations in that space are associated with the closest member of the point set. The result is a tessellation of the plane into a set of the regions associated with members of the point set (Okabe et al., 2000). A sample Voronoi diagram is illustrated in Fig. 1, which corresponds to the partition of the area of study for the case of the air quality monitoring network in Athens, which is used in this work for the evaluation of the interpolation methodologies in air pollution modeling. The nearest neighbor scheme utilizes the Voronoi diagram and for point predictions all estimations within a region are set as equal to the nearest point or to the average its surrounding points.

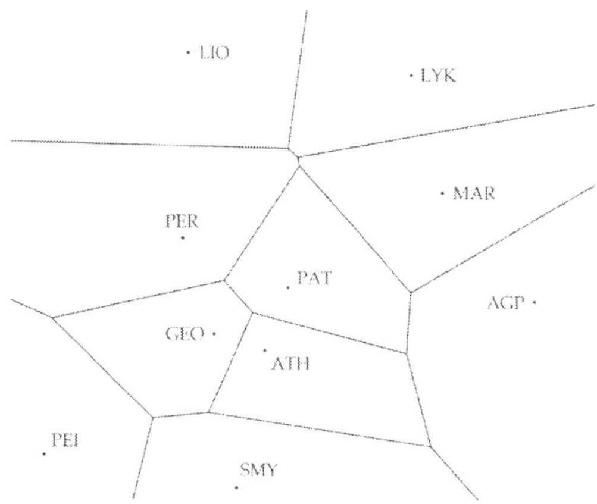

Fig. 1. Voronoi diagram for the air monitoring network of Athens, Greece

The nearest neighbor methodology is selected for its computational simplicity and its main weakness is that each prediction is based on just one measurement and therefore information from other neighboring points is ignored (Webster & Oliver, 2007). Furthermore, it may be more suitable for data that are more discrete than continuous in nature (Eberly et al., 2004). These drawbacks limit the use of the nearest neighbor scheme in air pollution modeling which has been used in spatial interpolation comparative studies (Deligiorgi et al., 2010; Marshal et al., 2008; Shao et al., 2007) and for air quality data imputation (Junninen et al., 2002).

2.2 Triangulated irregular network

The triangulated irregular network (polynomial interpolation) is a group of methodologies that utilize the triangulation of the input space for estimating the values of the variable under study in unsampled locations. Delaunay triangulation is typically employed, which assigns triangles in the dataset by employing the criterion that no data point is positioned inside the circumcircle of any other triangle (Watson & Philip, 1984). Delaunay triangulation maximizes the minimum angle of all the triangle angles and thus it avoids extremely sharp and obtuse angles. The Voronoi tessellation is the geometrical dual of the Delaunay triangulation, obtained by perpendicularly bisecting the edges of the triangulation. In the case of the air quality monitoring network in Athens, both diagrams are presented in Fig. 2. Triangulated irregular network estimates for a target point within one of the formed triangles is performed by fitting a polynomial function to the values at the vertices of the formed triangles. The predictive surface is typically generated by using a least-squares regression fit that minimizes squared differences between the surface and measured points (Eberly at al., 2004).

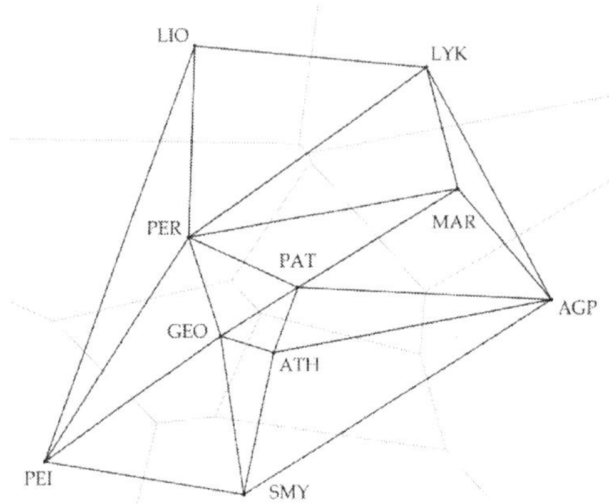

Fig. 2. Delaunay triangulation (black lines) and its dual Voronoi tessellation (gray lines) for the air quality monitoring station in Athens

2.3 Natural neighbor

The natural neighbor interpolation scheme is also based on the Voronoi tessellation and is a local interpolant, where the estimated function is a linear appropriately weighted average of the nearby data points. The selection of the nearby sites (natural neighbors) and the interpolation weights is based on a methodology proposed by R. Sibson (Sibson, 1981). Initially for a given set of irregular points $S = \{S_1 \ldots S_n\}$, the original Voronoi diagram is constructed (Fig. 1). Subsequently, the target interpolation point q is inserted in the diagram, altering the Voronoi tessellation, which for the case of the air monitoring network in Athens is illustrated in Fig. 3. The point q in the new Voronoi diagram is associated with the region $V(q)$ and the interpolation weights w_i are calculated from (Fan, 2005):

$$w_i = \frac{Area(V(q) \cap V(s_i))}{Area(V(q))} \tag{1}$$

where s_i the natural neighbors of q and $V(s_i)$ their associated regions in the original Voronoi diagram. The natural and the nearest neighbor schemes belong to the polygonal interpolation group of methodologies and are local, deterministic and abrupt schemes.

2.4 Inverse distance weighting

The Inverse distance weighting class of interpolation schemes are developed by D. Shepard and are based on the assumption that the degree of influence of the nearby sample points should be greater than the effect of more distant points. The interpolant is also a weighted average of the sample point values and the weights w_i are expressed as an inverse function of distance according to:

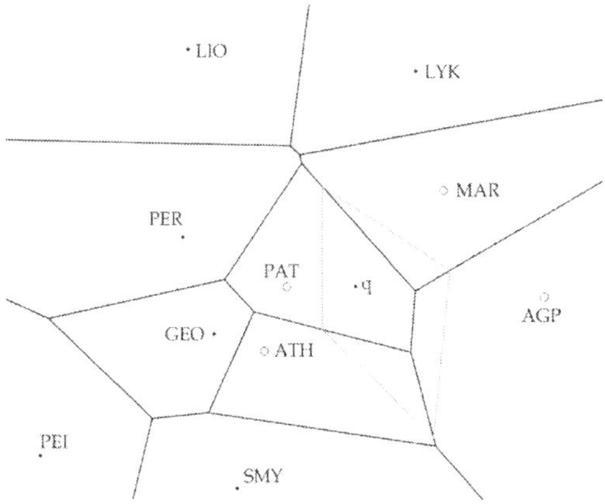

Fig. 3. Voronoi diagram of the air quality monitoring network in Athens, Greece (black lines), natural neighbors (open circle points) and the Voronoi region of q (defined from the gray lines)

$$w_i = \frac{1/d_i^{\,p}}{\sum_{i=1}^{N} 1/d_i^{\,p}} \tag{2}$$

where N, the number of the sample points, d_i the distance of the target point from each of the sample points and p the power parameter. The selection of p is arbitrary (typically equal to 2) and expresses the relative importance of the nearby points in relation to the distant points. As p becomes larger the weight assigned to points in large distances becomes smaller. Especially for $p \rightarrow 0$, the IDW estimate is equal to the spatial average of all sampled points, and when $p \rightarrow \infty$, the methodology is identical with the nearest neighbor scheme. The IDW schemes therefore can be global or local and are deterministic, exact and gradual interpolation methods. Although IDW methodology is one of the most widely used interpolation schemes, it has several limitations, mainly due to the fact that the spatial relationship between two locations is not simply a function of distance and in many cases this distance-decay relationship is not constant over space (Lu & Wong, 2008). In air pollution modeling the IDW method is a popular alternative to Kriging methodology in various scales. It is applied operationally by the Environmental Protection Agency (EPA) for generating real-time O_3, PM_{10} and Air Quality Index spatial predictions in nationwide scales (Eberly et al., 2004).

2.5 Radial Basis Functions

The use of Radial Basis Functions (RBF) as estimators for interpolating irregularly distributed data is introduced by R.L. Hardy (Hardy, 1971) and since then they have been widely used in various disciplines in multivariate scattered data interpolation. The method is based on the assumption that the estimated surface can be approximated to any degree of exactness by the summation of regular mathematically defined surfaces (Hardy, 1978). The RBF interpolation schemes are global, exact, deterministic interpolators. The RBF approximation function can be expressed by:

$$y(x) = \sum_{i=1}^{N} w_i \varphi(\|x - c_i\|) \tag{3}$$

where the estimation function y is a linear aggregate of N radially symmetric functions φ for a given norm $\|\cdot\|$, with Euclidean norm being the most common choice (Baxter, 1992). The data points are called the centers (c_i) of the RBF interpolant and each φ is associated with one of these centers. The coefficients w_i are chosen so that the interpolation conditions are satisfied and estimated by solving the system of linear equations (Rousos & Baxter, 2005):

$$Aw = f \tag{4}$$

where

$$A_{ij} = \varphi(\|x_i - x_j\|) \tag{5}$$

is the interpolation matrix, w the vector of coefficients and f the vector of function values. Some common φ functions are the linear and cubic:

$$\varphi(r) = r \tag{6}$$

$$\varphi(r) = r^3 \tag{7}$$

the Gaussian, the multiquadric and the inverse multiquadric::

$$\varphi(r) = e^{-\frac{r^2}{2c^2}} \tag{8}$$

$$\varphi(r) = (r^2 + c^2)^{0.5} \tag{9}$$

$$\varphi(r) = (r^2 + c^2)^{-0.5} \tag{10}$$

The parameter c, in the Gaussian, the multiquadric and the inverse multiquadric RBFs is called the shape parameter and can be chosen arbitrarily, with a reasonable choice being in the region of the average euclidean distance between the centers. In many cases the quality of approximation is very sensitive to the shape parameter. The main advantage of the RBF interpolation schemes is that they require generally few point observations but relatively well distributed centers in order to create a good approximation of the studied process. The method in general doesn't perform well for a process with abrupt changes in small distances and as an exact method, for processes associated with large observation errors.

2.6 Thin plate splines

Thin plate splines were initially introduced for interpolating meteorological fields (Wahba & Wendelberger, 1980) and can be considered as a special form of RBF based interpolation. The physical analog of the term 'thin plate spline', refers to the bending of a thin sheet of metal so it would pass through a predefined set of points. The Spline interpolation schemes in the case of scattered data interpolation is used to find the minimum curvature surface that passes through a set of irregularly spaced data points (Sandwell, 1987). The two fundamental conditions a thin plate spline satisfies are the exactness and smoothness (Myers, 1994):

$$\sum_{i=1}^{N}[f(x_i)-f^*(x_i)]^2=0 \tag{11}$$

$$\int_R [Df^*(x)]^2 dx \quad \text{is minimized} \tag{12}$$

where D is a second order differential operator and R the region of interest. An alternative restriction for a thin plate spline is the minimization of the quantity (Myers, 1994):

$$\sum_{i=1}^{N}[f(x_i)-f^*(x_i)]^2+\lambda\int_R [Df^*(x)]^2 dx \quad \text{is minimized} \tag{13}$$

This expression represents a tradeoff between the exactness and smoothness of the solution. The interpolant obtained from above restriction no longer interpolates the noisy data, but smooths it, with the smoothing parameter λ controlling the tradeoff between exactness and smoothness (Wahba, 1990).

2.7 Kriging

Kriging interpolation schemes are the fundamental tools in the field of spatial statistics, developed by the founder of geostatistics G. Matheron and named after D.G. Krige who proposed the technique for mining applications (Krige, 1951). The main difference of the Kriging schemes from the above non-geostatistical methodologies is the assumption that spatial correlation structure of the process under study is known and it is estimated from the observed data. The Kriging interpolation schemes are stochastic, local, gradual and exact interpolators. The Kriging methodology includes two stages: the analysis of the spatial variation and the estimation of the target variable, which is also based on the weighted average approach.

The analysis of the spatial variation is performed through the variogram assessment. The observed data are initially used for the estimation of the semivariance from:

$$\hat{\gamma}(h)=\frac{1}{2n(h)}\sum_{i=1}^{n}\left(z(x_i)-z(x_i+h)\right)^2 \tag{14}$$

where z, the variable under study, h the lag distance between the pre-ordered data and $n(h)$ the number of the paired data at distance h. The plot of the semivariance versus the lag distance is the empirical variogram, which is subsequently modeled by a parametric model. The most common models are the linear, the circular, the spherical, the exponential, the Gaussian or a nested model of one or more of the above. The characteristic parameters of the variogram are the sill, which is the semivariance value at the plateau of the variogram, representing the distance where there is no correlation between the observations, the range, which is the lag where the semivariance is equal to the sill and the nugget which represents theoretically the minimum variance and is the semivariance value at zero lag distance (Fig.4).

The Kriging scheme is also known as the best linear unbiased estimator, and its estimates are based on the variogram model and the values and location of the measured points (Eberly et al., 2004). The Kriging interpolation weights are chosen using the modeled variogram so that the estimate is unbiased and the estimation variance is less than any other linear combination of the observed values. The most common Kriging methods are the Simple Kriging, which assumes a known constant mean, Ordinary Kriging, which assumes that there is an unknown constant mean, estimated from the data and Universal Kriging which assumes that there is a trend in the surface that partly explains the data's variations (Eberly et al., 2004).

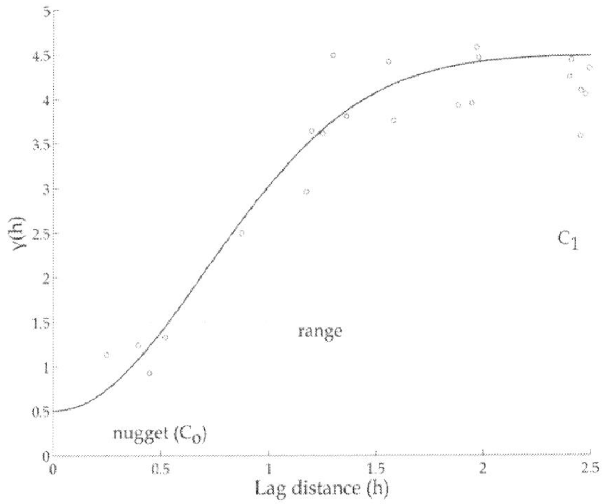

Fig. 4. Example of an empirical variogram with the fitted theoretical model (Gaussian)

2.8 Artificial Neural Networks

Recent advances in artificial intelligence propose an additional approach to the problem of point spatial interpolation of irregularly and gridded data An Artificial Neural Network (ANN) is a mathematical model inspired from the brain physiology to take advantage of their advanced problem solving abilities. Biological brains contain specialized cells called neurons, which interconnect and form highly complex networks. A biological neuron is consisted of a cell body (soma), the dendrites, which extend out of the cell body and an axon which is the transmitter of the neuron. In terms of information processing, the basic operation of an artificial neuron, is the summation of all the incoming signals through its dendrites. A pulse, in accordance with the biological functioning, is sent along the axon if sufficient input signal is received to stimulate the neuron to its threshold level. This simplified neuron operation (McCulloch-Pitts model) can be modeled by the data flow diagram presented in Fig. 5.

The output of the neuron in Fig. 5 is calculated using the following equation:

$$y = f(\sum_{i=1}^{n} x_i w_i - \theta) \tag{15}$$

where f is the binary threshold function. The abilities of the McCulloch-Pitts neuron model can be extended by replacing the activation functions and depending on the application, linear, sigmoid, tangent hyperbolic and other functions can be used.

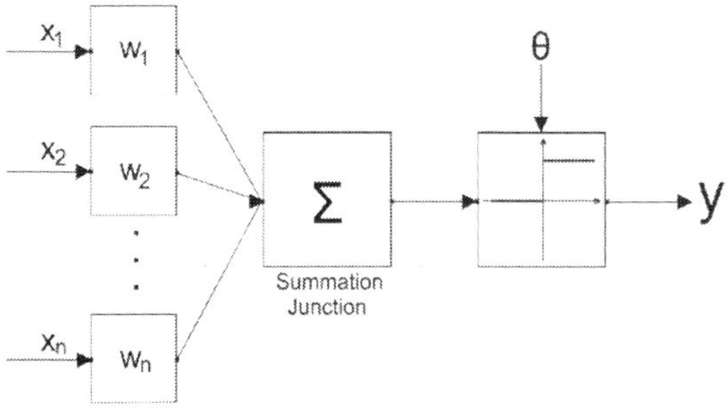

Fig. 5. Simplified McCulloch-Pitts model

An ANN is set of interconnected artificial neurons which can be classified according to their topology. The most commonly employed ANNs are the Feed Forward Networks, which have no feedback connections and their most important characteristic is the lack of memory. The output of such a network depends solely on its current inputs and weights and therefore the order in which the elements of a time series are presented to the network has no effect in its output. Feed forward ANN's are universal approximators, theoretically capable of estimating any measurable input-output function to any desired degree of accuracy (Hornik et al., 1989). In Fig. 6 their most common topology is presented, which contains three layers of neurons.

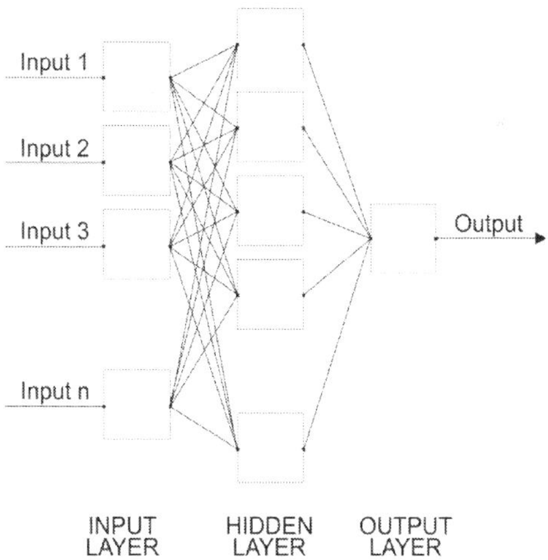

Fig. 6. Feed Forward Artificial Network

The input layer contains a set of artificial neurons with linear activation functions, while in the hidden layer, non-linear activation functions are used. The third layer is the output layer and depending on the application, linear or non-linear activation functions can be used. Besides feed forward ANNs, several topologies exist with feedback connections and their most important feature is the existence of memory. Such networks are called recurrent and their output depends on the current and previous inputs, since these define the state of its memory.

The behavior of an ANN, besides its topology, is also affected by the input neuron weights. The process of determining the weights is called training, which corresponds to the natural learning process in biological brains. Training requires the existence of a training data set, which is composed of input vectors along with the desired output from the network. In general, training can be considered as an optimization problem (selecting the set of weights that minimizes some error measure) and therefore any optimization algorithm can be applied, with backpropagation commonly employed. A successfully trained network is able to generalize, in terms of producing an output for an input that was not presented during training. Such an output can be considered as an interpolation or prediction, depending on the network input. Several issues need to be addressed during training in order to obtain an ANN with satisfactory generalization ability. A common problem, related to the entire family of gradient descent based training algorithms like backpropagation, is that a suboptimal set of weights can be initially determined. A technique to overcome this issue is the multiple training of the ANN, by altering the initial weights values. Another common problem related with training is directly related to the extreme flexibility of ANNs. Specifically, as the number of neurons increases so does the ability of the network to exploit patterns in the training set, irrelevant to the statistical dependence between input and output, such as noise (Blackwell et al., 2009). This effect is known as overtraining and the result is highly degraded generalization ability for the trained ANN. Overtraining can be resolved using early stopping, which requires the partition of the original training set to the normal training set and the validation set, which is continually used during training to evaluate the generalization ability of the ANN. A decrease in this ability will trigger a premature ending of the training process in order to avoid overtraining.

Besides these inherent drawbacks, the wider acceptance of ANNs for statistical interpolation and prediction is limited due to the requirement of a representative training set. This fact differentiates the ANNs from the previous interpolation methodologies.

3. Application of spatial interpolation methods

The accuracy assessment for the aforementioned spatial interpolation methodologies is performed for the greater area of metropolitan Athens in Greece. The area under study is the Athens basin, which is bounded by Mount Aegaleo to the west, Mount Parnitha to the north, Mount Pentelikon to the northeast, Mount Hymmetus to the east and the Saronic Gulf coastline to the southwest. These geophysical features along with the prevailing meteorological conditions, land-use types and pollution sources, create a complex pattern of air pollution spatial distribution. The main sources of pollution in the area are vehicular traffic, industrial activities and central heating emissions. The dispersion conditions at the Athens basin have been extensively studied focusing on air pollution episodes in relation to atmospheric conditions (Asimakopoulos et al., 1992; Lalas et al., 1983; Kassomenos et al. 1995; Güsten et al. 1988).

The accuracy assessment of the interpolation methodologies is carried out for Nitrogen Dioxide (NO_2) which is a primary pollutant and tropospheric Ozone (O_3) which is secondary photochemical pollutant. The analysis is carried out for their mean daily concentrations from the 1st of January 2009 to the 31st of December 2009. The air quality monitoring stations used for the purpose of this study are operated by the Hellenic Ministry of Environment, Energy and Climate Change (MEECC) and their locations are presented in Fig. 7.

The air quality data provided from the MEECC database are mean hourly concentrations of NO_2 and O_3, which are further averaged to provide the mean daily concentrations. Although the mean daily concentrations do not provide information regarding the diurnal pattern of each pollutant, a daily time scale is appropriate in epidemiological studies and for the assessment of the spatial interpolation methodologies. The hourly data availability and the characteristics of each station are presented in Table 1. Calendar days with missing mean daily observations from one or more stations are excluded from the dataset, resulting to a series of 227 and 269 days with concurrent observations for NO_2 and O_3 respectively. As it is observed the data availability for both pollutants is high for all stations except for Marousi and Peristeri and for NO_2 at Piraeus.

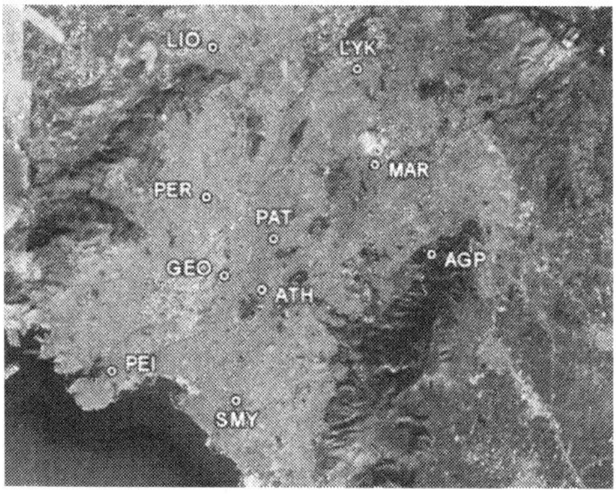

Fig. 7. Air quality monitoring network and experimental area

Station	ID	Station Type	Area Type	Data Completeness (%)	
				NO_2	O_3
Aghia Paraskevi	AGP	Background	Suburban	96.54	99.02
Athinas	ATH	Traffic	Urban	99.89	99.93
Geoponiki	GEO	Industrial	Suburban	97.73	91.67
Liosia	LIO	Background	Suburban	93.18	96.20
Lykovrisi	LYK	Background	Suburban	99.87	99.85
Marousi	MAR	Traffic	Urban	87.47	90.11
Patision	PAT	Traffic	Urban	98.74	98.47
Peristeri	PER	Background	Urban	89.26	89.19
Piraeus	PEI	Traffic	Urban	86.64	95.43
Nea Smyrni	SMY	Background	Urban	95.08	96.44

Table 1. Characteristics of the air quality monitoring stations and data availability

3.1 Descriptive statistical analysis

The analysis of the mean concentration levels for the examined pollutants reveals the increased spatial variability at the area of study. For NO_2 the maximum concentrations are observed for the urban station at Patision (91.23μgr/m^3) and the minimum for the suburban station at Aghia Paraskevi (18.31μgr/m^3). The spatial variability for O_3 is also high at the

Athens basin, but with the exact opposite distribution. Maximum concentrations are recorded for the suburban station of Aghia Paraskevi (84.91μgr/m^3) and minimum at Patision station (24.46μgr/m^3). The seasonal evolution of the studied pollutants is presented in Fig. 8 for the three stations situated at the urban core of metropolitan Athens (ATH, GEO, PAT) and for the suburban station at Marousi. It is clear that the values fall in different ranges and have different characteristics. Ozone maxima are observed at the suburban regions at the Athens basin, in locations where lower NO_2 concentrations are reported. Furthermore, O_3 exhibits a clear seasonal cycle, with maximum concentrations observed during the warm period of the year, while NO_2 concentrations remain relatively stable throughout the year, with a minimum at August, due to the decreased vehicular traffic.

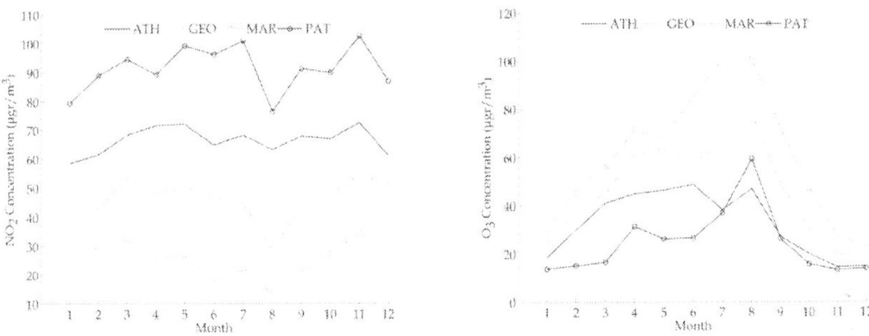

Fig. 8. Seasonal evolution of NO_2 and O_3 for the urban stations at Athinas, Geoponiki and Patision and for the suburban station at Marousi

The quality of the spatial prediction obtained by the various methods can be partly explained from the spatial correlation between the recorded time-series, which is presented in Table 2. The maximum correlation coefficient is observed between MAR and LYK for both pollutants.

	Nitrogen Dioxide (NO_2)					Ozone (O_3)				
	ATH	GEO	MAR	PAT	PER	ATH	GEO	MAR	PAT	PER
AGP	0.47	0.71	0.75	0.61	0.66	0.70	0.83	0.86	0.70	0.87
ATH	1.00	0.79	0.65	0.59	0.70	1.00	0.84	0.74	0.54	0.70
GEO	0.79	1.00	0.88	**0.84**	0.87	0.84	1.00	0.90	**0.81**	0.88
LIO	0.71	**0.90**	0.85	0.80	**0.89**	0.47	0.77	0.80	0.78	0.86
LYK	0.58	0.87	**0.92**	0.80	0.85	0.70	0.89	**0.94**	0.80	**0.94**
MAR	0.65	0.88	1.00	0.80	0.85	0.74	0.90	1.00	0.80	0.91
PAT	0.59	0.84	0.80	1.00	0.74	0.54	0.81	0.80	1.00	0.76
PER	0.70	0.87	0.85	0.74	1.00	0.70	0.88	0.91	0.76	1.00
PEI	**0.82**	0.58	0.38	0.37	0.50	0.82	0.68	0.54	0.42	0.55
SMY	0.77	0.85	0.78	0.68	0.84	**0.84**	**0.93**	0.90	0.76	0.91

Table 2. Correlation coefficient between the target and the remaining stations

3.2 Methodology

The assessment of the spatial interpolation methodologies for the greater area of metropolitan Athens includes thirteen schemes in total, which are presented in detail at

Table 3. The accuracy assessment of the evaluated interpolation methodologies is performed using the leave-one-out cross-validation methodology. According to this technique, one station is selected as the target station and an interpolation methodology is employed in order to estimate air pollution concentrations at its coordinates. The predictive accuracy of the interpolation scheme is evaluated against the observed concentrations. The leave-one-out cross-validation technique is applied multiple times for NO_2 and O_3 for the target stations situated at Athinas (ATH), Geoponiki (GEO), Marousi (MAR), Patision (PAT) and Peristeri (PER).

Interpolation Methodology	Abbreviation
Nearest Neighbor	NN
Triangulated Irregular Network	
• Linear Interpolation	Lin
• Cubic Interpolation	Cub
Natural Neighbor	NaN
Inverse Distance Weighted	
• Inverse distance weighted (p=1)	IDW
• Inverse distance squared weighted (p=2)	ID^2W
Kriging	
• Ordinary Kriging	OK
Radial Basis Functions Interpolation	
• Linear RBFs	RBFlin
• Gaussian RBFs	RBFg
• Multiquadric RBFs	RBFmq
Splines	
• Thin-plate splines	SPtp
• Biharmonic Splines	SPbh
Artificial Neural Networks	
• Feed Forward Neural Networks	ANN

Table 3. Spatial interpolation methodologies

The application of the NN, Lin, Cub, NaN, IDW, ID^2W, RBFlin, SPtp and SPbh schemes is straightforward, following the theoretical background presented at the second section of this chapter. Regarding the Ordinary Kriging methodology, an automated approach is adopted, where isotropy is assumed. During the phase of variogram modeling the circular, spherical, pentaspherical, exponential and Gaussian models are used. The best model for each day and station is selected using the coefficient of determination (R^2), and subsequently it is employed in the OK spatial interpolation phase. For the selected time frame, during the analysis of the spatial variation, in 56.04% of the days for NO_2 and in 56.13% of the days for O_3 the circular model is selected.

Regarding the ANN, the additional dataset required for training consists of daily NO_2 and O_3 concentrations from 2000 to 2008, acquired from the MEECC database. This dataset was randomly divided into training and validation sets, with the 80% of its values used for training and the remaining 20% for validation. The test set in this case, is also the mean daily NO_2 and O_3 concentrations from 2009. The validation set is used for avoiding overtraining and for selecting the optimum ANN architecture. Multiple ANNs are trained with varying numbers of hidden layer neurons (from 10-50) and the optimum architecture is the one that minimizes the Mean Absolute Error (MAE) on the validation set. In all cases for avoiding the inherent drawback of gradient decent algorithms, training is performed multiple times (50 repetitions). The selected architecture for each station and pollutant is presented in Table 4.

	ATH	GEO	MAR	PAT	PER
NO_2	9-32-1	9-34-1	9-32-1	9-31-1	9-38-1
O_3	9-36-1	9-26-1	9-37-1	9-37-1	9-33-1

Table 4. ANN architecture, number of input, hidden and output layer neurons

In the case of RBFg and RBFmq, the shape parameter was not chosen arbitrarily as the average Euclidean distance of all station pairs. A sensitivity assessment of the effect of the parameter c on the predictive accuracy of RBFg and RBFmq for each station and pollutant is performed using the additional dataset from 2000-2008. The optimum value of the parameter c is the one that minimizes the MAE and the results are presented in Table 5.

		ATH	GEO	MAR	PAT	PER
NO_2	RBFq	3	0.75	3.1	4.5	2
	RBFmq	6	4.5	1	6	2.8
O_3	RBFq	1.9	3.5	2	6	1.4
	RBFmq	4	4.5	2	10	2

Table 5. Values of shape parameter c in km

The performance evaluation of each interpolation scheme is based on a set of correlation and difference statistical measures (Willmott, 1982), along with the application of Wilcoxon's sign rank test. The proposed measures for assessing the accuracy of the interpolation schemes include the correlation coefficient (R) and coefficient of determination (R^2), the Mean Absolute Error (MAE), which is proposed as the optimal model performance error for dimensioned evaluations (Willmott & Matsuura, 2005), the Mean Absolute Percentage Error (MAPE), the Root Mean Square Error ($RMSE$), the Mean Bias Error (MBE) and the Index of agreement d (Willmott, 1982), which is a descriptive measure that scales with the magnitude of the variables, retains mean information and does not amplify outliers (Vicente-Serrano et al., 2003).
According to the above statistical criteria and in combination with data display graphics like scatterplots the best performing scheme is selected. In each case the best performing scheme is statistically tested for its equal predictive accuracy versus the rest interpolation schemes using the Wilcoxon's signed-rank test (Diebold & Mariano, 1995):

$$S_{3a} = \frac{\sum_{t=1}^{T} I_{+}(d_t)rank(|d_t|) - \frac{T(T+1)}{4}}{\sqrt{\frac{T(T+1)(2T+1)}{24}}} \tag{16}$$

where

$$d_t = |S_{it} - S'_{it}| - |S_{it} - S'_{itj}| \tag{17}$$

and

$$I_+(d_t) = \begin{cases} 1, d_t > 0 \\ 0, otherwise \end{cases} \tag{18}$$

where, T the number of the concurrent observations, S_{it} the concentration at the i station at time t, S'_{it} the prediction of the best performing interpolation scheme according to statistical measures, and S'_{itj} the prediction of the j scheme at station i and at time t. The null hypothesis of equal predictive accuracy between the methodologies is evaluated by comparing the test statistic (S_{3a}) against the t-distribution. Superior forecasts are indicated by the sign on the test statistic, which is determined by the order which the error terms are subtracted from each other in d_t (Snell et al., 2000).

3.3 Results

A general remark is that the performance of the various interpolation schemes varies significantly, depending on the target station and therefore a universal method cannot be selected for the entire field of interest as the optimum interpolation scheme. Furthermore, the air pollutant of concern affects the performance of spatial interpolation methods, attributed to the differentiations in transport mechanisms and chemical transformations.
Initially, the results for the primary pollutant NO_2 are discussed. The main performance statistics are the MAE and MAPE, which are presented in Table 6. A direct finding is that the ANN scheme for NO_2 consistently outperforms all interpolation methods for all target stations. Among these stations, the most difficult to be predicted is PAT station since all methods produce predictions with significant errors. However, even for this station, ANN predictions are distinctively better (MAPE = 15.67%) compared to other methods with MAPE values ranging from 28.22% for RBFg to 55.10% for IDW scheme. The best ANN predictions for NO_2 are obtained for GEO and MAR having MAPE values 5.9% and 6.1% respectively. The results of the other methods for these two stations are also superior compared to the rest of the target stations, although they cannot compare favorably against ANN predictions. The above results are in accordance with the rest statistical measures. The explained variation by ANN, as it expressed by R^2, has a mean 77.29% for all target stations, varying from 59.04% for PAT to 90.46% for GEO station. Furthermore, high d values (greater than 0.92) are reported for all stations except PAT. Regarding the average model bias, the low MBE values are obtained for ANNs (ranging from -1.34μgr/m^3 to 0.33μgr/m^3) provide strong indication of its superior predictive abilities compared to the other methods, which systematically underestimate or overestimate the observed concentrations.
Regarding O_3 predictions, ANN outperforms the rest interpolation schemes for three target stations (namely ATH, PAT, PER) although its predictive ability is generally lower compared to NO_2 predictions (slightly higher MAPE values). For the stations at GEO and MAR the best results are obtained from RBFg and NN respectively (Table 7). The unexpected superior performance of the simplistic NN scheme can be attributed to the extremely high spatial correlation (R = 0.94) between MAR and its nearest neighbor station, located at Lykovrisi. It should be noted that even for these two stations, ANN is among the best performing schemes. In detail, for ATH, PAT and PER the ANN's d values are greater

than 0.88 and the *RMSE* is lower than 13.89μgr/m^3. The higher prediction errors for all schemes are also observed for the urban PAT station, indicating the increased spatial variability of air pollution in the urban core of metropolitan Athens.

	ATH		GEO		MAR		PAT		PER	
	MAE	MAPE	MAE	MAPE	MAE	MAPE	MAE	MAPE	MAE	MAPE
NN	22.24	25.03	22.24	25.03	9.02	10.83	25.43	31.04	10.34	13.33
Lin	12.93	14.89	16.19	18.45	10.85	12.18	39.94	43.08	10.33	13.09
Cub	16.00	19.09	22.16	24.35	19.73	20.73	34.96	38.75	10.23	12.99
NaN	12.06	15.47	15.92	17.97	10.83	12.16	41.72	44.60	12.95	15.20
IDW	30.07	31.86	8.77	10.66	25.95	27.33	52.87	55.10	11.52	14.65
ID2W	32.72	34.47	12.56	14.50	27.89	29.70	52.52	54.88	12.29	16.05
OK	16.22	19.32	15.40	17.82	20.61	21.67	37.30	40.69	12.19	14.63
RBFlin	12.97	16.43	20.64	22.20	15.89	17.03	42.37	45.04	20.50	22.99
RBGg	13.20	16.38	15.47	18.58	8.37	9.65	22.53	28.22	13.69	16.03
RBFmq	12.40	15.04	15.77	18.92	22.45	23.73	23.02	28.87	17.15	21.31
SPtp	12.45	15.37	18.96	21.41	27.09	28.38	34.80	38.68	18.19	22.31
SPbh	12.48	15.41	19.33	21.74	33.39	34.80	34.51	38.41	20.00	24.34
ANN	**6.77**	**8.77**	**4.57**	**5.93**	**4.63**	**6.10**	**11.80**	**15.67**	**8.79**	**11.73**

Table 6. MAE and MAPE (%) performance statistics for NO_2

	ATH		GEO		MAR		PAT		PER	
	MAE	MAPE	MAE	MAPE	MAE	MAPE	MAE	MAPE	MAE	MAPE
NN	15.12	19.88	15.12	19.88	**7.42**	**11.04**	15.38	19.24	19.21	23.61
Lin	12.22	16.28	7.65	10.95	9.00	12.63	21.87	26.33	19.79	23.28
Cub	12.06	15.21	11.81	15.63	8.60	11.73	17.87	23.00	21.56	25.10
NaN	14.06	18.57	7.11	10.13	9.01	12.64	23.74	27.88	22.80	26.39
IDW	26.31	30.92	12.77	15.08	13.71	17.82	32.89	35.63	11.20	14.37
ID^{2}W	26.83	31.92	14.89	17.49	15.27	19.80	32.46	35.14	9.83	12.28
OK	13.81	18.41	6.98	9.97	11.99	15.02	26.12	30.14	19.56	24.01
RBFlin	12.93	17.43	9.11	12.27	8.42	11.90	24.22	28.23	30.26	33.68
RBGg	10.42	14.11	**6.15**	**7.78**	8.85	12.25	17.00	21.59	22.00	25.28
RBFmq	10.70	13.91	10.04	14.05	8.56	11.92	16.54	21.33	23.02	28.32
SPtp	10.90	14.87	9.52	13.32	9.91	12.89	18.66	23.78	26.50	31.00
SPbh	10.91	14.96	9.63	13.37	10.12	13.19	18.57	23.74	28.28	32.77
ANN	**8.07**	**10.42**	9.00	12.35	8.85	12.89	**10.47**	**13.89**	**9.13**	**12.27**

Table 7. MAE and MAPE (%) performance statistics for O_3

The scatter diagrams produced for each case are in accordance with the results from the statistical measures and some representative cases are presented in Fig. 9. Limited dispersion along the x=y axis is observed for the best performing schemes (for NO_2 at GEO for ANN, for O_3 at GEO for RBFg and for O_3 at MAR for NN) while systematic under-prediction is observed for the IDW scheme for NO_2 at GEO (MBE = -8.23μgr/m^3).
The methodologies based on the tessellation of the spatial field of interest (NN, Lin, Cub, NaN) along with the schemes that use simple functions of distance (IDW, ID^{2}W) for the

estimation of the spatial distribution of air pollution, in most cases fail to describe adequately the spatial variation of NO_2 and O_3. The spatial setting of the monitoring stations especially for the primary pollutant NO_2 is not dense enough to provide adequate information for the efficient modeling of the air pollution levels. Regarding O_3, especially for the MAR station an increased performance of the simplified schemes is observed attributed to the existence of a highly representative station in its proximate region (LYK).

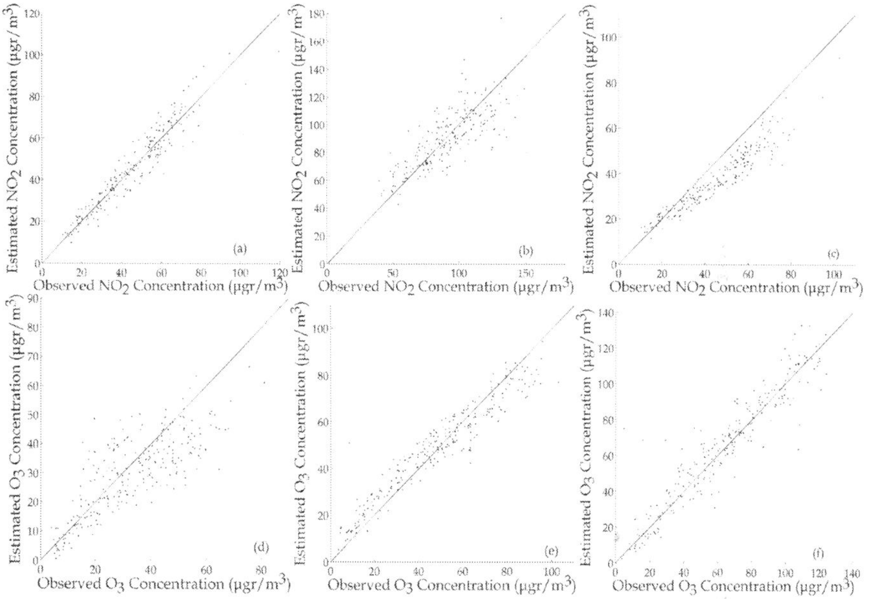

Fig. 9. Scatter diagrams of observed versus estimated concentrations for (a) NO_2 at GEO for ANN, (b) NO_2 at PAT for ANN, (c) NO_2 at GEO for IDW, (d) O_3 at ATH for ANN, (e) O_3 at GEO for RBFg and for (f) O_3 at MAR for NN.

The results of the OK methodology are greatly affected by the sampling configuration of the monitoring network. In most cases, except for O_3 at Geoponiki the OK prediction is comparable to more simplistic models, giving evidence that the scheme cannot give highly accurate results at every point of interest in the area under study. The performance of the methodology could however be increased taking into account, during the variogram analysis, the effect of meteorological factors and topography that induce the anisotropy in the spatial distribution of both primary and secondary pollutants. In our case anisotropism is ignored to simplify and automate the model fitting. The Gaussian scheme from the family of the RBF interpolants is found to be in average the best alternative scheme to ANNs with *d* values ranging from 0.62 at PAT to 0.89 at MAR and from 0.66 at PAT to 0.96 at GEO for NO_2 and O_3 respectively.

The results of the Wilcoxon's test statistic are presented in Table 8 and are used for assessing the statistical significance of the aforementioned results. Regarding NO_2 and for all stations the predictive ability of the ANN is assessed versus the alternative schemes, while for O_3 the

ANN is assessed for ATH, PAT and PER, the RBFg for GEO and the NN for MAR. The hypothesis of ANN's equal predictive accuracy for NO_2 is rejected in all cases at $p<0.05$ and in all cases except for NN, Lin and Cub for PER at $p<0.01$. The ANN predictive accuracy for O_3 is statistical significantly superior from the rest interpolation schemes at ATH, PAT and PER at $p<0.05$ and in all cases except for RBFg at ATH and PER at $p<0.01$. Regarding the RBFg at GEO and the NN at MAR, the null hypothesis of equal predictive accuracy is accepted at $p<0.05$ for the Lin and the OK models and for the RBFlin and ANN models respectively.

	Nitrogen Dioxide (NO_2)					Ozone (O_3)				
	ATH	GEO	MAR	PAT	PER	ATH	GEO	MAR	PAT	PER
NN	-11.72	-12.61	-9.01	-9.45	**-2.52**	-7.40	-9.48	--	-5.97	-7.40
Lin	-10.09	-11.66	-10.48	-12.51	**-2.49**	-4.80	**-2.33**	-3.33	-9.92	-4.80
Cub	-10.52	-12.61	-12.93	-12.01	**-2.43**	-5.90	-7.87	-3.36	-7.06	-5.90
NaN	-6.33	-11.70	-10.46	-12.61	-6.25	-6.25	-1.71	-3.36	-11.00	-6.25
IDW	-12.89	-8.62	-13.00	-13.03	-3.78	-12.44	-9.91	-7.89	-13.76	-12.44
ID²W	-12.96	-11.98	-12.97	-13.03	-4.34	-12.30	-11.06	-8.61	-13.72	-12.30
OK	-9.42	-11.41	-12.82	-12.30	-5.16	-5.83	**-1.93**	-7.97	-11.83	-5.83
RBFlin	-7.05	-12.67	-12.51	-12.64	-11.05	-2.89	-5.76	**-2.54**	-11.31	-2.89
RBFg	-7.50	-11.37	-8.63	-8.39	-6.82	**-2.46**	--	-3.23	-7.07	**-2.46**
RBFmq	-8.08	-11.42	-12.99	-8.62	-9.01	-3.53	-5.38	-2.96	-6.51	-3.53
SPtp	-7.58	-12.18	-13.06	-11.98	-9.59	-2.89	-4.85	-5.41	-7.58	-2.89
SPbh	-7.61	-12.25	-13.06	-11.94	-10.29	-2.81	-5.18	-5.47	-7.51	-2.81
ANN	--	--	--	--	--	--	-5.32	**-1.89**	--	--

Table 8. Wilcoxon's sign rank test statistics

4. Conclusion

The estimation of pollution fields, especially in densely populated areas, is an important application in the field of environmental science due to the significant effects of air pollution to public health. In this chapter the spatial interpolation methodologies, commonly employed for air pollution point estimations, are reviewed and their accuracy is assessed for five target sites located at the greater area of metropolitan Athens. In general, an optimal interpolation scheme that outperforms the rest methodologies at the area of study is not established. The errors associated with all interpolation schemes are related to the pollutant of concern and mainly to the spatial configuration of the monitoring network.

The Artificial Neural Network approach to point spatial interpolation is found to be specifically promising, compared to the traditional interpolation techniques. For the primary (NO_2) and secondary (O_3) pollutants addressed in this study, ANNs are found to be statistical significantly superior in all cases except for one case where their predications have equal predictive accuracy with the RBFg scheme and for O_3 at the Marousi station, where the simplistic Nearest Neighbor model outperforms the trained ANN scheme. In this context the ANNs are proposed as an alternative method to spatial interpolation, with increasingly important applications where point air quality estimations are prerequisite. The performance of the ANNs can be further improved, since there is no immediate restriction to the number of their inputs, by including parameters related to the atmospheric dispersion and the chemistry of the pollutant under consideration. The critical requirement of a representative dataset during training is the basis for their limited operational applications in air quality spatio-temporal forecasting.

5. Acknowledgements

This work has been partially funded by the KAPODISTRIAS Programme of the Special Account for Research Grants, National and Kapodistrian University of Athens.

6. References

Analitis, A.; Katsouyanni, K.; Dimakopoulou, K.; Samoli, E.; Nikoloulopoulos, A,K.; Petasakis, Y.; Touloumi, G.; Schwartz, J.; Anderson, H.R.; Cambra, K.; Forastiere, F.; Zmirou, D.; Vonk, J.M.; Clancy, L.; Kriz, B.; Bobvos, J. & Pekkanen, J. (2006). Short-term effects of ambient particles on cardiovascular and respiratory mortality, *Epidemiology*, Vol.17, No.2, (March2006), pp. 230-233, ISSN 1531-5487

Asimakopoulos, D.; Deligiorgi, D.; Drakopoulos, C.; Helmis, C.; Kokkori, K.; Lalas, D.; Sikiotis, D. & Varotsos, C. (1992). An experimental study of nighttime air-pollutant transport over complex terrain in Athens. *Atmospheric Environment*, Vol.26, No.1, (March 1992), pp. 59-71, ISSN 1352-2310

Baxter, B. (1992). *The interpolation theory of radial basis functions, Ph.D. Thesis*, Trinity College University of Cambridge, Cambridge, United Kingdom

Blackwell, W.J. & Chen F.W. (2009). *Neural Networks in Atmospheric Remote Sensing*, Massachusetts Institute of Technology, ISBN 978-1-59693-372-9, Lexington, USA

Deligiorgi, D.; Philippopoulos, K.; Thanou, L. & Karvounis, G. (2010). A comparative Study of Three Spatial Interpolation Methodologies for the Analysis of Air Pollution Concentrations in Athens, Greece, *7th International Conference of the Balkan Physical Union*, American Institute of Physics (AIP) Conference Proceedings Vol. 1203, pp. 445-450, ISBN 978-0-7354-0740-4

Diebold, F.X. & Mariano R.S. (1995).Comparing Predictive Accuracy, *Journal of Business & Economic Statistics*, Vol.13, No.3, (July 1995), pp. 253-263, ISSN 1537-2707

Eberly, S.; Swall, J.; Holland, D.; Cox, B.; Baldridge, E. (2004). *Developing Spatially Interpolatied Surfaces and Estimating Uncertainty*, United States Environmental Protection Agency

Fan, Q.; Efrat, A.; Koltun, V.; Krishnan, S. & Venkatasubramanian, S. (2005). Hardware-Assisted Natural Neighbor Interpolation, *Proceedings of ALENEX/ANALCO 2005 7th Workshop on Algorithm Engineering and Experiments*, pp. 111-120, ISBN 0-89871-596-2, Vancouver, Canada, 22 January, 2005

Gusten, H.; Heinrich, G.; Cvias, T.; Klasinc, L.; Ruscic, B.; Lalas, D. & Petrakis, D. (1988). Photochemical formation and transport of ozone in Athens, Greece. *Atmospheric Environment*, Vol.22, No.9, pp. 1855-1861, ISSN 1352-2310

Hardy, R.L. (1971). Multiquadric Equations of Topography and Other Irregular Surfaces, *Journal of Geophysical Research*, Vol.76, No.8, pp.1905-1915, ISSN 0148–0227

Hardy, R.L. (1978). *The Application of Multiquadric Equations and Point Mass Anomaly Models to Crustal Movement Studies*, National Oceanic and Atmospheric Administration

Hornik, K.; Stinchcombe, M. & White, H. (1989). Multilayer feedforward networks are universal approximators. *Neural Networks*, Vol.2, No.5, pp. 359-366, ISSN 0893-6080

Ito, K.; De Leon, S. F. & Lippmann, M. (2005). Associations Between Ozone and Daily Mortality: Analysis and Meta-Analysis, Epidemiology, Vol.16, No.4, (July 2005), pp. 446-457, ISSN 1531-5487

Junninen, H.; Niska, H.; Tuppurainen, K.; Ruuskanen, J. & Kolehmainen, M. (2004). Methods for imputation of missing values in air quality data sets. *Atmospheric Environment*, Vol.38, No.18, (June 2004), pp. 2895-2907, ISSN 1352-2310

Kassomenos, P.; Kotroni, V. & Kallos, G. (1995). Analysis of Climatological and Air Quality Observations from Greater Athens Area. *Atmospheric Environment*, Vol.29, No.24, (December 1995), pp.3671- 3688, ISSN 1352-2310

Krige, D.G. (1951). A statistical approach to some basic mine valuation problems on the Witwasersrand. *Journal of the Chemical, Metallurgical and Mining Society of South Africa*, Vol.52, No.6, pp. 119-139

Lalas, D.; Asimakopoulos, D. & Deligiorgi D. (1983). Sea-breeze circulation and photochemical pollution in Athens, Greece. *Atmospheric Environment*, Vol.17, No.9, pp. 6121-1632, ISSN 1352-2310

Li, J. & Heap A. D. (2008). *A Review of Spatial Interpolation Methods for Environmental Scientists*, Geoscience Australia, Retrieved from http://www.epa.gov/airtrends/studies.html

Lozano, A.; Usero, J.; Vanderlinden, E.; Raez, J.; Contreras, J.; Navarrete, B. & El Bakouri, H. (2009). Design of air quality monitoring networks and its application to NO_2 and O_3 in Cordova, Spain. *Microchemical Journal*, Vol. 93, No. 2, (November 2009), pp. 211-219, ISSN 0026-265X

Lu, G.Y. & Wong D.W. (2008). An adaptive inverse-distance weighting spatial interpolation technique, *Computers & Geosciences*, Vol.34, No.9, (September 2008), pp. 1044-1055, ISSN 0098-3004

Marshal, J.D.; Nethery, E. & Brauer, M. (2008). Within-urban variability in ambient air pollution: Comparison of estimation methods, *Atmospheric Environment*, Vol.42, No.6, (February 2008), pp. 1359-1369, ISSN 1352-2310

Mofarrah, A. & Husain, T. (2010). A holistic approach for optimal design of air quality monitoring network expansion in an urban area. *Atmospheric Environment*, Vol.44, No.3, (January 2010), pp. 432-440, ISSN 1352-2310

Myers D.E. (1994). Spatial interpolation: an overview. *Geoderma*, Vol.62, No.1-3, (March 1994), pp. 17-28, ISSN 0016-7061

Nejadkoorki, F.; Nicholson, K. & Hadad, K. (2011). The design of long-term air quality monitoring networks in urban areas using a spatiotemporal approach. *Environmental Monitoring and Assessment*, Vol.172, No.1-4 , (January 2011), pp. 215-223, ISSN 1573-2959

Okabe, A.; Boots, B.; Sugihara, K. & Chiu, S.N. (2000). *Spatial Tessellations Concepts and Applications of Voronoi Diagrams*, John Wiley & Sons Ltd, ISBN 0-471-986354, Chichester, England

Roussos, G. & Baxter, B. (2005). Rapid evaluation of radial basis functions. *Journal of Computational and Applied Mathematics*, Vol.34, No.9, (September 2008), pp. 1044-1055, ISSN 0098-3004

Samet, J.M.; Dominici, F.; Curriero, F.C.; Coursac, I. & Zeger, S.L. (2000). Fine particulate air pollution and mortality in 20 U.S. cities 1987-1994, *The New England Journal of Medicine*, Vol.343, No.24, (December 2000), pp. 1742-1749, ISSN 1533-4406

Sandwell, D. (1987). Biharmonic spline interpolation of GEOS-3 and SEASAT altimeter data. *Geophysical Research Letters*, Vol.14, No.2, (February 1987), pp. 139-142, ISSN 0094-8276

Shao, X.; Stein, M. & Ching, J. (2007). Statistical comparisons of methods for interpolating the output of a numerical air quality model. *Journal of Statistical Planning and Inference*, Vol.137, No.7, (July 2007), pp. 2277-2293, ISSN 0378-3758

Sibson, R. (1981). A brief description of natural neighbor interpolation, In: *Interpolating multivariate data*, Barnett V., pp. 21-36, John Wiley & Sons Ltd, ISBN 978-0471280392, Chichester, England

Snell, S.E.; Gopal, S. & Kaufmann R.K. (2000). Spatial Interpolation of Surface Air Temperatures Using Artificial Neural Networks: Evaluating Their Use for Downscaling GCMs, *Journal of Climate*, Vol.13, No.5, (March 2000), pp. 886-895, ISSN 1520-0442

Vicente-Serrano, S.M.; Saz-Sanchez, M.A. & Cuadrat J.M. (2003). Comparative analysis of interpolation methods in the middle Ebro Valley (Spain): application to annual precipitation and temperature, *Climate Research*, Vol.24, No.2, (July 2003), pp. 161-180, ISSN 1616-1572

Wahba G. (1990). *Spline Models for Observational Data*, Society for Industrial and Applied Mathematics, ISBN 978-0-898712-44-5, Philadelphia, USA

Wahba, G. & Wendelberger, J. (1980). Some new mathematical methods for variational objective analysis using spiline and cross validation. *Monthly Weather Review*, Vol.108, No. 8, (August 1980), pp. 1122-1143, ISSN 1520-0493

Watson, D.F. & Philip, G.M. (1984). Systematic Triangulations, *Computer Vision, Graphics, and Image Processing*, Vol.26, No.2, (May 1984), pp. 217-223, ISSN 0734-189X

Webster, R. & Oliver, M.A. (2007). *Geostatistics for Environmental Scientists*, John Wiley & Sons Ltd, ISBN 978-0-470-02858-2, Chichester, England

Willmott C.J. & Matsura K. (2005). Advantages of the mean absolute error (MAE) over the root mean square error (RMSE) in assessing average model performance. *Climate Research*, Vol.30, No.1, (December 2005), pp. 79-82, ISSN 1616-1572

Willmott C.J. (1982). Some Comments on the Evaluation of Model Performance. *Bulletin of the Maerican Meteorological Society*, Vol.63, No.11, (November 1982), pp. 1309-1313, ISSN 1520-0477

Part 4

New Techniques in Air Quality Management

20

The Electrical Conductivity as an Index of Air Pollution in the Atmosphere

Nagaraja Kamsali[1], B.S.N. Prasad[2] and Jayati Datta[3]
[1]*Department of Physics, Bangalore University, Bangalore*
[2]*University of Mysore, Mysore*
[3]*Indian Space Research Organization, Bangalore*
India

1. Introduction

Air is never perfectly clean. Many natural sources of air pollution have always existed. Ash from volcanic eruptions, salt particles from breaking waves, pollen and spores released by plants, smoke from forest and brushfires, and windblown dust are all examples of natural air pollution. Human activities, particularly since the industrial revolution, have added to the frequency and intensity of some of these natural pollutants.

Air pollution has considerable effects on many aspects of our environment: visually aesthetic resources, vegetation, animas, soils, water quality, natural and artificial structures, and human health. The effect of air pollution on vegetation include damage to leaf tissue, needles, or fruit, reduction in growth rates or suppression of growth, increased susceptibility to a variety of diseases, pests, and adverse weather. Air pollution can affect human health in several ways. The effects on an individual depend on the dose or concentration of exposure and other factors, including individual susceptibility. Some of the primary effects of air pollutants include toxic poisoning, causing cancer, birth defects, eye irritation and irritation of the respiratory system, viral infections causing pneumonia and bronchitis, heart diseases, chronic diseases, etc. The consequences of air pollution reach beyond health and agriculture, and they influence the weather as well. There is strong evidence that increased atmospheric contamination reduces visibility, modifies electrical conductivity, alters precipitation, and changes the radiation balance.

The very presence of a city affects the local climate, and as the city grows, so does its climate changes. Now-a-days, cities are warmer than the surrounding areas. The temperature increase is a result of the enhanced production of heat energy, may be the heat emitted from the burning of fossil fuels and other industrial, commercial, and residential sources, and the decreased rate of heat loss since the dust in the urban air traps and reflects back into the city as long wave radiation. In addition, particulates in the atmosphere over a city are often at least 10 times more abundant than in rural areas. Although the particulates tend to reduce incoming solar radiation by up to 30% and thus cool the city, this cooling effect of particulates is small in relation to the effect of processes that produce heat in the city.

Particulate matter encompasses the small particles of solid or liquid substances that are released into the atmosphere by many activities and are referred to as aerosols. Modern farming adds considerable amounts of particulate matter to the atmosphere, as do desertification and volcanic eruptions. Nearly all industrial processes, as well as the burning of fossil fuels, release particles into the atmosphere. Much particulate matter is easily visible as smoke, soot, and dust. Emission of aerosols to the atmosphere has increased significantly since the industrial revolution began. An aerosol is a particle with a diameter less than 10μm. Because the effects of collisions with air molecules dominate over gravity, the smaller aerosol particles tend to remain in the atmosphere for a long time. Bigger particles of diameter greater than 10μm drop out of the atmosphere faster because of gravity. Recent research has indicated that aerosols emitted from coal (sulfates) may contribute to global cooling because sulfates act as seeding for clouds, the aerosol particles provide surfaces for water to condense, forming clouds that reflect a significant amount of sunlight directly. The net cooling owing to sulfate aerosols may offset partially the global warming expected from the anthropogenic greenhouse effect.

The concern that man's activity in a world of rapidly growing population may lead to inadvertent modification of the climate on a local basis and even on a global scale, has led to the establishment of a worldwide network of air quality monitoring stations. An important parameter to be monitored is the particulate load of the atmosphere, which may have a direct impact on the climate change. Recently, atmospheric electric scientists have renewed their interest in the possibility of using atmospheric electrical parameters as indices of air pollution trends, despite the complexity of the relationship involved. Comparing fair weather conductivity recorded on the Carnegie expedition during 1967 revealed a decreasing trend of conductivity, which has dropped at least 30% in the North Atlantic. This decreasing trend in conductivity is attributed to a significant rise in the aerosol pollution over the Northern hemisphere.

An attempt has been made (Nagaraja Kamsali et al., 2009) to infer pollution trend from atmospheric electrical conductivity on a regional basis. There seems to be sufficient evidence that atmospheric electrical conductivity, properly deduced and analyzed, with due consideration of the local meteorology and air pollution climatology, may serve as a sensitive and practical tool, capable of documenting aerosol air pollution trends, especially secular changes, in largely populated urban area. Considering the relative simplicity and reliability of the electrical conductivity measurement, which is well suited for continuous and automatic recording, its inclusion in the measurement program of the regional air pollution may be preferred.

1.1 Radioactivity of the Earth's atmosphere

The rocks and soil in the solid Earth contain radioactive elements, ^{238}U and ^{232}Th in the form of minerals Uranite and Monozite; their concentrations vary from region to region. Decay of these elements release α, β and γ radiations along with daughter products, which themselves may or may not be radioactive. Some of the daughter elements are gases such as radon and thoron. These gases get either released into the atmosphere or trapped within the rock soil, depending on the permeability of the surrounding material. The radioactive gases released from the soil get mixed with the atmospheric air and contribute to ionization during their decay. The concentration of these gases varies from place to place, and with altitude at any given place. The distribution of these gases with altitude depends on the atmospheric stability. The decay products of these gases are mostly radioactive elements, and they also contribute to the ionization of air.

The origin of radon and thoron in the Earth's crust stems directly from the radium isotopes and their decay products distributed in minute quantities in the ground within few meters from the Earth's surface. Once the isotopes of the gas are produced they migrate to a significant distance from the site of generation, even during their brief half-lives. A fraction of the radon gas generated in soil near the Earth's surface enters into the atmosphere before undergoing into the radioactive decay. The amount of radon gas that escapes depends on the amount of ^{226}Ra and ^{232}Th in the ground, the type of soil, porosity, dampness and meteorological conditions. The amounts of ^{222}Rn (radon), ^{220}Rn (thoron) and ^{219}Rn (action) in the atmosphere depend primarily on the concentration of ^{238}U, ^{232}Th and ^{235}U. The relative abundances of the isotopes in natural uranium by weight are ^{238}U - 99.28%, ^{235}U - 0.71% and ^{234}U - 0.0054%. On the other hand, ^{232}Th is even more abundant than ^{238}U in the Earth's crust.

At present, twenty isotopes of radon are known. ^{222}Rn, from radium, has a halflife of 3.823 days; ^{220}Rn emanating naturally from thorium and called thoron has a halflife of 55.6s. ^{219}Rn emanating from actinium and called actinon has a halflife of 3.96s. It is estimated that every square mile of soil to a depth of 6 inches contains about 1g of radium, which releases radon in tiny amounts into the atmosphere.

1.2 Radon entry into the atmosphere

For a radon atom to escape from the mineral grain into the pore space, the decay must occur within the recoil distance of the grain surface. The recoil range for ^{222}Rn is 20-70 nm in common minerals, 100 nm in water, and 63 μm in air. Radon atoms entering the pore space are then transported by diffusion and advection through this space until they in turn decay or are gets released into the atmosphere. The escape of radon atoms from soil grains into the pore space is called radon emanation. The transportation of radon atoms from pore space by diffusion and advection through the space and gets released into the atmosphere is called radon exhalation. The amount of radon released per unit surface area per unit time is called radon exhalation rate. The rate of radon exhalation from soil depends on geophysical parameters such as geology of the region, soil porosity, soil texture, humidity and temperature of the soil, and also on meteorological parameters such as temperature, humidity, pressure etc. The influence of environmental parameters on the exhalation rate of ^{222}Rn from soil is not clearly understood. When the soil is frozen or covered with snow, the exhalation rate is reduced because diffusion processes are slowed down. It was observed by George (1980) that, the exhalation rate was linearly correlated with outdoor temperature. However, Arabzedegan et al., (1982) observed that the meteorological parameters had no significant influence on the value of ^{222}Rn exhalation rate. Raghavayya et al., (1982) have found that soil moisture had a strong influence on the ^{222}Rn exhalation rate. ^{222}Rn exhaled from the surface soil reaches the atmosphere and further decays to a series of radionuclides which are heavy metal radionuclides that attach to other atmospheric components to form aerosol particles.

1.3 Ionization in the atmosphere

Cosmic radiation is the primary source of ions over the oceans and above a couple of kilometers over the land (Hoppel et al., 1986). In the lower atmosphere, ions are predominantly produced by radiation emitted from radioactive materials in the soil, in building materials, and in the air (radon and its daughter products). The α and β particles and the γ rays from ^{222}Rn, ^{220}Rn and their decay products give up their energy by producing ions and raising atoms and molecules to excited states. An α particle from the decay of ^{222}Rn

atom (5.49 MeV) will produce about 150,000 ion pairs with the expenditure of about 34 eV per ion-pair along its trajectory through the surrounding air in the atmosphere (Israel, 1970). Cosmic radiation contributes about 10% to the ionization at ground level. However, at higher altitudes, the partitioning shifts drastically both because the radiation from the soil and airborne materials decreases whereas the intensity of the cosmic radiation increases.

The radioactive gases (^{222}Rn and ^{220}Rn) can diffuse from the ground into the atmosphere and contribute to the volume ionization. The concentration of these gases and the rate of ionization in the atmosphere depend upon the amount of uranium and thorium in the ground, and the temperature, dampness and porosity of the soil. Radon emanated from the Earths crust reaches the atmosphere and further decays to a series of short lived radionuclides namely, polonium (^{218}Po), lead (^{214}Pb) and bismuth (^{214}Bi). Radiation from radioactive gases exhaled from the ground and their daughter products causes ionization in the atmosphere. Radon is one of the relatively longer-lived gases. The principal decay modes and half-lives of ^{222}Rn and its short-lived daughters in order are ^{222}Rn - α, 3.82 days; ^{218}Po - α, 3.05 min; ^{214}Pb - β, 26.8 min; ^{214}Bi - β, 19.7 min; and ^{214}Po - α, 2×10^{-4} s. The radiations α , β and γ released during the decay of radon and its progeny cause ionization, and hence these are important in the study of atmospheric electricity. Ionization due to radioactive gases in the air is even more variable and depends not only on the amount exhaled from ground but also on atmospheric dispersion. Direct measurements of ionization due to radioactive gases in the atmosphere are difficult and have not generally been satisfactory. Estimation of ionization rate is therefore based on measurements of radioactive products in the air at that level, while the ionization due to cosmic radiation is almost constant with time. On cool nights with nocturnal temperature inversions the radioactive gases can be trapped in a concentrated layer close to the ground, whereas during unstable convective periods, the gases can be dispersed over an altitude of several kilometers. But during daytime due to convective processes, these gases disperse to higher altitudes. Because of this, the height distribution of radon in the atmosphere has also been used to determine the turbulent diffusion coefficient (Hoppel et al., 1986).

1.4 Influence of meteorological parameters on Ionization

In general, the ionization increases rapidly with height but shows certain changes in the troposphere and stratosphere that appear to be associated with the existence of aerosol layers (Israel, 1971). However, in the lowest region of the atmosphere the rate of ionization greatly depends upon the different meteorological parameters. The ionization rate near the Earth's surface is markedly influenced by variations in meteorological parameters such as wind and precipitation, in addition to the particles and air mass in the atmosphere. The wind speed which is produced under the influence of pressure gradient and Coriolis force is highly variable in time and space. It is an important factor in the transportation of atmospheric nuclei and ions in the atmosphere. It shows diurnal as well as seasonal variations. Its magnitude and direction vary with height and hence the density of ionization in the atmosphere is very much influenced by the wind. However, the precipitation is another important meteorological parameter which has a profound influence on the ion content of the atmosphere. It washes out both positive and negative ions, leaving the air fairly low in ion content. The degree of washout depends on the spectrum of the rain and its duration. Addition to this, the air mass also influences the ionization. The qualities of an air mass depend upon the region over which it stagnates for a sufficiently long time and acquires the characteristics of the environment. Air mass that lies over the densely populated areas and cities will be weak in the ion content because of the large quantity of particulate matter that goes into it from the city environment. Air mass that lies over the rural surroundings is mostly free from such particulate matter and hence it is rich in ion content. Thus the quality of an air mass influences the ion content and consequently columnar resistance of the local generator. Another major parameter is the suspended particles in the atmosphere. These particles always exist in the atmosphere but the number varies with the atmospheric conditions. The visibility of the atmosphere is a good indicator of the particle content in the atmosphere. It enhances the formation of fog and cloud in the atmosphere under certain favourable conditions. They promote attachment of ions to particles and reduce their mobility.

1.5 Particulate matter in the atmosphere

Particulate matter encompasses the small particles of solid or liquid substances that are released into the atmosphere by many activities and are referred to as aerosols. Modern farming adds considerable amounts of particulate matter to the atmosphere, as do desertification and volcanic eruptions. Nearly all industrial processes, as well as the burning of fossil fuels, release particles into the atmosphere. Much particulate matter is easily visible as smoke, soot, and dust. Emission of aerosols to the atmosphere has increased significantly since the industrial revolution began. An aerosol is a particle with a diameter less than 10μm. Because the effects of collisions with air molecules dominate over gravity, the smaller aerosol particles tend to remain in the atmosphere for a long time. Bigger particles of diameter greater than 10μm drop out of the atmosphere faster because of gravity. The aerosols emitted from coal (sulfates) may contribute to global cooling because sulfates act as seeding for clouds, the aerosol particles provide surfaces for water to condense on, forming clouds that reflect a significant amount of sunlight directly. The net cooling owing to sulfate aerosols may offset much of the global warming expected from the anthropogenic greenhouse effect.

1.6 Air pollution

Air is never perfectly clean. Many natural sources of air pollution have always existed. Ash from volcanic eruptions, salt particles from breaking waves, pollen and spores released by plants, smoke from forest and brushfires, and windblown dust are all examples of natural air pollution. Ever since people have been on earth, however, they have added to the frequency and intensity of some of these natural pollutants. As the faster moving fluid medium in the environment, the atmosphere has always been one of the most convenient places to dispose of unwanted materials. Ever since fire was first used by people, the atmosphere has been a sink for waste disposal. As long as a chemical is transported away or degraded relatively rapidly, there is no pollution problem. Chemical pollutants can be thought of as compounds that are in the wrong place or in the wrong concentrations at the wrong time. Pollutants that enter the atmosphere through natural or artificial emissions may be broken down or degraded not only within the atmosphere but also by natural processes in the hydrologic and geochemical cycles; pollutants that leave the atmosphere may become pollutants of water and of geological cycles.

Air pollution has considerable effects on many aspects of our environment: visually aesthetic resources, vegetation, animals, soils, water quality, natural and artificial structures, and human health. The effect of air pollution on vegetation include damage to leaf tissue, needles, or fruit, reduction in growth rates or suppression of growth, increased susceptibility to a variety of diseases, pests, and adverse weather. Air pollution can affect human health in several ways. The effects on an individual depend on the dose or concentration of exposure and other factors, including individual susceptibility. Some of the primary effects of air pollutants include toxic poisoning, causing cancer, birth defects, eye irritation and irritation of the respiratory system, viral infections, causing pneumonia and bronchitis, heart diseases, chronic diseases, etc. Wherever there are many sources emitting air pollutants over a wide area there is a potential for the development of smog. Whether air pollution develops depends on the topography and on weather conditions, because these factors determine the rate at which pollutants are transported away from their sores and converted to harmless compounds in the air. When the rate of production exceeds the rate of degradation and of transport, dangerous conditions may develop. Meteorological conditions can determine the fate of air pollution. Air pollution and meteorology are linked in two ways. One concerns the influence that weather conditions have on the dilution and dispersal of air pollutants. The second connection is the reverse and deals with the effect that air pollution has on weather and climate. Air pollution is a continuing threat to our health and welfare. The cleanliness of air, therefore, should certainly be as important to us as the cleanliness of food and water.

The concern that man's activity in a world of rapidly growing population may lead to inadvertent modification of the climate, on a local basis and even on a global scale, has led to the establishment of a worldwide network of air quality monitoring stations. An important parameter to be monitored is the particulate load of the atmosphere, which may have a direct impact on the climate change. Recently, atmospheric electric scientists have renewed their interest in the possibility of using atmospheric electrical parameters as indices of air pollution trends, despite the complexity of the relationship involved. Comparing fair weather conductivity recorded on the Carnegie expedition during 1967 revealed a decreasing trend of conductivity, which has dropped at least 30% in the North Atlantic. This decreasing trend in conductivity is attributed to a significant rise in the aerosol pollution over the Northern hemisphere. An attempt to infer pollution trend from atmospheric electrical conductivity on a regional basis, due to extensive urbanization has been made. There seems to be sufficient evidence that atmospheric electrical conductivity, properly reduced and analyzed, with due consideration of the local meteorology and air pollution climatology, may serve as a sensitive and practical tool, capable of documenting aerosol air pollution trends, especially secular changes, in largely populated areas.

1.7 Electrical nature of the Earth's atmosphere

The planetary boundary layer is that region of the lower atmosphere in which the influences of the earth's surface are directly felt. The primary influences of the surface are drag, heating (or cooling), and evaporation (or condensation). These processes cause vertical fluxes of momentum, sensible heat, and moisture, which penetrate into the lower atmosphere to a finite height. These fluxes, in turn, generate turbulence, ultimately controlling the mean profiles of wind speed, temperature, and water vapour in the planetary boundary layer.

Atmospheric electrical conductivity, ion mobility and small ion number density, etc. are important parameters for understanding the electrical nature of the atmosphere. The small ions consisting of aggregates of few molecules determine the electrical conductivity over the region (Israel, 1970). The number densities of these ions are controlled by ionizing mechanisms for the production of ions and electrons, and the loss processes for these charged species. The electrical conductivity of the air in an aerosol free atmosphere is mainly due to small ions. However, in a polluted atmosphere these ions soon get attached to the aerosol particles and form the intermediate and large ions. Since the mobility of small ions is at least two orders of magnitude more than that of large ions, the small ions are still considered to be, if not the only, the main contributors to the local electrical conductivity (Dhanorkar and Kamra, 1997). Since the presence of aerosol particles depletes the small ions, the electrical conductivity and aerosol concentration are generally considered to have an inverse relationship in the atmosphere and the electrical conductivity has been often proposed to act as an index of air pollution. The decrease in electrical conductivity with an increase in air pollution has been reported under different meteorological conditions such as in continental air (Misaki, 1964; Mani and Huddar, 1972), near coastlines (Misaki and Takeuti, 1970; Morita et al., 1972; Kamra and Deshpande, 1995), in marine air (Cobb and Wells, 1970), at the level of inversion (Rosen and Hofmann, 1981) and during volcanic eruption (Srinivas et al., 2001).

The electrical conductivity of the air is mainly due to the presence of highly mobile small ions produced in the atmosphere by cosmic rays and local radioactive sources (Retalis et al., 1991). A balance is maintained since the small ions are being removed from the atmosphere at the same rate that they are being produced. They are removed by recombination with oppositely charged small ions and by attachment to larger aerosol particles. These recombination and attachment processes thus determine the life times of small ions in the atmosphere and secondarily of the conductivity, since its value is formed predominantly by the small ion concentration in the air. Ion lifetimes near earth's surface vary from 20 sec in highly polluted air to 300 sec in very clean air (Cobb, 1973).

1.8 The electrical conductivity of the atmosphere

The atmospheric electrical conductivity depends on the existence of positive and negative ions. If a small potential is applied to two electrodes in an ionized gas, a week current flow is induced by the ions which flow in opposite directions to the electrodes where they deliver their charge. The current density (per unit cross section) is defined as the amount of electric charge which flows per unit time through a unit surface area perpendicular to the direction of flow. The current density is composed of two terms corresponding to the respective ion flow, and is given by the relation $i = e\left(n^+\mu^+ + n^-\mu^-\right)E = \sigma E$, where $\mu^+$and μ^- are the mobilities and n^+ and n^- are the ion concentrations of positive and negative polarity, and e is the electronic charge. The resulting conductivity can be expressed in terms of the number densities and mobilities of the individual species as $\sigma = e\left(n^+\mu^+ + n^-\mu^-\right)$. For several types of ions of different mobility are present, the above expression become $\sigma = e\sum_{i=1}^{n}\left(n_i^+\mu_i^+ + n_i^-\mu_i^-\right)$.

Near to the Earth's surface, electrical conductivity is not constant with time. Observations over long periods at various sites have shown that atmospheric conductivity has well defined diurnal and seasonal variations during fair weather periods. The exact nature of variation differs with stations, the mean values and amplitude being different. For a given station, the pattern of variation of conductivity at the surface normally remains more or less a constant during a year. The average behaviour at each station may therefore be characteristic of the 'atmospheric electric climate' of that station. But some similarities are seen in the diurnal behaviour of conductivity at most stations. For instance, conductivity is higher at night than during the day. This is typical for continental stations and makes it probable that, apart from all local and other influences, a more general factor which causes this general tendency in the diurnal variation must exist. This variation is possibly a reflection of the general rhythm of the atmosphere, since nights are usually calmer, with low winds and hardly any convective motion, and therefore conducive for accumulation of radon and other radioactive species near the surface. A sharp fall associated with sunrise is seen at many stations. This is believed to be due to an increase in the aerosol concentration due to onset of circulation and human activity. In marine air, the mean conductivity is slightly higher than that over land but the diurnal amplitude is very low. The average value on land is about 1.8×10^{-14} Ω^{-1} m^{-1} while over the ocean it is about 2.8×10^{-14} Ω^{-1} m^{-1} (Chalmers, 1967). And unlike over land, conductivity over ocean is minimum in the late afternoon or early evening hours, and remains low during night. It rises to a peak in the forenoon and is followed by a gradual decline. This is almost a mirror image of the pattern seen over continental stations. Further, while the positive conductivity displays a clear variation in the manner described, the variation in the negative conductivity has very low amplitude and is just discernible. The ratio of polar conductivities thus exhibits a clear pattern over ocean. From the life time it is well known that, the mobility of large or intermediate ions are a few orders of magnitude smaller than those of small ions. So, their contribution to the polar conductivity is relatively smaller than that of the small ions. Small ions of opposite polarity recombine and cause a decrease in the small ion concentration and consequently a reduction in the conductivity of air. There exist a close correlation between the concentration of small ions and polar conductivity near the surface of Earth. In the presence of aerosol particles losses in the small ion concentration are caused not only due to the ion-ion recombination process, but also due to attachment of small ions to the aerosol particles. Attachment of small ions to the aerosol particles makes them almost immobile and causes a further decrease in conductivity of the atmosphere. This develops inverse relationship between the aerosol concentration and the electrical conductivity (Cobb and Wells, 1970). Mani and Huddar (1972) observed the decrease in conductivity presumably caused due to increase in dust particle content at Pune, India over a period of 20 years. Hogan et al., (1973) pointed out that the conductivity inversely follows the product of aerosol size and number density better than the number density alone. Similar type of results was obtained by Srinivas et al., (2001) and they concluded that the conductivity of the atmosphere depends on the number density of aerosols rather than their concentration alone. Cobb (1973) suggested that this product is really close to what is meant by pollution. Due to this inverse relationship between the electrical conductivity and the pollution, the electrical conductivity has been proposed to act as a pollution index.

1.9 Need for theoretical estimation of electrical conductivity of the atmosphere

Measurements of atmospheric electrical conductivity are in general difficult to interpret because of large variety of influencing factors. Therefore, a thorough theoretical and experimental analysis is necessary in order to conduct research in atmospheric electricity. In the atmospheric surface layer, particularly the lowest few meters above the ground, large number of factors will dominate, for example, radioactive emanation from the ground; porosity, dampness and temperature of the soil; aerosol concentration; atmospheric electric field and mobility of small ions. The electrical conductivity is very sensitive to the presence of aerosols. Thus, the aerosol loading has a bearing on the conductivity of the atmosphere. The aerosols reduce the conductivity of the atmosphere by (i) converting the highly mobile small ions into less mobile aerosol ions through ion - aerosol attachment and (ii) neutralizing the small ions through the aerosol ion - small ion recombination. Another process that makes the ion-aerosol attachment rate faster is the charged aerosol - aerosol recombination. In the present communication ion - aerosol interactions are taken into account to develop a model for the electrical conductivity of the atmosphere. It is validated by comparing the computed values of conductivity with the experimental values.

Several model studies on the electrical conductivity of the atmosphere of higher altitude are available (Datta et al., 1987; Prasad et al., 1991; Srinivas and Prasad, 1993, 1996). Most of the model studies have considered the loss of small ions as solely due to a) ion - ion recombination b) ion - attachment to aerosols. The other two types of loss of small ions arise from a) the recombination of molecular ions with oppositely charged aerosol and b) charged aerosol - aerosol recombination, and hence further deplete the small ions. Obviously, addition of these two terms results in more realistic values (Srinivas et al., 2001). The equilibrium density of small ions is governed by the equations of continuity for the production and loss of these ions, where the gain and loss due to transport are negligible. The effective recombination of small ions is altered in the presence of aerosols, since these aerosols interact with the ions through various attachment and recombination processes. The attachment of small ions to neutral aerosols produce charged aerosols referred to as 'large ions', which are less mobile than the smaller molecular ions. The subsequent recombination of charged aerosols with ions as well as oppositely charged aerosols would result in the depletion of small ion concentration more rapidly than in the absence of aerosols. Therefore, the formation of less mobile aerosol ions and the reduction of more mobile molecular ions alter the electrical conductivity of the atmosphere. Thus, the pollution due to aerosols/dust in the atmosphere can considerably reduce the atmospheric conductivity. Hence, the atmospheric pollution level can be monitored through electrical conductivity measurements and conductivity is used as index of air pollution.

2. Instrumentation and methodology

2.1 Radon concentration in air

Concentration of radon in air at a height of 1 m above the surface is measured using the Low Level Radon Detection System. The procedure briefly consists of sampling the air in a collection chamber and exposing a circular metallic disc to the radon inside the collection chamber. A delay of at least 10-min is normally allowed for any thoron, which may be present in the chamber to decay completely. The positively charged ^{218}Po (RaA) atoms created in the chamber get collected on the metallic plate maintained at an optimum negative potential that should be sufficient to force all the RaA atoms onto the plate. The collection is carried out for an optimized period and thereafter the charged plate is removed from the chamber and alpha-counted. The concentration of radon, R_n, (in Bq m^{-3}) is calculated (Nagaraja et al., 2003a) with the expression: $Rn = 1000C/(EFVZ)$ where C is the total number of counts, E is the efficiency of alpha counting system, F is the efficiency of collection of RaA-atoms on the metallic disc and is empirically related to the relative humidity (H) by $F = 0.9 * [1 - \exp(0.039 * H - 4.118)]$, V is the volume of chamber, Z is the correction factor for build up and decay of radon daughter atoms on the metallic disc during the exposure and counting period.

2.2 Radon progeny concentration in air

An air flow meter kept at a height of 1 m above the surface is used to measure the radon progeny concentration. Air is drawn through a glass fiber filter paper by means of a suction pump at a known flow rate. The radon progeny in air sample are retained on the filter paper. The filter paper is then alpha-counted at any specific delay time. Total activity on the filter paper is measured at three different counting intervals of 2 - 5, 6 - 20 and 21 - 30 minutes. Activities of polonium (RaA), lead (RaB) and bismuth (RaC) (in Bq m^{-3}) are calculated using the modified equations given by Raghavayya (1998):

$$RaA = \frac{+4.249019 \times C_1 - 2.062417 \times C_2 + 1.949949 \times C_3}{V \times E}$$

$$RaB = \frac{-0.355129 \times C_1 + 0.006232 \times C_2 + 0.240618 \times C_3}{V \times E}$$

$$RaC = \frac{-0.215175 \times C_1 + 0.371319 \times C_2 - 0.502945 \times C_3}{V \times E}$$

$$R_d = \frac{+0.048445 \times C_1 - 0.019335 \times C_2 + 0.037053 \times C_3}{V \times E}$$

where C_1, C_2 and C_3 are the gross counts during the three counting intervals, E is the efficiency of alpha counting system, V is the sampling rate in liters per minute (LPM), R_d is the concentration of radon progeny (in Working Level)

2.3 Ion - pair production rate due to radioactivity

The ion-pair production rate due to radon and its progeny is calculated using the expression: $Q=\varepsilon/I$ and $\varepsilon = 5.49 \times 10^6 Rn + 6.00 \times 10^6 RaA + 0.85 \times 10^6 RaB + 7.69 \times 10^6 RaC$ where Rn is the concentration of radon (in Bq m^{-3}), RaA, RaB and RaC are the concentration of radon progenies i.e., polonium, lead and bismuth (in Bq m^{-3}), respectively, ε is the total energy (in eV) released by the decay of radon and its progeny, I is the energy required to produce one ion-pair = 32 eV, Q is the ion-pair production rate (ion-pairs $cm^{-3}s^{-1}$).

2.4 Aerosol concentration near the earth's surface

The concentrations of aerosols were measured with an Electrical Aerosol Analyzer in the size range from 3 to 750 nm diameter. The instrument works on the principle of diffusion charging-mobility analysis of particles. The ambient air is sampled at a rate of 50 litre per minute. The samples is first exposed to a Kr-85 radioactive neutralizer and then passed through a mobility analyzer that contains concentric cylindrical electrodes and a central collector rod. A pre-determined voltage is applied between the electrodes to produce an electric field in the condenser. The charged particles are deflected towards the collector rod by the electric field and the concentration of aerosols is measured in terms of mobility spectrum (Liu and Pui, 1975). A computer is interfaced with analyzer to measure the size distribution.

2.5 Electrical conductivity of the atmosphere

The Gerdien condenser is basically a cylindrical capacitor that collects atmospheric ions and provides air's electrical conductivity, which is shown in Fig. 1 and plate 1. It consists of two coaxial cylinders between which air is allowed to flow. A voltage is applied to one cylinder, known as the driving electrode, with respect to the other. This driving voltage repels ions of one polarity towards the other electrode where ions get collected. It is measured using a sensitive electrometer. For voltages that are sufficiently low, the collector current increases in proportion to the driving voltage. The collector current at these voltages is given by the expression: $I = \sigma CV/\varepsilon_0$. Gerdien condenser measurements have some sources of errors caused either due to non-fulfillment of some ideal conditions or due to improper adjustment of some important controlling factors such as potential, air flow rate etc and these factors have to be considered in the estimation of conductivities.

In case the humidity is very high, like during fog, it does not directly influence the proper function of the condenser. Apart from this, cobwebs, fragments of feathers, and accumulated dust, may, become conducting in humid air. In such a case the conductivity values increase drastically and reach unrealistic magnitudes. Similar effects can be observed when very small insects enter the inner cylinder. Hence, it is necessary to maintain the area surrounding the condenser free from the insects, cobwebs, dust, leaves etc.

The electric field inside the Gerdien condenser is assumed to be independent of axial distance. This is true only for an infinitely long condenser. For a condenser of finite length, the electric field distribution gets modified at the ends. This is called the 'edge effect'. At the ends of the condenser, some field lines originating at the outer surface of the driving electrode terminate at the collector. The field lines which are straight in the interior and perpendicular to the capacitor electrode are distorted at the edge. Because of this, some ions which are not present in the volume of air being sampled are also pulled in. Thus, the internal field does not end at the capacitor edge, but extends outside the region. Consequently, the ions experience an electric field even before entering the capacitor. Difficulties related to the edge effect can be reduced when the outer electrode is closer to the earth's potential. However, since it is easier to maintain the inner tube at a high level of insulation required for electrometer usage, the accelerating potential is most commonly applied to the outer tube and an electrometer is connected to inner cylinder. The diverging of the electric field around the entrance of the outer electrode may also be avoided with the aid of a grounded entrance cylinder. This and various other modifications suggested to homogenize the field at the capacitor inlet do not remove the field disturbance completely. Moreover, such arrangements introduce another source of error; the losses due to diffusion of ions to devices introduced to remove edge effects. Any such device is, therefore, not introduced in our condenser.

The atmospheric electrical conductivity of both positive and negative polarities is simultaneously measured with two Gerdien condensers housed separately in a single unit as in Fig. 1, locally fabricated at Pune, India. The apparatus consists of two identical cylindrical tubes of 10-cm diameter and 41 cm length joined by a U-shaped tube. The air is sucked through them with a single fan fixed at the end of U-shaped tube. The flow rate in each tube is about 19 liters per second. The inner co-axial aluminum electrodes in both the tubes are of 1-cm diameter and 20 cm length, and are fixed well inside the outer electrodes with Teflon insulation. The outer electrodes are shielded from any external electric field by two coaxial cylinders of 11 cm diameter and 35 cm length separated from the outer electrodes with Bakelite rings. Opposite but equal potentials of ± 35 V are applied to the

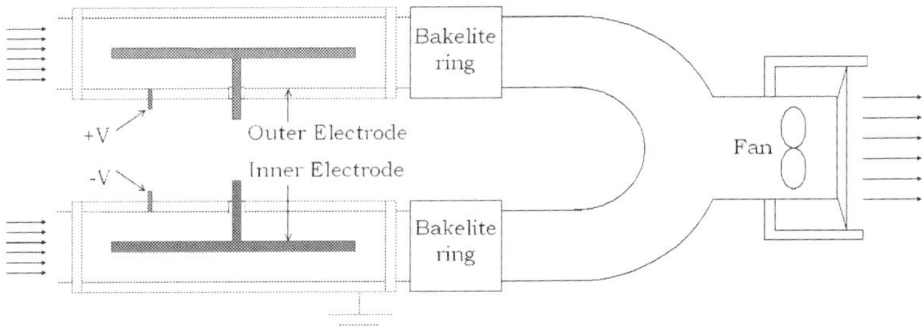

Fig. 1. Schematic representation of the Gerdien Condenser

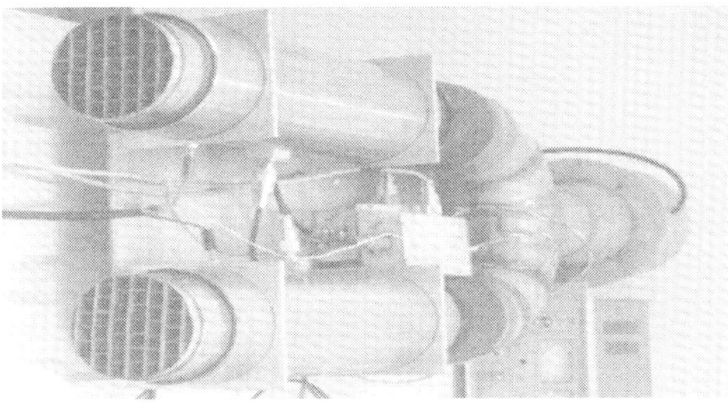

Plate 1. Gerdien Condenser

outer electrodes of the two condensers. The critical mobility of the instrument is greater than 2.93 x $10^{-4} m^2 V^{-1} s^{-1}$ (Dhanorkar and Kamra, 1993) and is capable of resolving the values of conductivity as small as 3 x 10^{-16} $\Omega^{-1} m^{-1}$ (Dhanorkar and Kamra, 1992). To avoid leakages along the insulators at higher humidity, Teflon is used to insulate all high impedance points in the apparatus because it maintains good insulation even at high humidity. The signals from the two condensers are amplified separately with two electrometer amplifiers AD 549J, and are carried through a separate coaxial cable and then recorded on a data logger card, which has a voltage range of ±15V with 10 mV resolutions. Though each sensor is scanned every second, the data logger card records averaged values of each parameter for every one-minute and then stores the values on a computer. Then it is averaged for any desired length of time with the help of an external program. In our case we have averaged this for every hour. The card has the capability of storing data at the maximum of 100 samples per second. In our case we made 60 samples per second and then it was averaged for a minute or for any desired length of time with the help of an external program. The diurnal and / or seasonal variations of radioactivity and the electrical conductivity of the atmosphere, and their dependence on meteorological parameters are presented for a continental station Pune (12 °N, 76 °E), India.

2.6 Meteorological parameters

The meteorological parameters such as temperature, pressure, wind speed and wind direction near the surface of the earth are obtained by automatic weather station installed near the observatory. Relative humidity is estimated from the wet and dry bulb thermometers.

3. Simple ion aerosol model for the estimation of electrical conductivity

There have been no reports on modeling study of the electrical conductivity in the lower part of the troposphere, in particular near the surface. Modeling for this region requires, ionic aerosols, in addition to the molecular ions. Based on this, a Simple Ion-Aerosol Model is proposed and the schematic diagram is shown in Fig. 2. It involves primary ion pair production rate due to surface radioactivity and cosmic rays, the small ion densities ($N_{\pm}$) and the aerosol number density. For the model prediction of equilibrium ion density and conductivity, it is necessary to use all the four loss processes of ions involved. The various recombination coefficients that enter into the model are: α_i due to the loss of oppositely charged small ions, α_a due to oppositely charged aerosol ions, α_s between small ions and aerosol ions, β due to attachment of small ions of similar polarities with aerosols.

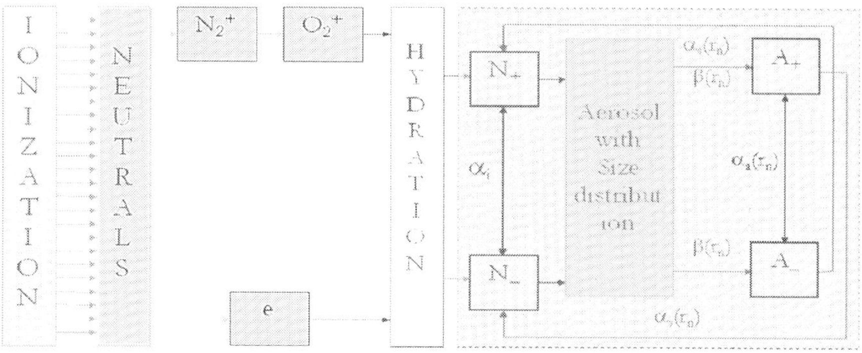

Fig. 2. Schematic representation of ion aerosol model

The temperature, pressure, relative humidity, ionization rate and aerosol number density along with the different attachment/recombination coefficients are the input parameters for the model. With these inputs the electrical conductivity is estimated. However, the model gives the total conductivity and does not make any distinction between positive and negative ion conductivities.

3.1 Theoretical description of the Model

The polar conductivity of air is defined as

$$\sigma = neb \tag{1}$$

where n is the concentration of ions of mobility b and having an elementary charge e. In the absence of aerosol particles, polar conductivity is mainly due to the high-mobility small ions. However, in the presence of aerosol particles, small ions get attached to the aerosol particles. In the Simple Ion - Aerosol Model, the detailed reaction paths for the formation of individual cluster ions are not considered. The total number densities of positive and negative ions are assumed to be equal from the charge neutrality criterion. The equilibrium ion densities are computed from the equations of continuity where the effect of transport is neglected. These equations include aforementioned four types of loss processes. The two values of β (for attachment of positive and negative ions with neutral aerosols) and α_s (for recombination of positive and negative ions with oppositely charged aerosols) are assumed to be equal.

The ion-aerosol interaction and involves primary ion pair production rate due to surface radioactivity and cosmic rays, the small ion densities ($N_\pm$) and the aerosol number density. For the model prediction of equilibrium ion density, it is necessary to use all the four loss processes of ions involved. The various recombination coefficients that enter into the model are: a) α_i - due to the loss of oppositely charged small ions, b) α_a - due to oppositely charged aerosol ions, c) α_s - is between small ions and aerosol ions and d) β - due to attachment of small ions of similar polarities with aerosols.

The equations of continuity (i) for small ions in the absence of aerosols and in the presence of aerosols and (ii) the charged aerosols are given by:

$$\frac{dN_0}{dt} = q - \alpha_i N_0^2 \quad (2)$$

$$\frac{dN_\pm}{dt} = q - \alpha_i N_\pm^2 - \beta Z N_\pm - \alpha_s A_\pm N_\mp \quad (3)$$

$$\frac{dA_\pm}{dt} = \beta Z N_\pm - \alpha_s A_\pm N_\mp - \alpha_a A_\pm^2 \quad (4)$$

where q is the ion production rate due to radioactivity and cosmic rays, N_0 is positive/negative ion density in the absence of aerosols, $N_\pm$ *and* $N_\mp$ are positive/negative ion densities in the presence of aerosols, $A_\pm$ is positively/negatively charged aerosol density, Z is neutral aerosol number density, β is the aerosol-ion attachment coefficient, α_i is the ion-ion recombination coefficient, α_s is the charged aerosol-ion recombination coefficient, α_a is the charged aerosol-aerosol recombination coefficient.

Under steady state conditions Eqs.(2), (3) and (4) reduce to:

$$q - \alpha_i N_0^2 = 0 \quad \Rightarrow \quad N_0 = \left(\frac{q}{\alpha_i}\right)^{\frac{1}{2}} \quad (5)$$

$$q - \alpha_i N_\pm^2 - \beta Z N_\pm - \alpha_s A_\pm N_\mp = 0 \quad (6)$$

$$\beta Z N_\pm - \alpha_s A_\pm N_\mp - \alpha_a A_\pm^2 = 0 \quad (7)$$

Solving Eqs. (5), (6) and (7) for $N_\pm$ with $A_\pm = N_0 - N_\pm$ we get:

$$N_\pm = \frac{(\alpha_a N_0 + \beta Z) \pm \sqrt{(\alpha_a N_0 + \beta Z)^2 - (\alpha_a - \alpha_i)(q + \alpha_a N_0^2)}}{(\alpha_a - \alpha_i)} \quad (8)$$

Solving Eqs. (6) and (7) simultaneously, one can get:

$$\beta Z = \left(\frac{q - \alpha_i N_{\pm}^2 + \alpha_a A_{\pm}^2}{2N_{\pm}} \right) \tag{9}$$

$$\alpha_s = \left(\frac{q - \alpha_i N_{\pm}^2 + \alpha_a A_{\pm}^2}{2A_{\pm} N_{\mp}} \right) \tag{10}$$

The ion depletion ΔN, i.e., loss of molecular ions in presence of aerosols is given by $\Delta N = N_0 - N_{\pm}$. Further, this ion depletion is equal to the charged aerosol concentration, $A_{\pm}$. Hence, the fractional depletion, η of small ions is used and computed as follows:

$$\eta = \left(\frac{\Delta N}{N_0} \right) = \left(\frac{A_{\pm}}{N_0} \right) \tag{11}$$

Using Eqs. (11) and (5), we can write Eqs.(9) and (10), respectively, as

$$\beta Z = \left(\frac{N_0 \eta [\alpha_i (2-\eta) + \alpha_a \eta]}{2(1-\eta)} \right) \tag{12}$$

$$\alpha_s = \left(\frac{\alpha_i (2-\eta) - \alpha_a \eta}{2(1-\eta)} \right) \tag{13}$$

The conductivity of the atmosphere in the absence and presence of aerosols, respectively, are given by:

$$\sigma_0 = N_0 e b_{\pm} \quad \& \quad \sigma_{\pm} = N_{\pm} e b_{\pm} \tag{14}$$

where e is the charge of an electron, $b_{\pm}$ is the ionic mobility, σ_0 is the conductivity of the atmosphere in the absence of aerosols and $\sigma_{\pm}$ is the conductivity of the atmosphere in the presence of aerosols.
The ionic mobility ($b_{\pm}$) is expressed in terms of the reduced mobility (b_0) at normal temperature (T_0) and pressure (P_0), and the ambient temperature (T) as well as pressure (P) as follows:

$$b_{\pm} = \frac{b_0 P_0 T}{T_0 P} \tag{15}$$

The reduction in conductivity (Δσ) due to depletion of small ions is given by:

$$\Delta \sigma = \sigma_0 - \sigma_{\pm} \tag{16}$$

The value of Δσ is computed in terms of different model parameters from Eqs. (5)-(16) and by following the method of Gringel et al. (1978), we get:

$$\Delta \sigma = \frac{\left[(2\alpha_i - \alpha_s)\sigma_0 + \beta Z e b_{\pm} \right] \mp \sqrt{\left[(2\alpha_i - \alpha_s)\sigma_0 + \beta Z b_{\pm} \right]^2 + 4(\alpha_s - \alpha_i)\sigma_0 \beta Z e b_{\pm}}}{2(\alpha_i - \alpha_s)} \tag{17}$$

3.2 Computation of conductivity

Modeling the conductivity of the atmosphere requires a knowledge of recombination coefficients α_i, α_a and α_s. The value of α_i is estimated from the parametric formula (Smith and Adams, 1982; Datta et al., 1987). From the theoretical considerations, Hoppel (1985) has shown that, for singly charged aerosols, the relative magnitudes of α_a and α_s are such that $\alpha_a \leq \alpha_s \leq \alpha_i$. Initially, with a suitable assumed value of α_a, the value of η is computed from Eq. (11), then the value of α_s is computed using Eq.(13). It is noted that, in this step, α_s becomes negative if the assumed value of α_a is unrealistically large. In the present computations $\alpha_a = 10^{-7}$ cm^3s^{-1} is found to be suitable. In presence of aerosols and depending on the magnitude of α_i, α_s and βZ, the realistic values of ΔN are obtained which lies between zero and N_0, and satisfies the condition $0 < N_\pm < N_0$. The values of βZ and α_s are obtained by solving Eqs.(9) and (12) simultaneously for βZ and Eqs.(10) and (13) for α_s through iterative procedure until the difference between the values gets minimized. These values are substituted in Eq.(8) to obtain $N_\pm$. By knowing the experimental values of q, Z, T, P and different attachment and recombination coefficients computed from the ion-aerosol balance equations, it is possible to estimate the values of N_0 and $N_\pm$ and hence, the values of σ_o, $\sigma_\pm$ and $\Delta\sigma_\pm$.

4. Results and discussion

In an experimental programme at Pune (18°32'N, 73°51'E, 559 m above mean sea level), India measurements of radon and its progeny, aerosol number density and atmospheric conductivities are measured. Using the ionization rate from radioactivity, ambient aerosol density and meteorological parameters such as temperature, pressure and relative humidity, the small ion concentration and hence the conductivity of the atmosphere is estimated from an Ion-aerosol model. Model computed conductivity of the atmosphere is used to validate the measured conductivity from a Gerdien condenser setup. Nagaraja et al. (2006) have described the experimental method for the measurement of ground level atmospheric electrical conductivity near the earth's surface and its validation using a simple Ion-aerosol model. The present model work is an extension of the earlier work for the estimation of percent reduction in conductivity for assumed higher ambient aerosol levels, which may be the result of increased pollution due to vehicular exhaust, industrial effluents, anthropogenic activity etc. Thus, monitoring of ground level atmospheric conductivity can be useful in air pollution studies (Nagaraja Kamsali et al., 2009).

The diurnal variation of radon, its progeny, aerosols, small ion concentration and conductivity for a typical day are described. Even though, the observations differ quantitatively, they show similar trends of variations over a day. Variability in the data of any of these parameters can be associated with different meteorological conditions. Moreover, simultaneous measurements of the diurnal variations of radon, its progeny, aerosols, conductivity, dry and wet bulb temperature, and wind speed and direction at the same observatory confirm that the day chosen is in no way electrically or meteorologically abnormal from other fair weather days. Thus, the observations described below can be reasonably generalized for other fair weather days as well. The results of collocated measurements of the parameters such as concentrations of radon and its progeny, ionization rate, aerosols, electrical conductivity and meteorological parameters such as temperature, relative humidity are examined for a typical day and are taken as representative.

4.1 Variation of Ionization rate due to radon and its progeny

The diurnal variations of activities of radon and its progeny near the earth's surface are depicted in Fig. 3(a). The radon concentration varies from 2.1 to 23.0 Bq m^{-3} with a median of 6.8 Bq m^{-3} showing a significant diurnal variation, whereas values of radon progeny vary from 0.5 to 6.4 Bq m^{-3} with a median of 2.0 Bq m^{-3}. It is seen from the figure that the concentrations are maxima during the early morning hours, generally between 0500 and 0700 hrs of Indian Standard Time (IST). The concentration decreases after sunrise, attaining minima during the afternoon, 1000 to 1800 hrs of IST. The variation of meteorological parameters such as temperature and relative humidity are shown in Fig. 3(b). It is observed that the concentration of radon and its progeny follow the trend of the relative humidity, in general. This is due to the fact that as temperature increases, the saturation vapour pressure increases so that the given air can take more water vapour. Consequently the relative humidity will decrease. The raise of temperature causes increased vertical mixing and raising of aerosols to the higher altitudes resulting in lower concentration of radon at the ground level (Hoppel et al., 1986). When the temperature decreases and relative humidity increases, the vertical mixing and raising of aerosols to the higher altitude reduces. As a consequence, the aerosol to which radon is attached, will be present at higher concentrations during night and in the early morning hours at ground level. This results in the increase of radon concentrations near the surface of the earth (Porstendörfer, 1994).

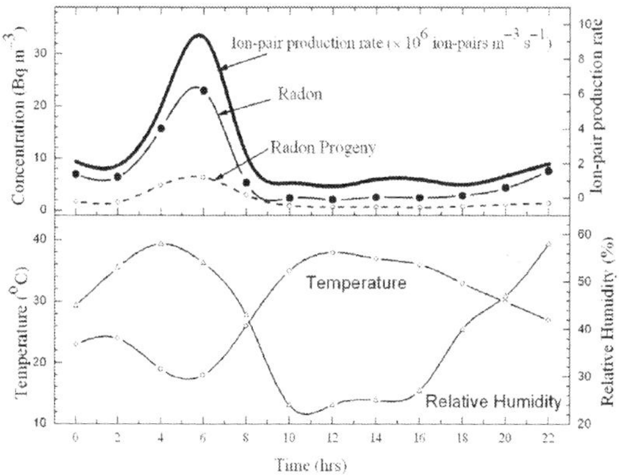

Fig. 3. Diurnal variation of a) radon, progeny and ion-pair production rate b) temperature and relative humidity

The ionization due to radioactivity also exhibits the diurnal variation as that of concentration of radon and its progeny and is shown in Fig. 3(a). At a height of 1 m above the surface ionization due to radioactive gases and their short-lived daughter products is predominantly caused by alpha particles. The rate of ionization due to radioactivity also exhibits the diurnal variation as that of concentration of radon and its progeny. The radon exhaled from the surface causes ionization of the atmosphere (Israel, 1970). It also shows a maximum in the early morning hours and a minimum in the afternoon (Dhanorkar and Kamra, 1994). The ion-pair production rate varies between 0.7 to 9.3 ion-pairs $cm^{-3}s^{-1}$ with a mean of 2.4 ion-pairs $cm^{-3}s^{-1}$. At night, ionization rates close to the ground are enhanced because of the accumulation of radioactive emanations from ground under temperature inversions and due to their lesser dispersion because of low winds (Hoppel et al., 1986). It also shows that the concentration of radon and its daughter products is low at noon when air is unstable and is highest during the night and early morning hours, when mixing is low. The radioactive emanations from the ground are trapped below inversions, and their accumulation causes an increase in ionization in the lower stable atmosphere during the nighttime. Large rates of ionization, produced due to these radioactive emanations in the presence of large aerosol concentration at night, produce large concentrations of ions of all categories. The accumulation of radioactive emanations will increase through the night until morning, when there are rapid changes in the stability of the atmosphere. In the morning, the vertical mixing caused due to increased eddy diffusivity will dilute the concentration of aerosols and radioactive emanations close to the ground. This may explain the occurrence of large concentrations at nighttime as compared to daytime (Dhanorkar and Kamra, 1993).

4.2 Variation of aerosol concentration and its size distribution

Figure 4(a) represents two hourly values of the aerosol concentration obtained for each of the eight size categories in the range of 13 to 750 nm. It shows a diurnal variation with the concentration showing a minimum during the early morning around 0600 hours IST and an increasing trend towards noon-hours. Early in the morning, due to anthropogenic activity and also due to the onset of convection resulting from the increase in atmospheric temperature, aerosols are pushed into the atmosphere resulting in an increased aerosol concentration (Israelsson et al., 1994; Sasikumar et al., 1995).

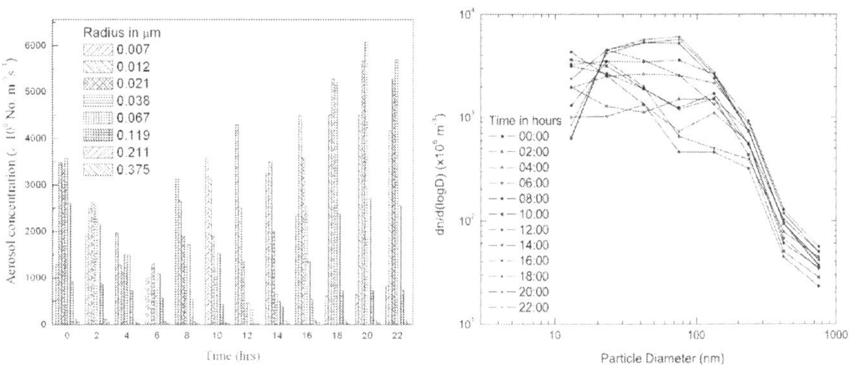

Fig. 4. Variation of aerosols a) over a day and b) its size distribution

The size distribution of aerosols is plotted in Fig. 4(b) are generally bimodal with their maxima at 75 and 23 nm during night (1800 to 0400 hrs). After 0400 hour with slight shift in maxima in the accumulation mode towards the higher size. During daytime, particularly in the afternoon, the shift in maxima in the accumulation mode to a higher diameter of 133 nm is distinct and the maxima in the nucleation mode seem to shift to smaller diameter. As a result of increase in the concentration of small particles, the size distribution curves during daytime are mostly open ended at the small particle side (Murugavel and Kamra, 1999). The magnitude of the peaks varies with time.

4.3 Variation of small ion concentration

The diurnal variation of estimated small ion concentration for a period of one week along with its mean values is shown in Fig. 5(a). Relatively a large day-to-day difference in the concentrations of ions that may arise from the difference in the accumulations of radioactive emanations and aerosol particles due to varying degrees of the lower atmospheric stability and/or due to varying intensities of advection. It is observed that the concentration show maxima in the early morning around 0600 hrs IST, it decreases after sunrise, and attains minimum in the afternoon. Radioactive emanations from ground are trapped below inversions, and their accumulations cause an increase in ionization and consequently increase in concentration in the lower atmosphere during the nighttime (Hoppel et al., 1986). Increased vertical mixing caused by increase in temperature and rising of aerosols to the higher altitudes results in the lower value of ionization rate at the ground level (Nagaraja et al., 2006). The concentration varies between 436 and 1663 ions cm^{-3} with a mean of 734 ions cm^{-3}.

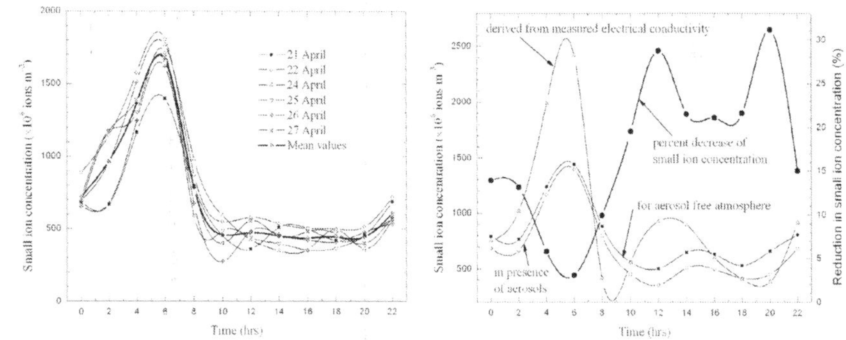

Fig. 5. Variation of a) small ion concentration b) small ion concentration in presence and absence of aerosols

The estimated small ion concentration shows a positive correlation with ionization rate due to surface radioactivity near to the Earth's surface. There is a correlation of 98.4% and suggests that the theoretical approach is best fitted and the correlation obtained indicates the correctness of ion-aerosol interaction considerations (Nagaraja Kamsali et al., 2009). Also shown in the figure is the estimated concentration of ions in the absence and in presence of aerosols using Eqs. (5) and (8), respectively. There is a fair agreement between theoretically estimated small ion concentration and experimentally derived values during daytime. However, there is a considerable difference during the morning hours. This difference is due to the large variability in measured conductivity. The percentage decrease in small ion concentration due to presence of aerosols is also shown in Fig. 5(b). The decrease is more during the noon-time when the aerosol concentration is high and less in the early morning when aerosol concentrations are minima. The maximum reduction in the estimated small ion concentration observed at around 2000 hrs coincides with the observed maximum concentration in aerosols. It is also seen that the existence of aerosols reduces the concentration and the percent reduction of small ion concentration is found to be minimum during early morning hours and starts increasing after sun rise and remains more or less same in the afternoon till sunset and reduces after words. This clearly shows that aerosols will have a control over the small ion concentration and hence the electrical structure of the atmosphere. A good correlation of 87.3% observed between the small ion concentration against derived values of small ion concentration from the measured electrical conductivity and suggests that the ion-aerosol balance equations considered are good enough to explain the interactions and are comparable with the experimental measurements of Dhanorkar and Kamra (1993).

4.4 Variation of conductivity

The diurnal behaviour of conductivity is found to be different at sites having different levels of natural radioactivity. Diurnal variations of experimental positive and negative ion conductivities for a continental station are shown in Fig. 6(a). The diurnal variation of conductivity of both polarities indicates that the pattern is representative of the regions. The positive and negative conductivities are approximately equal and their diurnal variations are generally mirror images to each other. It is observed that the conductivity of both polarities show maxima in the early morning hours between 0400 and 0700 hrs of IST, it decreases after sunrise, and attains minimum in the afternoon. This is due to the fact that the raise in temperature causes increased vertical mixing and rising of aerosols to the higher altitudes, which results in the lower value of ionization rate, and hence conductivity at the ground level (Dhanorkar et al., 1989; Nagaraja et al., 2006). The positive conductivity varies between 0.3 to 1.8 x 10^{-14} $\Omega^{-1}m^{-1}$ with a median of 0.8 x 10^{-14} $\Omega^{-1}m^{-1}$ and the negative conductivity varies between 0.2 to 2.0 x 10^{-14} $\Omega^{-1}m^{-1}$ with a median of 0.7 x 10^{-14} $\Omega^{-1}m^{-1}$.
During nighttime the atmosphere is relatively calm with low winds and little convective motion. The radon exhaled from the soil therefore accumulates near the ground and leads to increased ionization and higher conductivity. After sunrise, due to human activity and also due to the onset of convection resulting from the increase in atmospheric temperature, aerosols are pushed into the atmosphere. This causes a conversion of small ions, that are responsible for conductivity, into large ions through attachment, and an increase in the destruction of small ions through recombination with large ions of opposite polarity. The onset of circulation also removes radon from near the ground to higher altitude regions. These factors contribute to the observed reduction in conductivity in the afternoon (Subbaramu and Vohra, 1969). In the evening, with decreasing ground temperatures and also the anthropogenic activity, the aerosols that had been pushed to higher altitudes begin to settle down, and a greater fraction of small ions is lost through attachment. Conductivity therefore again starts raising after sunset. Finally, at nightfall, the aerosols settle down, and the conductivity recovers its normal night time high values. Since the major contribution to ionization at the surface comes from surface radioactivity and radon/thoron emanating from the soil, one would expect the diurnal variation pattern of conductivity to reflect that of the radon concentration near the surface. Radon and its short-lived daughters are understood to show a diurnal variation in concentration with a maximum early in the morning and a minimum in the afternoon (Nagaraja et al., 2003a). This is ascribed to the variations in the vertical mixing of air near the surface, which is controlled by atmospheric stability. As convection builds up with solar heating of the ground, radioactive gases are also transported upward, where they add to the ambient ionization. This leads to a reduction in the concentration of these gases, and consequently of ionization, near the surface (Nagaraja et al., 2003b). Therefore one should expect higher conductivities during

the period of atmospheric stability, with the values decreasing with the onset of turbulence (Sasikumar et al., 1995; Prasad et al., 2005). Wilkening and Romero (1981) reported the values of ionization rate, positive conductivity and negative conductivity, respectively, as 1.5 $cm^{-3}s^{-1}$, 1.4 x 10^{-14} $\Omega^{-1}m^{-1}$ and 1.4 x 10^{-14} $\Omega^{-1}m^{-1}$, for free atmosphere. These are of the same order as our measurements reported in this paper. The conductivity values presented here are of the same order of Dhanorkar and Kamra (1993). The total conductivity is the algebraic sum of the experimentally measured values of polar conductivities i.e., positive and negative conductivity. The variation of total conductivity is shown in Fig. 6(b). It varies between 0.6 to 3.9 x 10^{-14} $\Omega^{-1}m^{-1}$ with a median of 1.5 x 10^{-14} $\Omega^{-1}m^{-1}$. The increase of conductivity in the early morning hours is mainly because of the ionization produced by radioactive substances present in the atmosphere. The correlation coefficient between the ionization rate, obtained from the radioactivity, and the measured electrical conductivity is found to be 98% for a typical day and shows that the conductivity at the ground is mainly because of the ionization due to radioactivity near the surface.

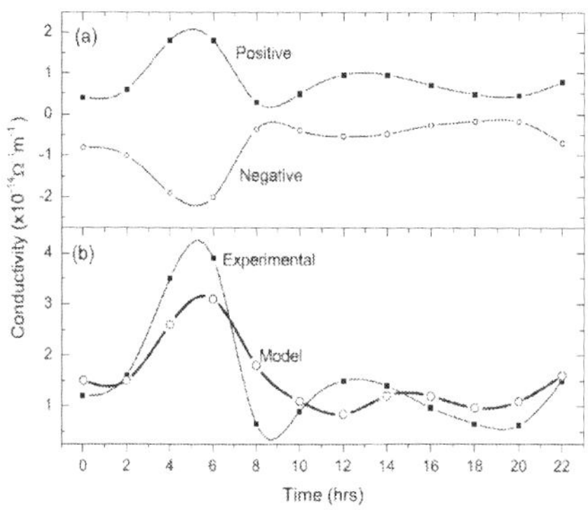

Fig. 6. Variation of a) positive and negative conductivity b) experimental and model conductivity

The experimental conductivity fluctuations in Fig. 6 is due to the similar variations in q which in turn is affected by the variations in meteorological parameters such as temperature, humidity, wind speed etc. The model of this study has reproduced well this trend of the diurnal variation in conductivity. It is seen that there is no exact one to one match between the experimental and model conductivities at certain points. However, there is a correlation coefficient of more than 90% between experimental and model derived conductivity values, which is an indication of correctness of the model results. In this study, the standard deviation in the measured ionization rates and aerosol concentration are both less than 10%. It is observed that the variations in the model conductivity values for the 10% variations in q and/or Z show variations of about 7% or less. It is evident that the simple Ion-aerosol model of this study has adequately reproduced the experimental conductivity.

4.5 Variation of conductivity under enhanced aerosols conditions

The conductivity of air is entirely due to small ions and depends upon the number of small ions present in the atmosphere. The conductivity decreases if the small ion concentration reduces. The small ion concentration is altered by the presence of aerosols due to attachment with aerosols, apart from the usual loss due to mutual recombination between the ions of opposite polarity. Hence, the conductivity alters due to the presence of aerosols in the atmosphere. Variation of conductivity due to enhanced (from the background) aerosols is plotted in Fig. 7(a). Three curves indicate that the conductivity decreases as the aerosol concentration increases. Fig. 7(b) shows the percent reduction in conductivity due to the presence of aerosols derived from the model for normal condition. In the early morning hours the concentration of aerosols is less and the conductivity is more (Houghton, 1985). Convective currents in the morning carry aerosol particles up from ground surface and the aerosols change the high mobility small ions into low mobility large ions. Thus the conductivity decreases after sunrise (Dhanorkar et al., 1989). As the day advances the aerosol concentration increases due to anthropogenic activities and conductivity decreases. The percent decrease in conductivity shows a diurnal variation with a minimum in the early morning hours when the aerosol concentration is minimum and reaches a maximum (30 - 32 %) in the afternoon when aerosol concentration also shows maximum. Due to pollution the concentration of aerosols increases and leads to decrease in conductivity of the atmosphere near the surface (Retalis et al., 1991; Dhanorkar and Kamra, 1993). Thus the measurement of atmospheric conductivity can be used as an index of pollution in the atmosphere. Also estimated from the Ion-Aerosol model is the reduction in small ion concentration and hence the conductivity of the atmosphere for varying levels of aerosols and as seen in Fig. 7(b). The results reveal that, for an increase of background aerosols by three-fold the percent reduction in conductivity is 7%, and for an increase of aerosols by six-fold the reduction in conductivity is 10% from the measured ambient aerosol level.

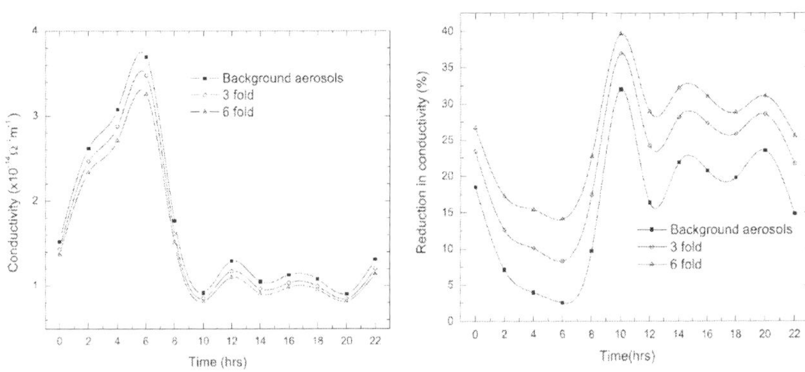

Fig. 7. Variation of conductivity a) in presence of aerosols and enhanced aerosols conditions b) percent decrease in conductivity for enhanced aerosol conditions

The diurnal, weekly and annual patterns of atmospheric electrical conductivity seem to be well correlated with the local aerosol air pollution patterns. The long-term decreasing trend seems to be affected by the air pollution build-up over metropolitan areas. Convective currents in the morning carry aerosol particles up into atmosphere; change the high mobility small ions into low mobility large ions. Thus the conductivity decreases after sunrise. As the day advances the aerosol concentration increases due to anthropogenic activities and conductivity decreases. The percent decrease in conductivity shows a diurnal variation with a minimum in the early morning hours when the aerosol concentration is minimum and reaches a maximum 32% for a typical day in the afternoon when aerosols also show maximum for background aerosols. Due to pollution the concentration of aerosols increases and leads to decrease in conductivity of the atmosphere near the surface. This gives an idea that the measurement of conductivity can be used as an index of pollution measurement. The result also reveals that, for an increase of aerosols by 3-fold the percent reduction in conductivity is 7%, and for an increase of aerosols by 6-fold the reduction in conductivity is 10% from the pollution free atmosphere.

5. Summary

The electrical conductivity of the atmosphere is mainly due to the radioactive substances in the earth's crust and their emanation near the surface. The radiations emitted by the radioactive substances ionize the atmospheric air and leads to conductivity. The conductivity gets altered because of the existence of aerosols in the atmosphere. A simple Ion aerosol model has been developed to deduce the conductivity theoretically by using ionization rate, aerosol concentration, and different attachment and recombination coefficients for the loss of small ions as input to the model. The model agrees well with the experimentally measured conductivity. In addition to this, it is seen from the model that the percent reduction in conductivity is 7% and 10% for an increase of ambient background aerosols by three-fold and six-fold, respectively.

6. Acknowledgements

The work was proposed under the financial support from the Department of Science and Technology (DST), Government of India. The authors (Nagaraja and Prasad) are thankful to the funding agency, and to Dr. A. K. Kamra and Dr. S. D. Pawar, Indian Institute of Tropical Meteorology, Pune, India for the fabrication of the Gerdien condenser and permitting us to carry out the experiments in Pune.

7. References

Chalmers, J.A. (1967). *Atmospheric Electricity*. Pergamon press, Oxford, London.

Cobb, W.E. (1973). Oceanic aerosol levels deduced from measurements of the electrical conductivity of the atmosphere, *Journal of Atmospheric Science*, 30, 101-106.

Cobb, W.E. & Wells, H.J. (1970). The electrical conductivity of oceanic air and its correlation to global atmospheric pollution. *Journal of Atmospheric Science*, 27, 814-819.

Datta, J., Chakravarty, S.C. & Mitra, A.P. (1987). A model for the cosmic ray produced ionization in the middle atmosphere. *Indian Journal of Radio and Space Physics*, 16, 257-266.

Dhanorkar, S. & Kamra, A. K. (1992). Relation between electrical conductivity and small ions in the presence of intermediate and large ions in the atmosphere, *Journal of Geophysical Research*, 97, 20345-20360.

Dhanorkar, S. & Kamra, A. K. (1993). Diurnal and seasonal variations of the small-, intermediate-, and large-ion concentrations and their contributions to polar conductivity, *Journal of Geophysical Research*, 98, 14895-14908.

Dhanorkar, S. & Kamra, A. K. (1994). Diurnal variation of ionization rate close to ground, *Journal of Geophysical Research*, 99, 18523-18526.

Dhanorkar, S. & Kamra, A. K. (1997). Calculation of electrical conductivity from ion-aerosol balance equations. *Journal of Geophysical Research*, 102, 30147-30159.

Dhanorkar, S., Deshpande, C. G. & Kamra, A. K. (1989). Observations of some atmospheric electrical parameters in the surface layer, *Atmospheric Environment*, 23, 839-841.

Gringel, W., Kaselan, K.H. & Muhleisen, R. (1978) Recombination rates of small ions and their attachment to aerosol particles. *Pure and Applied Geophysics*, 116, 1101–1113.

Hoppel, W.A. (1985). Ion-aerosol attachment coefficients, ion depletion, and the charge distribution on aerosols. *Journal of Geophysical Research*, 90, 5917-5923.

Hoppel, W.A., Anderson, R.V. & Willet, J.C. Atmospheric electricity in the planetary boundary layer, In: *The earth's electrical environment*, National Academy Press, Washington, D.C., USA, pp. 149-165, 1986.

Houghton, H.G. *Physical Meteorology*. MIT Publications, Cambridge, Massachusetts, England, 1–452, 1985.

Israel, H., *Atmospheric Electricity*, Vol.1, Israel Program for Scientific Translations, Jerusalem, 1-317, 1970.

Israelsson, S., Knudsen, E. & Anisimov, S.V. (1994). Vertical profiles of electrical conductivity in the lowermost part of the turbulent boundary layer over flat ground. *Journal of Atmospheric and Terrestrial Physics*, 56, 1545–1550.

Kamra, A.K. & Deshpande, C.G. (1995). Possible secular change and land-to ocean extension of air pollution from measurements of electrical conductivity over the Bay of Bengal. *Journal of Geophysical Research*, 100, 7105–7110, 1995.

Liu, B.Y.H. E. & Pui, D.Y.H. (1975). On the performance of the electrical aerosol analyzer. *Journal of Aerosol Science*, 6, 249-264.

Mani, A. & Huddar, B.B. (1972). Studies of the surface aerosols and their effect on atmospheric electric parameters. *Pure and Applied Geophysics*, 100, 154–166.

Misaki, M. & Takeuti, T. (1970). The extension of air pollution from land over ocean as related in the atmospheric electrical conductivity. *Journal of Meteorological Society of Japan*, 48, 263–269, 1970.

Misaki, M. (1964). Mobility spectrums of large ions in the New Mexico semidesert. *Journal of Geophysical Research*, 69, 3309–3318.

Morita, Y., Ishikawa, H. & Kanazawa, J. (1972). Atmospheric electrical conductivity measurements in the Pacific Ocean, exploring the background level of global pollution. *Journal of Meteorological Society of Japan*, 497–500, 1972.

Murugavel, P. & Kamra, A.K. (1999). Changes in the concentration and size-distribution of the sub-micron particles associated with the sea-and land-breezes at a coastal station. *Current Science*, 76, 994-997.

Nagaraja, K., Prasad, B. S. N., Madhava, M. S. & Paramesh, L., (2003a). Concentration of radon and its progeny near the surface of the earth at a continental station Pune (18 N, 74 E), *Indian Journal of Pure and Applied Physics,* 41, 562-569.

Nagaraja K., Prasad, B.S.N., Madhava, M.S., Chandrashekara, M.S., Paramesh, L., Sannappa, J., Pawar, S.D., Murugavel, P. & Kamra, A.K. (2003b). Radon and its short-lived progeny: Variations near the ground. *Radiation Measurement,* 36, 413-417.

Nagaraja K., Prasad, B.S.N., Srinivas, N. & Madhava, M.S. (2006). Electrical conductivity near the Earth's surface: Ion-aerosol model. *Journal of Meteorological Society of Japan,* 68, 757-768.

Nagaraja Kamsali., B.S.N. Prasad. & Jayati Datta. (2009). Atmospheric electrical conductivity measurements and modeling for application to air pollution studies. *Advances in Space Research,* 44, 1067-1078.

Porstendörfer, J. (1994). Properties and behaviour of radon and thoron and their decay products in the air, *Journal of Aerosol Science,* 25, 219-263.

Prasad, B.S.N., Nagaraja, K., Chandrashekara, M.S., Paramesh, L. & Madhava, M.S. (2005). Diurnal and seasonal variations of radioactivity and electrical conductivity near the surface for a continental location Mysore, India. *Atmospheric Research,* 76, 65-77.

Prasad, B.S.N., Srinivas, N. & Chandramma, S. (1991). A simplified ion-aerosol model for balloon measurements of ion conductivity and aerosol concentration. *Indian Journal of Pure and Applied Physics,* 20, 304-306.

Raghavayya, M. (1998). Modification of Kusnetz method for estimation of radon progeny concentration in air, *Radiation Protection and Environment,* 21, 127-132.

Retalis, D., Pitta, A. & Psallidas, P. (1991). The conductivity of the air and other electrical parameters in relation to meteorological elements and air pollution in Athens, *Meteorological Atmospheric Physics,* 46, 197-204.

Rosen, J.M. & Hofmann, D.J. (1981). Balloon-borne measurements of electrical conductivity, mobility and the recombination coefficients. *Journal of Geophysical Research,* 86, 7406–7410.

Sasikumar, V., Sampath, S., Muralidas, S. & Vijaykumar, K. (1995). Atmospheric electrical conductivity variations over different environments. *Geophysics Journal International,* 122, 89–96.

Smith, D. & Adams, N.G. (1982). Ionic recombination in the stratosphere. *Geophysical Research Letter,* 9, 1085-1087.

Srinivas, N. & Prasad, B.S.N. (1993). Seasonal and latitudinal variations of stratospheric small ion density and conductivity. *Indian Journal of Pure and Applied Physics,* 22, 122–127.

Srinivas, N. & Prasad, B.S.N. (1996). A detailed model study of stratospheric small ion density and conductivity. *Indian Journal of Pure and Applied Physics,* 25, 255-262.

Srinivas, N., Prasad, B.S.N. & Nagaraja, K. (2001). An ion-aerosol model study for the stratospheric conductivity under enhanced aerosol condition. *Indian Journal of Pure and Applied Physics,* 30, 31-35.

Subbaramu, M.C. & Vohra, K.G. (1969). Investigations on radioactive equilibrium in the lower atmosphere between radon and its short-lived decay products. *Tellus,* 21, 395–403.

Wilkening, M. & Romero, V. (1981). ^{222}Rn and atmospheric electrical parameters in the Carlsbad caverns. *Journal of Geophysical Research*, 86, 9911-9916.

21

Quick and Economic Spatial Assessment of Urban Air Quality

Panayotis C. Yannopoulos
University of Patras, Department of Civil Engineering, Environmental Engineering Laboratory
Greece

1. Introduction

The main sources of urban air pollution are transport and heating influenced by the background air pollution of the major area. Transport includes emissions from the road traffic, particle resuspension, petrol and gas stations, as well as emissions from ships and harbour traffic, in case of a city harbour. Heating includes emissions from oil or gas incinerators to provide central heating, which operate in houses and other buildings. Oncoming air pollution from any kind of emissions of the major area forms the background levels of the urban air pollution. These emissions may be originated from either anthropogenic or natural sources. Anthropogenic sources include activities of transport (airports, highways and ship emissions), power generation, oil refineries, waste incineration, industries and agriculture. Natural emissions come from area sources (sea, ocean, countryside land, forests) and point sources (volcanoes and territorial cracks). Air pollutants emitted from all these sectors include mainly the primary air pollutants of sulphur dioxide (SO_2), carbon monoxide (CO), oxides of nitrogen (NO_x), ammonia (NH_3), volatile organic compounds (VOC) and particulate matter (PM). Most of the traffic-related emissions are in the fine particulates range, i.e. $PM_{2.5}$ or of smaller size. Many of the primary air pollutants are deposited, but some of them interact with other air pollutants and form new secondary air pollutants, such as the photochemical air pollutants of ozone (O_3), aerosols, nitrogen dioxide (NO_2) and other. Although airborne particulates and ozone concern mostly human health, all aforementioned air pollutants (primary and secondary) affect adversely human health, ecosystems biodiversity, crops and forests, built environment, materials and cultural heritage. Besides the cost regarding general environmental damage, the death cost is huge. Several hundreds of thousands of premature deaths, increased hospital admissions, extra medication and millions of lost working days occur each year in Europe. Many developing countries, as China, are taking actions against air pollution. The Chinese policies to reduce SO_2 and NO_x emissions are similar to those of the European Union (EU). The developed countries, like USA and Japan, are not expected to cause any risk in competiveness to EU countries, because they have similar or more stringent air pollution policies than EU countries (Commission of the European Communities [CEC, 2005]). And of course there are other sources of potential pollutants, like noise causing annoyance to citizens, viruses causing human infections, and radioactive aerosols emitted due to probable failure episodes of nuclear reactors of power plants causing major impacts on major areas even globally

(Chernobyl of Ukraine in 1986 and, very recently in 2011, Fukushima of Japan). These pollutants need specific monitoring ways incompatible to the proposed methodology.
Cities of moderate or large population are facing air pollution problems and, therefore, a well organized monitoring program becomes mandatory to assess urban air quality and take decisions in the context of an air pollution abatement strategy. The first EU Council Directive regarding air pollution was the 80/779/EEC European Communities [EC, 1980]. That Directive in Article 6, partially kept in force up to 19 July 2001, had stated that SO_2 sampling stations might be located at sites where pollution is thought to be greatest and where the measured concentrations are representative of local conditions. The successor EU Council Directive 85/203/EEC (EC, 1985) in Annex III, which was kept in force up to 1 January 2010, gave among other specifications similar suggestions for the selection of NO_2 measurement points. A decade latter, the EU Council Directive 96/62/EC (EC, 1996) had enacted a common strategy on ambient air quality assessment and management, as well as measures and requirements to be adopted by Member States. The more recent EU Council Directive 1999/30/EC (EC, 1999), which has partially replaced the EC (1985), has established limit values and alert thresholds for concentrations of SO_2, nitrogen dioxide (NO_2) and NO_x, PM and lead (Pb) in ambient air. In addition, the latter EU Directive has specified common methods and criteria to assess concentrations for SO_2, NO_2 and NO_x, PM and Pb in ambient air and to obtain adequate information and maintain or improve ambient-air quality. It provides also lower and upper thresholds for determining the appropriate zones and the minimum number of sampling points for fixed measurement of the above pollutant concentrations. Interest is focused in zones where air pollution levels are higher than the lower thresholds or exceeding upper thresholds in contrast to zones where the levels are lower than the lower thresholds. The above directive forces Member States, which do not have representative measurements of the levels of pollutants for all zones, to undertake series of representative measurements, surveys or assessments in order to have the data available in time for implementation of the proposed legislation. For simplicity reasons, clarity and administrative efficiency, the five EU Directives 96/62/EC, 1999/30/EC, 2000/69/EC and 2002/3/EC, regarding air pollution issues have recently been replaced by the last EU Council Directive 2008/50/EC (EC, 2008). This Directive particularly seeks to promote the integration into the policies of the EU of a high level of environmental protection and the improvement of the environmental quality in accordance with the principle of sustainable development. This means that EC (2008) targets to reduce pollution to levels which minimise harmful effects on human health, paying particular attention to sensitive populations, and the environment as a whole, to improve the monitoring and assessment of air quality including the deposition of pollutants and to provide information to the public. Regarding urban air pollution monitoring, the Directive gives the criteria to establish an adequate network of pertinent monitoring stations installed at fixed locations where pollutants have to be measured, for taking either continuous measurements or measurements by random sampling. Managing a major city network, its high cost and need for reliable performance require convenient design and optimization based on a prior knowledge of the spatial and temporal concentration distributions of the city air pollution. Information from fixed measurements may be supplemented by modelling techniques and/or indicative measurements to enable spatial distribution of concentrations. In addition, the EC (2008) Directive allows the use of supplementary techniques of assessment, because they may reduce the required minimum number of fixed sampling points, contributing to economic monitoring networks and cost-effective air pollution abatement strategies.

In general, monitoring requires expert and assisting personnel, a network of rather permanent stations equipped by special analyzers and considerable funds on a yearly basis. Although such a network provides temporal concentration distributions of air pollutants at specified city locations, it is unable to give detailed spatial concentration distributions covering all the city areas. It is also unable to provide information on the apportionment of sources contribution to receptors. The latter information is necessary for taking appropriate measures in the application of the air pollution abatement strategy. However, knowledge of spatial concentration distributions of specific pollutants during rush hours may be necessary for the design of an optimised control of traffic (road and/or harbour traffic) and better the air quality of the city. It may be also important to obtain spatial concentration distributions for other atmospheric emissions coming from either urban sources (i.e. central heating) or rural and industrial sources; or it may be significant to identify the city locations and zones vulnerable to high pollution levels. Finally, short period measurements of spatial concentration distributions may contribute to the validation of models predicting urban air quality and assisting the apportionment of sources' contribution to the prescribed receptors.

This chapter describes an economic and quick methodology, which combines the determination of the spatial concentration distribution of air pollutants with suitable model predictions leading to source contribution for several urban receptors. For the determination of spatial concentration distribution, the bag sampling technique is used in conjunction with laboratory measurements performed by pertinent analysers (Yannopoulos, 2007a). The data obtained are then statistically analysed to get approximately spatial distribution of 1-h mean concentrations of specific air pollutants. For the model predictions of the same air pollutants, the USEPA integral dispersion models are employed (Yannopoulos, 2007b). In addition, this methodology may be employed to perform a preliminary spatial assessment of a city's air quality with the purpose to identify locations and zones susceptible to high pollution levels.

The advantages of the proposed methodology are the following:

a. Demand of limited personnel and small numbers of specialized sampling equipment and procedures. Only one analyzer per pollutant monitored may be used, since it involves the transfer of samples collected in the field to the laboratory for measurement and additional statistical treatment.
b. No need of permanent stations monitoring air pollution and, thus, flexibility in selecting the sampling sites.
c. Short sampling times in a rather restricted monitoring period lasting one to three months.
d. Satisfactory spatial coverage of urban air quality with measurements, which are adequate to further determine iso-concentration contours for each prescribed air pollutant.
e. Concentration results are characterized by mean probable errors less than around 25%.
f. Alternative usage of source emissions in predicting concentrations at the receptor locations and, thus, the ability to determine the contribution of each source. Note that this is unachievable by only using measurement data.
g. Possibility for a sporadic evaluation and harmonization of a network, which is used for continuous monitoring of the urban air quality, according to the needs of the air pollution abatement strategy of the city.
h. Possibility for prediction and assessment of the environmental impacts due to the city's development, which demands the installation of new infrastructure works or the

relocation of existing infrastructure (highways, bridges spanning rivers or sea narrow passages, underground railway, harbour, airport, industrial area e.t.c.).

While the proposed methodology has the following disadvantages:

a. The technique cannot be applied for all kinds of air pollutants, as for PM and aerosols that may sink during the time lag in the walls of the sampling bags. However, PM and aerosols are also difficult to predict using Gaussian models.
b. The time lag of air samples must be conveniently restricted and precautions must be taken, as the use of special bags (Teflon or Tedlar), depending on chemical properties of pollutants, and the daylight protection of samples, to avoid wall sinks and gas reactions. The conditions (a) and (b) bound the use of a laboratory for a prescribed city area satisfying limited time intervals to reduce the risk of pollutant alterations and reduction of their concentrations.
c. A small car in size or a monocycle must be available to enable visiting the city locations for air sampling and the quick transport of samples to the laboratory for immediate analysis.

The proposed methodology is applied for the air pollutant of SO_2. Mobile and stationary sources that use oil as a fuel (oil-engine cars, central heating and ships) constitute main sources of SO_2. For street level monitoring, SO_2 concentrations are mainly influenced by the traffic and therefore it is expected that the correlation coefficient should reflect the traffic-originated SO_2 emissions. In coastal cities where traffic presents important SO_2 emissions and the mixing height is reduced during rush hours due to the sea proximity, the street-level SO_2 concentrations monitored will be strongly correlated to traffic flow rates. This event obviously justifies the relatively high value 0.74 of the correlation coefficient reported by Jackson (2005). In contrast, in interior-land cities (without a harbour) the SO_2 concentrations are well correlated to particulate matter concentrations (Turalioğlu, 2005). Further on, cities are usually "hot spots" for certain atmospheric pollutants including SO_2 that may still affect large parts of forest ecosystems, particularly those under the immediate influence of pollutant sources (Smidt, 1998; Puxbaum et al., 1998; Herman et al., 2000). Therefore, it is interesting to give more light on spatial pollution levels in cities as the city sources may contribute significantly to increased regional levels. This implies that the reduction of emissions in city sources may effectively reduce the average of all emissions (Klemm and Lange, 1999).

2. Analytical description of methodology

The proposed methodology is a combination of the two alternatives already proposed by Yannopoulos (2007a, b), the air sampling bag alternative and the Gaussian modelling alternative. These are described step-by-step in the following.

2.1 Air sampling bag alternative

Plastic bags of various sizes (0.5 l to 100 l of volume) made by Tedlar, Teflon or aluminized Tedlar, have been used for air sampling, because they are simple in use and inexpensive. They can be easily cleaned by repeatedly filling the bag with pure nitrogen or zero air of high purity and evacuating it applying small under pressure. Leaving them filled with zero air overnight, they can be checked for leaks (Kumar & Viden, 2007). United States Environmental Protection Agency's [USEPA] Method 18 for the measurement of gaseous organic compound emissions by gas chromatography suggests the use of Tedlar bags

(USEPA, 1987). Some environmental agencies and specialized laboratories have established instrumented monitoring standards, which allow provision for bag sampling where direct measurement is difficult (Kennedy et al., 1995; USEPA, 1996; McGinley, 2002) or unsafe (Stewart, 1999). The air sampling bag method has been applied for both indoor and outdoor air quality measurements, regarding carbon dioxide (CO_2), methane (CH_4), non methane hydrocarbons (NMHC), ammonia (NH_3), nitrous oxide (N_2O), SO_2, NO, NO_2, CO and O_3 sampled with Tedlar sampling bags (Li et al., 2007; Lee, 1997; Lee and Chang, 2000; Li et al., 2001; Tsai et al., 2000). Although not valid in general, volatile organic compounds (VOC, Benzene) can be sampled and stored in Tedlar bags, because Tedlar bags present very little tendency to sorb organic compounds (Wang et al., 1996; Leong et al., 2002). Tedlar bags have been also used for sampling of odour characteristics and polycyclic aromatic hydrocarbons (PAH), analysing the samples within 24 h after sampling (Tsai et al., 2009). The losses from breath samples 52 h after filling the Tedlar bags were less than 10% (Steeghs et al., 2007). The losses from NO, NO_x and SO_2 air samples were about 15% 90 h after sampling, while CO didn't show any actual loss in Tedlar bags. The maximum losses have been observed from O_3 samples in Tedlar bags. They were more than 45% 90 h after sampling and more than 30% immediately after filling the bag (Chen, 2002). Alternatively, Teflon bags and sampling accessories are more inert than Tedlar, but they are more expensive and sensitive to punch and they need more care. Teflon bags can be used for air pollutant sampling and transport for analysis, in conjunction with all possible precautions taken to eliminate any potential errors in pollutant concentrations due to chemical transformations and wall sinks during the time lag in the sampling bags (Yannopoulos, 2007a; Kim et al., 2006).

2.1.1 Defining the study area

The first action to be made is the determination of the study area of the city. Information regarding population densities, sectors of the City Plan, iso-altitude contours, building heights, prevailing climatic conditions, location of air pollutant sources, dispersion conditions and previous measurements or indications of air pollution hot sports of the city under examination may be very useful in order to define the zones or agglomerations in conjunction with the legislation valid. It is worth to mention that each analyser in conjunction with the maximum time allowed from sampling and up to analysing the samples are the critical parameters to determine the size of the city zone, thus defining the general features of the area (number of maximum sampling points, personnel for sampling, necessary maximum time for the transportation of samples from the site to the laboratory for analysis).

2.1.2 Defining the project duration

The definition of the project timing and the whole project duration follows. This is based on available previous measurements, if any, or, if not, on measurements of similar air pollutants in other areas having similar sources, activities and climate. For primary air pollutants it is usual that the most convenient season is the winter period. This is because during the year the lowest mixing heights that worsen dispersion of air pollutants are observed in winter, while the highest mixing heights in the summer that favour dispersion and moderate heights are common during spring and autumn (Yannopoulos 2007b). Within the prescribed time interval of the year, the most susceptible hours for primary air pollutants are the morning hours or the evening hours of the working days, exempting the rainy workday hours. The determination of the project duration depends on the number of sampling points, the number of persons employed for simultaneous air sampling during each expedition and the statistically necessary revisits of each location for sampling in order to approximate hourly mean values. However, the maximum duration of the whole project should be less than the three months period, so as the project may be carried out entirely within a season of the year. If our purpose is the parallel run of a study of indicative measurements to a continuous monitoring network of stations, we have to design and apply four particular projects corresponding to the seasons of the year. According to the EC (2008) Directive, indicative measurements may have minimum time coverage 14% for air pollutants of SO_2, NO_2, NO_x, CO, benzene, $PM_{10}/PM_{2.5}$ and Pb, while more than 10% minimum time coverage during summer is required for O_3 and the related NO and NO_2. Therefore, the minimum duration of the whole project should be 13 working days for the winter months and 9 working days for the summer months, regarding the above air pollutants. A typically reasonable duration of a whole project of indicative air pollution measurements may be 1 to 2 months.

2.1.3 Defining the size and material of sampling bags

The size of the air sampling bags must be adequate to provide the necessary volume for analysing the sample by the analyser. The least volume of sample needed from the analyser to display the right value of the pollutant concentration depends on the rise time of the analyser and the pump discharge that the analyser uses. Therefore, the least sample volume can be calculated by multiplying the pump discharge by the rise time of the analyser employed. In case that more than one pollutant is going to be measured, the least sample volume must be the total sum of particular volumes needed for each analyser working independently. To save time during each sample analysis, all analysers have to be linked in parallel tubing with the bag. In such a case, the desired least volume should be calculated by multiplying the sum of discharges pumped from all analysers by the maximum rise time of the analysers employed. Indicative rise times of common analysers usually vary from 90 to 240 s, while their sampling rate varies from 0.2 to 1.0 l $minute^{-1}$. Sampling bags have to be filled up to 60-70% approximately for their protection by punch and leakage. This precaution is much more mandatory for Teflon bags than for Tedlar bags, because Teflon bags are more sensitive. Consequently, the least nominal capacity of each bag should be roughly 1.5 times larger than the volume of the sample needed for the analysis.

The bag material depends on the pollutant properties. For instance, even if a bag is made by a completely inert material, PM is adsorbed on the inner surface of bag in a more intensive way than a gaseous pollutant. O_3 is vulnerable to reactions with other air pollutants, thus it is reduced significantly during the time lag in the sampling bag, while CO indicates very good recovery rates when sampled in Tedlar bags. There are other air pollutants that may be absorbed by the wall material or may escape through the bag walls. All these restrictions bound the use of the bag sampling technique to only prescribed pollutants and bag materials.

Tedlar bags show low permeability in gases and they are chemically inert, causing minimal sample losses. They are also mechanically strong, reusable and cleanable. Teflon bags are ideal for unique applications, as they can withstand extreme temperature applications from -240°C up to 205°C. They are easily autoclavable and visually clear providing high transmittance to UV light. They are non-wettable, non-charring and showing no electrical tracking. They can be cleaned up easily and they have antistick and low frictional properties. They are physico-chemically inert, non contaminating samples, with low permeability to liquids, gases, moisture and organic vapours, as well as solvent resistant to virtually all chemicals, except molten alkali metals, gaseous fluorine, complex halogenated compounds at elevated pressure and temperature. In the market, the capacities available for Tedlar and Teflon bags range from 0.6 to 100 l. Teflon bags are more expensive than Tedlar bags.

For the requirements of an air pollution study of the present type, a concentration recovery rate of more than 85% may be acceptable for no more than 90 minutes time spent from the start of sampling and up to integrating the sample analysis, i.e. the required time for completing an expedition. Selecting bags made by the most inert materials (Tedlar or Teflon), the most common gasses that may be measured using the bag sampling technique are CO_2, CO, SO_2, NO, NO_2, N_2O, NH_3, CH_4, NMHC, total hydrocarbons (THC), as well as some VOCs, odorants and PAHs.

2.1.4 Defining the suitable point for sampling at each particular location

Normally, a typical city location for sampling is a cross road that has four corners, alike this one shown in Fig. 1. The suitable point mainly depends on the wind speed and direction. Thus, care will be paid in order to check for these factors before collecting the sample of each time. When the atmosphere is calm, these precautions are not too compulsory to follow, since the pollutants are diffused around with a uniform way. In case of a detectable local wind speed, the suitable point for carrying out sampling is the downwind point of the

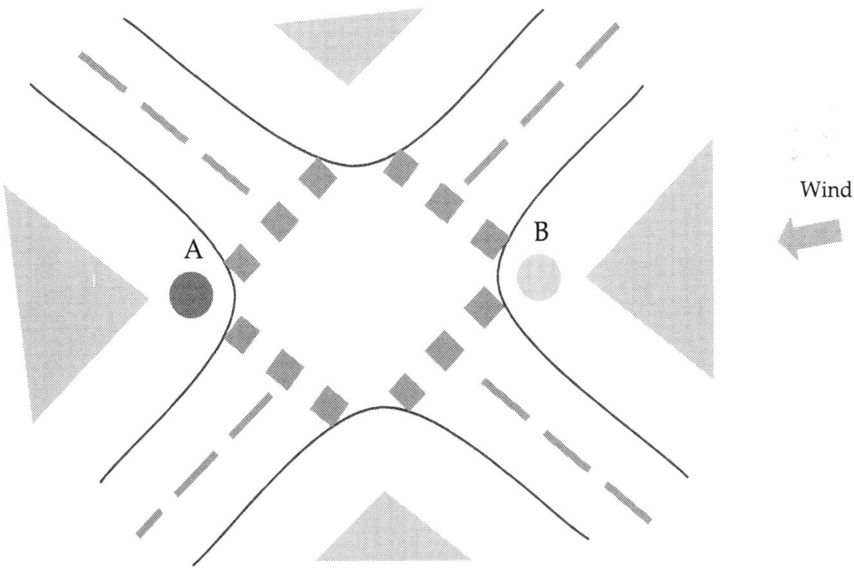

Fig. 1. A typical cross-road location where the downwind point A (red) and upwind point B (green) are shown

cross road with respect to traffic or other surrounding pollution sources. The data shown in Table 1 verify this argument. The results of two extra expeditions, including simultaneous sampling from two different positions of a junction, one upwind and another one downwind have been included in relation to the prevailing conditions. At each position five successive air samples were collected with 3-minutes duration per sample. Provided that the wind direction was detectable, data indicate that downwind points are mostly appropriate for peak concentration measurements, while this precaution becomes less important in a calm atmosphere. For instance, data given in Table 1 regarding Location 1 show that the wind conditions were calm to slight breeze with a rather well detectable direction. Thus, it is expected that the downwind point of the location would have higher impact from traffic than the upwind point, an event which is fully supported by the measurements presented in Table 1. In contrary, the measurements at Location 25 show approximately equal levels and this event is expectable due to unrecognised wind direction (calm atmosphere).

Point	A (red), Downwind						B (green), Upwind					
Sample No.	1	2	3	4	5	**Mean±σ_{n-1}**	1	2	3	4	5	**Mean±σ_{n-1}**
Location 1, Jan 7 1998, 10:27–10:45, clear sky, calm to slight breeze, traffic = 1658 vehicles/h												
SO_2	37	37	41	25	29	**34±7**	8	17	29	46	29	**26±14**
NO	326	237	455	227	268	**303±94**	62	75	72	102	265	**115±85**
NO_2	126	103	90	75	75	**94±22**	75	77	63	67	82	**73±8**
Location 25, Jan 7 1998, 12:20–12:40, clear sky, calm, traffic = 2011 vehicles/h												
SO_2	25	21	17	8	8	**16±8**	21	13	13	25	25	**19±6**
NO	266	175	291	137	131	**200±74**	168	40	26	195	203	**127±86**
NO_2	124	117	103	103	90	**108±13**	107	78	67	121	121	**99±25**

Table 1. Concentration measurements (μg m^{-3}) by simultaneous sampling at two points, one downwind (A) and another one upwind (B), taking five successive air samples of 3-minutes duration collected at each point

2.1.5 Estimating the maximum time lag of air samples in bags

For the pollutant under examination, estimation of the maximum time lag during its remaining in the bag should be made. The site locations in the city are initially selected. These locations should be distributed in areas, where the prescribed pollutant levels are expected to be moderate or high. The number of locations must be adequate for pertinent statistical analysis. Teflon or Tedlar bags of volume larger than the volume of the normal sampling are selected. The bags of 24×36-in size and 47-l maximum volume may be a good choice. Air sampling is performed following exactly the procedure for normal samples, but with a sampling duration of approximately 20 to 30 minutes, using an air pump convenient for sampling with such a flow rate as to fill the bag in the desired sampling duration time. Each sample collected in the bag is then transferred to the laboratory and measured without other delay. Then, the sample is preserved in its own bag enclosed in an opaque plastic sack in the room conditions. The measurement of the sample is repeated after two hours and then it may be useful to measure the sample after 12, 24, 48 and/or 72 hours. For all samples collected, the maximum losses per hour are calculated.

Having the possible recovery rate or the maximum rate of losses measured, the maximum allowed time lag for a smaller bag size may be estimated according to the following procedure (Yannopoulos, 2007a). It is considered that the wall losses are proportional to the internal wall surface of the sampling bag, S, or alternatively, the time lag is proportional to the air-sample height, h, defined as $h = V S^{-1}$, where V is the air volume sampled in the bag. For instance, taking that approximately 36-l and 3.6-l air volumes have been sampled using the 47-l and 4.7-l sampling bags, the corresponding air-sample heights are h = 3.23 and 1.94 cm. The hourly losses measured previously for 47-l bags correspond to losses during 36-minutes (1.94/3.23 h = 0.6 h) time lag for 4.7-l bags. Defining the allowed tolerance in recovery, which may be higher than 85%, it is then possible to determine the maximum time lag required for completing the entire process (sampling, transport and analysis) of each particular sample for normal sampling and analysis. Note that the time for the transport of samples depends on the distance between the city location and the laboratory in conjunction with the means for the transport. As mentioned previously, the use of small cars or motorcycles is preferable since they could be a good solution to minimise times, which may be critical to cost-effectively optimise the project area and site locations.

2.1.6 Estimating average air pollutant concentrations

The project may include measurements at the most critical hours of the day. According to previous experience described above, for measuring peaks of primary air pollution, the morning and evening hours from 08:00 to 10:00 and from 20:00 to 22:00, correspondingly, are the most suitable. The average concentration for all samples randomly picked at a particular location during morning or evening rush hours must be calculated separately for morning or evening hours and the average values are assigned initially at the nodes (sampling locations), using the point kriging method with a linear variogram model (Golden Software, 2002). All measurements used for the calculation of the above average concentration at each location have to pass the Q-test (Radojević and Bashkin, 1999). Thus, one grid matrix of average concentrations for each prescribed air pollutant for morning hours and another one for evening hours can be created. The next step is to filter each matrix using a low-pass linear convolution filter of 5-node averaging or, alternately, 9-node averaging with weights equal to 1 (moving average filter), as shown in Fig. 2. A number of four to five random measurements may be a good choice to carry out at each location, if a 9-node matrix for averaging is selected, while about the double random measurements are needed in case of selecting the 5-node averaging procedure for filtering. Thus, adopting the contribution of either 5-nodes or 9-nodes matrix in the calculation of the average value, which is assigned to the prescribed middle node of the group, the total number of samples is n, where $36 \leq n \leq 45$. After averaging the resulting matrices become smoother, as they are free of high frequency noise, and can be used to make the corresponding iso-concentration contour diagrams. The alternative use of Gaussian low-pass filter instead of the above will result in practically similar results with actually equal accuracy.

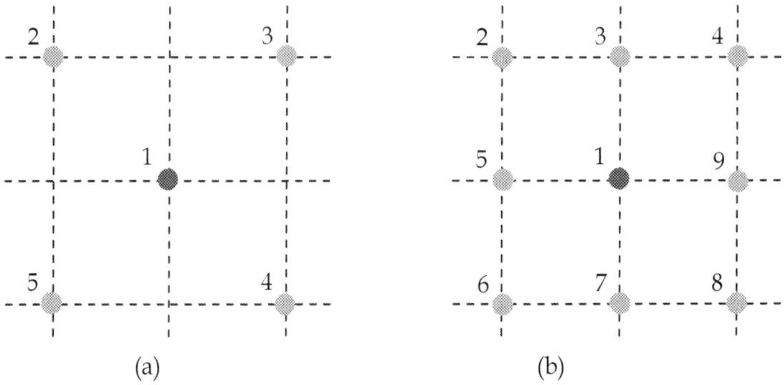

Fig. 2. Groups of matrix locations where averaging is applied: (a) Five-node matrix; (b) nine-node matrix

It can be shown that the final grid used above for contouring approximates mean concentrations of longer time interval at the specified locations. For this reason a confidence level 95% and the Student t-distribution are assumed. We define as probable error the percentage of the half-length of the confidence interval of the mean concentration for the specified level of confidence divided by the mean concentration at the particular location. This is expressed as a percentage of the mean concentration, i.e. $\pm 100 t_{n-1,\,a} n^{-1/2}$CV (%), where $t_{n-1,\,a}$ is the Student t-distribution parameter for n-1 degrees of freedom and a confidence level p; a=(1-p)/2; CV is the coefficient of variation defined as the standard deviation divided by the mean value. The mean probable error made for each final grid matrix of average concentrations can be calculated and tabulated. As we can see in Chapter 3, the maximum values of mean probable errors were around 25% and they depend on the parameter measured (Yannopoulos, 2007a).
The ratio of peak to mean concentrations at the ground level depends on the stability of the atmosphere and the type of terrain that the pollutant is passing over. Studies have also shown that the aforementioned ratio depends on the sampling time or averaging time, according to the following relationship (Turner, 1994; Schnelle and Dey, 2000)

$$C_t = C_s \left(\frac{s}{t} \right)^p \qquad (1)$$

where s, t are the shorter and longer averaging times (in minutes), for t longer than 10 minutes and up to 24 h; C_s, C_t are the mean concentrations for averaging times s, t; power p ranges from 0.17 to 0.75, but the most usual values are between 0.17 and 0.20 (Turner, 1994). Schnelle and Dey (2000) suggest the value p=0.17. Values of the ratio of peak to mean concentrations, C_t/C_s, for p=0.17 or 0.20, s=3 minutes and t=3 minutes, 10 minutes, 20 minutes, 1 h, 6 h and 24 h, are calculated using equation (1) and shown in Table 2. These values can be compared to the corresponding values coming from the measurements. For the locations designed, we get the maxima of concentrations for the typical sampling time and, for the average values calculated by applying the procedure described above, the ratio C_t/C_3 is computed. According to the procedure described above, the time t that corresponds to the common ratio C_t/C_3 reflects the approximate averaging time. As we can see below, this time approaches 1 h. In case that this procedure would give different averaging times than the desired time, the solution would be the revision of either the grid of sampling locations or the total number of visits for sampling at each location. For instance, making a denser grid, the averaging time would be decreased. The same result could be succeeded with the alternative procedure of collecting more samples randomly picked at each particular location during rush hours, with visits dispersed within workdays of the whole project period and using only 5-node averaging instead of 9-node averaging. A rarer grid of sampling locations, combined with increased number of nodes that participate in the group for the calculation of the average concentration of the middle node of the group, would lead to longer averaging times. Another way to get longer averaging times can be managed by increasing sampling time from 3 minutes to 5, 10, 15 or more minutes in relation to the desired result of the averaging time and simultaneously decreasing the total number of nodes or number of nodes participating in groups for the averaging procedure.

Averaging time or sampling time, t	C_t/C_3 (for p = 0.17)	C_t/C_3 (for p = 0.20)
3 minutes	1.000	1.000
10 minutes	0.815	0.786
20 minutes	0.724	0.684
30 minutes	0.676	0.631
1 h	0.601	0.549
6 h	0.443	0.384
24 h	0.350	0.291

Table 2. Values of the ratio of calculated concentration to 3-minutes concentration, C_t/C_3

If sampling intervals at the same location start at different times dispersed within the typical morning or evening sampling periods, it is expected that the aforementioned average concentrations will reflect mean levels of much greater time intervals than the 3 minutes, approaching the related typical morning or evening sampling periods. This assumption can be supported by additional expeditions organised within the program period, according to which simultaneous air sampling with 50-l and successive 5-l sampling bags at the same location must be performed, as indicated by the following case-study. Starting at the same time, the successive sampling of ten 5-l bags, each one filled with approximately 3.5-l air sample, lasted approximately 25 minutes, an equal time for a continuous collection of 30-40 l with a 50-l bag. The bag air samples were collected by four different locations of the central part of downtown Patras area and the concentrations measured for three air pollutants in regard to the conditions recorded are given in Table 3. The number of locations corresponds to that given in Fig. 5. A careful observation of data implies that the 25-minutes mean concentration (50-l bag) could be approximated fairly well by random picking of four to five 3-minutes concentration values (5-l bags).

Finally, the above procedure results in getting iso-concentration contour diagrams for average concentrations of a specified pollutant longer than the sampling time, within typical rush hours of the program period. These diagrams in conjunction with the probable errors encountered in iso-concentrations affect the determination of the boundaries of the high pollution zones and the mean displacements can be estimated.

Bag type	5 l											50 l
	1	2	3	4	5	6	7	8	9	10	**Mean**	**Mean**
	Location 2, Nov 26 1997, 10:35-11:00, cloudiness < 3/8, weak NE wind, traffic rate = 2682 veh h^{-1}											
SO_2	16	16	16	16	19	23	8	16	27	19	**17**	**24**
NO	192	202	172	134	201	157	137	134	242	193	**176**	**178**
NO_2	88	80	75	84	92	80	77	84	99	86	**85**	**104**
	Location 16, Dec 10 1997, 10:50-11:15, cloudiness < 3/8, calm, traffic rate = 1746 veh h^{-1}											
SO_2	8	8	12	32	12	16	20	20	8	4	**14**	**21**
NO	45	50	69	216	109	100	140	162	75	82	**105**	**100**
NO_2	61	67	71	134	78	73	82	84	67	61	**78**	**114**
	Location 22, Jan 14 1998, 10:45-11:10, cloudiness < 3/8, weak SW wind, traffic rate = 2328 veh h^{-1}											
SO_2	25	17	17	4	13	13	13	8	13	13	**13**	**20**
NO	147	82	106	24	55	21	49	17	64	40	**61**	**52**
NO_2	33	94	105	98	80	75	73	69	82	82	**79**	**106**
	Location 26, Jan 14 1998, 13:40-14:05, cloudiness < 3/8, weak SW wind, traffic rate = 1872 veh h^{-1}											
SO_2	8	8	8	4	4	8	8	8	4	13	**8**	**8**
NO	51	16	41	26	35	64	46	80	25	158	**54**	**25**
NO_2	77	59	55	65	55	75	75	78	59	94	**69**	**87**

Table 3. Concentration measurements (in μg m^{-3}) by simultaneous sampling at each selected location downtown of the city of Patras, using 10 successive 5-l and one 50-l Teflon air sampling bags

2.1.7 Correlating average air pollutant concentrations with traffic rates

Data regarding air pollution concentrations, measured within different rush-hour intervals of the project period, should present statistically insignificant concentration differences in order to be considered as belonging in a common group. If this statistical check is positive, then all concentration data can be considered suitable to be used for either calculating overall average pollutant concentrations or being correlated with traffic rates. Otherwise, the therapy would be to leave out concentrations measured at locations having a local effect that makes a different influence. Then, the rest data have to present statistically insignificant concentration differences and they can be used for the correlation with traffic rates. It must be noticed that data showing statistically insignificant concentration differences may be regarded as belonging in a common group and they can be used as an entity for the calculation overall average pollutant concentrations.

2.2 Gaussian modelling alternative

2.2.1 Gaussian model description

The USEPA Industrial Source Complex Long-Term model (ISCLT3) or/and Industrial Source Complex Sort-Term (ISCST3) model (http://sdi.odu.edu/air/isc3/) can be used to predict the dispersion of prescribed pollutants on a period basis in a city's atmosphere. These models provide options to model emissions from a wide range of sources. They use the straight-line, steady-state Gaussian sector-average plume equation for modelling pollutant emissions on either a long-term basis (month, season, and year) or a short-term basis (hour and day). Both models use the Gaussian plume equation with some modifications to account for simple point source emissions from stacks, emissions from stacks that experience the effects of aerodynamic downwash due to nearby buildings, isolated vents, multiple vents, storage piles, conveyor belts, and other sources of similar types. Emission sources are grouped into four basic types (point, volume, area, and open pit sources); the volume source and the area source may also be used to simulate line sources. The models are equipped with suitable algorithms to manipulate each source type and calculate dry or wet deposition for simple terrain (elevations below the release height). They use input meteorological data that have been summarized into joint frequencies of occurrence for particular wind speed classes, wind direction sectors, and stability categories (http://sdi.odu.edu/air/isc3/, USEPA 1995, Schnelle and Dey 2000, Turner 1994). In particular, the ISCLT3 model uses input meteorological data that have been summarized into joint frequencies of occurrence for particular wind speed classes, wind direction sectors, and stability categories. These summaries are called STAR summaries for STability ARray and may include frequency distributions over a monthly, seasonal or annual basis. The long term model can be used to calculate concentration or dry deposition values for each separate STAR summary input and/or for the combined period covered by all available STAR summaries. Since the wind direction input is the frequency of occurrence over a sector, with no information on the distribution of winds within the sector, the ISC long-term model uses a Gaussian sector-average plume equation as the basis for modelling pollutant emissions on a long-term basis. On the other hand, the ISCST3 model accepts hourly meteorological data records to define the conditions for plume rise, transport, diffusion, and deposition. The model estimates the concentration or deposition value for each source and receptor combination for each hour of input meteorology, and calculates user-selected short-term averages. For deposition values, either the dry deposition flux, the wet deposition flux, or the total deposition flux may be estimated. The total deposition flux is simply the sum of the dry and wet deposition fluxes at a particular receptor location. The user also has the option of selecting averages for the entire period of input meteorology.

The newer model systems AERMOD (http://www.epa.gov/scram001/dispersion_prefrec.htm) introduced by the American Meteorological Society/Environmental Protection Agency Regulatory Model Improvement Committee (AERMIC) can be also used to predict the dispersion of prescribed pollutants on a period basis in a city's atmosphere. The modelling system, AERMOD, has incorporated air dispersion based on planetary boundary layer turbulence structure and scaling concepts, including treatment of both surface and elevated sources, and both simple and complex terrain. There are two input data processors that are regulatory components of the AERMOD modelling system: AERMET, a meteorological data pre-processor that incorporates air dispersion based on planetary boundary layer turbulence structure and scaling concepts, and AERMAP, a terrain data pre-processor that incorporates complex terrain using USGS Digital Elevation Data. Other non-regulatory components of this system include: AERSCREEN, a screening version of AERMOD; AERSURFACE, a surface characteristics pre-processor, and BPIPPRIME, a multi-building dimensions program incorporating the GEP technical procedures for PRIME applications. The reader may visit the relevant USEPA site given above for further details on these model systems.

2.2.2 Defining model conditions

For the application of a model to predict the air quality levels on an area, the first action is the geographical definition of the area. Then, the characterisation of the area type (city, rural area, etc.), in conjunction with its morphological and physiological properties, as well as the area relations to the neighbouring areas and activities is required. It follows the collection of the necessary information from available sources regarding geographical data, meteorological data and information for all main sources of air pollution, which may affect the air quality of the area, either within or outside it. As the area that the model is going to be applied concerns a city, for the requirements of the present chapter, data regarding emissions from central heating, traffic and harbour are necessary. In many cases, it is important to carry out individual project studies in order to produce suitable data for the model input using the available raw information collected from the data sources. For instance, one may find available traffic rate measurements for some roads of the city in the municipal authorities, but they may realise that the data refer to periods other than the period of interest or they are not complete and they need to be completed and updated.

For air pollution studies, meteorological data concerning wind speed and direction are needed in order to group wind speed into categories. For the present work and concerning prescribed wind directions, the following seven classes of the wind speed measured in m s^{-1} are taken: WS-1 (0 - 1), WS-2 (1 - 2), WS-3 (2 - 3.5), WS-4 (3.5 - 5.5), WS-5 (5.5 - 6.5), WS-6 (6.5 - 8) and WS-7 (> 8). The number of prescribed wind directions may be at least four and at most sixteen depending on the study's requirements. Typical examples for classification and nomination of wind directions are shown in Fig. 3.

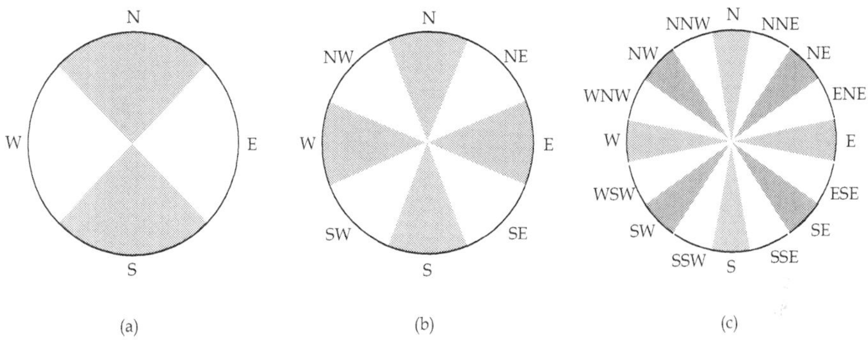

Fig. 3. Classification of wind directions in: (a) four sectors; (b) eight sectors; (c) sixteen sectors

The simplest four directions classification utilises the following sectors of equal angles, measured clockwise from the north direction in deg: N (-45 to 45), E (45 to 135), S (135 to 225) and W (225 to 315). The usual eight divisions for the classification of the wind direction include the following sectors: N (-22.5 to 22.5), NE (22.5 to 67.5), E (67.5 to 112.5), SE (112.5 to 157.5), S (157.5 to 202.5), SW (202.5 to 247.5), W (247.5 to 292.5) and NW (292.5 to 337.5). The most complete classification is in sixteen sectors as: N (-11.25 to 11.25), NNE (11.25 to 33.75), ENE (33.75 to 56.25), E (56.25 to 78.75), ESE (101.25 to 123.75), SE (123.75 to 146.25), SSE (146.25 to 168.75), S (168.75 to 191.25), SSW (191.25 to 213.75), SW (213.75 to 236.25), WSW (236.25 to 258.75), W (258.75 to 281.25), WNW (281.25 to 303.75), NW (303.75 to 326.25) and NNW (326.25 to 348.75).

In general, it must be noticed that available raw data regarding wind information for extended periods (years or decades) are very useful for the production of wind rose diagrams. These diagrams provide useful information for deciding and creating pertinent scenarios of prevailing winds, when air quality in adverse atmospheric conditions is claimed to predict. A very simple wind rose diagram is shown in Fig. 4. This diagram concerns a special period, which may be a month, year or any other defined period, and shows the wind occurrences per direction (thick line) for wind speeds within a prescribed range. Wind occurrences are given in frequency percentages 0, 5, 10 ... (%), which include calm occurrences shown in red colour in the rose centre.

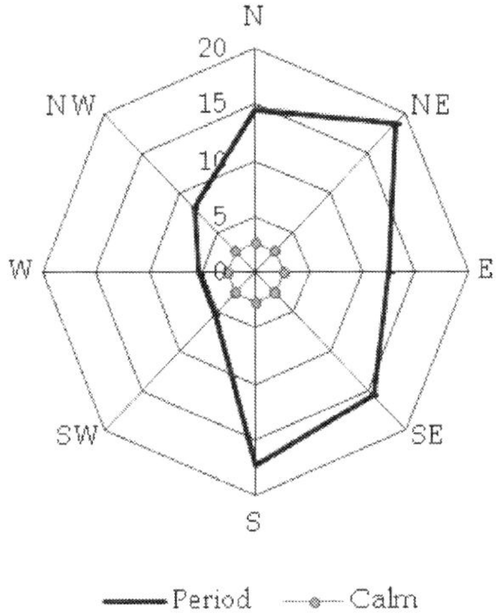

Fig. 4. A simple wind rose diagram showing the wind frequencies per direction for a prescribed speed range of winds

Another important magnitude to be defined is the atmospheric stability, which depends mainly on the convective and mechanical eddies. Classification is based on the assumption that stability at the near ground layers depends on the incoming solar radiation (insolation), which contributes to the production of convective eddies, and on wind speed, mainly responsible for the production of mechanical eddies. The high levels of incoming solar radiation in conjunction with low wind speed occurrences contribute to unstable atmospheric conditions. Conditions for a neutral atmosphere are probable when the sky is cloudy or with occurrences of high speed winds. The stable conditions occur when the net radiation index (NRI) gets negative values. The classification system followed herein is the Turner' system, which is based on hourly meteorological observations and Pasquill-Gifford's work (Turner 1994; Schnelle and Dey 2000). According to Schnelle and Dey (2000), Turner has classified atmospheric stability into the following seven categories:

1. Extremely unstable
2. Unstable
3. Slightly unstable
4. Neutral
5. Slightly stable

6. Stable
7. Extremely stable

The determination of the stability class is enabled by data given in Table 4, where the stability class number is given as a function of the class of the wind speed and the NRI. Schnelle and Dey (2000) has given some recipes regarding the correct choice of the stability class. The most important criteria are the following:

- Irrespective of day or night for complete cloudy sky (10/10) and ceiling less than 2000 m, the zero value for the net radiation index is suggested, NRI = 0.
- For night hours and small cloudy sky occurrences (less than 4/10), the value of NRI = -2 is recommended, while NRI = -1 for heavier cloudy sky occurrences (more than 4/10).
- For day hours, NRI is related to the insolation class number (ICN), which indicates the incoming net solar radiation. ICN depends on the angle that solar rays fall on the location of interest, called solar altitude a, as well as for the atmospheric conditions affecting insolation. The recommended values of ICN are:
  ICN = 1 for $a < 15$ deg (weak insolation)
  ICN = 2 for $15 < a < 35$ deg (slight insolation)
  ICN = 3 for $35 < a < 60$ deg (moderate insolation)
  ICN = 4 for $a > 60$ deg (strong insolation)
- For day hours and small to moderate cloudy sky occurrences (less than 5/10), the recommendation is to consider equality between NRI and ICN. Thus, for prescribed daylight conditions, the value of ICN can be determined as above and consequently the NRI value. Then, the value of Atmospheric Stability Class (ASC) can be determined from the data given in Table 4, using NRI = ICN and the class of the prevailing wind speed.
- For day hours and heavier cloudy sky occurrences (more than 5/10), NRI is determined by a modified ICN according to the procedure: NRI = ICN - 2 for ceiling less than 2000 m; NRI = ICN - 1 for ceiling between 2000 m and 5000 m; NRI = ICN - 1 for complete cloudy sky and ceiling higher than 2000 m.
- In cases when the modified ICN is found less than 1, then the value NRI = 1 is recommended.

Wind speed (WS) class	Net radiation index (NRI) of the area to the sunlight exposure (insolation)*						
	4 (strong)	3 (moderate)	2 (slight)	1 (weak)	0	-1	-2
WS-1	1	1	2	3	4	6	7
WS-2	1	2	2	3	4	6	7
WS-3	1	2	3	4	4	5	6
WS-4	2	3	3	4	4	4	5
WS-5	3	3	4	4	4	4	4
WS-6	3	4	4	4	4	4	4
WS-7	3	4	4	4	4	4	4

* Positive values indicate that radiation is directed toward the ground, while negative values indicate radiation directed away from the ground.

Table 4. Atmospheric stability class (ASC) as a function of the wind speed class and the net radiation index

It must be noticed that data in Table 4 have taken into consideration the effect of retaining heat by buildings and pavements after sunset. During the night hours this heat is re-radiated away from the earth. Due to more reflection conditions met in cities compared to rural areas, the city surfaces and buildings become warmer than the surfaces in rural areas and, therefore, they help the production of more convective eddies affecting ASC.

2.2.3 Determining source characteristics

The most significant sources for the study have to be defined. A source classification may help the knowledge of expected characteristics, using literature. Data regarding source location, emission inventories and other related information are very useful for producing emission loads or updating them, according to the prescribed procedure of the model. Always a pertinent preparation of data for the model input is requisite. It is needed either a slight data treatment or severe manipulation of data to become suitable for the model input. There are cases that input data has to be constructed by existing raw measurements. For instance, the central heating and ship emissions can be modelled as area sources, while traffic as a line source divided in segments of a prescribed length.
The calculation of the emission from each source category is based on the legislation in force and existing data. As an example, according to the USEPA data, the SO_2 central heating emissions are considered to be 852 mg l^{-1} per fuel mass consumed and varied with respect to the sector of the City Plan. The traffic emissions in a European country can be determined using the Computer Programme to calculate Emissions from Road Transport [COPERTIII] of the European Environmental Agency (http://vergina.eng.auth.gr/mech/lat/copert/copert.htm) or a newer version, in conjunction with the fuel quality characteristics for the period of interest. Traffic loads (cars per day or month), as well as the car types (conventional, catalytic passenger cars, light-duty and heavy-duty vehicles) and their percentages must be determined for the city's main streets and probably harbour and other significant areas. Both the average car speed and the average trip distance regarding the particular urban areas have to be defined. In addition, the daily or monthly ship and vehicle emissions due to a probable harbour, affecting the air quality of the city areas under examination, have to be calculated according to the model requirements. In most of cases, the raw data regarding traffic, central heating and harbour need laborious treatment efforts in order to create the suitable data for the model input.

2.2.4 Locating receptors

The city locations of special interest (downtown central points, traffic-jam street points, cross-roads with high traffic loads, parking lots, central squares etc.) may be defined as receptors. These locations may be any suspected points or points indicated by previous studies for probable high pollution levels. The locations, where measurements are available to be used for model validation or comparisons, should be also included in the list of receptors. Finally, receptors may be all locations where output values have to be provided by the model, for either direct comparison with existing measurements or plotting iso-concentration contour diagrams that can be used for several types of studies of the future development of the city.

3. Application and evaluation

3.1 The case-study area

The central part of the city of Patras is selected for the application of the methodology proposed. Patras, the third largest city of Greece, after Athens and Thessaloniki, is built on the NW coast of Peloponnese and today has a population of around 200,000 inhabitants. The population is further increased during summertime, festivals and working days, because of tourists and other visitors. Patras is the western gate of the country with a busy harbour due to many links to Ionian Greek Islands and Italy. The study area includes Sectors A, B, C, D, E and F of the City Plan, as shown in Fig. 5, which gives a general view of Patras highlighting areas of interest. The City Plan Sectors are included between the streets described in Table 5, where some important information is also given. For simplicity reasons, the town planning sectors included in the area of the present study have been grouped in regard to their common characteristics given in the City Plan of Patras: (a) Allowed Building Coefficient (ABC) defined as the area of all floors of a building that is allowed to be built, divided to the area of plot; (b) Allowed Cover Coefficient (ACC) defined as the maximum area of plot that is allowed to be covered by a building, divided to the area of plot. In addition, the storey number of buildings, an estimated range of building heights and useful remarks attaining in each sector part are given. Up to 1970's the coastal zone of Sectors A and D, in Akti Dymeon Street south of Papaflessa Street, constituted the old

City Plan Sector	ABC[a]	ACC[b]	Building Height (m)	Short description
A	2.4	0.70	15–28	Flat area, with most of the old buildings being rehabilitated. It has intensive traffic and business activities especially in main streets.
B	2.1	0.70	8–12	It includes the old flee Market and the area around the Odeum, built during the Roman Colony. The traffic and business activities are considerable mainly in the streets of boundaries.
C	1.6	0.60	8–21	Both parts are built in flat areas and there are intensive business activities and traffic, mainly in main streets.
D	1.2	0.70	8-24	It has parts of small plots in hilly old areas built rather disorderly and parts with plots having gardens of normal or big size. It has also a downgraded part with normal or big size plots and gardens. The traffic rates are intensive in main streets and low to moderate in secondary roads. The new harbour, which extends west of Sector D, is going soon to be in operation.
E	0.8	0.70	4–12	It is built on a hill having steep slopes and small plots, but it has a part built on a flat area with normal or big size plots and gardens. The traffic rates are low to moderate.
F	0.6			A small hilly area presenting moderate to high traffic rates in Alsiliou Str.

[a] Allowed Building Coefficient; it denotes the area of all floors of a building that is allowed to be built, divided by the plot area.
[b] Allowed Cover Coefficient; it denotes the maximum area of plot allowed to be covered by a building, divided by the plot area.

Table 5. Important details and short description of the City Plan Sectors shown with similar colours as in Fig. 5

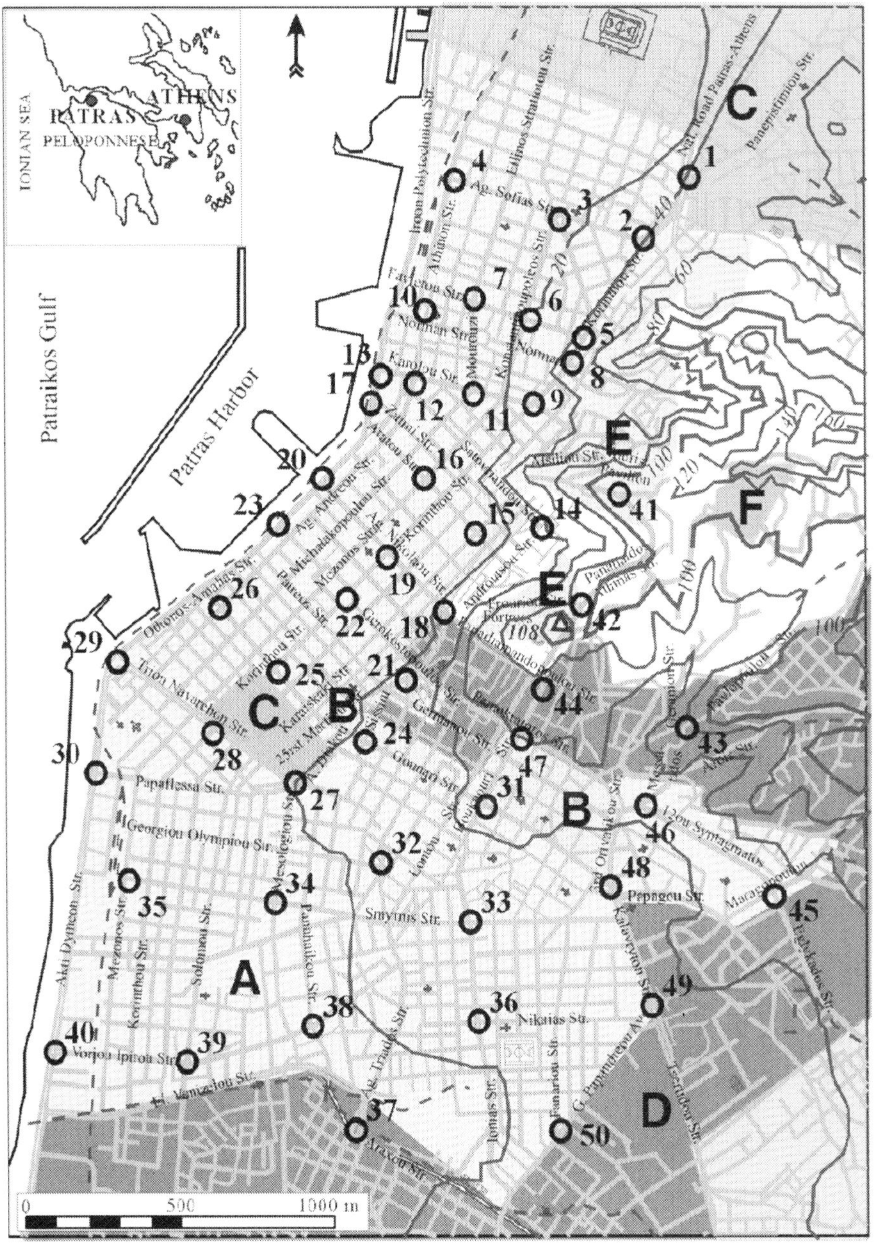

Fig. 5. A map showing interesting Patras areas, the City Plan Sectors A, B, C, D, E, F and the sampling locations 1-50

industrial area of Patras, but last years the industries - one by another - either ceased to operate or were removed in the new industrial area located at a distance of 17 km approximately towards SW of Patras. In now a day, some of the abandoned buildings and industrial structures have been modified to house municipal authorities or offer modern entertainment and shopping.

The marine Mediterranean climate is prevailing in the major Patras area, having relatively high humidity (annual mean 67%) and prolonged and sunny warm periods. As shown in Fig. 6, the NE and SW winds prevail during the actual program period with comparable frequencies (Yannopoulos and Skokaki, 2003). Frequencies (%) and directions of prevailing winds concern the period 1955-1988 and have been recorded by the station of the Hellenic National Meteorological Service (HNMS) at Akti Dymaion Str. of Patras. In addition, during the program periods, air temperatures varied between -6 and 31°C, with monthly mean values from 10 to 15°C. The annual mean precipitation is approximately 747 mm, with 52% occurrences during November, December and January; during these months, the monthly mean relative humidity is around 72 to 73%. During the program period a mean number of 45 cloudy days over 199 annually corresponds. Tables 6 and 7 include local climatic

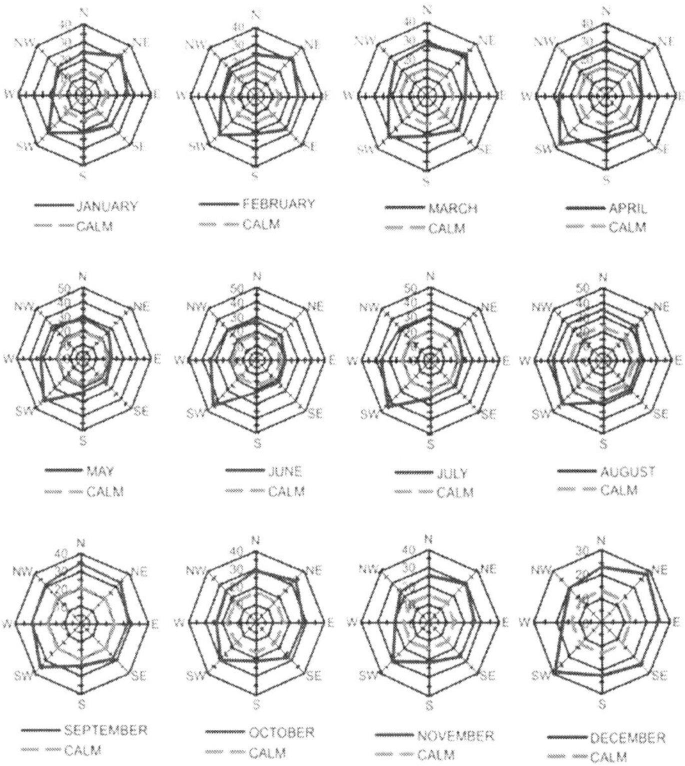

Fig. 6. Wind rose diagrams for Patras during months of a typical year

conditions recorded during the sampling period. Wind data in Table 7 show that calm or light winds were the most frequent weather conditions during sampling hours. Winds with very low velocity and rather detectable direction constitute suitable conditions for air sampling, since it is possible to define the downwind point for sampling performance. As for many other Mediterranean cities of moderate size and similar activities with Patras and apart of sporadic emissions due to industrial and service enterprises and background pollution due to probable remote sources, air pollution in Patras originates from the classic sources of harbour, traffic and central heating. More details and a thorough review for the actual air quality data for the city of Patras have been given by Yannopoulos and Skokaki (2003), Yannopoulos (2008) and Bloutsos & Yannopoulos (2011).

Month	Dates with cloudiness occurrences	Dates with rainfall occurrences
Nov 1997	13^{e}, 14, 17, 18, 19, 20^{e}, 21^{e}, 24^{m}, 25^{e}, 26, 27me, 28	13, 17, 18, 19, 20, 21, 24
Dec 1997	2, 4, 5, 8, 10, 11, 12, 16^{m}, 19me, 22^{m}	2, 4, 12, 16, 19, 22
Jan 1998	13^{e}, 14, 15, 16^{e}, 20, 22^{e}, 23	20, 23

Note: Superscript m denotes cloudiness more than 3/8 in the morning- and e in the evening-sampling period. The absence of an index denotes cloudiness less than 3/8 but not clear sky.

Table 6. Dates with cloudiness and rainfall occurrences during the project period

Date	Wind condition 08:30-09:30	Wind condition 20:30-21:30	Date	Wind condition 08:30-09:30	Wind condition 20:30-21:30	Date	Wind condition 08:30-09:30	Wind condition 20:30-21:30
Nov 13		S^{M}	Dec 2		NE	Jan 7	Calm	
14	SW	S	4		SW^{M}	8	Calm	Calm
17	Calm or S		5	Calm	NE	9	Calm	NE
18	Calm or N	Calm or NE	8	NE		12	Calm	
19	Calm or N		9	NE	Calm	13	Calm	Calm
20		Calm	10	Calm		14	Calm	
21		Calm	11	Calm	NE	15		Calm
24	NE^{G}		12	SE	SE	16		Calm
25	NE	NE	16	Calm	Calm	20		W
26	NE		19	Calm	NE^{M}	22		Calm
27	Calm	NE	22	SW^{M}		23		Calm
28	NE	NE						

Note: Superscript M denotes moderate- and G Gusty-wind speed. Index absence denotes slow wind, nearly calm.

Table 7. Local weather conditions prevailing during morning and evening sampling periods

3.2 Site dispersion conditions

Pollution rates and dispersion conditions present spatial and temporal differences. Dispersion conditions depend upon local and temporal characteristics and a measure of dispersion quality is the mixing height. In general, the mixing height over land is low during the morning, evening and night hours daily, when inversion effects occur, and increases significantly during the early afternoon hours (Schnelle and Dey, 2000). An indicator of its daily persistence is the percentage of calm frequency, which is 14% during program period and increases gradually from winter to summertime (Yannopoulos and Skokaki, 2003). Annually, the lowest mixing heights giving the worst dispersion conditions are observed in winter (usually in December), the highest with the best dispersion in summer (usually in July) and moderate values are common during spring and autumn. The diurnal variation of the mixing height depends mainly on the hour of day, the ground level conditions (roughness, thermal capacitance, reflection, radiation and altitude) and the weather conditions (cloudiness, prevailing winds, relative humidity e.t.c.). Coastal sites are more influenced by the higher thermal capacitance and the lower roughness height of the seawater than any other interior locations. Computations of the spatial variation of the mixing height over the whole Europe with land-sea distinction have shown that the mixing height becomes high (around 750 m) over Ionian Sea and small (around 50 m) over the land of Greece after midnight at 04:00 UT, while the opposite holds true in the afternoon at 16:00 UT (Delobbe et al., 2001). For Patras, a coastal city adjacent to Ionian Sea, it is evident that sea proximity will influence the actual mixing heights, which, over the sampling locations, are expected to increase with the distance from the shoreline (Hanna, 1987). Thus, in the typical hour of morning samplings, a small enough mixing height (around 50 m) prevailed over the sea and a higher one over the interior land, in contrast to the typical hour of evening samplings when the mixing heights over the sampling locations were decreased with the distance from the sea, up to a distance of ten times the interior land mixing height, i.e. around 500 m. Consequently, in the mornings, an improvement of the dispersion conditions occurred from sea towards interior locations, while the opposite in the evening (Yannopoulos, 2007b). It is evident that the influence of the harbour and the city emissions should be more important for sites located downwind. Therefore in the mornings, when breeze blows from sea towards land, a higher influence is expected to occur in the city locations downwind of the harbour, and a lower or none influence in the evenings when wind direction is reversed. Note that calm or winds of very low speed and rather detectable direction (light winds) were the most frequent during sampling hours, as shown in Table 7.

3.3 Air sampling bag alternative

3.3.1 Design characteristics of the study area

The study area of Patras has been defined to include the most critical parts of the city centre and its surroundings, as it is shown in Fig. 5. A number of 50 locations cover the most significant cross-road points and they are distributed over the area of interest as shown by numbered open circles. The program of air sampling and analysis started on the 13rd of November 1997 and finished on the 23rd of January 1998, in such a way to meet monthly peaks for primary air pollutant concentrations. It included 24 morning expeditions for sampling and an equal number of evening expeditions. Requiring getting nearly the critical situation for primary air pollutant concentrations, as for SO_2, morning and evening samplings were scheduled to be carried out during open market hours on working days, between 08:30 to 09:30 and 20:30 to 21:30, correspondingly. These hours of day were considered appropriate since SO_2 emissions are maximized due to peak-hour activities in conjunction with the low mixing heights (systematic daily radiation inversions). In order to minimise the duration of each expedition, a couple of students were used simultaneously, each one collecting 5 samples of short duration, of 3 minutes time approximately. Thus, ten air bag samples were normally collected, covering equal number of predefined city locations.

Sampling locations are shown in Fig. 5, where the City Plan Sectors A, B, C, D, E and F have been also plotted. The main characteristics of these sectors useful for model validations are given in Table 5 and Section 3.1.

3.3.2 Sampling equipment and procedure

Each air sample had a volume of about 3.5 to 4 l and was collected in a 5-l Teflon air-sampling bag (type 12×12in on/off) by using a small volume pump (model PM11/2 12VDC STD, Charles Austen Pumps Ltd.), which was operating for about 3 minutes. A small glass funnel (approximately 5-cm face diameter) fixed to face in front the pump followed by its standard filter resulted the pump intake. Sampling was performed 0.80 m above ground level on foot pavement, few meters far from the nearest kerbside and obstacles, depending on actual space availability. The knowledge of pollution levels at this altitude may be significant for assessing health effects on drivers and children since their nose is almost at 0.80 to 1.00 m above ground. Each sampling process was associated by a 3-minutes traffic recording and the recordings of useful local conditions (start time of sampling, wind and weather conditions as cloudiness and rainfall occurrences). Although not always feasible, care was paid for performing expeditions in nearly calm weather conditions and without drizzling or raining. The actual wind direction was considered in selecting the correct downwind position on the road junction with respect to the surrounding pollution sources, as justified above performing extra expeditions and the results are given in Tables 1 and 3.
For a typical expedition, the average time spent for the movement to the next sampling location was normally around 2 to 3 minutes and consequently the time needed to integrate 5-bags sampling was approximately 20 minutes. Finally, to protect bags against potential puncture and direct sunlight exposure and eliminate photochemical reactions and other sink losses on the sampling bag walls, the bags were placed immediately in thermally insulated and dark plastic sack. Then, they were transferred in the Environmental Engineering Laboratory (EEL) for the subsequent analysis without delay. A mean travel time of about 15 minutes was needed. The change of the pollutant concentration with time lag occurring in sampling bags was investigated. For seven locations distributed in downtown Patras with moderate to high pollution levels, specific air sampling using 50-l Teflon bags was made following exactly the procedure for normal samples. Subsequent analysis showed that negligible time lag effects encountered on the concentration measurement. The maximum SO_2 losses per hour were less than 0.5% and, since the maximum time lag required for completing the sampling and analysis process of each particular 5-l sample was always less than ¾ h, the procedure was proved appropriate. Note that each 50-l sample preserved in the laboratory in its Teflon bag entirely enclosed in a black sack.
The number of visits for sampling of each location was 4 to 5 mornings and equal evenings of different dates dispersed within the whole program period. The visits started at different times allocated within the typical morning or evening sampling hours. The target was to get an average pollutant concentration of the morning or evening measurements more or less unaffected of these conditions, demonstrating an average of the critical situation, as it is affected mainly by the sampling hour of day. Therefore, the sampling schedule proposed makes average concentrations rather free from local weather conditions, since they generally differ from sampling to sampling. In addition, as sampling intervals at the same location started at different times dispersed within the typical morning or evening sampling periods, the aforementioned average concentrations should reflect mean levels of much greater time

intervals than the 3 minutes, approaching the related typical morning or evening sampling periods. This assumption was supported by four additional expeditions within the program period, according to which simultaneous air sampling with 50-l and successive 5-l Teflon sampling bags at the same location were performed. The results are given in Table 3 along with the suitable justification and concluding remarks.

3.3.3 Laboratory equipment and measurement methods

The EEL is equipped with analyzers of continuous operation for SO_2 and other air pollutants. Regarding SO_2 the model 4108 of Dasibi Environmental Corp., USA, was available, designated by the USEPA as an equivalent method for the measurement of ambient air concentrations of SO_2. During the program period, indoor air temperatures varied from 13 to 24°C being within the range 5-40°C of the analyzers operating temperatures. The intake of SO_2 analyzer was linked through a "T" shaped connector to a common entrance pipe connector where the sampling bag was mounted; the length of the Tygon piping used was less than 1 m. The analyzer had its individual pumping system for air sampling and it was continuously operating during the whole program period. For the calibration, a standard mixture of 100 ppb SO_2 in pure Nitrogen (N_2) was used. An automatic zeroing of signals was systematically performed at 12-h intervals.

The analyzer uses the UV-fluorescence method with 1% linearity and 1% precision in the range 0-500 ppb. The sample flow rate is 0.5 l minute^{-1} and the rise time is 90-120 s with lower detectable limit less than 1 ppb in the range 0-1 ppm and noise at zero concentration less than 0.5 ppb. Therefore, the minimum air volume required for the sample analysis regarding SO_2 is approximately 1.0 l, which was less than the volume of each sample.

3.3.4 Treatment of measurements

For obtaining average concentration values at all locations of the study area of Patras, the procedure described above (Section 2.1.6) has been applied. All measurements at each location had passed the Q-test (Radojević and Bashkin, 1999). Thus, the average concentration of four or five 3-minutes samples randomly picked at a particular location during morning or evening rush hours was calculated separately for morning or evening hours and assigned initially at each node, using the point kriging method with a linear variogram model (Golden Software, 2002). Thus, two grid matrices of average concentrations for each air pollutant, one for morning hour and another for evening hour, were taken. Next, each of these matrices was filtered using a low-pass linear convolution filter of 9-node averaging with weights equal to 1 (moving average filter). The 9-node averaging is actually performed on a sample of n values, where $36 \leq n \leq 45$. For a confidence level 95% and using the Student t-distribution, the final grid used above for contouring is proved to approximate 1-h mean concentrations at the specified locations, as it will be shown in the following. The probable error defined in Section 2.1.6 is calculated and consequently the mean probable error made for each final grid matrix of average concentrations was determined and shown in Table 8. The mean error made for estimating SO_2 average concentrations was less than 23.2%, while the corresponding error for traffic rates was less than 1.4%.

For the dependence of concentration on averaging time, the relationship given above by Eq. 1 may be applied for p = 0.17 and s = 3 minutes and averaging times t = 25-60 minutes. The ratio of C_t/C_3 computed by the above equation varied in the range 0.601-0.697 with mean value 0.649. Using the data given in Table 3 concerning successive sampling with 5-l bags, $C_t/C_3 \pm\sigma_{n-1}$ is found to be 0.63±0.155. For examined Patras locations, the maxima for 3-

minutes and for morning or evening period averaging time that is roughly 1 h are given in Table 8 with respect to the location measured. Calculating again $C_t/C_3 \pm\sigma_{n-1}$, it is found to be 0.542 regarding the morning and evening SO_2 concentrations, while is 0.588±0.0562 based on an average value regarding SO_2, NO and NO_2 concentrations (Yannopoulos 2007a). All above findings support the conclusion that the morning- or evening-period average concentrations approximate 1-h mean values.

Pollutant	Averaging time	Mornings	Probable error	Evenings	Probable error
SO_2	3 minutes	81 (Location 30)*		108 (Location 25)	
	about 1 h	49 (Location 30)	±21.3%	56 (Location 19)	±23.2%
Traffic rate	3 minutes	3360 (Location 40)		721 (Location 26)	
	about 1 h	2920 (Location 2)	±1.4%	512 (Location 19)	±1.2%

* The locations where the peak value was observed are given in parentheses.

Table 8. Maximum values of 3-minutes and approximated 1-h average concentrations (in μg m^{-3}) for SO_2 and traffic rate (in veh h^{-1}) in the Patras area with regard to the locations where they have been measured

Finally, application of the procedure described above results in 1-h average values of SO_2 concentrations and traffic rates, within typical morning or evening hours of workdays during program period. Then, iso-concentration and iso-traffic contour diagrams are plotted and shown in Fig. 7 and 8, correspondingly. Each of these figures consists from two diagrams; (a) and (b) for morning and evening average values. It is noted that probable errors encountered in iso-concentrations may affect the determination of the boundaries of high pollution zones by mean displacements less than the mean spatial resolution (±360 m) regarding SO_2, approximately.

3.3.5 Evaluation of sulfur dioxide levels

According to Fig. 5 and 7(a), the 1-h SO_2 morning maxima found to be 40–45 μg m^{-3} and were located at the centre of downtown Patras around Vas. Georgiou Square (Locations 22, 23 and 25) and close to Ag. Andreas Church (Locations 29 and 30) on the main national road connecting Patras with the Western Peloponnese. Note that the numbered open circles in Fig. 5, 7 and 8 indicate locations. At Location 30, an extended municipal parking lot for citizen's cars was operating. These locations presented high rates of traffic (more than 2000 veh h^{-1}) as indicated in Fig. 8(a), while emission rates were augmented due to car stops controlled by traffic lights. The 1-h mean SO_2 concentrations are well correlated to the corresponding traffic rates. The correlation coefficient for all locations (n = 50) found to be r = 0.60, meaning that the SO_2 air pollution is a rather traffic oriented pollution. Note that Jackson (2005) has given r = 0.74. The fact that Locations 1, 2 and 49 were found less polluted by SO_2 concentrations, although they have comparable traffic rates, may be justified taking into consideration that the latter locations are far-off the centre of downtown Patras and belong in more open areas (near Sector A boundaries), which resemble more or less to suburbs with better dispersion conditions than other areas of downtown Patras. In addition, since these locations (being on the ground) are rather far from both harbour (traffic and ships) and central-heating (elevated) sources, more dilution is favoured compared to the nearby traffic sources, which were also free from long vehicles emitting SO_2, as such vehicles were enforced to drive along the seaside roads. It is noted that the peripheral diversion highway was then under construction, thus all vehicles coming from Pyrgos and going to Athens and vice-versa passed through the city of Patras.

According to Fig. 5 and 7(b), the 1-h SO_2 evening maxima found to be 45–50 μg m^{-3} and were located around Locations 19, 22 and 25, in the central area of downtown Patras. However, SO_2 concentrations equal to or higher than 40 μg m^{-3} covered almost all the central area of Sector A and the central part of Sector C. Figure 8(b) shows that traffic rates were more than 1250 veh h^{-1} in most of this area. For all locations (n = 50) and evening hours, a lower correlation coefficient (r = 0.48) than that for morning hours and enough lower than Jackson's (2005) was found. If evening SO_2 concentrations are compared to morning ones (Fig. 7a and b), the differences observed might not be considered statistically significant. The values of the correlation coefficient for either morning or evening observations remained actually unchanged. Since for most samplings the wind conditions recorded and given in Table 7 indicate calm or light NE winds, the harbour traffic and ship emissions might not significantly affect the study area compared to the effect by the local traffic and central heating. The lower value of r computed for evening hours may be attributed to more intense contribution of central heating emissions in conjunction with probably lower mixing heights during evening hours than morning hours. Applying the statistical test with a significance level 0.005, for the acceptance or not of the hypothesis that morning measurements at each location have equal mean values to the corresponding evening measurements, indicated that the above hypothesis is acceptable. Therefore the SO_2-concentration differences can be considered statistically insignificant. Further on, the low enough correlation coefficient (r = 0.25) found denotes that the aforementioned small differences are rather weakly correlated to the corresponding differences of traffic rates. Consequently, evening and morning concentrations could be statistically considered as coming from similar populations.

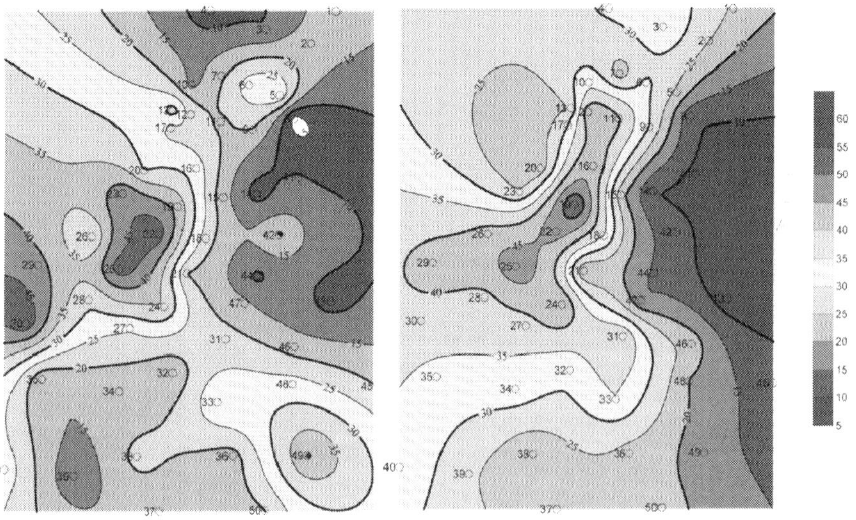

Fig. 7. Iso-concentration contours of 1-h mean SO_2 values (in μg m^{-3}) for (a) morning hours and (b) evening hours of a typical winter period (November 13 1997 - January 23 1998)

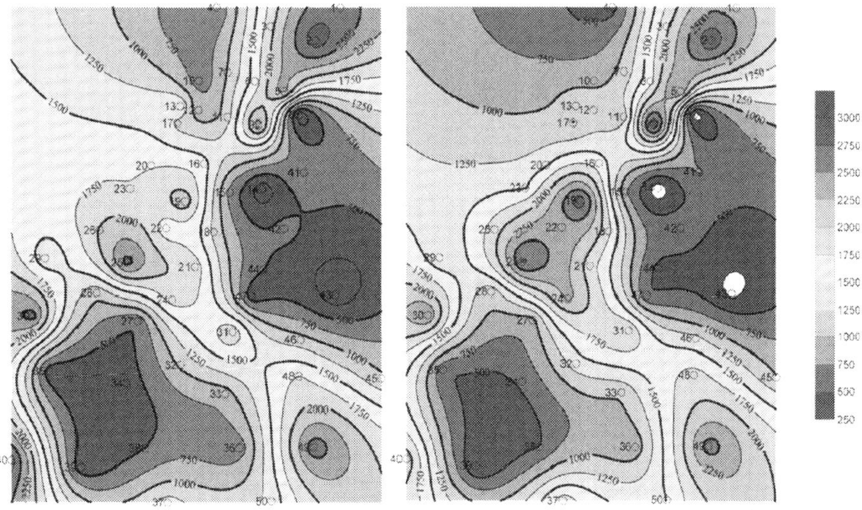

Fig. 8. Iso-traffic contours of 1-h mean traffic rates (in veh h^{-1}) for (a) morning hours and (b) evening hours of a typical winter period (November 13 1997 - January 23 1998)

In general, this implies relatively similar justifications for both morning and evening SO_2 concentrations. Therefore, the entire dataset of 1-h SO_2 morning and evening concentrations (n = 50×2) may be fairly well correlated to the corresponding traffic rates, while be weakly correlated to other factors through the prevailing wind and mixing height conditions. For traffic-explained SO_2 concentrations, the following relationship is taken (Yannopoulos 2007a):

$$SO_2\ (\mu g\ m^{-3}) = 0.0076 \times (veh\ /\ h) + 15.55,\ \text{with}\ r{=}0.54 \qquad (2)$$

The present findings can be verified once more by Yannopoulos and Skokaki's (2003) SO_2 concentration measurements. The measurements had been carried out for extended periods of year at Location 22 (former Location A), as well as at a location between the nearby Locations 12 and 17 (former Location B). The 1-h SO_2 levels determined in the present study at these locations during same year seasons and hours of the day were found close to the corresponding levels continuously monitored roughly two years earlier. Continuous measurements made by a Program of Achaia Prefecture during 1988-89, at a location between Locations 5 and 6 and at an elevation approximately 10 m above ground level, gave morning and evening maxima of 1-h mean SO_2 56 and 83 μg m^{-3} for the winter period (Yannopoulos and Amanatidis, 1993). According to Fig. 5, 7(a) and (b), the 1-h values of the present program for the above location were approximately 25 and 30 μg m^{-3}, which are less than the past concentrations. The higher levels recorded in 1988-89 than 1997-98 may be attributed mainly to the greater sulphur content of fuels in conjunction with the shorter distance of the monitoring site from central heating emission sources, since the fixed station had been installed on a building's roof. Yannopoulos (2007a) has given additional comparisons to available levels of other Greek and Mediterranean cities, as well as further details regarding NO, NO_2 and NO_x air pollutants and comparisons to air quality limits.

3.4 Gaussian modelling alternative

3.4.1 Application details of the model

The ISCLT3 model described above was used to simulate emissions from the sources affecting the Patras atmosphere. Besides background air pollution, these sources include mainly emissions from traffic, central heating and ships. The study area belonged in downtown Patras and had a magnitude of 3 km^2, approximately. It covered the central parts of Sector A, B and C of the Patras city plan. The main streets, which incorporated in this area, were Agias Sofias, Korinthou, Trion Navarhon, Agiou Andreou and Othonos Amalias Street (Fig. 5). These streets included in a polygon, which can be defined by Locations 2, 4, 29, 27, 31, 41 and 2. The necessary altitudes for the model input were available from the Municipality of Patras. In addition, the polygon included the two sites (receptors) where EEL had been carried out in the past continuous monitoring of air pollution. The site A coincides with the Location 22 at the Vas. Georgiou A' Square, while the site B is located in the junction of the streets of Ag. Andreou and Zaimi, between Locations 12 and 16.

The model input preparation was the first step. The central heating and ship emissions were modelled as area sources. Traffic is modelled as a line source and each main street was divided in 20-m length segments. Then, the emissions from the several types of vehicles were calculated according to globally acceptable methodologies and the European Union legislation. Primary data of traffic rates were available by previous EEL studies. To determine the emissions due to traffic, COPERTIII (http://vergina.eng.auth.gr/mech/lat/copert/copert.htm) was utilised. The fuel quality characteristics during the period of interest were considered according the European Union legislation. Due to the operation of the Patras harbour, increased traffic was observed in the main streets of each neighbourhood, i.e. Korinthou, Ag. Andreou and Othonos-Amalias. The monthly traffic loads applied were 524,960 vehicles. They consisted of 36% conventional type and 64% catalytic passenger cars. The light-duty vehicles were considered to be approximately equal to the heavy-duty vehicles and were taken 5% each. The traffic oil was considered to contain 0.3% sulphur by weight, while the petrol contained 0.2% and the oil consumed in central heating contained 0.2% by weight. To calculate traffic emissions, the average car speed was considered equal to 19 km h^{-1}, as usually taken for Greek urban areas, and applied to an average trip distance of 12 km. These assumptions imply that the final SO_2 emission load due to traffic was equal to 146 μg s^{-1}. In the neighbourhood of the harbour, the mean speed of vehicles was taken equal to 9 km h^{-1} with an average trip distance 1.5 km. Thus, the monthly emissions of ships and the vehicle traffic due to harbour are given in Table 9. The emissions of SO_2 due to central heating were based on the USEPA data. They were assumed to be 852 mg l^{-1} per fuel mass consumed, but varied with respect to the population density attained in the relevant sectors of the City Plan. Deposition of air pollutants and additional natural processes were not considered in this application.

As it was described above (Section 2.2.1), the wind speed was grouped in six categories and the atmospheric stability determined according to the Turner's classification system with respect to the incoming solar radiation and the wind speed class. During the winter, from October to March, 5 h of moderate insolation daily with slight insolation for the remaining 19 h was assumed. During summertime, from April to September, it was assumed that 5 h corresponded to strong insolation, 5 h to moderate insolation and 14 h to slight insolation.

The determination of the roughness height for downtown Patras and especially for the central parts of Sectors A and C was based on the average building height, which is among 15 to 28 m (Table 5). Thus, the roughness height was taken equal to 1.0 m.

One of the most important parameters for model calibration is the mixing height, which depends on the time during a day and the seasonal properties. The prevailing mixing height during a typical day for each season was determined on the basis of author's visual observations made during morning hours and agreed with predictions in coastal areas by applying the models proposed by Hanna (1987), Luhar (1998) and Delobbe et al. (2001), as well as according to computations made using the procedure provided by www.dar.csiro.au/pollution/MixHeight/botWindow.htm.

Month	Year 1994		Year 1995	
	Ships	Vehicles	Ships	Vehicles
January	12,311	63	11,303	55
February	10,571	55	9,102	61
March	12,362	63	9,255	61
April	12,126	63	10,739	58
May	12,043	63	11,229	65
June	14,690	78	14,609	62
July	13,308	71	14,932	72
August	15,136	78	14,780	51
September	15,157	78	14,634	56
October	14,102	71	13,541	75
November	11,978	63	11,975	76
December	12,268	63	12,248	76

Table 9. Monthly SO_2 emissions (kg month^{-1}) from ships and vehicles due to traffic in the Patras harbour

Then, according to the model requirements, the mixing height for the A stability class have to be equivalent to approximately 1.5 times the mixing height for stability classes B and C and double for the stability class D. Equal mixing heights have to be assigned for B and C classes of atmospheric stability. According to the analysis given above (Section 3.2), the lowest values of the mixing height assigned during the winter period and the highest values during summertime. For spring and autumn, the average of the corresponding values for winter and summer were taken.

3.4.2 Sensitivity analysis of the model

The most important feature of the model is the comprehension of its sensitivity with respect to input key parameters controlling the prediction behaviour. These parameters are the roughness height and the mixing height, which present in general the highest uncertainty. For the present case study, it was found that an increase of the roughness height from 1.0 to 3.0 m indicated insignificant effect (less than 1%) in concentrations and, therefore the mixing height of 1.0 m was used in the predictions. Since, the concentration is inversely related to the mixing height, a mixing height increase would cause a decrease of the concentrations. Giving a 10% increase in the mixing height of A-class and subsequent increase in the mixing heights of B-, C- and D-classes of atmospheric stability according to the aforementioned instructions, the average decrease in predicted concentrations was 2%. This test showed that the mixing height was the most critical parameter affecting the model predictions. Since neither measurements nor reliable estimations of the mixing height were available, the mixing height became the main key parameter for the model calibration aiming to obtain acceptable SO_2 predictions at the prescribed receptor locations. For simplicity reasons and a specified stability class, fixed values of mixing height were used during the same season, irrespective of the year with available concentration measurements. In Table 10 the values of mixing heights, which give the best model performance at the receptor Sites A and B, are shown. According to USEPA (1995), it is noted that the model assumes unlimited mixing heights for classes E and F of the atmospheric stability.

Class of Atmospheric stability	Mean seasonal mixing height (m)		
	Winter	Spring and Autumn	Summer
A	90	195	300
B, C	60	130	200
D	45	100	150

Table 10. Mean seasonal values of the mixing height (in m) for Patras city with respect to atmospheric stability class

3.4.3 Model predictions and discussion

The model ISCLT3 was applied to predict the mean monthly SO_2 concentrations in the area of downtown Patras described above considering only emissions from the predefined human activities of traffic, central heating and harbour activities including ship emissions and related harbour traffic. The results were calculated using the aforementioned input values of the several model parameters. The background influence is introduced by superimposing the predicted SO_2 values on the actual ambient background levels. The spatial distribution of the mean monthly SO_2 concentrations is shown in Fig. 9, including two iso-concentration contour diagrams for (a) December 1994 and (b) December 1995. Note, that December is a month of the winter period where maxima of concentrations have occurred. The mean monthly SO_2 concentrations predicted at the receptor Locations A and B are compared with the related measurements at these sites and they are shown in Fig. 10.

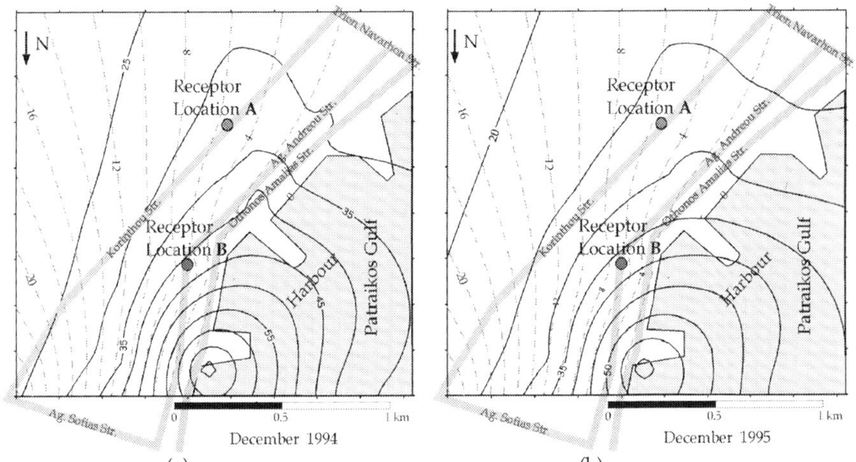

Fig. 9. Predicted mean monthly SO_2 iso-concentration contours (in $\mu g\ m^{-3}$) for (a) December 1994 and (b) December 1995

The SO_2 concentrations monitored at the W boundaries of the University of Patras Campus, a low activity outskirt, were considered equal to the ambient air background levels, which had been measured equal to 5.5 μg m^{-3}, according to Yannopoulos and Skokaki (2003). As shown in both diagrams of Fig. 9, the harbour areas that concentrate the most frequent ship calls are affected by high levels of pollution. Regarding the higher SO_2 concentration levels predicted in 1994 than 1995 (Fig. 9), it is believed that the concentration differences were due to somewhat different weather conditions and emission rates. As justified by the results shown in Fig. 10 and the following statistical analysis, the model performance is satisfactory in estimating both anthropogenic SO_2 levels and period pattern.

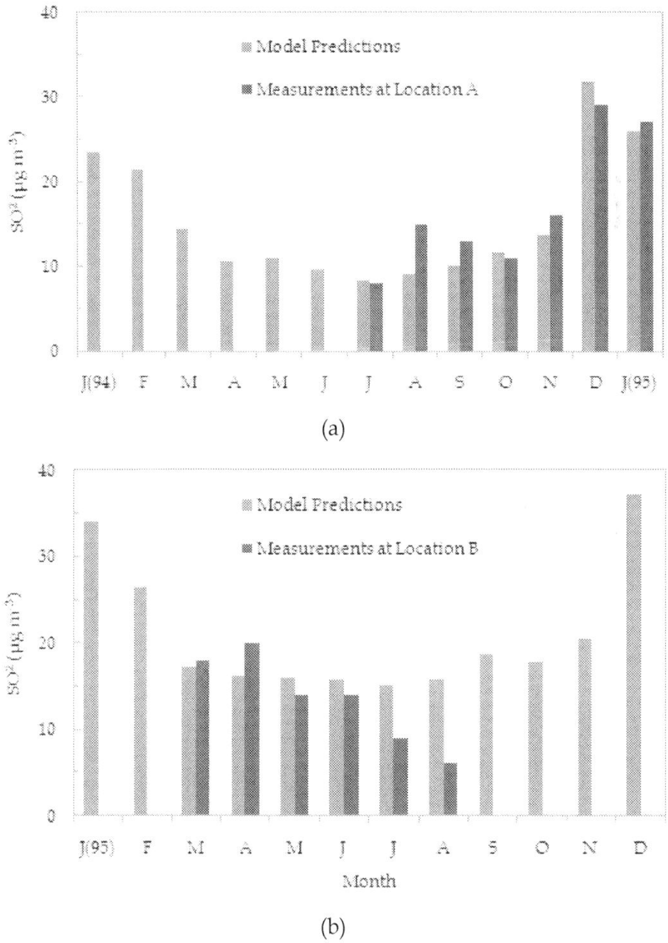

Fig. 10. Predicted and measured mean monthly SO_2 concentrations (in μg m^{-3}) at the receptor locations; (a) Location A and (b) Location B

For the statistical analysis, the paired t-test method is applicable (Kreyszig, 1999, p. 1125-1126), since two samples of the same size are available; the sample of observations and the sample of predictions. Each value of the sample of observations corresponds precisely to one value of the sample of predictions. Consequently, the sample of differences of corresponding values can be formed and tested. The test hypothesis is that the population corresponding to the differences should have a zero mean. As data lead definitely to the hypothesis acceptance with a significance level of 0.05 for a two-sided test, it is concluded that the two samples have statistically the same mean values.

Then, the contribution of the emissions from each source to the air pollution at the receptor Locations A and B has been computed and shown in Fig. 11 on a seasonal and an annual basis. It is obvious that the yearly average SO_2 contribution of ship emissions to both sites is around 60% and therefore significant, compared to all other sources, which show approximately equal contributions to each other, being around 20%. Although during winter the harbour contribution decreases to 39 and 57% at the receptor Locations A and B, respectively, it remains the dominant source for site B with central heating contributing to 22% of the SO_2 emissions. Central heating contributes 46% in SO_2 emissions at site A, with the harbour contributing less SO_2 than central heating. During summertime, the harbour contribution is maximised becoming 72 and 77% at the receptor locations A and B, respectively. An increased contribution of the traffic was found during summertime with the percentages of 27 and 28% than during winter, where contribution was 15 and 20% at the receptor Locations A and B, respectively.

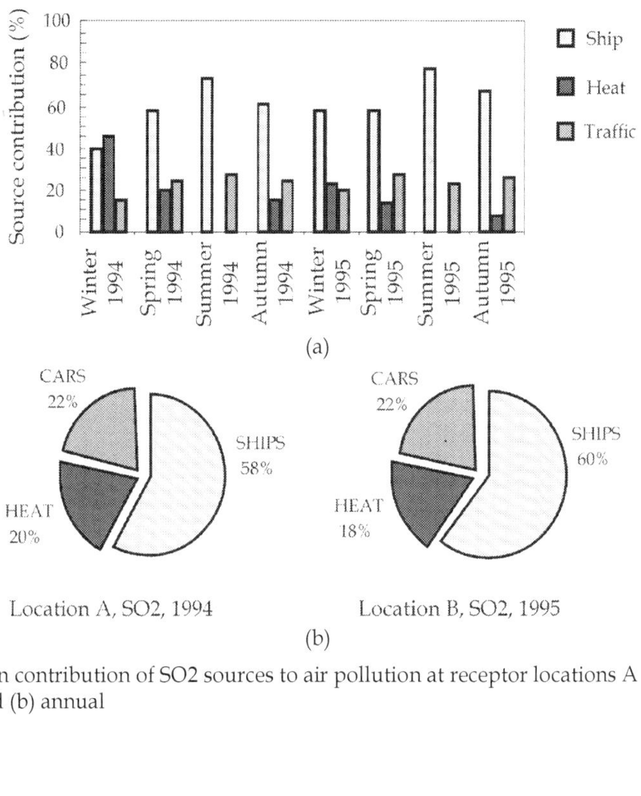

Fig. 11. Mean contribution of SO2 sources to air pollution at receptor locations A and B; (a) seasonal and (b) annual

3.5 Indicative cost analysis of methodology and comparison to custom monitoring

The cost of application of the proposed methodology in the city of Patras included the particular cost of the two alternatives increased by an extra cost for the evaluation and dissemination of results and preparation of the final report, as shown in Table 11. The Value Added Tax (VAT) is included in the particular item values. The cost of the air sampling bag technique can be analysed in the cost of suitable equipment, the necessary expenses for consumables, the labour cost of personnel needed for the collection, transfer and laboratory analysis of samples, as well as the transport and service expenses. The cost of equipment

Description	Unit	Quantity	Value per unit (Euros)	Partial cost (Euros)
A. Air sampling bag technique				
1. Equipment				
SO_2 analyser usage (1.5/36)×12,000.00	Item	1	500.00	500.00
Pump for air sampling (1.5/36)×900.00	Item	2	37.50	75.00
2. Consumables				
Calibration standard (100 ppb SO_2 in 10 l pure N_2 of 200 bar)	Item	1	200.00	200.00
Teflon air sampling bag of 5-l	Item	10	120.00	1,200.00
Teflon air sampling bag of 50-l	Item	1	300.00	300.00
Pump accessories (particulate filter, funnel, Tygon tubing, rechargeable battery and opaque plastic sack) and office consumables	Item	2	162.50	325.00
3. Labour				
Sampling and analysis of 450 samples	Person×months	3	2,000.00	6,000.00
Treatment of results	Person×months	0.5	3,000.00	1,500.00
4. Additional expenses				
Car or motorcycle rent for sample and personnel transport	Item	1	1,200.00	1,200.00
Calibration and service of the equipment	Item	1	500.00	500.00
Cost of A =				**11,800.00**
B. Gaussian modelling alternative				
1. Equipment				
Usage of a computer system (1.5/36)×4,800.00	Item	1	200.00	200.00
2. Labour				
Input data treatment, calculations and analysis of results	Person×months	2	3,000.00	6,000.00
Cost of B =				**6,200.00**
C. Evaluation and dissemination of results A and B and writing the final report	Person×months	1	3,000.00	**3,000.00**
Total cost of methodology application A+B+C=				**21,000.00**

Table 11. Indicative expenditure of methodology application in an urban area of Patras

included the cost proportion of the SO_2 analyser and the two sampling pumps to the 1.5-month project period, considering that the useful life of analyser and pump is 36 months. The cost of consumables included the SO_2 calibration standard for analyser calibration, 10 Teflon air sampling bags of 5-l nominal volume, 1 Teflon air sampling bag of 50-l nominal volume, accessories for two pumps (particulate filter, funnel, Tygon tubing, rechargeable batteries and opaque plastic sacks), as well as office consumables for preparing the report. The labour included the work for collecting the samples, transferring them and analysing in the laboratory, as well as the work for treatment of the results. There were additionally transport and service expenses, as the cost of renting a small car or motorcycle for the personnel and sample transport, as well as the analyser calibration and service expenses of the equipment. It is noted that the total distance covered for personnel and sample transport reached 2,200 km.
Since the total cost of methodology application was 21,000.00 Euros for 50 locations, the corresponding relative cost per location was **420,00** Euros.
The corresponding cost of measuring hourly mean concentrations of SO_2 using a permanent station, which constitutes the custom monitoring methodology, is difficult to account in a comparable way with the cost of the proposed methodology. The reason is the fussy determination of the area, where a permanent station could be considered representative for hourly mean concentration measurements. In my opinion, the best way to define this area is the consideration of a common mean spatial resolution. Thus, defining 360-m spatial resolution, comparable probable errors to the proposed methodology would be expected in plotting iso-concentration contours. Therefore, for the urban area of Patras of approximately

Description	Unit	Quantity	Value per unit (Euros)	Partial cost (Euros)
1. Equipment				
SO_2 analyser usage (1.5/36)×12,000.00	Item	1	500.00	500.00
Housing with network connections for power supply and communication (1.5/36)×6,000.00	Item	1	250,00	250.00
2. Consumables				
Calibration standard (100 ppb SO_2 in 10 l pure N_2 of 200 bar)	Item	0.1	200.00	20.00
Consumed electric power and office consumables	Item	0.02	500.00	10.00
3. Labour				
Treatment, evaluation and dissemination of results	Person×months	0.03	3,000.00	90.00
4. Additional expenses				
Calibration and service of the equipment	Item	0.02	6,500.00	130.00
Total cost of permanent station =				**1,000.00**

Table 12. Indicative expenditure of a permanent station in an urban area of Patras

6 km^2 covered by the 50 locations, a comparable number of permanent stations should be considered. Assuming that the corresponding cost of each permanent station for 1.5 months operation is as given in Table 12, it is evident that the proposed methodology has a considerable less cost than permanent station measurement. However, there is not a panache methodology that could substitute any other. The continuous monitoring using permanent stations is advantageous in getting direct mean values of any time interval at a location. Its disadvantage is the low flexibility in relocation of a permanent station, the difficulties encountered in finding suitable places for location with pertinent conditions being in agreement with installation standards, difficulties in the availability of power supply and network link, as well as the weakness in determining the contribution of a source to the pollution of a prescribed receptor location. As it has been already described, the proposed methodology has overcome all these disadvantages.

4. Conclusion

The proposed methodology combines the use of the air sampling bag technique and the Gaussian modelling alternative. The application in the city of Patras proved that the methodology is capable of evaluating cost effectively the spatial air pollution and detecting the city locations susceptible to high pollution levels needing more attention. The bag sampling technique allows measuring short time concentrations of specified air pollutants sampled at urban locations, using only one analyser per pollutant. The maximum mean errors made in the determination of hourly mean SO_2 levels were less than 25%. An acceptable prediction is also succeeded by employing the USEPA Gaussian models. The mean concentrations between predictions and measurements at two locations of Patras downtown showed statistically unimportant differences with a significance level of 0.05 for a two-sided test. Thus, the two samples have statistically the same mean values, a fact confirming the use of Gaussian modelling for air pollutant concentration predictions and therefore trusting the methodology application for estimating the seasonal and annual contribution of pollutant sources to prescribed receptor locations. The latter is extremely difficult or even impossible to be made using measurement stations only, because it is not practically feasible to stop emissions for long of a particular group of sources representing an activity, while monitoring continuously emissions from other activities. The disadvantages of the bag sampling technique concern air pollutants that show high enough chemical reactivity in the atmosphere and/or in the samples, like O_3, or they sink in the bag walls and generally present low enough recovery values, as for aerosols and particulates.
The future research may include use of available measurements and meteorological information for calibrating Gaussian models suitable for predicting the city air quality impact due to new activity installations and operations or activity relocations. The city of Patras needs such an air quality evaluation, because of the harbour relocation, which is going to start operating shortly, as well as an evaluation due to the construction and operation of the peripheral diversion highway and the Rion-Antirrion cable bridge. In general, the proposed methodology and its outcomes could be useful for the design and optimization of a city network of stations to monitor air quality, for environmental impact assessments, future reference and comparisons due to city development needs, and for validating dispersion models.

5. Acknowledgements

The author thanks the undergraduate students H. Marini, E. Nikolacopoulos, A. Bellou, K. Droulia, S. Krikeli, K. Liatsou, A. Mourtzoukou, M-N. Tzifa and M. Foteinopoulou, who had participated in the early EEL programs of sampling, measurement and estimate of source emissions.

6. References

Bloutsos, A. A. & Yannopoulos, P. C. (2011). Concentrations of selected toxic elements in airborne particulates of Patras, Greece. *Global NEST Journal* (in print).

Chen, C.-L., Tsuang, B.-J., Tu, C.-Y., Cheng, W.-L. & Lin, M.-D. (2002). Wintertime vertical profiles of air pollutants over a suburban area in central Taiwan. *Atmospheric Environment*, Vol.36, pp.2049-2059.

Commission of the European Communities (2005). *Commission Staff Working Paper, Annex to: The Communication on Thematic Strategy on Air Pollution and The Directive on "Ambient Air Quality and Cleaner Air for Europe" Impact Assessment.* Interinstitutional File: 2005/0183 (COD), 14335/05, ADD 1, Encl.: SEC(2005) 1133, Council of the European Union, DG I, Brussels, pp. 1-170. Retrieved from http://www.parlement.com/9353000/1/j4nvgs5kjg27kof_j9vvhy5i95k8zxl/vi7jgt1po0zi

Delobbe, L., Matthijsen, J. & Sauter, F. J. (2001): *Evaluation of Mixing Height Representations in the EUROS Model.* RIVM report 711002 001, National Institute of Public Health and the Environment: Bilthoven, Belgium.

EC (1980). Council Directive 80/779/EEC of July 15 1980: Air quality limit values and guide values for sulphur dioxide and suspended particulates. European Communities, OJ L 229.

EC (1985). Council Directive 85/203/EEC of 7 March 1985: Air quality standards for nitrogen dioxide. European Communities, OJ L 87.

EC (1996). Council Directive 96/62/EC of 27 September 1996: Ambient air quality assessment and management. European Communities, OJ L 296.

EC (1999). Council Directive 1999/30/EC of 22 April 1999: Limit values for sulphur dioxide, nitrogen dioxide and oxides of nitrogen, particulate matter and lead in ambient air. European Communities, OJ L 163.

EC (2008). Council Directive 2008/50/EC of 21 May 2008: On ambient air quality and cleaner air for Europe. European Union, OJ L 152.

Golden Software, Inc. (2002). *Surfer 8-Contouring and 3D surface mapping for scientists and engineers.* User's Guide, Golden Software, Inc., Colorado, USA.

Hanna, S. R. (1987). An Empirical Formula for the Height of the Coastal Internal Boundary Layer. *Boundary-Layer Meteorology* Vol.40, pp.205-207.

Herman, F., Smidt, S., Huber, S., Englisch, M. & Knoflacher, M. (2001): Evaluation of pollution-related stress factors for forest ecosystems in Central Europe. *ESPR – Environ Sci & Pollut Res*, Vol.8, pp.231-242.

Jackson, M. M. (2005). Roadside concentration of gaseous and particulate matter pollutants and risk assessment in Dar-Es-Salaam, Tanzania. *Environ Monit Assess*, Vol.104, pp.385-407.

Kennedy, E. R., Fischbach, T. J., Song, R., Eller, P. M. & Shulman, S. A. (1995). A NIOSH Technical Report: Guidelines for air sampling and analytical method development and evaluation. United States Department of Health and Human Services, National Institute for Occupational Safety and Health, Division of Physical Sciences and Engineering, Cincinnati, Ohio.

Kim, K.-H., Choi, G.-H., Choi, Y.-J., Song, H.-N., Yang, H.-S. & Oh, J.-M. (2006). The effects of sampling materials selection in the collection of reduced sulfur compounds in air. *Talanta,* Vol.68, pp.1713-1719.

Klemm, O. & Lange, H. (1999): Trends of air pollution in the Fichtelgebirge Mountains, Bavaria. *ESPR – Environ Sci & Pollut Res,* Vol.6, pp.193-199.

Kumar, A. & Viden, I. (2007). Volatile Organic Compounds: Sampling Methods and Their Worldwide Profile in Ambient Air. *Environ Monit Assess,* Vol.131, pp.301-321.

Lee, S.-C. (1997). Comparison of indoor and outdoor air quality at two staff quarters in Hong Kong. *Environ. Int.*, Vol.23, pp.791–797.

Lee, S. C. & Chang, M. (2000). Indoor and outdoor air quality investigation at schools in Hong Kong. *Chemosphere,* Vol.41, pp.109–113.

Leong, S.T., Muttamara, S. & Laortanakul, P. (2002). Influence of benzene emission from motorcycles on Bangkok air quality. *Atmospheric Environment,* Vol.36, pp.651-661.

Li, W.-M., Lee, S. C. & Chan, L. Y. (2001). Indoor ar quality at nine shopping malls in Hong Kong. *Sci Total Environ,* Vol.273, pp.27–40.

Li, X., Wang, S., Duan, L., Hao, J., Li, C., Chen, Y. & Yang, L. (2007). Particulate and Trace Gas Emissions from Open Burning of Wheat Straw and Corn Stover in Chine. *Environ. Sci. Technol.,* Vol.41, pp.6052-6058.

Luhar, A. K. (1998). An analytical slab model for the growth of the coastal thermal internal boundary layer under near-neutral onshore flow conditions. *Boundary-Layer Meteorology,* Vol.88, pp.103-120.

McGinley, C. M. (2002). Standardized odor measurement practices for air quality testing. In: *Air and Waste Management Association Symposium on Air Quality Measurement Methods and Technology -2002,* San Francisco, California, 13–15 Nov. 2002.

Puxbaum, H, Zambo, E. & Kalina, M. (1998): Assessment of wet, dry and occult deposition of sulfur and nitrogen at an Apline site. *ESPR – Environ Sci & Pollut Res,* Special Issue Vol.5, pp.53-58.

Radojević, M. & Bashkin, V. N. (1999). *Practical environmental analysis.* The Royal Society of Chemistry, ISBN 0-85404-594-5, Cambridge, UK.

Schnelle, K. B. Jr. & Dey, P. R. (2000). *Atmospheric dispersion modeling compliance guide.* McGraw-Hill, ISBN 0-07-058059-6, New York.

Smidt, S. (1998): Risk assessment of air pollutants for forested areas in Austria, Bavaria, Switzerland. *ESPR – Environ Sci & Pollut Res,* Special Issue Vol.5, pp.25-31.

Steeghs, M.M.L., Cristescu, S.M. & Harren, F.J.M. (2007). The suitability of Tedlar bags for breath sampling in medical diagnostic research. *Physiol. Meas.,* Vol.28, pp.73-84.

Stewart, R. (1999). *Background paper on offshore emission monitoring.* OG-IOR-4/16728904/4475/1192 /v3, AEA Technology plc, AEA Technology Environment, E5 Culham Science Centre, Abingdon, Oxfordshire.

Turalioğlu, F. S. (2005). An assessment on variation of sulphur dioxide and particulate matter in erzurum (Turkey). *Environ Monit Assess,* 104, 110–130.

Turner, D. B. (1994). *Workbook of Atmospheric Dispersion Estimates. An Introduction to Dispersion Modeling.* 2nd ed., Lewis Publidhers, CRC Press, ISBN 1-56670-023-X, USA.

Tsai, C.-J., Chen, M.-L., Chang, F.-K. & Mao, I.-F. (2009). The pollution characteristics of odor, volatile organochlorinated compounds and polycyclic aromatic hydrocarbons emitted from plastic waste recycling plants. *Chemosphere*, Vol.74, pp.1104–1110.

Tsai, J.-H., Hsu, Y.-C., Weng, H.-C., Lin, W.-Y. & Jeng, F.-T. (2000). Air pollutant emission factors from new and in-use motorcycles. *Atmospheric Environment,* Vol.34, pp.4747-4754.

U.S. Environmental Protection Agency (1987). *Method 18. Code of Federal Regulations,* Part 60, Title 40, Appendix A, U.S. GPO: Washington, DC, July 1, 1987.

U.S. Environmental Protection Agency (1995): *User's Guide for the Industrial Source Complex (ISC3) Dispersion Models. Vol. I – User Instructions, Vol II – Description of Model Algorithms.* EPA-454/B-95-003a, b, September 1995, Environmental Protection Agency, USA.

U.S. Environmental Protection Agency (1996). *Method 0040: Sampling of principal organic hazardous constituents from combustion sources using Tedlar bags.* United States Environmental Protection Agency, Test Methods, December.

Wang, Y., Raihala, T.S., Jackman, A.P. & John, R.ST. (1996). Use of Tedlar Bags in VOC Testing and Storage: Evidence of Significant VOC Losses. *Environ. Sci. Technol.,* Vol.30, pp.3115-3117.

Yannopoulos, P.C. (2007a). Spatial concentration distributions of sulfur dioxide and nitrogen oxides in Patras, Greece, in a winter period. *Environ Monit Assess,* Vol.135, pp.163-180.

Yannopoulos, P.C. (2007b). Sulfur dioxide dispersion and source contribution to receptors of downtown Patras, Greece. *Env Sci Pollut Res,* Vol.14, No.3, pp.172-175.

Yannopoulos, P.C. (2008). Long-term assessment of airborne particulate concentrations in Patras, Greece. *Fresenius Environmental Bulletin,* Vol.17, No.5, pp.608-616.

Yannopoulos, P. C. & Amanatidis, G. T. (1993). Air pollution levels in Patras, Greece. *Proceedings of International Conference on Environmental Pollution,* ISBN 0-9521673-1-X, Sitges, Barcelona, September 1993, Vol.2, pp.415–421.

Yannopoulos, P. C. & Skokaki, G. N. (2003). Particulate and sulfur dioxide concentration measurements in Patras, Greece. *J. Air & Waste Manage. Assoc.*, Vol.53, pp.957–970.

22

New Approaches for Urban and Regional Air Pollution Modelling and Management

Salvador Enrique Puliafito, David Allende, Rafael Fernández, Fernando Castro and Pablo Cremades

Grupo de Estudios Atmosféricos y Ambientales (GEAA)
Universidad Tecnológica Nacional - Facultad Regional Mendoza
Consejo Nacional de Investigaciones Científicas y Técnicas (CONICET)
Argentina

1. Introduction

Air pollution is a complex problem that plays a key role in human well-being, environment and climate change. Since cities are, by nature, concentrations of humans, materials and activities, air pollution is clearly a typical phenomenon associated with urban centres and industrialized regions (Fenger, 1999; de Leeuw et al., 2001). Since approximately half the population of the world lives in medium to large cities, it is essential to evaluate the air quality levels of the atmosphere in order to assess the possible health impact from exposition to pollutants (World Health Organization [WHO], 2002; Brunekreef & Holgate, 2002). Additionally, air pollution is not only a human health problem: the effects of pollution in ecosystems and materials are well identified and documented (Fowler et al., 2009); economic costs can also be associated with poor air quality, and with political/governmental measures taken in order to prevent or reduce pollution (Muller & Mendelsohn (2007)).

The simplest technique for evaluating patterns of local-scale urban air pollution concentration involves the interpolation of ambient concentrations from existing monitoring networks (Ballesta et al., 2008; Ferretti et al., 2008). However, the measured data from these stations are not necessarily representative of areas beyond their immediate vicinity, since concentrations of pollutants in urban areas may greatly vary on spatial scales that range from tens to hundreds of metres. At the same time, the temporal behaviour of primary and secondary pollutants changes considerably between day and night due to solar radiation, so that daily average measurements become unsatisfactory in determining or explaining high pollution episodes.

Air Quality Models (AQMs) are mathematical tools that simulate the physical and chemical processes that involve air pollutant dispersion and reaction in the atmosphere. Furthermore, they improve the limitations of monitoring networks by providing prediction of the temporal and spatial distribution of actual pollution levels. Modelling studies, in combination with air quality monitoring, are then essential and complementary tools for long and short term air pollution control strategies. A well calibrated model is a unique tool that allows the representation of the atmospheric dynamics and chemistry. Thus, AQMs have become a valid instrument for environmental managers in many activities, such as *a)* setting emission control regulations, *b)* testing the compliance of actual pollution levels, *c)* predicting the impact of new facilities on human health, *d)* selecting the best location for monitoring stations, and *e)* assessing the impact of different emissions scenarios on Global Climate Change.

This Chapter presents an overview of several techniques and available codes applied to urban modelling, plus describing three study cases from different approaches that we have selected according to the magnitude of the simulated problem.

2. Air Quality Models (AQMs)

The major elements of an AQM are depicted schematically in Fig. 1.

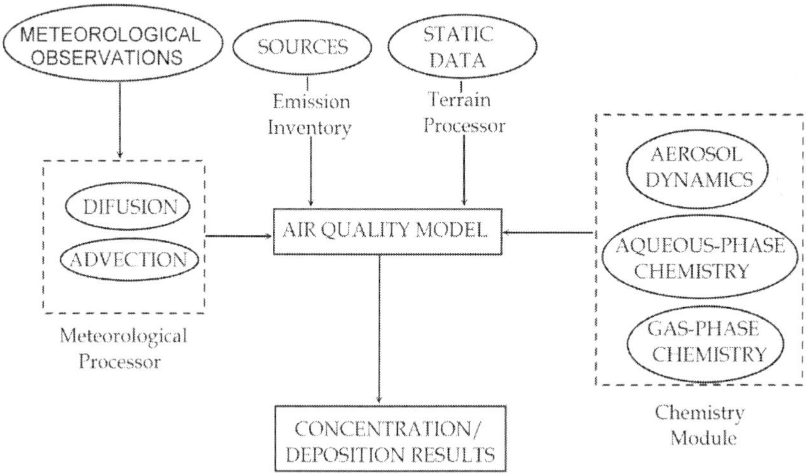

Fig. 1. Main components of an Air Quality Model and their interactions.

Basically, an air quality study implies the estimation of some pollutants concentration in a region of interest, during a finite period of time. The characteristics of each specific problem will define the physical and chemical processes involved, and consequently, the best model to use. The main criteria for choosing adequate software are: the dimension of the area under study, the number of pollution sources, the chemical species involved and the time scale of the episode.

2.1 Air quality and spatial scales

In a geophysical context, it is possible to define several superimposed scales for the air pollution phenomena, extending from local to global scales with several inputs and feedbacks at every level. A pollutant released from traffic or industries in a city may have a significant impact on the city itself or on a farther region, according to the properties and atmospheric lifetime of the pollutant. Consequently, local scale problems, such as the evaluation of the maximum ground level impact of primary pollutants released from traffic or industrial sources, cannot be treated similarly to long range problems due to the influence of regional scale processes, such as chemical transformation or deposition, which need not to be taken into account on a smaller scale. Therefore, following Zannetti (1990), five types of scales can be distinguished, depending on the distance from the emitting source: *a)* near-field phenomena (< 1 km from the source); *b)* short range transport (< 10 km from the source); *c)* intermediate range transport (between 10 and 100 km from the source); *d)* long range transport (> 100 km from the source); and *e)* global effects.

Recently, Air Quality Models have been extended to consider the interactions between processes at different scales, by including nesting capabilities to work with different grid resolutions simultaneously, or by integrating several models designed for a specific problem. Examples of the first approach are CMAQ (Binkowski & Roselle, 2003) and WRF/Chem (Grell et al., 2005) models. The BRUTAL model (Oxley et al., 2009) and the modelling system proposed by Jiménez-Guerrero et al. (2008) are examples of the second approach.

2.2 Emission inventories

Emission inventories have long been fundamental tools for air quality management. Emission estimates are important for developing emission control strategies, determining applicability to permit and control programs, ascertaining the effects of sources and appropriate mitigation strategies; and a number of other related applications by an array of users, including federal, state, and local agencies, consultants and industries.

Emission inventories are also a key input for AQMs and therefore must match their requirements, namely the spatial and temporal distribution of emissions in accordance with the model setup. Spatial and temporal emissions distribution of individual chemical compounds may not be available in emission inventories compiled at national, regional or local levels for regulatory purposes. Consequently, besides emission estimation itself, emission inventory preparation is a critical stage in air quality modelling. Many authors have clearly shown that emission input to AQMs is one of the main sources of uncertainty (Russell & Dennis, 2000; Hanna et al., 2001).

Data from source-specific emission tests or continuous emission monitors are usually preferred for estimating pollutant releases because those data provide the best representation of emissions from tested sources. However, test data from individual sources are not always available, and may not even reflect the variability of actual emissions over time. Thus, emission factors are frequently the best or only method available for estimating emissions, in spite of their limitations. The general methodology to estimate emissions from emission factors has the form:

$$E = A \times EF \times \left(1 - \frac{ER}{100}\right) \tag{1}$$

in which *E* represents the emissions (mass of pollutant), *A* is the activity rate in hours/year, *EF* is the emissions factor and *ER* defines the overall emission reduction efficiency, %. The emission factors are usually expressed as the weight of pollutant divided by a unit weight, volume, distance, or activity duration that releases the pollutant (e.g., kilograms of particulate emitted per megagram of coal burned). Such factors facilitate estimation of emissions from various sources of air pollution. In most cases, these factors are simply averages of all acceptable quality data available, and are generally assumed to be representative of long-term averages for all facilities in the source category (i.e., a population average).

The estimations of primary pollutants from numerous types of sources are generally performed by following well established methodologies, such as the Intergovernmental Panel on Climate Change (IPCC) Guidelines for National Greenhouse Inventories (IPCC, 2006), the EMEP Air Pollutant Emission Inventory Guidebook of the European Environmental Agency (EMEP/EEA, 2009), or the U.S. Environmental Protection Agency (U.S. EPA) AP-42 Compilation of Air Pollutant Emission Factors (U.S. EPA, 2010).
Even when the general methodology of estimation is similar, emissions from a specific type of sources have their own characteristics and must be treated by AQMs in a different way. Emissions from point sources are estimated for individual sources. Besides the amount of pollutants emitted (usually in g/s) and a precise geographic location, a few physical parameters are used in AQMs to characterise the emission, such as stack height and diameter, emission escape velocity and exit temperature.
Area (e.g., due to residential heating) or mobile sources are not treated individually and their emissions can be estimated by following one of two major approaches, namely top-down and bottom-up. The first one uses aggregated activity data and general *EF* to calculate the amount of emissions from all sources, often giving little spatial and temporal detail. Bottom-up approaches use detailed activity and source-specific data to calculate the emissions with high temporal and spatial resolution. Top-down approach is usually chosen because it needs a reduced amount of input parameters compared to bottom-up methods, therefore being more cost-efficient and easier to implement.
Emissions from area and mobiles sources are incorporated into an AQM as a regular grid with a resolution that depends on the modelling setup and the scope of the study. In order to spatially distribute emissions obtained through a top-down method, a series of simplified approaches of spatial disaggregation based on surrogate data (e.g. population density, principal road network density) and Geographical Information System (GIS) tools are generally used (Puliafito et al, 2003; Tuia et al, 2007). Fig. 2 shows a scheme for the spatial distribution of traffic emissions obtained from different approaches.

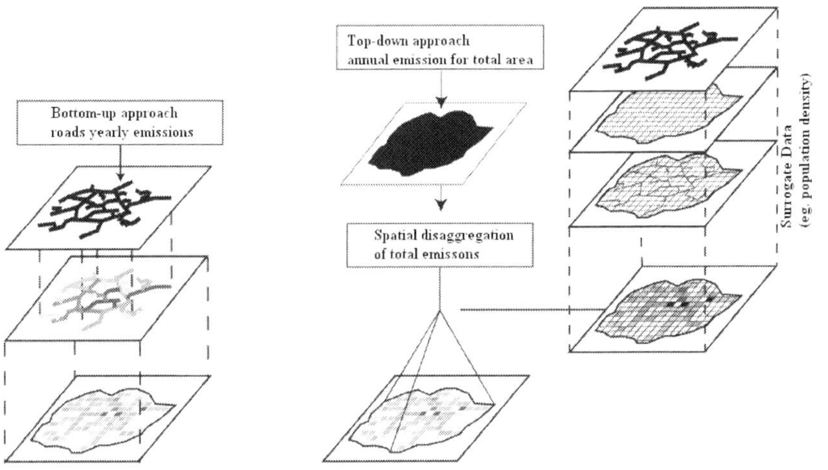

Fig. 2. Distribution of mobile emissions with bottom-up (left) and top-down (right) approach (adapted from Tuia et al, 2007).

Due to the intrinsic complexity of some sources, emission models are generally used in order to obtain total emissions and to evaluate different emission scenarios. For example, mobile emission can be estimated by using several models like COPERT (Ntziachristos et al, 2009), MOVES (U.S. EPA, 2009a) and IVE (Davis et al., 2005).

Temporal distribution of emission is another issue that needs to be addressed in order to generate acceptable inputs to AQMs. Time-resolved emissions are needed to properly simulate air quality through deterministic Eulerian models like WRF/Chem. Generally, emissions are expressed in annual or daily bases and temporal profiles are used to break them down to the temporal resolution expected by AQMs (usually 1 hour). These profiles are obtained by evaluating the temporal activity patterns from each type of source.

Emission inventories are built and reported for a variety of compounds or chemical classes such as CO, NO_X, NMVOC, PM_{10}, and SO_2. However, photochemical mechanisms, included in some AQMs like WRF/Chem, contain a simplified set of equations that use representative "model species" to simulate atmospheric chemistry (Dodge, 2000). Consequently, when chemical transformations are going to be considered in the AQMs, "inventory species" must be converted to "model species" through the use of speciation profiles. Among the most popular chemical mechanisms are CBIV (Gery et al., 1989), SAPRC-99 (Carter et al., 2000) and RADM-2 (Stockwell et al., 1990). Most of the speciation profiles are derived from North-American references (SPECIATE database, U.S. EPA, 2009b).

As described above, compiling and adapting emission inventories to meet the requirements of AQMs are demanding tasks. In order to simplify this process, a series of emission processing systems have been developed. These are capable of processing national or local inventories and global emission databases, as GFED (van der Werf et al., 2010), GEIA (GEIA/ACCENT, 2005) or RETRO (Schultz et al., 2008). For biogenic or mobile sources, some processing systems can be used to directly estimate the emissions. SMOKE is one of them and has been designed to create gridded, speciated and hourly disaggregated emissions inputs for a variety of AQMs in the United States. Adaptations of this model to non U.S. cases have been made for Spain (Borge et al., 2008) and for Europe (Bieser et al., 2010). Another example of emission pre-processor is the *Prep_chem_sources* code (Freitas et al., 2010) designed to work mainly with WRF/Chem, but adaptable to other AQMs.

2.3 Selection of the proper AQM

The complexity of the study case (meteorological effects, terrain interactions) and the scale and significance of potential effects (sensitivity of the receiving environment, human health) must be considered in order to determine which modelling tool is more appropriate for a particular problem. For short-range transport, models with a Gaussian approach like ISC3 and AERMOD, are used to study the interaction of emissions and concentrations of criteria pollutants in urban streets and surrounding urban neighbourhoods. CALPUFF, a non steady-state puff dispersion model, is best fitted for urban scales, with meteorological inputs provided by either on-site measurements and/or a separate meteorological model simulation. The development of the new WRF/Chem model (WRF with Chemistry) constitutes an adaptable and useful tool intended to perform the "on-line" modelling of the chemistry and meteorology over a wide range of scales. Its main applications are the study of secondary pollutants and formation of aerosols in an urban and regional context.

2.3.1 Short-range transport models

ISC3 (Industrial Source Complex; U.S. EPA, 1995) is a steady-state Gaussian plume model appropriate for assessing gas and particle pollutant concentrations from a wide variety of sources associated with an industrial zone. The model includes point, area, line, and volume sources; it can account for settling and dry deposition of particles, building downwash, and includes a limited terrain adjustment. ISC3 operates in both long-term and short-term modes (ISCLT3, ISCST3). The U.S. EPA, recently changed the model status to "alternative model" for regulatory applications, and it suggests using the AERMOD model instead.

AERMOD (American Meteorology Society/EPA Regulatory Model; Cimorelli et al., 2003) is a steady-state plume model. The concentration distribution function differs for the Stable Boundary Layer (SBL) and the Convective Boundary Layer (CBL): a vertical and horizontal Gaussian concentration distribution is considered into the SBL, while a vertical bi-Gaussian probability density function is included for the CBL. The model includes a boundary-layer similarity theory to define turbulence, including a variable altitude treatment for emissions released at different heights. Dispersion coefficients are treated as a continuum rather than as a discrete set of stability classes, including a non-Gaussian representation for unstable conditions, like those observed close to a stack under convective conditions. Variation of turbulence with height allows a better treatment of dispersion for different release heights. The model also uses a simple approach to incorporate terrain interactions. The modelling system consists of one main program (AERMOD) and two pre-processors (AERMET and AERMAP). The main purpose of AERMET is to incorporate all meteorological observations (from surface and upper air stations) and to calculate boundary layer parameters to be used by AERMOD. AERMAP is a terrain pre-processor used to create receptor grids and to generate gridded terrain data from several Digital Elevation Models datasets.

In general, ISC3 and AERMOD do not require significant computational resources, they are easy to use and have simple meteorological requirements. ISC3 is principally designed for calculating impacts on flat terrain regions. The more advanced AERMOD is designed for use with a complex terrain. In any case, Gaussian-plume models are best used for near-field applications where the steady-state meteorology assumption is most likely to apply.

2.3.2 CALPUFF

The CALPUFF modelling system (Scire et al., 2000a) is a multi-layer, multi-species non-steady-state Gaussian puff dispersion model capable of simulating time and space effects of varying meteorological conditions on pollutant transport. The model accounts for a variety of effects such as spatial variability of meteorological conditions, causality effects, dry deposition and dispersion over different types of land surfaces, plume fumigation, low wind-speed dispersion, primary pollutant transformation and wet removal. Recently, an improved chemical module has been included but it is still under evaluation. CALPUFF has various algorithms to include the use of turbulence-based dispersion coefficients derived from a similarity theory or observations. The system has three components: CALMET, a meteorological model that generates three dimensional hourly gridded fields of winds and temperatures, and two dimensional parameters such as mixing height, surface characteristics and dispersion parameters. CALPUFF is the transport and dispersion model that computes the advection of "puffs" containing the material emitted from modelled sources, using the fields generated from the meteorological model. CALPOST is the post-processing tool designed to summarize the results of the simulations (hourly concentrations or deposition fluxes) of CALPUFF.

The model is usually recommended by the U. S. EPA to simulate the effects of pollutant dispersion in long range transport, typically between 50 and 200 km, but contains algorithms which apply to much shorter distances (U.S. EPA, 2008). Advanced models like CALPUFF are more sophisticated than Gaussian models and are aimed at producing more realistic results. However, the potential benefits of using this advanced tool must be compared to the costs of producing more detailed meteorology, complete source inventories and terrain datasets, configuring and running the code.

2.3.4 WRF/Chem

The modelling of the chemical processes in AQMs is usually performed after and independently of the meteorological modelling. This is often called "off-line" integration of the chemical mechanisms, and computes the reactivity differential equations into an external grid containing the information of mass transport and meteorological fields. The development of the novel model WRF/Chem (WRF with Chemistry) allows performing a coupled modelling of the chemistry and the meteorology into a unique coordinate system (Grell et al., 2005; Wang et al., 2010). In this way, a wide range of chemical and physical parameterizations can be used without the need of interpolating them into different spatial and temporal domains. Of all WRF dynamic cores, the mass coordinate version, named Advanced Research WRF (ARW; Wang et al., 2009) possesses the ideal properties to perform the "on-line" modelling of the atmospheric chemistry. The WRF/Chem model has a modular structure which allows considering a variety of coupled physical-chemistry processes such as: transport, deposition, emission, homogeneous and heterogeneous chemistry, aerosol interaction, photolysis, long-wave and short-wave radiative transfer, etc. (Peckham et al., 2010). WRF/Chem application in different fields can explain past episodes, evaluate the potential effects of emission reduction strategies and perform air quality forecasting, always considering the interaction between high resolution chemistry and 3-D meteorology.

Besides the meteorological physical parameterizations (see Section 2.4.3), WRF/Chem model offers a wide range of chemical parameterizations which must be selected in order to perform a forecasting simulation. These include: *i)* the chemical mechanism and the chemical speciation; *ii)* the type of photolysis rate constant computation; *iii)* the aerosol optical properties, *iv)* the biogenic emissions; *v)* the inclusion of biomass and/or plume rise options; *vi)* the dust, gas and aerosol initial and boundary conditions (Grell et al., 2005; Peckham et al., 2010).

One of the most user-dependent features of WRF/Chem is the inclusion of pollutant emission inventories: because of its wide applicability, a time-dependent and high-resolution local inventory must be constructed and adapted by the user to fit the area of interest. By default, the WRF/Chem global configuration includes the implementation of the RETRO (Schultz, 2008) and EDGAR (Oliver & Berdowsky, 2001) emission inventories, which is performed by using the *Prep_chem_sources* application (see Section 2.2). This methodology is limited to very low resolution databases ($\sim 1^{\circ} \times 1^{\circ} \approx 100$ km $\times$ 100 km) with monthly average values of pollutants. Then, the spatial and temporal resolution must be improved if a local or regional air quality forecasting system is to be implemented. Only in the case of spatial domains located inside the U. S., the National Emissions Inventory (NEI) can be used to include high resolution pollutant emissions by means of the *emiss_v3* application. For all other countries, a detailed emission inventory must be developed as described in section 2.2.

2.4 Meteorological data

Meteorological data is a critical input for AQMs, as it is necessary to obtain accurate description of winds, turbulence fields and radiation in order to correctly describe transport, dispersion, deposition and chemical reactions of a released pollutant (Seaman, 2003; Gilliland et al., 2008; Schürmann et al., 2009; Demuzere et al., 2009; Pearce et al., 2011). The meteorological data requirements for local scale models (i.e., steady-state Gaussian, Puff-models) and more complex models vary considerably.

2.4.1 Meteorological data for Gaussian models

Steady-state Gaussian plume models, like ISCST3 (U.S. EPA, 1995) and AERMOD (Cimorelli et al., 2003), need data only from a single station, since they assume that meteorological conditions do not vary throughout the domain up to the top of the boundary layer. Although both codes can be used with screening options generally available from the model website, this is only suggested to generate first conservative guess estimates of ground level concentrations. The most appropriate practice is the use of existing data sets available in the study area providing enough accuracy to meet the local environmental authority criteria. National and local weather services and private consultants usually produce site-specific meteorological data sets and provide advice on its applicability for specific modelling studies. Also, the existing data are supplemented by new measurements from meteorological automatic stations collected on the site of interest. This approach is fully covered in U.S. EPA (2000).

2.4.2 Meteorological data for advanced models

More advanced models (both puff and grid models), allow meteorological conditions to vary across the modelling domain and up through the atmosphere, requiring thus more complex meteorological data. An example is the CALPUFF modelling system (Scire et al., 2000a). Since there are no meteorological measurements at every point of the domain, these models use pre-processed data for analysis. The CALMET meteorological model (Scire et al., 2000b) is a diagnostic model developed as a component of the CALPUFF modelling system for its use in air quality applications. CALMET in its basic form is designed to produce hourly fields of three-dimensional winds and various micro-meteorological variables based on the input of routinely available surface and upper air meteorological observations.

2.4.3 Mesoscale meteorological models: WRF

All prognostic models contain realistic dynamic and physical formulations, and can produce the most realistic meteorological simulations for regions where data are sparse or non existent. If local meteorological data is unavailable, the use of a prognostic modelling system could be a sensible option as part of a regulatory assessment.

The Weather Research and Forecasting Model, WRF (Skamarock et al., 2008) is now the most commonly used prognostic meteorological model. The WRF model is a state-of-the-art model developed and maintained by NCAR, NOAA/NCEP and other research centers (Michalakes et al., 2005). It is an open source code originally conceived as a next-generation mesoscale model oriented to cover both atmospheric research studies and operational forecasting. The model can be run in a nested way with the outer domain on a regional scale, covering distances usually in the order of 500–1,000 km. All domains are initialised using analyses from global models, like the NCEP Global Forecasting System (NCEP-GFS), which describe the three-dimensional fields of temperature, wind speed and direction, and moisture in global 1.0 x 1.0 degree grids prepared operationally every six hours for the whole world.

The WRF model can be easily configured to select and define the size and resolution of the computational domain. Moreover, it allows introducing "nested" domains, where smaller or child domains with higher resolution are inserted into bigger or parent domains. The resolution ratio between child and parent domains is usually a factor of three, which is a compromise between improving the spatial resolution and increasing the computing time.

In order to configure a WRF simulation, it is necessary to define the modeled domain and its properties, as well as to determine the initial and boundary conditions for the selected case. This is usually performed by the WRF Pre-processing System (WPS). This tool allows setting up the model by including: *a)* static databases like terrain elevation (with different spatial resolutions), Land Use Land Cover (LULC) fields and surface levels; and *b)* initial values for boundary meteorological conditions (with temporal resolution), the number of vertical levels, among others.

Before the forward forecasting is performed, a number of physical and dynamic parameterization must be selected. This includes a wide range of options from simple and efficient modules (used for operational purposes) to sophisticated and computationally expensive routines (for atmospheric research purposes). The physical parameterizations are classified in: *i)* micro-physic schemes; *ii)* long-wave and short-wave radiative modules; *iii)* surface layer models; *iv)* land surface models; *v)* Planetary Boundary Layer (PBL) schemes; and *vi)* cumulus parameterizations; among others.

2.4.4 CALWRF

When modelling a large domain with complex geography or when there is a lack of meteorological data, it is possible to use the outputs of any mesoscale meteorological model as input to the AQM. That is the case of CALWRF, an external tool developed to perform a CALPUFF simulation including WRF output data. CALWRF is basically a data filter and format converter code, which produces an intermediate three dimensional data file which is used by CALMET as a "first-guess" meteorological field. Then, CALMET merges these initial and boundary conditions with the terrain elevation data and land-use, and CALPUFF is run in its usual diagnostic mode. This approach not only increases the horizontal resolution of the meteorological fields, but also reduces the computational time that running the WRF/Chem model with a greater resolution would require.

2.5 Geophysical data

Terrain features around a pollutant source can significantly affect the pattern of dispersion. Steady-state Gaussian models like ISC3 contain limited algorithms to include terrain effects. Advanced models like CALPUFF contain more sophisticated procedures for modelling the effects of terrain, with a correspondingly greater effort required by the user to specify the static data. Since terrain data will be required for every receptor on the grid, there are several pre-processing tools that extract and format the Digital Elevation Model (DEM) data. The most common global data sets are: the United States Geological Service (USGS) GTOPO30 with a horizontal grid spacing of 30 arc-seconds (approximately 1km); USGS SRTM30, with the same horizontal grid spacing, but covering the globe only from 60° N latitude to 56° S latitude, with a seamless and uniform representation; and SRTM3 data with a horizontal grid spacing of 3 arc-seconds (about 90 m). Advanced models also need surface parameters, generally as gridded fields, to compute properly the dispersion of pollutants. These include surface roughness length, albedo, Bowen ratio, soil heat flux parameter, vegetation leaf area index and anthropogenic heat flux. Land Use and Land Cover (LULC) data are also available from the USGS, at the 1:250,000 scale, or in some cases at the 1:100,000 scale. The USGS Global Land Cover Characterization (GLCC) Database (GLCC; U.S. Geological Survey (USGS), 2010) is developed in a continent basis for land use, while land cover maps are classified into 37 categories, with a spatial resolution of 1 km.

2.6 Model evaluation

Evaluation of an AQM is the process of assessing its performance in simulating spatial-temporal features embedded in the air quality observations. The Atmospheric Modelling and Analysis Division of U.S. EPA classifies the different aspects of model evaluation under four general categories: operational, dynamic, diagnostic, and probabilistic: *a)* Operational performance evaluation is accomplished by comparing model simulated values, against observed data. The concept applies not only to air quality parameters, but also to meteorological variables in the case of advanced coupled models like WRF; *b)* Dynamic evaluation focuses on assessing the AQM response to changes in emissions and meteorology, which is central to applications in air quality management. This type of analysis can help in determining the key factors governing air pollution; *c)* Diagnostic evaluation investigates the processes and input drivers that affect model performance in order to find possible improvements to the algorithms; and *d)* Probabilistic evaluation generally requires a set of tools that helps addressing the model response to statistically varying inputs, without having to run the simulation for every case, which is very time consuming.

When evaluating air quality management strategies, policy-makers need information about relative risk and likelihood of success of different options. In these cases, a range of values reflecting the model uncertainties is more important than the model best guess, or actual output. End users are more likely to work with operational and dynamic evaluation tools, while the other two categories of evaluation are more related to model development.

The kind of data needed for verifying model output, will depend on the model itself and the user's needs. For models with meteorological pre-processors, like CALMET, or coupled meteorological/chemical models like WRF/Chem, atmospheric variables observation in some points of the domain would be required in order to validate results. Observations can be made at ground level or with a vertical profile, in the case of three dimensional simulations. In the case of chemical species concentration, monitoring stations could supply data needed to check model results. Some ground or satellite instruments can also provide vertical profile for chemical species (Martin, 2008). In any case, a consistent procedure should be applied in order to evaluate the model performance.

The most usual practice is to use the information content shown between the observed and the model-predicted values. In this respect, Willmott (1982) and Seigneur et al. (2000) propose some statistical performance measures. Table 1 summarizes these commonly used statistical tools for analysing and comparing model results versus measurements.

Indicator	Formula	Range	Ideal value	Comments
Correlation coefficient	$r = \frac{\sum_{i=1}^{N}\left[(C_{oi} - \bar{C}_o)(C_{pi} - \bar{C}_p)\right]}{\sqrt{\sigma_o \sigma_p}}$	[0,1]	1.0	C_o and C_p are observed and predicted concentrations.
Mean Bias	$MBE = \frac{1}{N}\sum_{i=1}^{N}(C_{pi} - C_{oi})$	n.a.	0.0	C_{pi} and C_{oi} are observed and predicted concentrations for N cases.
Fractional Bias	$FB = \frac{\bar{C}_o - \bar{C}_p}{0.5(\bar{C}_o + \bar{C}_p)}$	[-2,2]	0.0	$\bar{C}_O$ and $\bar{C}_P$ are the averages of observed and predicted concentrations.
Root mean square error	$RMSE = \sqrt{\frac{1}{N}\sum_{i=1}^{N}(C_{oi} - C_{pi})^2}$	n.a.	0.0	
Normalized mean square error	$NMSE = \frac{\overline{(C_o - C_p)^2}}{\overline{C_o C_p}}$	n.a.	0.0	
Index of agreement	$d = 1 - \frac{\sum_{i=1}^{N}(C_{pi} - C_{oi})^2}{\sum_{i=1}^{N}\left(\lvert C_{pi} - \bar{C}_o\rvert + \lvert C_{oi} - \bar{C}_p\rvert\right)^2}$	[0,1]	1.0	
Geometric Mean	$MG = \exp\left(\ln\overline{(C_o)} - \ln\overline{(C_p)}\right)$	> 0	1.0	
Geometric variance	$VG = \exp\left[\left(\ln\overline{(C_o)} - \ln\overline{(C_p)}\right)^2\right]$	> 0	1.0	

Table 1. Statistical indicators for the evaluation of AQMs.

3. Air pollution modelling in Argentina: results of several study cases

The following study cases were selected with the aim of describing the simulation process, from the selection of the appropriate model until the evaluation of final results.

3.1 Urban and regional modelling in Mendoza: inventories and monitoring

The extended or Gran Mendoza urban area is located in the north-central part of the Province of Mendoza (32.8° S; 68.8° W), Argentina, in a region of foothills and high plains, on the eastern side of the Andes Range. The metropolitan population is around 1,000,000 inhabitants, making this city the fourth largest metropolitan area in the country. The area is located in an arid to semi-arid zone of low rainfall: 120–400 mm/yr, mean 230 mm/yr, which occurs especially during the summer months (November to March).The closeness of the Andes Mountains has a strong influence on local meteorology and, therefore, on air quality.

Air quality in the area under study is affected by intensive and intermediate industrial activities, emissions due to transportation, residential sources and, to a lesser degree, agriculture and animal husbandry (Puliafito et al., 2003). Food and wine industry are very low energy consumers, but alloys, cement and petrochemical industries are relevant to energy consumption and their contribution to total emissions is important. Transportation is the second most important source of pollutants in the emission inventory, but the main concern in the downtown area. The emissions by the residential sector are mainly due to the use of natural gas for heating. Animal husbandry is concentrated in the South East of the Province and its contribution to total emission is low.

3.1.1 Modelling approaches

The impact of all anthropogenic sources in the air quality of the urban centre was evaluated using two different approaches: *a)* to reproduce the pollutant dispersion in the metropolitan area, we chose CALPUFF modelling system, setting the modelling domain in the urban area of Gran Mendoza; and *b)* to evaluate the impact of regional circulation in tropospheric chemistry, we used WRF/Chem in a greater modelling domain.

3.1.2 Urban pollutant dispersion

The modelling domain covers an area of 41×36 km^2 (32.8° S to 33.1° S and 68.8° W to 69.0° W) including the urban zone and part of the Andes. Because of the complex topography, detailed terrain features were incorporated using the USGS global 3 arc-sec SRTM3 data (~ 90m resolution). The LULC data was obtained from the Global Land Cover Characterization (GLCC) database captured by a 1-km resolution. Finer features of land use were obtained using the soil classification from the EcoAtlas Program from Mendoza's Rural Development Institute (IDR), which includes the interpretation and classification of Landsat images during the year 2006. Surface and upper meteorological data was obtained from the National Weather Service at the local airport station El Plumerillo (32.78° S, 68.78° W, 704 m above sea level) located at 6 km North-East of the urban center. Fig. 3 shows a view of the modelling domain. Industrial sources are located in two industrial areas in the periphery of the city. The position and emission rates for the industrial stacks are known, measured and compiled by the local environmental authority. Twenty one fixed (point) sources at constant rates for the modelling period were considered. Emissions from residential and commercial sources were estimated using data of natural gas consumption in the city and spatially distributed according to the land-use maps of the urban center.
A top-down approach (macro scale) associated to GIS was used to calculate mobile source emissions and their spatial distribution. The COPERT III model (Ntziachristos et al., 2000) was selected to estimate road transport emissions because of the similarities between the Argentinean and the European fleets (D'Angiola et al., 2010). Emission factors for Natural Gas Vehicles (NGV) were derived from local studies (ARPEL, 2005). The vehicle fleet was arranged into 4 vehicle classes and 28 technology categories depending on fuel type, vehicle size, fuel delivery system and exhaust control system; 19 of these categories, belong to light duty vehicles and passenger cars, 6 to heavy duty vehicles and 3 to urban buses.
Information from the National Vehicle Registration Directory (DNRPA) and the Automobile Manufacturers Association (ADEFA) was used to determine the fleet composition. To derive

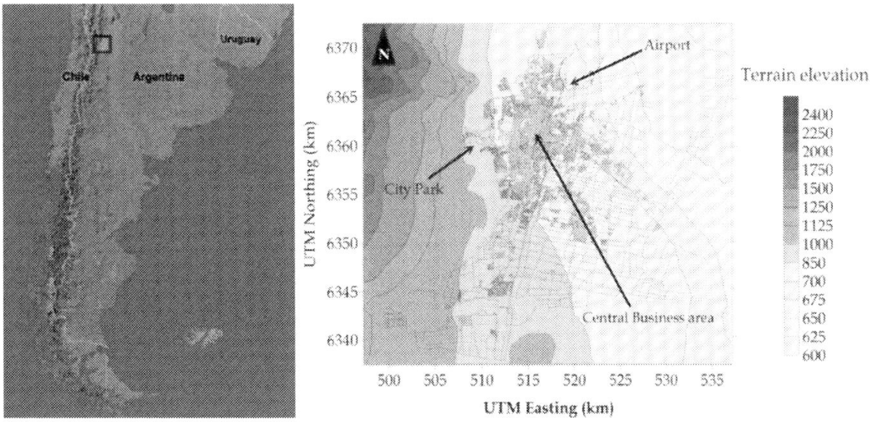

Fig. 3. Geographical location of the City of Mendoza (left) and terrain elevations (metres a.s.l.) in the modelling domain (right).

emission rates, the model considers different average speeds for different types of driving conditions. Three driving conditions were associated with specific road hierarchies: *a)* Highways: roads characterized as main intra-county, suburban areas or inter-regional highways connecting main town poles in the metropolitan area, with high traffic, no traffic lights and high average speed (70-100 km/h); *b)* Primary roads: roads that are main streets connecting important urban districts, with a high vehicle density, with most intersections regulated by traffic lights, and a low average speed (20-30 km/h); *c)* Secondary roads: roads that are mainly residential streets with a low vehicle density, very few or no traffic lights regulating intersections, although, in some cases, with the presence of speed limiters and a low-medium average speed (25-35 km/h). Average speed on each road hierarchy and annual average kilometres traveled (VKT) were estimated from information collected on a set of vehicles equipped with a Global Positioning Satellite (GPS) unit. The total amount of emissions estimated for highways, primary roads and secondary roads was distributed in the road network proportionally to the length of segments. Then, these emissions were geographically distributed into a gridded map with cells of 500 m x 500 m (Fig. 4, left).

Fig. 4 (right) depicts the daily mean ambient PM_{10} concentration modeled with CALPUFF. It shows concentrations of (70−80 $\mu g/m^3$) in downtown area, mainly produced by mobile sources. These values reduce towards the periphery of the city. The south-western area of the modelling domain is greatly influenced by the industrial complex located in the south western area of the city. PM_{10} emissions from dust sources were estimated according to OAQPS (1977) and modeled as area sources in CALPUFF. The emissions were incorporated as a discontinuous process when wind speeds were above a threshold of 6.8 m/s. Fig. 5 shows a time series comparison of the modeled results versus the PM_{10} measurements at the monitoring station in downtown for June 2009. Despite some peak values due to high traffic density the results show a good correlation between modelled concentration and measured data. A more detailed analysis of pollutant distribution in the metropolitan area of Mendoza can be found in Puliafito & Allende (2007) and Puliafito et al. (2011).

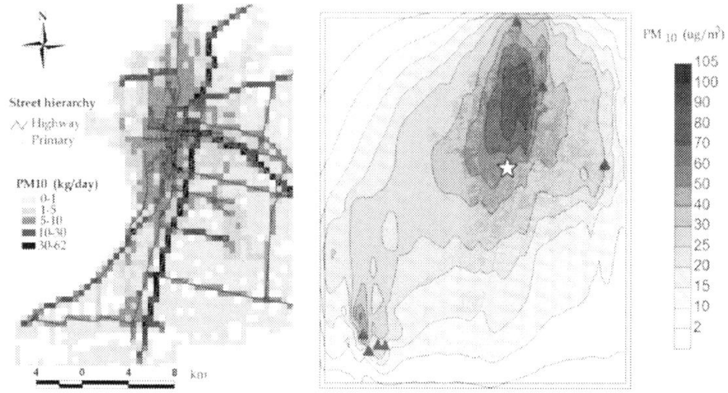

Fig. 4. Left panel: distribution of PM_{10} emission from mobile sources (kg/day). Right panel: spatial distribution of PM_{10} daily mean concentration values. Black triangles show the location of the main industrial sources. The white star indicates the location of a monitoring site.

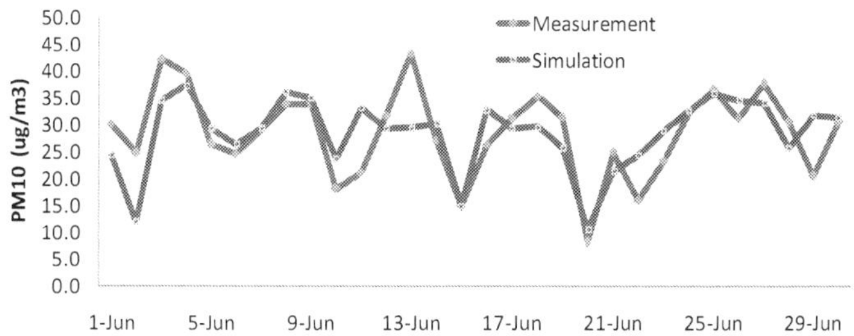

Fig. 5. Comparison between measured and simulated PM_{10} concentrations at the measurement site, indicated in Fig. 4.

3.1.3 Regional modelling

The mountain-valley circulation dominates the medium-range-transport in the area, so not only urban but also regional air pollution levels are influenced by these synoptic scale weather patterns. Then, in order to study the temporal and spatial distribution of primary and secondary pollutants in the northern region of the Province of Mendoza, the WRF/Chem model was used. Three nested domains with 28 vertical altitude levels were defined to downscale the chemical and physical properties of the atmosphere. The smaller domain covers the northern region of Mendoza Province (4 km resolution, 200 km N/S x 160 km E/W), including the urban center.

In order to capture the complexity of the terrain features in the study area, the WRF Preprocessing System (WPS) was modified to introduce the Shuttle Radar Topography Mission (SRTM3) data, increasing standard resolution over 10 times. Also, LULC maps developed by several local institutions and agencies were adapted to the WPS module using a GIS system (Fernandez et al., 2010). The emission inventory for Mendoza was developed to meet WRF/Chem requirements by means of adapting the *emiss_v3* routine, originally developed to process the U.S. National Emissions Inventory database (NEI; U.S. EPA, 2005). Boundary and initial conditions for concentrations were taken from global chemical model MOZART (Emmons et al., 2009). A 24 hour spin up was used for the inner domain. Default parameterization schemes were used in the model runs (see Peckham et al. (2010) for detailed descriptions of the model configuration).

Fig. 6 shows the simulated NO_2 surface concentration in p.p.m. during the night (left) and day (right) for a typical spring day in Mendoza. The contour lines represent the altitude of the calculated PBL within the WRF/Chem model. The model captures the behavior of the PBL height: at night, it is higher over the mountains (west) than over the valley, while during the day this effect is the opposite, surpassing 1,500 m right over the foothills. The NO_2 plume direction follows the prevailing wind patterns in the region, i.e., during the day pollutants emitted in the city are transported towards the South-West, while during the night the same air parcels return to the city in the opposite direction. Absolute values are greater during the night, mainly because of the smaller dilution volume limited by the PBL altitude. Here, the reader should recognize that, because of the local topography of Mendoza, the mountain-valley circulation is not exactly West-East, but shifted SW-NE.

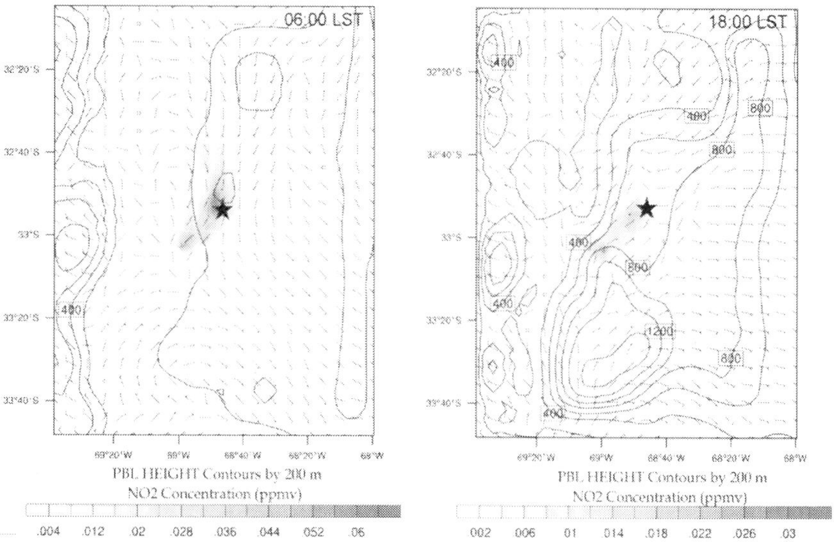

Fig. 6. Regional NO_2 concentrations during the night (left) and day (right). The star indicates the location of Mendoza City. The surface plots represent the altitude of the PBL in m. Wind barbs are superimposed in the figure.

3.2 Study case of Bahía Blanca

The district of Bahía Blanca (62.2° W; 38.5° S) is located on the southern Atlantic coast of the Province of Buenos Aires, in the boundary between the southern Pampas and the northern Patagonia. Bahía Blanca and its surroundings concentrate a population of 310,000 inhabitants in the urban area. The landscape is mainly flat with terrain elevations between

200 m and 500 m, except for the low mountain system of "Sierra de la Ventana", which reaches an altitude of 1,200 m. The area is located in a temperate zone with warm continental type weather. Its annual precipitation is about 600 mm, with high monthly variations. March, September, October and November are the months with higher probability of rainfall episodes. Winds are generally moderate, with predominant direction from North or Northwest. Mean wind speed is 5.3 m/s, with 3 % calms.
Bahía Blanca has an important seaport system handling intense industrial and commercial activities. The city has a highly developed oil refinery and petrochemical industries, producing ethanol, naphtha, gasoline, asphalt GLP, ethylene, PVC, polyethylene, urea, ammonia, chlorine and caustic soda. An important power station is located outside the industrial zone, SW of the urban centre, in the port area, which may operate with natural gas or heavy-oil, producing the main contribution of SO_2 emissions in the zone. Along the north-eastern shore of the bay, there are several small seaports dedicated to grains and containers. The ambient concentration of gases like NO_x, CO, NH_3 and SO_2 and concentrations of particulate matter have been measured by the local environmental authority since 1997.

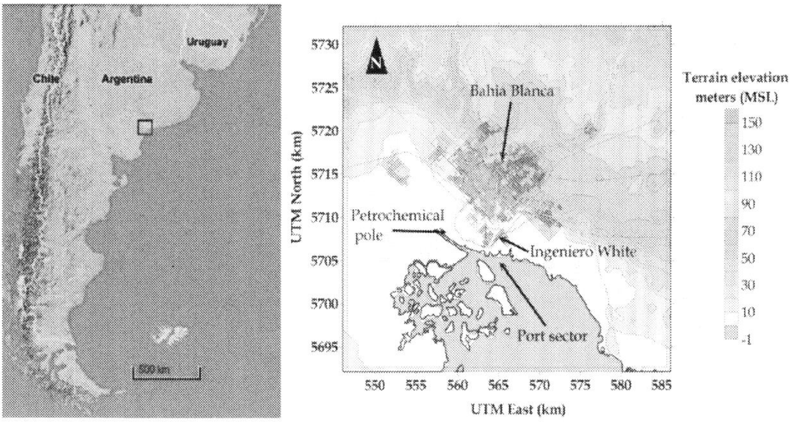

Fig. 7. Map of Argentina showing location of the modelling domain. To the right, a detail of the urban area of Bahia Blanca, major highways, the industrial complex, and the port area.

In order to complement the information provided by the monitoring data, pollutant dispersion was simulated with the CALPUFF modeling system. The modelling domain covered an area of 1,600 km^2 between 38.5° and 39.0° S latitude and 62.0° and 62.5 W longitude in a flat region with elevations up to 150 m, increasing to the NE. A view of the modelling domain is presented in Fig. 7. All industrial sources are located on a Petrochemical Pole, an Industrial Park and the port sector and were treated as point sources in CALPUFF.
Emissions of NO_x, SO_2 and NH_3 due to transportation (including road, railway and ships) were estimated following the top-down emission model COPERT III similarly to the Mendoza study case. Cargo oriented railway transportation emissions were estimated using specific fuel consumption factors and a number of train operations, while maritime activity emissions were calculated using a number of operations, port waiting time and bulk

emission factors. Both of them were calculated with the simplest CORINAIR methodology (EMEP/EEA, 2009). The sum of all emissions from mobile sources was spatially distributed using Geographical Information System (GIS) tools in a regular grid of 500 m x 500 m, covering the whole modelling domain. The transportation emissions were included in the dispersion model as area sources. Residential and fugitive sources were also included in the simulation, estimated with natural gas consumption statistics, as in the Mendoza study case.
Meteorological data was obtained from the local airport station Comandante Espora (38° 42′ S, 62° 09′ W, 75 m above sea level) located 12 km East of the urban center. The model runs were conducted over one year, and five chemical species (NO_x, SO_2, NH_3, SO_4^{2-} and NO_3^-) were simulated. The model includes the MESOPUFF II scheme to incorporate chemical transformation of SO_2 and NO_x. Monthly background concentrations for ammonia and ozone were provided by the environmental authority monitoring data over one year period. Fig. 8 shows the contribution of each source type to the total emissions of CO, SO_2, NO_x, PM_{10} and NH_3 for Bahia Blanca (left) and the average 24-h concentrations for NO_x (right).
Finally, the 24 h concentrations obtained with CALPUFF were compared to the measured ones in the monitoring station. Fig. 9 shows the comparisons between the modelling results

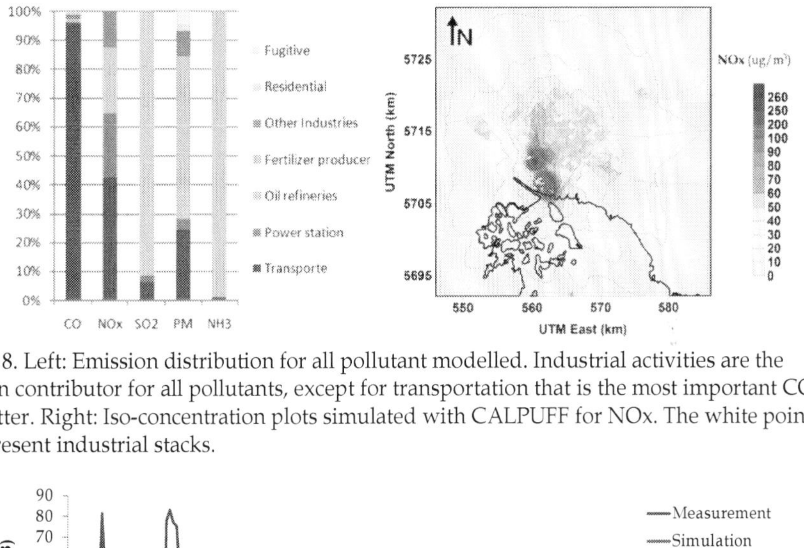

Fig. 8. Left: Emission distribution for all pollutant modelled. Industrial activities are the main contributor for all pollutants, except for transportation that is the most important CO emitter. Right: Iso-concentration plots simulated with CALPUFF for NOx. The white points represent industrial stacks.

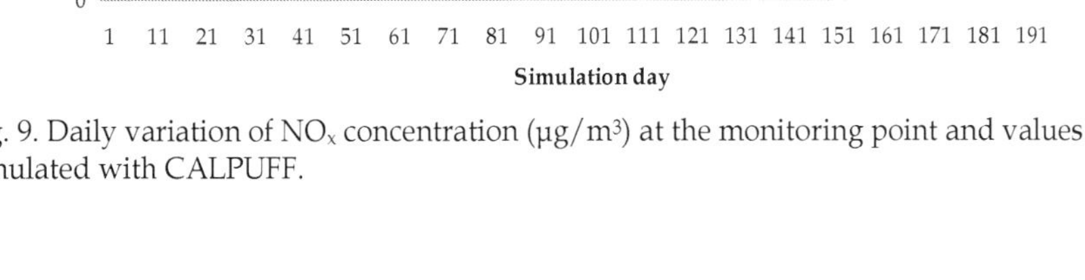

Fig. 9. Daily variation of NO_x concentration (μg/m³) at the monitoring point and values simulated with CALPUFF.

and the observations near the industrial complex. Although the model is unable to reproduce the concentration peaks exactly, the values of the statistical measures (MBE: -2.02; FB: -0.04; Willmott's d: 0.97) suggest that the dispersion simulation approaches the concentrations patterns with acceptable accuracy. The modelling results showed that the most polluted region is the town of Ingeniero White, located to the SE of the metropolitan area, and very close to the Petrochemical Pole (see detailed analysis in Puliafito & Allende (2007) and Allende et al. (2010a)). The fair agreement between concentrations measured by the environmental authority over one year and the simulated ones with CALPUFF validated the proposed emissions and dispersion models.

3.3 The Buenos Aires case

The City of Buenos Aires (34.6° S, 58.3° W) is the Capital and largest city of Argentina, and the second largest metropolitan area (around 13 million inhabitants), in South America after São Paulo, Brazil. It is located on the western shore of the Río de la Plata estuary, on the south-eastern coast of the South American continent. Buenos Aires has a humid subtropical climate (Köppen classification) with four distinct seasons, January being the warmest month and July the coldest. Relative humidity tends to be high throughout the whole year (~ 72%). Winds that blow clean air from Río de La Plata towards the city have an annual frequency of 58.0%, and there are 3.8% calms. At present, in spite of its extension and its high population density, Buenos Aires City has only a small air quality monitoring network, mainly dedicated to capture traffic pollution in 4 points of the city. Besides this fact, there have been several measurement campaigns at different sites in the city (Arkouli et al., 2010; Bogo et al., 1999; Mazzeo et al., 2005; Reich et al., 2006). Even though there are no large industries inside the city, a great industrial complex is located in the southeast. Also, four large partly gas and heavy-oil fired power stations are located close to the river shore.

For the air pollution evaluation in the area we used the CALPUFF model in a 25 km x 25 km modelling domain, extending from 34.5° S to 34.7° S and 58.3°S W to 58.6° W in a flat region with elevations up to 40 m increasing to the SW (Fig. 10). Surface meteorological observations were obtained from the local airport meteorological station in Aeroparque (34°34′ S, 58°30′ W). Since CALMET requires twice daily upper air meteorological

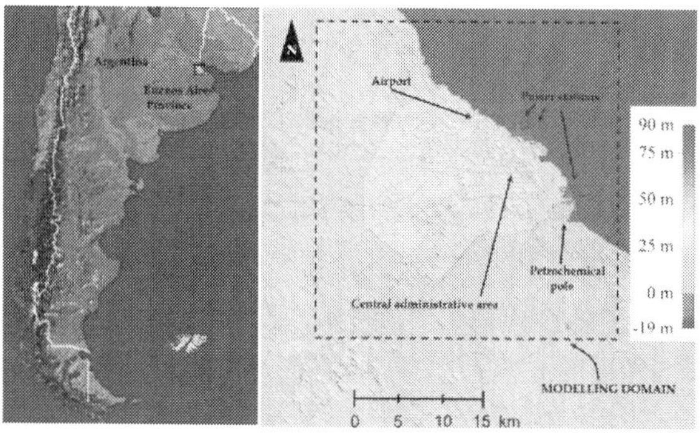

Fig. 10. Map of Argentina (left) and location of the modelling domain (right). Terrain elevations are in metres. In red are depicted major highways and the triangles indicate the location of the power stations.

observations for dispersion simulating, and since observations with such frequency are not available for this site, the required meteorological parameters were derived from the Weather Research and Forecasting model. The CALWRF off-line preprocessor was used to generate an intermediate three dimensional data file to be used by CALMET as a "first-guess" meteorological field. The WRF model was run with nested domains, reaching a resolution of 3 km, high enough to resolve the important meteorological features which can only be simulated by the prognostic WRF model, such as land and sea breezes.

Emissions from the power stations were estimated using emission factors of CORINAIR (EMEP/EEA, 2009) for most pollutants and IPCC (IPCC, 2006) for greenhouse gases, according to their thermal capacity and use of fuel. The emissions from oil refineries, chemical industries and storage tanks located very close to the city, near the SE border, were previously measured and included directly in the CALPUFF model. All these emissions were associated to 54 point sources. Temporal allocation of emissions was estimated using statistical data from the local Electricity Regulation Office and gross domestic product evolution for the industrial sector.

Emissions of NO_x, SO_2 and CO due to transportation were estimated following the top-down emission model COPERT III similar to the study cases of Mendoza and Bahía Blanca (Perez Gunella et al., 2009). Daily and monthly variation of road traffic in the city is often measured, with traffic rush hours during working days being from 08:00 to 09:00 a.m. and from 07:00 to 09:00 p.m. These emissions were geographically distributed into a gridded map with cells of 300 m x 300 m (Fig. 11, right).

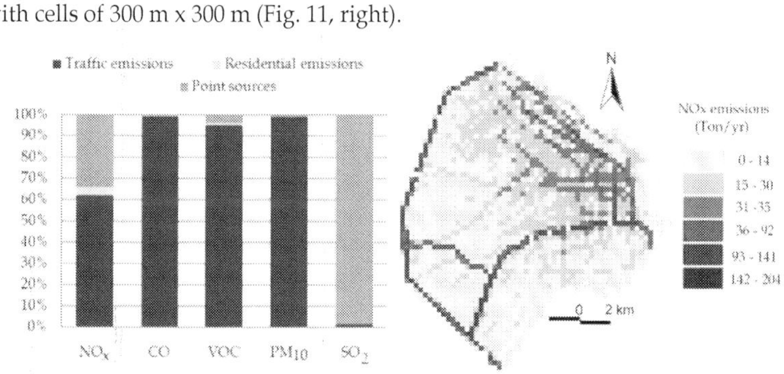

Fig. 11. Left: Emission contributions for all pollutants simulated in the study area. Right: Spatial allocation for NO_x traffic emissions in the city of Buenos Aires.

Residential emissions were associated with natural gas and LPG consumption, and estimated using emission factors from CORINAIR. Spatial allocation of the emissions was made proportional to the population density in the city districts and further weighted with the Unsatisfied Basic Needs index (NBI) to account for the population socioeconomic levels. The emission distribution from all sources is shown in Fig. 11 (left), and concentration patterns obtained with CALPUFF are shown in Fig. 12.

SO_2 concentration maxima are clearly located near the power stations. The road traffic appears to be responsible for high NO_x values near major roadways (Puliafito et al., 2010; Allende et al., 2010b). Monitoring values obtained in a toll gate located in the south of the city were compared to simulated ones. For CO, the comparison is shown in Fig. 13.

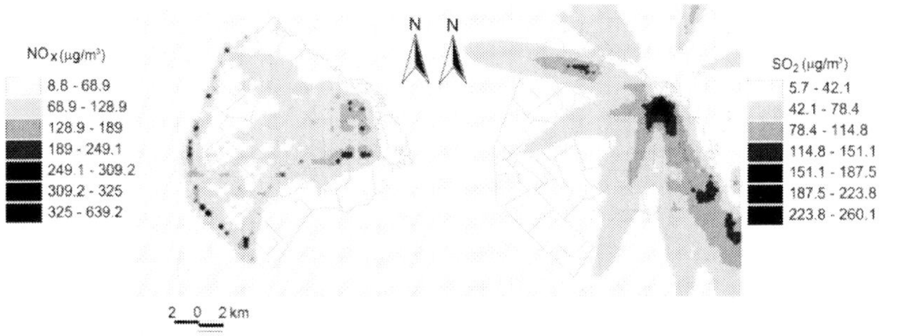

Fig. 12. 24 h average iso-concentration plots simulated with CALPUFF for NOx and SO_2.

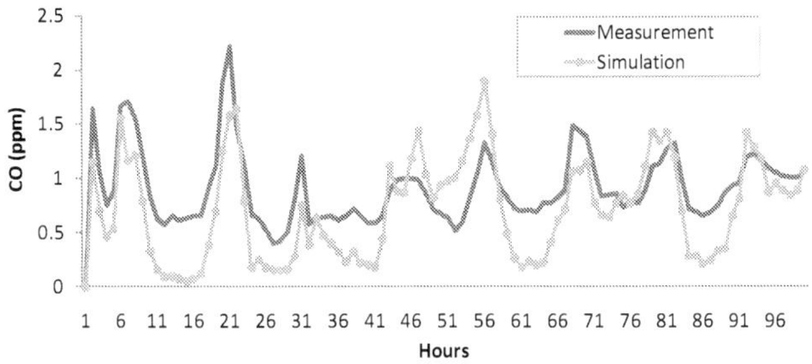

Fig. 13. Hourly variation of CO concentration (p.p.m.) at the monitoring point and values simulated with CALPUFF.

4. Summary and concluding remarks

The purpose of this chapter was to provide an overview of several modelling tools available for air quality evaluation, providing general and prescriptive but also flexible recommendations toward the adoption of best practice. Air Quality Models (AQMs) are mathematical tools that simulate the physical and chemical processes that involve air pollutants as they disperse and react in the atmosphere. The use of AQMs improve the limitations faced by the use of the monitoring network approach by providing prediction of the temporal and spatial distribution of actual pollution levels. Modelling studies, in combination with air quality monitoring are essential and complementary tools for long and short term air pollution control strategies. A well calibrated model is a unique tool that allows the representation of the atmosphere dynamics and chemistry, including real conditions and atmospheric disturbances.

From a theoretical standpoint, AQMs were divided in short-range Gaussian models (ISC3, AERMOD), advanced urban models (CALPUFF) and complex Chemical Transport Models coupled to mesoscale atmospheric models (WRF/Chem). The greatest differences between the model types are the meteorological input required, computer resources and user experience. For each of these models, we have discussed the configurations, data requirements, applicability and some physical and chemical formulations. We have identified several important aspects for selecting the proper AQM model: *a)* identification of the proper scale problem; *b)* the availability of meteorological data; *c)* the preparation of emission inventories; *d)* the inclusion of land use and topographical information; *e)* the integration of results in an adequate geographical information system; and finally *f)* a comparison of the simulated data to existing monitoring information, if any. From a practical point of view, and following the above scheme, the three study cases presented at the end of this chapter give practical implications for the selection of an AQM applied to different data availability, scale, topography and meteorology. We have described how to prepare the emission inventories for each case, and used local and regional AQMs to calculate the pollutant concentration. The acceptable accordance between the simulated concentrations and the monitored data confirms the proposed methodology described in this chapter.

Finally, we hope, this chapter encourages readers to use "state of the art" models since they provide the most realistic representation of the phenomena involved in the fate of pollutants released to the atmosphere.

5. Acknowledgements

We thank the authorities of the National Technological University of Argentina (UTN) and the Argentine Research Council (CONICET) for their support in our research activities. We further thank the Argentine National Weather Service (Servicio Meteorológico Nacional, SMN) for the availability of the meteorological data. This research was supported partially by grant PICT2005 # 23-32686 from the National Science Agency (Agencia Nacional de Promoción Científica y Tecnológica, ANPCyT).

6. References

Allende, D. G., Castro, F. H., & Puliafito, S. E. (2010a). Air Pollution Characterization and Modeling of an Industrial Intermediate City. *International Journal of Applied Environmental Sciences*, 5(2), 275-296.

Allende, D. G., Cremades, P., Puliafito, Enrique, Fernandez, R. P., & Perez Gunella, F. (2010b). Estimación de un Factor de Riesgo de Exposición a la Contaminación Urbana par ala población de la Ciudad de Buenos Aires. *Avances en Energías Renovables y Medio Ambiente*, 14, 127-134.

Arkouli, M., Ulke, A. G., Endlicher, W., Baumbach, G., Schultz, Eckart, Vogt, U., et al. (2010). Distribution and temporal behavior of particulate matter over the urban area of Buenos Aires. *Atmospheric Pollution Research*, 1(1), 1-8.

ARPEL. (2005). Medición de emisiones de vehículos en servicio en San Pablo, Santiago y Buenos Aires. *Informe Ambiental ARPEL* Número 25-2005.

Ballesta, P. P., Field, R. A., Fernandez-Patier, R., Madruga, D. G., Connolly, R., Caracena, A. B., et al. (2008). An approach for the evaluation of exposure patterns of urban populations to air pollution. *Atmospheric Environment*, 42(21), 5350-5364.

Bieser,J., Aulinger,A., Matthias,V., Quante, M., & Builtjes,P. (2010). SMOKE for Europe - adaptation, modification and evaluation of a comprehensive emission model for Europe. *Geoscientific Model Development Discussions.*, 3, 949-1007.

Binkowski, F. S., & Roselle, S. J. (2003). Models-3 Community Multiscale Air Quality (CMAQ) model aerosol component 1. Model description. *Journal of Geophysical Research - Atmospheres*, 108(D6), 4183.

Bogo, H., Negri, R. M., & Roman, E. S. (1999). Continuous measurement of gaseous pollutants in Buenos Aires city. *Atmospheric Environment*, 33(16), 2587-2598.

Borge, R., Lumbreras, J., & Rodriguez, E., 2008, Development of a high-resolution emission inventory for Spain using the SMOKE modelling system: A case study for the years 2000 and 2010. *Environmental Modelling and Software*, 23, 1026–1044.

Brunekreef, B., & Holgate, S. T. (2002). Air pollution and health. *Lancet*, 360(9341), 1233-42.

Carter, W. P. L. (2000).Documentation of the SAPRC-99 Chemical Mechanism for VOC Reactivity Assessment. *Report to the California Air Resources Board*, Available at http://cert.ucr.edu/~carter/absts.htm#saprc99

Cimorelli, A. J., Venkatram, A., Weil, J. C., Paine, R., Wilson, R., Lee, R., et al. (2003). AERMOD : Description of model formulation. Environmental Protection Agency.

Davis, N., Lents, J., Osses, M., Nikkila, N., & Barth, M. (2005). Development and Application of an International Vehicle Emissions Model. *Transportation Research Record*, 51-59.

Demuzere, M., Trigo, R. M., Arellano, J. V.-guerau D., & Lipzig, N. P. M. V. (2009). The impact of weather and atmospheric circulation on O3 and PM10 levels at a rural mid-latitude site. Atmospheric Chemistry and Physics, 9, 2695-2714.

Dodge, M. C. (2000). Chemical oxidant mechanisms for air quality modeling: critical review. *Atmospheric Environment*, 34(12-14), 2103-2130.

D'Angiola, A., Dawidowski, L. E., Gomez, D. R., & Osses, M. (2010). On-road traffic emissions in a megacity. *Atmospheric Environment*, 44(4), 483-493.

EMEP/EEA. (2009). EMEP/EEA Air Pollutant Emission Inventory Guidebook – 2009. Technical guidance to prepare national emission inventories. Copenhagen. Available from: www.eea.europa.eu/help/infocentre/enquiries

Emmons, L., Walters, S., Hess, P. G., Lamarque, J., Pfister, G. G., Fillmore, D., et al. (2009). Description and evaluation of the Model for Ozone and Related chemical Tracers, version 4 (MOZART-4). *Geosci. Model Dev. Discuss.*, 2, 1157-1213. Max-Planck-Institute for Meteorology, Hamburg, Germany.

Endlicher, W., Zahnen, B., Schultz, E., Alessandro, M., Mikkan, R., & Polimeni, M. (1998). Clima Urbano y Contaminación Atmosférica en Mendoza/Argentina. In : *Clima y Ambiente Urbano en Ciudades Ibéricas e Iberoamericanas* (pp. 115-134). Madrid.

Fenger, J. (1999). Urban air quality. *Atmospheric Environment*, 33(29), 4877-4900.

Fernandez, R., Allende, D., Castro, F., Cremades, P., Puliafito, E. (2010). Modelado regional de la calidad de aire utilizando el modelo WRF/Chem : implementación de datos globales y locales para Mendoza. *Avances en Energías Renovables y Medio Ambiente*, 14, 01.43-01.50.

Ferretti, M., Andrei, S., Caldini, G., Grechi, D., Mazzali, C., Galanti, E., et al. (2008). Integrating monitoring networks to obtain estimates of ground-level ozone concentrations -- A proof of concept in Tuscany (central Italy). *Science of the Total Environment*, 396(2-3), 180-192.

Fowler, D., Pilegaard, K., Sutton, M. A., Ambus, P., Raivonen, M., Duyzer, J., et al. (2009). Atmospheric composition change: Ecosystems-Atmosphere interactions. *Atmospheric Environment*, 43(33), 5193-5267.

Freitas, S.R., Longo, K.M., Alonso,M.F., Pirre, M., Marecal, V., Grell, G., Stockler,R., Mello, R.F., & Sánchez Gácita,M.. (2010). A pre-processor of trace gases and aerosols emission fields for regional and global atmospheric chemistry models. *Geoscientific Model Development Discussions*, 3, 855-888.

GEIA/ACCENT. (2005). GEIA-ACCENT database, an international cooperative activity of AIMES/IGBP. ACCENT EU Network of Excellence. http://www.geiacenter.org

Gery, M.W., Whitten, G.Z., Killus, J.P. & Dodge, M.C. (1989). A photochemical kinetics mechanism for urban and regional scale computer modeling. *Journal of Geophysical Research*, 94, 1292-12956.

Gilliland, A. B., Hogrefe, C., Pinder, R. W., Godowitch, J. M., Foley, K. L., & Rao, S. T. (2008). Dynamic evaluation of regional air quality models: Assessing changes in O_3 stemming from changes in emissions and meteorology. *Atmospheric Environment*, 42(20), 5110-5123.

Grell, G. A, Peckham, S. E., Schmitz, Rainer, McKeen, S. a, Frost, G., Skamarock, W. C., et al. (2005). Fully coupled "online" chemistry within the WRF model. *Atmospheric Environment*, 39(37), 6957-6975.

Hanna, S. R., Lu, Z., Frey, H. C., Wheeler, N., Vukovich, J., Arunachalam, S., et al. (2001). Uncertainties in predicted ozone concentrations due to input uncertainties for the UAM-V photochemical grid model applied to the July 1995 OTAG domain. *Atmospheric Environment*, 35(5), 891-903.

IPCC. (2006). 2006 IPCC Guidelines for National Greenhouse Gas Inventories. Hayama, Kanagawa, Japan. Available from http://ipcc-nggip.iges.or.jp

Jiménez-Guerrero, P., Jorba, O., Baldasano, J. M., & Gassó, S. (2008). The use of a modelling system as a tool for air quality management: annual high-resolution simulations and evaluation. *The Science of the Total Environment*, 390(2-3), 323-40.

Leeuw, F. A. A. M. de, Moussiopoulos, N., Sahm, P., & Bartonova, A. (2001). Urban air quality in larger conurbations in the European Union. *Environmental Modelling & Software*, 16(4), 399-414.

Martin, R. V. (2008). Satellite remote sensing of surface air quality. *Atmospheric Environment*, 42(34), 7823-7843.

Mazzeo, N. A., Venegas, L. E., & Choren, H. (2005). Analysis of NO, NO_2, O_3 and NO_x concentrations measured at a green area of Buenos Aires City during wintertime. *Atmospheric Environment*, 39(17), 3055-3068.

Michalakes, J., Dudhia, J., Gill, D., Henderson, T., Klemp, J., Skamarock, W., et al. (2005). the Weather Research and Forecast Model: Software Architecture and Performance. *Use of High Performance Computing in Meteorology - Proceedings of the Eleventh ECMWF Workshop*, (June), 156-168. Singapore: World Scientific Publishing Co.

Muller, N. Z., & Mendelsohn, R. (2007). Measuring the damages of air pollution in the United States. *Journal of Environmental Economics and Management*, 54(1), 1-14.

Ntziachristos, L., Samaras, Z., & Kouridis, C. (2000). COPERT III Computer programme to calculate emissions from road transport. Copenhagen.

OAQPS. (1977). Guideline for development of control strategies in areas with fugitive dust problems. EPA-405/2- 77-029.

Olivier, J. G. J., & Berdowski, J. J. M. (2001). Global emissions sources and sinks. *Berdowski, J., Guicherit, R. and B.J. Heij (eds.) "The Climate System"* (pp. 33-78). Lisse, The Netherlands. A.A. Balkema Publishers/Swets & Zeitlinger Publishers.

Oxley, T., Valiantis, M., Elshkaki, a, & ApSimon, H. M. (2009). Background, Road and Urban Transport modelling of Air quality Limit values (The BRUTAL model). *Environmental Modelling & Software*, 24(9), 1036-1050.

Pearce, J. L., Beringer, J., Nicholls, N., Hyndman, R. J., Uotila, P., & Tapper, N. J. (2011). Investigating the influence of synoptic-scale meteorology on air quality using self-organizing maps and generalized additive modelling. *Atmospheric Environment*, 45(1), 128-136.

Peckham, S., Grell, G., McKeen, S., Fast, J., Gustafson, W., Ghan, S., et al. (2010). WRF/Chem Version 3.2 User's Guide. Available from http://ruc.noaa.gov/wrf/WG11/Users_guide.pdf

Perez Gunella, F., Puliafito, S.E., & Pirani, K. (2009). Calculo de las emisiones del transporte para la Ciudad de Buenos Aires usando un Sistema de Informacion Geografico. *Avances en Energías Renovables y Medio Ambiente*, 13, 57-64.

Puliafito, E, Guevara, M., & Puliafito, C. (2003). Characterization of urban air quality using GIS as a management system. *Environmental pollution*, 122(1), 105-17.

Puliafito, E., & Allende, D. G. (2007). Emission patterns of urban air pollution. *Revista Facultad de Ingeniería Universidad de Antioquia*, 42, 38-56.

Puliafito, E., Castro, F., & Allende, D. G. (2011). Air quality impact of PM_{10} emission in urban centers. *International Journal of Environment and Pollution*, (in press).

Puliafito, E., Perez Gunella, F., & Allende, D. G. (2010). Modelo de Calidad del Aire para la Ciudad de Buenos Aires. *World Congress & Exhibition ENGINEERING 2010-ARGENTINA*. Buenos Aires, Argentina.

Reich, S., Magallanes, J., Dawidowski, L., Gómez, D., Grošelj, N., & Zupan, J. (2006). An Analysis of Secondary Pollutants in Buenos Aires City. *Environmental Monitoring and Assessment*, 119(1), 441-457. Springer Netherlands.

Russell, A., & Dennis, R. (2000). NARSTO critical review of photochemical models and modeling. *Atmospheric Environment*, 34(12-14), 2283-2324.

Schlink, U., Herbarth, O., Richter, M., Rehwagen, M., Puliafito, J. L., Puliafito, S. E., et al. (1999). Ozone-monitoring in Mendoza, Argentina : Initial results. *Journal of the Air & Waste Management Association*, 49(1), 82-87.

Schürmann, G. J., Algieri, A., Hedgecock, I. M., Manna, G., Pirrone, N., & Sprovieri, F. (2009). Modelling local and synoptic scale influences on ozone concentrations in a topographically complex region of Southern Italy. *Atmospheric Environment*, 43(29), 4424-4434.

Schultz, M. G., Backman, L., Balkanski, Y., Bjoerndalsaeter, S., Brand, R., Burrows, J. P., et al. (2008). REanalysis of the TROpospheric chemical composition over the past 40 years. A long-term global modeling study of tropospheric chemistry funded under the 5th EU framework programme. *Final Report*. (Vol. 30, pp. 281-8). Hamburg.

Scire, J. S., Strimaitis, D. G., & Yamartino, R. J. (2000a). A User's Guide for the CALPUFF Dispersion Model. Concord, MA.

Scire, J. S., Robe, F. R., Fernau, M. E., & Yamartino, R. J. (2000b). A User's Guide for the CALMET Meteorological Model. Concord, MA.

Seaman, N. L. (2003). Future directions of meteorology related to air-quality research. *Environment International*, 29(2-3), 245-252.

Seigneur, C., Pun, B., Pai, P., Louis, J., Solomon, P. A., Emery, C., et al. (2000). Guidance for the Performance Evaluation of Three-Dimensional Air Quality Modeling Systems for Particulate Matter and Visibility. *Journal of the Air & Waste Management Association*, 50(4), 588-599.

Skamarock, W. C., Klemp, J. B., Gill, D. O., Barker, D. M., Wang, W., & Powers, J. G. (2008). A Description of the Advanced Research WRF Version 3. Mesoscale and Microscale Meteorology Division, National Center for Atmospheric Research.

Stockwell, W. R., Middleton, P., & Chang, J. S. (1990). The second generation regional acid deposition model chemical mechanism for regional air quality modeling. *Journal of Geophysical Research, 95*, 16343-16367.

Stockwell, William R, Kirchner, F., Kuhn, M., & Seefeld, S. (1997). A new mechanism for regional atmospheric chemistry modeling. *Journal of Geophysical Research., 102*(D22), 25847-25879.

Tuia D., Ossés de Eicker M., Zah R., Osses M., Zarate E., Clappier A. (200). Evaluation of a simplified top-down model for the spatial assessment of hot traffic emissions in mid-sized cities. *Atmospheric Environment*, 41, 3658–3671.

U.S. EPA. (1995). User's Guide for the Industrial Source Complex (ISC3) Dispersion Model. Description of model algorithms. Research Triangle Park, NC 27711.

U.S. EPA. (2000). Meteorological Monitoring Guidance for Regulatory Modeling Applications (p. 171). Research Triangle Park, NC 27711.

U.S. EPA. (2005). 2005 National Emissions Inventory Data & Documentation. Retrieved January 1, 2010, from http://www.epa.gov/ttn/chief/net/2005inventory.html

U.S. EPA. (2008). Technical Issues Related to use of the CALPUFF Modeling System for Near-field Applications. Research Triangle Park, NC 27711.

U.S. EPA. (2009a). Motor Vehicle Emission Simulator (MOVES) 2010 User Guide (p. 150).

U.S. EPA. (2009a). SPECIATE 4.2 Speciation Database Development Documentation. EPA600-R-09/038. Available at: http:/www.epa.gov/ttn/chief/software/speciate

U.S. EPA. (2010). AP 42, Fifth Edition Compilation of Air Pollutant Emission Factors. Retrieved January 1, 2010, from http://www.epa.gov/ttnchie1/ap42/

U.S. Geological Survey (USGS). (2010). The Global Land Cover Characterization (GLCC) Database. Available from: http://www.src.com/datasets/GLCC_Info_Page.html

Van der Werf, G. R., Randerson, J. T., Giglio, L., Collatz, G. J., Mu, M., Kasibhatla, P. S., Morton, D. C., DeFries, R. S., Jin, Y., & van Leeuwen, T. T. (2010). Global fire emissions and the contribution of deforestation, savanna, forest, agricultural, and peat fires (1997–2009), *Atmospheric Chemistry and Physics Discussions.*, 10, 11707-11735.

Wang, Xuemei, Wu, Z., & Liang, G. (2009). WRF/CHEM modeling of impacts of weather conditions modified by urban expansion on secondary organic aerosol formation over Pearl River Delta. *Particuology*, 7(5), 384-391.

Wang, Xueyuan, Liang, X.-Z., Jiang, W., Tao, Z., Wang, J. X. L., Liu, H., et al. (2010). WRF-Chem simulation of East Asian air quality: Sensitivity to temporal and vertical emissions distributions. *Atmospheric Environment*, 44(5), 660-669.

WHO. (2002). *The World Health Report 2002: Reducing Risks, Promoting Healthy Life*. World Health (p. 13). Geneva.

Willmott, C. (1982). Some Comments on the Evaluation of Model Performance. *Bulletin of the American Meteorological Society*, 63, 1309-1369.

Zannetti, P. (1990). *Air Pollution Modelling*. Van Nostrand Reinhold. New York.

23

A New Air Quality Index for Cities

Lígia T. Silva and José F. G. Mendes
University of Minho
Portugal

1. Introduction

Global population growth has led to increased populations living in urban areas. Often, this enhances stresses on space, ecosystems, infrastructures, facilities and personal lifestyles. Problems related to quality of life in cities are increasingly relevant, especially with regard to environmental issues.

Due to a generalised increase of mobility and road traffic in urban areas, the total emissions from road traffic have risen significantly, assuming the main responsibility for the disregard of air quality standards. In urban environment the typical anthropogenic sources are mainly the road traffic and, when existing, the industrial activity.

The quantitative evaluation of traffic air pollution levels is the basis on which air pollution control policies stand. The evaluation of air quality may be occasional or long-term. Occasional evaluation is useful in the context of information and alert systems for the population, working normally in real or almost-real time. Data is acquired through measurements made on an hourly or daily average basis and concentration episodes are evaluated and reported. When long-term data is considered, then we talk about long-term trend analysis, this kind of approach can be adequate for identifying the major emission source contributors to urban pollution (Butterwick et al. 1991).

In order to find an air quality index, the pollutant concentrations are combined through a classification scale anchored on the legal limits and, on the other side, on the impacts over human health. Typically these classification models consider only the worse pollutant, i.e. the one which concentration is higher given a certain scale. Two air quality evaluation models are referred, both working in real time: a Canadian and a Portuguese experience.

The objective of this chapter is to present a new air quality index, cityAIR, developed for urban contexts. The mathematical formulation of cityAIR stands on two logics: whenever at least one of the pollutants considered overcomes the legal limits for the concentration, this will be the only relevant one for the index calculation, and the value will be the minimum of the scale (zero or red); when there is no limit violation, then all the pollutants are considered for the overall air quality, which is calculated through a multi-criteria combination of the concentrations, where trade-off is allowed.

A case study is presented for Viana do Castelo, a mid-sized Portuguese city, in which cityAIR values were calculated in consideration of concentrations of CO, NO_2, O_3, C_6H_6 and PM10.

2. Urban air pollution

Urban air pollution became one of the main factors of degradation of the quality of life in cities. This problem tends to worsen due to the unbalanced development of urban spaces and the significant increase of mobility and road traffic. As a consequence, the total emissions from road traffic have risen significantly, assuming the main responsibility for the disregard of air quality standard (Butterwick, L. et al., 1991).

The atmospheric pollutants are emitted from existent sources and, subsequently, transported and dispersed several times in the atmosphere before reaching receptors through wet deposition (rainout and washout by rain and snow) or dry deposition (particle adsorption). In an urban environment, typical anthropogenic sources are mainly the road traffic and, when existing, the industrial activity. Emissions from mobile sources contribute to primary and secondary air pollution that can threaten human health, damage ecosystems and influence climate (Sharma et al. 2010; Nagurney et al. 2010). Traffic patterns, vehicle characteristics, and street configurations have a cumulative effect on exhaust emissions (Pandian et al. 2009).

The combustion of hydrocarbon fuel in the air generates mainly carbon dioxide (CO_2) and water (H_2O). However, the combustion engines are not totally efficient, which means that the fuel is not totally burned. In this process the product of the combustion is more complex and could be constituted by hydrocarbons and other organic compounds as well as benzene (C_6H_6), carbon monoxide (CO) and particles (PM) that contain carbon and other pollutants. On the other hand, the combustion conditions - high pressures and temperatures - originate partial oxidation of the nitrogen present in the air and in the fuel, forming oxides of nitrogen (mainly nitric oxide and some nitrogen dioxides) conventionally designated by NO_x.

Traffic-related air pollution levels can be evaluated by either direct measurements or predictive models. The direct measurement method is only feasible for evaluating actual situations; predictive methods can be applied throughout the planning process from the initial concept to the final detailed design of air pollution abatement measures. However measurements provide essential information to validate the predictive methods.

Numerous available dispersion models represent an important set of tools for simulating air pollution scenarios. The model adopted for this research was developed by Cambridge Environmental Research Consultants (CERC) in the United Kingdom.

This model has been used by local authorities all over Europe for urban air quality forecasting (Carruthers et al., 1997, 1998, 2003; Timmis et al, 2000; McHugh et al., 1997).It uses a parameterisation of boundary layer physics in terms of boundary layer depth and Monin-Obukhov length, and it applies a skewed-Gaussian concentration profile for convective meteorological conditions. For stable and neutral meteorological conditions, the model assumes a Gaussian plume for the concentration profile distribution with reflection at the ground and in the inversion layer.

The dispersion model has a meteorological processor for input variables, which typically include day of the year, time of day, cloud cover, wind direction and speed and temperature. These variables are used to calculate model parameters such as boundary layer depth and Monin-Obukhov length. The model does not account for anthropogenic heat sources.

An additional and important feature that makes this dispersion model suitable for modelling the urban environment is a chemistry scheme that facilitates the calculation of chemical reactions between nitric oxide, nitrogen dioxide, ozone and volatile organic compounds in the atmosphere.

2.1 Existing air quality evaluation models

The evaluation of air quality may be occasional or long-term. Occasional evaluation is useful in the context of information and alert systems for the population, working normally in real or almost-real time. Data is acquired through measurements made on an hourly or daily average basis and concentration episodes are evaluated and reported. When long-term data (6-month or yearly evaluations) is considered, than we talk about long-term trend analysis.
In the following subchapters two air quality evaluation models are referred, both working in real time: a Canadian and a Portuguese experience.

2.1.1 AQI, Canada

Integrated in a public information system of Vancouver, the FPCAP (Federal-Provincial Committee on Air Pollution) provides the information on pollution levels in form of an Air Quality Index (AQI). The AQI is based on measurements taken throughout the region of Greater Vancouver (Butterwick et al., 1991).
The AQI is expressed as a single value taking into consideration the concentrations of five major air pollutants (CO, NO_2, O_3, SO_2, PM). The index is based on the pollutant with the highest concentration relative to Federal and Provincial air criteria. This pollutant is called the Index Pollutant. The values of the other four pollutants are then disregarded.
The numeric value of the Air Quality Index is correlated to a classification system. For each category of air quality, information is provided on the associated general health effects and recommended precautionary action. Table 1 summarizes this information.

AQI	Air Quality	General Health Effects	Cautionary Statements
0 - 25	Good	No measured effects are associated	No precautions are necessary
26 - 50	Fair	Is adequate protection against effects on general population	No precautions are necessary
51 - 100	Poor	Short-term exposure may result in irritation or mild aggravation of symptoms in sensitive persons.	Persons with heart or respiratory ailments should reduce physical action and outdoor activity
Over 100	Very poor	Significant aggravation of persons with heart and lung disease. Many people may notice symptoms.	Persons with respiratory and cardiovascular diseases should stay indoors and minimize physical activity.

Table 1. Great Vancouver Air Quality Index. Source: (Butterwick et al., 1991)

2.1.2 QualAr, Portugal

The APA (Agência Portuguesa do Ambiente) of the Ministry of Environment of Portugal provides public information on pollution levels based on measurements taken through a pollution monitoring network. The information on pollutant levels is presented as an index called "Índice de Qualidade do Ar" (QualAr) (APA, 2011). The QualAr is based on 24-hour average concentrations, and therefore does not reflect short term peak levels.

The QualAr is expressed as a single value taking into consideration the concentrations of five major air pollutants (CO, NO_2, O_3, SO_2, PM). The index is based on the pollutant with the highest concentration relative to the Portuguese annual limit values for the protection of human health. The values of the other four pollutants are then disregarded. The calculation of QualAr takes into account the following averages:

- Nitrogen Dioxide (NO_2) - hourly average
- Sulphur Dioxide (SO_2) - hourly average
- Ozone (O_3) - hourly average
- Carbon Monoxide (CO) - 8-hour average
- Suspended Particulates (PM10) - daily average

The air quality assumes the classification from Poor to Good according to a classification system summarized in the Table 2.

Pollutant	CO ($\mu g/m^3$)		NO_2($\mu g/m^3$)		O_3($\mu g/m^3$)		PM10($\mu g/m^3$)		SO_2($\mu g/m^3$)	
Classification	Min.	Max.	Min.	Max.	Min.	Max.	Min.	Max.	Min.	Max.
Very Poor	10000	-----	400	-----	240	-----	120	-----	500	-----
Poor	8500	9999	200	399	180	239	50	119	350	499
Fair	7000	8499	140	199	120	179	35	49	210	349
Good	5000	6999	100	139	60	119	20	34	140	209
Very Good	0	4999	0	99	0	59	0	19	0	139

Table 2. Classification of QualAr for 2010.Source: (APA, 2010)

The classification of the air quality is based on the pollutant with the highest concentration relative to the Portuguese annual limit values for the protection of human health (Decreto-Lei 102/2010), [i.e. for an atmosphere with pollutants levels SO_2 - 35 $\mu g/m^3$ (very good), NO_2 - 180 $\mu g/m^3$ (fair); CO - 6000 $\mu g/m^3$ (good), PM10 - 15 $\mu g/m^3$ (very good) and O_3 - 365 µg/m3 (very poor): Air Quality was Very Poor due to Ozone].

3. The cityAIR index

Both models presented above are approaches which prevent trade-off between pollutant concentrations because they are based on the pollutant with the highest concentration relative to the legal limits. For situations where the concentrations are below the legal limit, i.e. when there is no limit violation, a model integrating all the pollutants could offer a more complete evaluation of the air quality. Such a model requires that whenever at least one of the pollutants considered overcomes the legal limits for the concentration (or any other limit assumed for this purpose), this one will be the only relevant for the index calculation, and the value will be the minimum of the scale.

A multicriteria air quality index is proposed, which allows for trade-off between pollutants whenever concentration values stay under the considered limits.

The cityAIR model proposed stands on the combination of long-term concentrations, which may result from past measurements or, differently, from mathematical simulation models providing in this case a prospective view of air quality.

When air pollution concentrations are computer-simulated for a city, the values for each point or area considered are compared to a standard (in this paper the legal limit). This comparison generates a dummy variable: zero if the standard is exceeded and one if it is not.

The cityAIR index results from the weighted linear combination of normalised concentration values, which are subjected to the product of the dummy variables (eqn 1).

$$cityAIR = \sum_i w_i\, c_i \prod_i v_i \qquad (1)$$

Where:
w_i is the relative weight of the pollutant i;
c_i is the normalised concentration of the pollutant i;
v_i is the dummy variable of the legal limit violation L_i of pollutant i, defined as follows:

$v_i = 1$ when $c_i \leq L_i$

$v_i = 0$ when $c_i > L_i$

The proposed model makes use of multi-criteria techniques for combining, aggregating and standardising pollutant concentration data.

3.1 Pollutants and weights

The selection of pollutants to be included in the cityAIR index may vary according to the type of sources or even the data availability. For the purpose of this paper we present the pollutants considered in the case study, which are typically result from road traffic:
CO: Carbon Monoxide
NO_2: Nitrogen Dioxide
PM10: Particulate< 10 µm
C_6H_6: Benzene
O_3: Ozone
Equal weights were considered, which means 0.2 for each of the five pollutants.

3.2 Normalization of concentrations

Because of the different scales upon which concentrations are measured, it is necessary to standardize them before aggregation. The process of standardisation is essentially identical to that of fuzzification in fuzzy sets. Standardisation is intended to transform any scale into a normalised range (i.e. zero to one). In our case, the results express a membership grade that ranges from 0.0 to 1.0, representing a continuous spectrum from non-membership (bad air quality) to complete membership (very good air quality), on the basis of the criterion (pollutant concentration) being fuzzified.
For the standardization a sigmoidal function has been adopted (eqn 2).

$$score = \frac{1}{\sin^2 \alpha} \cos^2 \alpha \qquad (2)$$

Where,

$$\alpha = \left[{(x - x_a)}/{(x_a - x_b)} \right] \times {\pi}/{2} \qquad (3)$$

Where x is the concentration value being normalized, and x_a and x_b are control points in the function. Figures 1a to 1e present this function graphically for each of the five pollutants. The control points adopted (a and b) are listed in Table 3.

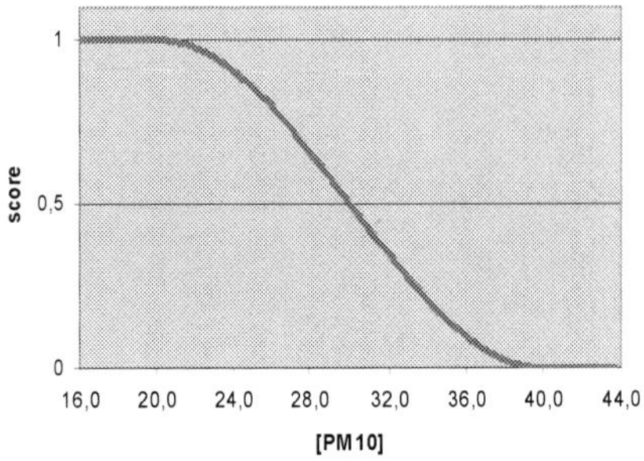

a) Fuzzy function of PM10

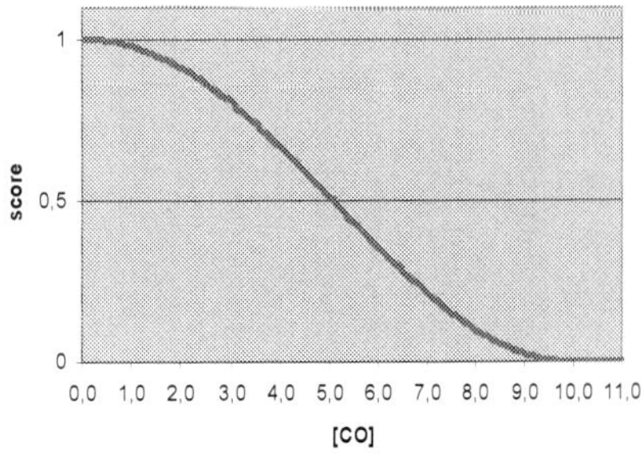

b) Fuzzy function of CO

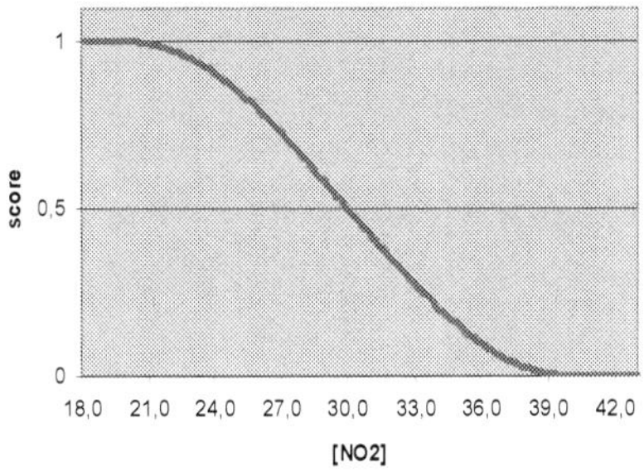

c) Fuzzy function of NO_2

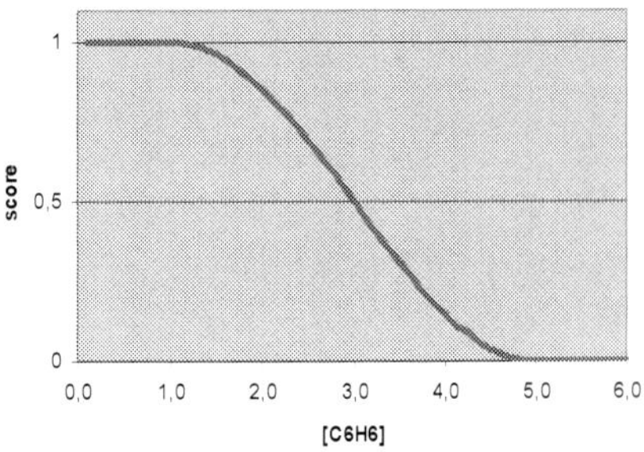

d) Fuzzy function of C_6H_6

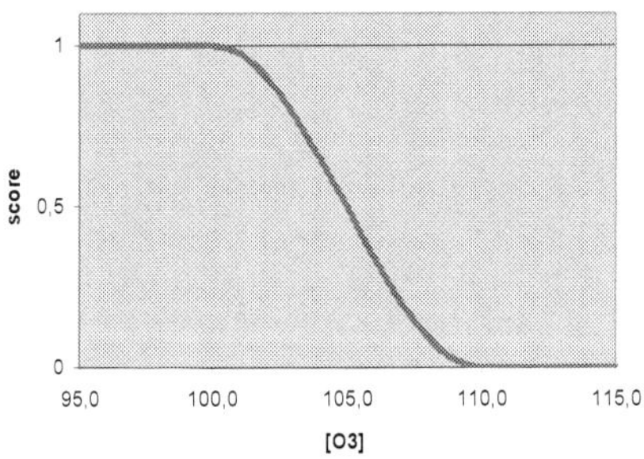

e) Fuzzy function of O_3

Fig. 1. Normalisation functions

Control points of the sigmoidal functions were selected according to the following criteria: score = 0 for the concentration limit values considered in the Portuguese legislation for human health protection and score = 1 for the concentration guidance values recommended by the World Health Organisation (WHO, 2005)(NO2 and CO values represented a non-polluted atmosphere (Seinfeld, 1997). Table 3 presents the adopted values.

Pollutants	Score = 0	Score = 1	Averaging period
CO	[CO] > 10.0 mg/m^3	[CO] ≤ 0.140 mg/m^3	8 hours (rolling average) for calendar year
PM	[PM10] > 40.0 µg/m^3	[PM10] ≤ 20.0 µg/m3	Calendar year
NO_2	[NO_2] > 40.0 µg/m^3	[NO_2] ≤ 20.0 µg/m^3	Calendar year
O_3	[O_3] > 110.0 µg/m^3	[O_3] ≤ 100.0 µg/m^3	8-hour average for calendar year
C_6H_6	[C$_6$H6] > 5.0 µg/m^3	[C_6H_6] ≤ 1.0 µg/m^3	Calendaryear

Table 3. Control points of the fuzzy functions

3.3 Dummy variables

Dummy variables switch from zero to one at the concentration limits mentioned above (third column of Table 3). Figures 2a to 2e show a graphical view of the dummy variable functions.

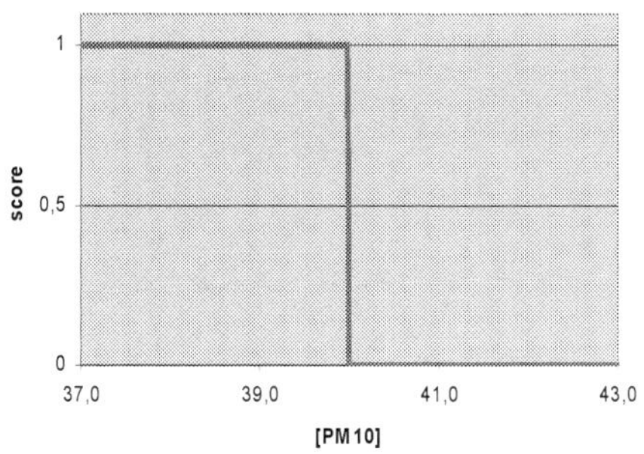

Where
score = 1 when [PM10]≤ 40 µg/m^3
score = 0 when [PM10] > 40 µg/m^3

[PM10] = calendar-year average concentration, expressed in µg/m^3.

a) Dummy variable function of PM10

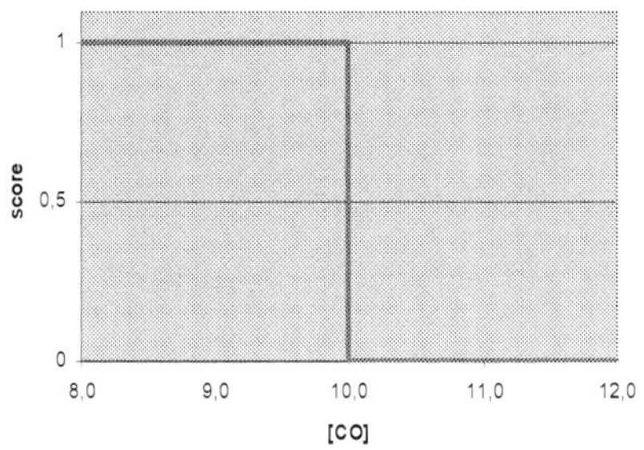

Where
score = 1 when [CO]≤ 10.0 mg/m^3
score = 0 when [CO] > 10.0 mg/m^3

[CO] = 8-hour rolling average concentration for the calendar year, expressed in mg/m^3.

b) Dummy variable function of CO

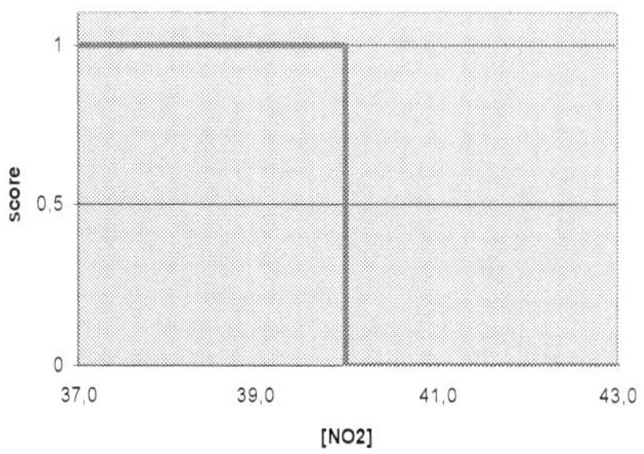

c) Dummy variable function of NO_2

Where
score = 1 when $[NO_2] \leq 40\ \mu g/m^3$
score = 0 when $[NO_2] > 40\ \mu g/m^3$

[NO2] = calendar-year average concentration, expressed in $\mu g/m^3$.

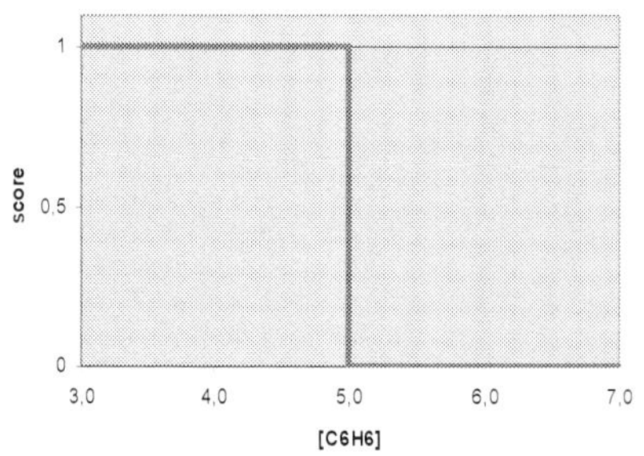

d) Dummy variable function of C_6H_6

Where
score = 1 when $[C_6H_6] \leq 5.0\ \mu g/m^3$
score = 0 when $[C_6H_6] > 5.0\ \mu g/m^3$

$[C_6H_6]$ = calendar-year average concentration, expressed in $\mu g/m^3$.

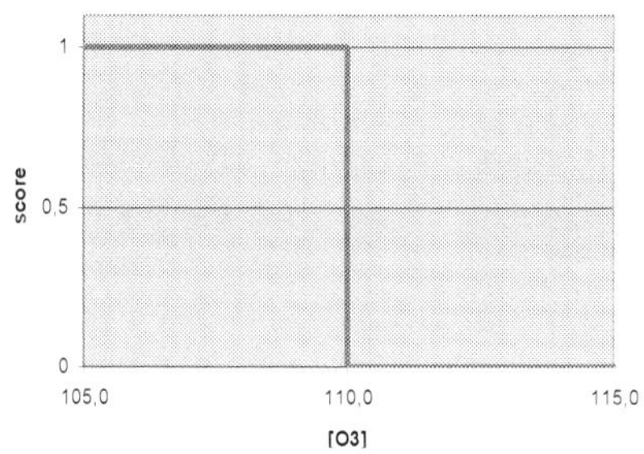

e) Dummy variable function of O3

Where
score = 1 when $[O_3] \leq 110\ \mu g/m^3$
score = 0 when $[O_3] > 110\ \mu g/m^3$

[O3] = 8-hour rolling average concentration for the calendar year, expressed in mg/m^3.

Fig. 2. Dummy variables

4. Case Study: Air quality index of one mid-sized city

A case study was undertaken to evaluate urban environmental quality in the Portuguese city of Viana do Castelo, which is located on the north-western seaside.
This mid-sized city has a population of around 36,000 in an overall area of 37 km2. The most notable source of noise and air pollution is a main road (Avenida 25 de Abril) that crosses the city and divides it into two parts.
Based on traffic data and the physical characteristics of the area, horizontal concentration maps were created for five main pollutants: CO, NO_2, C_6H_6, PM10 and O_3. A range of numerical models were used to produce results.
The ADMS-Urban model was used for pollutant dispersion. The Hills model was used to calculate air flow and turbulence over complex terrain and to account for the effects of variable surface roughness (CERC, 2001). The COPERT4 model (COPERT4), which is based on CORINAIR v.5 (CORINAIR, 2006), was used to estimate traffic emissions.

4.1 Air pollution of Viana do Castelo

The sources characterization data, and considering that Viana do Castelo is a touristic seaside city, two traffic counting campaigns were carried out, one in winter time and another one in summer time, of which resulted the data for two scenarios. Each campaign included most of the city streets and traffic was counted round-the-clock in a typical week day.
Main and secondary roads were modelled explicitly, as were one pulp and paper mill located in the vicinity of the city. The factory was modelled as one point source that represents the stack.
One single profile was developed to represent the hourly variation of traffics flows on all the roads. A full survey, including topographic characteristics, surface roughness and the specification of the emission sources, cross and longitudinal profiles (for canyon roads) was carried out for the whole city.

4.2 Validation

There is no direct technique for determining if a model is good or bad because model performance depends on so many factors. These are related with model input data, model set-up parameters and model algorithms. Besides model performance depends on the averaging time for the pollutant concentration, the pollutant itself and the monitoring sites locations. Much research has gone into prepare acceptable validations techniques. The usually used BOOT statistics approach derives from that of Hanna and Paine (Hanna & Paine, 1989) and employs a series of statistical measures comprising the mean, correlation, normal mean square error and fractional bias. The methodology adopted was based in BOOT statistical approach.
For the validation process it was guaranteed the same meteorological conditions, the same geographical base and the same reading points (coordinates x, y, z). Pollutant concentrations are predicted for each hour of the monitoring period. For hours with inadequate met data predictions are not made and the corresponding measured values are neglected.
The following simplifications were assumed:
The same flow and composition of traffic and the same traffic daily profile in both periods (measurement and modelling period);

- The validation process was developed at two levels:

- averaging the data in order to obtain daily concentrations profiles, both for monitored and predicted data;
- for each monitoring site comparison of the averaged daily concentrations profiles by the BOOT statistical methodology.

- The pollutant used in the validating process was CO, a primary and typical road traffic pollutant.

Pollutants were measured at three monitoring sites in the city (Fig. 3) during the monitoring periods shown in Table 4.

Monitoring sites	Monitoring periods
A1	0h00 19.Jan. to 24h00 21.Jan
A2	0h00 23.Jan. to 24h00 25.Jan
A3	0h00 30.Jan. to 24h00 1.Feb

Table 4. Monitoring periods

Point A1 is located at the Largo João Tomás da Costa, next to the River Garden. This site is particularly influenced by the road traffic that circulates near the garden.
Point A2 is located in the Campo do Castelo, a large square where some outdoor activities take place.
The third point, A3, is located in Rio Lima's Street, at the South edge of the City's Park and close to highway A28.

Fig. 3. Automatic monitoring sites in Viana do Castelo

The statistics calculated include Average, Standard Deviation, Normalised Mean Square Error (NMSE), Fractional Bias (FB) and the FAC2. The data format was hour by hour for the measured concentrations and predicted concentration. A perfect model would have FAC2=1.0, NMSE=0.0 and FB=0.0.
The Table 5 present statistics of comparisons between measured concentrations and the ADMS-Urban calculations. Statistics have been calculated based on hourly comparisons for each site (A1, A2 and A3) and for overall statistics (Ov.S.).
Figures 4 to 6 compare the predicted (ADMS-Urban calculations) and observed (measured) average daily concentrations of CO at the monitoring sites A1, A2 and A3.

Monitorin g Sites	Average		Standard deviation		FAC2 (objective 1)	NMSE (objective 0)	FB (Objective 0)
	Monit.	Pred.	Monit	Pred.			
A1	0.34	0.42	0.0888	0.0233	1.28	0.172	-0.20
A2	0.71	0.67	0.3184	0.0956	1.14	0.172	0.06
A3	0.31	0.33	0.0942	0.0550	1.10	0.049	-0.05
Ov.S.	0.46	0.47	0.1671	0.1322	1.17	0.131	-0.06

Table 5. Monitored and predicted CO concentrations (µg/m3)

The comparison of the output of ADMS-Urban with pollutant concentrations measured in the control points has confirmed the generally good performance of the model.
The variations of the mean concentrations along the day, shown in Figures 4 to 6, reveal a quite fair agreement between predicted and measured values.

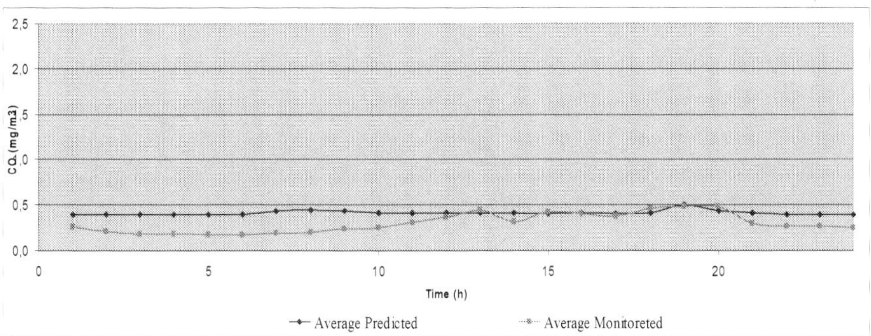

Fig. 4. Temporal variation at A1

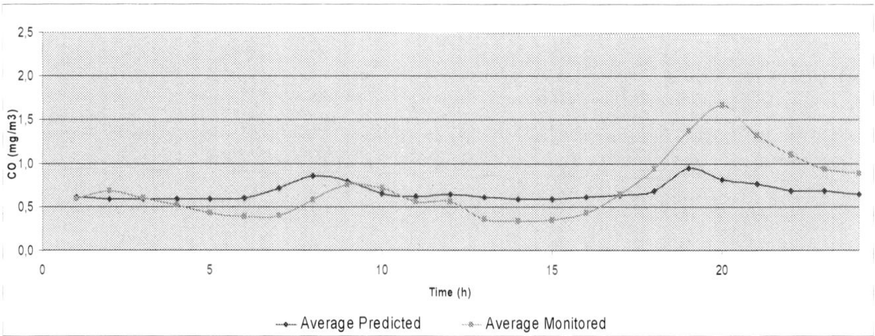

Fig. 5. Temporal variation at A2

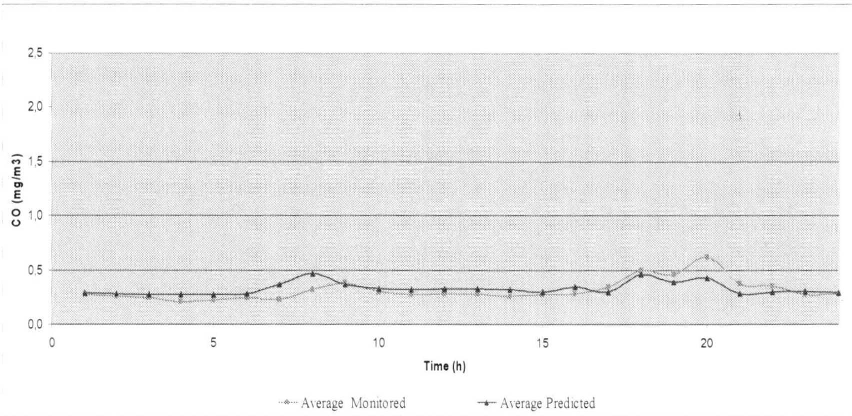

Fig. 6. Temporal variation at A3

4.3 Air pollutant maps

Horizontal concentration maps were created using ADMS-Urban model. These maps represent the average atmospheric pollution situation in one year. The following calculation parameters were adopted:

- Grid spacing: variable grid spacing (less than 10 meters);
- Height of the map: 1.20 m;
- Meteorological conditions: Data gathered at the automatic monitoring sites for one year (hourly);
- Monin-Obukhov length: 30 m ;
- Surface roughness: 0.5 m;
- Emissions inventory: database prepared for Viana do Castelo including road sources and industrial sources;
- Background file: annual average background concentration of NO2, CO, PM10and O3 at background monitoring sites (Silva, 2008);
- Output: hourly average CO [mg/m^3], NO_2[µg/m^3],PM10[µg/m^3], C_6H_6, O_3[µg/m^3];

- Average speed: variable.

This article presents results for the summer scenario, the most critical (Figures 7a to 7e).

a) Carbon monoxide, CO

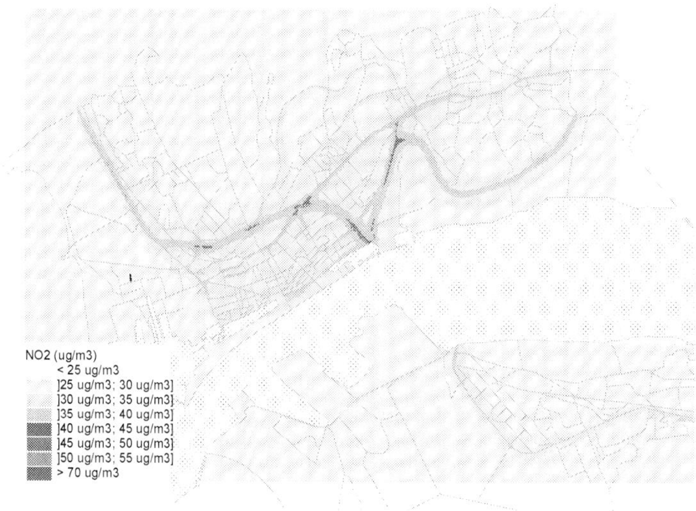

b) Nitrogen dioxide, NO_2

c) Benzene, C_6H_6

d) Particulate matter, PM10

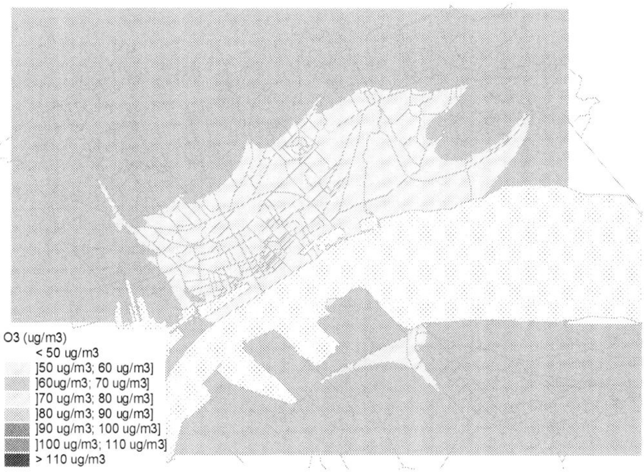

c) Ozone, O_3

Fig. 7. Air pollution maps, summer scenario

4.4 CityAIR of Viana do Castelo

The combination of the concentration maps, according to eqn.(1), results in an overall air qualitymap (Fig. 8).

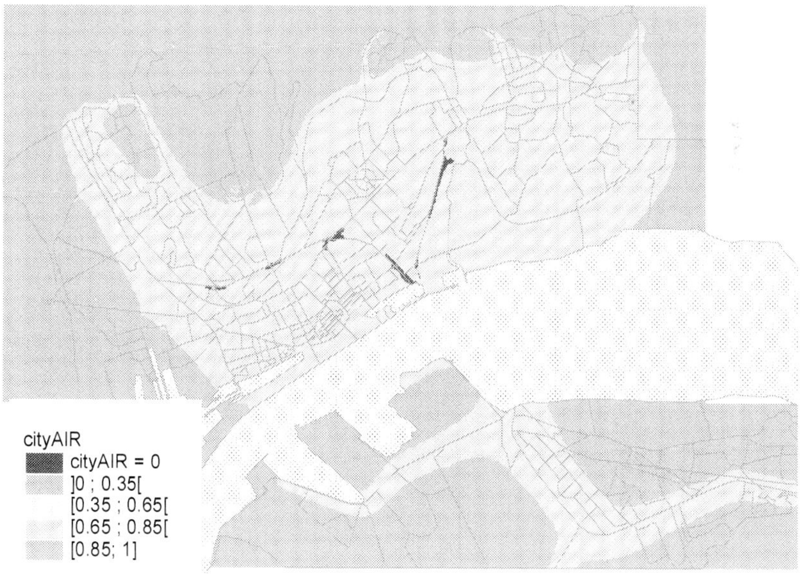

Fig. 8. CityAIR, summer scenario

Model results were overlain with a population GIS layer to estimate the affected population. Table 6 presents a synthesis of the areas and populations affected by air pollution in the city.

cityAIR	Population		Area	
	hab	%	m^2	%
= 0	69	0.2%	26332	0.2%
[0 ; 0.35[	0	0.0%	0	0.0%
[0.35 ; 0.65[	9	0.0%	3296	0.0%
[0.65; 0.85[	20477	71.7%	5152484	47.3%
[0.85; 1.0]	8002	28.0%	5711768	52.4%
Total	28557	100%	10893880	100%

Table 6. Areas and populations affected by air pollution

4.5 Analysis

The cityAIR index was developed to quantify city air quality. It depends on the concentrations of the five major urban pollutants: CO, NO_2, O_3, C_6H_6 and PM10. When none of the concentrations of these pollutants exceed the legal limit, cityAIR is calculated by combining the concentrations of different species. The cityAIR index is zero for areas with at least one pollutant concentration above the limit. Each area was also described as a binary data set, whereby a value of one signified that all concentrations were below the limit and a value of zero signified a threshold violation.

In combination with a Geographic Information System platform, the model and technologies used in this study proved to be useful for evaluating urban environmental quality, allowing the calculation of the areas and populations affected by air pollution in the city.

Although the inventory of air pollutant emission sources in the city of Viana do Castelo included one integrated pulp and paper mill, the road traffic sources made the greatest contribution to air pollution in the city. A dispersion model was used tocalculate air pollution in a continuous space of urban pollutant concentrations in the city. Horizontal maps were created for major air pollutants in urban areas (CO_2, C_6H_6, CO, NO_2, PM10 and O_3). The results demonstrated that the highest concentrations of primary air pollutants were in areas adjacent to the major roads. The obtained results are in agreement with field measurements and expected values. The highest pollutant concentrations were found in areas with a greater traffic flow or on roads with channel characteristics. Additionally, the weather conditions in the summer scenario were relatively unfavourable for dispersion and natural pollutant removal.

The cityAIR index defines air quality within a range from zero (poor air quality) to one (good air quality). Applying this model to the city of Viana do Castelo proved very useful for comparing the concentrations of major air pollutant species to their standards. This model generates a summary index of air quality that is easy to understand and intuitive for the general public.

Of the species studied, only NO_2 was found to be above legal limits. Air pollution maps reveal that the concentrations of PM10, NO_2, CO, CO_2 and C_6H_6 are higher in areas that are adjacent to high-traffic roads running through and around the city.

Because ozone is a secondary pollutant, its highest concentrations are not found near emission sources. Horizontal maps of O_3 show that the maximum ozone concentrations do not exceed the legal limit. However, in outlying areas, ozone concentrations reach the threshold for vegetation protection (65 µg/m3).
The distribution of the calculated cityAIR index over Viana do Castelo revealed that air quality is globally acceptable in this city (Figure 8). Nonetheless, the dearth of small zones on this map may be problematic. Small zones, including Av. 25 de Abril (the ramp to the bridge), the roundabouts of the hospital and football field and the access to the main road (IC1), have high levels of NO_2 over the legal limit.
Based on an analysis of the Table 6, we can conclude that only 0.2% of the population (69 pop.) in summer is exposed to a cityAIR index of zero, during which time 71.7% and 28.0% of the population benefit a cityAIR index of the population benefit a cityAIR index ranging between [0.65 ; 0.85[and above 0.85, respectively.

5. Conclusion

The urban air index used in Viana do Castelo, cityAIR, aggregates data for the air quality of a city and presents results in the context of standardised legal limits for air pollution.
Based on cityAIR, several priority areas for future mitigation and monitoring are proposed: Avenida 25 de Abril, Avenida Gaspar de Castro, access to west IC1 and access to north and south EN13.
When the results of this analysis were presented to the Municipality of Viana do Castelo cityAIR index proved to be easily understood and quite intuitive. For the problematic areas, a Monitoring & Mitigation Plan is being prepared.
In order to estimate pollutant emissions to the atmosphere from vehicular traffic, which is essential for evaluating both local and global air quality, emission factors were obtained for vehicles in Viana do Castelo and combined with the calculated vehicle circulation in that city. Thus, it is now generally possible to estimate the contribution of car traffic to climate change due to CO_2 emissions.
The model for air pollutant dispersion is consistent with field measurements. The model calculates concentrations of PM10, a component of secondary emission that originates under typical city conditions. The developed database of secondary particle concentrations in the city facilitates simulations of PM10 concentrations.
The cityAIR index developed here is transparent, simple and easy to understand. Indicator weights and dimensions used in the cityAIR calculation depend on the control points defined during the normalisation process. A variety of options can be used during the cityAIR calculation to focus index results on different dimensions and indicators of overall urban environmental quality.

6. Acknowledgment

This work was partially supported by Department of Civil Engineering, University of Minho and the Municipality of Viana do Castelo. The authors would like to thank the anonymous reviewers for their valuable comments and suggestions to improve the quality of the paper.

7. References

APA (2010) on line: www.iambiente.pt

Butterwick, L., Harrison, Roy. e Merritt Q.1991. Handbook for Urban Air Improvement. Commission of the European Communities.

Carruthers, D.J., Edmunds, H. A., M. Bennett, P.T. Woods, M.J.T. Milton, R. Robison, B.Y. Underwoon and FranklinC.J. 1997. Validation of ADMS dispersion model and assessment of its performance relative to R-19 and ISC using archived LIDAR data. Int. J. Environment and Pollution, Vol. 8, 264-278.

Carruthers, D.J., Edmunds, H. A., Lester, A.E., McHugh, C. A. and Singles,. R. J. 1998. Use and Validation of ADMS-Urban in Contrasting Urban and Industrial Locations. Int. J. Environment and Pollution, Vol. 14, 364-374.

Carruthers, D.J.,J.W. Blair and K.L. Johnson, 2003. Validation and Sensivity Study of the ADMS-Urban, Cambridge EnvironmentalResearch Consultants (2003) TR-0191.

CERC 2001. ADMS-Urban User Guide, Version 1.6.

COPERT4, Acessed 2008. Available from: http://lat.eng.auth.gr/copert

CORINAIR 2006. Emission Inventory Guidebook, Version 5.0, September.

Diário da República 2010. Dec. Lei n°102/10, 23 de Set. 2010. I Série-A, N.° 186, Lisboa.

Hanna, Steven R., Robert J. Paine. 1989. Hybrid Plume Dispersion Model (HPDM) Development and Evaluation. J. Appl. Meteor., 28, 206–224.

McHugh, C.A., Carruthers, D.J., Edmunds, H.A. 1997. ADMS-Urban: an air quality management system for traffic, domestic and industrial pollution, International Journal of Environment and Pollution, pp. 666-675.

Nagurney, A., Qiang, Q., Nagurney, L. 2010. Environmental Impact Assessment of Transportation Networks with Degradable Links in an Era of Climate Change. International Journal of Sustainable Transportation, 4: 3, 154 — 171.

Pandian S., Gokhale S., Ghoshal A. K. 2009. Evaluating effects of traffic and vehicle characteristics on vehicular emissions near traffic intersections. Transportation Research Part D 14, 180–196.

Seinfeld, J. H., and Padis, S. N. 1997. Atmospheric Chemistry and Physics From Air Pollution To Climate Change, John Wiley Sons.

Sharma A.R., Kharol S.K., Badarinath K.V.S. 2010. Influence of vehicular traffic on urban air quality - A case study of Hyderabad, India. Transp. Research Part D 15, 154–159.

Silva, L.T. 2008. Avaliação da Qualidade Ambiental Urbana. PhDThesis, Universidade do Minho, Braga, Portugal.

Timmis, R., Wilkinson, S., Carruthers, D. and McHugh C. 2000. Recent studies to validate and compare atmospheric dispersion models for regulatory purposes in the U.K., International Journal of Environment and Pollution Vol.14, p. 431-442.

WHO 2005. Air quality guidelines global update 2005 meeting report. Report on a working group meeting. WHO, Bonn, Germany.

An Analytical Application for the Determination of Metals in PM_{10}

Tony Byrd[2,3], Mary Stack[3] and Ambrose Furey[1,2]
[1]*PROTEOBIO, Mass Spectrometry Centre for Proteomics and Biotoxin Research, Cork Institute of Technology, Bishopstown, Cork*
[2]*Team Elucidate, Department of Chemistry, Cork Institute of Technology, Bishopstown, Cork*
[3]*Cork County Council, Environment Directorate, County Cork, Ireland*

1. Introduction

Air quality standards for Ireland European Council, 2008: Ambient Air Quality and Clean Air for Europe Directive 2008/50/EC [1], require that the annual limit for particulate matter ten microns (PM_{10}) per defined agglomeration zone within Ireland has an annual limit of 40 μgm^{-3}, the annual limit for $PM_{2.5}$ from January 1st 2010 is 25 μgm^{-3}. A 24 h PM_{10} mean per defined agglomeration zone of 50 μgm^{-3} is not to be exceeded more than thirty five times in one year. This directive came into effect in 2010 and has replaced air quality standards European Council, 1999. Air Quality Daughter Directive 1999/30/EC [2], (which) was implemented in 2005 under Air Quality Standard Regulations 2002 S.I. No. 271 of 2002 [3].

Particulate matter ten microns (PM_{10}) is the subject of international publicity, particularly with reference to its associated health impacts [4, 5]. Research in the field has varied from the development of analytical applications for the determination of particulate matter and its chemical components, to epidemiological studies, to the impacts of PM on human health. A specific investigation of particulate matter exposure has also presented a link to mortality rates [6]. Epidemiological studies should consider the determination of the metal content in particulate matter due to their potential negative effects upon human health [7].

PM_{10} is classified into two modes; coarse particles and fine particles. Coarse particles arise from such primary sources as the re-suspension of road dust, agricultural activity, sea spray, concrete production and other industrial processes [8]. Fine particles can result from secondary inorganic aerosols and volatile organic compounds, which arise from anthropogenic activity like combustion and traffic emissions [9]. Anthropogenic activities can cause atmospheric chemical reactions and produce such pollutants as nitrous oxides, sulphur dioxide, nitrate and ammonium along with soluble metallic substances like iron and chromium [10]. It has been determined that the calcium portion of particulate matter can be found with sulphate, as calcium sulphate and chloride can be found as sodium chloride [11].

Atmospheric particulate matter is often non spherical with a range of densities [12]. Some particles are liquid, some solid and others have solid cores surrounded by liquid. Particles may be hygroscopic, contain particle bound water, inorganic ions, metallic compounds and elemental carbon. It is impossible to assess fully all components of PM_{10} due to the significant portion of particle bound water, which may not be completely removed by pre-sample filter conditioning. Furthermore, the removal of particle bound water may result in loss of semi-volatile particulate matter [12]. Studies have shown variable losses of ammonium nitrate at certain temperatures from quartz filters while, the loss of semi-volatile particles can occur during transport and storage of filters [13].

The European Union has designated EN 14902:2005 [14] as the standard method for the analysis of the metallic component of PM_{10}, utilising Inductively Coupled Plasma Mass Spectroscopy (ICP-MS) and microwave digestions with nitric acid and hydrogen peroxide extraction. The standard states that, demonstration of validation can determine the equivalency of an alternative test. The limitation of resources by accredited laboratories means that the use of alternative instrumentation as Inductively Coupled Plasma Optical Emission Spectroscopy (ICP-OES) can be used in determining PM_{10} metallic fractions.

For Ireland, the climate is influenced by the Atlantic Ocean with prevailing south westerly-to-westerly winds and little significant variation in temperatures [15]. This research study details the investigation of validation of particulate matter metal analysis and investigates the impact of the metallic portion of PM_{10} levels at a sampling site in County Cork, Ireland in 2005 and examines the impact of source emissions on ambient air quality following EU Directive 1999/30/EC [1]. The study site is located in an area, which includes an industrial zone.

2. Methodology

2.1 Sampling sites

The determination of the PM_{10} fraction of suspended particulate matter was measured following the European reference method EN 12341 European Union 1998 [16]. The site selected is located in an area, which includes an industrial zone with a mineral extraction quarry and asphalt production plant. During the study time period, a landfill was operating within the study site. The road network in the area suffered much wear and tear during the study due to heavy vehicles and machinery utilising the small local access roads. Housing development became more frequent in this area during this time and the site location is also in close proximity to the coast with a large water body near-by with a potential for high sea salt contribution to particulate matter.

2.2 Sampling and measurements

The sampling of atmospheric particulate matter was carried out by means of a 'TCR TECORA' PM_{10} Analyser "Sky post PM/HV" sampler equipped with PM_{10} inlet and quartz fibre filters (47mm Schleicher and Schuell). Daily sampling was performed from 18 March to 19 May 2005.

Once gravimetric analysis and filter mass loading had been computed, sample and control filters were digested following the USEPA [17] method described in

Byrd et al., (2010) [18]. The solutions obtained were analysed by inductively coupled argon plasma optical emission spectrophotometer. Settings for the ICP-OES are outlined in Table 1.

Forward rf power	1300 W
Frequency of rf generator	40 MHz (axial view)
Coolant gas flow rate	15 l/min
Auxiliary gas flow rate	0.2 l/min
Sample gas flow rate	0.8 l/min
Solution delivery	1.0 ml/min
Automatic sampler	Perkin-Elmer AS900
Nebulizer	Cross-flow with Scott type expansion chamber
Polychromator	Echelle grating, cross-dispersed wavelength range (nm): 165–403
Detector	Segmented-array charge coupled device
Measurement mode	Continuous nebulization
Signal processing mode	3 pixels peak area
Background correction	2 points
Wavelengths (nm)	Al, 308.215; Cd, 228.802; Cr, 267.716; Cu, 324.742; Fe, 238.204; Mn, 257.610; Ni, 232.003; Pb, 217.000; Zn, 206.200; Ca, 317.949; K, 404.721; Na, 330.236; Mg, 279.075.

Table 1. Instrument characteristics and settings for ICP-OES

2.3 Standards and controls

Single and mixed metallic standards are commercially available for use with Inductively Coupled Plasma Spectroscopy (ICP-OES). The National Institute of Standards and Technology (NIST) standard reference materials for calibration (1 - 30 μg/ml) were prepared in dilute digestion acid solution. Validation parameters were assessed [19] using these standards. Control samples were also used to determine detection limits. Control samples were pre-conditioned quartz fibre filters and these were placed in the PM_{10} Analyser "Sky post PM/HV", but not exposed to ambient air.

3. Validation

Low volume daily PM_{10} sampling involves the exposure of quartz fibre filters to a known quantity of ambient air over a twenty four hour period in which each filter paper exposed and collected represents one daily sample. Some sample collection equipment provides for multiple samples per day, which is useful when attempting to determine the impact of local activity at various times. However, each individual sample filter is different in composition from the next, as it represents a different time period.

Splitting a sample is not possible, as material will not necessarily collect on the filter in a uniform manner [20, 21]; therefore it is necessary to use the complete filter in the treatment process for chemical characterisation. The extraction of quartz fibre filters results in the total destruction of the filter and so the performance of a series of standard additions to a single paper is not possible. The solution to this was to add NIST certified reference standard to blank filters which have undergone the entire sample conditioning process. Calibration and validation of this methodology was performed in this way. The sample matrix changed as the test progressed with a solid filter being incurred with a known concentration of liquid analyte before conditioning and extraction of analyte to a complex liquid matrix. Standard addition of mixed heavy metals in nitric acid was performed. The final volume of standard was 20ml.

Determination of method performance parameters was carried out through revision of expected analyte concentrations in a PM_{10} sample. Limits of Detection (LOD) and limits of Quantitation (LOQ) were assessed through the use of blank filters in accordance with IUPAC rules. From the series of standards devised, the lowest standard took account of blank concentrations and instrumentation sensitivity and was used in the determination of linearity where possible. Blank materials used in the determination of trace metal analysis all contain some level of analyte of interest in the form of contamination. These materials were used to establish a baseline concentration level and results obtained were subtracted from standards and sample values.

The impact of interference is worth considering prior to sample collection. Positive and negative interference can result during transportation of samples to the laboratory; loss of semi-volatile components represents the main volume of such losses through the loss of ammonium nitrate. The loss of particle bound water can result in significant deterioration of PM_{10} sample concentration [22]. Conditioning of filters by humidification standardises the preparation of the test filter so as to ensure traceability and accuracy of the complete analytical process.

3.1 Filter paper humidification assessment

A critical step in the process is the conditioning of filter papers at a stable low relative humidity. Variable humidity causes changes in the size of hygroscopic particles. This results in the loss of volatile ammonium nitrate and semi volatile organic components by evaporation [12]. This step in the procedure was developed to determine its influence upon analyte recovery. First, a standard addition series was devised (Table 2). Each standard contained a mixture of all elements to be analysed.

Second, a batch of blank filter papers was prepared by spiking a mixture of standards of decreasing concentration (Table 2). This batch consisted of 6x10 filter papers. The standard elemental mix was incurred onto 10 filter papers over 6 standard concentrations [Table 2]. This batch of filter papers was subjected to the full procedure described in Byrd et al., 2010 [18] including the treatment of filter papers before standards were added and the treatment

Element	Blank	Std 1	Std 2	Std 3	Std 4	Std 5
Al	0	0.05	0.2	0.5	1	1.5
Ba	0	0.05	0.2	0.5	1	1.5
Cd	0	0.05	0.2	0.5	1	1.5
Co	0	0.05	0.2	0.5	1	1.5
Cr	0	0.05	0.2	0.5	1	1.5
Cu	0	0.05	0.2	0.5	1	1.5
Fe	0	0.05	0.2	0.5	1	1.5
Mn	0	0.05	0.2	0.5	1	1.5
Ni	0	0.05	0.2	0.5	1	1.5
Pb	0	0.05	0.2	0.5	1	1.5
Zn	0	0.05	0.2	0.5	1	1.5
Ca	0	1	5	10	50	100
Mg	0	5	2.5	5	25	50
K	0	0.1	0.5	1	5	10
Na	0	0.5	2.5	5	25	50

Table 2. Elements and standard concentrations (μg/ml) applied to filter papers during validation studies.

of filter papers after the standard was added in a desiccator at fixed relative humidity and temperature [Humidification study: Batch A]. A second batch of blank filter papers was prepared consisting of 6x10 filter papers. These filters were treated in the same way as the filters in batch A without desiccating after standard addition [Non-humidification study: Batch B]. Spiked filter papers for both batches were weighed using a semi-micro balance prior to and following conditioning.

3.2 ICP-OES validation results

Average concentrations (n=10) for blank filters in Batches A and B were determined. These results were used to determine performance characteristics such as LOD and LOQ (Table 3 and 4). Limit of detection is derived from the standard deviation of the mean blank result obtained. LOD is the lowest concentration of analyte that can be detected and reliably distinguished from zero [23]. LOD can be expressed as:

$$LOD = Y_{blank} + 3s_{blank} \quad (1)$$

Y_{blank} is the average value of the blank signal and s_{blank} is the corresponding standard deviation of that average blank signal. Limit of Quantitation is the lowest concentration of analyte that can be determined quantitatively with an acceptable level of precision [23]. LOQ can be expressed as:

$$LOQ = Y_{blank} + 10s_{blank} \quad (2)$$

LOD values for humidified filters (Batch A) of Al, Ba, Cd, Co, Cr, Cu, Fe, Mn, Ni, Pb, Zn ranged from 0.001-0.038 µg/ml, LOQ ranged from 0.008-0.21 µg/ml (Table 3). LOD values for Ca, Mg, K, and Na ranged from 0.06-1.81 µg/ml with the LOQ ranging from 0.32-1.32 µg/ml (Table 3). LOD values for non-humidified filters (Batch B) of Al, Ba, Cd, Cr, Co, Cu, Fe, Mn, Ni, Pb, and Zn ranged from 0.001-0.056 µg/ml with the LOQ ranging from 0.008-0.31 µg/ml (Table 4). LOD values for Ca, Mg, K, and Na ranged from 0.003-3.91 µg/ml and the LOQ ranged from 0.18-21.41 µg/ml (Table 4).

The elements Ba, Cu, Fe, Mn, Ni, K and Na all presented favourable LOD and LOQ values for Batch A compared with Batch B. Further comparison of the data presented (Table 3 and 4) shows that LOD, LOQ and average value of blank signal for the elements Al, Cr, Zn, Ca and Mg are of greater concentration on those filters that have been humidified as part of the conditioning treatment process. This can be attributed to the impact of the desiccant on the humidification process and on the impurities in the digestion acid mixture. This presents a challenge to the linearity range obtainable for those elements, which inevitably will have higher LOQ values.

LOD and LOQ values obtained for cobalt in those filter papers that have been humidified (Batch A) was 0 µg/ml. This is because the standard deviation of these 10 filters was 2.25.x10-19 µg/ml.

In filters that have been humidified (Batch A), Al and Fe had LOQ values greater than the lowest standard concentration used (0.05 µg/ml). The linear range of these elements therefore does not include this standard.

In Batch A (Table 5), accepted relative standard deviation values (%RSD) ranged from 0.9-15.1% (Limit <20%) with acceptable recovery values ranging from 80.3-118.8% (Limits 80-120%). These results correlate with the precision values obtained (Table 5). There are some

Element	Humidification Process (Batch A)						
	Mean ng/m^3	LOD soln µg/ml	LOD air ng/m^3	LOQ soln µg/ml	LOQ air ng/m^3	IDL soln µg/ml	IDL air ng/m^3
Al	23.4	0.014	11.75	0.077	64.15	0.008	6.42
Ba	5.3	0.005	4.00	0.026	21.66	0.003	2.17
Cd	0.65	0.001	1.21	0.008	6.67	0.001	0.67
Co	0.83	0.000	0.00	0.000	0.00	0.000	0.00
Cr	4.67	0.004	3.25	0.021	17.50	0.002	1.75
Cu	1.92	0.002	1.25	0.008	6.67	0.001	0.67
Fe	50.32	0.038	31.49	0.207	172.47	0.021	17.25
Mn	1.08	0.002	1.25	0.008	6.67	0.001	0.67
Ni	3.33	0.004	3.25	0.021	17.50	0.002	1.75
Pb	1.82	0.007	6.17	0.041	34.16	0.004	3.42
Zn	23.66	0.017	14.00	0.093	77.48	0.009	7.75
Ca	230.87	0.240	199.96	1.316	1096.45	0.132	109.64
K	17.5	0.058	47.99	0.316	263.28	0.032	26.33
Mg	33.58	0.070	58.24	0.383	319.10	0.038	31.91
Na	1560.5	1.810	1507.70	0.992	826.50	0.099	82.65

Table 3. Average concentrations (n=10) for blank filters in Batch A (Humidified) with limits of detection, limits of quantitation and instrument detection limits reported.

Element	Non-Humidification Process (Batch B)						
	Mean ng/m^3	LOD soln µg/ml	LOD air ng/m^3	LOQ soln µg/ml	LOQ air ng/m^3	IDL soln µg/ml	IDL air ng/m^3
Al	15.83	0.004	3.73	0.025	20.83	0.003	2.08
Ba	4.42	0.005	4.09	0.027	22.50	0.003	2.25
Cd	0.58	0.001	1.21	0.008	6.67	0.001	0.67
Co	0.58	0.001	1.21	0.008	6.67	0.001	0.67
Cr	2.58	0.002	1.84	0.012	10.00	0.001	1.00
Cu	2.25	0.004	3.34	0.022	18.33	0.002	1.83
Fe	60.82	0.056	46.70	0.307	255.78	0.031	25.58
Mn	1.17	0.002	1.29	0.009	7.50	0.001	0.75
Ni	3.08	0.005	3.92	0.026	21.66	0.003	2.17
Pb	1.75	0.007	6.17	0.041	34.16	0.004	3.42
Zn	26.33	0.015	12.48	0.082	68.32	0.008	6.83
Ca	139.56	0.145	120.48	0.794	661.53	0.079	66.15
K	42.08	0.075	62.08	0.409	340.77	0.041	34.08
Mg	33.58	0.003	2.65	0.175	145.80	0.018	14.58
Na	1780.98	3.905	3253.43	21.411	17838.93	2.141	1783.89

Table 4. Average concentrations (n=10) for blank filters in Batch B (Non-humidified) with limits of detection, limits of quantitation and instrument detection limits reported.

Element µg/ml		Mean µg/ml	SD	RSD %	Recovery %	Precision	Bias
Al	0.05	0.056	0.013	**23.4**	105.5	0.012	0.003
	0.2	0.196	0.013	6.8	93.2	0.019	-0.014
	0.5	0.442	0.021	4.8	84.0	0.087	-0.080
	1	0.845	0.053	6.3	80.3	0.214	-0.200
	1.5	1.308	0.033	2.5	82.8	0.280	-0.260
Ba	0.05	0.055	0.007	13.2	104.7	0.007	0.002
	0.2	0.200	0.009	4.5	95.0	0.061	-0.050
	0.5	0.489	0.023	4.7	92.8	0.044	-0.036
	1	0.862	0.022	2.6	88.8	0.190	-0.180
	1.5	1.307	0.043	3.3	82.8	0.274	-0.260
Cd	0.05	0.064	0.004	6.2	**122.0**	0.012	0.011
	0.2	0.208	0.010	4.8	98.6	0.043	-0.039
	0.5	0.493	0.025	5.1	93.7	0.041	-0.032
	1	0.882	0.014	1.6	83.8	0.170	-0.160
	1.5	1.365	0.044	3.2	86.4	0.220	-0.200
Co	0.05	0.063	0.004	6.4	118.8	0.011	0.009
	0.2	0.201	0.002	1.0	95.7	0.061	-0.051
	0.5	0.492	0.022	4.5	93.5	0.040	-0.033
	1	0.887	0.020	2.3	84.3	0.170	-0.160
	1.5	1.354	0.020	1.5	85.8	0.230	-0.214
Cr	0.05	0.064	0.004	6.3	**120.7**	0.012	0.010
	0.2	0.206	0.005	2.4	97.7	0.010	-0.005
	0.5	0.462	0.024	5.2	87.8	0.068	-0.061
	1	0.862	0.020	2.3	81.9	0.190	-0.180
	1.5	1.348	0.040	3.0	85.4	0.230	-0.220
Cu	0.05	0.053	0.008	15.1	98.6	0.008	-0.001
	0.2	0.198	0.005	2.5	96.0	0.012	-0.008
	0.5	0.479	0.018	3.8	93.3	0.039	-0.034
	1	0.885	0.030	3.4	82.2	0.190	-0.180
	1.5	1.387	0.042	3.0	84.2	0.250	-0.240
Fe	0.05	0.029	0.026	**91.9**	**54.2**	0.035	-0.023
	0.2	0.184	0.015	8.1	87.5	0.027	-0.025
	0.5	0.444	0.042	9.5	84.4	0.091	-0.078
	1	0.851	0.040	4.7	80.8	0.210	-0.190
	1.5	1.305	0.060	4.6	82.6	0.275	0.261

Element µg/ml		Mean µg/ml	SD	RSD %	Recovery %	Precision	Bias
Mn	0.05	0.061	0.005	8.2	115.9	0.010	0.008
	0.20	0.202	0.004	2.2	95.9	0.009	-0.008
	0.50	0.491	0.020	4.1	93.3	0.039	-0.034
	1.00	0.867	0.030	3.5	82.4	0.188	-0.180
	1.50	1.348	0.022	1.6	85.4	0.232	-0.220
Ni	0.05	0.057	0.003	4.5	111.9	0.007	0.006
	0.20	0.206	0.002	1.1	93.9	0.015	-0.012
	0.50	0.459	0.020	4.4	91.0	0.051	-0.045
	1.00	0.876	0.030	3.4	81.6	0.196	-0.184
	1.50	1.421	0.014	1.0	84.5	0.246	-0.233
Pb	0.05	0.063	0.008	12.7	128.8	0.017	0.014
	0.20	0.216	0.011	5.1	100.2	0.003	0.001
	0.50	0.475	0.025	5.3	94.1	0.039	-0.029
	1.00	0.914	0.030	3.3	85.3	0.160	-0.150
	1.50	1.435	0.040	2.8	86.3	0.216	-0.210
Zn	0.05	0.049	0.014	**28.7**	105.8	0.013	0.003
	0.20	0.213	0.002	1.1	101.5	0.014	0.003
	0.50	0.492	0.024	4.9	92.0	0.048	-0.040
	1.00	0.899	0.017	1.9	82.9	0.180	-0.170
	1.50	1.498	0.022	1.5	87.7	0.200	-0.185
Ca	1.00	0.671	0.024	3.6	**63.7**	0.383	-0.363
	5.00	5.438	0.484	8.9	103.3	0.492	1.660
	10.00	9.590	0.672	7.0	91.1	1.134	-0.890
	50.00	50.883	2.018	4.0	96.7	2.597	-1.661
	100.00	105.723	3.197	3.0	100.4	3.072	0.437
K	0.10	0.039	0.013	**33.0**	**37.4**	0.067	-0.063
	0.50	0.464	0.016	3.4	88.1	0.065	-0.060
	1.00	0.885	0.059	6.7	84.1	0.177	-0.159
	5.00	4.812	0.176	3.7	91.4	0.482	-0.429
	10.00	10.332	0.336	3.3	98.2	0.374	-0.185
Mg	0.50	0.463	0.012	2.6	87.9	0.065	-0.060
	2.50	2.606	0.116	4.4	99.0	0.113	-0.025
	5.00	4.793	0.242	5.0	91.1	0.524	-0.447
	25.00	25.440	0.994	3.9	96.7	1.289	-0.832
	50.00	50.240	1.738	3.5	95.5	2.909	-2.272
Na	5.00	3.947	1.477	**37.4**	**75.0**	1.925	-1.250
	25.00	23.847	1.324	5.6	90.6	2.774	-2.345
	50.00	49.727	1.471	3.0	94.5	3.227	-2.759

Table 5. Batch A. Filters have been conditioned at constant relative humidity and temperature following addition of standard and prior to analysis by ICP-OES. Bold values represent outliers.

exceptions in the relative standard deviation values and recovery values obtained. The lowest standard (0.05 µg/ml) for Al, Zn and Fe displayed relative standard deviations greater than 20%. On further examination, the precision values obtained for Al and Zn show that there is no significant deviation from the true value of this standard when compared with other standards. The reason for these high %RSD values is that at such a low concentration, a slight variation has a big impact upon relative standard deviation. In the case of Fe, the mean concentration value obtained for this standard is a half of the true value spiked on the filter. The LOQ value for Fe (Table 3) is greater than the 0.05 µg/ml spiked standard. The recovery values for Cd, Cr and Pb in standard one (0.05 µg/ml) are greater than the upper percentage limit of 120% (Table 5). The average values obtained for these standards are greater than the true concentration. There is a positive bias observed in these standards (Table 5). With reference to Ca, Na, Mg and K; Na (0.5, 2.5 µg/ml) presented no elemental reading due to instrument sensitivity constraints. The result for standard 5 µg/ml is that the recovery value is less than the 80% limit (Table 5).

Ca (1.0 µg/ml) and K (0.1 µg/ml) also presented recovery values less than the lower limit for recovery (Table 5). These results can be attributed to the high LOQ obtained for these elements (Table 3). For Mg, all standards are within the required limits for relative standard deviations and recovery.

In Batch B (Table 6), acceptable relative standard deviation values (%RSD) range from 0.876-18.527% (Limit <20%). Acceptable recovery values range from 80.1-119.9% (Limits 80-120%), (Table 6). However, the following standards: Al, Fe (0.05 µg/ml), present relative standard deviation values above the limit of 20% RSD. On further evaluation, recovery values (Table 6) for the elements Al and Fe are less than the lower limit for recovery of 80%.

These values can be attributed to the loss of particle bound water from the standards due to the lack of humidity and temperature control. The elements Na, Ca and K at concentrations Na (0.5, 2.5, 5.0 µg/ml) Mg (0.5, 2.5 µg/ml), Ca (1.0, 5 µg/ml), K (0.1, 0.5 µg/ml) gave no detectable reading (Table 6). These results can be attributed to the high LOQ results obtained for these elements (Table 4).

Element µg/ml		Mean µg/ml	SD	% RSD	Recovery %	Precision	Bias
Al	0.05	0.012	0.022	**179.8**	**23.6**	0.046	-0.038
	0.20	0.162	0.007	4.1	**77.1**	0.049	-0.046
	0.50	0.417	0.038	9.1	**79.2**	0.116	-0.104
	1.00	0.822	0.011	1.4	**78.1**	0.231	-0.219
	1.50	1.355	0.044	3.2	85.8	0.234	-0.213
Ba	0.05	0.052	0.007	13.3	99.0	0.007	-0.001
	0.20	0.206	0.007	3.3	97.6	0.008	-0.005
	0.50	0.474	0.032	6.8	90.0	0.061	-0.050
	1.00	0.879	0.021	2.4	83.5	0.176	-0.165
	1.50	1.402	0.043	3.1	88.8	0.188	-0.169
Cd	0.05	0.063	0.002	3.8	119.5	0.011	0.010
	0.20	0.216	0.006	2.6	102.4	0.007	0.005
	0.50	0.486	0.015	3.0	92.3	0.043	-0.039
	1.00	0.895	0.019	2.1	85.0	0.159	-0.150
	1.50	1.453	0.024	1.7	92.0	0.145	-0.120
Co	0.05	0.060	0.002	3.2	114.6	0.008	0.007
	0.20	0.210	0.007	3.4	99.8	0.007	0.000
	0.50	0.473	0.031	6.6	89.8	0.061	-0.051
	1.00	0.902	0.008	0.9	85.7	0.151	-0.014
	1.50	1.433	0.034	2.4	90.8	0.157	-0.138
Cr	0.05	0.058	0.004	6.4	110.0	0.006	0.005
	0.20	0.206	0.008	3.6	97.9	0.008	-0.004
	0.50	0.449	0.017	3.9	85.2	0.080	-0.074
	1.00	0.862	0.018	2.0	82.8	0.183	-0.172
	1.50	1.408	0.032	2.3	89.2	0.182	-0.162
Cu	0.05	0.053	0.005	9.8	100.5	0.005	0.000
	0.20	0.198	0.011	5.5	93.9	0.017	-0.012
	0.50	0.479	0.014	2.8	91.1	0.049	-0.045
	1.00	0.885	0.011	1.2	84.1	0.168	-0.159
	1.50	1.387	0.028	2.0	87.9	0.200	-0.182
Fe	0.05	0.040	0.127	**316.7**	**76.4**	0.122	-0.012
	0.20	0.157	0.017	10.8	**74.5**	0.056	-0.051
	0.50	0.422	0.033	7.8	80.1	0.109	-0.100
	1.00	0.864	0.013	1.5	82.1	0.189	-0.179
	1.50	1.404	0.016	1.1	88.9	0.188	-0.167

Element µg/ml		Mean µg/ml	SD	% RSD	Recovery %	Precision	Bias
Mn	0.05	0.058	0.002	3.8	110.6	0.006	0.005
	0.20	0.208	0.007	3.5	98.8	0.007	-0.003
	0.50	0.487	0.013	2.7	92.5	0.042	-0.038
	1.00	0.894	0.099	11.1	84.9	0.159	-0.151
	1.50	1.426	0.034	2.4	90.3	0.163	-0.145
Ni	0.05	0.057	0.002	3.8	108.7	0.005	0.004
	0.20	0.206	0.010	4.8	97.9	0.010	-0.004
	0.50	0.459	0.030	6.6	87.3	0.073	-0.064
	1.00	0.876	0.021	2.4	83.2	0.179	-0.168
	1.50	1.421	0.033	2.3	90.0	0.166	-0.151
Pb	0.05	0.063	0.006	8.9	119.9	0.012	0.010
	0.20	0.216	0.008	3.7	102.4	0.009	0.005
	0.50	0.475	0.035	7.4	90.2	0.061	-0.049
	1.00	0.914	0.024	2.6	86.9	0.141	-0.132
	1.50	1.435	0.030	2.1	90.9	0.155	-0.136
Zn	0.05	0.045	0.008	18.5	85.1	0.011	-0.007
	0.20	0.213	0.008	3.6	101.0	0.008	0.002
	0.50	0.492	0.007	1.3	93.5	0.035	-0.033
	1.00	0.899	0.028	3.1	85.4	0.156	-0.146
	1.50	1.498	0.027	1.8	94.9	0.099	-0.077
Ca	10.00	11.133	0.823	7.4	105.8	0.990	0.576
	50.00	52.163	4.590	8.8	99.1	4.385	-0.446
	100.00	105.233	2.170	2.1	100.0	2.060	-0.030
K	1.00	0.894	0.050	5.6	84.9	0.170	-0.150
	5.00	5.137	0.514	10.0	97.6	0.500	-0.120
	10.00	10.600	0.196	1.8	100.7	0.200	0.070
Mg	5.00	5.125	0.356	6.9	97.4	0.356	-0.132
	25.00	26.260	2.150	8.2	99.8	2.040	-0.050
	50.00	52.010	0.944	1.8	98.8	1.100	-0.590
Na	5.00	N. R.	N. R.	N. R.	N. R.	N. R.	N. R.
	25.00	24.212	1.960	8.1	92.0	2.810	-1.998
	50.00	49.625	1.560	3.1	94.6	3.190	-2.700

Table 6. Batch B. Filters have not been conditioned at constant relative humidity and temperature following addition of standard and prior to analysis by ICP-OES. Bold values represent outliers.

A linearity study was conducted on Batch A (Humidification). Good linearity was obtained across the accepted ranges (Table 7). High limit of quantitation values (LOQ), which led to poor recovery values for elements at certain low standard concentrations have already been discussed. For this reason the linear range of some elements was reduced.
The Pearson coefficient (R^2) was calculated for each element for those filters that were humidified (Batch A) (Table 7). R^2 ranges from 0.997-0.999.

Element	Equation of the Line	Linear Range	R^2	P-Value
Al	y=0.8528x+0.0149	0.05-1.5	0.9992	0.824
Ba	y=0.8502x+0.0297	0.05-1.5	0.9983	0.852
Cd	y=0.8831x+0.0283	0.05-1.5	0.9985	0.8968
Co	y=0.8807x+0.0269	0.05-1.5	0.999	0.8906
Cr	y=0.8725x+0.211	0.05-1.5	0.986	0.8662
Cu	y=0.8649x+0.0257	0.05-1.5	0.9981	0.8649
Fe	y=0.8566x+0.0106	0.2-1.5	0.9995	0.7913
Mn	y=0.8727x+0.0265	0.05-1.5	0.9981	0.878
Ni	y=0.8662x+0.0228	0.05-1.5	0.9985	0.8606
Pb	y=0.883x+0.0331	0.05-1.5	0.9992	0.907
Zn	y=0.8948x+0.0204	0.05-1.5	0.9975	0.8971
Ca	y=1.0586x-0.7582	5-100	0.9996	0.9604
K	y=1.0406x-0.175	0.5-10	0.9989	0.9982
Mg	y=1.0408x-0.2198	0.5-50	0.9999	0.9765
Na	y=1.02x-1.3319	5.0-50	0.9999	0.9678

Table 7. Batch A. Regression statistics assigned to standards from ANOVA analysis.

The values are acceptable considering the matrix involved and the impact of the conditioning process (Fig. 1). The data was analysed by ANOVA.
The p values obtained show insignificant statistical variation within the linear ranges (Table 7). The linear range for the elements Al, Ba, Cd, Co, Cr, Co, Cu, Mn, Ni, Pb, and Zn are 0.05-1.5 µg/ml. For Fe, the linear range was 0.2-1.5 µg/ml, Ca 5-10 µg/ml, K 0.5-10 µg/ml, Mg 0.5-50 µg/ml and Na 5-50 µg/ml.
Repeat analysis of Batch A was performed at a time delay of three days after the initial analysis. In all elements analysed, the lowest standard (0.05 µg/ml) did not produce an elemental reading. Furthermore, standards Al (0.2, 1.0 µg/ml) Cr (0.2 µg/ml), Fe (0.2 µg/ml) presented recovery values less than the lower recovery limit of 80% (Table 8). This can be attributed to the deterioration of standards over time following digestion. Therefore, analysis should be performed immediately following digestive treatment of filter papers.

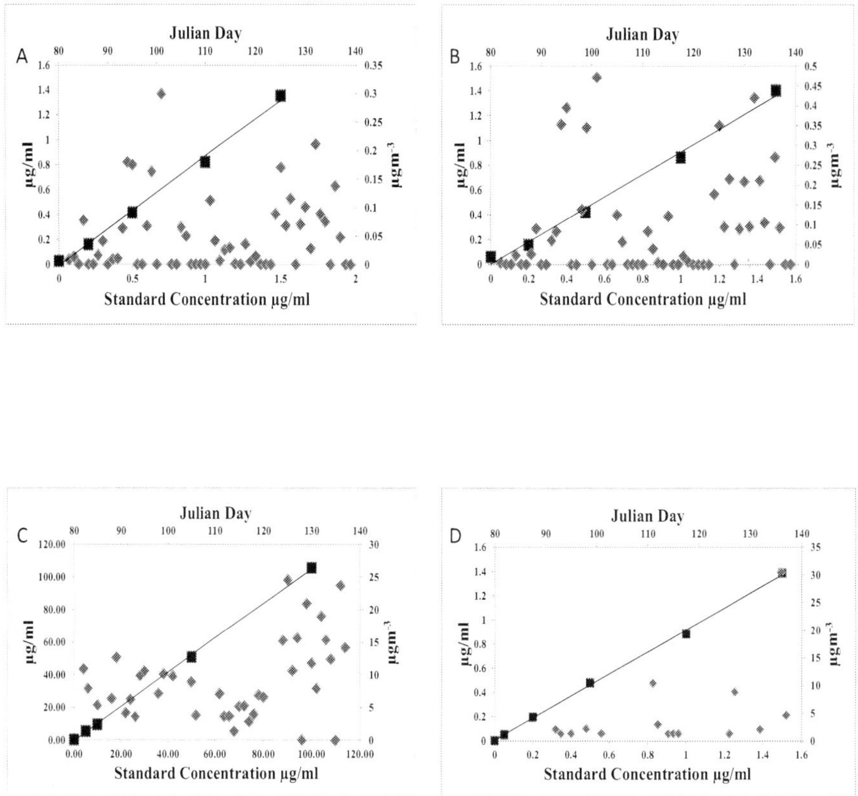

Fig. 1. Overlay of the standard calibration curves (µg/ml) against the actual recorded ICP readings for the elements from sample filters (n=39; blue diamond symbol). Curve A represents Aluminium, with a linear regression of R^2=0.9956 for spiked standards; Curve B: Iron, R^2=0.9973; Curve C: Calcium, R^2=0.9996; Curve D: Copper, R^2=0.9986.

Element µg/ml		Mean µg/ml	SD	RSD %	Recovery %	Precision	Bias
Al	0.2	0.131	0.013	9.7	**62.2**	0.081	-0.076
	0.5	0.468	0.033	7.1	88.9	0.066	-0.055
	1	0.831	0.010	1.2	**78.9**	0.222	-0.211
	1.5	1.295	0.062	4.8	82.0	0.290	-0.270
Ba	0.2	0.185	0.009	4.7	87.7	0.027	-0.025
	0.5	0.495	0.031	6.3	94.0	0.043	-0.030
	1	0.889	0.019	2.2	84.4	0.165	-0.156
	1.5	1.351	0.059	4.4	85.5	0.236	-0.217
Cd	0.2	0.190	0.007	3.6	90.2	0.022	-0.020
	0.5	0.512	0.036	7.0	97.3	0.037	0.013
	1	0.930	0.029	3.1	88.3	0.126	-0.117
	1.5	1.421	0.020	1.4	90.0	0.159	-0.150
Co	0.2	0.193	0.007	3.7	91.5	0.019	-0.017
	0.5	0.501	0.031	6.2	95.2	0.039	-0.024
	1	0.914	0.018	2.0	86.8	0.140	-0.132
	1.5	1.385	0.041	3.0	87.7	0.199	-0.185
Cr	0.2	0.146	0.008	5.3	**69.5**	0.065	-0.061
	0.5	0.499	0.032	6.4	94.8	0.041	-0.026
	1	0.908	0.019	2.0	86.3	0.146	-0.137
	1.5	1.364	0.048	3.5	86.4	0.220	-0.205
Cu	0.2	0.189	0.010	5.4	90.0	0.023	-0.020
	0.5	0.483	0.034	7.0	91.9	0.054	-0.041
	1	0.875	0.016	1.9	83.2	0.178	-0.169
	1.5	1.337	0.031	2.3	84.7	0.244	-0.230
Fe	0.2	0.147	0.023	15.8	**69.9**	0.067	-0.060
	0.5	0.478	0.030	6.3	90.9	0.056	-0.046
	1.0	0.861	0.019	2.2	81.8	0.193	-0.183
	1.5	1.346	0.056	4.2	85.3	0.239	-0.221
Mn	0.2	0.191	0.007	3.6	90.8	0.020	-0.018
	0.5	0.503	0.031	6.1	95.6	0.038	-0.022
	1.0	0.912	0.017	1.9	86.7	0.142	-0.134
	1.5	1.375	0.038	2.8	87.1	0.207	-0.193
Ni	0.2	0.188	0.007	3.8	89.2	0.024	-0.022
	0.5	0.491	0.035	7.0	93.2	0.049	-0.034
	1.0	0.914	0.032	3.5	86.8	0.142	-0.132
	1.5	1.376	0.036	2.6	87.1	0.207	-0.193
Pb	0.2	0.203	0.010	4.9	96.4	0.012	-0.007
	0.5	0.512	0.033	6.5	97.4	0.034	-0.013
	1.0	0.946	0.033	3.5	89.8	0.112	-0.102
	1.5	1.390	0.043	3.1	88.0	0.194	-0.180
Zn	0.2	0.179	0.011	6.0	84.9	0.033	-0.030
	0.5	0.536	0.033	6.2	101.8	0.033	0.009
	1.0	0.963	0.030	3.1	91.5	0.094	-0.085
	1.5	1.450	0.020	1.4	91.8	0.141	-0.123

Table 8. Batch A. repeat analysis to determine stability of filters which have been conditioned at constant relative humidity and temperature following addition of standard and prior to analysis by ICP-OES. Bold values indicate outliers.

Significance testing was performed on Batch A against Batch B to determine the statistical significance of the removal of the conditioning step in Batch B. A test was also applied to data from batch A and the repeat analysis of Batch A to determine the statistical significance of delay in analysis following digestion.
Significant statistical differences were observed between Batch A repeat (N=2) (Table 9). This was expected due to the deterioration of standards as previously discussed. No statistical differences were observed between Batch A and Batch B (Table 9). This was despite the fact that poor recovery values were obtained in Batch B.

Element	Batch	F-Values						Fcrit
	Compared	Standard Conc.		μg/ml				
		0	0.05	0.2	0.5	1	1.5	3.0204
Al	A1vsA2	0.3434		0.2342	1.8520	1.8182	0.7384	
	A1vsB1	0.3286	0.1063	0.1571	0.5140	0.9996	1.0358	
Cu	A1vsA2	**4.8889**		**5.9221**	1.9381	1.6138	2.2133	
	A1vsB1	0.6481	1.6438	0.9650	0.7862	0.5859	0.8649	
Fe	A1vsA2	0.2544		2.6744	1.1611	2.1950	4.2165	
	A1vsB1	0.8155	0.0001	0.1629	0.7884	1.1858	0.3047	
Mn	A1vsA2	2.1523		0.6792	1.5032	1.2188	3.0147	
	A1vsB1	0.9111	0.9714	1.1028	1.0376	0.7959	0.3581	
Ni	A1vsA2	0.6221		0.6958	2.9769	1.0100	2.4429	
	A1vsB1	1.2095	1.6625	0.9714	0.4916	1.2891	0.3237	
Pb	A1vsA2	0.8889		1.2557	2.6848	1.9018	0.0196	
	A1vsB1	0.6162	2.1552	0.5370	0.4081	1.2534	0.5629	
Ba	A1vsA2	**93.4444**		0.3876	**7.6962**	2.6462	**5.1197**	
	A1vsB1	2.4873	1.5556	0.4276	0.6406	0.8042	0.3436	
Zn	A1vsA2	**143.8148**		0.2590	0.4886	0.5395	0.0104	
	A1vsB1	0.8896	1.2818	1.6404	0.8925	0.4590	0.2799	

Table 9. Results of significance testing of Humidified filters (Batch A) versus Humidified filters repeat analysis (Batch A Repeat) and Humidified filters (Batch A) versus Non-Humidified filters (Batch B). Bold values indicate significant statistical differences.

4. Case study

In the spring-summer of 2005 fifty-nine daily samples were taken and thirty-nine of these daily samples were analysed for metals by ICP-OES. Average daily PM_{10} values were 34 μgm^{-3}, with a high value of 89 μgm^{-3} PM_{10} (Table 10). Metal values displayed (Table 10), show the constituent portions for each element on the day this maximum value was observed. The median PM_{10} value observed for the period was 29 μgm^{-3} (Table 10). Wind speed was a significant contributor to high particulates when coupled with low precipitation, particularly when the weather system was originating in the Atlantic Ocean.

Monitoring Period	Type	Samples Total Filters	PM_{10} μgm^{-3}	Al ngm^{-3}	Fe ngm^{-3}	Ca μgm^{-3}	K ngm^{-3}	Mg ngm^{-3}	Na μgm^{-3}	Pb ngm^{-3}	Zn ngm^{-3}	Cu ngm^{-3}	Mn ngm^{-3}
Spring-Summer 2005	Mean	59	34	82	147	9.9	185	164	0.124	9	72	5	14
	Max	Daily	89	299	471	24.5	604	464	2.18	22	146	31	31
59 daily samples	Max	Daily	%	0.46	0.72	27.84	1.31	0.71	5.89	0.04	0.32	0.04	0.03
	Max	Daily	PM_{10}	65	65	88	46	65	37	49	45	65	89
	Median		29	65	92	8.4	167	124	0	8	68	2	11
	Mean		%	0.24	0.43	29.12	0.54	0.48	0.36	0.03	0.21	0.02	0.04

Table 10. Collection Site PM_{10} data and selected chemical analysis results

Back trajectories arriving at this observation site were obtained from the web version of the Hybrid Single Particle Langrangian Integrated Trajectory (HYSPLIT) model [24] from March 15th to May 22nd 2005 (Fig. 2 and 3). NOAA's Air Resources Laboratory (ARL) archives were used to establish back trajectories [25].

These back trajectories show the air mass over the observation site during the study period. Air mass was primarily emanating from a west-south westerly direction during the study period bringing Atlantic Air. This air mass is associated with higher wind speed causing the increase in concentration of coarse particulate matter (2.5-10 μm) due to earth crust and sea-salt sources [26]. Brief periods also prevailed when the air mass originated from the United Kingdom (Fig. 2 C), Northern European countries such as Poland (Fig. 2 D) and the Arctic regions (Fig. 3 A). This air mass is associated with anthropogenic pollutants and is considered part of the fine particulate matter fraction (0-2.5 μm) [15].

4.1 Principal component analysis

Principle Component Analysis (PCA) was applied to data using STATISTICA software (StatSoft, Tulsa, OK, USA) upon those samples obtained which exceeded the daily PM_{10} limit of 50 μgm^{-3}. The most related groups were (Mg^{2+}, Zn^{2+}, Cu^{+}, Ca^{2+}), (Al^{+}, Mn^{+}, Fe^{2+}), (K^{+}, Pb^{2+}).

VARIMAX rotation of the data presents three factors that were observed to be associated with particulate matter concentrations during these daily exceedances accounting for 84.3% of total variation (Table 11). The first factor has a high loading for Cu^{+}, Ca^{2+} Mg^{2+} whilst also observing significant loading of Zn^{+} (Table 11) and identifies crustal material and road dust re-suspension as the source [26]. During the study period, local access roads in the area were in need of repair. The influence of heavy industrial traffic upon these roads was significant. The high concentration of quarrying extractive industry in the area led to much deposition of dust along access roads, which was re-suspended on days of high wind speed, leading to an increase in the concentration of particulate matter. The second factor presented high loading for Al^{+}, Mn^{+} and Fe^{2+} associated with traffic with extractive industry in the area (Table 11). Na^{+} was not found to be present in any of the samples, which exceeded the daily limits and therefore could not have a principal component analysis applied to it in this case. The third factor is indicative of differing types of industrial activity in the locality. K and Pb^{2+} are indicative of industrial factors [27].

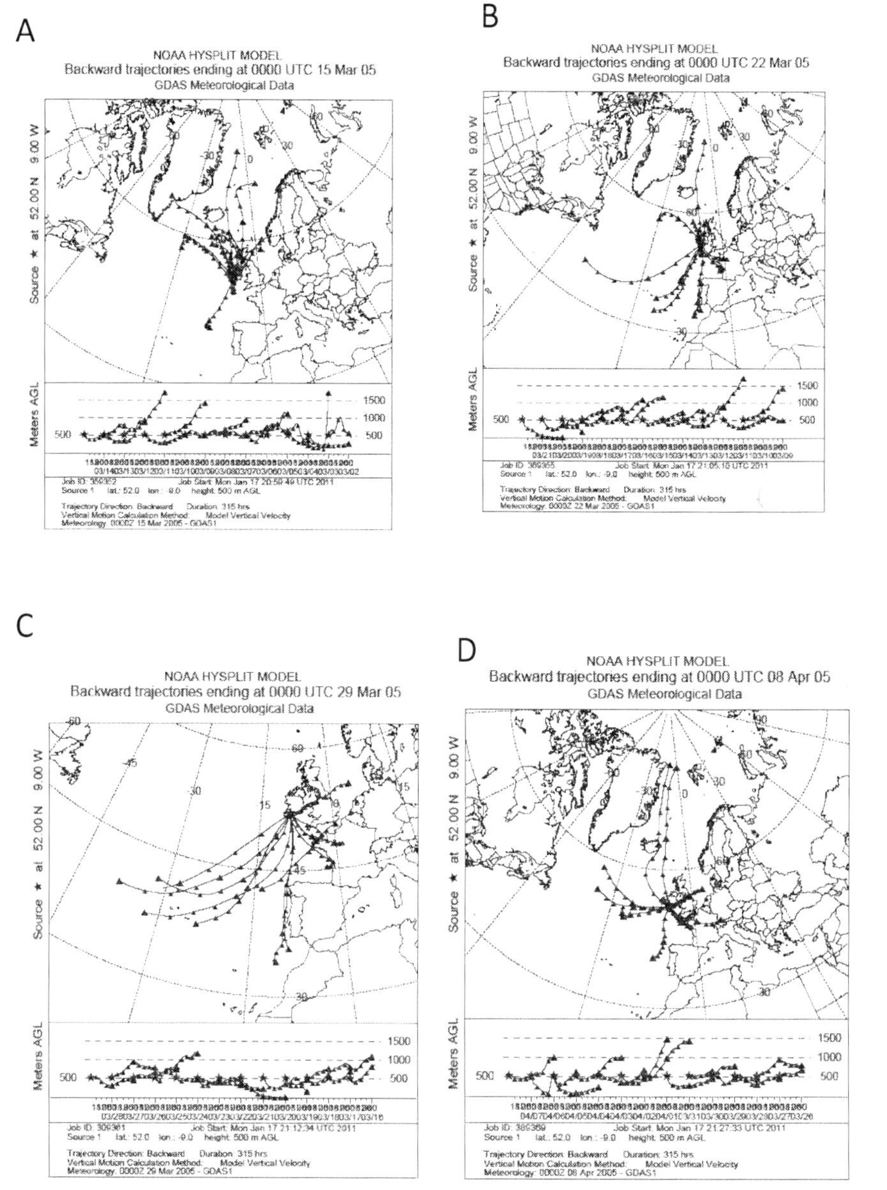

Fig. 2. Back trajectories of air mass flow over the observation site during study period obtained from NOAA HYSPLIT web based model; (A) March 15th 2005; (B) March 22nd 2005; (C) March 29th 2005; (D) April 8th 2005.

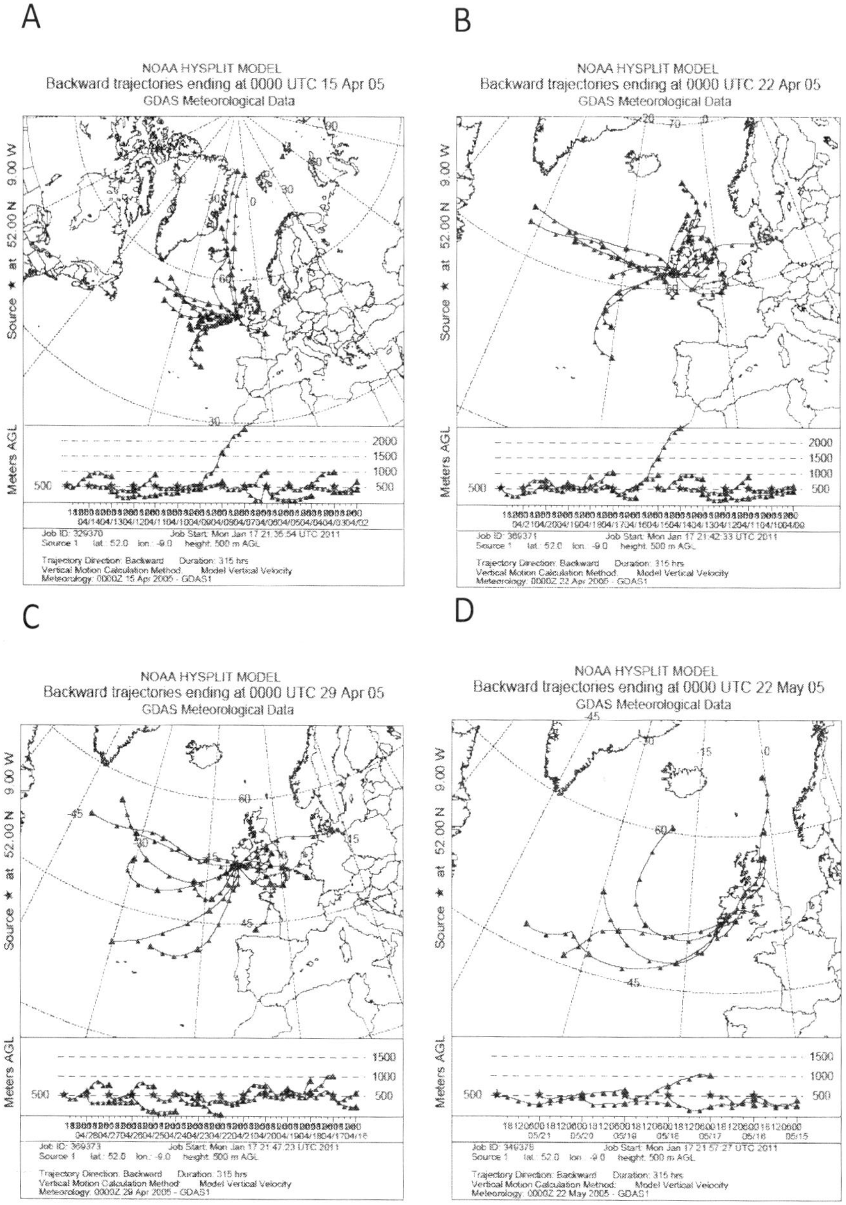

Fig. 3. Back trajectories of air mass flow over the observation site during study period obtained from NOAA HYSPLIT web based model; (A) April 15th 2005; (B) April 22nd 2005; (C) April 29th 2005; (D) May 22nd 2005.

	PM_{10}	Spring 2005		
Element	Factor 1	Factor 2	Factor 3	Communality
Al	0.01	**0.84**	-0.07	0.7
Zn	**-0.85**	-0.14	0.09	0.75
Cu	**0.87**	-0.06	0.39	0.9
Ca	**0.71**	0.45	0.22	0.75
Fe	0.03	**0.89**	0.16	0.82
K	0.41	0.48	**0.71**	0.91
Mg	**0.92**	0.17	0.06	0.87
Mn	0.37	**0.85**	-0.22	0.91
Pb	0.05	-0.17	**0.97**	0.98
Eigenvalues	4.2	2.06	1.33	
Variance %	46.72	22.84	14.74	
Cumulative %	46.72	69.56	84.3	
Origin	Crustal-Road dust	Traffic	Industrial	

Table 11. Principal Component Factor analysis on PM_{10} element concentration data set. Bold values represent strong correlation to specific factors.

Principal Component plots (95% confidence level) indicate a strong correlation of elements associated with coarse particles (2.5-10 μm) and PM_{10} concentration values obtained. Al (r-0.6717) and Fe (r-0.741) are traffic in origin and Mg (r-0.59) is associated with high wind speed causing the re-suspension of road dust (Fig. 4 A, C, D). Ca (r-0.399) is associated with crustal origin and the re-suspension of road dust, though does not correlate as strongly with particulate matter concentration (Fig. 4. B). K (r-0.37) is associated with industrial combustion (Fig. 5. C). Mn (r-0.49) in this case is associated with traffic sources (Fig. 5. A). Cu (r-0.054) and Zn (r-0.046) are associated with wear and tear from tyres of vehicles. These elements display no correlation with PM_{10} concentration during the study (Fig. 5. B, D). Principal component scatter plots were used to determine the correlation between PM_{10} and its constituent elements in a study in Argentina [21] with significant correlations highlighted and discussed.

4.2 Enrichment factors

The anthropogenic contribution to the elemental composition of PM_{10} in samples, which exceeded the daily limit of 50 μgm^{-3} was assessed by calculation of Enrichment Factor (EF) for each element using Al as a reference, utilising the crustal composition equation as given by Mason and Moore 1966 [28].

$$EF_{Al}(X) = (X/Al)_{air} / (X/Al)_{crust} \quad (3)$$

Enrichment factors (Fig. 6) show Fe, K, Mg <10 EF attributable to natural sources. Fe is assigned to traffic factor, Mg is assigned to re-suspension of road dust and K, assigned to an industrial factor is attributable to heavy extractive industry in the area. Ca and Cu show slightly enriched levels correlating with the crustal material factor assigned under PCA. Mn, Pb and Zn are all enriched to an elevated level. Zn and Mn are assigned to traffic factor; Pb is assigned to industrial activity.

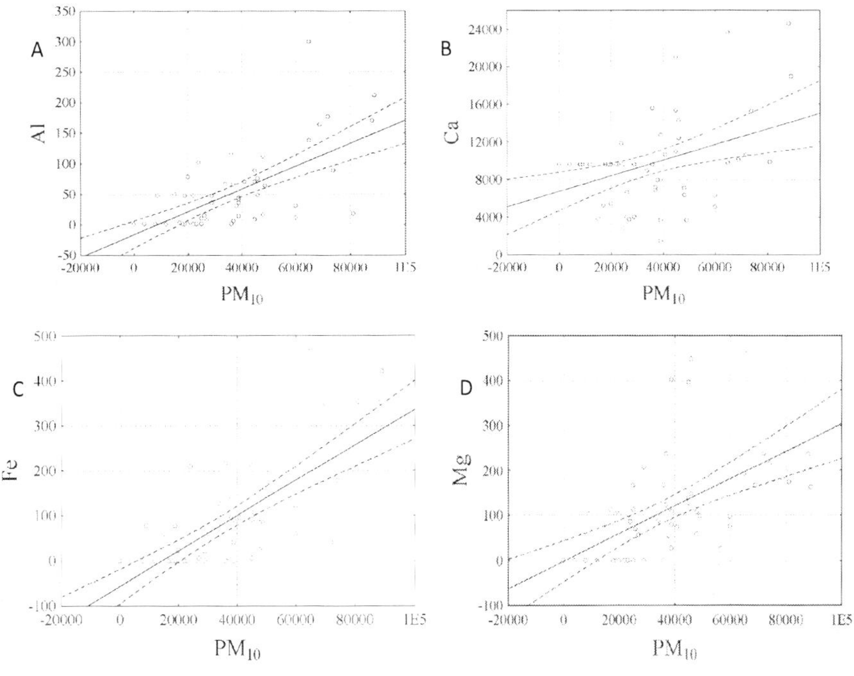

A; PM_{10} μgm^{-3} vs. Al ngm^{-3}; Al = -16.16 + 0.00187 * PM_{10}; Correlation: r = 0.67167; 0.95, Confidence Interval

B; PM_{10} μgm^{-3} vs. Ca ngm^{-3}; Ca= 6754.5 + 0.08292 * PM_{10}; Correlation: r = 0.39914; 0.95, Confidence Interval

C; PM_{10} μgm^{-3} vs. Fe ngm^{-3}; Fe = -56.61 + 0.00393 * PM_{10}; Correlation: r = 0.74102; 0.95, Confidence Interval

D; PM_{10} μgm^{-3} vs. Mg ngm^{-3}; Mg = -1.679 + 0.00306 * PM_{10}; Correlation: r = 0.58740; 0.95, Confidence Interval

Fig. 4. Correlation of PM_{10} concentration versus concentration of individual metallic elements as a constituent component of particulate matter for the assessment site. PM_{10} Concentration as the line graph while elemental readings are illustrated as points

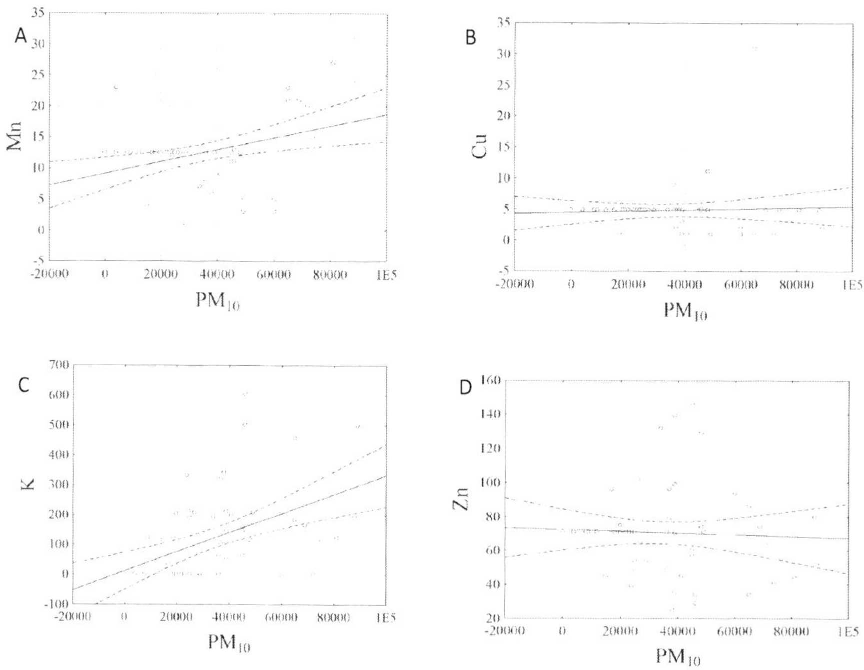

A; PM_{10} μgm^{-3} vs. Mn ngm^{-3}; Mg= 9.1546 + 0.00010 * PM_{10}; Correlation: r = 0.37044; 0.95, Confidence Interval

B; PM_{10} μgm^{-3} vs. Cu ngm^{-3}; Cu= 4.4761 + 0.00001 * PM_{10}; Correlation: r = 0.05387; 0.95, Confidence Interval

C; PM_{10} μgm^{-3} vs. K ngm^{-3}; K = 12.317 + 0.00321 * PM_{10}; Correlation: r = 0.48838; 0.95, Confidence Interval

D; PM_{10} μgm^{-3} vs. Zn ngm^{-3}; Zn = 72.388 - 0.0001 * PM_{10}; Correlation: r = -0.0459; 0.95, Confidence Interval

Fig. 5. Correlation of PM_{10} concentration versus concentration of individual metallic elements as a constituent component of particulate matter for the assessment site. PM_{10} Concentration as the line graph while elemental readings are illustrated as points

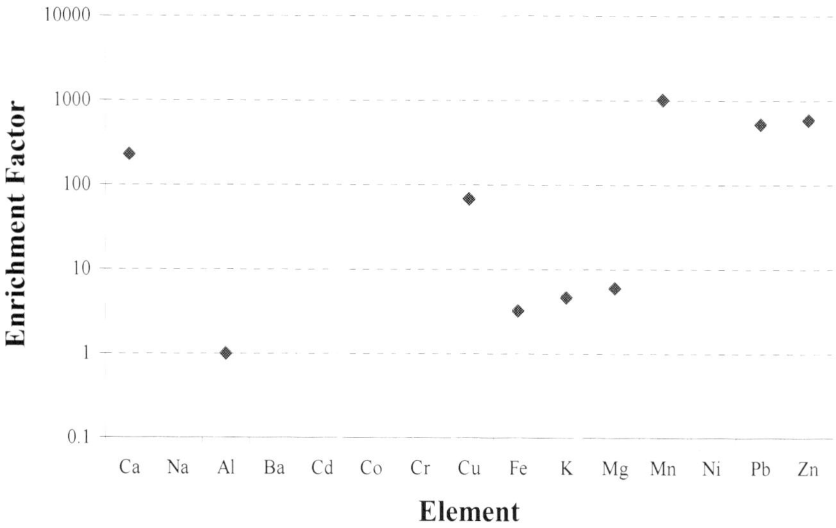

Fig. 6. Enrichment factors for elements for the data set of samples that exceeded daily particulate matter limits at the sampling site.

5. Comparative study

Other validation studies of the elemental composition of PM_{10} have been presented with the goal of achieving a validated, robust working method for the analysis of particulate matter. The use of extractive techniques is commonplace with various extractive materials used [29]. Extractive material used will depend upon the instrumentation involved, in order to present the analyte in the appropriate form for analysis.

In this study, blank filter papers were used for the determination of the average value of blank signal (n=10). Standard deviations from these blank filters were used to determine LOD and LOQ values [21]. The average concentration of blank values, were presented in a specific study ranging from 1.6-40 ngm^{-3} [20]. Comparable results 1.92-50 ngm^{-3}, were obtained in this study (Table 3). LOD values in this study, not including those elements previously mentioned as having some level of interference, are comparable with previous similar studies [30]. Previous studies presented LOQ values ranging from 0.97-124 μgl^{-1} [7]. Our values present favourably, ranging from 0.008-1.32 μg/ml. The differences in this case are due to the sensitivity of the instrument used in each study and the differing matrices involved.

Determination of the percentage recovery of analyte is expressed in the majority of studies [7, 21, 31] involving the validation of methodologies. Acceptable recovery values in this study 80.3-118.8% (Table 5) are a good expression of extraction efficiency. These results are comparable with other recovery studies undertaken [21, 31]. At low standard concentration levels, certain elements in our study (Fe, Zn, K, Na) produced poor recovery values attributable to LOQ values greater then the concentrations applied in those standards. A previous validation study highlighted poor recovery for Al, Fe and Zn, which were assigned to metal oxides and interference from filter holders [31]. Linearity studies undertaken have presented ranges of R^2 0.62-1.0 [32] and R^2 0.965-0.996 [7]. In this study, linearity is comparable across the accepted linear range, R^2 ranged from 0.986-0.999.

6. Conclusion

The determination of the metallic components of PM_{10} by Inductively Coupled Plasma Optical Emission Spectroscopy (ICP-OES) is confirmed to be suitable for application on individual ambient air samples with good analytical performance.

Limits of Detection (LOD); 0.001-0.038 µg/ml equivalent to (1.21 - 31.49 ngm^{-3}) for heavy metals and 0.06-1.81 µg/ml equivalent to (47.99 – 1507.7 ngm^{-3}) for Na, Ca, K, Mg. Limits of Quantitation (LOQ) ranged from 0.008-0.21 µg/ml for heavy metals and 0.32-1.32 µg/ml for Na, Ca, K, Mg. These results are considered satisfactory. % RSD ranged from 0.9-15.1%, while recoveries were 80.3-122.0% for the majority of elements, with exceptions highlighted in the study. The accepted linear range for the method of analysis was established with good Pearson coefficient values (R^2-0.986-0.999) obtained. Precision and trueness values were also established in the validation study.

The validation study gives information on the effects of the conditioning process upon filter papers, standards and samples and provides information on the limitations involved with the determination of certain metals at low concentrations. The conditioning study also confirms that the requirement for the humidification of filter papers is necessary to ensure the loss of analyte due to particle bound water is minimised. Assessment of filter papers not subjected to the humidification step in the procedure presents LOD values 0.001-0.056 µg/ml equivalent to (1.21-46.7 ngm^{-3}) for heavy metals, LOQ 0.008-0.31 µg/ml and 0.003-3.91 µg/ml equivalent to (2.65-3253 ngm^{-3}) for Na, Ca, K, Mg, LOQ 0.18-21.41 µg/ml. Acceptable % RSD are 0.88-18.5%. Recovery values ranged between 23.6 and 19.9%.

This paper also provides information about the nature and composition of particulate matter in samples where the daily limit for PM_{10} of 50 μgm^{-3} was exceeded. Principal component analysis (PCA) on these results shows the most related groups were (Mg^{2+}, Zn^{2+}, Cu^{+}, Ca^{2+}), (Al^{+}, Mn^{+}, Fe^{2+}), (K^{+}, Pb^{2+}). VARIMAX rotation presents three associated factors (traffic, road dust and crustal material, industrial) accounting for 84% of total variation. The Atlantic air mass has a major influence on Ireland with high wind speed causing the re-suspension of road dust. Anthropogenic influences may be local in origin or due to the influence of air masses originating over Northern or Central Europe. Principal component plots were presented showing the correlation between particulate matter concentration and individual constituent elements. Enrichment factors calculated for the data set indicate primary sources were the significant influence over particulate matter concentrations.

7. Acknowledgements

The authors gratefully acknowledge the financial support and provision of instrumentation from Cork County Council and the NOAA Air Resources Laboratory (ARL) for the provision of the HYSPLIT transport and dispersion model (http://www.arl.noaa.gov/ready.html) used in this publication. The Science Foundation Ireland Basic Research Grant Programme 2004 awarded to Dr A. Furey is acknowledged for funding the formation of the 'Team Elucidate' research group and the Higher Education Authority (Programme for Research in Third-Level Institutions 4; PRTLI IV) is also acknowledged.

8. References

[1] European Council, 2008. Ambient Air Quality and Clean Air for Europe Directive 2008/50/EC.

[2] European Council, 1999. Air Quality Daughter Directive 1999/30/EC.

[3] Minister for the Environment and Local Government, Ireland. Air Quality Standard Regulations 2002 S.I No. 271 of 2002.

[4] D. W. Dockery and C. A. III Pope, "Acute Respiratory Effects of Particulate Air Pollution", *Annual Review of Public Health*, 15, 107-132, 1994.

[5] D. Dockery and C.A Pope, "Epidemiology of acute health effects": Summary of time series studies in: R. Wilson, J. Spengler (Eds), Particles in our air: Concentrations and health effects, *Harvard Univ. press*, Cambridge, MA, USA, pp. 123-147, 1996.

[6] R. Van Dingenen, F. Raes, J-P. Putaud, U. Baltensperger, A. Charron, M. C. Facchini, S. Decesari, S. Fuzzi, R. Gehrig, H. C. Hansson, R. M. Harrison, C. Hüglin, A. M. Jones, P. Laj, G. Lorbeer, W. Maenhaut, F. Palmgren, X. Querol, S. Rodriguez, J. Schneider, H. ten Brink, P. Tunved, K. Tørseth, B. Wehner, E. Weingartner, A. Wiedensohler, P. Wåhlin, "A European aerosol phenomenology—1: physical characteristics of particulate matter at kerbside, urban, rural and background sites in Europe". *Atmospheric Environment, 38*, 2561-2577, 2004.

[7] F. Bianchi, M. Maffini, A. Mangia, E. Marengo, C. Mucchino, "Experimental design optimisation for the ICP-AES determination of Li, Na, K, Al, Fe, Mn and Zn in human serum", *Journal of pharmaceutical and biomedical analysis*, 43, 659-665, 2007.

[8] X. Querol, A. Alastuey, S. Rodríguez, F. Plana, E. Mantilla, C. R. Ruiz, "Monitoring of PM_{10} and $PM_{2.5}$ around primary particulate anthropogenic emission sources", *Atmospheric Environment 35*, 845-858, 2001.

[9] G. M. Marcazzan, S. Vaccaro, G. Valli, R. Vecchi, "Characterisation of PM_{10} and $PM_{2.5}$ particulate matter in the ambient air of Milan (Italy)", *Atmospheric Environment, 35*, 4639-4650, 2001.

[10] J. D. Whyatt, S. E. Metcalfe, J. Nicholson, R. G. Derwent, T. Page, J. R. Stedman, "Regional scale modelling of particulate matter in the UK, source attribution and an assessment of uncertainties", *Atmospheric Environment*, 41, 3315-3327, 2007.

[11] T. Moreno, W. Gibbons, T. Jones, R. Richards, "The geology of ambient aerosols: characterising urban and rural/coastal silicate $PM_{10-2.5}$ and $PM_{2.5}$ using high-volume cascade collection and scanning electron microscopy", *Atmospheric Environment, 37*, 4265-4276, 2003.

[12] W. E. Wilson, J. C. Chow, C. Claiborn, W. Fusheng, J. Engelbrecht, J. G. Watson, "Monitoring of particulate matter outdoors", *Chemospere 49*, 1009-1043, 2002.

[13] J-P. Putaud, F. Raes, R. Van Dingenen, E. Bruggemann, M-C. Facchini, S. Decesari, S. Fuzzi, R. Gehrig, C. Hüglin, P. Laj, G. Lorbeer, W. Maenhaut, N. Mihalopoulos, K. Muller, X. Querol, S. Rodriguez, J. Schneider, G. Spindler, H. ten Brink, K. Tørseth, A. Wiedensohler, "A European aerosol phenomenology—2: Chemical characteristics of particulate matter at kerbside, urban, rural and background sites in Europe", *Atmospheric Environment, 38*, 2579-2595, 2004.

[14] European Standard EN 14902:2005, Ambient air quality-Standard method for the measurement of Pb, Cd, as and Ni in the PM10 fraction of suspended particulate matter, Brussels.

[15] J. Yin, A. G. Allen, R. M. Harrison, S. J. Jennings, E. Wright, M. Fitzpatrick, T. Healy, E. Barry, D. Ceburnis, D. McCusker, "Major component composition of urban PM_{10} and $PM_{2.5}$ in Ireland", *Atmospheric Research, 78*, 149-165, 2005.

[16] European Standard EN 12341, 1998. Air quality-Determination of the PM_{10} fraction of suspended particulate matter-Reference method and field test procedure to demonstrate reference equivalence of measurement methods.

[17] Centre for Environmental Research Information. USEPA (1999) Office of Research and Development. Compendium of methods for the determination of inorganic compounds in ambient air. Method IO-3.1. Selection, preparation and extraction of filter material.

[18] T. Byrd, M. Stack, A. Furey, "The assessment of the presence and main constituents of particulate matter ten microns (PM_{10}) in Irish, rural and urban air", *Atmospheric Environment, 44,* 75-87, 2010.

[19] UNEP, Nairobi, 1997. *GEMS/Air Methodology reviews,* Vol3: Measurements of Particulate Matter in Ambient Air.

[20] J. Marrero, R. J. Rebagliati, D. Gómez, P. Smichowski, "A study of uniformity of elements deposition on glass fiber filters after collection of airborne particulate matter (PM-10), using a high-volume sampler", *Talanta, 68,* 442-447, 2005.

[21] P. Smichowski, J. Marrero, D. Gómez, "Inductively Coupled Plasma Optical Emission Spectrophotometric determination of trace elements in PM_{10} airborne particulate matter collected in an industrial area of Argentina", *Microchemical Journal, 80,* 9-17, 2005.

[22] A. Charron, R. M. Harrison, S. Moorcroft, J. Booker, "Quantitative Interpretation of divergence between PM_{10} and $PM_{2.5}$ mass measurement by TEOM and Gravimetric (Partisol) instruments", *Atmospheric Environment,* 38, 415-423, 2004.

[23] F. LT. Goncalves, S. Braun, P. L. Silva Dias, R. Sharovsky, "Influences of the weather and air pollutants on cardiovascular disease in the metropolitan area of Sao Paulo", *Environmental Research,* 104, 275-281, 2007.

[24] R. R. Draxler, and G. D. Rolph, 2003. HYSPLIT (Hybrid Single-Particle Langrangian Integrated Trajectory) Model access via NOAA ARL READY Website (http://www.arl.noaa.gov/ ready/hysplit4.html). NOAA Air Resources Laboratory, Silver Spring, MD.

[25] G. D. Rolph, 2003. Real-time Environmental Applications and Display sYstem (READY) Website (http://www.arl.noaa.gov/ready/hysplit4.html). NOAA Air Resources Laboratory, Silver Spring, MD.

[26] S. G. Jennings, D. Ceburnis, A. G. Allen, J. Yin, R. M. Harrison, M. Fitzpatrick, E. Wright, J. Wenger, J. Moriarty, J. R. Sodeau, E. Barry, "Air Pollution - Nature and Origin of PM_{10} and Smaller Particulate Matter in Urban Air", Published by the Environmental Protection Agency, Ireland, 2006.

[27] M. Viana, X. Querol, T. Götschi, A. Alastuey, J. Sunyerj, B. Forsberg, J. Heinrich, D. Norbäck, F. Payo, J. A. Maldonado, N. Künzli, "Source apportionment of ambient $PM_{2.5}$ at five Spanish centres of the European community respiratory health survey (ECRHS II)", *Atmospheric Environment, 41,* 1395-1406, 2007.

[28] B. Mason, C. Moore, *Principles of Geochemistry,* Fourth ed. Wiley, New York, 1996.

[29] M. R. Heal, L. R. Hibbs, R. M. Agius, I. J. Beverland, "Total and water-soluble trace metal content of urban background PM10, PM2.5 and black smoke in Edinburgh, UK", *Atmospheric Environment,* 39, 1417-1430, 2005.

[30] M. Dos Santos, D. Gomez, L. Dawidowski, P. Smichowski, "Determination of water-soluble and insoluble compounds in size classified airborne particulate matter", *Microchemical Journal,* 91, 133-139, 2009.

[31] S. Canepari, E. Cardarelli, A. Pietrodangelo, M. Strincone, "Determination of metals, metalloids and non-volatile ions in airborne particulate matter by a new two-step sequential leaching procedure. Part B: Validation on equivalent real samples", *Talanta* 69, 588-595, 2006.

25

Artificial Neural Networks - a Useful Tool in Air Pollution and Meteorological Modelling

Primož Mlakar and Marija Zlata Božnar
MEIS environmental consulting d.o.o.
Slovenia

1. Introduction

Artificial neural networks have become a widely used tool in several air pollution and meteorological applications. Yi and Prybutok (1996) used MPNN for surface ozone predictions, as well as Comrie (1997). Several prediction models were also made for other pollutants; for instance for SO_2 (Božnar et al., 1993) and for CO (Moseholm et al., 1996). Marzban & Stumpf (1996) used MPNN for predicting the existence of tornadoes.

A review article by Gardner (1998) described a variety of applications, mainly in the field of air pollution forecasting and pattern classification. Though the number of applications is growing, especially in recent years, no special attention has been paid to the principles of artificial neural network usage in environmental applications.

Our group first established a method for short term forecasting of SO_2 concentrations on the basis of a multilayer perceptron neural network (Božnar et al, 1993), but in the following years we use an artificial neural networks in several other applications that differ very much each another.

In this article we intend to show examples of a variety of applications of artificial neural networks in air pollution and the meteorological field. Examples are taken from our past experience, extending over a decade.

Several applications in this field start from fundamentals and too much attention is paid to optimization and speeding up of the learning algorithms. From our experience this should be a minor problem for an environmental modeller and does not significantly affect the final model quality if modern tools are used. In the process of model construction other factors are much more crucial - such as feature determination, pattern selection, and learning process optimization. These are the methods that are derived from the basic principle of artificial neural networks - that is the ability to learn information from given examples.

In this article we intend to show some solutions for the effective transformation of measured information into air pollution and meteorological models. We hope that the variety of examples will inspire new applications and methods that will serve the air pollution modelling community. The mystique of artificial neural networks, derived directly from their name, prevents many modellers from using them. It is the purpose of this article to demystify this useful mathematical tool and in this way encourage its usage.

2. Artificial neural networks – several types for different purposes

Artificial neural networks can be divided into several groups according to their topology. The tool was firstly widely used in the pattern recognition field. The topologies vary from feed forward neural networks with several hidden layers, to topologies with backward loops that make the result sequence dependent, to fuzzy logic and several automatic sorting tools. A detailed explanation of this groups is far beyond the scope of this article. The reader interested in this issue can get information from several books (Lawrence, 1991).

In this article we focus on two main "species" of artificial neural networks that can cover a huge variety of air pollution and meteorological modelling applications. The two selected are the Multilayer Perceptron artificial Neural Network (MPNN) and the Kohonen neural network (KNN). Both can be replaced by other artificial neural networks for the same purpose, but this does not change the method of using these tools. In this article MPNN and KNN can both be treated as one of the best possible solutions. The authors of this article have no intention to argue about the qualifications of other topologies.

In this article it will be shown what the most suitable applications of MPNN and KNN are. The latter is not so widely used although it has great potential in environmental problems.

MPNN is mathematically speaking a universal approximator (Hornik, 1991; Kurkova, 1992). It can reconstruct arbitrary multivariable and highly non-linear functions. Therefore it is a suitable tool for modelling atmospheric phenomena whose behaviour has not yet been described by formulas but is only known from measured examples.

KNN, on the other hand, is a structure capable of sorting a multitude of multivariable samples or patterns into groups of similar ones. It is important that it can find these groups without a teacher - so-called unsupervised learning. This ability becomes extremely important when dealing with multivariable patterns where similarity rules are not obvious.

3. Multilayer perceptron artificial neural network (MPNN)

The structure of MPNN was introduced by Rumelhart (1986). It is one of the basic neural network structures from which several others were derived.

The basic element of the MPNN is a neuron. Several neurons are organized into layers - input, hidden (one or more) and output layer. Each neuron has a simple structure that mimics the functionality of the neuron found in animals and the whole structure of layers mimics the brain structure. This similarity gives rise to the name. Each neuron firstly summarizes the weighted input values and then passes the sum through the transfer function. If the transfer function is nonlinear, such as a basic sigmoid function or hyperbolic tangent, then the whole structure acquires its great ability as an universal approximator.

The neurons in the input layer take the values from the model input variables and pass the values to the neurons in the hidden layer, the hidden layer neurons pass the values to the higher hidden layers and finally to the output layer that gives the model output value. The output of each neuron is passed to the input of all neurons in the next higher layer. All the connections between neurons are weighted. These interconnection weights are the basic parameters of the model that are adjusted during the learning process.

Model inputs take their values from the input features - measured parameters that determine the output of the model. Model output(s) represents the phenomenon that is being reconstructed (approximated). Outputs are called output features.

The values of one particular realization of all inputs is called the input vector, and the model outputs values form the output vector. Both vectors together form a pattern. A pattern is therefore like one dot in the multivariable space lying on the surface of the function the model is approximating.

The whole idea of constructing a model to approximate a multivariable function is the following: Firstly enough patterns should be available (for instance from the measurements) with known input and output features. These patterns should be uniformly spread over the whole investigated domain. Then the model topology is designed according to the number of input and output features. The model learning stage consists of several adjustments of model interconnection weights - in order to minimize the average error between the actual measured output values and the output values that are produced by the neural network. One of the algorithms that can be used for this purpose is the backpropagation algorithm. In the process of learning the MPNN takes the information (about phenomenon under investigation) that is available in the learning patterns and when learning is completed (the model constructed) it can give the results for previously unknown patterns - where only input values are presented to the network. This is possible if there were similar patterns (to the unknown pattern) in the learning set. This is the so-called generalizing capability of the MPNN. The similarity is mathematically speaking the distance between two patterns.

The basic rule of MPNN model construction is therefore to provide information rich learning patterns.

There are some basic steps and methods that should be used in the model construction process to obtain effective models. These steps will be summarized in the following paragraphs and their practical use is shown in the exemplary applications that follow this section.

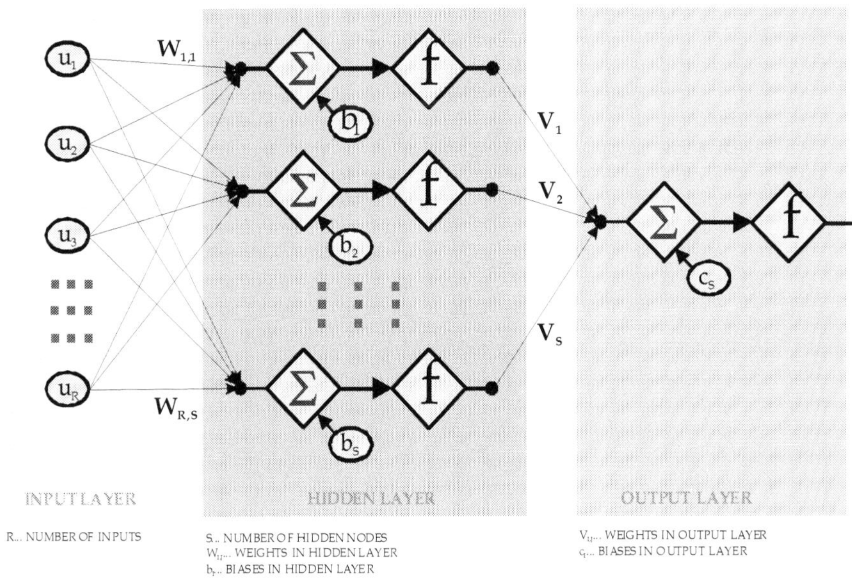

Fig. 1. The structure of a feedforward multilayer perceprtron neural network

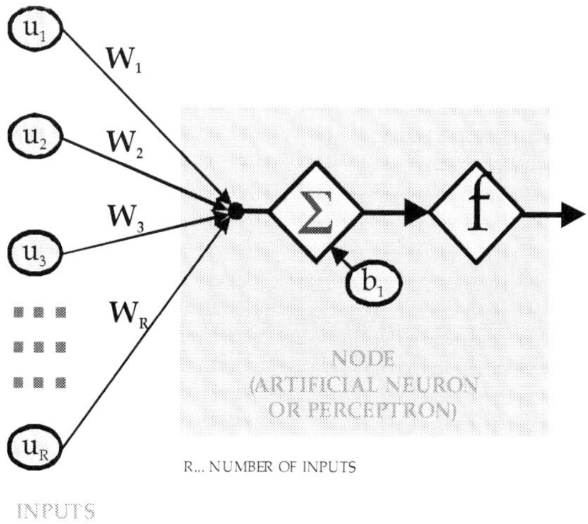

Fig. 2. Node (artificial neuron or perceptron)

3.1 Feature determination

Feature determination should be done in order to properly define the modelled domain (independent variables), to enable all important information to be captured, to simplify MPNN and therefore achieve more effective learning, to reduce the number of learning patterns needed and to increase the probability of finding the global minimum of the error function during learning.

Firstly the modeller should determine what the desired output of the model is. This can be one or several parameters that can be measured or calculated. These are the output features. For several output features it is usually more effective to establish one model for each feature than one model for all. Then the input features should be determined from several other measured parameters that represent the possible variables that cause or influence the output parameter. Input features are the ones that have significant influence on the outputs. Feature determination can be done heuristically (using expert knowledge about the phenomenon under investigation) or using other methods (feature reduction that can be extraction or selection (Devijver, 1982); examples of selection are contribution factors or Saliency metrics techniques). In both the latter methods basically the model is firstly trained with all available features and the higher absolute values of interconnection weights reveal the more important input features.

It is extremely important that the feature determination process should not be based on a linear method. Most of the processes in the atmosphere over complex terrain are not linearly dependent on each other. Therefore if the input features are chosen from the possible input measurements by a linear criterion (for instance calculation of the linear correlation factor between the examined input measurement and output modelled parameter), then most probably the important ones are rejected. An MPNN has the very important ability of being able to simulate highly non-linear dependencies and the modeller should obtain the most advantage out of this. The above mentioned contribution factors and Saliency metrics techniques both allow highly-nonlinear relationships to be found.

3.2 Model construction

The data base of the measurements (values of input and output features for several situations - for instance for several measuring intervals) form the data base of patterns. It should be divided into several sets (training, testing, production, on-line, remaining). The training set is used to adjust the interconnection weights of the MPNN model. The testing set is used periodically during the learning process to test the model's generalizing capabilities - this is optimization during learning. The final model is the one that gives the best results on the testing set. In this way we prevent the model from becoming too dependent on known patterns and therefore losing the generalizing capabilities. The training and testing sets together form the learning set. A third set different from the previous ones is the production set. This set is used for model verification to determine its expected error. All three sets should have known input and output vectors. When the model has been tained, it can be used on patterns with unknown output values. This set of patterns is the on-line set - when a newly measured situation arises, the model gives us an answer.

3.3 Pattern selection

Only patterns with valuable information should be put into the learning set, while others are rejected and form the remaining set. The pattern selection can be done either heuristically or a Kohonen neural network can be used to sort patterns into groups and in this way the KNN shows which ones are more important. The main goal of pattern selection techniques is to select patterns over the whole of the modelled domain. These patterns should contain all the information about the studied phenomenon. Patterns selected for the training and testing set should represent all important but usually rare situations that may appear during the further use of the model. Just having a lot of patterns that are the most common, but do not represent the rare complicated situations, is certainly not enough for an effective model.

3.4 Network topology determination

The topology (number of neurons in the input, hidden and output layers) is determined from the number of features and the number of patterns. Input and output features determine the number of neurons in the input and output layers. The number of neurons in the hidden layer(s) is usually determined as the number of inputs divided by two plus the square root of the number of patterns. There is no rule for a perfect solution - the user should acquire some experience.

3.5 Training and testing process

After the topology has been determined and the patterns prepared, a training algorithm (for instance a backpropagation algorithm) should be used to determine the model's interconnection weights. Basically the algorithms have parameters that determine the speed of learning. Learning is a process of finding the global minimum of the error function. If during the learning process we move in big steps, the model cannot reach the bottom of the minimum function, but escapes quickly to other local minima. If the steps are too small, the model can be stuck in a local minimum far from the global one. During the learning process, the network should be periodically tested on the testing set (not included in the training set) to prevent overtraining. At the end the model is the network giving the best results on the optimization - testing set. This is an optimizing process that finds the network with best generalizing capabilities instead of the best memorizing capabilities. Learning speed determination and optimization are usually far more important for successful learning than having a slightly better or worse algorithm.

3.6 Model verification

When the model is trained, it should always be validated on the production set to determine the expected error in further on-line use. To obtain a fair judgement of the model's abilities, the patterns that form the production set for validation should not be presented to the model in the training or testing set at all.
The training, testing and production sets should reflect all the situations that can arise in the on-line use of the model.
Feature determination and pattern selection are therefore the most crucial steps in model construction and usually determine the model's abilities.

4. Short term forecasting of ambient SO_2 concentrations using MPNN

First let us use the MPNN as a basis for short term ambient SO_2 concentration forecasting. As an example for study the area around the Šoštanj Thermal Power Plant in Slovenia was used. The studied domain of 30 by 30 km with the TPP in the centre lies in very complex terrain - a basin surrounded by several hills that are cut by valleys. The area is characterized by very low wind speeds, frequent calm situations and thermal inversions in winter that cause severe air pollution. The whole of the studied area is covered by 6 ambient automatic measuring stations (measuring basic meteorological parameters like wind, air temperature, relative humidity and precipitation and pollutant concentrations) and emission stations in the TPP. All the stations collect data every half hour.
The idea was to test the forecasting abilities of the new MPNN tool. Low winds and quick wind changes in the area cause severe air pollution peaks of very short duration (only a few intervals). We tried to establish a model that would forecast the SO_2 concentration for the following half hour from the data available for present or past intervals (air pollution and meteorological measurements). The task was a difficult one, because the work was concentrated on rapid warning of short but severe SO_2 peaks and on not causing false alarms.
The data base of measurements was huge in all dimensions. There were over 50 parameters that were measured every half hour and several years of data were available for analysis (one year consists of over 17000 half hour intervals). It is obvious that all the data could not be simply used together because of the computational space and time problems (this was at the beginning of the PC era) and more importantly because the patterns with less information would prevail over the sparse patterns carrying crucial information. The same is valid for the different measured parameters that are the possible inputs to the model. This huge data base forced us to establish methods for feature determination and pattern selection. The idea was to find patterns that carry most of the available information and to determine which measurements influence the modeled ambient SO_2 concentration at a chosen station. It is very important to stress that we were seeking for highly non-linear dependencies that the MPNN is able to model.
The whole procedure of feature determination and pattern selection techniques is explained in detail in several our publications (Božnar, 1997; Mlakar, 1997; Mlakar & Božnar, 1996; Božnar et. al, 1993; Božnar & Mlakar, 2001).

This approach resulted in a model for a chosen station that used around 15 input measurements from that and other stations to forecast the local SO_2 air pollution. The model was trained with small (in comparison to the huge data base available) data sets of chosen patterns. This resulted in a significant improvement of model forecasting ability.

It is also important that the usual cost functions (linear correlation coefficient, mean square error, ...) are not suitable for forecasting problems where most of the time nothing out of the ordinary is happening, but when the peak of concentration comes, it is severe and short. It was very easy to obtain very good values of the above mentioned cost functions - but this does not tell anything about the real model capabilities (if it really correctly and on-time predicts the coming SO_2 peak). Therefore we defined a new cost function termed p6 (Mlakar, 1997). This is the probability of successful forecasting of a high concentration without causing false alarms. It is a very sharp cost function that clearly distinguishes good models from the ones that are in the range of naive predictors.

In the process of SO_2 modelling it was clearly proven that feature determination and pattern selection techniques influence the final model performance much more than the training algorithms and other details of the establishment of the model. This is caused by the fact that the information carried in the features and patterns of the available data set should be presented to the model in the learning phase in a "model understandable" way. To generalize this principle it can be stated that an understandable way is similar to a humanly understandable way. People also cannot learn effectively if the informative and key examples are hidden in large quantity of useless examples.

5. Daily ozone peak forecasting for a semi-urban area

A model for ozone forecasting was established for the city of Nova Gorica in Slovenia close to the Adriatic sea (Grašič, 2006). During the hot summer period high ozone episodes are often recorded. The idea of constructing the model is to have information about the ozone pollution peak of the following day already available in the evening of the day before 19:00. That would allow the population sensitive to ozone to plan their activities for the following day.

Slovene legislation defines warning values for a one hour average ozone concentration and for eight hour moving average values. We concentrated our research on determination of the maximum hourly value of ozone concentration of the following day. Ozone peaks usually occur during the midday period, therefore the task deals with forecasting cca 17 hours in advance.

The available data were measurements from a local air pollution measuring station (SO_2, O_3, NO, NO_2, CO, VOC) that also measures ground level meteorological parameters (wind, air temperature, relative humidity, air pressure and global solar radiation). In principle in the evening meteorological forcasts are available for the city of Nova Gorica. Of these values two are more reliable - the maximum daily air temperature and the average wind speed and direction for the following midday. For the purpose of establishing the model from the historical data base, actual measurements of these two parameters on the following days were taken instead of prognostic values.

A two year data base was available for model construction and verification. In this case only one pattern per day is available. Therefore two years data give cca 700 patterns only. Out of this data base one winter and one summer month were excluded (were not used in the learning process at all) for independent model verification.

Because of the small data base available for learning it was only divided into a randomly taken group of 10% for testing (optimising) and the remaining 90% used for adjustment of the model's weights (training). No other pattern selection was performed.

Feature selection was done in two steps. Firstly a wide selection of possible input features was made using chemical knowledge about ozone formation and other related processes. Then this wide range was narrowed using contribution factors. The finally selected input features were air temperature, global solar radiation, NO, NO_2, NO_x, CO, O_3 all as 24h average values calculated at 19:00 on the previous day, prognostic vector wind speed, sine of wind direction and maximal hourly air temperature for the day of prediction (all three prognostic values were taken from the available measured data base).
The verification of the model for approximately two months not used in the learning process showed that the model has a good performance. For final judgement, a longer verification period would be necessary. It is also expected that its performace would be slightly worse if actual meteorological prognostic model predictions were taken instead of real measurements (for the last three features).

6. Ground level wind reconstruction over complex terrain

Air pollution prediction was the first but not the only field where we successfully constructed MPNN based models.
Recently we encountered the problem of missing ground level wind data on the location of a planned industrial plant. The time available for the task was short and therefore it was not possible to perform one year of measurements, and only 6 weeks of measurements were available. The location was again in the complex terrain of Zasavje, Slovenia. Study of the winds in the area clearly show that ground level wind reconstruction from global prognostic meteorological models would not be useful because of the orographic complexity of the area.
But there are six existing meteorological stations in the area on sited from 2 to 10 km from the planned location. None of these locations has the same characteristics as the new location, so their data could not be used directly.
Our idea was to reconstruct one year of ground level wind data on the new location from one year of wind data at the old station locations. This is a very suitable task for a MPNN based model. The six weeks data base when wind measurements were available at both old and new locations was used to train and verify the model.
In contrast to the SO2 forecasting problem, this problem again has a small data base consisting of 6 weeks of half hour average values of wind speed and direction measurements at 7 locations. Therefore only the last week of measurements was reserved for final model verification and was not used for model learning. The remaining five week data base was again divided into a randomly taken 10% test set for optimization and 90% for training.
For every station vector and scalar half hour average values and maximum values of wind speed were available, as well as wind direction. The vectors were also decomposed into cosine and sine components. The decomposition into cosine and sine components is a trick that should be used whenever we have a measurement of circular nature (such as azimuth angle or hour within a day). All these measurements and their combination at the old stations locations are candidates for model input features.
Firstly a heuristic feature selection was performed by simply comparing the similarity of wind roses for the new and old locations. Then the final feature selection was repeated using the contribution factors technique.
The results of the model verification show better results than expected, considering the high complexity of the area studied.
The reconstructed wind speed mean absolute error at the new location was less than 0.4 m/s, the mean squared error 0.45 m/s and the linear correlation coefficient 0.84. The

average absolute error for the wind direction was 35 degrees over the whole verification data set (which contains a lot of very low wind speeds and calms) and as little as 15.5 degrees if only cases with a wind speed over 3 m/s were examined.

7. Other meteorological applications of MPNN

We successfully applied MPNN in the following meteorological problems that will be only shortly explained:

- reconstruction of SODAR measurements,
- short term forecasting of ground level wind,
- reconstruction of diffuse solar radiation,
- correction of long wave solar radiation measurements.

7.1 Reconstruction of SODAR measurements

SODAR measurements are crucial for modern numerical Lagrangean particle models used for short scale air pollution reconstruction over complex terrain. But SODAR measurements are not always available. SO_2 air pollution was studied in detail (fourth chapter of this article) in the Šoštanj area of Slovenia. In the Šoštanj basin SODAR measurements were available only for an aproximately two month period during a measuring campaign (Elisei et. al, 1992). The area of the basin and surrounding hills is well covered with ground level wind measuring stations.

We made a MPNN based model to see whether it was possible to reconstruct SODAR upper layer (not ground level) measurements from the measurements at other stations. A test model was made for the level 50m above the ground. The results were quite good (comparable to the Trbovlje wind reconstruction). Some details can be found in paper by Božnar and Mlakar (1995).

7.2 Short term forecasting of ground level wind

In the same area around Šoštanj short term ground level wind forecasts would also be very useful as an input to an SO_2 concentration forecasting model. Forecasts of wind changes for the next few half hour intervals are more dependent on local thermal and solar radiation changes than on the movement of global fronts. Due to terrain complexity again such forecasts cannot be derived from regional prognostic meteorological models, because they operate in too sparse (time and space) coordinates.

We constructed a model for ground level wind forecasting for one of the stations in the Šoštanj region. The forecast was made for one averaging interval in advance. The input features were ground level wind measurements from the studied station and from two other stations for the current time interval. For wind speed one interval in the past was also used. The results were very good for wind speed and acceptable for wind direction prediction. Some details can be found in paper by Božnar and Mlakar (1995).

7.3 Reconstruction of diffuse solar radiation measurements and correction of long wave solar radiation measurements

Our colleagues from Sao Paulo, Brazil made extensive research on the measurement of and construction of correlation based models for diffuse solar radiation in the Sao Paulo urban area (Oliveira et. al, 2002). The diffuse solar radiation component requires expensive measuring procedures in comparison to other basic meteorological measurements,

including global and long wave solar radiation. Therefore it would be useful for many purposes if the diffuse solar radiation component could be reconstructed from other simpler meteorological measurements. An MPNN-based model was constructed for this purpose that gives significantly better results than previously available models. Details can be found in paper by Soares et. al, (2004).
Another problem arose from this work - correction of long wave measurements according to the Fairall formula (Fairall et al, 1998). This correction requires additional measurements of the temperature of the long wave sensor's dome and base. There exist several years of long wave solar radiation measurements for Sao Paulo but without the required additional measurements for correction. We solved the problem by several months measurements of the missing parameters and then establishing a MPNN-based model for reconstruction of the Fairall correction from the basic meteorological measurements that are available for several years (Oliveira, 2006). The model again gave very good results.
In both the above explained models, feature determination and pattern selection techniques were applied in the model construction phase.

8. Kohonen neural network (KNN)

The Kohonen neural network (KNN) (Kohonen, 1995) differs significantly from the MPNN. The main purpose of KNN is to sort multivariable patterns into groups (clusters) of similar ones. It is important that the grouping criteria need not be known - therefore this is unsupervised learning.
KNN is a very practical and effective tool for finding groups of similar patterns in data sets where it is not known in advance (through some other available knowledge) what their natural division into groups of similar patterns is.
The sorting principle is as follows: firstly the user prepares a data set of multivariable patterns that should be searched for groups of similar ones. The pattern consists of input features (the same definition as in MPNN). The output feature is the number of the cluster that the pattern belongs to. The quantity of clusters should be determined by the user. The natural number of clusters (the number of clusters that best fits the examined problem) cannot be determined automatically. But there is a relatively simple way of finding it. The process of dividing data set into groups is repeated for several different quantities of groups. For each division the average standard deviation of the distance of all patterns from the corresponding centre of the group should be calculated. On increasing the number of groups, the standard deviation decreases rapidly until the natural number of groups is reached. After that, if we divide these groups into more groups, the standard deviation decreases significantly slower than before. Using this rule, the "natural" number of groups can be easily derived from a graph of the average standard deviation of the distance versus the number of groups.
The crucial part of sorting is selection of the measure of distance appropriate to the problem examined. In most cases the Euclidean distance between two vectors can be used. But it should be noted that if the components of the vector represent measurements of different natural processes, then each process should be normalized. If this is not done, some components may prevail over others. Beside Euclidean distance, many other distance measures that are known from pattern recognition theory can also be used.
In the iterative process when KNN sorts the available data set of patterns into a chosen number of groups, it actually puts together patterns that are close one to another in terms of the distance function used. The algorithm is again an iterative one and the user can stop the process of division when the groups become stable.

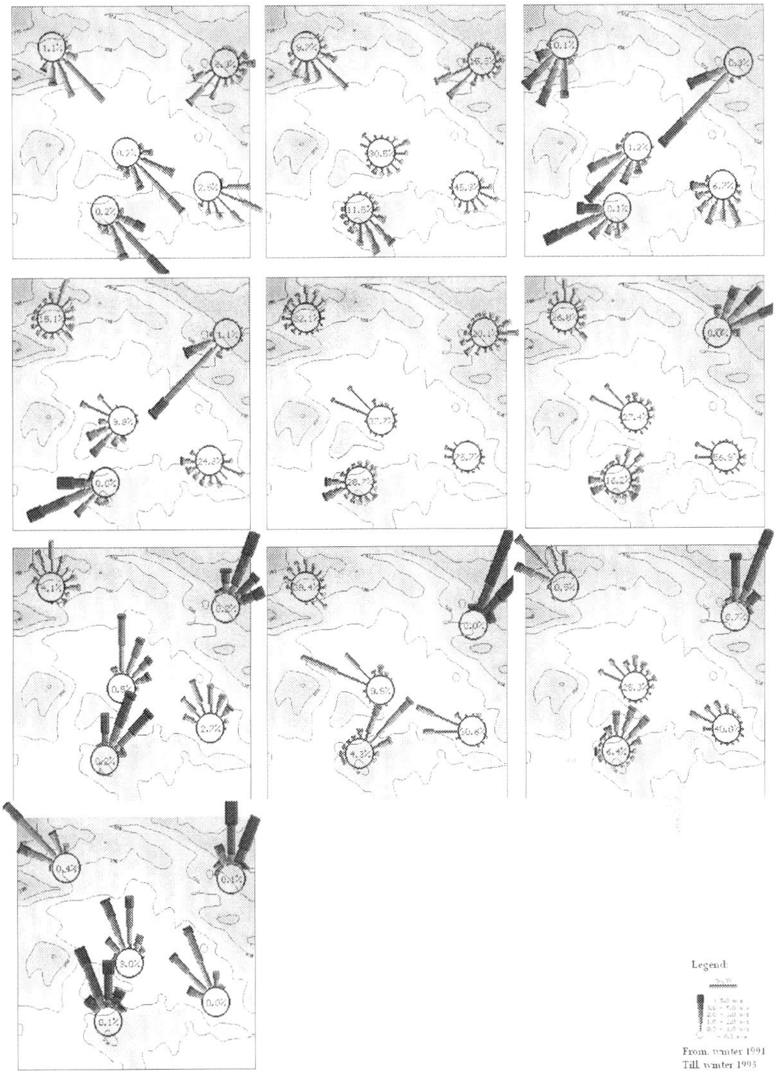

Fig. 3. Presentation of wind roses for all clusters for the division into 10 clusters

Feature determination is also an important process when using KNN. In this case feature selection means that the user should find the inputs that can provide some information about how a particular pattern differs from other ones. With KNN, feature determination is mostly done heuristically according to the user's knowledge about the examined phenomenon.

KNN is an extremely useful sorting tool for problems dealing with huge data bases and multivariable patterns.

In the following paragraphs two successful problems that we solved using KNN will be presented.

9. Sorting of ground level wind fields using KNN

The Šoštanj area (a basin in complex terrain, explained in previous sections) has very colorful ground level wind field patterns due to the fact that the very low winds that prevail there meander in the basin and follow the shape of nearby lying valleys and become stronger over the hills and passes. Hence the wind roses of six ambient automatic measuring stations (that were examined for SO_2 forecasting) look totally different, in spite of the fact that they are only a few kilometers away one from another. This fact illustrates the complexity of the ground level winds in the area.

We examined the following problem (Mlakar & Božnar, 1996): is it possible to find groups of similar wind fields (a wind field in this case is represented by a packet of simultaneously measured wind data at all stations) occurring in this area, or is the problem too stochastic to be grouped? If there are groups, what is the "natural" number of groups there?

To answer these questions we examined the wind data for five stations. One station was excluded because its location was inappropriate for representative wind measurements.

The data base of over 26000 half hour intervals was examined when wind measurements were available for all five stations. Due to the complexity of the wind roses for the whole data set it was expected that the natural number of groups would be very large. Several divisions from 10 to 100 groups were tested. As a measure of division quality a special index was defined - the weighted sum of the standard deviation of wind speed and the standard deviation of wind direction within the obtained groups. The natural number of groups was found to be around 32.

The quality of the division of the 26000 wind patterns into 32 groups was easily controlled by plotting wind roses for the new groups for each station. The new wind roses were not similar to the wind roses composed of the whole data set for each station. And also the wind roses of different groups (and the pattern for five stations) were different one from another. The pictures of the wind roses for 32 groups at five stations proved very obviously that the sorting process was done in a very effective and successful way.

This example of wind data sorting is a very persuasive one to convince the user about the effectiveness of KNN. This is due to the fact that wind roses are a graphical presentation that can be easily comprehended and the differences or similarities visualized. And on the other hand, there is no way (because of the area complexity) to do this sorting manually - only on the basis of some meteorological knowledge.

10. KNN as a tool for pattern selection techniques for MPNN based models

Another very successful application where we used KNN was as a pattern selection technique. When establishing methods for pattern selection that would not need user knowledge about meteorological phenomenon, we used KNN to sort a huge data base of air pollution patterns into natural groups of patterns. When the groups are obtained it is easy to construct a training and testing set to effectively train the MPNN.

The method was developed for the case of SO_2 prediction in the Šoštanj area (Božnar, 1997).

The method of pattern selection using KNN showed the same improvement in the MPNN model effectiveness as the method of using all the available detailed expert knowledge about air pollution in the area. Therefore KNN pattern selection techniques are particularly suitable for problems where detailed expert knowledge is not available.

11. Conclusion

Two types of artificial neural networks were shown to be useful tools for environmental modelling: the multilayer perceptron neural network MPNN and the Kohonen neural network KNN. MPNN is an universal approximator. Therefore it can be used for modeling phenomena where reconstruction or prediction of one (or several) parameters is required on the basis of other measured parameters. In the model construction phase there are two important steps that are often neglected. These are feature determination and pattern selection techniques. The methods that we suggest can be used in very different applications. They contribute much to the final model performance. Their contribution can be described as extraction of useful information from the available data base of measurements and presentation of this information to the neural network during the learning process in the most plausible way. The variety of presented examples from air pollution prediction to meteorological applications shows how flexible MPNN models can be. Meteorological applications especially demonstrate that MPNN models can be a useful additional tool in the field of meteorological preprocessors for modern air pollution models. The variety of examples presented also proves that feature determination and pattern selection techniques are more or less universal.

KNN is not used so widely as MPNN in atmospheric research. The examples presented here prove that KNN is a very effective tool for sorting problems. It actually performs very well also in cases where there is no a priori knowledge about similarities at all.

We hope that the given examples of successful use of artificial neural networks will inspire other applications in atmospheric research.

12. Acknowledgment

The study was partially financed by the Slovenian Research Agency, Project No. L1-2082.

13. References

Božnar, M. (1997). Pattern selection strategies for a neural network - based short term air pollution prediction model. Adeli, H. (ed.). *Intelligent Information Systems IIS'97*, Grand Bahama Island, Bahamas, December 8-10, Proceedings. Los Alamitos, California[etc.]: IEEE Computer Society, pp. 340-344

Božnar, M., Lesjak, M., & Mlakar, P. (1993). A neural network-based method for short-term predictions of ambient SO2 concentrations in highly polluted industrial areas of complex terrain. *Atmospheric Environment*, B 27(2), pp. 221-230.

Božnar, M. & Mlakar, P. (1995). Neural networks - a new mathematical tool for air pollution modelling., Editors: Power, H., Moussiopoulos, N., Brebbia, C. A. *Air pollution III*. Vol. 1, Air pollution theory and simulation. Southampton; Boston: Computational Publications, pp. 259-266

Božnar, M. & Mlakar, P. (2001). Artificial neural network-based environmental models. *Air pollution modeling and its application XIV*, Kluwer Academic: Plenum Publishers, New York, pp. 483-490

Comrie, A.C. (1997). Comparing neural networks and resression models for ozone forecasting. *Journal of Air and Waste Management* 47, pp. 653-663.

Devijver, J.K., (1982). *Pattern recognition: A statistical Approach.* Englewood Cliffs, Prentice-Hall, Inc.

Elisei et. al (1992). *Experimental Campaign for the Environmental Impact Evaluation of Sostanj Thermal Power Plant, Progress Report,* CISE, Milano

Fairall, C.W., Persson. P.O.G., Bradley, E.F., Payne, R.E., & Anderson, S.P. (1998). A New Look at Calibration and Use of Eppley Precision Infrared Radiometers. Part I: Theory and Application. *Journal of Atmospheric and Oceanic Technology*; 15, pp. 1229-1242

Gardner, M.W., & Dorling S.R. (1998). Artificial neural networks (the multilayer perceptron) - a review of applications in the atmospheric sciences. *Atmospheric Environment* 32 (14/15), pp. 2627-2636.

Grašič, B., Mlakar, P., & Božnar, M. Z. (2006). Ozone prediction based on neural networks and Gaussian processes. *Il Nuovo Cimento C*, Vol. 29, Issue 6, pp. 651-661

Hornik, K. (1991). Approximation capabilities of multilayer feedforward networks. *Neural Networks 4*, pp. 251-257

Kohonen, T. (1995). *Self-organizing maps*. Springer, Berlin

Kurkova, V. (1992). Kolmogorov's Theorem and Multilayer Neural Networks, *Neural Networks*, 5, pp. 501-506

Lawrence, J. (1991). *Introduction to Neural Networks*, California Scientific Software, Grass Valley

Marzban, C., & Strumpf, G.J. (1996). A neural network for tornado prediction based on doppler radar derivd attributes. *Journal of Applied Meteorology* 35, pp. 617-626

Mlakar, P. (1997). Determination of features for air pollution forecasting models. Adeli, H. (ed.). *Intelligent Information Systems IIS'97,* Grand Bahama Island, Bahamas, December -10, Proceedings. Los Alamitos, California, USA: IEEE Computer Society 1997, pp. 350-354.

Mlakar, P. & Božnar, M. (1996). Analysis Of Winds And SO2 Concentrations In Complex Terrain. Editors: Caussade, B., Power, H., Brebbia, C.A.. *Air Pollution IV: Monitoring, Simulation And Control.* Southampton; Boston: Computational Mechanics Publications, Cop., pp. 455-464.

Moseholm, L., Silva, J., & Larson, T. (1996). Forecasting carbon monoxide concentrations near a sheltered intersection using video traffic surveillance and neural networks. *Transport Research* 1D (1), pp. 15-28

Oliveira, A.P., Escobedo J.F., Machado A.J., & Soares J. (2002). Correlation models of diffuse solar radiation applied to the City of São Paulo (Brazil). *Applied Energy* 71(1), pp. 59-73.

Oliveira, A. P. De, Soares, J. R., Božnar, M., Mlakar, P., Escobedo, J. F. (2006). An application of neural network technology to corrent the dome temperature effects on pyrgeometer measurements. *Journal of Atmospheric and Oceanic Technolog*, vol. 23, pp. 80-89.

Rumelhart, D.E., Hilton, G.E., Williams, R.J. (1986). Learning internal representation by error propagation, in *Parallel Distributed Processing: Explorations in the Nicrostructure of Cognition*. eds. DE Rumelhart and JL McClelland, Vol 1, MIT Press, Cambridge, MA

Soares, J., Oliveira, A.P., Božnar, M.Z., Mlakar, P., Escobedo, J.F., & Machado, A.J., (2004). Modeling hurly diffuse solar-radiation in the city of São Paulo using a neural'network technique. *Applied Energy* 79(1), pp. 201-214.

Yi, J., & Prybutok, R. (1996). A neural network model forecasting for prediction of daily maximum ozone concentration in an industrialised urban area. *Environmental Pollution* 92 (3), pp. 349-357.

26

Indoor Air Control by Microplasma

Kazuo Shimizu
Shizuoka University
Japan

1. Introduction

Sick-Building Syndrome (SBS) has become an environmental issue worldwide in recent decades (P. Burge, 2004). Much research on this syndrome has been carried out. Generally, it is known that since buildings became more airtight for improvement of air-conditioning and heating, the effect of Volatile Organic Compounds (VOCs) diffusions from building materials has increased, causing many symptoms of SBS. Control of these indoor air pollutants is necessary to maintain Indoor Air Quality (IAQ). Recently, Indoor Air Quality (IAQ) is recognized as an important factor of home and building construction (P. Wolkoff, G. D.Nielsen, 2001).

Formaldehyde (HCHO) is one of the most common indoor VOCs. This substance is emitted from resins, plastics and often building materials, such as plywood, chipboard, and paneling, and is one of the main causes of SBS (H. Yoshino, 2004). Indoor air contaminants which worsen IAQ are characterized by the presence of not only VOCs, but also of fungus and bacteria, various malodorous substances and Environmental Tobacco Smoke (ETS) etc. VOCs such as formaldehyde (HCHO) are well known as a cause of sick building syndrome (A. Seki et al., 2007). Indoor air concentration of HCHO is regulated at 0.08 ppm by Ministry of Health, Labor and Welfare (MHLW) in Japan. Decompositions of VOCs by nonthermal plasma technique have been researched recently (A. Koutsospyros et al., 2004; D. Li et al., 2002; K. N. Faungnawakji et al., 2004; K. Urashima and J. S. Chang, 2000; T. Oda et al., 2004; Y.-H. Song et al., 2002).

In addition, infective diseases such as new influenza strains caused by pathogenic organisms have been spread worldwide. Recently this causes serious problems in schools and a lack of vaccines in many countries. In this context, home appliance manufactures have developed air purifiers for the market in Japan, but not in the market worldwide (Sharp, Panasonic, Daikin). Research on air purification; both odor treatment and bacteria, viruses treatment based on the nonthermal plasma technique has also been intensively studied in universities as well as by home appliance manufacturers (A. Sakudo and H. Shintani, 2010; F. J. Trompeter et al., 2002; H. Ghomi et al., 2005; K. Kitano et al., 2006; M. Laroussi et al., 2004; M. Nagatsu et al., 2003; T. C. Montie et al., 2000, N. Hayashi et al., 2006; N. S. Panikov et al., 2002, T. Ito et al., 1992.)

These nonthermal plasma techniques were carried out with rather "high voltage" region (5-15 kV, sometimes more than 20 kV)(A. Mizuno et al., 1995). It is expected that they can be alternatives to other simple plasma techniques to purify indoor air. In this chapter, a technique for indoor air control by microplasma will be introduced (K. Shimizu et al., 2008, 2009, and 2010 a, b, and c).

2. About microplasma

Atmospheric microplasma is a type of dielectric barrier discharge (DBD) (K. Tachibana, 2006; L. A. Rosocha et al., 1993; S. Agnihotri et al., 2004; U. Kogelschatz, 2007). The discharge gap is set to an order of micrometers which is extremely narrow, enabling the plasma to generate at a discharge voltage of around 600 V. Streamers between the electrodes are also very small (in the order of micrometers), resulting in a relatively compact and dense plasma. Fig. 1 is an image of the microplasma during discharge. Streamers were generated not only between the electrodes but also around the holes of the electrodes.

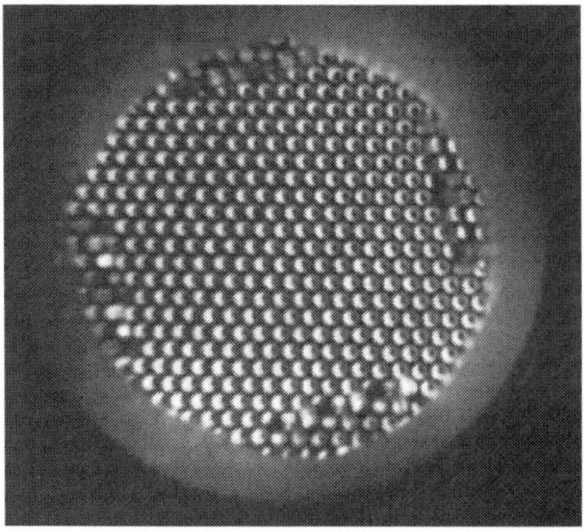

Fig. 1. Image of the microplasma electrode (Vd=1 kV, front view). A photo taken by a digital camera with 5 seconds of shutter opening.

Discharge gap was set based on Paschen's law, which indicates the minimum sparking voltage and discharge gap for various gases at atmospheric pressure. High reduced electric fields were readily obtainable with such small discharge gaps, resulting in a reduction of low energy electrons (1-2 eV), which dissociate ozone (B. Eliasson et al., 1987; J. Kitayama and M. Kuzumoto, 1997). This microplasma electrode has the advantage of generating a high concentration of ozone with low discharge voltage and power.
Fig. 2 shows the schematic image of microplasma electrodes for odor treatment (K. Shimizu et al., 2010b).

3. Experimental setup

3.1 Microplasma electrode and reactors

Fig. 2 is a schematic image of the microplasma electrode. Two perforated metal plates covered with dielectric materials were faced together, and an alternating voltage (about 25 kHz, 1 kV) was applied. Pulse power supply can also be used, a technique detailed in 3.2, 3.3, and 3.4. Innumerable streamers generated between the electrodes, which could excite various radicals (O*, N*, etc) and generate ozone. These radicals could react with the flowed gas and detoxify it (H. X. Din et al., 2005). The electrode had a diameter of 45 to 60 mm and a thickness of 1 mm. Since this electrode had a large aperture (aperture ratio: about 30 to 40%), the pressure loss through the electrode was extremely low (less than 5 mmH_2O at gas flow rate 10 L/min). This could enable large volume gas treatment passes.

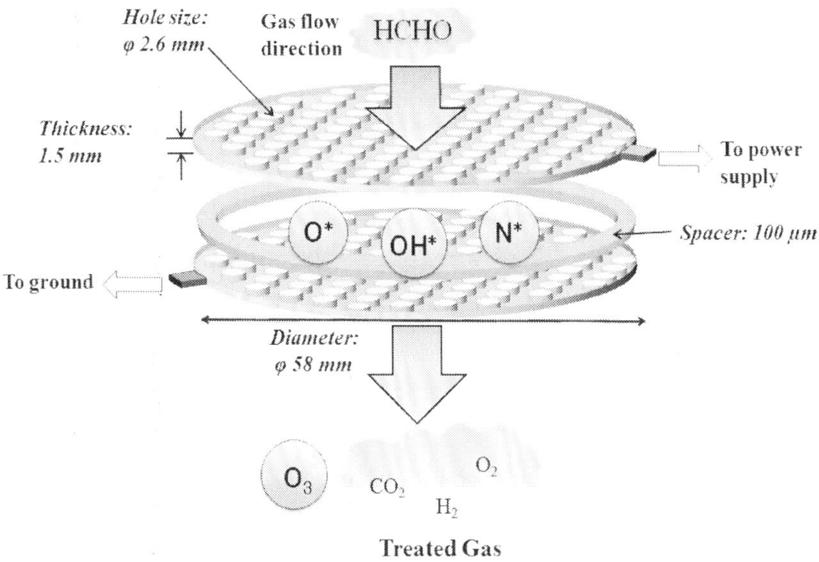

Fig. 2. Schematic image of microplasma electrodes for treating odor molecules. The pressure loss between the electrodes is very small.

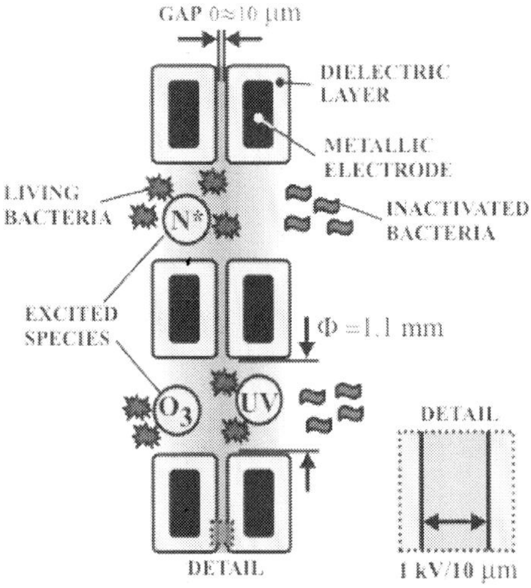

Fig. 3. Schematic image of the sterilization process of microplasma between the electrodes.

Fig. 3 shows a schematic image of the active species generated by the microplasma between the electrodes and the exposure of the living bacteria to the microplasma causing sterilization. A suspension of living bacteria in water was sprayed perpendicular to the electrode. The living bacteria were carried through the holes by the carrier gas. Due to the small discharge gap between electrodes (0–10 μm) and the direction of the gas flow, few of the bacteria enter the discharge gap between electrodes, and most pass through the holes. The few bacteria that enter the narrow zone between the electrodes attach to the surface of the electrodes and are lost from the flow of the carrier gas. The bacteria and their colonies that were counted on the nutrient medium were exclusively those bacteria which passed through the holes.

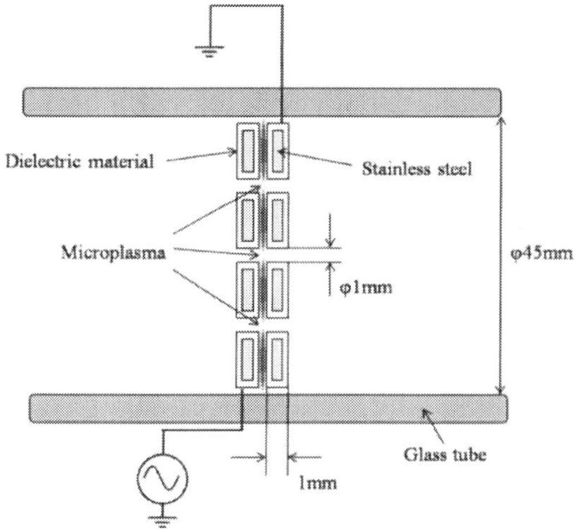

Fig. 4. Schematic image of the microplasma reactor (side view).The discharge gap is about 10 to 100 μm with or without a spacer.

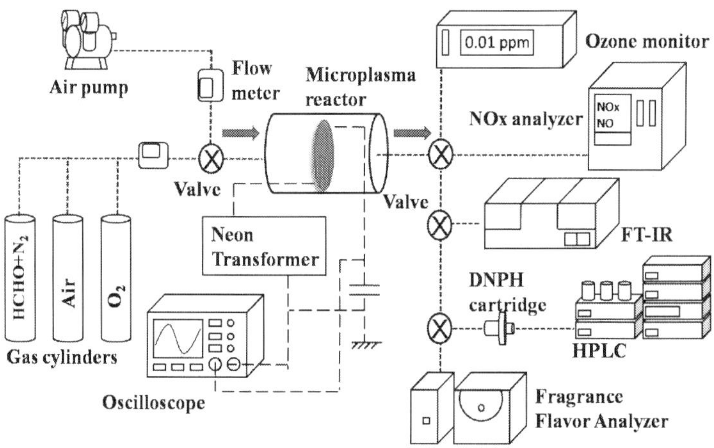

Fig. 5. An experimental setup for one pass process of formaldehyde removal.

Fig. 4 shows the microplasma reactor, and the experimental setup for VOCs removal is presented in Fig. 5. In this experiment, formaldehyde was used as VOCs. Air was flowed into diluted formalin solution by an air pump or gas cylinders, supplying a constant low concentration of formaldehyde to the microplasma reactor.

The treated gas was then sent to an ozone monitor (Ebara Jitsugyo, EG-2001B), NOx analyzer (Shimadzu, NOA-7000), FTIR (Shimadzu, IRPrestige-21), HPLC (Agilent, 1100 series) and Fragrance Flavor Analyzer (Shimadzu, FF-2A) for investigation of the gas composition change, identification and quantity analysis of by-products, and distinction of the gas smell.

Also, an oscilloscope (Tektronix, TDS 3014) was used to measure the discharge voltage, current, and power. Lissajous figures were used for the estimation of discharge power.

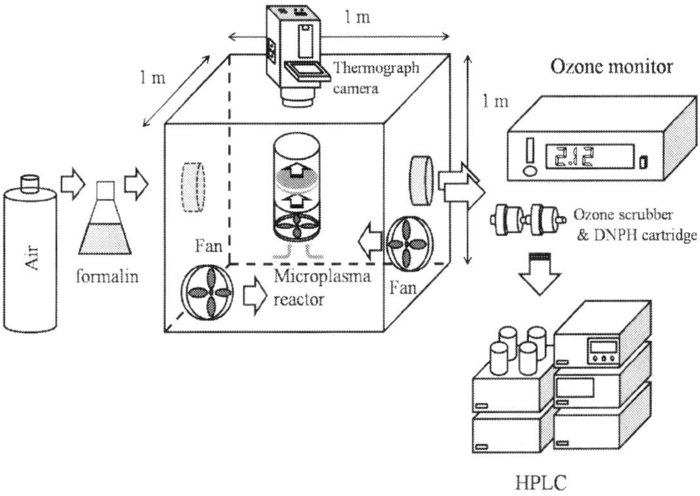

Fig. 6. An experimental setup for large volume process of VOCs removal.

The experimental setup for large volume treatment of VOCs is shown in Fig. 6. Formalin solution (abt. HCHO 37 %) was used to generate 0.5 ppm in 1 m^3 acrylic chamber. The inside gas was treated by microplasma reactor at a gas flow rate of 1,500 L/min. Before the microplasma treatment a fan was used to equalize the concentration of HCHO in the whole volume. Treatment time was 60 minutes; HCHO and O_3 concentration was measured each 12 minutes by ozone monitor and HPLC. The discharge voltage of an HV amplifier and Marx Generator circuit was set at 1.2 kV, and frequency of both power sources was set at 1.5 kHz. These power supplies are explained in the next section.

Sterilization effect of bacteria at low voltage by using atmospheric microplasma was also investigated. *Escherichia coli* JCM20135 and *Bacillus subtilis* JCB20036 were used as the target to be sterilized. An experimental setup for sterilization of bacteria is shown in Fig. 7.

This Bacteria sterilization described here by microplasma was a one pass process. The liquid culture medium was introduced in the microplasma reactor and sprayed at a gas flow rate of 3.5 L/min by use of a medical nebulizer, through the electrode against a petri dish with culture medium. Carrier gas was also introduced in the reactor at a gas flow rate of 5 L/min. After the experiment, the laboratory dishes were incubated in the incubator at 37°C for 15 hours after microplasma treatment. Sterilization effect of microplasma was inspected by comparing the number of colonies with and without microplasma treatment.

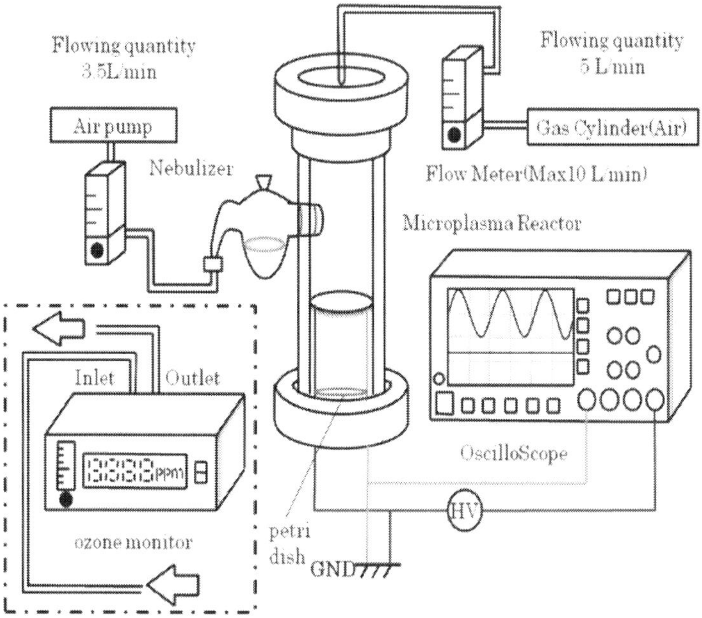

Fig. 7. An experimental setup for bacteria sterilizaion.

3.2 Power supply

In another series of experiments, high frequency AC neon transformer (LECIP, M-1H shown in Fig. 8), a pulse power supply consisting of a function generator (Tektronix, AFG3021B) and high voltage amplifier (Trek, MODEL 5/80) were also used as power supplies to energize the electrodes.

A Marx Generator with MOSFET switches was developed as a pulse power supply. It generates negative pulses up to - 1.6 kV, rise time 100 ns and pulse width 1.5 μs. The electronic circuit of Marx Generator with 2 stages is shown in Fig. 9.

When the MOSFET switches are opened, the capacitors are linked in parallel connection and charged at a value V. By turning on the MOSFET switches, the capacitors discharge in a series connection. Thus the output voltage has the value V multiplied with the number of capacitors, in this case 2V (M. Blajan et al., 2010). The DC power source which charges the capacitors has a maximum output of 0.8 kV, thus the Marx Generator can generate negative pulses up to -1.6 kV.

It is worth mentioning that the power supply was small and easy to handle, generating microplasma as shown in Figs. 8 and 10.

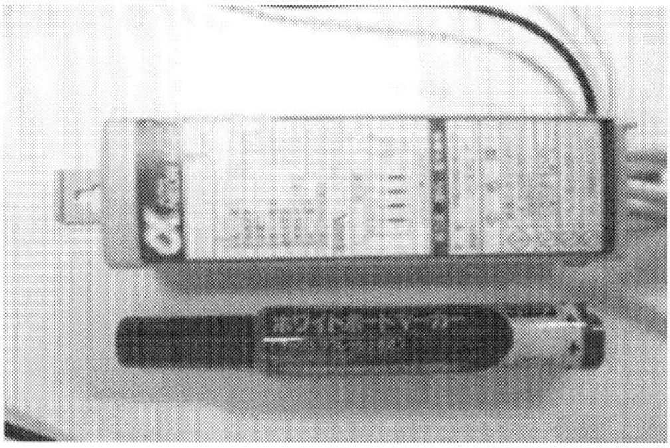

Fig. 8. A small AC power supply.

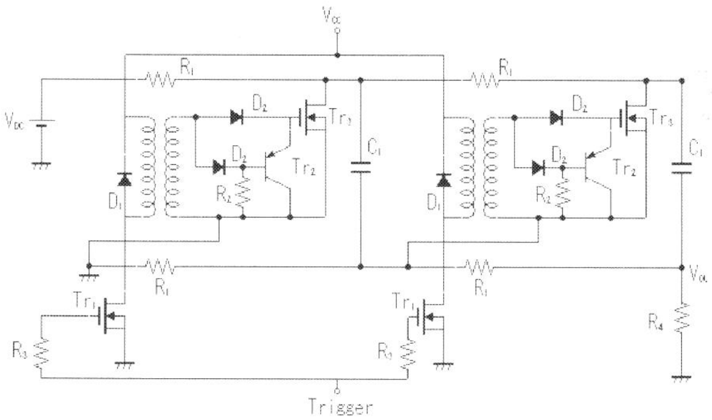

Fig. 9. Example of a negative pulse Marx Generator circuit.

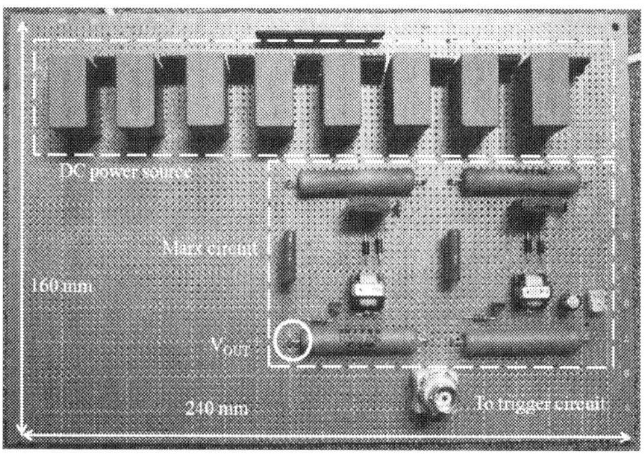

Fig. 10. Image of a Marx Generator circuit.

3.3 Waveforms for generating microplasma

Figure 11 shows the typical waveform of discharge voltage and corresponding discharge current generating microplasma at about 1 kV. This waveform shows alternate current and frequency is about 25 kHz. Corresponding discharge current showed a typical waveform of dielectric barrier discharge. The microplasma reactor can generate atmospheric plasma at about 1 kV, since its discharge gap was narrow (about 50 μm) (K. Shimizu et al., 2008, 2009, and 2010a and b).

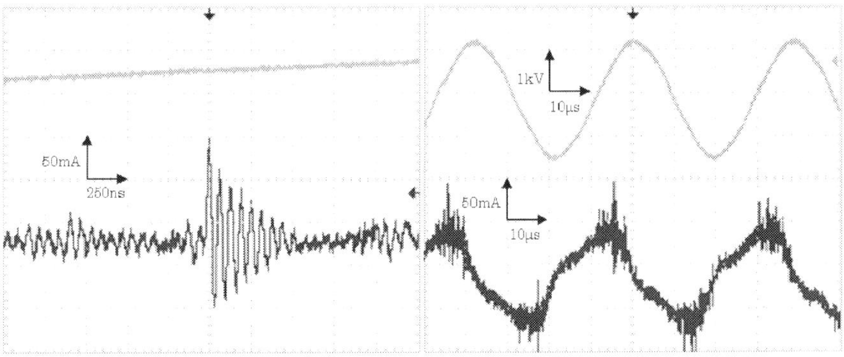

Fig. 11. Typical waveform of microplasma with AC neon transformer; applied voltage (above) and corresponding discharge current (below). Time scale; 10 μs (left), 250 ns (right).

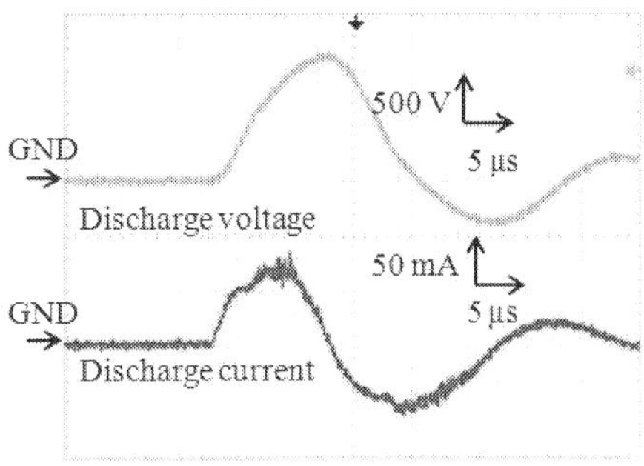

Fig. 12. Discharge voltage and corresponding discharge current by a high voltage amplifier.

Pulse voltage was also applied to generate microplasma as shown in Figs. 12 and 13. Repetition frequency of both pulses was fixed to 1 kHz. The waveforms of pulse voltage generated by an HV amplifier are shown in Fig. 12. Spike currents occurred due to the streamers, convoluted on the current waveform, and were observed in addition to the capacitive current at the steepest slopes of the waveform.
The waveforms of pulse voltage by a Marx Generator circuit are shown in Fig. 13. Rise time of discharge voltage by a Marx Generator circuit was 100 ns, and fall time was 4 μs. Sharp discharge current was observed.
Waveforms are different from Figs. 11 to 13, and the microplasma generation was confirmed with all the applied voltage, since the microplasma was generated by typical dielectric barrier discharge. The indoor air control devices in the market are usually driven by weak corona discharge ranged from 5 to 10 kV without dielectric barriers on their electrodes. This is a significant difference between the microplasma technique under discussion here, and

the technique employed by indoor air control devices on the market now which generate ions, according to the manufacturers claims.

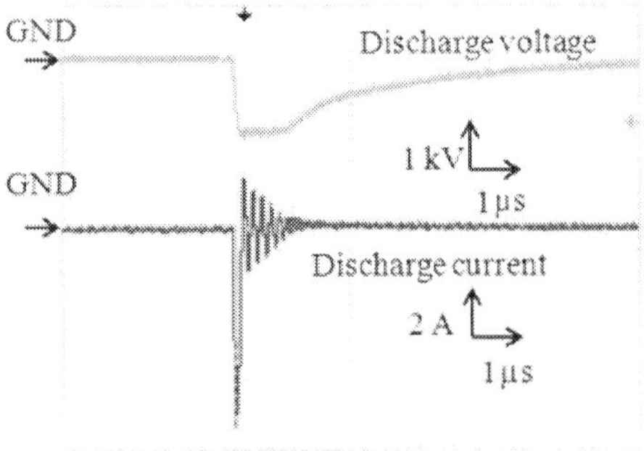

Fig. 13. Discharge voltage and corresponding discharge current by a Marx Generator.

3.4 Power consumption

The characteristic of discharge power versus discharge voltage is shown in Fig. 14. Discharge power was estimated by integrating the waveform of voltage and current product in time using an oscilloscope and dividing it with time in order to obtain the power for one cycle.

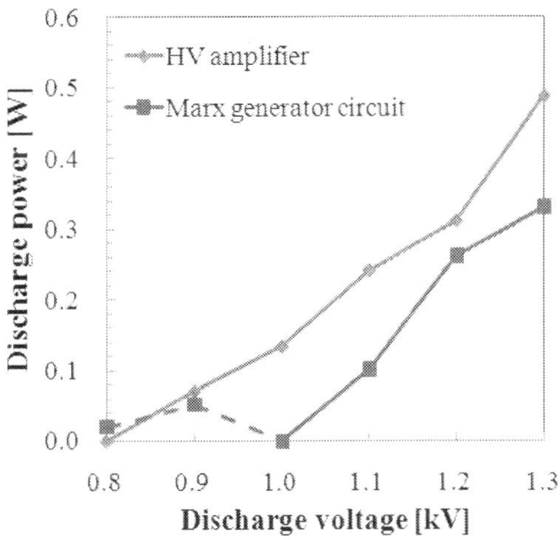

Fig. 14. Characteristic of discharge power by both power sources.

Discharge power of Marx Generator at 1.0 kV was lower than that at 0.9 kV. This could be explained by the decrease of transient form of the discharge current when microplasma discharge started to occur at 1 kV.

As the discharge voltage increases, the discharge power increases. Discharge power of an HV amplifier was 0.49 W at 1.3 kV for one cycle, and discharge power of Marx Generator was 0.33 W. Discharge power could attributed to the ozone generation which will discuss in the next part.

4. Results and discussion

In this part, experimental results for indoor air control by microplasma is presented. First, control of VOCs in simulated indoor air will be introduced with a small quantity of ozone generation. After that byproduct analysis, chemical reactions will be discussed. Finally, bacteria in simulated indoor air will be presented. Mechanisms of sterilization process are also discussed along with data such as UV light emission by microplasma.

4.1 Removal of HCHO and ozone generation in air

The experimental results of HCHO removal in one pass process using a microplasma reactor (shown in Fig. 5)powered by a HV amplifier are shown in Fig. 15. Discharge voltage was set at 0.8 ~ 1.3 kV. The microplasma discharge started at 0.9 kV. Thus no HCHO removal was measured at 0.8 kV (initial concentration of HCHO: 0.707 ppm). The removal ratio of HCHO at 0.9 kV was 35.4 %. At this discharge voltage, ozone generation was initiated (about 0.2 ppm). As discharge voltage increased, the HCHO removal ratio and generated ozone concentration also increases. The HCHO removal ratio reached 95.8 % at the discharge voltage of 1.1 kV, and 3.46 ppm of ozone was generated at this voltage. The removal ratio of HCHO was saturated at around 96 % when the voltage was increased above 1.1 kV. Generated ozone concentration increased with discharge voltage. Thus the maximum value of ozone concentration was 12.1 ppm at 1.3 kV.

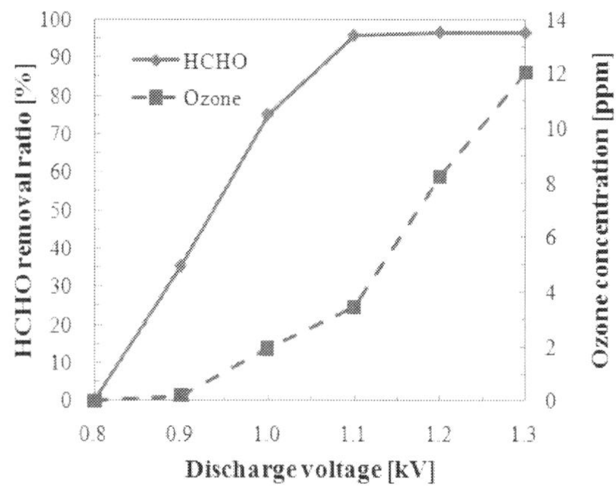

Fig. 15. HCHO and O_3 concentration in one pass process by an HV amplifier.

The optimal value for the discharge voltage, in the HCHO removal process when an HV amplifier was used, could be considered to be 0.9 kV due to the relatively high removal ratio and low generated ozone concentration.

The experimental results of HCHO removal in one pass process using a Marx Generator are shown in Fig. 16. Initial concentration of HCHO was set at 0.762 ppm. The removal ratio of

HCHO was 30.7% at 1.1 kV. At this discharge voltage, generation of ozone was confirmed (0.16 ppm), and was lower than with an HV amplifier. The HCHO removal ratio reached 96.2 % when discharge voltage increased to 1.3 kV. Generated ozone concentration of 3.81 ppm could be considered low compared to an HV amplifier results.

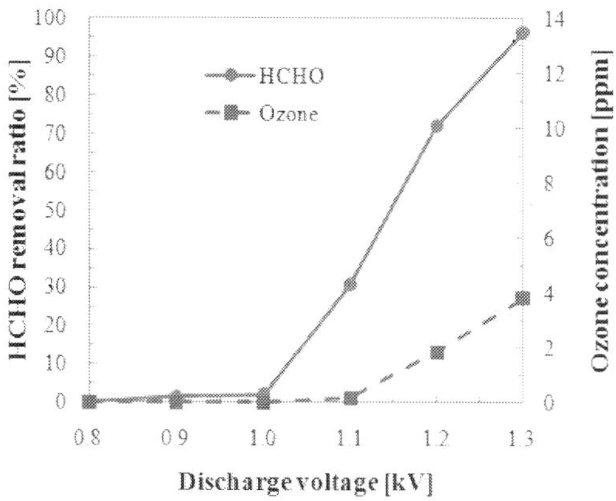

Fig. 16. HCHO and O_3 concentration in one pass treatment by Marx Generator circuit.

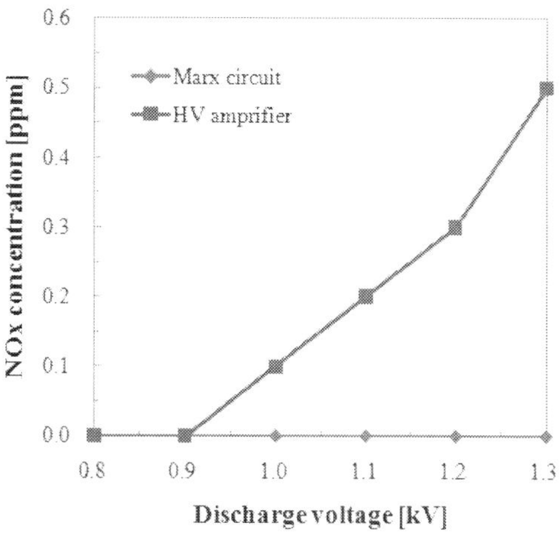

Fig. 17. NOx generation difference by the pulse power supplies.

In the one pass HCHO removal process by microplasma, mixing HCHO from a gas cylinder with air from another gas cylinder was used as simulated indoor air. Thus the humidity of the mixed gas was 0%. Reaction process by microplasma with humidity will be presented bellow (Figs.19 and 20). This process could have OH radicals which play a role in leading oxidization reactions.

NOx generation characteristics by both power sources is shown in Fig. 17. When an HV amplifier was used, NOx started to be generated at 1.0 kV. After increasing the discharge voltage, the NOx concentration increased to 0.5 ppm at 1.3 kV. On the contrary, NOx generation was not observed when using a Marx Generator as the pulse power supply.

Atmospheric microplasma leads to the dissociation of nitrogen molecules to generate nitrogen atoms or meta stable state $N_2(A)$ molecules to produce NO as shown in equations (1) and (2) (A. Rousseau et al., 2005). Since NO production depends on the pulse duty cycle ratio, which is proportional to the pulse power, microplasma powered by short pulses could not be responsible for the NOx generation because of low pulse duty cycle ratio, even at the voltage as high as 1.3 kV.

$$N_2(A^3\Sigma_u^+) + O \rightarrow NO + N(^2D) \qquad k_1=0.4\times10^{-11}\ [cm^3s^{-1}] \tag{1}$$

$$N(^2D) + O_2 \rightarrow NO + O \qquad k_2=9.7\times10^{-12}e^{-185/T}\ [cm^3s^{-1}] \tag{2}$$

Considering that NOx are well known for their toxicity, in a further implementation of microplasma as a technology for cleaning the room air, a Marx Generator could be favorable to use as the solution for the power supply.

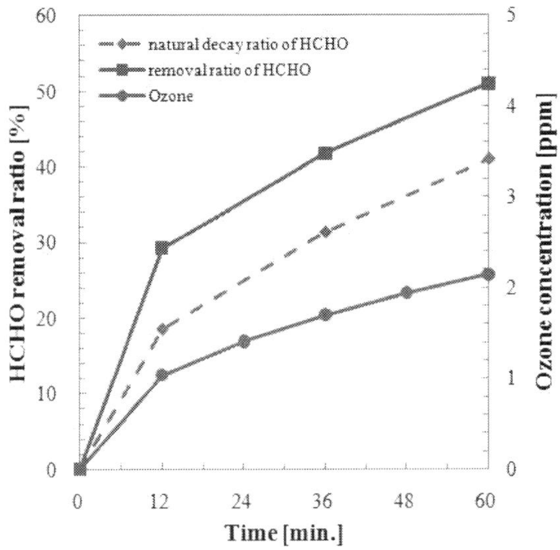

Fig. 18. Large volume HCHO treatment by an HV amplifier.

The results of large volume HCHO treatment using 1 m^3 chamber (shown in Fig. 6, gas flow rate: 1,500 L/min.) with an HV amplifier as a power supply are shown in Fig. 18. In this case, the sample HCHO gas was generated from the evaporation of formalin in room air. The gas humidity was equivalent to room air (abt. 60 %).

The removal ratio and generated ozone concentration increased with the treatment time. A natural decay of HCHO concentration was measured (Fig. 18). The natural decay after 60 minutes was 41 % and the removal ratio of HCHO with microplasma reached 51 %. Initial concentration of HCHO was set at 0.537 ppm and after 60 minutes of microplasma treatment 0.05 ppm of HCHO was removed, also taking into consideration the effect of natural decay. The generated ozone concentration was 2.14 ppm in the chamber after 60 minutes of microplasma treatment.

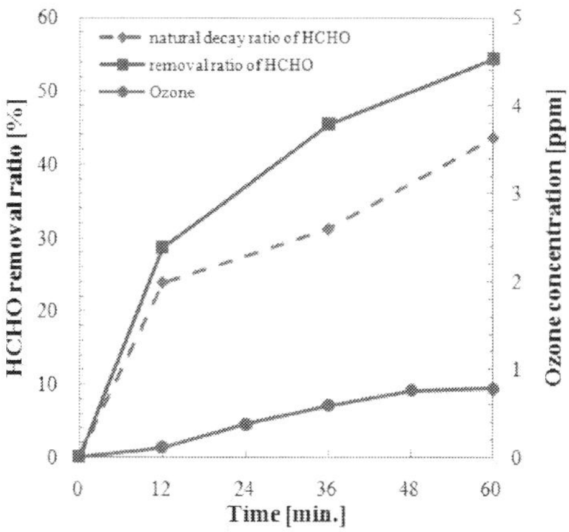

Fig. 19. Large volume HCHO treatment by Marx Generator.

The results of large volume HCHO treatment using a Marx Generator to energize the microplasma electrodes are shown in Fig. 19. The natural decay ratio was 43.6 %, and the removal ratio of HCHO reached 54.4 % after 60 minutes of treatment. Generated ozone concentration was 0.78 ppm after the same period of time. HCHO removal ratio was comparable with the removal ratio obtained with an HV amplifier as a power supply; however, the generated ozone concentration was lower.

The results obtained with a Marx Generator are similar in both treatment cases (with one pass reactor and large reactor) from the point of view of HCHO removal ratio, but the NOx and ozone generation are lower. Thus it can be concluded that the use of a Marx Generator as a power supply has advantages over than that of an HV amplifier.

HCHO concentration of actual houses is usually below 0.1 ppm. Removal performance of HCHO by microplasma treatment at large gas flow rate was about 0.05 ppm. This recommends microplasma technology as a solution for the treatment of indoor air contaminants.

4.2 Byproduct analysis and reaction process by microplasma

There are many reports on treating hazardous substances, such as cigarette smoke and VOCs (H. Yoshida et al., 1989, T. Kuroki et al., 2001, R.Atkinson et al., 2004, 2006). Most of these methods require high voltages and have difficulty handling large volume. Therefore, we have carried out HCHO removal in simlated indoor air experimentally using a microplasma reactor, which is more compact, and has low energy consumption. Major plasma chemical reactions are known in the treatment of HCHO are as follows (D. G. Storch and M. J. Krushner, 1993):

$$HCHO+O \rightarrow HCO+OH \qquad k_3=1.75\times10^{-13}\ [cm^3s^{-1}] \qquad (3)$$

$$HCHO + OH \rightarrow HCO + H_2O \qquad k_4=1.11\times10^{-11}\ [cm^3s^{-1}] \qquad (4)$$

$HCO + O \rightarrow CO_2 + H$ $k_5=5.00\times10^{-11}$ [cm^3s^{-1}] (5)

$HCO + O \rightarrow CO + OH$ $k_6=5.00\times10^{-11}$ [cm^3s^{-1}] (6)

$HCO + OH \rightarrow H_2O + CO$ $k_7=5.00\times10^{-11}$ [cm^3s^{-1}] (7)

Fig. 20 and 21 are the analysis results obtained by the FTIR. In this case, the initial concentration of HCHO was set to 10 ppm to observe the spectrum difference clearly, and the discharge voltage was set to 600, 800, and 1000 V with an AC neon transformer. The gas flow rate was set to 2.0 L/min and humidity conditions were changed.

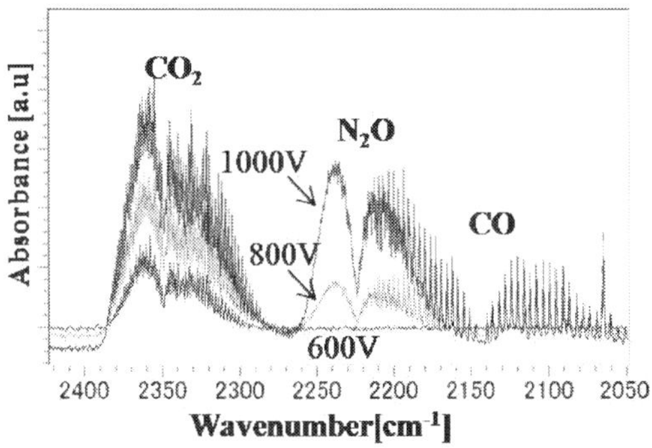

Fig. 20. FT-IR analysis (2000-2450 cm^{-1}) after formaldehyde treatment without humidity (0%) in the flowing air.

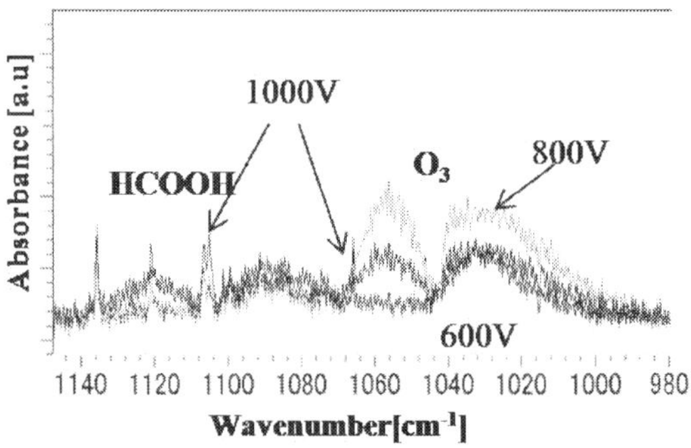

Fig. 21. FT-IR analysis afterformaldehyde treatment with humidity (60%) in the flowing air.

From the analysis results, carbon dioxide (CO_2: 2250-2400 [cm^{-1}]), carbon monoxide (CO: 2000-2250 [cm^{-1}]) and nitrous oxide (N_2O: 2175-2250 [cm^{-1}]) were discovered as byproducts in zero humidity air conditions. CO_2 and CO could be a result of equations (5-7). The generation of N_2O according to (8) is realized by the reaction of excited species of N_2 with O_2. The presence of N_2 excited species in microplasma was demonstrated by the measurement of N_2 second positive band system and N_2^+ first negative band system for microplasma discharge in air (M. Blajan et al., 2009). N_2O was a byproduct which derives also from reaction (9).

$$N_2^* + O_2 \rightarrow N_2O + O \qquad k_8=3.00\times10^{-14}\ [cm^3s^{-1}] \qquad (8)$$

$$NO_2 + N \rightarrow N_2O + O \qquad k_9=3.00\times10^{-12}\ [cm^3s^{-1}] \qquad (9)$$

When there was humidity in the flowing air, formic acid (HCOOH: 1100 [cm^{-1}]) is also confirmed as a byproduct (Fig.19). Electron impact dissociation of H_2O leads to the production of H and OH radicals and also to excited state $O(^1D)$ dissociated H_2O to generate OH a process described bellow (4.5). Thus formaldehyde reacts with generated OH radicals, and HCOOH was formed from the following equation:

$$HCHO + OH \rightarrow HCOOH + H \qquad k_{10}=2.00\times10^{-13}\ [cm^3s^{-1}] \qquad (10)$$

It is not desirable to produce byproducts which are harmful to humans. Therefore, there should either be a second treatment process for these byproducts, or the electron energy should be controlled in order not to dissociate N_2 and CO_2. These issues will be investigated and could be controlled by a power supply.

4.3 Smell analysis

HCHO is known to have a sweet smell, and can be detected when there is 0.08 ppm or more in the air. After microplasma treatment, a few byproducts were confirmed. This fact suggests the smell of the treated gas could also be changed. A smell analysis was carried out to confirm the difference of smell before and after the microplasma treatment.

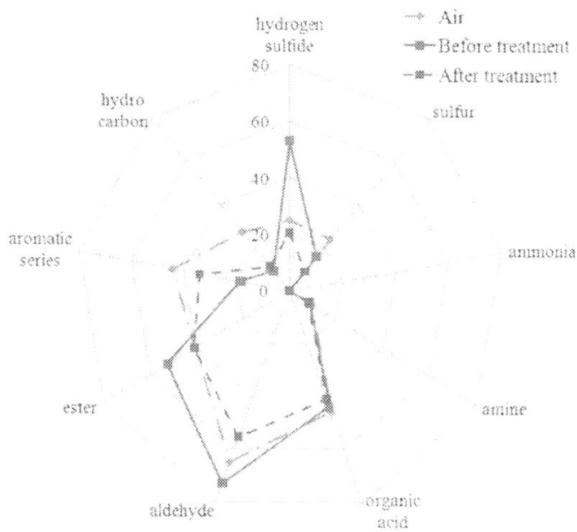

Fig. 22. Smell similarity before/after HCHO treatment and room air.

Sample gas was collected and analyzed by a Fragrance Flavor Analyzer, which has ten different semiconductor sensors with different sensitivities. Nine standard gases, which each have a different kind of smell, are input in this device and the analyzed sample gas is compared with them, providing data on the similarity against each standard gas. It is able to compare the change of the smell. The numbers in the radar graph stand for the similarity percentages against the nine standard gases(K. Shimizu et al., 2008).

Fig. 22 shows the smell similarity of simulated indoor air contained HCHO before and after the microplasma treatment for one pass process. Initial concentration of HCHO was set at 1ppm.

The smell of HCHO had a strong similarity with hydrogen sulfide, ester, and aldehyde. After microplasma treatment, the similarities of these three contents decreased, and the similarities with hydrocarbons and aromatics increased. It was observed that the smell of the sample gas changed to that of air after microplasma treatment.

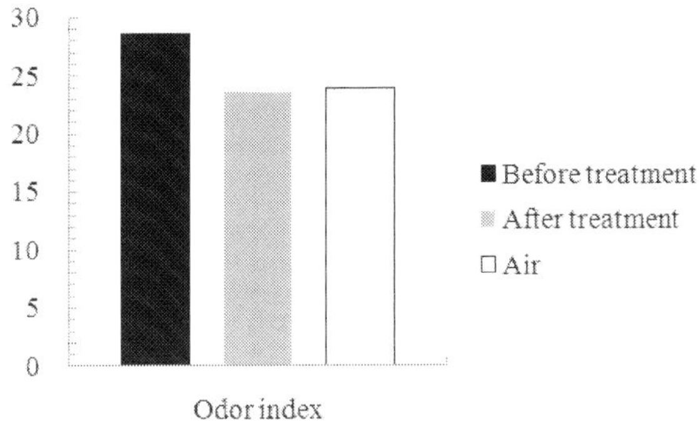

Fig. 23. The odor index before/after HCHO treatment and room air.

Fig. 23. shows the odor index before and after the microplasma one pass treatment, compared to that of room air. Odor index was about 29 before the microplasma treatment (Vd=1.24 kV), and after the microplasma treatment, it decreased to about 24. This value was slightly lower than the room air.

4.4 Sterilization of bacteria by microplasma

Virus or bacteria contained in tiny water droplets suspended in room air could cause serious illnesses such as influenza. One pass treatment of microplasma (shown in Fig. 7) can be an effective sterilization method for indoor air.

The diameters of the water particles formed by a medical nebulizer were measured by a laser particle counter (Kanomax, 3886). The gas flow rate was restricted to the range of 5-7 L/min because the pressure loss of the nebulizer was high. As shown in Fig. 24, a particle counter observed particles with diameters of 0.3 and 0.5 μm without water in the nebulizer. These particles could be the dust in the room air. With water added to the nebulizer, water droplets with diameters ranging from 0.5 to 5 μm were generated at the applied gas flow rates of 5, 6, and 7 L/min. Smaller water droplets did not coalesce to form larger droplets because the number of 0.5 μm droplets exceeds the number of 1.0 μm droplets. Since the sizes of colon bacilli *Escherichia coli* JCM20135 and *Bacillus subtilis* JCB 20036 are about 0.5 to 2.5 μm, they were contained within the water droplets generated by the nebulizer.

Inactivation of *Escherichia coli* and *Bacillus subtilis* were experimentally investigated at a total gas flow rate of 8.5 L/min by using microplasma electrodes. Ambient air and nitrogen were used to compare the effect of the oxidization effect of ozone and to confirm the effect of high electric field and UV radiation from microplasma.

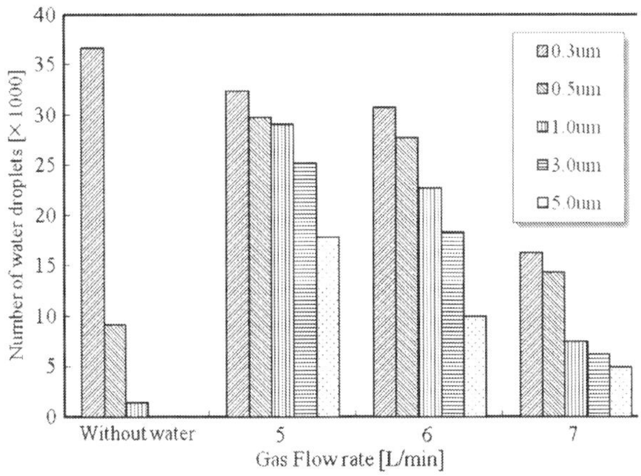

Fig. 24. Size distribution of water particles generated in by the nebulizer at gas flow rates of 5, 6, and 7 L/min.

Photographs of petri dishes with *E. coli*, before and after the microplasma treatment with air and nitrogen as the carrier gases, are shown in Fig. 25 and 26, respectively. *E. coli* was incubated at 37 degrees for 15 h after the microplasma treatment. Decrease of the colonies were observed both for air plasma and nitrogen plasma.

(a) Before treatment. (b) Discharge voltage 1.0 kV. (c) Discharge voltage 1.4 kV.

Fig. 25. Images of the *E.coli* samples before and after the air plasma treatment.

From these photos, air-plasma had better results to sterilize *E.coli* than that with nitrogen plasma. In the case of nitrogen as carrier gas, the presence of oxidation species such as ozone was not confirmed. This could explain the difference between air and nitrogen.

Photographs of petri dishes with *B.subtilis*, before and after the microplasma treatment with air and nitrogen as the carrier gases, are shown in Fig. 27 and 28, respectively. *B.subtilis* was incubated at 30 degrees for 18 h after the microplasma treatment. A decrease of the number of colonies was observed when the discharge voltage increased in both air plasma and nitrogen plasma. Decrease of the colonies was rather low compared to the of *E.coli* results. The inactivation process for bacteria may occur between the electrodes that generate microplasma or in the space near the electrodes after passing through the holes of the electrodes.

(a) Before treatment. (b) Discharge voltage 1.0 kV. (c) Discharge voltage 1.4 kV

Fig. 26. Images of the *E.coli* samples before and after the nitrogen plasma treatment.

(a) Before treatment. (b) Discharge voltage 1.0 kV. (c) Discharge voltage 1.4 kV.

Fig. 27. Images of the *B. subtilis* samples before and after the air plasma treatment.

(a) Before treatment. (b) Discharge voltage 1.0 kV. (c) Discharge voltage 1.4 kV.

Fig. 28. Images of the *B. subtilis* samples before and after the nitrogen plasma treatment.

As evident in these photos, more effective results were obtained for *E. coli* (gram-negative bacteria) than that of *B. subtilis*. Lower sterilization of *B. subtilis* (gram-positive bacteria) could be caused by its relatively impermeable cell walls, which have a thickness in the range of 22 to 25 nm. The cell wall of gram-positive bacteria is composed of peptidoglycan and secondary polymers. Gram-negative bacteria have thin peptidoglycan layers (2-3 nm) plus an overlying lipid-protein bilayer (7-8 nm) known as the outer membrane (R. Stainer et al., 1986, T. Beveridge, 2001).

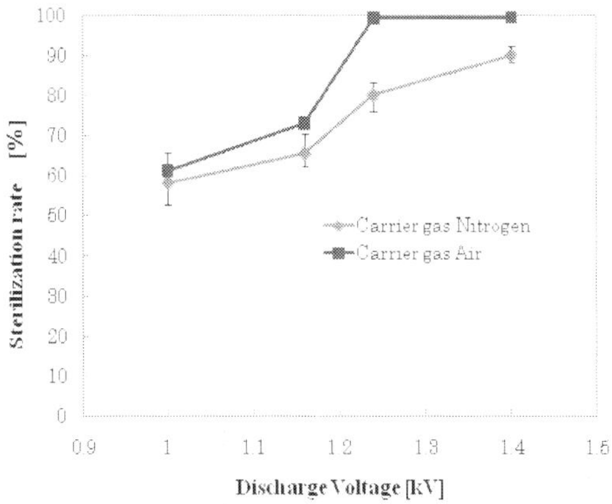

Fig. 29. Comparison of sterilization rate of *E.coli* with air and nitrogen plasma.

Sterilization rate of *E. coli* versus discharge voltages by air plasma and nitrogen plasma is shown in Figure 29. 100% sterilization of *E. coli* was accomplished with air as carrier gas for one pass treatment by microplasma as shown in Fig. 29. When nitrogen was the carrier gas, the sterilization rate surpassed 90% corresponding to a discharge voltage of 1.4 kV. Ozone was not formed during the discharge in the presence of nitrogen, and the sterilization of *E. coli* could be considered to be due to the effects of high electric field, excited nitrogen ions, active species such as OH, and UV radiation by microplasma, as described below in section 4.5.

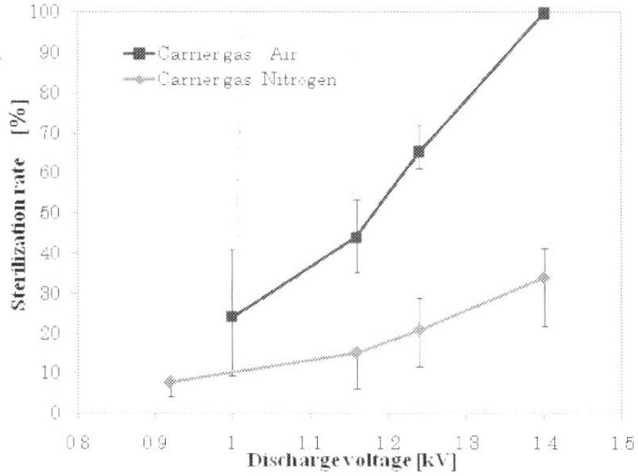

Fig. 30. Comparison of sterilization rate of *B. subtilis* with air and nitrogen plasma.

Sterilization rate of *B. subtilis* versus discharge voltage by air plasma and nitrogen plasma is shown in Fig. 30.

When air was used as carrier gas, maximum concentration of ozone was 22 ppm in the reactor. The effective reactor volume was 0.2 L, and the gas residence time of the reactor was about 1.4 seconds. When air was used as carrier gas, a near 100% sterilization rate of *B. subtilis* was achieved at discharge voltage of 1.4 kV. With nitrogen as the carrier gas, a sterilizaion rate of about 30% was achieved for *B. subtilis*. From this result, the sterilization process could be considered a synergetic effect of UV radiation, high electric field, (not only oxidative) radicals, and ozone. Various invesitigators show the sterilization effect (not by microplasma but) by atmospheric plasma or plasma jet (G. Fridman et al., 2008, M. G. Kong et al., 2009).

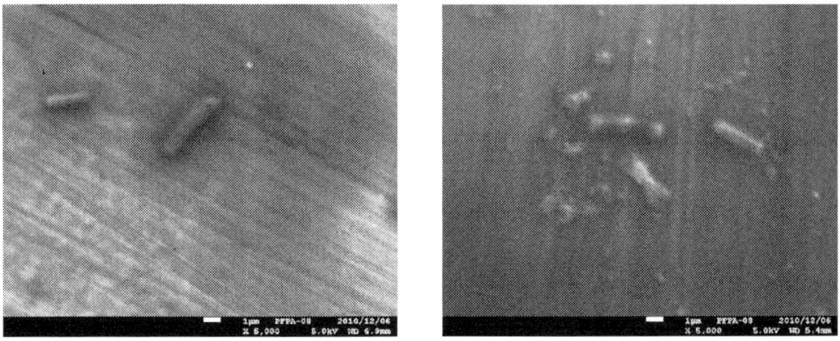

(a) before treatment (x 5000) (b) after treatment (Vd= 1.0 kV, x 5000)

Fig. 31. Image of *B.subtilis* before and after the microplasma treatmnet.

Fig. 31 shows the images of *B. subtilis* before and after sterilization by microplasma. The images were taken by a Field Emission Scanning Electron Microscope (JOEL, JSM-7001F). Image (b) shows how the bacteria was affected by the microplasma discharge. The shape of the *B. subtilis* was changed and torn to pieces after the microplasma treatment. During the microplasma treatment process, every bacteria was exposed to a high electric field and UV

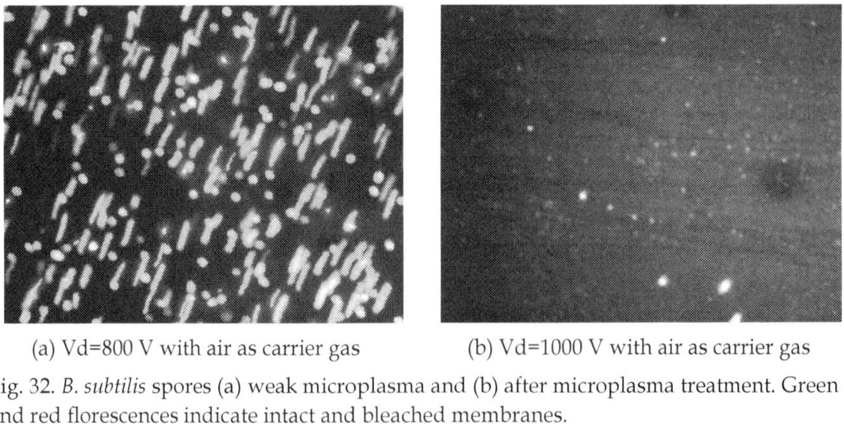

(a) Vd=800 V with air as carrier gas (b) Vd=1000 V with air as carrier gas

Fig. 32. *B. subtilis* spores (a) weak microplasma and (b) after microplasma treatment. Green and red florescences indicate intact and bleached membranes.

radiation while passing through the electrodes' holes generating microplasma. This could cause physical damage to the bacteria and affect their shape. In particular, active radical species had an etching effect to break cell walls, and UV affects the DNA directly to sever their structure (R. Stainer et al., 1986). Fig. 32 shows the bacteria (*B. subtilis*) damage by microplasma, where LIVE/DEAD BacLight Bacterial Viability Kits was used to identify intact and bleached membranes of *B. subtilis* spores with green and red florescence, respectively.

4.5 UV light emission by microplasma

UV light emissions from microplasma were observed to confirm the effect of UV light on the bacteria sterilization process (K. Shimizu et al., 2010). The emission spectra were measured by an intensified charge-coupled device (ICCD) camera (Ryoushi-giken, SMCP-ICCD 1024 HAM-NDS/UEmV), a spectrometer (Ryoushi-giken, VIS 351), and by a photomultiplier tube (Hamamatsu Photonics, R3896). A pulse generator (Tektronix, AFG 3021B) was used to trigger the ICCD camera and the Marx Generator consisting of semiconductor switches. A Marx Generator with four-stage MOSFET switches was used as the power supply. The spectrum was observed at -1.4 kV with a pulsewidth of 500 ns and a frequency of 1 kHz. The gas flow rate of dry nitrogen was set at 5 L/min. Data obtained from the ICCD camera were transferred to a computer for analysis.

Fig. 33 shows the emission spectrum of the microplasma discharge in N_2. Larger peaks indicate the N_2 second positive band system (N_2 SPS) and smaller peaks indicate the N_2 first negative band system (N_2 FNS). The spectrum indicates the generation of active molecular nitrogen species in the microplasma discharge (Z. Machala et al., 2007).

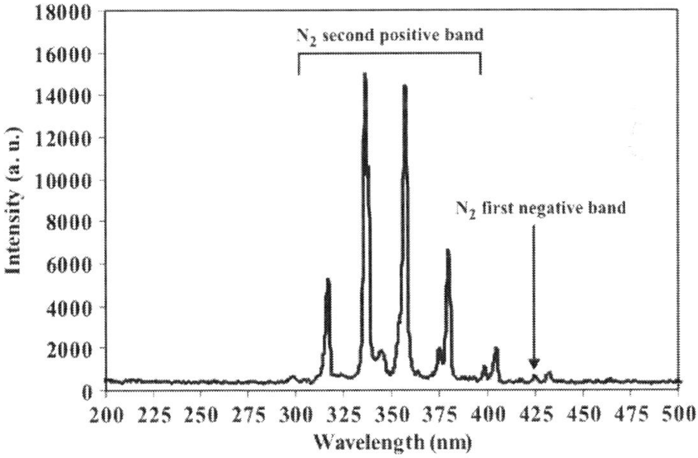

Fig. 33. Spectrum of N_2 second positive band system and N_2 first negative band system by microplasma discharge in nitrogen. Gas flow rate was set at 5 L/min.

The elementary processes (11) and (12) describe the radiation kinetics for the N_2 SPS with a wavelength of 337.1 nm and at atmospheric pressure (K. V. Kozlov and H.-E. Wagner, 2007).

The excitation of nitrogen molecules in the ground state by direct electron impact is described by

$$e + N_2\,(X^1\Sigma_g{}^+)_{v=0} \rightarrow N_2(C^3\pi_u)_{v'=0} + e\ (\Delta E = 11\ \text{eV}) \tag{11}$$

The spontaneous radiation of nitrogen in the excited state is described by

$$N_2(C^3\pi_u)_{v'=0} \rightarrow N_2(B^3\pi_g)_{v''=0} + h\upsilon\ (\tau_0{}^C = 40\ \text{ns}) \tag{12}$$

Water droplets from the nebulizer were entrained in the nitrogen gas that entered the chamber to confirm another active species in the microplasma discharge. Fig. 34 shows the emission spectrum of nitrogen with entrained water droplets.

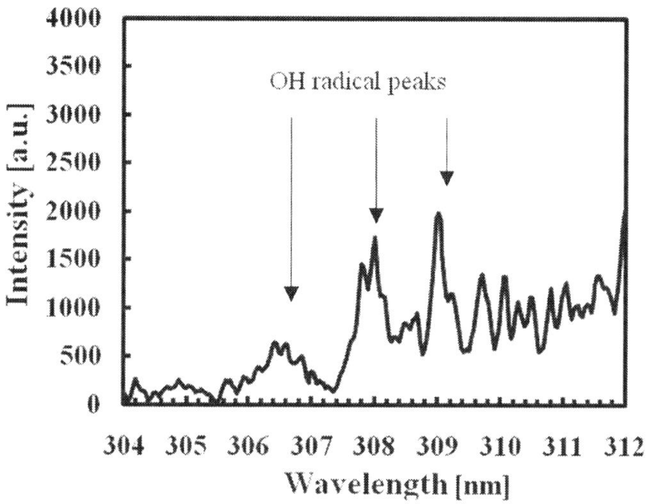

Fig. 34. OH radical peaks of emission spectrum of microplasma discharge in nitrogen with H_2O by an HV amplifier at 1.44 kV. Gas flow rate was set at 5 L/min.

The UV light emission confirmed active species such as OH radicals measured at 306.4 nm, 307.8 nm and 308.9 nm (Z. Machala et al., 2007, F. Liu et al., 2007). UV light emissions from 316 nm and higher wavelengths were also observed, which affect the sterilization process of bacteria (J. M. C. Robertson et al., 2005, A. K. Benabbou et al., 2007). OH radicals are generated via electron impact dissociation of H_2O which leads to the production of H and OH radicals:

$$e + H_2O \rightarrow e + H + OH \tag{13}$$

Also, the excited state $O(^1D)$ dissociated H_2O to generate OH:

$$O(^1D) + H_2O \rightarrow 2\,OH \tag{14}$$

Thus OH radical peak was obtained in the presence of H_2O in air or in nitrogen. The combination of UV light and active species could have contributed to the sterilization process of bacteria in air.

5. Conclusions and future work

Indoor air control a keen issue worldwide, since pandemics such as new influenza strains are now serious problems in every country. People are also worrying about smells or odours in homes, offices, hospitals, hotels, on trains etc. Various devices which claim to generate ozone or ions to control indoor air are on the market now. But their performances are insufficient, especially for air pollutant removal (A. Nozaki et al., 2010). Mechanisms for the removal of air pollutants or bacteria are still not proven, especially for reactions with coagulated aerosols in the air in so called "clustered ions" emitted by indoor air control devices (T. Yamauchi et al., 2007). Such mechanisms require more research to improve the performances of indoor air control devices.

In this chapter, a technique for indoor air control by microplasma was presented with the various data for VOCs removal with low concentration of ozone and bacteria sterilization in one pass treatment. Smell analysis and byproduct analysis were also presented to evaluate the microplasma process. Also, various power supplies including a self-made Marx Generator are presented resulting in low ozone concentration in output gas. It is also important to treat indoor air in such a way as to not generate gas harmful to human beings. Moreover, microplasma has various features as discussed in "2. About Microplasma". Discharge voltage is lower than that of other nonthermal plasma methods, or other indoor air control devices on the market. This means microplasma devices will possibly be made small, light and cost effective for commercial production. Air control devices generate ions to treat room air; in contrast, microplasma can treat room air, while room air flows into the microplasma electrode. There could be UV light emission, active species, strong electrical field between the electrodes. Odor gas molecules, bacteria or viruses could be decomposed or sterilized here.

Mechanisms of VOCs removal or sterilization process have also been demonstrated with chemical reactions and emission spectra which could attribute emitting active species such as OH radicals from microplasma. But this data is still insufficient to explain these mechanisms.

In this chapter, there is no information about ion density generated by microplasma. Ions could be generated during the microplasma process. However evidence supporting claims that ions contribute to indoor air treatment processes is still not overwhelmingly conclusive. Further study is required to show these reaction processes or sterilization processes by microplasma are "effective", and how safe this process is for practical use in home appliances.

The author would like to thank for the financial support for part of this project the Japanese Ministry of Education, Sciences, Sports and Culture, Grant-in-Aid. The author would also like to thank Dr. Blajan of Shizuoka University for fruitful discussions.

6. References

A. Mizuno, K. Shimizu, A. Chakrabarti, L. Dascalescu, and S. Furuta. (1995). NOx Removal Process Using Pulsed Discharge Plasma, *IEEE Trans. on IAS*, Vol. 31, No. 5, pp. 957-963

A. Nozaki, Y. Ichijo, and Y. Narita. (2010). A Study on the Indoor Air Pollutant Removal Performance of Room Air Cleaners, *Proc. Int'l. Symp. On Contamination Control 2010*, Tokyo, pp. 505-510

A. K. Benabbou, Z. Derriche, C. Felix, P. Lejeune, and C. Guillard. (2007). Photocatalytic inactivation of Escherischia coli: Effect of concentration of TiO2 and microorganism, nature, and intensity of UV irradiation, *Appl. Catalysis B, Environ.*, vol. 76, no. 3/4, pp. 257-263, Nov, 2007

A. Koutsospyros, S.-M. Yin, C. Christodoulatos and K. Becker. (2004). Destruction of Hydrocarbons in Non-thermal, Ambient-Pressure, Capillary Discharge Plasmas, Int. J. Mass. Spectrum., vol. 233, pp. 305-315

A. Sakudo, H. Shintani. (2010). *Sterlization and Disinfection by Plasma,* Nova Science Publishers, Inc., New York.

A. Seki, T. Takigawa, R. Kishi, K. Sakabe, S. Torii, M. Tanaka, T. Yoshimura, K. Morimoto, T. Katoh, S. Kira and Y. Aizawa. (2007). Review of 'Sick House Syndrome', *Jpn. J. Hyg,* Vol. 62, pp. 939-948

A. Rousseau , L. V. Gatilova, J. Röpcke, A. V. Meshchanov and Y. Z. Ionikh. (2005). NO and NO_2 production in pulsed low pressure dc discharge, *Appl. Phys. Let.,* vol.86, pp. 2115011-2115013

B. Eliasson, M. Hirth and U. Kogelschatz. (1987). Ozone synthesis from oxygen in dielectric barrier discharge, *J. Phys. D: Appl. Phys.,* vol. 20, pp. 1421-1437

D. li, D. Yakushiji, S. Kanazawa, T. Ohkubo and Y. Nomoto.(2002). Decomposition of Toluene by Streamer Corona Discharge with Catalyst, *J. Electrost.,* vol. 55, pp. 311-319

Daikin company website: http://www.daikin.com/global_ac/products/residential/airpurifier_mc707/mechanism.html

D. G. Storch and M. J. Krushner. (1993). Destruction mechanisms for formaldehyde in atmospheric pressure low temerpature plasmas, *J. Appl. Phys,* vol. 73, no. 1

F. J. Trompeter, W. J. Neff, O. Franken, M. Heise, M. Neiger, S. Liu, G. J. Pietsch, and A. B. Saveljew. (2002). Reduction of Bacillus Subtilis and Aspergillus Niger Spores Using Nonthermal Atmospheric Gas Discharges, *IEEE Trans. on plasma science,* Vol.30, No. 4, pp. 1416-1423

F. Liu, W. Wang, S. Wang, W. Zheng, and Y. Wang. (2007). Diagnosis of OH radical by optical emission spectroscopy in a wire-plate bi-directional pulsed corona discharge, *J. Electrostat.,* vol. 65, no. 7, pp. 445–451, Jun, 2007

G. Fridman, G. Friedman, Al. Gutsol, A. B. Shekhter, V. N. Vasilets, A. Fridman. (2008). Applied Plasma Medicine, *Plasma Process. Polym.,* Vol. 5, 503–533

H. Ghomi, M. S. Rahman, P. R. Chalise, Y. Hayashi, M. Watanabe, A. Okino, T. Ano, M. Shoda and E. Hotta. (2005). Experimental Investigation of Effect of Low-Energy Pulsed Atmospheric Electron Beam on Bacterial Cells, *Jpn. J. Appl. Phys,* Vol. 44, No. 12, pp. 8698-8701

H. X. Ding, A. M. Z., X. F. Yang, C. H. Li, and Y. Xu. (2005). Removal of formaldehyde form gas streames via packed-bed dielectric barrier discharge plasmas, *J. Phys. D: Appl. Phys,* vol. 38, pp. 4160-4167

H. Yoshida, Z. Marui, M. Aoyama, J. Sugiura and A. Mizuno. (1989). Removal of Ordor Gas Component Utilizing Plasma Chemical Reactions Promoted by the Partial Discharge in a Ferroelectric Pellet Layer, *J. IEJ,* vol. 13, no. 5

J. Kitayama and M. Kuzumoto. (1997). Theoretical and experimental study on ozone generation charcteristics of an oxygen-fed ozone generator in silent discharge, *J. Phys. D:Appl. Phys.,* vol. 30, pp. 2453-2461

J. M. C. Robertson, P. K. J. Robertson, and L. A. Lawton. (2005). A comparison of the effectiveness of TiO_2 photocatalysis and UVA photolysis for the destruction of three pathogenic micro-organisms, *J. Photochem. Photobiol. A, Chem.,* vol. 175, no. 1, pp. 51–56

K. Kitano, H. Aoki, S. Hamaguchi. (2006). Radio-Frequency-Driven Atmospheric-Pressure Plasmas in Contact with Liquid Water, *Jpn. J. Appl. Phys*, Vol. 45, No. 10, pp. 8294-8297

K. N. Faungnawakji, D. Sano, T. Yamamoto, T. Kanki, T. Charinpanitkul and W. Tanthapanichakoon. (2004). Removal of Acetaldehyde in Air Using a Wetted-Wall Corona Discharge reactor, *Chem.Eng.J.*, vol.103, pp.115-122

K. Shimizu, M. Yamada, M. Kanamori, and M. Blajan. (2010a) Basic Study of Bacteria Inactivation at Low Discharge Voltage by Using Microplasmas, *IEEE Trans. on IAS*, Vol. 46, No. 2, pp. 641-649

K. Shimizu, M. Kanamori, and M. Blajan. (2010b). Application of Atmospheric Microplasma for Indoor Air Treatment, *IJPEST*, Vol. 4, No. 1, pp. 45-51

K. Shimizu, T. Ishii, and M. Blajan. (2010c). Emission Spectroscopy of Pulse Power Microplasma for Atmospheric Pollution Control, *IEEE Trans. on IAS*, Vol. 46, No. 3, pp. 1125-1131

K. Shimizu, T. Sugiyama, M. Nishamani L. S. and M. Kanamori. (2009). Application of Micro Plasma for NOx Removal, *IEEE Trans. on IAS*, Vol. 45, No. 4, pp. 1506-1512

K. Shimizu, T. Sugiyama, and L. S. Manisha Nishamani. (2008). Study of Air Pollution Control by Using Micro Plasma Filter, *IEEE Trans. on IAS*, Vol. 44, No. 2, pp. 506-511

K. Shimizu, T. Sugiyama and M. Kanamori. (2008). Application of microplasma for ozone generation and environmental protection, *IJPEST*, vol. 2, no. 1, pp.38-43

K. Tachibana. (2006). Current Status of Microplasma Research, *IEEJ Trans.*, Vol. 1, pp. 145-155

K. Urashima and J. S. Chang. (2000). Removal of Volatile Organic Compounds from Air Streams and Industrial Flue Gases by Non-thermal Plasma Technology, *IEEE Dielec. Elect. Insul.*, vol. 7, pp. 602-613

K. V. Kozlov and H.-E. Wagner. (2007). Progress in spectroscopic diagnostics of barrier discharges, *Contrib. Plasma Phys.*, vol. 47, no. 1/2, pp. 26–33, Feb, 2007

L. A. Rosocha, G. K. Anderson, L. A. Bechtold, J. J.Coogan, H. G. Heck, M. Kang, W. H. McCulla, R. A. Tennant, and P. J. Wantuck. (1993). Treatment of Hazardous Organic Wastes Using Silent Discharge Plasmas, *Non-thermal Plasma Techniques for Pollution Control*, NATO ASI Ser. 34, Part B, Springer-Verlag Pub., Oxford, pp. 281-308

M. Blajan, S. Muramatsu, T. Ishii, H. Mimura, and K. Shimizu. (2010). Emission Spectroscopy of Microplasma Driven by a Pulsed Power Supply, *Journal of the Institute of Electrostatics Japan*, Vol. 34, No. 2, pp. 99-104, 2010, ISSN 0386-2550

M. Blajan, A. Umeda, S. Muramatsu, K. Shimizu. (2010). Emission spectroscopy of pulsed powered microplasma for surface treatment of PEN film, *Proc. of 2010 IEEE Industry Applications Society Annual Meeting*, CD-ROM, 3-7 October 2010 Houston, Texas, USA, Digital Object Identifier: 10.1109/IAS.2010.5614482

M. Blajan, T. Ishii, H. Mimura, K. Shimizu. (2009). Emission spectrometry of microplasma for NOx removal process, *Book of abstracts ISPC 19 International Symposium on Plasma Chemistry*, Ruhr-University Bochum, Germany, July 27 - 31 (2009) 398

M. Laroussi, G. S. Sayler, B. B. Glascock, B. McCurdy, M. E. Pearce, N. G. Bright, and C. M. Malott. (2004). Images of Biological Samples Undergoing Sterilization by a Glow Discharge at Atmospheric Pressure, *IEEE Transactions on plasma science*, Vol.27, No. 1, pp. 34-35

M. G. Kong, G Kroesen, G Morfill, T Nosenko, T Shimizu,J van Dijk and J L Zimmermann. (2009). Plasma medicine: an introductory review, *New Journal of Physics*, Vol. 11, pp. 1-35

M. Nagatsu, F. Terashita and Y. Koide. (2003). "Low-Temperature Sterilization with Surface-Wave-Excited Oxygen Plasma", Jpn. J. Appl. Phys, Vol. 42, No. 7B, pp. 856-859

N. Hayashi, W. Guan, S. Tsutsui, T. Tomari and Y. Hanada. (2006). Sterilization of Medical Equipment Using Radicals Produced by Oxygen/Water vapor RF Plasma, *Jpn. J. Appl. Phys,* Vol. 45, No. 10B, pp. 8358-8363

N. S. Panikov, S. Paduraru, R. Crowe, P. J. Ricatto, C. Christodoulatos, and K. Becker. (2002) Destruction of Bacillus Subtilis Cells Using an Atmospheric-Pressure Capillary Plasma Electrode Discharge, *IEEE Trans. on plasma science,* Vol. 30, No. 4, pp. 1424-1428

P. Wolkoff, G. D.Nielsen. (2001). Organic compounds in indoor air- their relevance for perceived indoor air quality?, *Atmospheric Environment,* Vol. 35, pp. 4407 - 4417

Panasonic company website: http://www.peshk.panasonic.hk/ap/index.html

R. Atkinson, D. L. Baulch, R. A.Cox, J. N. Crowley, R. F. Hampson, R. G. Hynes, M. E. Jenkin, M. J. Rossi, and J. Troe. (2004). Evaluated kinetic and photochemical data for atmospheric chemistry: Volume 1 - gas phase reactions of Ox, HOx NOx and SOx species, *Atomos. Chem. Phys.,* Vol.4, pp. 1461 - 1738

R. Atkinson, D. L. Baulch, R. A.Cox, J. N. Crowley, R. F. Hampson, R. G. Hynes, M. E. Jenkin, M. J. Rossi, and J. Troe. (2006). Evaluated kinetic and photochemical data for atmospheric chemistry: Volume 2 - gas phase reactions of organic species, *Atmos. Chem. Phys.,* Vol.6, pp. 3625 - 4055

R. Stainer, J. Ingraham,M.Wheelis, and P. Painter. (1986). *The Microbial World.,* 5th ed. New York: Prentice-Hall

S. Agnihotri, M. P. Cal, and J. Prien. (2004). Destruction of 1,1,1-Trichloroethane Using Dielectric Barrier Discharge Nonthermal Plasma, *J. Env. Engg.,* pp. 349-354

Sharp company website: http://www.sharp.ca/products/ion/plasma_home.asp

T. Beveridge. (2001). "Use of the Gram stain in microbiology," *Biotech Histochem.,* vol. 76, no. 3, pp. 111-118, May, 2001

T. Ito, Y. Murayama, M. Suzuki, N. Yoshimura, K. Iwano and K. Kudo. (1992). Evidence for sterilization of Saccharomyces Cerevisiae K 7 by an External Magnetic Flux, *Jpn. J. Appl. Phys,* Vol. 31, No. 6A, pp. 676-678

T. Kuroki, M. Okubo and T. Yamamoto. (2001). Indoor Air Cleaning Technology Using Non-equilibrium Plasma, *JSME. Trans. B,* vol. 67, no. 658, Jun, Japan

T. Oda, K. Yamaji, and T. Takahashi. (2004). Decomposition of Dilute Trichloroethylene by Nonthermal Plasma Processing ~Gas Flow Rate, Catalyst, and Ozone Effect, *IEEE Trans. Ind. Appl.,* vol. 40, pp. 430-436

T.C. Montie, K. Kelly-Wintenberg, J. R. Roth. (2000). An overview of research using the one atmosphere uniform glow discharge plasma (OAUGDP) for sterilization of surfaces and materials, *IEEE Trans. Plasma Sci,* Vol. 28, pp. 41-50

T. Yamauchi, H. Suda, and Y. Matsui. (2007). Development of Home Appliances Using Electrostatic Atomization, *J. Aerosol Res.,* vol. 22, pp. 5-10

U. Kogelschatz. (2007). Applications of Microplasmas and Microreactor Technology, *Contrib. Plasma Phys.* Vol. 47, No. 1-2, pp. 80-88

Y.-H. Song, S.-J. Kim, K.-I. Choi and T. Yamamoto. (2002). Effect of adsorption and temperature on a nonthermal plasma plasma process for removing VOCs, *J. Electrost.,* vol. 55, pp. 189-201

Z. Machala, M. Janda, K. Hensel, I. Jedlovský, L. Leštinská, V. Foltin, V. Martišovitš, and M. Morvová. (2007). Emission spectroscopy of atmospheric pressure plasmas for bio-medical and environmental applications, *J. Mol. Spectrosc.,* vol. 243, no. 2, pp. 194–201

Method for Validation of Lagrangian Particle Air Pollution Dispersion Model Based on Experimental Field Data Set from Complex Terrain

Boštjan Grašič, Primož Mlakar and Marija Zlata Božnar
MEIS environmental consulting d.o.o.
Slovenia

1. Introduction

Validation of air pollution dispersion model is very important process. It determines performances and efficiency of model in well defined conditions. Conditions consist of type of terrain orography (flat or complex), size of domain (local, regional, continental, global), number of grid cells in domain, meteorological conditions (strong or weak winds, etc.) and emission types (stacks, traffic, domestic heating). Results of validation give good guidelines how, where and when model can be successfully applied.

Validation is especially important when model is used for regulatory purposes. FAIRMODE European guidelines for air pollution modelling explicitly require that modeling tool must be successfully validated in similar environment (FAIRMODE, 2010). Slovenian legislation (Ur.l. RS, št. 31/2007, 2007) that is following European Council Directive of 28th June 1984 on combating air pollution from industrial plants (EUR-Lex 84/360/EEC, 1984) requires that the modeling tool for reconstructions of air pollution around stationary industry sources meet the requirements of complex terrain because most of Slovenian industry is located in the bottom of basins, river canyons and valleys. Complex terrain defines a set of specific atmospheric conditions: low wind speeds, temperature inversions, flow over topography, presence of terrain obstacles or discontinuities (land-sea, urban-rural environment), etc. Lagrangian particle dispersion model is the only air pollution model at the moment that is successfully achieving these requirements (Wilson and Sawford, 1996, Schwere et al., 2002). It has significantly evolved in last years and moved from research usage to usage for operational regulatory purposes (Tinarelli et al., 2000, Graff, 2002).

Validations over complex terrain are still very rare. They are very important for research community and governmental environment agencies. Research community use the results for further developments and improvements of modeling techniques and environment agencies for setting up and implementation of regulatory policies.

A study has been made to improve traditional air pollution model validation methodology. It is upgraded to estimate inaccuracy in position and time of the Lagrangian particle air pollution dispersion model. New validation methodology has been demonstrated on a field from a very complex terrain from Šaleška region (Slovenia). For validation Lagrangian particle air pollution dispersion model *SPRAY* produced by ARIANET Srl from Milano, Italy is selected. It has been chosen for validation because it follows Slovenian legislation about air pollution modeling over complex terrain. Validation is performed on one very complex terrain air pollution situation that is very difficult for reconstruction and includes phenomenon of air pollution accumulation. Traditional statistical indexes are determined at four locations in different directions from the point of view of air pollution source. To estimate model's inaccuracy in position and time new enhanced validation methodology is demonstrated and described in details. Results of this validation will serve for future improvements of selected air pollution dispersion model.

2. Methodology

2.1 Traditional validation methodology

Traditional validation methodology for air pollution modeling is based on statistical comparison between measured and reconstructed data about air pollution concentrations in environment. It is well described in model validation framework named "Model evaluation toolkit" that has been established and maintained by Olesen (1996).

Measured data are collected from automatic environmental measuring stations located on the area of interest (domain) usually around sources of air pollution. Reconstructed concentrations are obtained from the air pollution modeling simulation.

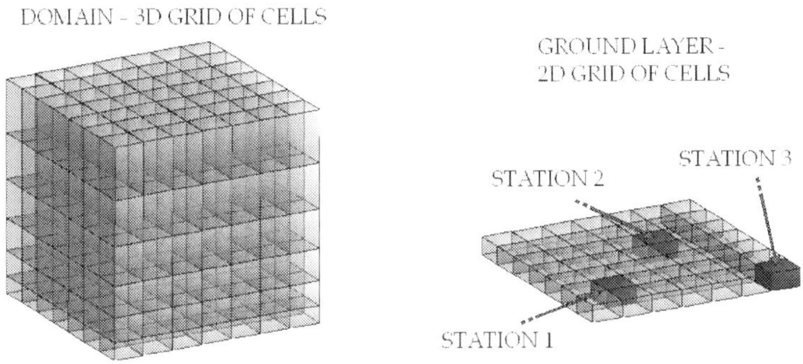

Fig. 1. The domain split in 3D grid of cells is presented on the left side where ground layer is colored in green; on the right side only ground layer is presented where the cells where stations are located are highlighted in red color.

In the air pollution model usually area of interest consists of a grid of cells where each cell describes average air pollution situation in certain part of the domain (i.e. in presented study case in next chapter domain is split into 100 x 100 cells in horizontal and in 20 layers in vertical which give 200 000 cells for the domain). For the comparison reconstructed average concentration from the ground cell where measuring station is located is taken. An example is presented on Figure 1.

Statistical analysis of data is performed for selected time interval where measured and reconstructed data are available. For this time interval a set of data patterns must be prepared. Each data pattern from this set consists of a pair of measured and reconstructed concentraion obtained at time step *t* as presented in equation (1).

$$\{C_{meas}(t), C_{recon}(t)\} \tag{1}$$

Using traditional validation methodology most often three statistical indexes are determined:

- the correlation coeficient (CR):

$$CR = \frac{\frac{1}{T}\sum_{t=0}^{T}\left(C_{meas}(t) - \hat{C}_{meas}\right)\left(C_{recon}(t) - \hat{C}_{recon}\right)}{\sigma_{C_{meas}}\sigma_{C_{recon}}} \tag{2}$$

- the normalized mean square error (NMSE):

$$NMSE = \frac{\frac{1}{T}\sum_{t=0}^{T}\left(C_{meas}(t) - C_{recon}(t)\right)^2}{\hat{C}_{meas}\cdot\hat{C}_{recon}} \tag{3}$$

- the fractional bias (FB):

$$FB = 2\frac{\hat{C}_{meas} - \hat{C}_{recon}}{\hat{C}_{meas} + \hat{C}_{recon}} \tag{4}$$

Definitions of variables and functions for determination of statistical indexes:

$C_{meas}(t)$ ...measured concentration at time step t
$C_{recon}(t)$ ...reconstructed concentration at time step t
$\hat{C}_{meas}$ ...average measured concentration
$\hat{C}_{recon}$ ...average reconstructed concentration
σ_C ...standard deviation of (measured or reconstructed) concentrations
t ...time step index
T ...length of full time interval (number of measured concentrations)

2.2 Enhanced validation methodology

In the model validation framework named "Model evaluation toolkit" maintained by Olesen (1996) difficulties that can arise in model validation are outlined. Differences between measured and reconstructed concentrations are caused by measuring errors, inherent uncertainty, input uncertainty and model formulation error. In the paper by Grašič et al. (2007) it has been determined that inaccuracy in position and time exists in the model. To estimate these inaccuracies enhanced validation methodology is presented. It is based on methodology where additionally reconstructed ground level concentrations in neighboring cells of the cell where station is located are also used in validation. Each measured value is during enhanced validation compared with one reconstructed concentration selected from a set of reconstructed concentrations. Set of this reconstructed concentrations NC as described in equation (5) consists of average concentration in the cell where station is located and neighboring cells. Neighborhood is defined in position (space) (i.e. for neighborhood of 1 cells in position we create a set of 9 cells as presented on Figure 2 and equation (6)) and in time scale (i.e. neighborhood of 1 time interval consist of 3 time intervals as presented on Figure 3 and equation (7)).

$$NC(t,m,n)=\begin{Bmatrix} C_{recon}(t,m,n); \\ t-\Delta T<t<t+\Delta T; \\ m-\Delta H<m<m+\Delta H; \\ n-\Delta H<n<n+\Delta H \end{Bmatrix} \tag{5}$$

Definitions of variables for determination of set of neighborhood concentrations *NC*:

NC ...set of reconstructed concentrations in the station's neighborhood
t ...time step index
ΔT ...length of neighborhood in time scale (number of time steps)
m ...index (number) of cell in east-west direction
n ...index (number) of cell in east-west direction
ΔH ... length of neighborhood in position (space) (number of cells)

$$NC_{position}(t,m,n)=\begin{Bmatrix} C_{recon}(t,m,n); \\ m-\Delta H<m<m+\Delta H; \\ n-\Delta H<n<n+\Delta H \end{Bmatrix} \tag{6}$$

$$NC_{time}(t,m,n)=\begin{Bmatrix} C_{recon}(t,m,n); \\ t-\Delta T<t<t+\Delta T \end{Bmatrix} \tag{7}$$

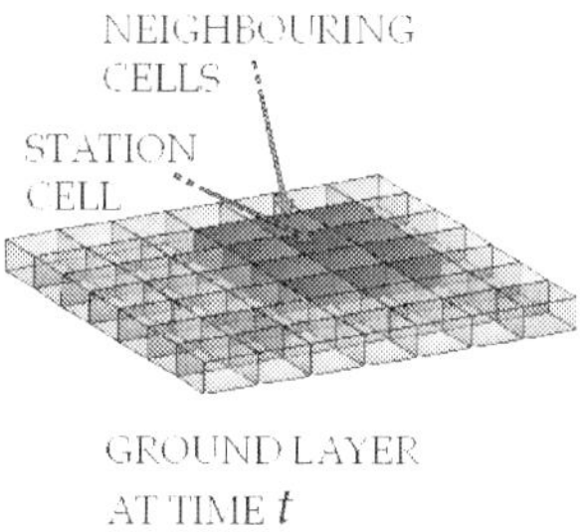

Fig. 2. Example of neighboring cells in position ($\Delta H=1$) where set of neighborhood concentrations *NC* consists of 9 cells

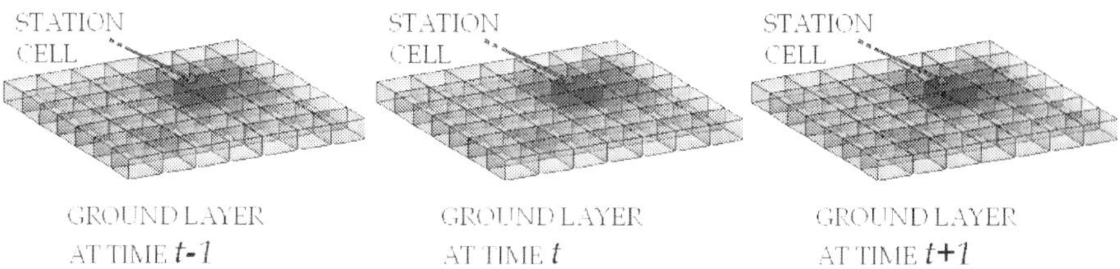

Fig. 3. Example of neighboring cells in time ($\Delta T=1$) where set of neighborhood concentrations *NC* consists of 3 cells

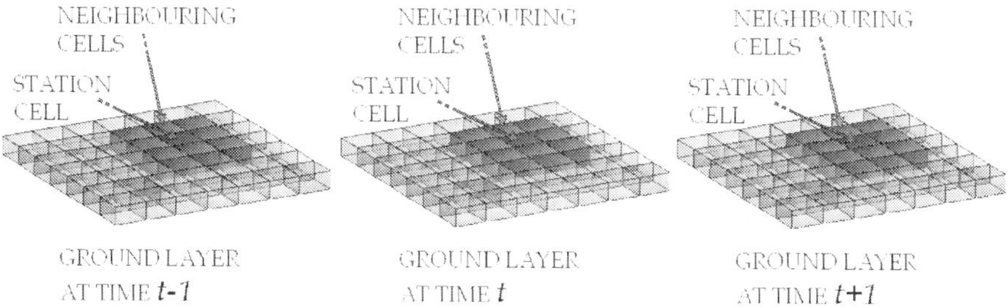

Fig. 4. Example of neighboring cells in position and time *(ΔH=1* and *ΔT=1)* where set of neighborhood concentrations *NC* consists of 27 cells

Finally in enhanced validation methodology each measured value is compared with one reconstructed concentration selected from a set of neighborhood concentrations *NC*. From this set of reconstructed concentrations one concentration $C_{BMrecon}$ is selected using best matching function according to measured concentration as described in equation (8). Best matching function selects one element from *NC* set where difference of this element and measured concentration in lowest compared to other elements in *NC* set.

$$C_{BMrecon}(t,m,n) = BM(NC(t,m,n)) \tag{8}$$

3. Study case

Presented method is demonstrated on a field data set from a complex terrain. In the following sub-chapters field data set from Šaleška region (Slovenia) is described. Field data set from Šaleška region has been chosen for several reasons:

- The first reason is the complex terrain of the region where all typical complex terrain meteorological conditions occur (Grašič, 2007, Blumen et al, 1990).
- The second reason are high emissions from thermal power plant which were about 100000 tons of sulphur dioxide SO_2 and 12400 tons of nitrogen oxides NO_x per year (Elisei, 1991) because no desulphurization plant has been installed at that time. These high emissions represented the main air-pollution source in the region where ambient SO_2 concentrations higher than 1 mg/m³ were measured at surrounding automatic environmental measuring stations. All other local source of air pollution can be practically neglected for this reason. Experimental campaign had been therefore organized as a tracer experiment.
- The third reason is the availability of all measured data from environmental automatic measuring stations and emission station for the whole period of measuring campaign. Complete database is available in final report (Elisei et al., 1991) and also on the internet web page (Šoštanj91 Campaign home page, 2007).
- And the fourth reason is that database obtained during the campaign had been used to validate several available air pollution models (Grašič, 2007).

Study case continues with description of air pollution modelling and comparison of validation results using standard and new presented method.

3.1 "Šoštanj91" field data set description

An experimental measuring campaign named had been performed in spring of year 1991 from 15th of March till 5th of April 1991 in surrounding of Thermal power plant Šoštanj (TPPŠ). Main purpose of the campaign was determination of environmental impact of the air pollution from the three stacks of thermal power plant. The emphasis has been on the meteorological conditions that cause severe air pollution episodes.

TPPŠ is located in the centre Šaleška valley as presented on Figure 5. In the central part of Šaleška valley there is a plain located north of Paka river. Average altitude of the valley is three hundred meters above sea level. Valley is surrounded by hills on the south side and by high mountains Karavanke Alps on the west, north and east side. There are two towns and several small villages in the valley and its surrounding where approximately 36000 people lived in the time when campaign had been performed (Elisei et al., 1991). Map on Figure 6 shows the location of Šaleška valley in the north-eastern part of Slovenia.

The experimental campaign had been performed by researchers from three research institutions: ENEL-CRAM and CISE, Milano, Italy and Jozef Stefan Institute, Ljubljana, Slovenia. Data obtained during the campaign had been used to validate several available air pollution models: standard and advanced Gaussian models, Gaussian puff model and Lagrangian particle dispersion model (Brusasca et al., 1992, Božnar et al., 1993, Božnar et al., 1994). Final results of this studies proved that the Lagrangian particle dispersion model is the most effective tool for air pollution modelling in very complex terrain. Campaign was described in details in a final report (Elisei et al., 1991) where also all measured data is available. Database consists of measurements from different measuring systems: automatic measuring stations of Environmental Information System (EIS) maintained by TPPŠ, automatic mobile laboratory, one mobile Doppler SODAR and DIAL. Pictures of some of equipment are presented on Figure 7.

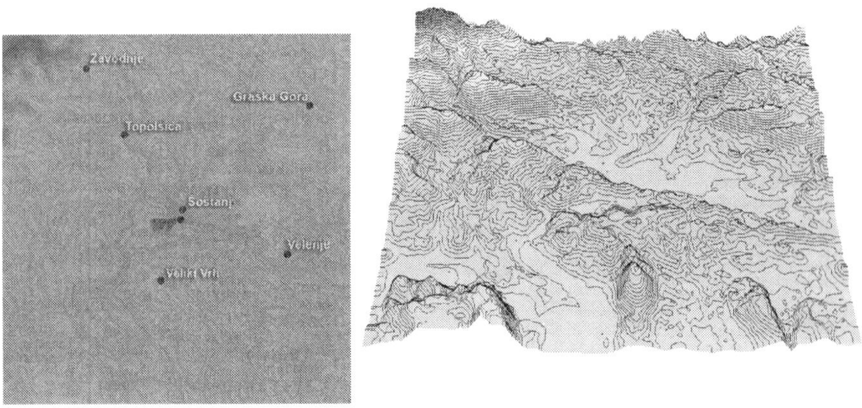

Fig. 5. Map of Šaleška region with locations of automatic environmental stations and location of the Thermal power plant Šoštanj in the centre (left picture) and the topography of the region (right picture)

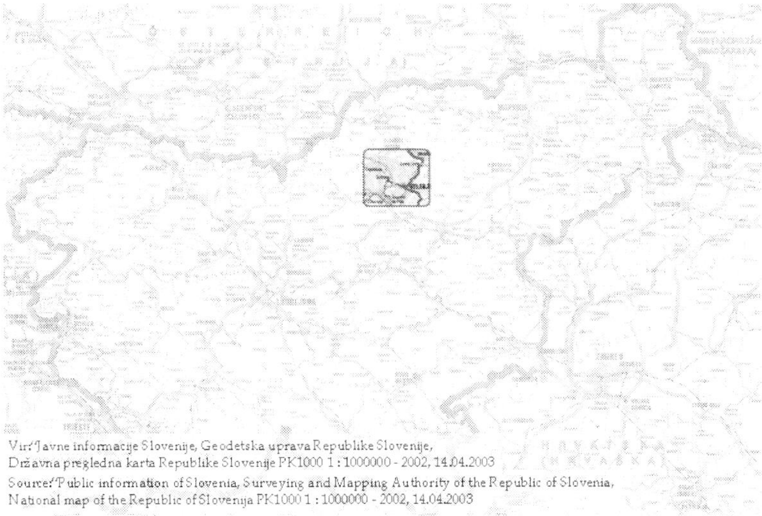

Fig. 6. Location of Saleška region in the north-eastern part of Slovenia

Fig. 7. Pictures of some of equipment used in campaign in spring 1991: environmental automatic measuring station (left), mobile SODAR (right - upper) and DIAL (right -lower)

Environmental Information System of TPPŠ consisted of six stationary automatic measuring stations and one mobile station. Locations of the stations are presented on Figure 5. Environmental parameters measured on stations are presented in Table 1.

Station name / Parameter name	Zavodnje	Graška Gora	Topolšica	Veliki Vrh	Šoštanj	Velenje
Air temperature	x	x	x	x	x	x
Relative humidity	x	x	x	x	x	x
Global solar radiation						x
Precipitation					x	
Air pressure					x	
Wind	x	x	x	x	x	x
SO2	x	x	x	x	x	x
NO	x					
NOx	x					
O3	x					

Table 1. List of parameters measured at automatic environmental stations (x denotes that parameter is measured at certain station)

TPPŠ had during the campaign three operating stacks of different heights: 100m, 150 m and 230 m. Neither of the stacks had installed desulphurization plant during the experimental campaign. Measured emissions are presented in Table 2 where static and dynamic parameters are given. Emissions from generators Block 1, Block 2 and Block 3 are emitted from one stack named Block 1,2,3. Picture of TPP Šoštanj is presented on Figure 7.

Stack name	Height	Diameter	Location
Block 1,2,3	100 m	6.50 m	46.373N 15.052E
Block 4	150 m	6.34 m	46.372N 15.053E
Block 5	230 m	6.20 m	46.371N 15.055E

Stack name	Emission rate	Exit temperature	Exit velocity
Block 1,2,3	0.01 ÷ 0.24 kg/s	155 ÷ 171°C	0.7 ÷ 2.9 m/s
Block 4	0.87 ÷ 2.05 kg/s	155 ÷ 183°C	8.8 ÷ 12.3 m/s
Block 5	0.53 ÷ 2.46 kg/s	172 ÷ 202°C	8.6 ÷ 12.7 m/s

Table 2. List of emission parameters for all TPP Šoštanj stacks that were operating during experimental campaign in spring 1991, upper table presents static and lower dynamic parameters.

Fig. 8. Thermal power plant Šoštanj

3.2 Air pollution modelling

In this study case the Lagrangian particle air pollution dispersion model has been chosen for validation. The name of the model is SPRAY and its detailed description is given in papers by its authors (Brusasca et al, 1992, Tinarelli et al., 2000).

Model has been chosen for validation due to several reasons:

- First version of the model has already been validated on the "Šoštanj91" field data set (Brusasca et al., 1992, Božnar et al., 1993, Božnar et al., 1994). Model has significantly evolved in last years. It has moved from research usage to usage for operational regulatory purposes (Tinarelli et al., 2000).
- Model follows new Slovenian legislation where for complex terrain it is required to use Lagrangian particle dispersion model. Model is coupled with corresponding meteorological pre-processor module which is able to reconstruct three dimensional diagnostic non-divergent wind fields.
- Most of Slovenian industry is located in the complex terrain at the bottom of valleys, river canyons or basins. The results of validation can therefore be applied also on the other similar cases in Slovenia or anywhere else where complex terrain is present.
- Latest version has also been recently validated on "Šoštanj91" field data set (Grašič, 2007). The validation results in this study are extended and focused on the validation method rather than on the validation of the model.

Detailed description of model parameters and settings for this study are described in paper about latest validation of the model (Grašič, 2007). For input into meteorological pre-processor measured data mobile Doppler SODAR and from automatic environmental measuring stations as described in Table 1 has been used. All measured data are available in half-hour intervals. Mobile Doppler SODAR has been located in the centre of the domain.

Meteorological fields have been reconstructed at 150 m horizontal resolution. The same resolution has also been used to describe the complex topography (i.e. orography, Corine land use, etc.). Given all this topography and local meteorological data three dimensional mass consistent wind fields have been generated and used in Lagrangian particle dispersion model for air pollution reconstruction.

Lagrangian particle dispersion model has been generating half-hour average ground concentration fields at the same resolution 150 m as meteorological pre-processor. It has been using Thomson's 1987 scheme with Gaussian random forcing (Thomson, 1987). The number of emitted virtual particles has been set in order to assure minimum resolution for ground level concentrations less than 1 $\mu g/m^3$. Anfossi's formulation (Anfossi, 1993) has been used for plume rise of hot stack plumes where horizontal and vertical variations of both mean wind and atmospheric stability had been taken into consideration.

Air pollution reconstruction has been made for the full duration of the experimental campaign: from 15th of March till 5th of April 1991. Results from simulation are available in half-hour intervals. Each half-hour result represents average air pollution situation over complete domain for one half-hour interval. This result is a three-dimensional (3D) concentration field describing concentrations for each cell of the domain. Domain consists of 100x100 grid cells in horizontal and of 20 layers in vertical that is 200000 grid cells in one 3D concentration field. For validation of the model only two-dimensional (2D) concentration field is relevant representing ground level concentrations. This ground-level concentration field consists of 100x100 cells from ground layer where each cell size is 150 m x 150 m in horizontal and 10 m in vertical.

For demonstration of new validation method only one very complex air pollution situation has been selected. It is a typical complex terrain situation, very difficult for reconstruction and still represents greatest challenge to all available air pollution dispersion models. The situation is described in details in paper by Grašič et al. (2007). It lasted from 1st of April 1991 at 20:00 until 2nd of April 1991 at 20:00.

Spreading of reconstructed plume in three-dimensional domain is presented on Figure 9 where it is shown that plume has been spreading in all directions over domain during a relatively short period of time. This is also seen from the Doppler SODAR measurements presented on Figure 10. This graph represents measurements from SODAR for each half-hour time interval at different heights. Each arrow on the graph represents direction of horizontal wind component at certain height. The length of the arrow represents the magnitude of horizontal wind speed component.

Air pollution spreading in all directions is also proven by measurements of half-hour average SO_2 concentrations at four environmental stations at different directions from TPPŠ as presented on Figures 11, 12, 13, 14.

In the paper by Grašič et al. (2007) it is also reported that during this selected period the phenomenon of air pollution accumulation occurred. Very stable meteorological situation was main cause for very slow mixing of plume with air. Pollution plume was moving very slowly according to measured average wind speed and direction. At the beginning of this situation the air pollution from the point of view of a measuring station came from the direction of the source. But when the main wind changed its direction to opposite direction, also the air pollution cloud changed its direction. From now on from the point of view of measuring station it appeared that the air pollution cloud is coming from the virtual emission source located on the other side. In our case selected domain was not wide enough to capture this phenomenon by Lagrangian particle dispersion model. Part of the air pollution cloud has been lost out of domain which should be taken into account when model is being validated. Lagrangian particle model could reproduce this phenomenon correctly if the domain would be widened but in this case we would have to decrease the final resolution of the domain due to computational limits of the model. Decrease of the resolution (i.e. from 150 m grid cell to 500 m grid cell) would result in more coarse results and also some local complex terrain effects could be lost.

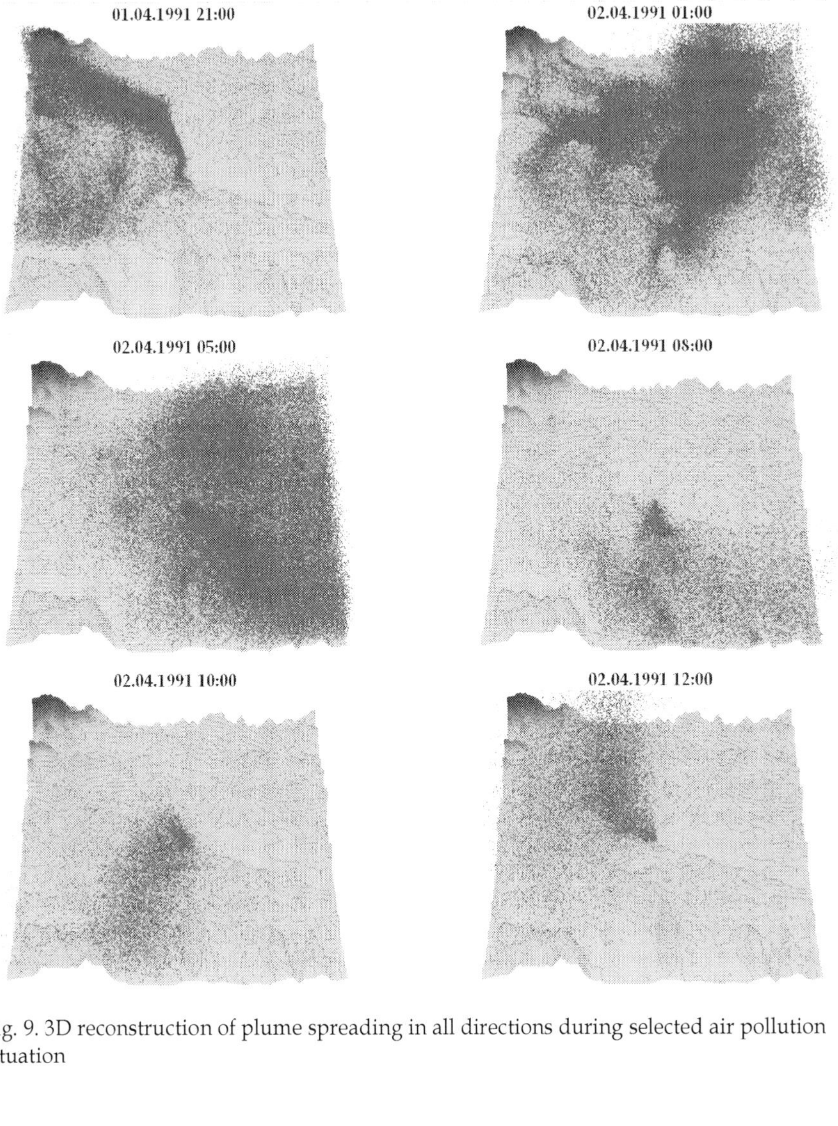

Fig. 9. 3D reconstruction of plume spreading in all directions during selected air pollution situation

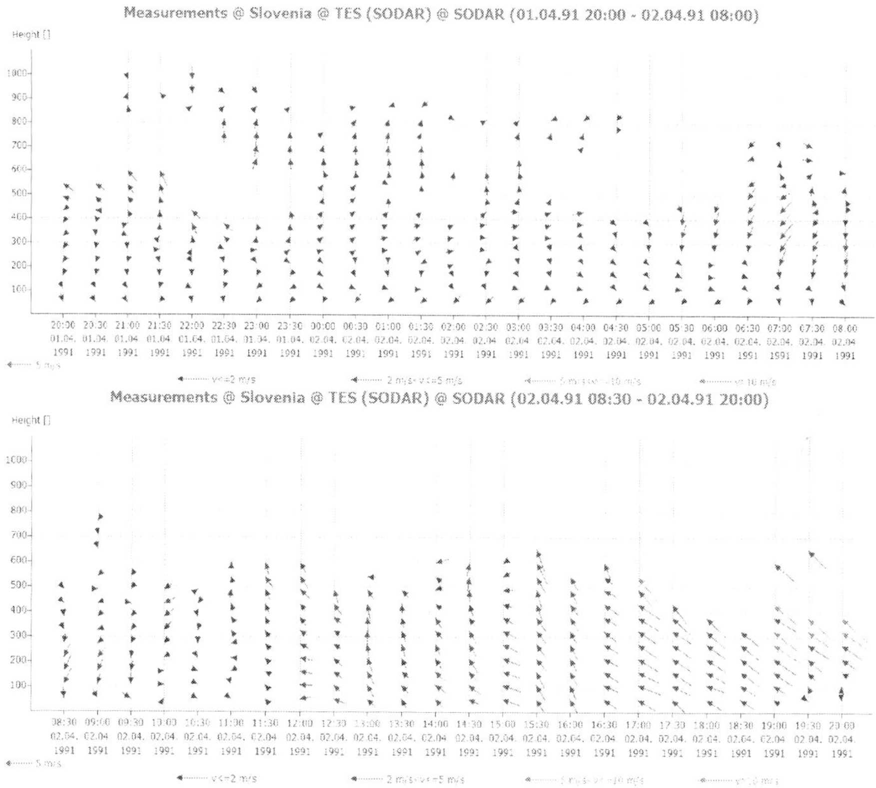

Fig. 10. SODAR measurements during selected air pollution situation from 1st of April 1991 at 20:00 until 2nd of April 1991 at 20:00 when air pollution accumulation occurred; direction of arrows presents horizontal wind direction at certain height; length and color of arrow presents horizontal wind speed at certain height

3.2 Validation results

Validation of modelling results is performed at four stations located in different directions from the point of view of thermal power plant Šoštanj. Four locations are selected according to positions of four environmental automatic measuring stations as presented on Figure 5: Graška Gora, Šoštanj, Veliki Vrh and Zavodnje. From all these stations measurements of half-hour average SO_2 concentrations are available for selected air pollution situation from 1st of April 1991 at 20:00 till 2nd of April 1991 at 20:00.

As presented on Figure 11 measured SO_2 concentration was increased due to wind change at the beginning of selected air pollution situation. Wind at approximate height 250 m changed its direction from north-west to south-east. Next wind change was toward the south which caused an increase of SO_2 concentrations at Šoštanj (Figure 12) and Veliki Vrh (Figure 13) stations. Figure 14 presents measured SO_2 concentrations at Zavodnje station which is the most distant station from the TPPŠ. This result is interesting especially because

of the measured SO_2 concentration peak at the ending of air pollution situation. This peak was caused by air pollution accumulation phenomenon as describe in previous sub-chapter about air pollution modelling. Because the station is located near the border of domain (Figure 5) it is expected that the model results will be underestimated in this case.
In the following sub-chapters a comparison between measured and reconstructed SO_2 concentrations at the locations of presented four stations is made using traditional and enhanced validation methods. Within validation using traditional validation methodology modelling problems will be described that cause under or over estimations of reconstructed concentrations. And the sub-chapter using enhanced validation methodology is presenting different comparison results which can be used as a good estimation of model's inaccuracy of position and time.

3.2.1 Traditional validation results

Figure 11 shows comparison between measured and reconstructed SO_2 concentrations at station Graška Gora. Reconstructed concentrations agree very well with measured ones. Also comparison using traditional statistical indexes for complete duration of experimental measuring campaign from 15th of March till 5th of April 1991presented in Table 3 shows good correlation where correlation reaches value higher than 0.3.
Same comparison of SO_2 concentrations at station Šoštanj is presented on Figure 12. Comparison on the graph shows underestimation of reconstructed concentration values. The first reconstructed peak at 11:30 hour is underestimated due to model's inaccuracy of position. In the paper by Grašič et al. (2007) it is shown that correct peak has been reconstructed just two cells away from the station. The second underestimated concentration peak is caused by short distance between station and stacks (approximately 500 m). There are two effects that are not well captured due to this short distance. First is the stack tip down-wash effect. And the second is the combination of low-wind speed in direction towards the station and convective turbulences (Grašič et al., 2007). Comparison using traditional statistical indexes presented in Table 3 shows almost no correlation and medium underestimation of reconstructed concentrations
The comparison of SO_2 concentrations at Veliki Vrh station are presented on Figure 13. During the air pollution situation two concentration peaks have been reconstructed (from 00:00 till 04:00 and from 06:00 till 12:00). Both peaks are not correctly reconstructed due to inaccuracy of the model in position. Such peaks can appear in real situation just few meters from the measuring station without being detected (Grašič et al., 2007). Comparison presented in Table 3 shows poor correlation between measured and reconstructed concentrations.
Even more obvious phenomenon of model's inaccuracy in position is presented on Figure 14 where comparison between measured and reconstructed SO_2 concentrations at the location of Zavodnje station is made. The phenomenon is more expressed because of the long distance between the station and thermal power plant. It generated first reconstructed peak in time interval from 00:00 till 04:00 hour. The second measured concentration peak has been underestimated due to air pollution accumulation that has been lost because the domain was not wide enough to capture the phenomenon. This event occurred at the end of air pollution situation when the wind changed direction from south back to north-west direction (Grašič et al., 2007). Comparison presented in Table 3 shows none correlation between measured and reconstructed concentrations and very high normalized mean square error.

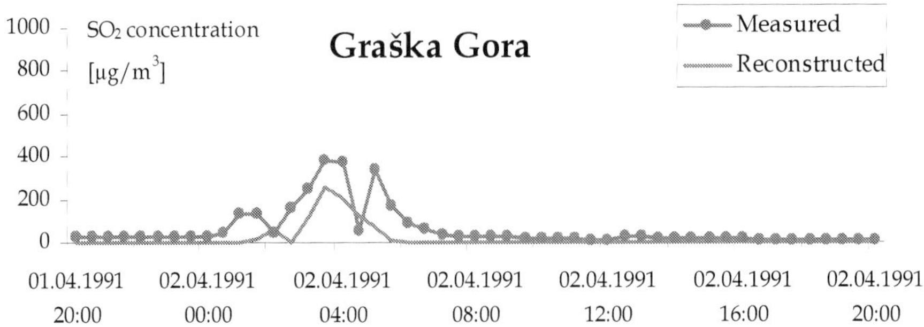

Fig. 11. Comparison of measured and reconstructed ambient SO_2 concentrations at Graška Gora station

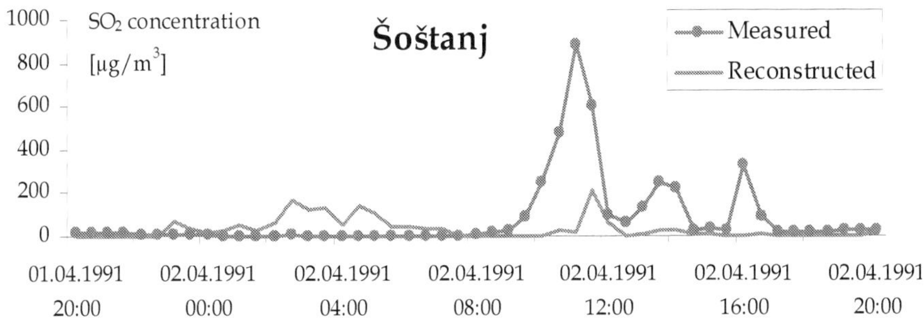

Fig. 12. Comparison of measured and reconstructed ambient SO_2 concentrations at Šoštanj station

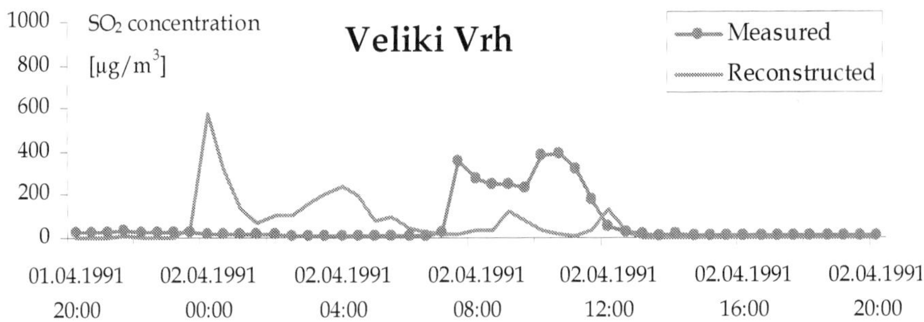

Fig. 13. Comparison of measured and reconstructed ambient SO_2 concentrations at Veliki Vrh station

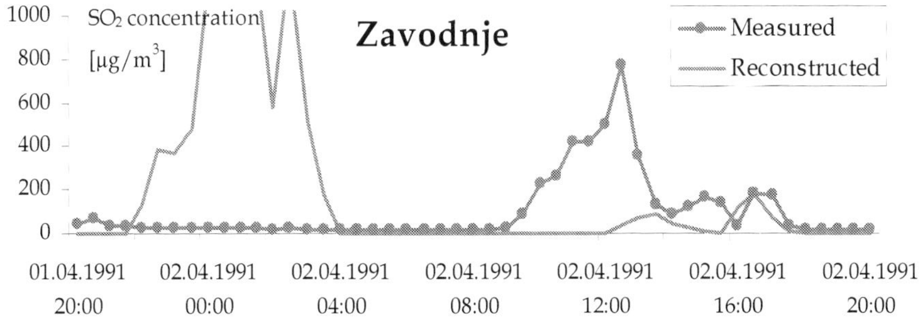

Fig. 14. Comparison of measured and reconstructed ambient SO_2 concentrations at Zavodnje station

Table 3 presents results of comparison between measured and reconstructed SO_2 concentrations using statistical indexes for complete duration of experimental measuring campaign in spring of year 1991 from 15th of March till 5th of April 1991. In this case traditional point-to-point comparison has been made. From the results seems that only the reconstructed concentrations at Graška Gora are satisfactory comparing to results of many authors in published papers (Ferrero et al., 1996, Rizza et al., 1996, Kaasik, 2005) which were also participating in model validation framework named "Model evaluation toolkit" that has been established and maintained by Olesen (1996). Within this research Olesen aslo outlined difficulties that can arise in model validation: differences between measured and reconstructed concentrations are caused by measuring errors, inherent uncertainty, input uncertainty and model formulation error.

Detailed analysis of selected air pollution situation (Grašič et al., 2007) determined that inaccuracy in position and time of reconstructed concentrations have been caused mostly by model's sensitivity to wind speeds and directions measured at different stations and by SODAR. Model's sensitivity strongly depends on the complexity of the terrain which is in our case highly complex.

STATION NAME	NUMBER OF PAIRS	CORRELATION COEFICIENT (CR)	NORMALIZED MEAN SQUARE ERROR (NMSE)	FRACTIONAL BIAS (FB)
Graška Gora	884	0,34	40,42	1,60
Šoštanj	881	0,02	17,32	0,37
Veliki Vrh	839	0,13	8,70	0,09
Zavodnje	858	-0,004	38,35	0,10

Table 3. Statistical indexes of comparison using traditional methodology for complete duration of experimental measuring campaign from 15th of March till 5th of April 1991

3.2.2 Enhanced validation results

Figure 15, 16, 17 and 18 shows comparison between measured and reconstructed SO_2 concentrations at stations Graška Gora, Šoštanj, Velikih Vrh and Zavodnje. There are three types of reconstructed concentrations with different size of neighborhood as described in section 3.2 *Enhanced validation methodology*:

- Recon. (ΔH=0,ΔT=0) - size of neighborhood is 0, only 1 cell where station is located is used for comparison, results are identical to traditional validation method
- Recon. (ΔH=1,ΔT=1) - size of neighborhood is 27 cells (9 cells in horizontal scale and 3 cells in time scale)
- Recon. (ΔH=2,ΔT=2) - size of neighborhood is 125 cells (9 cells in horizontal scale and 3 cells in time scale)

Agreement between measured and reconstructed concentrations is significantly improving when neighborhood is expanding. Similar result is obtained within comparison using traditional statistical indexes for complete duration of experimental measuring campaign from 15th of March till 5th of April 1991 presented in Tables 4 and 5. Comparison of results presented in Tables 3, 4 and 5 show significant improvement of all statistical indexes.

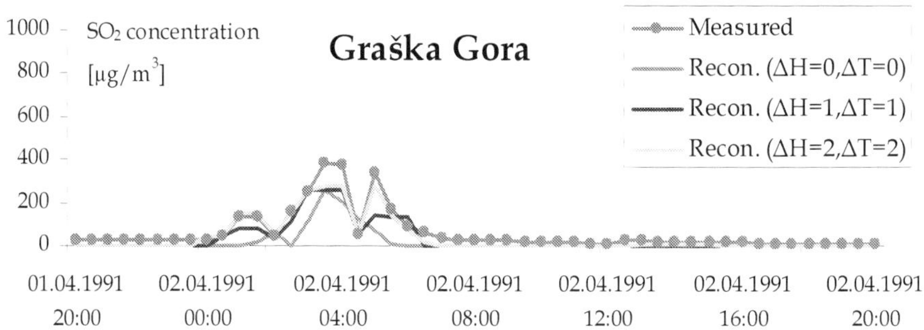

Fig. 15. Comparison of measured and reconstructed ambient SO_2 concentrations at Graška Gora station using different sizes of neighborhood

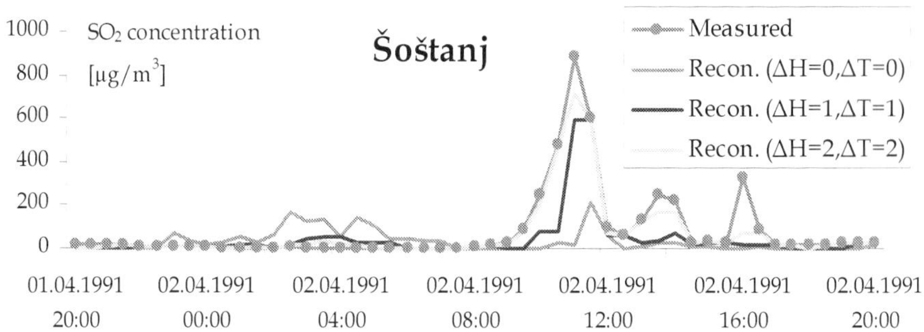

Fig. 16. Comparison of measured and reconstructed ambient SO_2 concentrations at Šoštjan station using different sizes of neighborhood

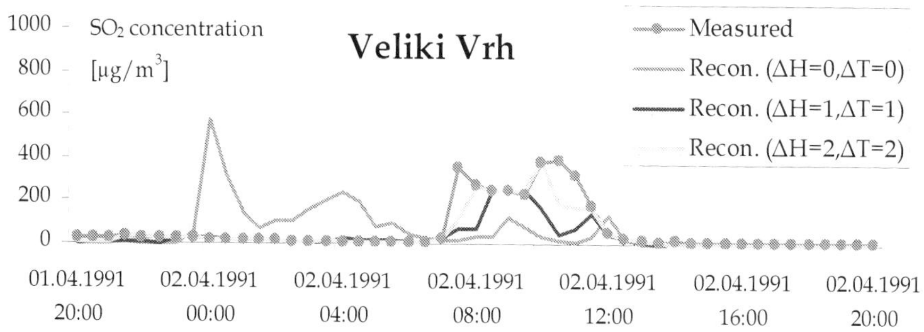

Fig. 17. Comparison of measured and reconstructed ambient SO_2 concentrations at Veliki Vrh station using different sizes of neighborhood

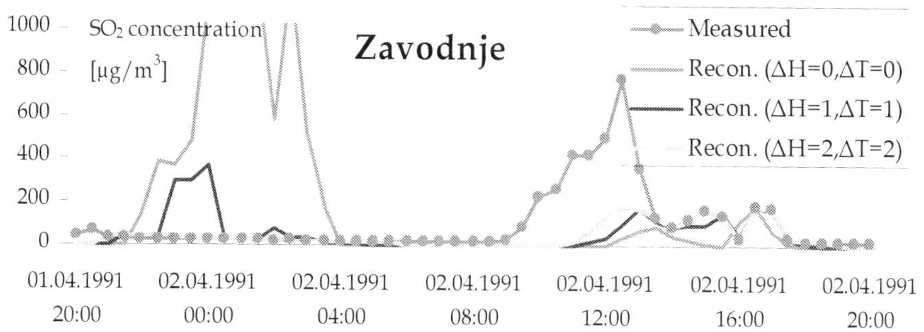

Fig. 18. Comparison of measured and reconstructed ambient SO_2 concentrations at Zavodnje station using different sizes of neighborhood

STATION NAME	NUMBER OF PAIRS	CORRELATION COEFICIENT (CR)	NORMALIZED MEAN SQUARE ERROR (NMSE)	FRACTIONAL BIAS (FB)
Graška Gora	881	0,70	1,08	9,07
Šoštanj	836	0,38	0,93	19,72
Veliki Vrh	878	0,75	0,61	3,10
Zavodnje	855	0,39	0,78	6,02

Table 4. Statistical indexes of comparison using enhanced validation methodology for complete duration of experimental measuring campaign from 15th of March till 5th of April 1991 where size of neighborhood consists of 27 cells ($\Delta H=1$ and $\Delta T=1$)

STATION NAME	NUMBER OF PAIRS	CORRELATION COEFICIENT (CR)	NORMALIZED MEAN SQUARE ERROR (NMSE)	FRACTIONAL BIAS (FB)
Graška Gora	879	0,76	0,83	5,22
Šoštanj	876	0,47	0,79	14,92
Veliki Vrh	834	0,88	0,38	1,30
Zavodnje	853	0,67	0,65	3,38

Table 5. Statistical indexes of comparison using enhanced validation methodology for complete duration of experimental measuring campaign from 15th of March till 5th of April 1991 where size of neighborhood consists of 125 cells ($\Delta H=2$ and $\Delta T=2$)

4. Further improvements of enhanced validation methodology

In the paper by Grašič et al. (2007) validation has been performed using enhanced validation methodology as explained in chapter *3.2 Enhanced validation methodology*. For this validation neighborhood of 27 cells (one cell in each horizontal direction $\Delta H=1$ and one time step on time scale $\Delta T=1$) has been used. Figures 19, 20, 21 and 22 present comparisons of the results obtained in paper by Grašič et al. (2007) and results presented in previous sub-chapter *3.2.2 Enhanced validation results* where also neighborhood of 27 cells has been used.
This comparison shows slightly better results for recent study than for the previous study. This is also apparent from statistical indexes presented in Table 6. Main difference between previous and recent study is in removing of used reconstructed concentrations for further comparison in the old method. Main idea of enhanced validation methodology is to assign each measured concentration one reconstructed concentration from the neighborhood. Focusing on the time scale this means that one the same reconstructed concentration can be assigned to three measured concentrations when size of neighborhood is one time interval $\Delta T=1$. To avoid this in the previous study to each measured concentration only one unique reconstructed concentration has been assigned that appeared firstly in the set. The set was processed in time order from the oldest to youngest measured concentration. If processing had been performed in opposite order for the youngest to oldest the results would be slightly different. To find out the best order how to process this set very advanced optimizing algorithm performing in three dimensions would have to be developed which will be our main task in the future. Another task that will have to be solved in parallel will also be determination of appropriate criteria function to measure success of this optimization algorithm.

STATION NAME	NUMBER OF PAIRS	CORRELATION COEFICIENT (CR)	NORMALIZED MEAN SQUARE ERROR (NMSE)	FRACTIONAL BIAS (FB)
Graška Gora	881	0,69	1,14	10,38
Šoštanj	836	0,36	0,95	20,64
Veliki Vrh	878	0,74	0,61	3,26
Zavodnje	855	0,37	0,79	6,30

Table 6. Statistical indexes of comparison using enhanced validation methodology for complete duration of experimental measuring campaign from 15th of March till 5th of April 1991 where size of neighborhood consists of 27 cells ($\Delta H=1$ and $\Delta T=1$) and unique reconstructed concentrations are used

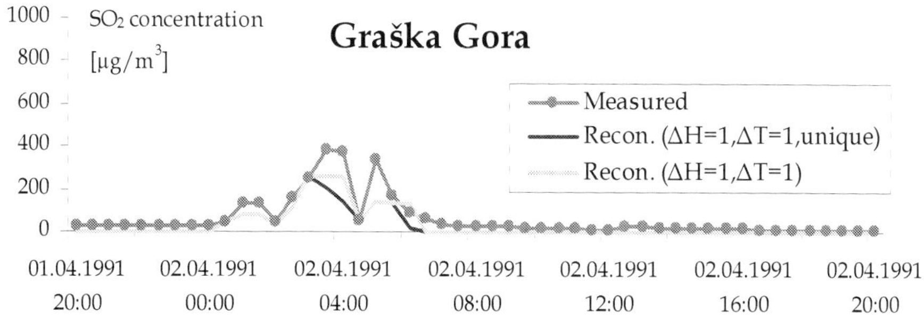

Fig. 19. Comparison of measured and reconstructed ambient SO_2 concentrations at Graška Gora station where for first comparison unique reconstructed concentrations are used

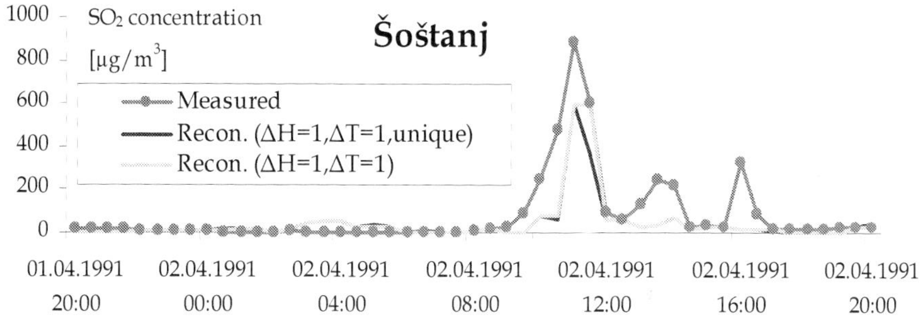

Fig. 20. Comparison of measured and reconstructed ambient SO_2 concentrations at Šoštanj station where for first comparison unique reconstructed concentrations are used

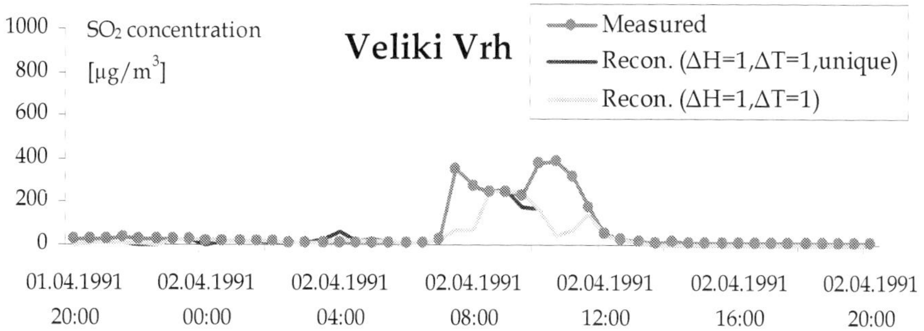

Fig. 21. Comparison of measured and reconstructed ambient SO_2 concentrations at Veliki Vrh station where for first comparison unique reconstructed concentrations are used

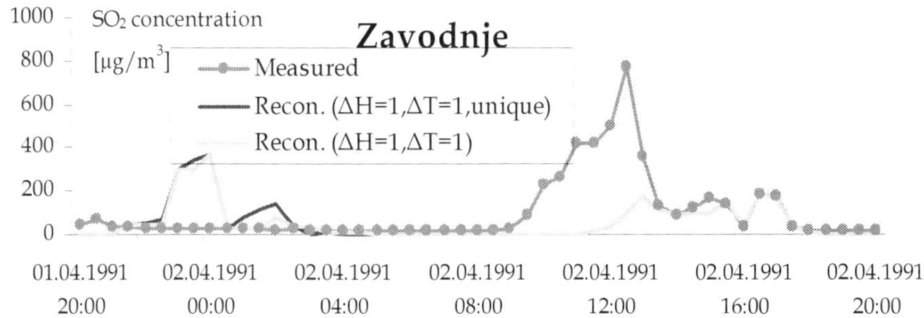

Fig. 22. Comparison of measured and reconstructed ambient SO_2 concentrations at Zavodnje station where for first comparison unique reconstructed concentrations are used

5. Conclusion

Traditional air pollution model validation methodology has been extended in this paper. It is based on statistical comparison between measured and reconstructed air pollution concentrations in the environment where different statistical indexes are determined. The method been upgraded to estimate inaccuracy in position and time of the Lagrangian particle air pollution dispersion model. To obtain these inaccuracies additional reconstructed air pollution concentrations from the neighborhood are used. Neighborhood is defined in spatial and time scale.

Enhanced validation methodology has been demonstrated on a field data set »Šoštanj91« from a very complex terrain from Šaleška region (Slovenia). Field data set is described in details and it has been selected mainly due to high emissions during experimental campaign where SO_2 air pollution situation behaved as tracer experiment.

Air pollution modeling has been made using Lagrangian particle air pollution dispersion model *SPRAY* produced by ARIANET Srl from Milano, Italy. This model has been chosen for validation because it follows Slovenian legislation about air pollution modeling over complex terrain where most of Slovenian industry is located.

For validation of the model only one very complex air pollution situation has been selected. It is a typical complex terrain situation, very difficult for reconstruction and includes phenomenon of air pollution accumulation and convective mixing afterwards. Validation using standard statistical indexes has been made at four locations in different directions from the point of view of air pollution source.

Validation begins using traditional validation methodology. Comparison between measured and reconstructed SO_2 concentrations gives relatively poor results. Only reconstructed concentrations at one station are satisfactory. It has been determined that these results are caused by model's sensitivity to measured wind speeds and directions.

To "measure" this model's inaccuracies in position and time enhanced validation methodology is demonstrated. It gives more satisfactory results at location of all stations and it also estimates inaccuracies. It has been estimated that model's inaccuracy in position is about +-300 m and in time +-1 hour which is indeed excellent result for such a complex terrain. These results give good information for future improvement of air pollution dispersion model.

On the other hand also inaccuracies of measurements should be taken into account during the validation process. It is very important to be aware that the measurements are made at certain location. In certain meteorological conditions these measurements are not representative for the nearest neighborhood. This effect is even more obvious for the complex terrain where the air pollution plume can be present in the nearest neighborhood of the station but it is not detected due to certain local phenomenon.

6. Acknowledgment

The study was partially financed by the Slovenian Research Agency, Project No. L1-2082.

7. References

Anfossi, D., Ferrero, E., Brusasca, G., Marzorati, A., & Tinarelli, G. (1993). A simple way of computing buoyant plume rise in Lagrangian stochastic dispersion models. *Atmospheric Environment* 27A, pp. 1443-1451.

Blumen, W., Banta, R. M., Berri, G., Carruthers, D. J., Dalu, G. A., Durran, D. R., Egger, J., Garratt, J. R., Hanna, S. R., Hunt, J. C. R., Meroney, R. N., Miller, W., Neff, W. D., Nicolini, M., Paegle, J., Pielke, R. A., Smith, R. B., Strimaitis, D. G., Vukicevic, T., & Whiteman, C. D. (1990). Atmospheric processes over complex terrain. *Meteorological monographs*, Volume 23, Number 45, American Meteorological Society, Boston, USA

Božnar, M., Brusasca, G., Cavicchioli, C., Faggian, P., Finardi, S., Minella, M., Mlakar, P., Morselli, M. G., & Sozzi, R. (1993). Model evaluation and application of advanced and traditional gaussian models on the experimental Šoštanj (Slovenia, 1991) campaign. Editor: Cuvelier, C., *Intercomparison of Advances Practical Short-Range Atmospheric Dispersion Models* : Proceedings of the Workshop : August 30 - September 3, 1993, Manno-Switzerland, (Joint Research centre, EUR 15603 EN). Brussels: ECSC-EEC-EAEC, 1994, pp. 112-121.

Božnar, M., Brusasca, G., Cavicchioli, C., Faggian, P., Finardi, S., Mlakar, P., Morselli, M. G., Sozzi, R., & Tinarelli, G. (1994). Application of advanced and traditional diffusion models to en experimental campaign in complex terrain. Editor: Baldasano, J. M., *Second International Conference on Air Pollution*, Barcelona, Spain, 1994. Air Pollution II. Volume 1, Computer simulation. Southampton; Boston: Computational Mechanics Publications, 1994, pp. 159-166.

Brusasca, G., Tinarelli, G., & Anfossi, D. (1992). Particle model simulation of diffusion in low windspeed stable conditions. *Atmospheric Environment* Vol. 26, pp. 707-723

Elisei, G., Bistacchi, S., Bocchiola, G., Brusasca, G., Marcacci, P., Marzorati, A., Morselli, M. G., Tinarelli, G., Catenacci, G., Corio, V., Daino, G., Era, A., Finardi, S., Foggi, G., Negri, A., Piazza, G., Villa, R., Lesjak, M., Božnar, M., Mlakar, P., & Slavic, F. (1991). *Experimental campaign for the environmental impact evaluation of Sostanj thermal power plant, Progress Report,* ENEL S.p.A, CRAM-Servizio Ambiente, Milano, Italy, C.I.S.E. Tecnologie Innovative S.p.A, Milano, Italy, Institute Jozef Stefan, Ljubljana, Slovenia

EUR-Lex (2007). Council Directive 84/360/EEC of 28 June 1984 on the combating of air pollution from industrial plants. Available from: http://eur-lex.europa.eu/LexUriServ/LexUriServ.do?uri=CELEX:31984L0360:SL:NOT, 08.10.2007

FAIRMODE (2010). Guidance on the use of models for the European Air Quality Directive, working document of the Forum for Air Quality Modelling in Europe FAIRMODE ETC/ACC report Version 6.2, Editor: Bruce Denby, In: *FAIRMODE*. Available from: http://fairmode.ew.eea.europa.eu/fol429189/forums-guidance/model_guidance_document_v6_2.pdf, 08.03.2011

Ferrero, E., Anfossi, D., Brusasca, G., Tinarelli, G., Alessandrini, S., & Trini Castelli, S. (1993). Simulation of atmospheric dispersion in convective boundary layer: comparison between different Lagrangian particle models. *4th Workshop on Harmonisation within Atmospheric Dispersion Modelling for Regulatory Purposes*, Oostende, 6-9 May 1996, 67-74. International Journal of Environment and Pollution, Vol 8, Nos. 3-6, pp. 315-323.

Government of Republic Slovenia (October 2007). Decree on the emission of substances into the atmosphere from stationary sources of pollution, In: *Slovenian legislation register Ur.l. RS, št. 31/2007*, Available from: http://zakonodaja.gov.si/rpsi/r06/predpis_URED4056.html, 08.10.2007

Graff, A. (2002). The new German regulatory model - a Lagrangian particle dispersion model. *8th International Conference on Harmonisation within Atmospheric Dispersion Modelling for Regulatory Purposes*, October 14-17, Sofia, Bulgaria, pp. 153-158

Grašič, B., Božnar, M. Z., & Mlakar, P. (2007). Re-evaluation of the Lagrangian particle modelling system on an experimental campaign in complex terrain. *Il Nuovo Cimento C*, Vol. 30, No. 6, pp. 19-

Kaasik, M. (2005). Validation of the AEROPOL model against the Kincaid data set. *10th International Conference on Harmonisation within Atmospheric Dispersion Modelling for Regulatory Purposes*, October 17-20 2005, Sissi, Crete, pp. 327-331

Olesen, H.R. (1996). Toward the establishment of a common framework for model evaluation. *Air Pollution Modeling and Its Application XI*, Edited by S-E. Gryning and F. Schiermeier, Plenum Press, New York, pp. 519-528

Rizza, U., Mangia, C., & Tirabassi, T. (1996). Validation of an operational advanced Gaussian model with Copenhagen and Kincaid datasets. *4th Workshop on Harmonisation within Atmospheric Dispersion Modelling for Regulatory Purposes*, Oostende, 6-9 May 1996, 67-74. International Journal of Environment and Pollution, Vol 8, Nos. 3-6, pp. 41-48.

Schwere S., Stohl A., Rotach M. W. (2002). Practical considerations to speed up Lagrangian stochastic particle models. *Computers & Geosciences* 28, pp. 143-154

Šoštanj91 Campaign home page (2007). Experimental Campaign for the Environmental Impact Evaluation of Šoštanj Thermal Power Plant. In: *Šoštanj91 home page*. Available from: http://193.77.212.133/tes-campaign91/indexe.html

Thomson, D.J. (1987). Criteria for the selection of stochastic models of particle trajectories in turbulent flows. *Journal of Fluid Mechanics*, Vol. 180, pp. 529-556

Tinarelli, G., Anfossi, D., Bider, M., Ferrero, E., & Trini Casteli, S. (2000). A new high performance version of Lagrangian particle dispersion model SPRAY, some case studies. *Air pollution modelling and its Applications XIII*, S. E. Gryning and E. Batchvarova eds., Kluwer Academic / Plenum Press, New York, pp. 499-507

Wilson, J. D., & Sawford, B. L. (1996). Review of Lagrangian stochastic models for trajectories in the turbulent atmosphere. *Boundary-Layer Meteorology* 78, pp. 191-210

28

Air Quality and Bioclimatic Conditions within the Greater Athens Area, Greece - Development and Applications of Artificial Neural Networks

Panagiotis Nastos[1], Konstantinos Moustris[2], Ioanna Larissi[3] and Athanasios Paliatsos[4]
[1]*Laboratory of Climatology and Atmospheric Environment, Faculty of Geology and Geoenvironment, University of Athens,*
[2]*Department of Mechanical Engineering, Technological Educational Institute of Piraeus,*
[3]*Department of Electronic-Computer Systems Engineering, Technological Educational Institute of Piraeus,*
[4]*General Department of Mathematics, Technological Educational Institute of Piraeus, Greece*

1. Introduction

Over the past few decades the phenomenon of urbanization resulted in severe problems. The quality of human life has been deteriorated in the megacities around the world. This chapter deal with the Artificial Networks (ANNs) forecasting ability in predicting the air quality as well as the bioclimatic conditions in an urban environment.

For this purpose, different ANNs are demonstrated in this chapter. These ANNs have been developed in order to predict the air quality as well as the bioclimatic conditions within the Greater Athens Area (GAA), Greece. The prognosis for both air quality and bioclimatic conditions within GAA concerns the next three days (24 to 72 hours prediction).For the proper ANNs training for both air quality and bioclimatic conditions, hourly values of specific meteorological parameters such as air temperature, relative humidity, wind speed, wind direction, air pressure, sunshine and solar radiation, as well as hourly values of air pollutants concentrations have been used. These hourly data have been recorded in many different sites within GAA from the network of the Greek Ministry of Environment Energy and Climatic Change (GMEECC) during the period 2001-2005. Hourly values of barometric pressure and total solar irradiance for the same time period were acquired from the National Observatory of Athens (NOA).

This chapter is divided into nine sections. The first section is brief introduction concerning ANNs. The second section presents air quality indices that have been used in this work in order to describe the air quality within GAA. The third section presents bioclimatic indices, which describe the human thermal comfort-discomfort due to meteorological conditions. The fourth section presents statistical performance indices that have been used in order to investigate the predictive ability and reliability of the developed ANNs models. The fifth

section demonstrates the examined sites within the GAA and the data/methodology used in this study.
The sixth section presents the ANNs that were developed in order to predict the maximum daily value of the air pollution indices as well as the persistence of the phenomenon, namely the number of consecutive hours within the day with high/strong air pollution. The seventh section presents the ANNs that were developed in order to predict the daily values of the bioclimatic indices as well as the number of consecutive hours within the day with dangerous bioclimatic conditions for humans' health. The eighth section includes the spatial variation for both air quality levels and human comfort/discomfort levels within GAA. The ninth is the last section summarizing briefly the extracted results by the performed analysis and how these results can contribute positively to the economy, energy, environment and quality of human life in general.
Finally the results of this work have shown that the ANNs could give an adequate forecast for both air quality and bioclimatic conditions within the urban environment of the GAA for the next three days at a statistical significant level of $p<0.01$.

2. Artificial Neural Networks

Artificial Neural Networks (ANNs) are a branch of artificial intelligence developed in the 1950s aiming at imitating the biological brain architecture. They are an approach to the description of functioning of human nervous system through mathematical functions. Typical ANNs use very simple models of neurons. These artificial neurons models retain only very rough characteristics of biological neurons of the human brain (McCulloh & Pitts, 1943). ANNs are parallel-distributed systems made of many interconnected non-linear processing elements (PEs), called neurons (Hecht-Nielsen, 1990). A renewal of scientific interest has grown exponentially since the last decade, mainly due to the availability of appropriate hardware that has made them convenient for fast data analysis and information processing (Viotti et al., 2002).
Figure 2.1 presents the structure of a biological neuron (upper graph) as well as the structure of an artificial neural (lower graph).
ANNs have been applied in time series prediction (Lapedes & Farber, 1987; Werbos, 1988). Although their behaviour has been related to non-linear statistical regression (Bishop, 1995), the big difference is that ANNs seem naturally suited for problems that show a large dimensionality of data, such as the task of identification for systems with great number of state variables. Over the last years, black box approaches have been recognized to constitute a viable alternative to conceptual models for input-output simulation and forecasting and also to allow shortening the time required for the model development. In particular, ANNs concentrated a general consensus in predicting different pollutants time series, as shown by the review of Gardner & Dorling (1998a, 1998b).
Many ANNs were developed for very different environmental purposes. Heymans & Baird (2000) have used network analysis to evaluate the carbon flow model built for the northern Benguela upwelling ecosystem in Namibia. Antonic et al. (2001) have estimated the forest survival after building the hydroelectric power plant on the Drava River, Croatia by means of GIS constructed database and a neural network. Karul et al. (2000) used a three-layer Levenberg-Marquardt feedforward neural network to model the eutrophication process in three water bodies in Turkey. Besides, Moustris et al. (2011) used ANNs for long term precipitation forecast, using long-term monthly precipitation time series of four meteorological stations in Greece.

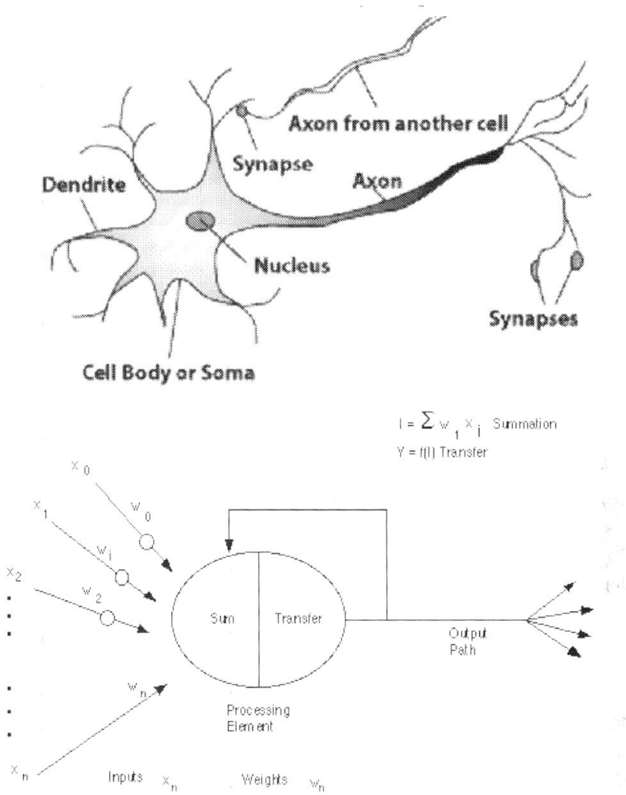

Fig. 2.1. Biological (upper graph) and artificial (lower graph) neuron structure.

Viotti et al. (2002) used ANNs to forecast short and middle long-term concentration levels for some of the well-known pollutants at the urban area of Perugia, Italy. The ANNs approach proved to be viable also for O_3, PM_{10}, NO_2, NO_x forecasting, outperforming alternative techniques in different case studies (Nunnari et al., 1998; Prybutok et al., 2000; Kolehmainen et al., 2001; Balaguer Ballester et al., 2002; Schlink et al., 2003; Corani, 2005; Slini et al., 2006; Dutot et al., 2007; Papanastasiou et al., 2007).

2.1 Multi-Layer Perceptron and feed-forward ANNs

The Multi-Layer Perceptron (MLP) is the most commonly used type of ANNs. Its structure consists of Processing Elements (PEs) and connections (Hecht-Nielsen, 1991). PEs, which are called neurons, are arranged in layers. The first layer is the input layer, one or more hidden layers follow and the final layer is the output layer. An input layer serves as buffer that distributes input signals to the next layer, which is a hidden layer. Each neuron of the hidden layer communicates with all the neurons of the next hidden layer, if any, having in each connection a typical weight factor. So, each unit-artificial neuron in the hidden layer sums its input, processes it with a transfer function and distributes the result to the output layer. It is also possible that there are several hidden layers connected in the same fashion. The units-artificial neurons in the output layer compute their output in a similar manner. Finally, the signal reaches the output layer, where the output value from the ANN is compared to the target value and an error is estimated. Thus, the values of weight factors are amended appropriately and the training cycle repeats until the error is acceptable, depending on the application.

Since data flow within the artificial neural network from a layer to the next one without any return path, such kind of ANNs are defined as feed-forward ANNs. The structure of a feed-forward Multi-Layer Perceptron artificial neural network can be represented as in Figure 2.1.1.

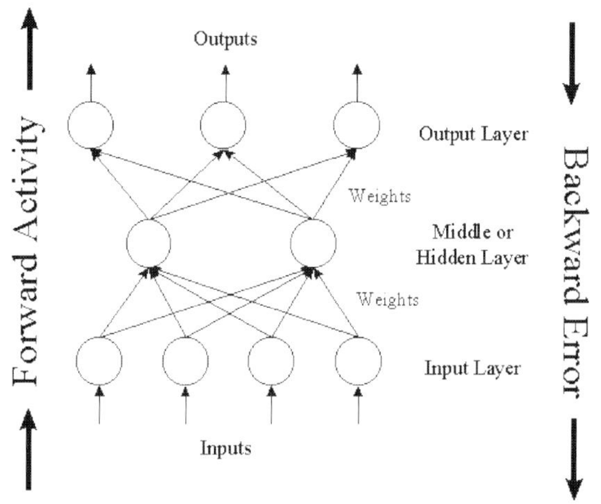

Fig. 2.1.1. Typical MLP feed-forward Artificial Neural Network Structure (Caudill & Butler, 1992).

2.2 Feed-forward ANNs training and the Back-propagation training algorithm

The training-learning process of ANNs can be far from the ensemble optimum in some cases, and the problem can be solved only with a very good database, a best choice of the input configuration for training, or using most powerful learning algorithms (Viotti et al., 2002).

The back-propagation learning algorithm consists of two steps of computation: a forward pass and a backward pass. In the forward pass, an input pattern vector is applied to the sensory nodes of the network, i.e. to the units in the input layer. The signals from the input layer are propagated to the units in the first layer and each unit produces an output. The outputs of these units are propagated to the units in the subsequent layers and this process continues until, finally, the signals reach the output layer, where the actual response of the network to the input vector is obtained (Figure 2.1.1).

During the forward pass, the synaptic weights of the network are fixed. During the backward pass, on the other hand, the synaptic weights are all adjusted in accordance with an error signal, which is propagated backward through the network against the direction of synaptic connections.

The mathematical analysis of the algorithm is as follows (Viotti et al., 2002). In the forward pass, given an input pattern vector $y^{(p)}$, each hidden node-neuron j receives a net input:

$$x_j^{(p)} = \sum_k w_{jk} y_k^{(p)}$$

where w_{jk} represents the weight between the hidden neuron j and the input neuron k. Thus, the hidden neuron j produces an output:

$$y_j^{(p)} = f\left(x_j^{(p)}\right) = f\left[\sum_k w_{jk} y_k^{(p)}\right]$$

where f(x) is the activation faction of the hidden layer. Different kinds of activation functions are referenced in the literature, such as linear, sigmoid, hyperbolic tangent, logistic, etc. (Norgaard et al., 2000). In the following, we consider a hyperbolic tangent activation function for the neurons in the hidden layer, hence, the value returned by the activation function of neuron j of the hidden layer is:

$$f(x_j) = \frac{e^{2x_j} - 1}{e^{2x_j} + 1}$$

Each output neuron receives the input from the preceding hidden layer by the forecasted value, so that the entry to the output neuron can be written as:

$$x^{(p)} = \sum_j w_j y_j^{(p)} = \sum_j w_j f\left[\sum_k w_{jk} y_k^{(p)}\right]$$

where w_j represents the weight between the output neuron and the hidden neuron j. It therefore produces the final output:

$$y^{(p)} = f\left(x^{(p)}\right) = f\left(\sum_j w_j y_j^{(p)}\right) = f\left[\sum_j w_j f\left(\sum_k w_{jk} y_k^{(p)}\right)\right]$$

The presentation of all the patterns is usually called *epoch*. Many epochs are generally needed before the error becomes acceptably small. In the batch neuron the error signal is calculated for each input pattern but the weights are modified only when the input patterns have been presented. The error function is calculated referring to the Mean Square Error (MSE) and the weights are modified accordingly:

$$E = \frac{1}{2}(d-y)^2 = \frac{1}{2}\left\{d - f\left[\sum_j w_j f\left(\sum_k w_{jk} y_k\right)\right]\right\}^2$$

where d is the desired or real output (monitored variable value) and y is the ANN output or the forecasted value. In the batch mode, E is equal to the sum of all MSEs on all the patterns of the training set. E is obviously a differentiable function of all weights (and thresholds) and therefore we can apply the gradient descent method. For the hidden to output connections the gradient descent rule gives:

$$\Delta w_j = -\eta \frac{\partial E}{\partial w_j}$$

where η is a number called *learning rate*. The learning rate is a parameter that determines the size of the weights adjustment each time the weights are changed during the training process. Small values for the learning rate cause small weight changes and large values cause large changes (Attoh-Okine, 1999). The best learning rate is not obvious. If the learning rate is 0.0, the network will never learn.

Refenes et al. (1994) reported that one and tow layered network with a learning rate of η=0.2 and a momentum rate of $0.3<\alpha<0.5$ yield the best combination of convergence. The momentum term is a factor used to speed network training. It adds a proportion of the previous weight changes to the current weight changes.

Using the chain rule it can be written as:

$$\frac{\partial E}{\partial w_j} = \frac{\partial E}{\partial y^{(p)}} \frac{\partial y^{(p)}}{\partial x^{(p)}} \frac{\partial x^{(p)}}{\partial w_j} = -\sum_p \left(d^{(p)} - y^{(p)}\right) f\left(x^{(p)}\right) y_j^{(p)}$$

Thus, the hidden to output connections are updated according to the following equation:

$$\Delta w_j = \eta \sum_p \left(d^{(p)} - y^{(p)}\right) f\left(x^{(p)}\right) y_j^{(p)} = \eta \sum_p \delta^{(p)} y_j^{(p)}$$

where $\delta^{(p)} = \left(d^{(p)} - y^{(p)}\right) f\left(x^{(p)}\right)$

For the input to hidden layer connections the gradient descent rule is:

$$\Delta w_{jk} = -\eta \frac{\partial E}{\partial w_{jk}}$$

Then using the chain rule, we obtain:

$$\frac{\partial E}{\partial w_{jk}} = \frac{\partial E}{\partial y_j^{(p)}} \frac{\partial y_j^{(p)}}{\partial x_j^{(p)}} \frac{\partial x_j^{(p)}}{\partial w_{jk}} = \frac{\partial E}{\partial y_j^{(p)}} f\left(x_j^{(p)}\right) y_k^{(p)}$$

Particularly, with reference to $\frac{\partial E}{\partial y_j^{(p)}}$, it can be written as:

$$\frac{\partial E}{\partial y_j^{(p)}} = -\sum_p \left(d^{(p)} - y^{(p)}\right) \frac{\partial \left[f\left(x^{(p)}\right)\right]}{\partial y_j^{(p)}}$$

and after simple passages, we obtain:

$$\frac{\partial E}{\partial y_j^{(p)}} = -\sum_p \left(d^{(p)} - y^{(p)}\right) f\left(x^{(p)}\right) w_j$$

Therefore, with reference to $\frac{\partial E}{\partial w_{jk}}$, it can be written as:

$$\frac{\partial E}{\partial w_{jk}} = -\sum_{p}\left(d^{(p)} - y^{(p)}\right)f\left(x^{(p)}\right)w_j f\left(x_j^{(p)}\right)y_k^{(p)}$$

from which the input to hidden connections updating is obtained as:

$$\Delta w_{jk} = \eta\sum_{p}\left(d^{(p)} - y^{(p)}\right)f\left(x^{(p)}\right)w_j f\left(x_j^{(p)}\right)y_k^{(p)}$$

and finally we get:

$$\Delta w_{jk} = \eta\sum_{p}\delta^{(p)}w_j f\left(x^{(p)}\right)y_k^{(p)} = \eta\sum_{p}\delta_j^{(p)}y_k^{(p)}$$

with $\delta_j^{(p)} = \delta^{(p)}w_j f\left(x^{(p)}\right)$

It is worthwhile noting that, a network architecture having just one hidden layer, and activation functions arranged as described above, constitutes a universal predictor and it can theoretically approximate any continuous function to any degree of accuracy. In practice, such degree of flexibility is not achievable because parameters must be estimated from sample data, which are both finite and noisy (Barazzetta & Corani, 2004).

The ANNs work on a matrix containing more patterns. Particularly, the patterns represent the rows while the variables are the columns. This data set is a sample. To be more precise, giving the ANN three different subsets of the available sample we can get the forecasting model; the three subsets concern the training, the validation and the test subsets. These subsets are briefly described:

- *Training subset,* the group of data with which we train-educate the network according to the gradient descent for the error function algorithm, in order to reach the best fitting of the nonlinear function representing the phenomenon.
- *Validation subset,* the group of data, given to the network still in the learning phase, by which the error evaluation is verified, in order to update the best thresholds and weights effectively.
- *Test subset,* one or more sets of new and unknown data for the ANN, which are used to evaluate ANN generalization, i.e. to evaluate whether the model has effectively approximated the general function representative of the phenomenon, instead of learning the parameters uniquely.

3. Air quality indices

Urban air pollution is a growing problem in big cities with large urbanization, where adverse health effects have been established. Bad city design combined with specific topographical and meteorological conditions allowing poor circulation, are associated with frequent episodes of critically high atmospheric pollution, enforcing in some cases extreme actions by the authorities, such as restriction of motor vehicles circulation within large area of the city.

For a better and more effective monitoring and analysis of air quality in big cities, air pollution indices are often used. Most of them have resulted after a series of epidemiological studies, which investigated the impact of air pollution on public health. In this work, two air pollution indices are presented and applied in order to forecast the air quality within GAA using ANNs.

3.1 Description of the European Regional Pollution Index (ERPI)

The European Regional Pollution Index (ERPI) has been proposed and developed by Moustris (2009). This air quality index is based on the air pollution index that is known as Regional Pollution Index (RPI). The New South Wales government in Sydney, Australia used RPI since the mid 1990s (NSW-EPA 1998, 2006).

The calculation of ERPI was performed using the thresholds prescribed by the European Community (EC) based on the framework directive 1996/62/EC and the three affiliated directives 1999/30/EC, 2000/69/EC, and 2002/3/EC (Table 3.1.1). Due to the way of calculation of ERPI, based on EC air pollution thresholds, the Australian RPI was renamed as European Regional Pollution Index (ERPI).

In this work, ERPI was calculated for five main air pollutants. Concretely, the air pollutants concern nitrogen dioxide (NO_2), sulfur dioxide (SO_2), carbon monoxide (CO), ozone (O_3) and particulate matter with aerodynamic diameter less than or equal to 10 μm (PM_{10}). For any observed concentration C_i, the value of the sub-index I_i is given by:

$$I_i = \frac{C_i}{Limit_i} \times 50$$

Air Pollutant	Limit values
NO_2	**Hourly value: 200 μg/m³**
SO_2	**Hourly value: 350 μg/m³**
CO	**Maximum daily mean value for 8 hours: 10 mg/m³**
O_3	**Maximum daily mean value for 8 hours: 120 μg/m³**
PM_{10}	**Mean daily value: 50 μg/m³**

Table 3.1.1. Limit concentration values of ambient air pollutants according to EC directives.

Once a sub-index I_i is obtained for each air pollutant (Table 3.1.1), the overall ERPI is simply taken as the maximum of all the I_i values according the formula:

$$ERPI = \max\{I_1, I_2, I_3, I_4, I_5\}$$

where I_1, I_2, I_3, I_4, and I_5 are the sub-indices whose values are defined by the NO_2, SO_2, CO, O_3 and PM_{10}, respectively. If ERPI ≥ 50 this means that at least one of the pollutants is over its limit value (Table 3.1.1). Table 3.1.2 presents the classification of air quality according to ERPI values (Moustris, 2009; Moustris et al., 2010).

ERPI	ERPI Class	Classification
0 - 2	1	Very Good
2 - 21	2	Good
21 - 40	3	Satisfactory
40 - 60	4	Sufficient
60 - 79	5	Poor
≥79	6	Very Poor

Table 3.1.2. Air quality classification according to ERPI values.

3.2 Description of Daily Air Quality Index (DAQx)

A new impact-related air quality index obtained on a daily basis and abbreviated as DAQx (Daily Air Quality Index) has been recently developed and tested by the Meteorological Institute of Freiburg, Germany, and the Research and Advisory Institute for Hazardous Substances, Freiburg, Germany (Mayer et al., 2002a, 2002b; Makra et al., 2003). DAQx considers the air Pollutants SO_2, CO, NO_2, O_3 and PM_{10}. To enable a linear interpolation between index classes, DAQx is calculated for each pollutant by:

$$DAQx = \left[\left(\frac{DAQx_{up} - DAQx_{low}}{C_{up} - C_{low}} \right) \times \left(C_{inst} - C_{low} \right) \right] + DAQx_{low}$$

with C_{inst}: highest daily 1 hour concentration of SO_2, NO_2, and O_3, highest daily running 8 hours concentration of CO, and mean daily concentration of PM_{10}. C_{up} is the upper threshold of specific air pollutant concentration range; C_{low} is the lower threshold of specific air pollutant concentration range; $DAQx_{up}$ is the value of DAQx according to C_{up}; $DAQx_{low}$ is the value of DAQx according to C_{low} (Table 3.2.1).

The daily value of DAQx is considered the highest value extracted by the calculated values for each pollutant.

SO_2 (μg/m³)	CO (mg/m³)	NO_2 (μg/m³)	O_3 (μg/m³)	PM_{10} (μg/m³)	DAQx value	DAQx Class	Classification
0-24	0.0-0.9	0-24	0-32	0.0-9.9	0.5-1.4	1	Very Good
25-49	1.0-1.9	25-49	33-64	10.0-19.9	1.5-2.4	2	Good
50-119	2.0-3.9	50-99	65-119	20.0-34.9	2.5-3.4	3	Satisfactory
120-349	4.0-9.9	100-199	120-179	35.0-49.9	3.5-4.4	4	Sufficient
350-999	10.0-29.9	200-499	180-239	50.0-99.9	4.5-5.4	5	Poor
≥1000	≥30.0	≥500	≥240	≥100	≥5.5	6	Very Poor

Table 3.2.1. Upper and lower limits for air pollutant concentrations and DAQx values, DAQx classes and classification according to Mayer et al. (2002a, 2002b).

4. Bioclimatic indices

The growth of the city of Athens during the last decades and the phenomenon of urbanization (Philandras et al., 1999) have established the well known Urban Heat Island (UHI) at a great areal extent of the city, resulting in explicit effects on human thermal comfort-discomfort. Thermal comfort is defined as the condition of mind, which expresses satisfaction with the thermal environment, absence of thermal discomfort, or conditions in which 80% or 90% of humans do not express dissatisfaction (Givoni, 1998).

Several indices, which describe the human thermal comfort-discomfort, have been developed worldwide. In this chapter three bioclimatic indices will be presented. The Discomfort Index (DI), the Cooling Power index (CP) and the Physiologically Equivalent Temperature (PET). In the process, these indices are briefly described.

4.1 Discomfort Index (DI)

The Discomfort Index (DI) was originally developed by Thom (Thom, 1959) and was supported by later works (Clarke & Bach, 1971; Giles et al., 1990). This index describes the degree of thermal load under various meteorological conditions, suitable for both outdoor and indoor environments. It is useful to evaluate how current temperature and relative humidity can affect the sultriness or discomfort sensation and cause health danger in the population.

DI (°C)	Classification of human comfort-discomfort sensation
DI<21	No discomfort feeling
21≤DI<24	Less than 50% of the total population feels discomfort
24≤DI<27	More than 50% of the total population feels discomfort
27≤DI<29	Most of the population feels discomfort
29≤DI<32	The discomfort is very strong and dangerous
DI≥32	State of medical emergency

Table 4.1.1. Classification of human comfort-discomfort sensation for DI.

Several formulas of the index have been proposed for use along with tables of boundary values that indicate degrees of comfort-discomfort. In the present work we used the following formula of DI, calculated as a combination of air temperature T (°C) and relative humidity RH (Giles et al., 1990):

$$DI = T - (0.55 \times 0.0055 \times RH) \times (T - 14.5), \qquad (^{o}C)$$

The classification of the DI values with the equivalent feeling of thermal comfort-discomfort is given in Table 4.1.1 (Giles et al., 1990).

4.2 Cooling Power index (CP)

The Cooling Power Index (CP) was developed by Siple & Passel (1945) and describes the loss of energy, per unit of time and body surface, which a human organism can tolerate. The CP index, in contrast to the DI index, takes into consideration the wind speed instead of relative humidity. It describes the heat flux per surface unit of the human body towards the environment and the vice versa. For the calculation of the CP index hourly values of air temperature (T, °C) and wind speed (V, m/sec) were used. The calculation of CP is based on the following formula (Tzenkova et al., 2003):

$$CP = 1.163 \times (10.45 + 10 \times V^{0.5} - V) \times (33 - T), \qquad (W/m^2)$$

The classification of the CP index values modified by Besancenot et al. (1978), with the equivalent feeling of thermal comfort-discomfort is given in Table 4.2.1.

CP (W/m²)	Classification of human comfort
CP<0	Endothermal - very hot discomfort
0<CP≤174	Atonic - hot discomfort
175≤CP≤349	Hypotonic - hot sub comfort
350≤CP≤699	Neutral - comfort
700≤CP≤1049	Tonic - cold sub comfort
CP≥1050	Cold discomfort

Table 4.2.1. Classification of human comfort-discomfort sensation for CP.

4.3 Physiologically Equivalent Temperature (PET)

The thermal index Physiologically Equivalent Temperature (PET) is based on the total energy balance of the human body. PET values were evaluated (Mayer & Höppe, 1987; Höppe, 1999), in order to interpret the grade of the thermophysiological stress (Table 4.3.1). It describes the effect of the thermal environment as a temperature value (°C) and can be quantified easier for non specialists in this topic. For night time situation, air temperature corresponds very close to the PET value. It has been applied in heat waves and climatic variability studies (Nastos & Matzarakis 2008, Matzarakis & Nastos 2010) and weather impacts on health (Nastos & Matzarakis, 2006).

The PET analysis was performed by the use of the radiation and bioclimate model, RayMan, which is well-suited to calculate radiation fluxes and human biometeorological indices (Matzarakis et al., 1999, 2010) and was chosen for all our calculations of mean radiant temperature and PET. The RayMan model, developed according to the Guideline 3787 of the German Engineering Society (VDI, 1998) calculates the radiation flux in easy and complex environments on the basis of various parameters, such as air temperature, air humidity, degree of cloud cover, time of day and year, albedo of the surrounding surfaces and their solid-angle proportions (Matzarakis et al., 2010).

PET (°C)	Thermal sensation	Physiological stress level
< 4	very cold	extreme cold stress
	....	....
8	cold	strong cold stress
	....	....
13	cool	moderate cold stress
	....	....
18	slightly cool	slight cold stress
	....	....
23	comfortable	no thermal stress
	....	....
29	slightly warm	slight heat stress
	....	....
35	warm	moderate heat stress
	....	....
41	hot	strong heat stress
	....	....
> 41	very hot	extreme heat stress

Table 4.3.1. Physiologically Equivalent Temperature (PET) for different grades of thermal sensation and physiological stress on human beings (during standard conditions: heat transfer resistance of clothing: 0.9 clo, internal heat production: 80 W) (Matzarakis & Mayer, 1996)

5. Statistical performance indices

The quality and reliability of the developed ANNs, concerning their ability to forecast both air quality and bioclimatic conditions within GAA, were tested using several statistics indices that have already been applied in similar studies (Moustris et al., 2010). The statistical performance indices that used in this work are presented and described briefly:

$$\text{Mean Bias Error: } MBE = \frac{1}{N}\sum_{i=1}^{N}(P_i - O_i)$$

where N is the number of the data points, O_i is the observed data and P_i is the predicted data. The MBE represents the degree of correspondence between the mean forecast (P_i) and the mean observation (O_i). MBE is used to describe how much the model underestimates or overestimates the observed data. Positive/negative values indicate over estimated/under estimated prediction.

$$\text{Root Mean Square Error: } RMSE = \sqrt{\frac{1}{N}\sum_{i=1}^{N}(P_i - O_i)^2}$$

RMSE provides a measure of how well future outcomes are likely to be predicted by the model.
The coefficient of determination (R^2) indicates how much of the observed variability is accounted by the estimated model (Kolehmainen et al., 2001). The coefficient of determination is a number between 0 and +1 and measures the degree of association between two variables. The coefficient of determination is calculated according to the equation (Comrie, 1997):

$$R^2 = \frac{\sum_{i=1}^{N}(P_i - O_{iave})^2}{\sum_{i=1}^{N}(O_i - O_{iave})^2}$$

where O_{iave} is the average of the observed data.
A relative measure of error, called the index of agreement (IA), is also discussed in Willmott et al. (1985). Index of agreement is calculated according to the formula:

$$IA = 1 - \frac{\sum_{i=1}^{N}(P_i - O_i)^2}{\sum_{i=1}^{N}(|P_i - O_{iave}| + |O_i - O_{iave}|)^2}$$

where O_{iave} is the average of the observed data. This is a dimensionless measure that is limited to the range of 0-1. If IA=0, that means no agreement between prediction and observation and if IA=1, that means perfect agreement between prediction and observation.

6. Data and methodology

For the calculation of the bioclimatic indices as well as the air quality indices, appropriate meteorological data in hourly basis were used. More specifically, hourly values of air temperature (°C), relative humidity (%) and wind speed (m/s were used for DI and CP

calculation. In addition to the aforementioned meteorological parameters, total cloudiness cover (octas) was taken into consideration for PET calculation, using the RayMan model (Matzarakis et al., 1999, 2010). The appropriate meteorological parameters used as inputs in the RayMan model were acquired from the National Observatory of Athens, for the period 2001-2004. Besides, hourly values of air pollutants concentrations (NO_2, SO_2, CO, O_3 and PM_{10}) were used in order to estimate the two air quality indices ERPI and DAQx. All the above datasets have been recorded by the network of the GMEECC covering the period 2001- 2005 and concern nine (9) different regions within the GAA, namely the regions: Agia Paraskevi, Thrakomakedones, Lykovrissi, Maroussi, Liossia, Galatsi, Patission, Aristotelous, and Geoponiki (Fig. 6.1). For a better surveillance, the examined regions-stations listed below with the following abbreviations: Agia Paraskevi (APA), Galatsi (GAL), Liossia (LIO), Maroussi (MAR), Patission (PAT), Aristotelous (ARI), Thrakomakedones (THR), Lykovrissi (LYK). The hourly values of air barometric pressure and total solar irradiance for the same time period were obtained from the National Observatory of Athens.
To describe the air quality within the GAA the values of the air quality indices ERPI and DAQx were calculated on an hourly basis in seven different regions-stations (APA, THR, LYK, MAR, LIO, GAL and PAT) with respect to the pollutants NO_2, SO_2, CO, and O_3 and in five different regions-stations (APA, THR, LYK, MAR and ARI) with respect to the particulate matter PM_{10}. The maximum value of the 24 hourly values was considered as the daily value for each one of the two air quality indices. Thus, for each one station-region two daily values for each one of the two examined air quality indices were calculated. The first daily value concerns the air pollutants NO_2, SO_2, CO, and O_3 and the second concerns the particulate matter PM_{10}. This happened because the daily concentrations of particulate matter PM_{10} as well as the daily concentrations of ozone are both high enough. If only one daily value for each of the two air quality indices was calculated, then, we will not be able to know if that value is due to ozone or PM_{10}. Thereafter, an appropriate number of ANN models were developed and trained in order to predict for the next three days the daily value for each one of the two air quality indices as well as the number of consecutive hours during the day where the value of the index is greater than a threshold value.

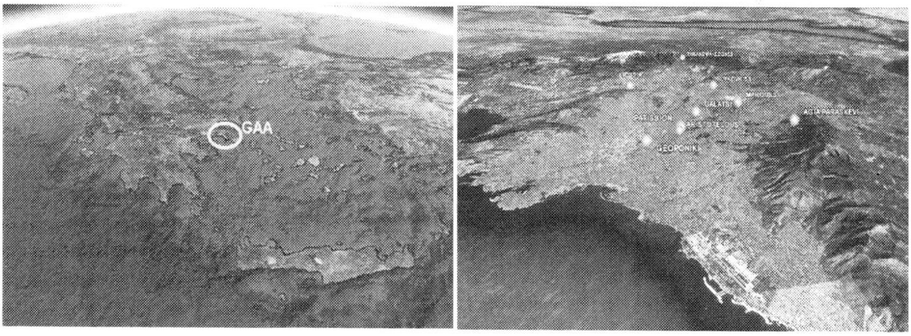

Fig. 6.1 Map of Greece (left graph) and spatial distribution of the GMEECC network's stations within GAA (right graph).

The bioclimatic conditions within the GAA are interpreted by the use of the bioclimatic indices DI and CP, which were calculated on hourly basis for eight different regions-stations (APA, THR, LYK, MAR, LIO, GAL, GEO and PAT). In the process, the daily value for each index in each region-station was calculated. The calculation was carried out only during the warm period of the year (May-September) in order to describe the human discomfort due to heat stress weather conditions. Then, an appropriate number of ANNs were developed and trained in order to predict for the next three days the daily value for each one of the two bioclimatic indices as well as the number of consecutive hours during the day, where the value of the index is greater than a threshold value (DI ≥ 24 °C) or less than a threshold value (CP ≤ 174 W/m^2). Furthermore, the mean daily values of PET index were estimated only for the National Observatory of Athens, because of the availability of the total parameters needed as inputs in RayMan model. Thereafter, the developed ANN was evaluated in forecasting PET for the next three days.

7. Air quality forecasting using ANNs

7.1 ANNs description

Six different ANNs were developed in order to forecast the air quality levels within the GAA. The first one (ANN#1) was trained in order to forecast the daily value of ERPI (for the pollutants CO, NO_2, SO_2 and O_3) for the next day at seven different areas of GAA (APA, THR, LYK, MAR, LIO, GAL and PAT). The second one (ANN#2) was trained in order to forecast the daily value of DAQx (for the pollutants CO, NO_2, SO_2 and O_3) for the next day at the above seven different areas within the GAA. The third one (ANN#3) was trained in order to forecast the daily number of the consecutive hours for the next day, with at least one of the pollutants concentrations (CO, NO_2, SO_2 and O_3) above a threshold according to directives of European Community, for each one of the seven examined stations within the GAA. The fourth (ANN#4) was trained in order to forecast the daily value of ERPI (with respect to PM_{10}) for the next day, at five different areas of GAA (APA, THR, LYK, MAR, and ARI). The fifth (ANN#5) was trained in order to forecast the daily value of DAQx (with respect to PM_{10}) for the next day, at the mentioned five different areas within the GAA. Finally the sixth (ANN#6) was trained in order to forecast the daily number of the consecutive hours for the next day with the PM_{10} concentrations above a threshold according to EC directives, for each one of the five examined stations within the GAA.

In each case, the group of data defined as "the training set", used for ANNs training, concerns the time period 2001-2004. The group of data defined as "the validation set", given to the network still in the learning phase, accounts 20% of 'the training set" for each one of the developed ANN models. Finally "the test set" refers to the year 2005. The year 2005 is absolutely unknown to the models, in order to reveal the models forecasting ability. Table 6.1 presents the input and output data for the six developed ANN models.

The combination of selected data for the appropriate ANN models training was done after a series of several tests (trial and error method). At the end, the combination that gave the best forecasting result in each case was selected (Table 7.1.1).

In this point, we have to mention that for all the constructed ANN models we have used as input data, in addition to other parameters, the maximum and minimum air temperature, the maximum and minimum wind speed for the next day as well as the mean daily air barometric pressure and the mode daily wind direction for the next day. This may produce a limitation in the forecasting attempt, but it is easy to have access to these forecasted values through the network of the Hellenic National Meteorological Service (HNMS).

INPUT DATA (input layer)	ANN#1	ANN#2	ANN#3	ANN#4	ANN#5	ANN#6
Stations' number (1,2,3,4,5,6,7)	√	√	√	√	√	√
Month (1,2,3,…,12)	√	√	√	√	√	√
Mean daily air pressure (mbar) for the six previous days	√	√	√	√	√	√
Daily sum of the global solar irradiance for the six previous days (W/m^2)	√	√	√	√	√	√
Maximum (T_{max}) and minimum (T_{min}) daily temperature (^0C) for the six previous days	√	√	√	√	√	√
Maximum (WS_{max}) and minimum (WS_{min}) daily wind speed (m/sec) for the six previous days	√	√	√	√	√	√
Maximum ($RH\%_{max}$) and minimum ($RH\%_{min}$) daily relative humidity for the six previous days		√			√	
Cosine and sine of the mode daily wind direction for the six previous days	√	√	√	√	√	√
ERPI daily value for the six previous days	√		√	√		√
DAQx daily value for the six previous days		√			√	
The number of consecutive hours during the day with ERPI≥50 for the six previous days			√			√
Mean daily air pressure (hPa) one day ahead	√	√	√	√	√	√
Maximum (T_{max}) and minimum (T_{min}) daily temperature (^0C) one day ahead	√	√	√	√	√	√
Maximum ($RH\%_{max}$) and minimum ($RH\%_{min}$) daily relative humidity one day ahead		√			√	
Maximum (WS_{max}) and minimum (WS_{min}) daily wind speed (m/sec) one day ahead	√	√	√	√	√	√
Cosine and sine of the mode daily wind direction one day ahead	√	√	√	√	√	√
OUTPUT DATA (output layer)						
ERPI daily value for the next day	√			√		
DAQx daily value for the next day		√			√	
The number of consecutive hours with ERPI≥50 for the next day			√			√

Table 7.1.1. Input and output data for the appropriate training of the six developed ANN models.

7.2 Forecasting of daily ERPI and DAQx values for the next day

The global fit agreement statistical indices as well as the excess statistical indices for the observed and predicted ERPI and DAQx values were calculated and demonstrated for the eight examined stations respectively. More specifically, O_{ave}, P_{ave}, MBE, RMSE, IA and R^2 values for ERPI index are presented in Table 7.2.1.

	ANN#1						ANN#2						ANN#3					
	O_{ave}	P_{ave}	MBE	RMSE	IA	R^2	O_{ave}	P_{ave}	MBE	RMSE	IA	R^2	O_{ave}	P_{ave}	MBE (hours)	RMSE (hours)	IA	R^2
APA	44.0	42.5	-1.460	6.629	0.925	0.752	3.4	3.4	0.013	0.396	0.853	0.557	2.8	3.2	0.374	3.015	0.877	0.605
GAL	35.6	35.7	0.118	0.644	0.903	0.697	3.3	3.3	0.060	0.471	0.751	0.378	0.5	0.7	0.199	1.415	0.657	0.231
LIO	37.7	37.8	0.114	6.081	0.919	0.726	3.3	3.3	0.032	0.463	0.746	0.358	0.7	0.9	0.168	1.761	0.671	0.238
MAR	38.3	37.1	-1.272	5.852	0.920	0.738	3.3	3.3	0.000	0.428	0.795	0.446	0.8	0.8	-0.008	1.825	0.708	0.292
PAT	33.1	32.2	-0.944	9.233	0.717	0.381	3.7	3.7	-0.024	0.343	0.791	0.442	0.3	0.3	0.042	1.365	0.299	0.017
THR	42.4	40.6	-1.810	7.290	0.922	0.760	3.3	3.3	-0.002	0.450	0.876	0.637	3.9	3.7	-0.208	4.924	0.829	0.516
LYK	38.9	38.6	-0.317	7.317	0.937	0.826	3.3	3.3	0.021	0.431	0.889	0.686	2.0	1.4	-0.590	3.092	0.742	0.401
	ANN#4						ANN#5						ANN#6					
	O_{ave}	P_{ave}	MBE	RMSE	IA	R^2	O_{ave}	P_{ave}	MBE	RMSE	IA	R^2	O_{ave}	P_{ave}	MBE (hours)	RMSE (hours)	IA	R^2
APA	40.0	40.0	0.356	18.718	0.674	0.266	3.7	3.6	-0.006	0.639	0.911	0.673	6.0	6.0	-0.310	4.259	0.764	0.377
MAR	46.0	46.0	-0.225	19.441	0.699	0.306	3.9	3.9	0.000	0.646	0.930	0.629	8.0	7.0	-0.141	4.908	0.791	0.425
THR	29.0	31.0	2.026	13.285	0.792	0.487	3.0	3.1	0.076	0.636	0.895	0.689	2.0	3.0	0.485	4.002	0.690	0.361
LYK	53.0	53.0	-0.195	19.927	0.729	0.381	4.2	4.2	-0.066	0.634	0.965	0.611	10.0	10.0	-0.303	5.594	0.816	0.492
ARI	53.0	53.0	0.292	18.352	0.721	0.295	4.3	4.3	0.023	0.528	0.779	0.578	10.0	9.0	-1.084	5.845	0.785	0.430

Table 7.2.1. Global fit agreement indices for ERPI predicted values for the next day.

Concerning the pollutants CO, NO_2, SO_2 and O_3, the R^2 values show a very satisfactory prediction for ERPI-ANN#1 ($0.381 \leq R^2 \leq 0.826$) as well as for the DAQx-ANN#2 ($0.378 \leq R^2 \leq 0.686$) during the test year 2005. Besides, IA values show also a very good prediction for ERPI-ANN#1 ($0.717 \leq IA \leq 0.937$) and the DAQx-ANN#2 ($0.746 \leq IA \leq 0.889$). In all cases, it seems that the prediction for the pollutants CO, NO_2, SO_2 and O_3 is much more successful using the ERPI, which is according to the European Community directives, instead of the DAQx. But using both predictions we can have a better and safe "picture" about air quality one day ahead within the GAA. As far as the air pollution persistence (for the pollutants CO, NO_2, SO_2 and O_3) is concerned, it seems that ANN#3 gives an adequate prediction. The R^2 values range between 0.017 and 0.605 while IA range between 0.299 and 0.877.

Finally, the worst prediction with respect to the air quality index ERPI appears for the region-station PAT (city centre) against the region-station LIO (urban area) concerning the air quality index DAQx. Generally, it seems that the prediction for the stations, which are closer to the GAA's downtown, is not so good compared to the prediction of the peripheral regions-stations. This is likely due to the traffic load and the bad air circulation within the city's centre, meaning that, more relevant data, associated with the above mentioned factors, are needed for a better ANNs training.

Figure 7.2.1 presents the best prediction (LYK) and the worst prediction (PAT) for ERPI concerning the pollutants CO, NO_2, SO_2 and O_3, while the best prediction (LYK) and the worst prediction (LIO) for DAQx concerning the same pollutants are depicted in Figure 7.2.2 Accordingly, Figure 7.2.3 presents the best prediction (THR) and the worst prediction (APA) for ERPI concerning the pollutant PM_{10}, and Figure 7.2.4 shows the best prediction (LYK) and the worst prediction (ARI) for DAQx with respect to the pollutant PM_{10}. During the warm period of the year (May-September) the values of ERPI (Figure 7.2.1) are greater than 50, meaning that at least one pollutant's concentration is above its threshold according to the EC directives. In most cases (more than 90%) the corresponding pollutant for these high values of ERPI is the ozone. The same results revealed from Figure 7.2.2 regarding DAQx, where during the warm period of the year the daily values of DAQx are greater than 3.5, meaning that a bad air quality exist in most cases. As far as the PM_{10} concentrations are concerned (Figures 7.2.3 and 7.2.4), it is shown that, for almost half of the days throughout the year are above the threshold concentration value, indicating bad air quality in the most of the examined stations-regions.

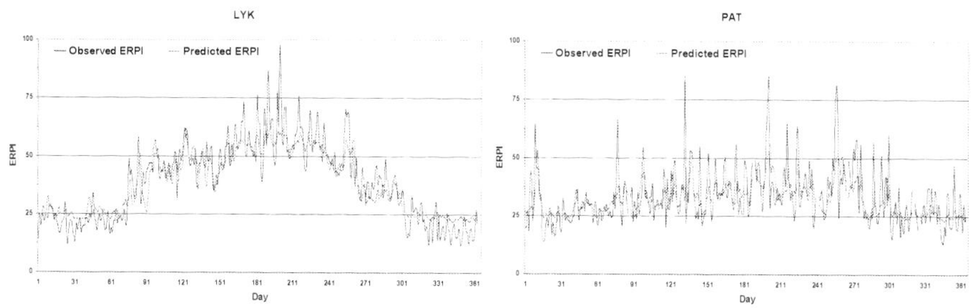

Fig. 7.2.1. Predicted vs. observed ERPI values for the CO, NO_2, SO_2 and O_3, pollutants during the test year 2005.

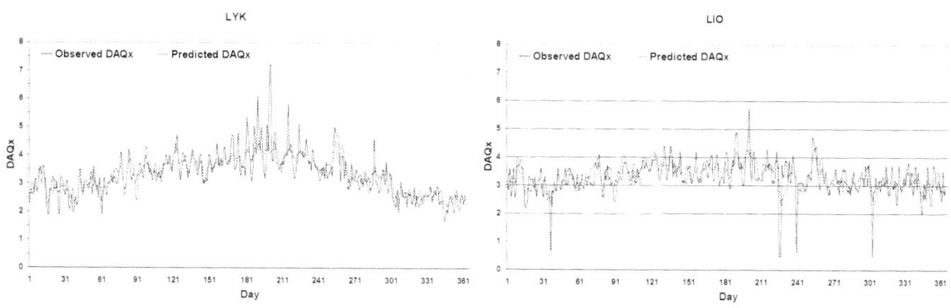

Fig. 7.2.2. Predicted vs. observed DAQx values for the CO, NO_2, SO_2 and O_3, pollutants during the test year 2005.

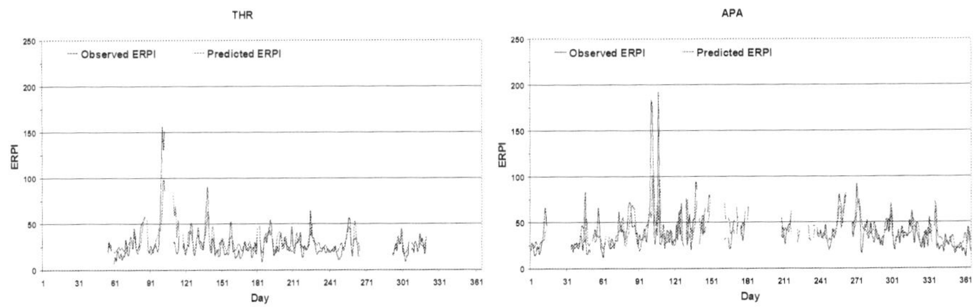

Fig. 7.2.3. Predicted vs. observed ERPI values for the PM_{10}, pollutant during the test year 2005.

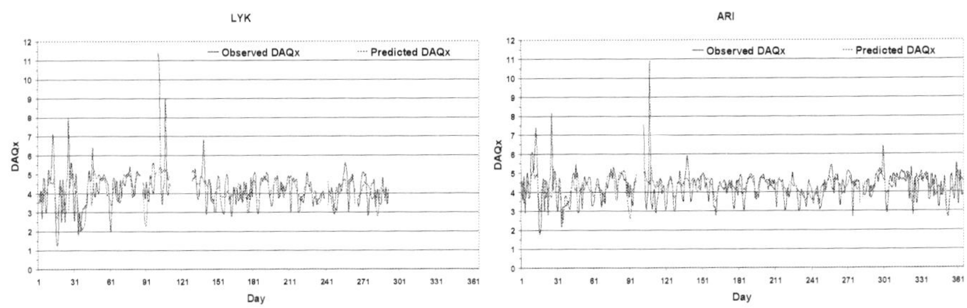

Fig. 7.2.4. Predicted vs. observed DAQx values for the PM_{10}, pollutant during the test year 2005

8. Bioclimatic conditions forecasting using ANNs

8.1 ANNs description for DI and CP forecasting

Four different ANN models were developed in order to forecast the bioclimatic conditions within the GAA during the warm period of the year (May-September). The first one (ANN#7) was trained in order to forecast the daily value of Thom's DI index for the next day at eight different areas of GAA (APA, THR, LYK, MAR, LIO, GAL, GEO and PAT). The second one (ANN#8) was trained in order to forecast the daily value of CP index for the next day at the above mentioned eight different areas within the GAA. The third one (ANN#9) was trained in order to forecast the daily number of the consecutive hours with DI ≥ 24 °C for the next day at each one of the eight examined stations within the GAA. Finally, the fourth (ANN#10) was trained in order to forecast the daily number of the consecutive hours with CP ≤ 174 W/m² for the next day at each one of the eight examined stations within the GAA.

In each case the group of data named as "the training set" used for ANNs training concerns the time period 2001-2004. The group of data named as "the validation set" given to the network still in the learning phase accounts 20% of the training set for each one of the above ANNs. Finally "the test set" refers to the year 2005, which is absolutely unknown to the models in order to reveal the models forecasting ability. Table 8.1.1 presents the input and output data for the four developed ANNs. The combination of selected data for the appropriate ANN models training was done after a series of several tests (trial and error method). At the end, the combination that gave the best forecasting result in each case was selected (Table 8.1.1).

INPUT DATA (input layer)	ANN#7	ANN#8	ANN#9	ANN#10
Stations' number (1,2,3,4,5,6,7)	√	√	√	√
Month (5,6,7,8,9)	√	√	√	√
The maximum (T_{max}) daily temperature for the six previous days.	√	√	√	
The maximum (RH_{max}) daily relative humidity for the six previous days.	√		√	
The maximum (DI_{max}) daily value of DI for the six previous days.	√		√	
The daily number of consecutive hours with DI≥24 ^{0}C for the six previous days.	√		√	
The maximum (V_{max}) daily wind speed for the six previous days.		√		
The minimum (CP_{min}) daily value of CP for the six previous days.		√		
The daily number of consecutive hours with CP≤174 W/m^2 for the six previous days.		√		√
The maximum (T_{max}) and minimum (T_{min}) daily temperature for the six previous days.				√
The maximum (V_{max}) and minimum (V_{min}) daily wind speed for the six previous days.				√
The maximum (CP_{max}) and minimum (CP_{min}) daily value of CP for the six previous days.				√
OUTPUT DATA (output layer)				
The maximum (DI_{max}) daily value of DI for the next day.	√			
The minimum (CP_{min}) daily value of CP for the next day.		√		
The daily number of consecutive hours with DI≥24 ^{0}C for the next day.			√	
The daily number of consecutive hours with CP≤174 W/m^2 for the next day.				√

Table 8.1.1. Input and output data for the appropriate training of the four developed ANNs.

8.2 DI and CP daily value forecasting for the next day

The global fit agreement statistical indices as well as the excess statistical indices for the observed and predicted values were calculated and demonstrated for the eight examined stations respectively. More specifically, O_{ave}, P_{ave}, MBE, RMSE, IA and R^2 values for DI are presented in Table 8.2.1.

The R^2 values show a very satisfactory prediction for DI-ANN#7 ($0.676 \le R^2 \le 0.841$) during the test year 2005 as well as for the CP-ANN#8 ($0.591 \le R^2 \le 0.814$). Concerning the IA values, a very satisfactory prediction for DI-ANN#7 ($0.849 \le IA \le 0.956$) as well as for the CP-ANN#8 ($0.813 \le IA \le 0.948$) appears. Taking into consideration the persistence of the phenomenon with respect to the daily number of consecutive hours with high discomfort conditions, due to strong heat stress, it seems that ANN#9 and ANN#10 give an adequate prediction. Additionally, the R^2 values show a very satisfactory prediction for ANN#9 (0.140

	ANN#7						ANN#8					
	O_{ave}	P_{ave}	MBE (°C)	RMSE (°C)	IA	R^2	O_{ave}	P_{ave}	MBE (W/m²)	RMSE (W/m²)	IA	R^2
APA	23.2	23.3	0.082	0.972	0.942	0.794	141	135	-6.196	49.754	0.920	0.731
GAL	24.1	24.0	-0.057	0.889	0.952	0.824	98	97	-1.137	39.863	0.948	0.814
GEO	23.2	23.3	0.111	0.831	0.956	0.841	143	130	-12.739	44.964	0.942	0.806
LIO	23.5	23.4	-0.069	0.929	0.944	0.801	126	126	0.145	48.369	0.930	0.764
MAR	24.0	23.9	-0.145	1.046	0.934	0.773	115	113	-2.154	48.893	0.926	0.743
PAT	25.0	25.1	0.090	0.889	0.953	0.832	82	78	-4.338	46.999	0.936	0.792
THR	19.9	20.7	0.779	1.584	0.849	0.676	292	246	-45.137	80.978	0.813	0.591
LYK	22.5	22.7	0.247	1.051	0.929	0.771	167	160	-6.791	55.554	0.878	0.664
	ANN#9						ANN#10					
	O_{ave}	P_{ave}	MBE (hours)	RMSE (hours)	IA	R^2	O_{ave}	P_{ave}	MBE (hours)	RMSE (hours)	IA	R^2
APA	3.6	4.2	0.654	2.610	0.930	0.760	7.0	8.0	1.333	4.372	0.907	0.702
GAL	6.6	6.4	-0.170	3.135	0.951	0.832	12.0	12.0	0.255	4.158	0.946	0.812
GEO	4.0	4.3	0.301	2.526	0.946	0.810	8.0	9.0	0.667	4.292	0.931	0.762
LIO	4.1	4.3	0.204	2.475	0.943	0.795	8.0	9.0	0.849	3.959	0.932	0.764
MAR	6.1	6.0	-0.112	3.254	0.943	0.797	10.0	10.0	0.430	4.436	0.934	0.766
PAT	10.9	11.0	0.103	3.601	0.901	0.861	13.0	14.0	0.524	4.923	0.928	0.751
THR	0.1	0.6	0.541	1.126	0.368	0.104	1.0	2.0	1.281	2.716	0.750	0.443
LYK	2.0	2.6	0.669	2.083	0.897	0.680	5.0	6.0	1.118	3.836	0.903	0.689

Table 8.2.1. Global fit agreement indices for DI predicted values for the next day.

$\leq R^2 \leq 0.832$) as well as for the CP-ANN#10 ($0.443 \leq R^2 \leq 0.812$) during the test year 2005. Besides, the IA values, show a very satisfactory prediction regarding ANN#9 ($0.368 \leq IA \leq 0.951$) and ANN#10 ($0.750 \leq IA \leq 0.946$). The worst prediction for the daily number of consecutive hours with high discomfort conditions, due to strong heat stress, refers to the region-station of THR (suburban region-station). This may be attributed to the fact that, in this suburban region (Thrakomakedones) the bioclimatic conditions are better than all the other examined regions within the GAA due to lower temperature values. Both discomfort indices, DI and CP, present daily values over their thresholds for a short period of time during the examined period. Thus, there is not a "memory-experience" of the persistence in THR, so the developed ANN models cannot have the appropriate training in order to forecast the number of consecutive hours with strong discomfort.

Figure 8.2.1 reveals that within the city's centre (PAT), the strong discomfort conditions (DI ≥ 24 °C) appear from the end of June to the first half of September. At the suburban station (THR) there is not a significant discomfort, according to DI values. Just a few days during the warm period of the year appear to be over the threshold of DI ≥ 24 °C; namely at least 50% of the population feels discomfort due to heat stress.

Figure 8.2.2 illustrates that close to the city's center (urban area of Galatsi), the hot sub comfort conditions according to CP values (CP ≤ 174 W/m²) appear from the middle of June until the first half of September. At the suburban station (THR), the discomfort due to heat stress conditions starts at the beginning of July until the middle of August. In all the above cases it seems that the prediction of bioclimatic conditions one day ahead with the use of ANN models is very satisfactory and realizable.

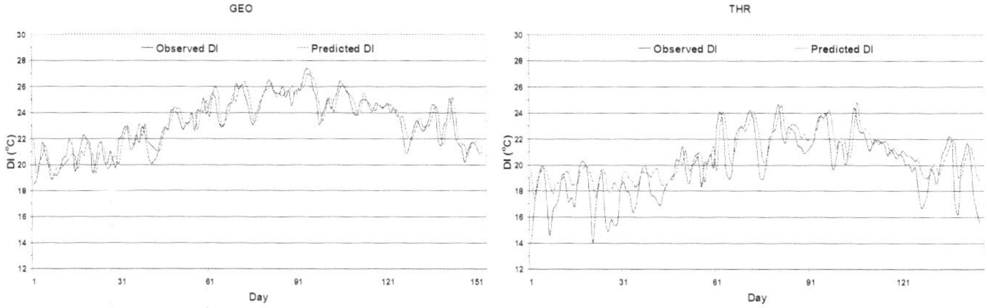

Fig. 8.2.1. Predicted vs. observed values of DI for the next day, concerning the best (GEO) and the worst (THR) prediction for DI daily maximum value, during the warm period of the test year 2005.

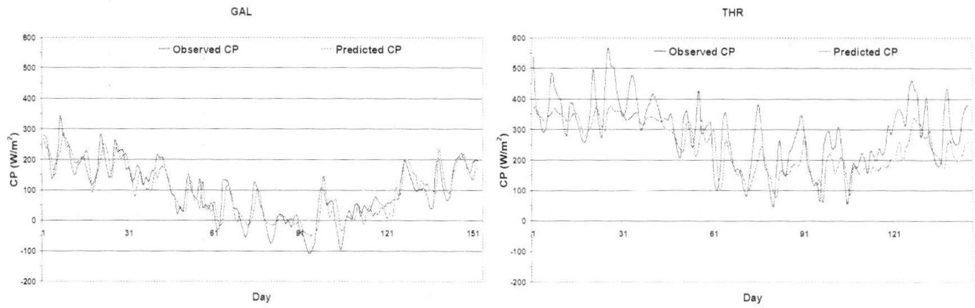

Fig. 8.2.2. Predicted vs. Observed values of CP for the next day, concerning the best (GAL) and the worst (THR) prediction for CP daily minimum value, during the warm period of the test year 2005.

8.3 ANNs description for PET forecasting

Three developed ANNs were trained using back-propagation algorithm to forecast the mean daily PET value for the next day (ANN#11), two next days (ANN#12) and three next days (ANN#13). The training dataset concern the period 2001-2003, while the validation dataset concern the year 2004, which was absolutely unknown to the constructed model, in order to test the predictive ability of the model. Superposed epoch analysis on the training datasets indicated that three days before the incidence of strong heat/cold stress are adequate to forecast PET value for the next days. Thus, the input data (Table 8.3.1) which were taken for ANNs training concern the mean daily air temperature, relative humidity, wind speed and sunshine for the previous three days from the National Observatory of Athens.

Table 8.3.2 presents the fit agreement indices between the observed and the predicted PET values, for the validation year 2004. It is remarkable the high values of IA and R^2, which indicate that the constructed ANNs have an excellent forecasting ability of PET for the next three days. This gives evidence that the developed ANNs, taking into account simple meteorological parameters recorded in the previous three days, are capable to predict a bioclimatic index, which is not easily calculated (PET was estimated using the RayMan model), while the most remarkable finding is that of pronounced agreement between observed and predicted PET values. Figure 8.3.1 depicts the predicted and observed mean daily PET time series for the next day (a), the next two days (b) and the next three days (c), along with the respective scatter plots.

INPUT DATA (input layer)	ANN#11	ANN#12	ANN#13
Mean daily air temperature (^0C) for the three previous days	√	√	√
Mean daily wind speed (m/s) for the three previous days	√	√	√
Mean daily relative humidity (RH%) for the three previous days	√		
The sunshine duration (hours) for the three previous days	√	√	√
OUTPUT DATA (output layer)			
Mean daily PET value for the next day	√		
4Mean daily PET value for the next two days		√	
Mean daily PET value for the next three days			√

Table 8.3.1. Input and output data for the appropriate training of the developed ANN#11.

	MBE	RMSE	IA	R^2
Mean daily PET value for the next day (ANN#11)	+0.5	2.8	0.982	0.933
Mean daily PET value for the next two days (ANN#12)	+0.5	3.8	0.966	0.874
Mean daily PET value for the next three days (ANN#13)	+0.4	4.3	0.956	0.839

Table 8.3.2. Global fit agreement indices for PET predicted values for the next one, two and three days.

9. Spatial distribution of air quality and bioclimatic conditions in the GAA

9.1 Spatial variation of air quality within GAA

The mean annual value for both air quality indices ERPI and DQAx was calculated at all the examined regions within GAA, during the time period 2001-2005. Figure 9.1.1 shows the spatial variation of air quality levels within GAA. As far as the air quality index ERPI is concerned, only the station THR appears a satisfactory air quality level in annual basis (ERPI < 40). The stations MAR, APA and GAL appear a tolerable air quality level (ERPI < 50). Moreover, the air quality levels at LYK, LIO and PAT stations are very close to the limit value of ERPI ≥ 50. Finally, the air quality level appears to be poor in the city centre station ARI. This may be attributed to the high PM_{10} concentration levels almost during the whole year. In this point, we have to mention that the station PAT is also in the centre of the city and very close to the ARI station, but unfortunately for this station we don't have any PM_{10} observations. Similar conclusions are extracted with respect to the air quality index DAQx. The only exception is the LIO station in which the air quality levels seems to be much closer to the stations GAL, MAR and APA.

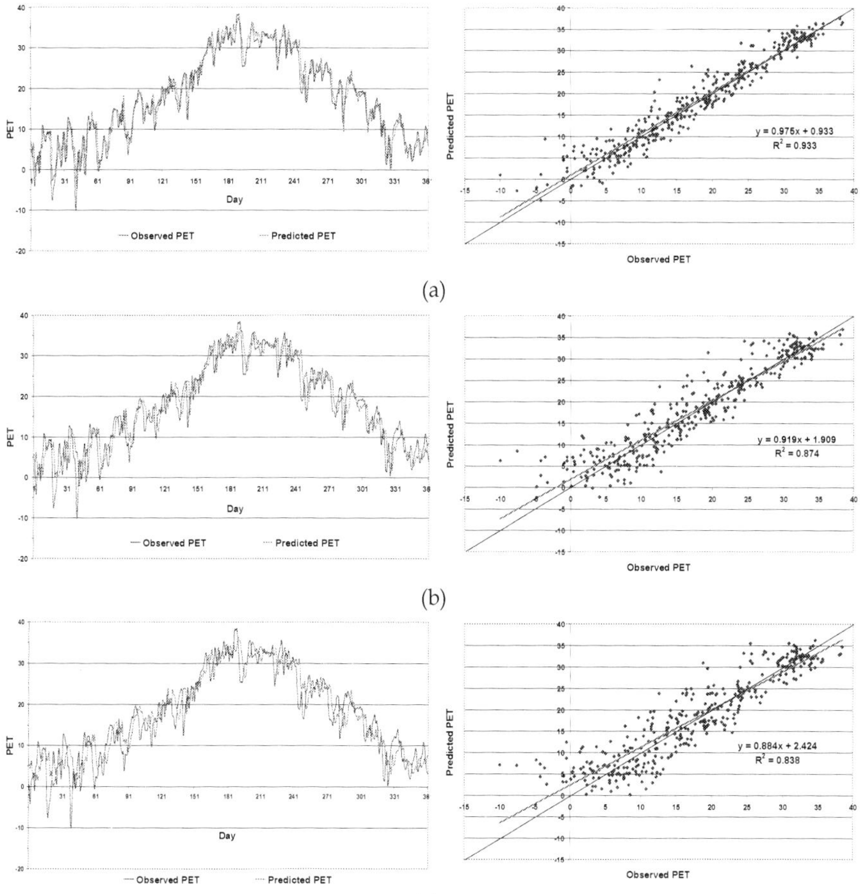

Fig. 8.3.1. Mean daily PET predicted values vs. observed values, for the next day (a), next two days (b) and next three days (c) three days, for the test year 2004.

9.2 Spatial variation of bioclimatic conditions within GAA

During the period 2001-2005, the mean annual value for both bioclimatic indices DI and CP was calculated at all the examined regions within the GAA. Figure 9.2.1 depicts the spatial variation of the bioclimatic conditions within the GAA during the warm period of the year (May-September), where three different bioclimatic zones appear. The first zone is the north suburban zone (THR), which can be characterized as a comfortable zone. The second zone extends peripherally the city's center (LIO, LYK, MAR and APA) and can be marginally characterized as a comfortable zone or warm zone. Finally, the third zone concerns the city's center (GAL, PAT and GEO), which can be characterized as an uncomfortable zone or a strong heat stress zone.

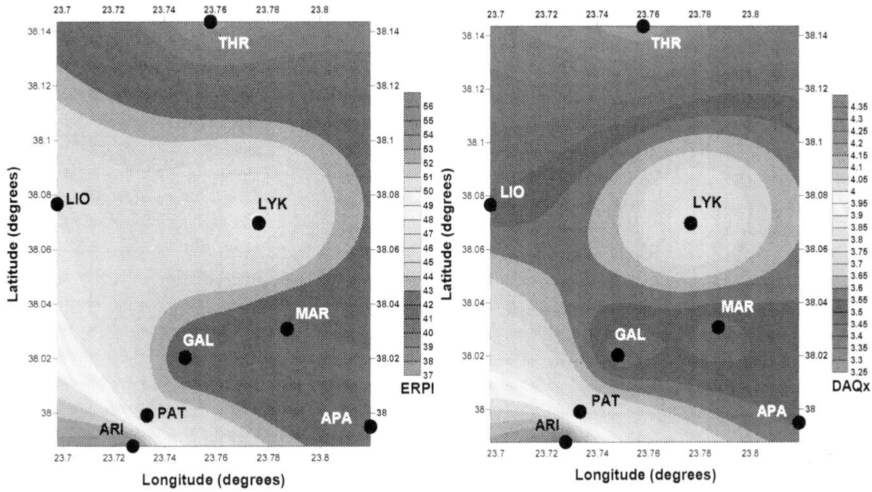

Fig. 9.1.1. Spatial variation of air quality levels within GAA for the pollutants NO_2, SO_2, CO, O_3 and PM_{10}. Mean annual values of ERPI (left graph) and DAQx (right graph), during 2001-2005.

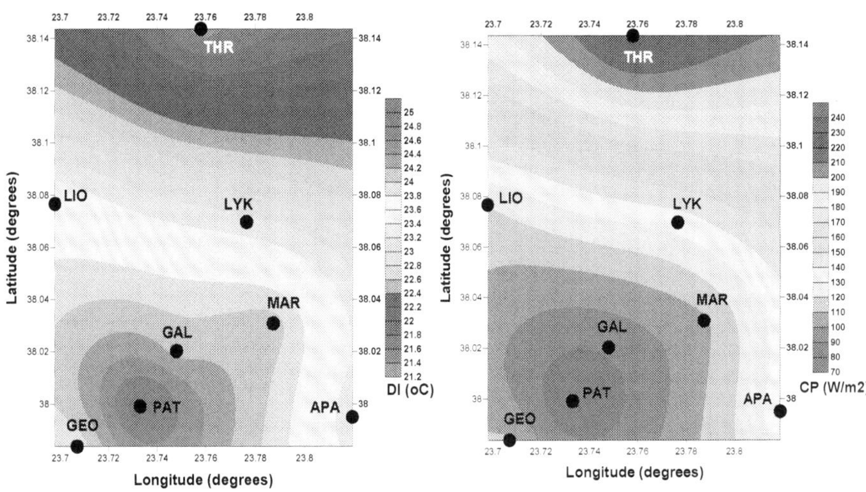

Fig. 9.2.1. Spatial variation of bioclimatic conditions within the GAA. Mean annual values of DI (left graph) and CP (right panel), for the warm period of the year during 2001-2005.

As far as the persistence of discomfort during the examined period 2001-2005 is concerned, the mean seasonal number of consecutive hours during the day with high levels of human discomfort appears in the station PAT; 11.3 and 13.6 consecutive hours with respect to DI and CP, respectively, against 1.0 and 2.7 consecutive hours at the station THR, respectively. All the other examined regions-stations within the GAA present a bioclimatic behavior between PAT and THR. This means that for a given building within the city's center region (PAT), we need 5 to 11 times more energy for cooling during the warm period of the year than the energy for cooling at the north suburban area (THR).

10. Conclusions

In this study an application, which concerns the development and the use of ANN models on environmental issues and generally in environmental management, is presented. A number of ANN models have been developed and trained in order to forecast the air quality levels, as well as the bioclimatic conditions in different regions within the GAA. The findings of this work appoint the ANN models forecasting capacity.

The Results showed that the use of ANN models as forecasting tool is realizable and satisfactory at a statistically significant level of $p<0.01$. In particular for the air quality forecasting for the next day, the R^2 values ranged between 0.381 and 0.826 (ERPI) and between 0.378 and 0.686 (DAQx). Besides, the IA index between the predicted and observed values ranged between 0.717and 0.937 for ERPI forecasting, while it ranged between 0.746 and 0.889 for DAQx forecasting. It seems that in all cases, the air quality forecasting is more sufficient using the ERPI air quality index than the DAQx. In this point we have to mention that the ERPI is according to the European Community directives for the air quality levels. The same results are extracted regarding the forecasting of the persistence of air pollution episodes and especially the number of consecutive hours during the day with poor air quality.

Concerning the forecasting of bioclimatic conditions for the next day, the R^2 values ranged between 0.676 and 0.841 for DI and between 0.591 and 0.814 for CP. The IA values ranged between 0.849 and 0.956 for DI and between 0.813 and 0.948 for CP. Taking into account the persistence of the phenomenon (the number of consecutive hours during the day with high discomfort conditions due to strong heat stress), it seems that ANN#9 (consecutive discomfort hours according to DI values) and ANN#10 (consecutive discomfort hours according to CP values) give an adequate prediction.

A remarkable finding of this research is that the high values of IA (0.956 - 0.982) and R^2 (0.839 - 0.933) with respect to PET forecasting for the next three days indicate that the constructed ANNs have an excellent forecasting ability of PET, a more complex bioclimatic index based on the human energy balance. This gives evidence that the developed ANNs, taking into account simple meteorological parameters recorded in the previous three days, are capable to predict a bioclimatic index, which is not easily calculated (PET was estimated using the RayMan model).

11. References

Antonic, O., Hatic, D., Krian, J. & Bukocev, D. (2001). Modeling groundwater regime acceptable for the forest survival after the building of the hydro-electric power plant. *Ecol. Model.*, Vol. 138, pp. 277-288.

Attoh-Okine, N.O. (1999). Analysis of learning rate and momentum term in backpropagation neural network algorithm trained to predict pavement performance. *Adv. Eng. Softw.*, Vol. 30, No. 4, pp. 291-302.

Balaguer Ballester, E., Valls, G., Carrasco-Rodriguez, J., Soria Oliva, E., Valle-Tascon, S. (2002). Effective 1-day ahead prediction of hourly surface ozone concentrations in eastern Spain using linear models and neural networks. *Ecol. Model.*, Vol. 156, pp. 27-41.

Barazzetta, S., Corani, G. (2004). First results in the prediction of PM10 in Milan: the Air Sentinel project. *Proceedings of the 9th International Conference on Harmonisation within Atmospheric Dispersion Modelling for Regulatory Purposes*, Garmisch, May 2004.

Besancenot, J.P. (1978). Le bioclimat humain de Rio. In: *Recherches de Climatologie en Milieu Tropical et Mediterranean*, Suchel, J.B., Altes, E., Besancenot, J.P. & Maheras, P. (Editors), Cahier No. 6 du Centre de Recherches de Climatologie, Universite de Dijon, Dijon, France.

Bishop, C.M. (1995). *Neural Networks for Pattern Recognition*, Oxford University Press, ISBN 019853864, Oxford, U.K.

Caudill, M. & Butler, C. (1992). Understanding Neural Networks; Computer Explorations, MIT Press-Cambridge, ISBN 02625309996, MA, USA.

Clarke, J.F., Bach, W. (1971). Comparison of the comfort conditions in different urban and suburban microenvironments. *Int. J. Biometeorol.*, Vol. 15, pp. 41–54.

Comrie, A.C. (1997). Comparing neural networks and regression models for ozone forecasting. *J. Air Waste Manage.*, Vol.. 47, pp. 653-663.

Corani, G. (2005). Air quality prediction in Milan: feed-forward neural networks, pruned neural networks and lazy learning. *Ecol. Model.*, Vol. 185, pp. 513-529.

Council Directive 96/92/EC (1996). *On ambient air quality assessment and management.* Official Journal of the European Communities, L296, 21.11.1996, pp. 55-63.

Council Directive 1999/30/EC (1999). *Limit values of sulphur dioxide, nitrogen dioxide and oxides of nitrogen, particulate matter and lead in ambient air.* Official Journal of the European Communities, L163, 29.6.1999, pp. 41-60.

Directive 2000/69/EC of the European Parliament and the Council (2000). *Limit values for benzene and carbon monoxide in ambient air.* Official Journal of the European Communities, L313, 13.12.2000, pp. 12-22.

Directive 2002/3/EC of the European Parliament and the Council (2002). *Ozone in ambient air.* Official Journal of the European Communities, L67, 9.3.2002, pp. 14-31.

Dutot, A.L., Rynkiewicz, J., Steiner, F.E. & Rude, J. (2007). A 24-h forecast of ozone peaks and exceedance levels using neural classifiers and weather predictions. *Environ. Modell. & Softw.*, Vol. 22, No. 9, pp. 1261-1269.

Gardner, M.W., Dorling, S.R. (1998a). Artificial Neural Networks, the multilayer perceptron. A review of applications in the atmospheric sciences. *Atmos. Environ.*, Vol. 32, pp. 2627-2636.

Gardner, M.W., Dorling, S.R. (1998b). Neural network modeling and prediction of hourly NOx and NO2 concentrations in urban air in London. *Atmos. Environ.*, Vol. 33, pp. 709-719.

Giles, B.D., Balafoutis, C.H., Maheras, P. (1990). Too hot for comfort: the heatwaves in Greece in 1987 and 1988. *International Journal of Biometeorology*, Vol. 34, pp. 98-104.

Givoni, B. (1998). *Climatic considerations in building and urban design*, John Wiley & Sons, ISBN 0471291773, New York.

Hecht-Nielsen, R. (1990). *Neurocomputing*, Addison-Wesley, ISBN 0201093553, Reading, M.A.

Heymans, J.J., Baird, A. (2000). A carbon flow model and network analysis of the northern Benguela upwelling system, Namibia. *Ecol. Model.*, Vol. 126, pp. 9-32.

Höppe, P.R. (1999). The physiological equivalent temperature–a universal index for the biometeorological assessment of the thermal environment. *Int. J. Biometeorol.*, Vol. 43, pp. 71–75.

Karul, C., Soyupak, S., Cilesiz, A.F., Akbay, N. & Germen, E. (2000). Case studies on the use of neural networks in eutrophication modeling. *Ecol. Model.*, Vol. 164, pp. 145-152.

Kolehmainen, M., Martikainen, H., Ruuskanen, J. (2001). Neural networks and periodic components used in air quality forecasting. *Atmos. Environ.*, Vol. 35, pp. 815-825.

Lapedes, A. & Farber, R. (1987). *Non-linear signal processing using neural networks*. Technical Report no. LA-UR-2662 Los Alamos National Laboratory.

Makra, L., Mayer, H., Beczi, R. & Borsos, E. (2003). Evaluation of the air quality of Szeged with some assessment methods. *Acta Climatologica et Chorologica*, Universitatis Szegediensis, Tom. 36-37, pp. 85-93.

Matzarakis, A., Mayer, H. (1996). *Another kind of environmental stress: Thermal stress*. WHO Collaborating Centre for Air Quality Management and Air Pollution Control, NEWSLETTERS No. 18, pp. 7-10.

Matzarakis, A., Mayer, H., Iziomon, M. (1999). Applications of a universal thermal index: physiological equivalent temperature. *Int. J. Biometeorol.*, Vol. 43, pp. 76-84.

Matzarakis, A., Rutz, F., Mayer, H. (2010). Modelling Radiation fluxes in simple and complex environments - Basics of the RayMan model. *Int. J. Biometeorol.*, Vol. 54, pp. 131-139.

Mayer, H., Höppe, P. (1987). Thermal comfort of man in different urban environments. *Theor. Appl. Climatol.*, Vol. 38, pp. 3-49.

Mayer, H., Kalberlah, F. & Ahrens, D. (2002a). TLQ-Am impact related air quality index obtained on a daily basis. *Proceedings of the fourth Symposium on the Urban Environment*, Norfolk, Virginia (USA), 20-24 May 2002, pp. 80-81.

Mayer, H., Kalberlah, F., Ahrens, D. & Reuter, U. (2002b). Analysis of indices for the assessment of the air (in German). *Gefahrstoffe-Reinhaltung der Luft*, Vol. 62, pp. 177-183.

McCulloh, W.S., Pitts W. (1943). A logical calculus of ideas immanent in Nervous activity. *Bulletin of Mathematical Biophysics*, Vol. 5, (1943), pp.115-133.

Moustris, K.P. (2009). *Air quality forecasting with the use of Artificial Neural Networks in the greater Athens area*. PhD Thesis, School of Chemical Engineering, National Technical University of Athens, Athens, Greece.

Moustris, K.P., Ziomas, I.C., Paliatsos, A.G. (2010). 3-Day-Ahead Forecasting of Regional Pollution Index for the Pollutants NO_2, CO, SO_2, and O_3 Using Artificial Neural Networks in Athens, Greece. *Water Air Soil Poll.*, Vol. 200, pp. 29–43.

Moustris, K.P., Larissi, I.K., Nastos, P.T., Paliatsos, A.G. (2011). Precipitation forecast using artificial neural networks in specific regions of Greece. *Water Resour. Manag.*, DOI 10.1007/s11269-011-9790.

Nastos, P.T., Matzarakis, A. (2006). Weather impacts on respiratory infections in Athens, Greece. *Int. J. Biometeorol.*, Vol. 50, pp. 358-369.

Nastos, P.T., Matzarakis, A. (2008). Variability of tropical days over Greece within the second half of the twentieth century. *Theor. Appl. Climatol.*, Vol. 93, pp. 75-89.

Matzarakis, A., Nastos, P.T. (2010). Human-Biometeorological assessment of heat waves in Athens. *Theor. Appl. Climatol.*, DOI 10.1007/s00704-010-0379-3.

Norgaard, M., Ravn, O., Poulsen, N.K., Hansen, L.K. (2000). *Neural Networks for Modelling and Control of Dynamic Systems*, Springer-Verlag, ISBN 1852332271, London.

Nunnari, G., Nucifora, M., Randieri, C. (1998). The application of neural techniques to the modeling of time-series of atmospheric pollution data. *Ecol. Model.*, Vol. 111, pp. 187-205.

Papanastasiou, D.K., Melas, D., Kioutsioukis, I. (2007). Development and Assessment of Neural Network and Multiple Regression Models in Order to Predict PM10 Levels in a Medium-sized Mediterranean City. *Water Air Soil Poll.*, Vol. 182, pp. 325-334.

Philandras, C.M., Metaxas, D.A., Nastos, P.T. (1999). Climate variability and urbanization in Athens. *Theor. Appl. Climatol.*, Vol. 63, No. 1-2, pp. 65-72.

Prybutok, R., Junsub, Y., Mitchell, D. (2000). Comparison of neural network models with ARIMA and regression models for prediction of Huston's daily maximum ozone concentrations. *Eur. J. Oper. Res.*, Vol. 122, pp. 31-40.

Refenes, A.N., Zapranis, A. & Francis, G. (1994). Stock performance modeling using neural networks: A comparative study with regression models. *Neural Networks*, Vol. 7, No. 2, pp. 375-388.

Schlink, U., Dorling, S., Pelikan, E., Nunnari, G., Cawley, G., Junninen, H., Greig, A., Foxall, R., Eben, K., Chatterton, T., Vondracek, J., Richter, M., Dostal, M., Bertucco, L., Kolehmainen, M., Doyle, M. (2003). A rigorous inter-comparison of ground-level ozone predictions. *Atmos. Environ.*, Vol. 37, pp. 3237-3253.

Siple, P.A., Passel, C.F. (1945). Measurements of dry atmospheric cooling in subfreezing temperatures. *Proceedings of the American Philosophical Society*, Vol. 89, No. 1, pp. 177–199.

Slini, T., Kaprara, A., Karatzas, K. & Moussiopoulos, N. (2006). PM10 Forecasting for Thessaloniki, Greece. *Environ. Model. Softw.*, Vol. 21, No. 4, pp. 559-565.

Thom, E.C. (1959). The discomfort index. *Weatherwise*, Vol. 12, pp. 57–60.

VDI (1998). VDI 3787, Part I: Environmental Meteorology, Methods for the human biometeorological evaluation of climate and air quality for the urban and regional planning at regional level. Part I: Climate, Beuth, Berlin.

Viotti, P., Liuti, G. & Di Genova, P. (2002). Atmospheric urban pollution: applications of an artificial neural network (ANN) to the city of Perugia. *Ecol. Model.*, Vol. 148, No. 1, pp. 27-46.

Werbos, P. (1988). Generalization of Backpropagation with application to a recurrent gas market model. *Neural Networks*, Vol. 1, pp. 339-356.

Willmott, C.J., Ackleson, S.G., Davis, R.E., Feddema, J.J., Klink, K.M., Legates, D.R., O'Donnell, J. & Rowe, C. (1985). Statistics for the evaluation and comparison of models. *J. Geophys. Res.*, Vol. 90, pp. 8995-9005.